Quantity	SI Equivalence	English Equivalence
Work or energy	$1 \text{ in.} - \text{lb}_f = 0.1129848 \text{ J}$ $1 \text{ ft} - \text{lb}_f = 1.355818 \text{ J}$ $1 \text{ Btu} = 1055.056 \text{ J}$ $1 \text{ kWh} = 3.6 \times 10^6 \text{ J}$	$1 \text{ J} = 8.850744 \text{ in.} - \text{lb}_f$ $1 \text{ J} = 0.737562 \text{ ft} - \text{lb}_f$ $\quad = 0.947817 \times 10^{-3} \text{ Btu}$ $\quad = 0.277778 \text{ kWh}$
Power	$1 \text{ in} - \text{lb}_f/\text{sec} = 0.1129848 \text{ W}$ $1 \text{ ft} - \text{lb}_f/\text{sec} = 1.355818 \text{ W}$ $\quad = 0.0018182 \text{ hp}$ $1 \text{ hp} = 745.7 \text{ W}$	$1 \text{ W} = 8.850744 \text{ in.} - \text{lb}_f/\text{sec}$ $1 \text{ W} = 0.737562 \text{ ft} - \text{lb}_f/\text{sec}$ $\quad = 1.34102 \times 10^{-3} \text{ hp}$
Area moment of inertia or second moment of area	$1 \text{ in}^4 = 41.6231 \times 10^{-8} \text{ m}^4$ $1 \text{ ft}^4 = 86.3097 \times 10^{-4} \text{ m}^4$	$1 \text{ m}^4 = 240.251 \times 10^4 \text{ in}^4$ $\quad = 115.862 \text{ ft}^4$
Mass moment of inertia	$1 \text{ in} - \text{lb}_f - \text{sec}^2 = 0.1129848 \text{ m}^2 \cdot \text{kg}$	$1 \text{ m}^2 \cdot \text{kg} = 8.850744 \text{ in.} - \text{lb}_f - \text{sec}^2$

Spring constant:
  translatory
 	
  torsional
	$1 \text{ lb}_f/\text{in.} = 175.1268 \text{ N/m}$
	$1 \text{ lb}_f/\text{ft} = 14.5939 \text{ N/m}$
	$1 \text{ in.} - \text{lb}_f/\text{rad} = 0.1129848 \text{ m} \cdot \text{N/rad}$
		$1 \text{ N/m} = 5.71017 \times 10^{-3} \text{ lb}_f/\text{in.}$
		$\quad = 68.5221 \times 10^{-3} \text{ lb}_f/\text{ft}$
		$1 \text{ m} \cdot \text{N/rad} = 8.850744 \text{ in-lb}_f/\text{rad}$
		$\quad = 0.737562 \text{ lb}_f - \text{ft/rad}$

Damping constant:
  translatory
  torsional
	$1 \text{ lb}_f - \text{sec/in} = 175.1268 \text{ N} \cdot \text{s/m}$
	$1 \text{ in} - \text{lb}_f - \text{sec/rad}$
	$\quad = 0.1129848 \text{ m} \cdot \text{N} \cdot \text{s/rad}$
		$1 \text{ N} \cdot \text{s/m} = 5.71017 \times 10^{-3} \text{ lb}_f - \text{sec/in.}$
		$1 \text{ m} \cdot \text{N} \cdot \text{s/rad} = 8.850744 \text{ lb}_f - \text{in} - \text{sec/rad}$

Angles	$1 \text{ rad} = 57.295754 \text{ degrees};\qquad 1 \text{ degree} = 0.0174533 \text{ rad};$ $1 \text{ rpm} = 0.166667 \text{ rev/sec} = 0.104720 \text{ rad/sec};\qquad 1 \text{ rad/sec} = 9.54909 \text{ rpm}$	

errors

Mechanical Vibrations

Second Edition

Singiresu S. Rao
Purdue University

Addison-Wesley Publishing Company
Reading, Massachusetts · Menlo Park, California · New York
Don Mills, Ontario · Wokingham, England · Amsterdam ·
Bonn · Sydney · Singapore · Tokyo · Madrid ·
San Juan

This book is in the Addison-Wesley Series in Mechanical Engineering

Library of Congress Cataloging-in-Publication Data

Rao, S. S.
    Mechanical vibrations/Singiresu S. Rao —2nd ed.
       p.   cm.
    Bibliography: p.
    Includes index.
    ISBN 0-201-50156-2
    1. Vibration.   I. Title.
TA355.R37 1990
620.3—dc20

Reprinted with corrections June, 1990.

CDEFGHIJ-DO-943210

To Lord Sri Venkateswara

Preface

This text is intended for use as an introduction to the subject of vibration engineering at the undergraduate level. The style of presentation from the prior edition, of presenting the theory, computational aspects, and applications of vibrations in a manner as simple as possible is retained. As in the first edition, computer techniques of analysis are emphasized. Expanded explanations of the fundamentals, emphasizing physical significance and interpretation that build upon previous experiences in undergraduate mechanics are given. Numerous examples and problems are used to illustrate principles and concepts.

This book was first published in 1986. Favorable reactions and friendly encouragements from professors, students and my publisher have provided me with the impetus to come out with a new edition. In this second edition several new chapters have been added. Modifications and corrections to many topics have been made. Most of the additions to the first edition were suggested by those who have used the text and by numerous reviewers.

Some of the important changes in this edition are:

* Approximately forty percent of the problems are new.
* Design type problems, identified by asterisks, are included in various chapters.
* Project type problems are added at the end of several chapters.
* The section on vibration measuring instruments has been expanded into a full chapter entitled, "Vibration Measurement."
* The chapter on further topics in vibration is now deleted.
* New chapters on finite element method, nonlinear vibration, and random vibration are added.
* All the examples in the book have been presented in a new format. Following the statement of each example, the known information, the quantities to be determined, and the approach to be used are first identified and then the detailed solution is given.

FEATURES

Each topic in *Mechanical Vibrations* is self-contained. All the concepts are explained fully and the derivations are presented with complete details for the benefit of the reader. The computational aspects are emphasized throughout the book. Several Fortran computer programs, most of them in the form of general purpose subroutines, are given at the end of the chapters. These programs are given for use by the students. Although the programs have been tested, no warranty is implied as to their accuracy.

Problems, which are based on the use/development of computer programs, are given at the end of each chapter. It is highly desirable that students solve these problems to obtain exposure to many important computational and programming details.

Some subjects are presented in a somewhat unconventional manner. The topics of Chapters 9, 10, and 11 fall in this category. Most textbooks discuss the topics of isolators, absorbers, and balancing at different places. Since one of the main purposes of the study of vibrations is to control vibration response, all the topics directly related to vibration control are given in Chapter 9. The vibration measuring instruments, along with vibration exciters and signal analysis procedures, are presented in Chapter 10. Similarly, all the numerical integration methods applicable to single- and multidegree-of-freedom systems, as well as continuous systems, are unified in Chapter 11.

Specific features include:

* 23 Computer programs to aid the student in the numerical implementation of the methods discussed in the text.
* Nearly 100 illustrative examples following the presentation of most of the topics.
* More than 250 review questions to help students in reviewing and testing their understanding of the text material.
* Over 600 problems, with solutions in the instructor's manual.
* More than 290 references to lead the reader to specialized and advanced literature.
* Biographical information about scientists and engineers, who contributed to the development of the theory of vibrations, is given on the opening pages of chapters and appendixes.

NOTATION AND UNITS

Both the SI and the English system of units have been used in the examples and problems. A list of symbols, along with the associated units in SI and English systems, is given at the beginning of the book. A brief discussion of SI units as they apply to the field of vibration is given in Appendix C. Arrows are used over symbols to denote column vectors and square brackets are used to indicate matrices.

CONTENTS

Mechanical Vibrations is organized into 14 chapters and 3 appendixes. The material of the book provides flexible options for different types of vibration courses. For a one-semester senior or dual-level course, Chapters 1 through 5, portions of Chapters 6, 7, 8 and 10, and Chapter 9 may be used. The course can be given a computer orientation by including Chapter 11 in place of Chapter 8. Alternatively, with Chapters 12, 13 and 14, the text has sufficient material for a one-year sequence at the senior level. For shorter courses, the instructor can select the topics depending on the level and orientation of the course. It is hoped that the relative simplicity with which the various topics are presented makes the book useful to students as well as practicing engineers for purposes of self-study and as a source of references and computer programs.

Chapter 1 starts with a brief discussion of the history and importance of vibrations. The basic concepts and terminology used in vibration analysis are introduced. The free vibration analysis of single-degree-of-freedom undamped translational and torsional systems is given in Chapter 2. The effects of viscous, Coulomb and hysteretic damping are also discussed. The harmonic response of single-degree-of-freedom systems is considered in Chapter 3. Chapter 4 is concerned with the response of a single-degree-of-freedom system under general forcing conditions. The roles of convolution integral, Laplace transformation, and numerical methods are discussed. The concept of response spectrum is also introduced in this chapter. The free and forced vibration of two-degree-of-freedom systems is considered in Chapter 5. The self-excited vibration and stability of the system are discussed. Chapter 6 presents the vibration analysis of multidegree-of-freedom systems. Matrix methods of analysis are used for the presentation of the theory. The modal analysis procedure is described for the solution of forced vibration problems. Several methods of determining the natural frequencies of discrete systems are outlined in Chapter 7. Dunkerley's, Rayleigh's, Holzer's, matrix iteration, and Jacobi's methods are discussed.

The vibration analysis of continuous systems including strings, bars, shafts, beams, and membranes is given in Chapter 8. The Rayleigh and Rayleigh-Ritz methods of finding the approximate natural frequencies are also described. Chapter 9 discusses the various aspects of vibration control including the problems of elimination, isolation and absorption. The balancing of rotating and reciprocating machines and whirling of shafts are also considered. The vibration measuring instruments, vibration exciters and signal analysis are the topics of Chapter 10. Chapter 11 presents several numerical integration techniques for finding the dynamic response of discrete and continuous systems. The central difference, Runge-Kutta, Houbolt, Wilson, and Newmark methods are summarized and illustrated. The finite element analysis, with applications involving one dimensional elements, is given in Chapter 12. An introductory treatment of nonlinear vibration, including a discussion of subharmonic and superharmonic oscillations, limit cycles, and systems with time dependent coefficients, is given in Chapter 13. The random vibration of linear vibration systems is considered in Chapter 14. Appendixes A, B, and C outline the basic relations of matrices, Laplace transforms, and SI units, respectively.

ACKNOWLEDGMENTS

I would like to express my appreciation to the many students and faculty whose comments have helped me improve the presentation. I am most grateful to Professors F. P. J. Rimrott of the University of Toronto and Raymond M. Brach of the University of Notre Dame for pointing out some of the errors in the previous edition and offering suggestions for improvement. The comments and suggestions made by the following reviewers have been of great help in revising the book: Michael K. Wells, Montana State University; Pinhas Barak, GMI Engineering & Management Institute; Donald A. Grant, University of Maine; Tom Burton, Washington State University; and Ramesh S. Guttalu, University of Southern California. It was gratifying throughout to work with the staff of Addison-Wesley on this revision. In particular, the help of Tom Robbins, Executive Editor, has been most valuable. Helen Wythe, Production Supervisor, handled the task of incorporating my corrections and revisions very efficiently. I would like to thank Purdue University for granting me permission to use the Boilermaker Special in Problem 2.49. Finally, I wish to thank my wife Kamala and daughters Sridevi and Shobha without whose patience, encouragement, and support this revised edition might never have been completed.

S. S. Rao

Contents

CHAPTER 2
Free Vibration of Single Degree
of Freedom Systems

CHAPTER 3
Harmonically Excited Vibration

CHAPTER 4
Vibration under General Forcing Conditions

CHAPTER 5
Two Degree of Freedom Systems

CHAPTER 6
Multidegree of Freedom Systems

CHAPTER 7
Determination of Natural Frequencies and Mode Shapes

CHAPTER 10
Vibration Measurement

CHAPTER 11

Numerical Integration Methods in Vibration Analysis

CHAPTER 12

Finite Element Method

CHAPTER 13
Nonlinear Vibration

CHAPTER 14

Random Vibration

APPENDIX A

Matrices   693

LIST OF SYMBOLS

Symbol	Meaning	English Units	SI Units
$a, a_0, a_1, a_2, \ldots$	constants, lengths		
a_{ij}	flexibility coefficient	in./lb	m/N
$[a]$	flexibility matrix	in./lb	m/N
A	area	in^2	m^2
$A, A_0, A_1, \ldots$	constants		
$b, b_1, b_2, \ldots$	constants, lengths		
$B, B_1, B_2, \ldots$	constants		
$\vec{B}$	balancing weight	lb	N
$c, \underline{c}$	viscous damping coefficient	lb-sec/in.	N · s/m
$c, c_0, c_1, c_2, \ldots$	constants		
c	wave velocity	in./sec	m/s
c_c	critical viscous damping constant	lb-sec/in.	N · s/m
c_i	damping constant of ith damper	lb-sec/in.	N · s/m
c_{ij}	damping coefficient	lb-sec/in.	N · s/m
$[c]$	damping matrix	lb-sec/in.	N · s/m
C, C_1, C_2, C_1', C_2'	constants		
d	diameter, dimension	in.	m
D	diameter	in.	m
$[D]$	dynamical matrix	sec^2	s^2
e	base of natural logarithms		
e	eccentricity	in.	m
$\vec{e}_x, \vec{e}_y$	unit vectors parallel to x and y directions		
E	Young's modulus	lb/in^2	Pa
$E[x]$	expected value of x		
f	linear frequency	Hz	Hz
f	force per unit length	lb/in.	N/m
$\underset{\sim}{f}$	unit impulse	lb-sec	N · s
F, F_d	force	lb	N
F_0	amplitude of force $F(t)$	lb	N
F_t, F_T	force transmitted	lb	N
F_i	force acting on ith mass	lb	N
$\vec{F}$	force vector	lb	N
$\underset{\sim}{F}$	impulse	lb-sec	N · s
g	acceleration due to gravity	in./sec^2	m/s^2
$g(t)$	impulse response function		

Symbol	Meaning	English Units	SI Units
G	shear modulus	lb/in^2	N/m^2
h	hysteresis damping constant	lb/in	N/m
$H(i\omega)$	frequency response function		
i	$\sqrt{-1}$		
I	area moment of inertia	in^4	m^4
$[I]$	identity matrix		
$\text{Im}(\ )$	imaginary part of ( )		
j	integer		
J	polar moment of inertia	in^4	m^4
$J_0, J_1, J_2, \ldots$	mass moment of inertia	lb-in./sec^2	$\text{kg} \cdot \text{m}^2$
$k, \underset{\sim}{k}$	spring constant	lb/in.	N/m
k_i	spring constant of ith spring	lb/in.	N/m
k_t	torsional spring constant	lb-in/rad	N-m/rad
k_{ij}	stiffness coefficient	lb/in.	N/m
$[k]$	stiffness matrix	lb/in.	N/m
l, l_i	length	in.	m
$m, \underset{\sim}{m}$	mass	$\text{lb-sec}^2/\text{in.}$	kg
m_i	ith mass	$\text{lb-sec}^2/\text{in.}$	kg
m_{ij}	mass coefficient	$\text{lb-sec}^2/\text{in.}$	kg
$[m]$	mass matrix	$\text{lb-sec}^2/\text{in.}$	kg
M	mass	$\text{lb-sec}^2/\text{in.}$	kg
M	bending moment	lb-in.	$\text{N} \cdot \text{m}$
$M_t, M_{t1}, M_{t2}, \ldots$	torque	lb-in.	$\text{N} \cdot \text{m}$
M_{t0}	amplitude of $M_t(t)$	lb-in.	$\text{N} \cdot \text{m}$
n	an integer		
n	number of degrees of freedom		
N	normal force	lb	N
N	total number of time steps		
p	pressure	lb/in^2	N/m^2
$p(x)$	probability density function of x		
$P(x)$	probability distribution function of x		
P	force, tension	lb	N
q_j	jth generalized coordinate		
$\vec{q}$	vector of generalized displacements		

LIST OF SYMBOLS (*continued*)

Symbol	Meaning	English Units	SI Units
$\dot{\vec{q}}$	vector of generalized velocities		
Q_j	jth generalized force		
r	frequency ratio $= \omega/\omega_n$		
$\vec{r}$	radius vector	in.	m
Re( )	real part of ( )		
$R(\tau)$	autocorrelation function		
R	electrical resistance	ohm	ohm
R	Rayleigh's dissipation function	lb-in/sec	N · m/s
R	Rayleigh's quotient	$1/\text{sec}^2$	$1/s^2$
s	exponential coefficient, root of equation		
S_a, S_d, S_v	acceleration, displacement, velocity spectrum		
$S_x(\omega)$	spectrum of x		
t	time	sec	s
t_i	ith time station	sec	s
T	torque	lb-in	N-m
T	kinetic energy	in.-lb	J
T_i	kinetic energy of ith mass	in.-lb	J
T_r	transmissibility ratio		
u_{ij}	an element of matrix $[U]$		
U, U_i	axial displacement	in.	m
U	potential energy	in.-lb	J
$\vec{U}$	unbalanced weight	lb	N
$[U]$	upper triangular matrix		
v, v_0	linear velocity	in./sec	m/s
V	shear force	lb	N
V	potential energy	in.-lb	J
V_i	potential energy of ith spring	in.-lb	J
w, w_1, w_2, ω_i	transverse deflections	in.	m
w_0	value of w at $t = 0$	in.	m
$\dot{w}_0$	value of $\dot{w}$ at $t = 0$	in./sec	m/s
w_n	nth mode of vibration		
W	weight of a mass	lb	N
W	total energy	in.-lb	J

Symbol	Meaning	English Units	SI Units
W	transverse deflection	in.	m
W_i	value of W at $t = t_i$	in.	m
$W(x)$	a function of x		
x, y, z	cartesian coordinates, displacements	in.	m
$x_0, x(0)$	value of x at $t = 0$	in.	m
$\dot{x}_0, \dot{x}(0)$	value of $\dot{x}$ at $t = 0$	in./sec	m/s
x_j	displacement of jth mass	in.	m
x_j	value of x at $t = t_j$	in.	m
$\dot{x}_j$	value of $\dot{x}$ at $t = t_j$	in./sec	m/s
x_h	homogeneous part of $x(t)$	in.	m
x_p	particular part of $x(t)$	in.	m
$\vec{x}$	vector of displacements	in.	m
$\vec{x}_i$	value of $\vec{x}$ at $t = t_i$	in.	m
$\dot{\vec{x}}_i$	value of $\dot{\vec{x}}$ at $t = t_i$	in./sec	m/s
$\ddot{\vec{x}}_i$	value of $\ddot{\vec{x}}$ at $t = t_i$	in./sec^2	m/s^2
$\vec{x}^{(i)}(t)$	ith mode		
X	amplitude of $x(t)$	in.	m
X_j	amplitude of $x_j(t)$	in.	m
$\vec{X}^{(i)}$	ith modal vector	in.	m
$X_i^{(j)}$	ith component of jth mode	in.	m
$[X]$	modal matrix	in.	m
$\vec{X}_r$	rth approximation to a mode shape		
y	base displacement	in.	m
Y	amplitude of $y(t)$	in.	m
z	relative displacement, $x - y$	in.	m
Z	amplitude of $z(t)$	in.	m
$Z(i\omega)$	mechanical impedance	lb/in.	N/m
α	angle, constant		
β	angle, constant		
β	hysteresis damping constant		
γ	specific weight	lb/in^3	N/m^3
δ	logarithmic decrement		
$\delta_1, \delta_2, \dots$	deflections	in.	m
δ_{st}	static deflection	in.	m
δ_{ij}	Kronecker delta		

LIST OF SYMBOLS (*continued*)

Symbol	Meaning	English Units	SI Units
Δ	determinant		
ΔF	increment in F	lb	N
Δx	increment in x	in.	m
Δt	increment in time t	sec	s
ΔW	energy dissipated in a cycle	in.-lb	J
ε	a small quantity		
ε	strain		
ζ	damping ratio		
θ	constant, angular displacement		
θ_i	ith angular displacement	rad	rad
θ_0	value of θ at $t = 0$	rad	rad
$\dot{\theta}_0$	value of $\dot{\theta}$ at $t = 0$	rad/sec	rad/s
Θ	amplitude of $\theta(t)$	rad	rad
Θ_i	amplitude of $\theta_i(t)$	rad	rad
λ	eigenvalue $= 1/\omega^2$	sec^2	s^2
$[\lambda]$	transformation matrix		
μ	viscosity of a fluid	lb-sec/in^2	kg/m $\cdot$ s
μ	coefficient of friction		
μ_x	expected value of x		
ρ	mass density	lb-sec^2/in^4	kg/m^3
η	loss factor		
σ_x	standard deviation of x		
σ	stress	lb/in^2	N/m^2
τ	period of oscillation, time	sec	s
τ	shear stress	lb/in^2	N/m^2
ϕ	angle, phase angle	rad	rad
ϕ_i	phase angle in ith mode	rad	rad
ω	frequency of oscillation	rad/sec	rad/s
ω_i	ith natural frequency	rad/sec	rad/s
ω_n	natural frequency	rad/sec	rad/s
ω_d	frequency of damped vibration	rad/sec	rad/s

SUBSCRIPTS

cri	critical value		
eq	equivalent value		
i	ith value		

Symbol	Meaning	English Units	SI Units
L	left plane		
max	maximum value		
n	corresponding to natural frequency		
R	right plane		
0	specific or reference value		
t	torsional		

OPERATIONS

Symbol	Meaning	English Units	SI Units
$(\dot{\ })$	$\dfrac{d(\ )}{dt}$		
$(\ddot{\ })$	$\dfrac{d^2(\ )}{dt^2}$		
$(\vec{\ })$	column vector ( )		
[ ]	matrix		
$[\ ]^{-1}$	inverse of [ ]		
$[\ ]^{T}$	transpose of [ ]		
$\Delta(\ )$	increment in ( )		
$\mathscr{L}(\ )$	Laplace transform of ( )		
$\mathscr{L}^{-1}(\ )$	inverse Laplace transform of ( )		

CHAPTER 1

Fundamentals of Vibration

Galileo Galilei (1564 – 1642), an Italian astronomer, philosopher, and professor of mathematics at the Universities of Pisa and Padua, in 1609 became the first man to point a telescope to the sky. He wrote the first treatise on modern dynamics in 1590. His works on the oscillations of a simple pendulum and the vibration of strings are of fundamental significance in the theory of vibrations. (Courtesy of the Granger Collection)

1.1 PRELIMINARY REMARKS

This chapter introduces the subject of vibrations in a relatively simple manner. The chapter begins with a brief history of the subject and continues with an examination of its importance. The various steps involved in vibration analysis of an engineering system are outlined, and essential definitions and concepts of vibration are introduced. There follows a presentation of the concept of harmonic analysis, which can be used for the analysis of general periodic motions. No attempt at exhaustive treatment is made in Chapter 1; subsequent chapters will develop many of the ideas in more detail.

1.2 BRIEF HISTORY OF VIBRATION

People became interested in vibration when the first musical instruments, probably whistles or drums, were discovered. Since then, people have applied ingenuity and critical investigation to study the phenomenon of vibration. Galileo discovered the relationship between the length of a pendulum and its frequency and observed the resonance of two bodies that were connected by some energy transfer medium and tuned to the same natural frequency. Further, he observed the interrelationships of the density, tension, length, and frequency of a vibrating string [1.1]. Although it had long been understood that sound was related to the vibration of a mechanical system, it was not clear that pitch is determined by the frequency of vibration until Galileo found the result. At about the same time as Galileo, Hooke showed the relationship between frequency and pitch.

Among mathematicians, Taylor, Bernoulli, D'Alembert, Euler, Lagrange, and Fourier made valuable contributions to the development of vibration theory. Wallis and Sauveur observed, independently, the phenomenon of mode shapes (with stationary points, called nodes) in vibrating strings. They also established that the frequency of the second mode is twice that of the first and the frequency of the third mode three times that of the first. Sauveur is credited with coining the term *fundamental* for the lowest frequency and *harmonics* for the others. Bernoulli first proposed the principle of linear superposition of harmonics: Any general configuration of free vibration is made up of the configurations of individual harmonics, acting independently in varying strengths [1.2].

After the enunciation of Hooke's law of elasticity in 1676, Euler (1744) and Bernoulli (1751) derived the differential equation governing the lateral vibration of prismatic bars and investigated its solution for the case of small deflections. In 1784, Coulomb did both theoretical and experimental studies of the torsional oscillations of a metal cylinder suspended by a wire.

There is an interesting story related to the development of the theory of vibration of plates [1.3]. In 1802, Chladni developed the method of placing sand on a vibrating plate to find its mode shapes and observed the beauty and intricacy of the modal patterns of the vibrating plates. In 1809, the French Academy invited Chladni to give a demonstration of his experiments. Napoleon Bonaparte, who attended the meeting, was very impressed and presented a sum of 3000 francs to the Academy, to be awarded to the first person to give a satisfactory mathematical theory of the vibration of plates. By the closing date of the competition in October, 1811, only one candidate, Sophie Germain, had entered the contest. But Lagrange, who was one of the judges, noticed an error in the derivation of her differential equation of motion. The Academy opened the competition again, with a new closing date of October, 1813. Sophie Germain again entered the contest, presenting the correct form of the differential equation. However, the Academy did not award the prize to her because the judges wanted physical justification of the assumptions made in her derivation. The competition was opened once more. In her third attempt, Sophie Germain was finally awarded the prize in 1816, although the judges were not completely satisfied with her theory. In fact, it was later found that her differential equation was correct but that the boundary conditions were erroneous. The correct boundary conditions for the vibration of plates were given in 1850 by Kirchhoff.

After this, vibration studies were done on a number of practical mechanical and structural systems. In 1877, Lord Rayleigh published his book on the theory of sound [1.4]; it is considered a classic on the subject of vibrations even today. Notable among the many contributions of Rayleigh is the method of finding the fundamental frequency of vibration of a conservative system by making use of the principle of conservation of energy—now known as Rayleigh's method [1.5]. In 1902, Frahm investigated the importance of torsional vibration study in the design of propeller shafts of steamships. The dynamic vibration absorber, which involves the addition of a secondary spring-mass system to eliminate the vibrations of a main system, was also proposed by Frahm in 1909. Among the modern contributors to the theory of vibrations, the names of Stodola, Timoshenko, and Mindlin are notable. Stodola's method of analyzing vibrating beams is also applicable to turbine blades. The works of Timoshenko and Mindlin resulted in improved theories of vibration of beams and plates.

It has long been recognized that many basic problems of mechanics, including those of vibrations, are nonlinear. Although the linear treatments commonly adopted are quite satisfactory for most purposes, they are not adequate in all cases. In nonlinear systems, there often occur phenomena that are theoretically impossible in linear systems. The mathematical theory of nonlinear vibrations began to develop in the works of Poincaré and Lyapunov at the end of the last century. After 1920, studies undertaken by Duffing and van der Pol brought the first definite solutions into the theory of nonlinear vibrations and drew attention to its importance in engineering. In the last 20 years, authors like Minorsky and Stoker have endeavored to collect the main results concerning nonlinear vibrations in the form of monographs [1.6, 1.7].

Random characteristics are present in diverse phenomena such as earthquakes, winds, transportation of goods on wheeled vehicles, and rocket and jet engine noise. It became necessary to devise concepts and methods of vibration analysis for these random effects. Although Einstein considered Brownian movement, a particular type of random vibration, as long ago as 1905, no applications were investigated until 1930. The introduction of the correlation function by Taylor in 1920 and of the spectral density by Wiener and Khinchin in the early 1930s opened new prospects for progress in the theory of random vibrations. Papers by Lin and Rice, published between 1943 and 1945, paved the way for the application of random vibrations to practical engineering problems. The monographs of Crandall and Mark, and Robson systematized the existing knowledge in the theory of random vibrations [1.8, 1.9].

Until about 25 years ago, vibration studies, even those dealing with complex engineering systems, were done by using gross models, with only a few degrees of freedom. However, the advent of high-speed digital computers in the 1950s made it possible to treat moderately complex systems and to generate approximate solutions in semi-closed form, relying on classical solution methods but using numerical evaluation of certain terms that cannot be expressed in closed form. The simultaneous development of the finite element method enabled engineers to use digital computers to conduct numerically detailed vibration analysis of complex mechanical, vehicular, and structural systems displaying thousands of degrees of freedom [1.10, 1.11]. Figure 1.1 shows the finite element idealization of the body of a bus [1.12].

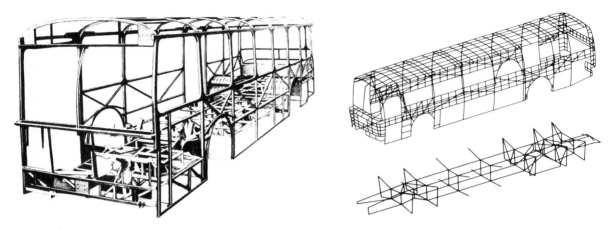

Figure 1.1   Finite element idealization of the body of a bus [1.12]. *(Reprinted with permission © 1974 Society of Automotive Engineers, Inc.)*

1.3 IMPORTANCE OF THE STUDY OF VIBRATION

Most human activities involve vibration in one form or other. For example, we hear because our eardrums vibrate and see because light waves undergo vibration. Breathing is associated with the vibration of lungs and walking involves (periodic) oscillatory motion of legs and hands. We speak due to the oscillatory motion of larynges (tongue) [1.13]. Early scholars in the field of vibration concentrated their efforts on understanding the natural phenomena and developing mathematical theories to describe the vibration of physical systems. In recent times, many investigations have been motivated by the engineering applications of vibration, such as the design of machines, foundations, structures, engines, turbines, and control systems.

Most prime movers have vibrational problems due to the inherent unbalance in the engines. The unbalance may be due to faulty design or poor manufacture. Imbalance in diesel engines, for example, can cause ground waves sufficiently powerful to create a nuisance in urban areas. The wheels of some locomotives can rise more than a centimeter off the track at high speeds due to unbalance. In turbines, vibrations cause spectacular mechanical failures. Engineers have not yet been able to prevent the failures that result from blade and disk vibrations in turbines. Naturally, the structures designed to support heavy centrifugal machines, like motors and turbines, or reciprocating machines, like steam and gas engines and reciprocating pumps, are also subjected to vibration. In all these situations, the structure or machine component subjected to vibration can fail because of material fatigue resulting from the cyclic variation of the induced stress. Furthermore, the vibration causes more rapid wear of machine parts such as bearings and gears and

Figure 1.2 Tacoma Narrows bridge during wind-induced vibration. The bridge opened on 1 July 1940 and collapsed on 7 November 1940. *(Farquharson photo, Historical Photography, Collection, University of Washington Libraries.)*

also creates excessive noise. In machines, vibration causes fasteners such as nuts to become loose. In metal cutting processes, vibration can cause chatter, which leads to a poor surface finish.

Whenever the natural frequency of vibration of a machine or structure coincides with the frequency of the external excitation, there occurs a phenomenon known as *resonance*, which leads to excessive deflections and failure. The literature is full of accounts of system failures brought about by resonance and excessive vibration of components and systems (see Fig. 1.2). Because of the devastating effects that vibrations can have on machines and structures, vibration testing [1.14] has become a standard procedure in the design and development of most engineering systems (see Fig. 1.3).

In many engineering systems, a human being acts as an integral part of the system. The transmission of vibration to human beings results in discomfort and loss of efficiency. Vibration of instrument panels can cause their malfunction or difficulty in reading the meters [1.15]. Thus one of the important purposes of vibration study is to reduce vibration through proper design of machines and their mountings. In this connection, the mechanical engineer tries to design the engine or machine so as to minimize unbalance, while the structural engineer tries to design the supporting structure so as to ensure that the effect of the imbalance will not be harmful [1.16].

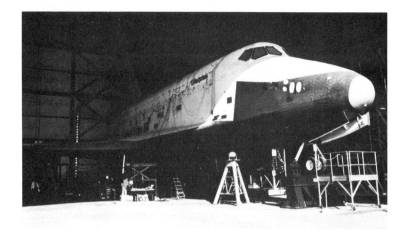

Figure 1.3   Vibration testing of the space shuttle Enterprise. *(Courtesy of NASA.)*

In spite of its detrimental effects, vibration can be utilized profitably in several industrial applications. In fact, the applications of vibratory equipment have increased considerably in recent years [1.17]. For example, vibration is put to work in vibratory conveyors, hoppers, sieves, washing machines and compactors. Vibration is also used in pile driving, vibratory testing of materials, vibratory finishing processes, and electronic circuits to filter out the unwanted frequencies (see Fig. 1.4). Vibration has been found to improve the efficiency of certain machining, casting, forging, and welding processes. It is employed to simulate earthquakes for geological research and also to conduct studies in the design of nuclear reactors.

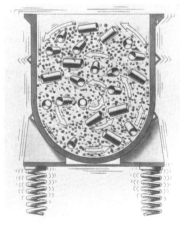

Figure 1.4   Vibratory finishing process. *(Reprinted courtesy of the Society of Manufacturing Engineers, © 1964 The Tool and Manufacturing Engineer.)*

1.4 BASIC CONCEPTS OF VIBRATION

1.4.1
Vibration

Any motion that repeats itself after an interval of time is called *vibration* or *oscillation*. The swinging of a pendulum and the motion of a plucked string are typical examples of vibration. The theory of vibration deals with the study of oscillatory motions of bodies and the forces associated with them.

1.4.2
Elementary Parts
of Vibrating
Systems

A vibratory system, in general, includes a means for storing potential energy (spring or elasticity), a means for storing kinetic energy (mass or inertia), and a means by which energy is gradually lost (damper).

The vibration of a system involves the transfer of its potential energy to kinetic energy and kinetic energy to potential energy, alternately. If the system is damped, some energy is dissipated in each cycle of vibration and must be replaced by an external source if a state of steady vibration is to be maintained.

As an example, consider the vibration of the simple pendulum shown in Fig. 1.5. Let the bob of mass m be released after giving it an angular displacement θ. At position 1 the velocity of the bob and hence its kinetic energy is zero. But it has a potential energy of magnitude $mgl(1 - \cos \theta)$ with respect to the datum position 2. Since the gravitational force mg induces a torque $mgl \sin \theta$ about the point O, the bob starts swinging to the left from position 1. This gives the bob certain angular acceleration in the clockwise direction, and by the time it reaches position 2, all of its potential energy will be converted into kinetic energy. Hence the bob will not stop in position 2, but will continue to swing to position 3. However, as it passes the mean position 2, a counterclockwise torque starts acting on the bob due to gravity and causes the bob to decelerate. The velocity of the bob reduces to zero at the left extreme position. By this time, all the kinetic energy of the bob will be converted to potential energy. Again due to the gravity torque, the bob continues to attain a counterclockwise velocity. Hence the bob starts swinging back with progressively

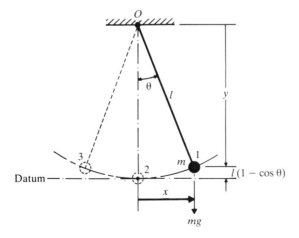

Figure 1.5 A simple pendulum.

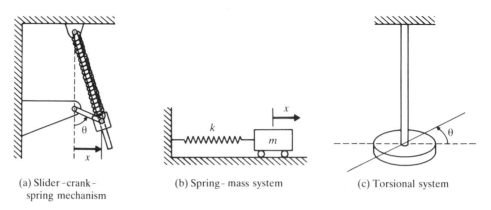

(a) Slider-crank-
   spring mechanism

(b) Spring-mass system

(c) Torsional system

Figure 1.6   Single degree of freedom systems.

increasing velocity and passes the mean position again. This process keeps on repeating, and the pendulum will have oscillatory motion. However, in practice, the magnitude of oscillation (θ) gradually decreases and the pendulum ultimately stops due to the resistance (damping) offered by the surrounding medium (air). This means that some energy is dissipated in each cycle of vibration due to damping by the air.

**1.4.3
Degree
of Freedom**

The minimum number of independent coordinates required to determine completely the positions of all parts of a system at any instant of time defines the degree of freedom of the system. The simple pendulum shown in Fig. 1.5, as well as each of the systems shown in Fig. 1.6, represents a single degree of freedom system. For example, the motion of the simple pendulum (Fig. 1.5) can be stated either in terms of the angle θ or in terms of the cartesian coordinates x and y. If the coordinates x and y are used to describe the motion, it must be recognized that these coordinates are not independent. They are related to each other through the relation $x^2 + y^2 = l^2$, where l is the constant length of the pendulum. Thus any one coordinate can describe the motion of the pendulum. In this example, we find that the choice of θ as the independent coordinate will be more convenient than the choice of x or y. For the slider shown in Fig. 1.6(a), either the angular coordinate θ or the coordinate x can be used to describe the motion. In Fig. 1.6(b), the linear coordinate x can be used to specify the motion. For the torsional system (long bar with a heavy disk at the end) shown in Fig. 1.6(c), the angular coordinate θ can be used to describe the motion.

Some examples of two and three degree of freedom systems are shown in Figs. 1.7 and 1.8, respectively. Figure 1.7(a) shows a two mass–two spring system that is described by the two linear coordinates x_1 and x_2. Figure 1.7(b) denotes a two rotor system whose motion can be specified in terms of θ_1 and θ_2. The motion of the system shown in Fig. 1.7(c) can be described completely either by X and θ or by x, y, and X. In the latter case, x and y are constrained as $x^2 + y^2 = l^2$ where l is a constant.

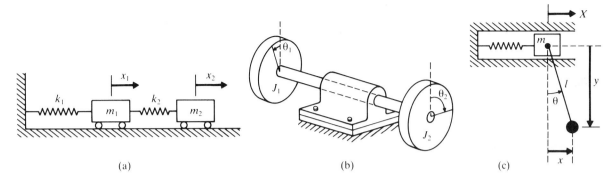

Figure 1.7 Two degree of freedom systems.

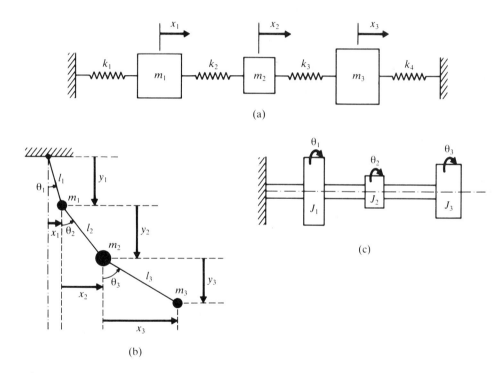

Figure 1.8 Three degree of freedom systems.

For the systems shown in Figs. 1.8(a) and 1.8(c), the coordinates x_i ($i = 1, 2, 3$) and θ_i ($i = 1, 2, 3$) can be used, respectively, to describe the motion. In the case of the system shown in Fig. 1.8(b), θ_i ($i = 1, 2, 3$) specifies the positions of the masses m_i ($i = 1, 2, 3$). An alternate method of describing this system is in terms of x_i and y_i ($i = 1, 2, 3$); but in this case the constraints $x_i^2 + y_i^2 = l_i^2$ ($i = 1, 2, 3$) have to be considered.

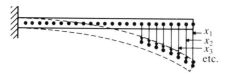

Figure 1.9 A cantilever beam (an infinite number of degrees of freedom system).

The coordinates necessary to describe the motion of a system constitute a set of *generalized coordinates*. The generalized coordinates are usually denoted as $q_1, q_2, \ldots$ and may represent cartesian and/or noncartesian coordinates.

1.4.4 Discrete and Continuous Systems

A large number of practical systems can be described using a finite number of degrees of freedom, such as the simple systems shown in Figs. 1.5 to 1.8. Some systems, especially those involving continuous elastic members, have an infinite number of degrees of freedom. As a simple example, consider the cantilever beam shown in Fig. 1.9. Since the beam has an infinite number of mass points, we need an infinite number of coordinates to specify its deflected configuration. The infinite number of coordinates defines its elastic deflection curve. Thus the cantilever beam has an infinite number of degrees of freedom. Most structural and machine systems have deformable (elastic) members and therefore have an infinite number of degrees of freedom.

Systems with a finite number of degrees of freedom are called *discrete* or *lumped parameter* systems, and those with an infinite number of degrees of freedom are called *continuous* or *distributed* systems.

Most of the time, continuous systems are approximated as discrete systems, and solutions are obtained in a simpler manner. Although treatment of a system as continuous gives exact results, the analysis methods available for dealing with continuous systems are limited to a narrow selection of problems, such as uniform beams, slender rods, and thin plates. Hence most of the practical systems are studied by treating them as finite lumped masses, springs, and dampers. In general, more accurate results are obtained by increasing the number of masses, springs, and dampers—that is, increasing the number of degrees of freedom.

1.5 CLASSIFICATION OF VIBRATION

Vibration can be classified in several ways. Some of the important classifications are as follows.

1.5.1 Free and Forced Vibration

Free Vibration. If a system, after an initial disturbance, is left to vibrate on its own, the ensuing vibration is known as free vibration. No external force acts on the system. The oscillation of a simple pendulum is an example of free vibration.

Forced Vibration. If a system is subjected to an external force (often, a repeating type of force), the resulting vibration is known as *forced vibration*, The oscillation that arises in machines such as diesel engines is an example of forced vibration.

If the frequency of the external force coincides with one of the natural frequencies of the system, a condition known as *resonance* occurs, and the system undergoes dangerously large oscillations. Failures of such structures as buildings, bridges, turbines, and airplane wings have been associated with the occurrence of resonance.

1.5.2 Undamped and Damped Vibration

If no energy is lost or dissipated in friction or other resistance during oscillation, the vibration is known as *undamped vibration*. If any energy is lost in this way, on the other hand, it is called *damped vibration*. In many physical systems, the amount of damping is so small that it can be disregarded for most engineering purposes. However, consideration of damping becomes extremely important in analyzing vibratory systems near resonance.

1.5.3 Linear and Nonlinear Vibration

If all the basic components of a vibratory system—the spring, the mass, and the damper—behave linearly, the resulting vibration is known as *linear vibration*. On the other hand, if any of the basic components behave nonlinearly, the vibration is called *nonlinear vibration*. The differential equations that govern the behavior of linear and nonlinear vibratory systems are linear and nonlinear, respectively. If the vibration is linear, the principle of superposition holds, and the mathematical techniques of analysis are well developed. For nonlinear vibration, the superposition principle is not valid, and techniques of analysis are less well known. Since all vibratory systems tend to behave nonlinearly with increasing amplitude of oscillation, a knowledge of nonlinear vibration is desirable in dealing with practical vibratory systems.

1.5.4 Deterministic and Random Vibration

If the value of the excitation (force or motion) acting on a vibratory system is known at any given time, the excitation is called *deterministic*. The resulting vibration is known as *deterministic vibration*.

In some cases, the excitation is *nondeterministic* or *random*; the value of the excitation at a given time cannot be predicted. In these cases, a large collection of records of the excitation may exhibit some statistical regularity. It is possible to estimate averages such as the mean and mean square values of the excitation.

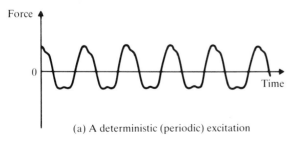

(a) A deterministic (periodic) excitation

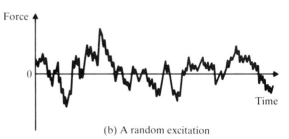

(b) A random excitation

Figure 1.10

Examples of random excitations are wind velocity, road roughness, and ground motion during earthquakes. If the excitation is random, the resulting vibration is called *random vibration*. In the case of random vibration, the vibratory response of the system is also random; it can be described only in terms of statistical quantities. Figure 1.10 shows examples of deterministic and random excitations.

1.6 VIBRATION ANALYSIS PROCEDURE

A vibratory system is a dynamic system for which the variables such as the excitations (inputs) and responses (outputs) are time-dependent. The response of a vibrating system generally depends on the initial conditions as well as the external excitations. The analysis of a vibrating system usually involves mathematical modeling, derivation of the governing equations, solution of the equations, and interpretation of the results.

Step 1: Mathematical Modeling. The purpose of mathematical modeling is to represent all the important features of the system for the purpose of deriving the mathematical (or analytical) equations governing the behavior of the system. The mathematical model should include enough details to be able to describe the system in terms of equations without making it too complex. The mathematical model may be linear or nonlinear depending on the behavior of the components of the system. Linear models permit quick solutions and are simple to handle; however, nonlinear models sometimes reveal certain characteristics of the system that cannot be predicted using linear models. Thus a great deal of engineering judgment is needed to come up with a suitable mathematical model of a vibrating system.

Sometimes the mathematical model is gradually improved to obtain more accurate results. In this approach, first a very crude or elementary model is used to get a quick insight into the overall behavior of the system. Subsequently, the model is refined by including more components and/or details so that the behavior of the system can be observed in more detail. To illustrate the procedure of refinement used in mathematical modeling, consider the forging hammer shown in Fig. 1.11(a). The forging hammer consists of a frame, a falling weight known as the tup, an anvil, and a foundation block. The anvil is a massive steel block on which material is forged into desired shape by the repeated blows of the tup. The anvil is usually mounted on an elastic pad to reduce the transmission of vibration to the foundation block and the frame [1.18]. For a first approximation, the frame, anvil, elastic pad, foundation block, and the soil are modeled as a single degree of freedom system as shown in Fig. 1.11(b). For a refined approximation, the weights of the frame and anvil and the foundation block are represented separately with a two degree of freedom model as shown in Fig. 1.11(c). Further refinement of the model can be made by considering eccentric impacts of the tup, which cause each of the masses shown in Fig. 1.11(c) to have both vertical and rocking (rotation) motions in the plane of the paper.

Step 2: Derivation of Governing Equations. Once the mathematical model is available, we use the principles of dynamics and derive the equations that describe the

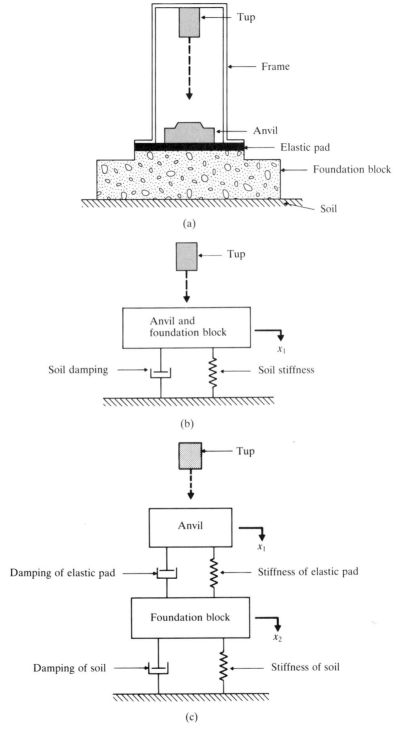

Figure 1.11   Modeling of a forging hammer.

vibration of the system. The equations are usually in the form of a set of ordinary differential equations for a discrete system and partial differential equations for a continuous system. The equations may be linear or nonlinear depending on the behavior of the components of the system. Several approaches are commonly used to derive the governing equations. Among them are Newton's second law of motion, d'Alembert's principle, and the principle of conservation of energy.

Step 3: Solution of the Governing Equations. The equations of motion must be solved to find the response of the vibrating system. Depending on the nature of the problem, we can use one of the following techniques for finding the solution: standard methods of solving differential equations, Laplace transformation methods, matrix methods,* and numerical methods. If the governing equations are nonlinear, they can seldom be solved in closed form. Further, the solution of partial differential equations is far more involved than that of ordinary differential equations. Numerical methods, using computers, can be used to solve the equations. However, it will be difficult to draw general conclusions about the behavior of the system using computer results.

Step 4: Interpretation of the Results. The solution of the governing equations gives the displacements, velocities, and accelerations of the various masses of the system. These results must be interpreted with a clear view of the purpose of the analysis and the possible design implications of the results.

1.7 SPRING ELEMENTS

A linear spring is a type of mechanical link which is generally assumed to have negligible mass and damping. A force is developed in the spring whenever there is relative motion between the two ends of the spring. The spring force is proportional to the amount of deformation and is given by

$$F = kx \qquad (1.1)$$

where F is the spring force, x is the deformation (displacement of one end with respect to the other), and k is the *spring stiffness* or *spring constant*. If we plot a graph between F and x, the result is a straight line according to Eq. (1.1). The work done in deforming a spring is stored as strain or potential energy in the spring.

---

* The basic definitions and operations of matrix theory are given in Appendix A.

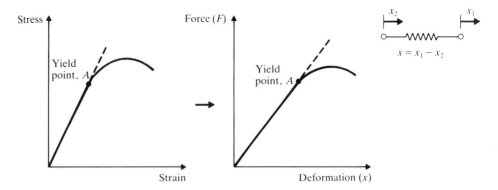

Figure 1.12 Nonlinearity beyond proportionality limit.

Actual springs are nonlinear and follow Eq. (1.1) only up to a certain deformation. Beyond a certain value of deformation (after point A in Fig. 1.12), the stress exceeds the yield point of the material and the force-deformation relation becomes nonlinear [1.20, 1.39]. In many practical applications we assume the deflections to be small and make use of the linear relation in Eq. (1.1). Even if the force-deflection relation of a spring is nonlinear, as shown in Fig. 1.13, we often approximate it as a linear one by using a linearization process [1.19, 1.20]. To illustrate the linearization process, let the static equilibrium load F acting on the spring cause a deflection of x^*. If an incremental force ΔF is added to F, the spring deflects by an additional quantity Δx. The new spring force $F + \Delta F$ can be expressed using Taylor's series expansion about the static equilibrium position x^* as

$$F + \Delta F = F(x^* + \Delta x) = F(x^*) + \frac{dF}{dx}\bigg|_{x^*} (\Delta x) + \frac{1}{2!} \frac{d^2F}{dx^2}\bigg|_{x^*} (\Delta x)^2 + \cdots \quad (1.2)$$

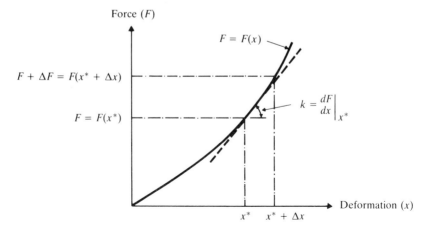

Figure 1.13 Linearization process.

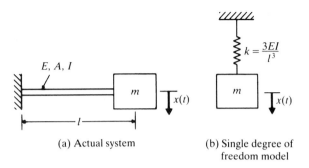

(a) Actual system   (b) Single degree of
                               freedom model

Figure 1.14   Cantilever with end mass.

For small values of Δx, the higher order derivative terms can be neglected to obtain

$$F + \Delta F = F(x^*) + \frac{dF}{dx}\bigg|_{x^*}(\Delta x) \tag{1.3}$$

Since $F = F(x^*)$, we can express ΔF as

$$\Delta F = k\Delta x \tag{1.4}$$

where k is the linearized spring constant at x^* given by

$$k = \frac{dF}{dx}\bigg|_{x^*}$$

We may use Eq. (1.4) for simplicity, but sometimes the error involved in the approximation may be very large.

Elastic elements like beams also behave as springs. For example, consider a cantilever beam with an end mass m, as shown in Fig. 1.14. We assume, for simplicity, that the mass of the beam is negligible in comparison with the mass m. From strength of materials [1.21], we know that the static deflection of the beam at the free end is given by

1ST Moment Area Theorem
$$\delta_{st} = \frac{Wl^3}{3EI} \tag{1.5}$$

where $W = mg$ is the weight of the mass m, E is Young's modulus and, I is the moment of inertia of the cross section of the beam. Hence the spring constant is

$$k = \frac{W}{\delta_{st}} = \frac{3EI}{l^3} \tag{1.6}$$

Similar results can be obtained for beams with different end conditions.

**1.7.1
Combination
of Springs**

In many practical applications, several linear springs are used in combination. These springs can be combined into a single equivalent spring as indicated below.

Case (i): Springs in Parallel. Let the springs be parallel as shown in Fig. 1.15(a). If W is the weight of mass m, we have for equilibrium

$$W = k_1\delta_{st} + k_2\delta_{st} \tag{1.7}$$

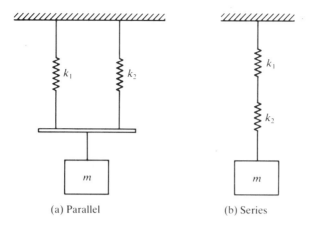

(a) Parallel   (b) Series

Figure 1.15 Combination of springs.

where δ_{st} is the static deflection of the mass m. If k_{eq} denotes the equivalent spring constant of the combination of the two springs, then for the same static deflection δ_{st}, we have

$$W = k_{eq}\delta_{st} \tag{1.8}$$

Equations (1.7) and (1.8) give

$$k_{eq} = k_1 + k_2 \tag{1.9}$$

In general, if we have n springs with spring constants $k_1, k_2, \ldots, k_n$ in parallel, then the equivalent spring constant k_{eq} can be obtained:

$$k_{eq} = k_1 + k_2 + \cdots + k_n \tag{1.10}$$

Case (ii): Springs in Series. Next we consider two springs connected in series, as shown in Fig. 1.15(b). Since both the springs are subjected to the same force W, we have for equilibrium

$$\left. \begin{array}{l} W = k_1\delta_1 \\ W = k_2\delta_2 \end{array} \right\} \tag{1.11}$$

where δ_1 and δ_2 are the elongations of springs 1 and 2, respectively. As the total elongation is equal to the static deflection δ_{st},

$$\delta_1 + \delta_2 = \delta_{st} \tag{1.12}$$

If k_{eq} denotes the equivalent spring constant, then for the same static deflection,

$$W = k_{eq}\delta_{st} \tag{1.13}$$

Equations (1.11) and (1.13) give

$$k_1\delta_1 = k_2\delta_2 = k_{eq}\delta_{st}$$

or

$$\delta_1 = \frac{k_{eq}\delta_{st}}{k_1} \quad \text{and} \quad \delta_2 = \frac{k_{eq}\delta_{st}}{k_2} \tag{1.14}$$

Substituting these values of δ_1 and δ_2 into Eq. (1.12), we obtain

$$\frac{k_{eq}\delta_{st}}{k_1} + \frac{k_{eq}\delta_{st}}{k_2} = \delta_{st}$$

that is,

$$\frac{1}{k_{eq}} = \frac{1}{k_1} + \frac{1}{k_2} \tag{1.15}$$

Equation (1.15) can be generalized to the case of n springs in series:

$$\frac{1}{k_{eq}} = \frac{1}{k_1} + \frac{1}{k_2} + \cdots + \frac{1}{k_n} \tag{1.16}$$

In certain applications, springs are connected to rigid components such as pulleys, levers, and gears. In such cases, an equivalent spring constant can be found using energy equivalence, as illustrated in Example 1.2.

EXAMPLE 1.1   Equivalent *k* of Hoisting Drum

A hoisting drum, carrying a steel wire rope, is mounted at the end of a cantilever beam as shown in Fig. 1.16(a). Determine the equivalent spring constant of the system when the suspended length of the wire rope is l. Assume the net cross-sectional diameter of the wire rope as d and the Young's modulus of the beam and the wire rope as E.

Given: Dimensions of the cantilever beam: length = b, width = a, and thickness = t. Young's modulus of the beam = E. Wire rope: length = l, diameter = d, and Young's modulus = E.

Find: Equivalent spring constant of the system.

Approach: Series springs.

Solution. The spring constant of the cantilever beam is given by

$$k_b = \frac{3EI}{b^3} = \frac{3E}{b^3}\left(\frac{1}{12}at^3\right) = \frac{Eat^3}{4b^3} \tag{E.1}$$

The stiffness of the wire rope subjected to axial loading is

$$k_r = \frac{AE}{l} = \frac{\pi d^2 E}{4l} \tag{E.2}$$

The cantilever beam and the wire rope can be considered as series springs (Fig. 1.16b) whose equivalent spring constant k_{eq} is given by

$$\frac{1}{k_{eq}} = \frac{1}{k_b} + \frac{1}{k_r} = \frac{4b^3}{Eat^3} + \frac{4l}{\pi d^2 E}$$

or

$$k_{eq} = \frac{E}{4}\left(\frac{\pi a t^3 d^2}{\pi d^2 b^3 + l a t^3}\right) \tag{E.3}$$

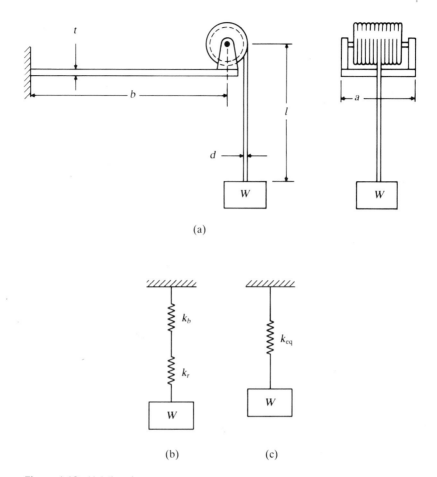

(a)

(b)   (c)

Figure 1.16 Hoisting drum.

EXAMPLE 1.2   **Equivalent _k_ of a Crane**

The boom _AB_ of the crane shown in Fig. 1.17(a) is a uniform steel bar of length 10 m and area of cross-section 2500 mm². A 1000 kg mass is suspended while the crane is stationary. The cable _CDEBF_ is made of steel and has a cross-sectional area of 100 mm². Neglecting the effect of the cable _CDEB_, find the equivalent spring constant of the system in the vertical direction.

Given: Steel boom: length = 10 m, cross-sectional area = 2500 mm², and material = steel. Cable _FB_: material = steel and cross-sectional area = 100 mm². Base: _FA_ = 3 m.

Find: Equivalent spring constant of the system.

Approach: Equivalence of potential energy.

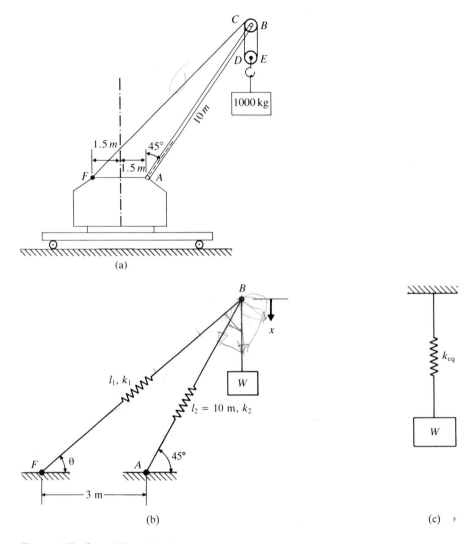

Figure 1.17 Crane lifting a load.

Solution. A vertical displacement x of point B will cause the spring k_2 (boom) to deform by an amount $x_2 = x \cos 45°$ and the spring k_1 (cable) to deform by an amount $x_1 = x \cos(90° - \theta)$. The length of the cable FB, l_1 is given by (Fig. 1.17b)

$$l_1^2 = 3^2 + 10^2 - 2(3)(10)\cos 135° = 151.426, \qquad l_1 = 12.3055 \ m$$

The angle θ satisfies the relation

$$l_1^2 + 3^2 - 2(l_1)(3)\cos \theta = 10^2, \qquad \cos \theta = 0.8184, \qquad \theta = 35.0736°$$

The total potential energy (U) stored in the springs k_1 and k_2 is given by

$$U = \tfrac{1}{2}k_1(x\cos 45°)^2 + \tfrac{1}{2}k_2[x\cos(90° - \theta)]^2 \tag{E.1}$$

where

$$k_1 = \frac{A_1 E_1}{l_1} = \frac{(100 \times 10^{-6})(207 \times 10^9)}{12.3055} = 1.6822 \times 10^6 \text{ N/m}$$

and

$$k_2 = \frac{A_2 E_2}{l_2} = \frac{(2500 \times 10^{-6})(207 \times 10^9)}{10} = 5.1750 \times 10^7 \text{ N/m}$$

Since the equivalent spring in the vertical direction undergoes a deformation x, the potential energy of the equivalent spring (U_{eq}) is given by

$$U_{eq} = \tfrac{1}{2}k_{eq}x^2 \tag{E.2}$$

By setting $U = U_{eq}$, we obtain the equivalent spring constant of the system as

$$k_{eq} = 26.4304 \times 10^6 \text{ N/m}$$

1.8 MASS OR INERTIA ELEMENTS

The mass or inertia element is assumed to be a rigid body; it can gain or lose kinetic energy whenever the velocity of the body changes. From Newton's second law of motion, the product of the mass and its acceleration is equal to the force applied to the mass. Work is equal to the force multiplied by the displacement in the direction of the force and the work done on a mass is stored in the form of kinetic energy of the mass.

In most cases, we must use a mathematical model to represent the actual vibrating system, and there are often several possible models. The purpose of the analysis often determines which mathematical model is appropriate. Once the model is chosen, the mass or inertia elements of the system can be easily identified. For example, consider the cantilever beam with a tip mass shown in Fig. 1.14(a). For a quick and reasonably accurate analysis, the mass and damping of the beam can be disregarded; the system can be modeled as a spring-mass system, as shown in Fig. 1.14(b). The tip mass m represents the mass element, and the elasticity of the beam denotes the stiffness of the spring. Next, consider a multistory building subjected to an earthquake. Assuming that the mass of the frame is negligible compared to the masses of the floors, the building can be modeled as a multidegree of freedom system, as shown in Fig. 1.18. The masses at the various floor levels represent the mass elements, and the elasticities of the vertical members denote the spring elements.

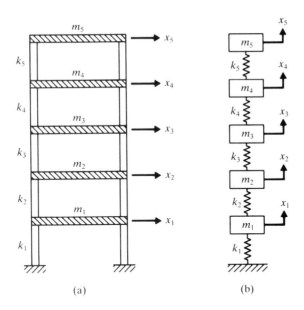

Figure 1.18 Idealization of a multistory building as a multidegree of freedom system.

**1.8.1
Combination
of Masses**

In many practical applications, several masses appear in combination. For a simple analysis, we can replace these masses by a single equivalent mass, as indicated below [1.22].

Case (i): Translational Masses Connected by a Rigid Bar. Let the masses be attached to a rigid bar that is pivoted at one end, as shown in Fig. 1.19(a). The equivalent mass can be assumed to be located at any point along the bar. To be specific, we assume the location of the equivalent mass to be that of mass m_1. The

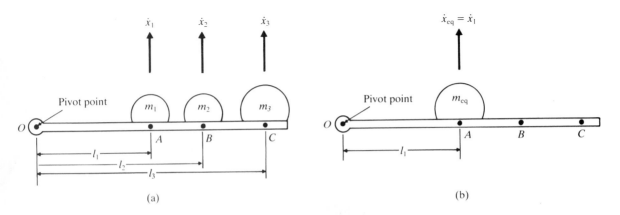

Figure 1.19 Translational masses connected by a rigid bar.

velocities of masses m_2 ($\dot{x}_2$) and m_3 ($\dot{x}_3$) can be expressed in terms of the velocity of mass m_1 ($\dot{x}_1$), by assuming small angular displacements for the bar, as

$$\dot{x}_2 = \frac{l_2}{l_1}\dot{x}_1, \qquad \dot{x}_3 = \frac{l_3}{l_1}\dot{x}_1 \tag{1.17}$$

and

$$\dot{x}_{eq} = \dot{x}_1 \tag{1.18}$$

By equating the kinetic energy of the three mass system to that of the equivalent mass system, we obtain

$$\frac{1}{2}m_1\dot{x}_1^2 + \frac{1}{2}m_2\dot{x}_2^2 + \frac{1}{2}m_3\dot{x}_3^2 = \frac{1}{2}m_{eq}\dot{x}_{eq}^2 \tag{1.19}$$

This equation gives, in view of Eqs. (1.17) and (1.18),

$$m_{eq} = m_1 + \left(\frac{l_2}{l_1}\right)^2 m_2 + \left(\frac{l_3}{l_1}\right)^2 m_3 \tag{1.20}$$

Case (ii): Translational and Rotational Masses Coupled Together. Let a mass m having a translational velocity $\dot{x}$ be coupled to another mass (of mass moment of inertia J_0) having a rotational velocity $\dot{\theta}$, as in the rack and pinion arrangement shown in Fig. 1.20. These two masses can be combined to obtain either 1) a single equivalent translational mass m_{eq} or 2) a single equivalent rotational mass J_{eq}, as shown below.

1. Equivalent translational mass. The kinetic energy of the two masses is given by

$$T = \frac{1}{2}m\dot{x}^2 + \frac{1}{2}J_0\dot{\theta}^2 \tag{1.21}$$

and the kinetic energy of the equivalent mass can be expressed as

$$T_{eq} = \frac{1}{2}m_{eq}\dot{x}_{eq}^2 \tag{1.22}$$

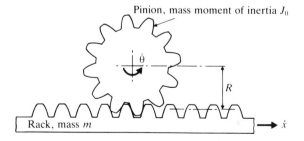

Pinion, mass moment of inertia J_0

$\dot{\theta}$

R

Rack, mass m

$\dot{x}$

Figure 1.20   Translational and rotational masses in a rack and pinion arrangement.

Since $\dot{x}_{eq} = \dot{x}$ and $\dot{\theta} = \dot{x}/R$, the equivalence of T and T_{eq} gives

$$\frac{1}{2} m_{eq} \dot{x}^2 = \frac{1}{2} m \dot{x}^2 + \frac{1}{2} J_0 \left(\frac{\dot{x}}{R} \right)^2$$

that is,

$$m_{eq} = m + \frac{J_0}{R^2} \tag{1.23}$$

2. *Equivalent rotational mass.* Here $\dot{\theta}_{eq} = \dot{\theta}$ and $\dot{x} = \dot{\theta}R$, and the equivalence of T and T_{eq} leads to

$$\frac{1}{2} J_{eq} \dot{\theta}^2 = \frac{1}{2} m (\dot{\theta}R)^2 + \frac{1}{2} J_0 \dot{\theta}^2$$

or

$$J_{eq} = J_0 + mR^2 \tag{1.24}$$

EXAMPLE 1.3   **Cam-Follower Mechanism**

A cam-follower mechanism (Fig. 1.21) is used to convert the rotary motion of a shaft into the oscillating or reciprocating motion of a valve. The follower system consists of a pushrod of mass m_p, a rocker arm of mass m_r, and mass moment of inertia J_r about its C.G., a valve of mass m_v, and a valve spring of negligible mass [1.23, 1.24, 1.38]. Find the equivalent mass

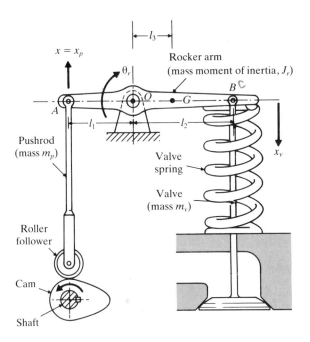

Figure 1.21   Cam-follower system.

(m_{eq}) of this cam-follower system by assuming the location of m_{eq} as (i) point A and (ii) point C.

Given: Mass of pushrod = m_p, mass of rocker arm = m_r, mass moment of inertia of rocker arm = J_r, and mass of valve = m_v. Linear displacement of pushrod = x_p.

Find: Equivalent mass of the cam-follower system (i) at point A, (ii) at point C.

Approach: Equivalence of kinetic energy.

Solution. Due to a vertical displacement x of the pushrod, the rocker arm rotates by an angle $\theta_r = x/l_1$ about the pivot point, the valve moves downward by $x_v = \theta_r l_2 = xl_2/l_1$ and the C.G. of the rocker arm moves downward by $x_r = \theta_r l_3 = xl_3/l_1$. The kinetic energy of the system (T) can be expressed as[†]

$$T = \frac{1}{2} m_p \dot{x}_p^2 + \frac{1}{2} m_v \dot{x}_v^2 + \frac{1}{2} J_r \dot{\theta}_r^2 + \frac{1}{2} m_r \dot{x}_r^2 \qquad \text{(E.1)}$$

where $\dot{x}_p$, $\dot{x}_r$, and $\dot{x}_v$ are the linear velocities of the pushrod, C.G. of the rocker arm and the valve, respectively, and $\dot{\theta}_r$ is the angular velocity of the rocker arm.

(i) If m_{eq} denotes the equivalent mass placed at point A, with $\dot{x}_{eq} = \dot{x}$, the kinetic energy of the equivalent mass system T_{eq} is given by

$$T_{eq} = \frac{1}{2} m_{eq} \dot{x}_{eq}^2 \qquad \text{(E.2)}$$

By equating T and T_{eq}, and noting that

$$\dot{x}_p = \dot{x}, \qquad \dot{x}_v = \frac{\dot{x}l_2}{l_1}, \qquad \dot{x}_r = \frac{\dot{x}l_3}{l_1}, \qquad \text{and} \qquad \dot{\theta}_r = \frac{\dot{x}}{l_1}$$

we obtain

$$m_{eq} = m_p + \frac{J_r}{l_1^2} + m_v \frac{l_2^2}{l_1^2} + m_r \frac{l_3^2}{l_1^2} \qquad \text{(E.3)}$$

(ii) Similarly, if the equivalent mass is located at point C, $\dot{x}_{eq} = \dot{x}_v$ and

$$T_{eq} = \frac{1}{2} m_{eq} \dot{x}_{eq}^2 = \frac{1}{2} m_{eq} \dot{x}_v^2 \qquad \text{(E.4)}$$

Equating (E.4) and (E.1) gives

$$m_{eq} = m_v + \frac{J_r}{l_2^2} + m_p \left(\frac{l_1}{l_2} \right)^2 + m_r \left(\frac{l_3}{l_2} \right)^2 \qquad \text{(E.5)}$$

1.9 DAMPING ELEMENTS

In many practical systems, the vibrational energy is gradually converted to heat or sound. Due to the reduction in the energy, the response, such as the displacement of the system gradually decreases. The mechanism by which the vibrational energy is gradually converted into heat or sound is known as damping. Although the amount

---

[†] If the valve spring has a mass m_s, then its equivalent mass will be $\frac{1}{3} m_s$ (see Example 2.5). Thus the kinetic energy of the valve spring will be $\frac{1}{2} (\frac{1}{3} m_s) \dot{x}_v^2$.

of energy converted into heat or sound is relatively small, the consideration of damping becomes important for an accurate prediction of the vibration response of a system. A damper is assumed to have neither mass nor elasticity, and damping force exists only if there is relative velocity between the two ends of the damper. It is difficult to determine the causes of damping in practical systems. Hence damping is modeled as one or more of the following types [1.25].

Viscous Damping. Viscous damping is the most commonly used damping mechanism in vibration analysis. When mechanical systems vibrate in a fluid medium such as air, gas, water, and oil, the resistance offered by the fluid to the moving body causes energy to be dissipated. In this case, the amount of dissipated energy depends on many factors, such as the size and shape of the vibrating body, the viscosity of the fluid, the frequency of vibration, and the velocity of the vibrating body. In viscous damping, the damping force is proportional to the velocity of the vibrating body. Typical examples of viscous damping include (1) fluid film between sliding surfaces, (2) fluid flow around a piston in a cylinder, (3) fluid flow through an orifice, and (4) fluid film around a journal in a bearing.

Coulomb or Dry Friction Damping. Here the damping force is constant in magnitude but opposite in direction to that of the motion of the vibrating body. It is caused due to friction between rubbing surfaces that are either dry or have insufficient lubrication.

Material or Solid or Hysteretic Damping. When materials are deformed, energy is absorbed and dissipated by the material [1.26]. The effect is due to friction between the internal planes, which slip or slide as the deformations take place. When a body having material damping is subjected to vibration, the stress-strain diagram shows a hysteresis loop as indicated in Fig. 1.22. The area of this loop denotes the energy lost per cycle due to damping.

1.9.1
Construction of Viscous Dampers

A viscous damper can be constructed using two parallel plates separated by a distance h, with a fluid of viscosity μ between the plates (see Fig. 1.23). Let one plate be fixed and the other plate be moved with a velocity v in its own plane. The fluid layers in contact with the moving plate move with a velocity v, while those in contact with the fixed plate do not move. The velocities of intermediate fluid layers are assumed to vary linearly between 0 and v as shown in Fig. 1.23. According to Newton's law of viscous flow, the shear stress (τ) developed in the fluid layer at a distance y from the fixed plate is given by

$$\tau = \mu \frac{du}{dy} \tag{1.25}$$

where $du/dy = v/h$ is the velocity gradient. The shear or resisting force (F) developed at the bottom surface of the moving plate is

$$F = \tau A = \frac{\mu A v}{h} = cv \tag{1.26}$$

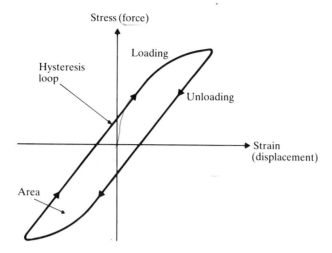

Figure 1.22. Hysteresis loop for elastic materials.

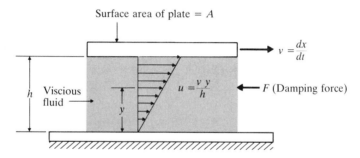

Figure 1.23 Parallel plates with a viscous fluid in between.

where A is the surface area of the moving plate and

$$c = \frac{\mu A}{h} \tag{1.27}$$

is called the damping constant.

If a damper is nonlinear, a linearization procedure is generally used about the operating velocity (v^*), as in the case of a nonlinear spring. The linearization process gives the equivalent damping constant as

$$c = \frac{dF}{dv}\Big|_{v^*}$$

1.9.2 Combination of Dampers

When dampers appear in combination, they can be replaced by an equivalent damper by adopting a procedure similar to the one described in Secs. 1.7 and 1.8 (see Problem 1.20).

EXAMPLE 1.4   **Piston-Cylinder Type Dashpot**

Develop an expression for the damping constant of the dashpot shown in Fig. 1.24(a).

Given: Diameter of cylinder = $D + 2d$, diameter of piston = D, velocity of piston = v_0, axial length of piston = l, and viscosity of fluid = μ.

Find: Damping constant of the dashpot.

Approach: Shear stress equation for viscous fluid flow. Rate of fluid flow equation.

Solution. As shown in Fig. 1.24(a), the dashpot consists of a piston of diameter D, and length l, moving with velocity v_0 in a cylinder filled with a liquid of viscosity μ [1.20, 1.27]. Let the clearance between the piston and the cylinder wall be d. At a distance y from the moving surface, let the velocity and shear stress be v and τ, and at a distance $(y + dy)$ let the velocity and shear stress be $(v - dv)$ and $(\tau + d\tau)$, respectively (see Fig. 1.24(b)). The negative sign for dv shows that the velocity decreases as we move toward the cylinder wall. The viscous force on this annular ring is equal to

$$F = \pi D l\, d\tau = \pi D l \frac{d\tau}{dy}\, dy \tag{E.1}$$

But the shear stress is given by

$$\tau = -\mu \frac{dv}{dy} \tag{E.2}$$

where the negative sign is consistent with a decreasing velocity gradient [1.28]. Using Eq.

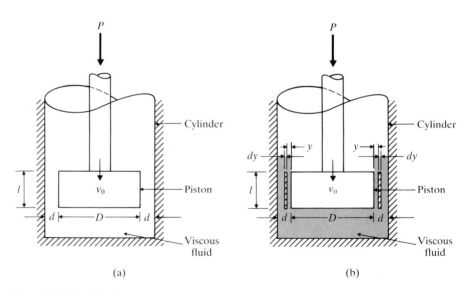

(a)   (b)

Figure 1.24   A dashpot.

(E.2) in Eq. (E.1), we obtain

$$F = -\pi Dl\, dy\, \mu \frac{d^2v}{dy^2} \tag{E.3}$$

The force on the piston will cause a pressure difference on the ends of the element, given by

$$p = \frac{P}{\left(\dfrac{\pi D^2}{4}\right)} = \frac{4P}{\pi D^2} \tag{E.4}$$

Thus the pressure force on the end of the element is

$$p(\pi D\, dy) = \frac{4P}{D}\, dy \tag{E.5}$$

Where $(\pi D\, dy)$ denotes the annular area between y and $(y + dy)$. If we assume uniform mean velocity in the direction of motion of the fluid, the forces given in Eqs. (E.3) and (E.5) must be equal. Thus we get

$$\frac{4P}{D}\, dy = -\pi Dl\, dy\, \mu \frac{d^2v}{dy^2}$$

or

$$\frac{d^2v}{dy^2} = -\frac{4P}{\pi D^2 l\mu} \tag{E.6}$$

Integrating this equation twice and using the boundary conditions $v = -v_0$ at $y = 0$ and $v = 0$ at $y = d$, we obtain

$$v = \frac{2P}{\pi D^2 l\mu}(yd - y^2) - v_0\left(1 - \frac{y}{d}\right) \tag{E.7}$$

The rate of flow through the clearance space can be obtained by integrating the rate of flow through an element between the limits $y = 0$ and $y = d$:

$$Q = \int_0^d v\pi D\, dy = \pi D\left[\frac{2Pd^3}{6\pi D^2 l\mu} - \frac{1}{2}v_0 d\right] \tag{E.8}$$

The volume of the liquid flowing through the clearance space per second must be equal to the volume per second displaced by the piston. Hence the velocity of the piston will be equal to this rate of flow divided by the piston area. This gives

$$v_0 = \frac{Q}{\left(\dfrac{\pi}{4}D^2\right)} \tag{E.9}$$

Equations (E.9) and (E.8) lead to

$$P = \left[\frac{3\pi D^3 l\left(1 + \dfrac{2d}{D}\right)}{4d^3}\right]\mu v_0 \tag{E.10}$$

By writing the force as $P = cv_0$, the damping constant c can be found as

$$c = \mu\left[\frac{3\pi D^3 l}{4d^3}\left(1 + \frac{2d}{D}\right)\right] \tag{E.11}$$

1.10 HARMONIC MOTION

Oscillatory motion may repeat itself regularly, as in the case of a simple pendulum, or may display considerable irregularity, as in the case of ground motion during an earthquake. If the motion is repeated after equal intervals of time, it is called *periodic motion*. The simplest type of periodic motion is *harmonic motion*. The motion imparted to the mass m due to the Scotch yoke mechanism shown in Fig. 1.25 is an example of simple harmonic motion [1.29, 1.30, 1.20]. In this system, a crank of radius A rotates about the point O. The other end of the crank P slides in

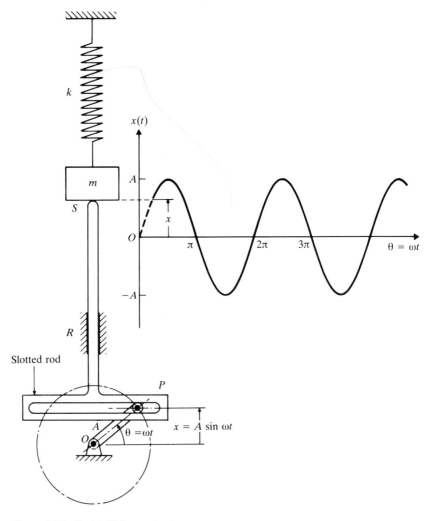

Figure 1.25   Scotch Yoke mechanism.

a slotted rod, which reciprocates in the vertical guide R. When the crank rotates at an angular velocity ω, the end point S of the slotted link and hence the mass m of the spring-mass system are displaced from their middle positions by an amount x (in time t) given by

$$x = A \sin \theta = A \sin \omega t \qquad (1.28)$$

This motion is shown by the sinusoidal curve in Fig. 1.25. The velocity of the mass m at time t is given by

$$\frac{dx}{dt} = \omega A \cos \omega t \qquad (1.29)$$

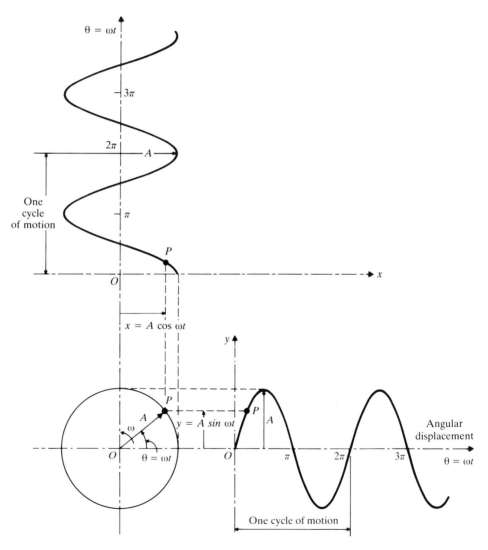

Figure 1.26 Harmonic motion as the projection of the end of a rotating vector.

and the acceleration by

$$\frac{d^2x}{dt^2} = -\omega^2 A \sin \omega t = -\omega^2 x \qquad (1.30)$$

It can be seen that the acceleration is directly proportional to the displacement. Such a vibration, with the acceleration proportional to the displacement and directed towards the mean position, is known as *simple harmonic motion*. The motion given by $x = A \cos \omega t$ is another example of a simple harmonic motion. Figure 1.25 clearly shows the similarity between cyclic (harmonic) motion and sinusoidal motion.

**1.10.1
Vectorial
Representation
of Harmonic
Motion**

Harmonic motion can be represented conveniently by means of a vector $\overrightarrow{OP}$ of magnitude A rotating at a constant angular velocity ω. In Fig. 1.26, the projection of the tip of the vector $\vec{X} = \overrightarrow{OP}$ on the vertical axis is given by

$$y = A \sin \omega t \qquad (1.31)$$

and its projection on the horizontal axis by

$$x = A \cos \omega t \qquad (1.32)$$

**1.10.2
Complex Number
Representation
of Harmonic
Motion**

Any vector $\vec{X}$ in the xy plane can be represented as a complex number:

$$\vec{X} = a + ib \qquad (1.33)$$

where $i = \sqrt{-1}$ and a and b denote the x and y components of $\vec{X}$, respectively (see Fig. 1.27). Components a and b are also called the *real* and *imaginary* parts of the vector $\vec{X}$. If A denotes the modulus or absolute value of the vector $\vec{X}$, and ϕ represents the argument or the angle between the vector and the x-axis, then $\vec{X}$ can also be expressed as

$$\vec{X} = A \cos \phi + iA \sin \phi = Ae^{i\phi} \qquad (1.34)$$

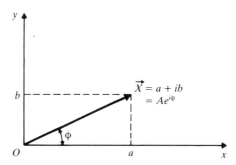

Figure 1.27

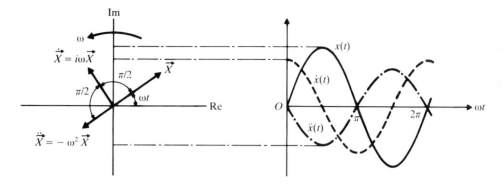

Figure 1.28 Displacement, velocity, and accelerations as rotating vectors.

with

$$A = (a^2 + b^2)^{1/2} \qquad (1.35)$$

and

$$\phi = \tan^{-1} \frac{b}{a} \qquad (1.36)$$

The rotating vector $\vec{X}$ of Fig. 1.26 can be represented as a complex number:

$$\vec{X} = Ae^{i\omega t} \qquad (1.37)$$

The differentiation of Eq. (1.37) with respect to time gives

$$\frac{d\vec{X}}{dt} = \frac{d}{dt}(Ae^{i\omega t}) = i\omega Ae^{i\omega t} = i\omega \vec{X} \qquad (1.38)$$

$$\frac{d^2\vec{X}}{dt^2} = \frac{d}{dt}(i\omega Ae^{i\omega t}) = -\omega^2 Ae^{i\omega t} = -\omega^2 \vec{X} \qquad (1.39)$$

Thus the displacement, velocity, and acceleration can be expressed as*

$$\text{displacement} = \quad \text{Re}[Ae^{i\omega t}] = A\cos\omega t \qquad (1.40)$$

$$\text{velocity} = \quad \text{Re}[i\omega Ae^{i\omega t}] = -\omega A \sin\omega t$$
$$= \omega A \cos(\omega t + 90°) \qquad (1.41)$$

$$\text{acceleration} = \text{Re}[-\omega^2 Ae^{i\omega t}] = -\omega^2 A \cos\omega t$$
$$= \omega^2 A \cos(\omega t + 180°) \qquad (1.42)$$

---

*If the harmonic displacement is originally given as $x(t) = A\sin\omega t$, then we have

$$\text{displacement} = \text{Im}[Ae^{i\omega t}] = A\sin\omega t$$
$$\text{velocity} = \text{Im}[i\omega Ae^{i\omega t}] = \omega A \sin(\omega t + 90°)$$
$$\text{acceleration} = \text{Im}[-\omega^2 Ae^{i\omega t}] = \omega^2 A \sin(\omega t + 180°)$$

where Im denotes the imaginary part.

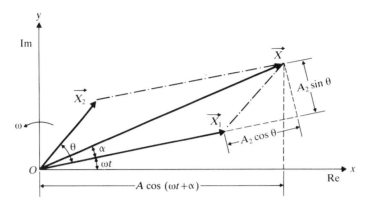

Figure 1.29   Vectorial addition of harmonic functions.

where Re denotes the real part. These quantities are shown as rotating vectors in Fig. 1.28. It can be seen that the acceleration vector leads the velocity vector by 90°, and the latter leads the displacement vector by 90°.

Harmonic functions can be added vectorially, as shown in Fig. 1.29. If $\text{Re}(\vec{X_1}) = A_1 \cos \omega t$ and $\text{Re}(\vec{X_2}) = A_2 \cos(\omega t + \theta)$, then the magnitude of the resultant vector $\vec{X}$ is given by

$$A = \sqrt{(A_1 + A_2 \cos \theta)^2 + (A_2 \sin \theta)^2} \tag{1.43}$$

and the angle α by

$$\alpha = \tan^{-1}\left(\frac{A_2 \sin \theta}{A_1 + A_2 \cos \theta}\right) \tag{1.44}$$

Since the original functions are given as real components, the sum $\vec{X_1} + \vec{X_2}$ is given by $\text{Re}(\vec{X}) = A \cos(\omega t + \alpha)$. The sum of $\vec{X_1}$ and $\vec{X_2}$ can also be found using complex numbers:

$$\begin{aligned}
\vec{X} = \vec{X_1} + \vec{X_2} &= A_1 e^{i\omega t} + A_2 e^{i(\omega t + \theta)} = \left(A_1 + A_2 e^{i\theta}\right)e^{i\omega t} \\
&= \left(A_1 + A_2 \cos \theta + iA_2 \sin \theta\right)e^{i\omega t} \\
&= Ae^{i\alpha}e^{i\omega t} = Ae^{i(\omega t + \alpha)}
\end{aligned} \tag{1.45}$$

where A and α are given by Eqs. (1.43) and (1.44).

EXAMPLE 1.5   Addition of Harmonic Motions

Find the sum of the two harmonic motions $x_1(t) = 10 \cos \omega t$ and $x_2(t) = 15 \cos(\omega t + 2)$.

Given: Two harmonic motions, $x_1(t) = 10 \cos \omega t$ and $x_2(t) = 15 \cos(\omega t + 2)$.

Find: Sum of harmonic motions.

Approach: Equation for sum of trigonometric terms. Principle of vector addition.

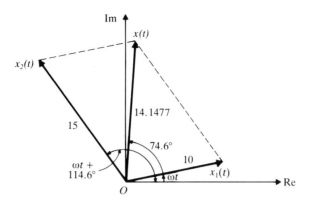

Figure 1.30

Solution

Method 1: By using trigonometric relations: Since the circular frequency is the same for both $x_1(t)$ and $x_2(t)$, we express the sum as

$$x(t) = A \cos(\omega t + \alpha) = x_1(t) + x_2(t) \tag{E.1}$$

that is,

$$A(\cos \omega t \cos \alpha - \sin \omega t \sin \alpha) = 10 \cos \omega t + 15 \cos(\omega t + 2)$$
$$= 10 \cos \omega t + 15(\cos \omega t \cos 2 - \sin \omega t \sin 2) \tag{E.2}$$

that is,

$$\cos \omega t (A \cos \alpha) - \sin \omega t (A \sin \alpha) = \cos \omega t (10 + 15 \cos 2) - \sin \omega t (15 \sin 2) \tag{E.3}$$

By equating the corresponding coefficients of $\cos \omega t$ and $\sin \omega t$ on both sides, we obtain

$$A \cos \alpha = 10 + 15 \cos 2$$
$$A \sin \alpha = 15 \sin 2$$
$$A = \sqrt{(10 + 15 \cos 2)^2 + (15 \sin 2)^2}$$
$$= 14.1477 \tag{E.4}$$

and

$$\alpha = \tan^{-1}\left(\frac{15 \sin 2}{10 + 15 \cos 2}\right) = 74.5963° \tag{E.5}$$

Method 2: By using vectors: For an arbitrary value of ωt, the harmonic motions $x_1(t)$ and $x_2(t)$ can be denoted graphically as shown in Fig. 1.30. By adding them vectorially, the resultant vector $x(t)$ can be found to be

$$x(t) = 14.1477 \cos(\omega t + 74.5963°) \tag{E.6}$$

Method 3: By using complex number representation: The two harmonic motions can be denoted in terms of complex numbers:

$$x_1(t) = \text{Re}\left[A_1 e^{i\omega t} \right] \equiv \text{Re}[10 e^{i\omega t}]$$
$$x_2(t) = \text{Re}\left[A_2 e^{i(\omega t + 2)} \right] \equiv \text{Re}[15 e^{i(\omega t + 2)}] \tag{E.7}$$

The sum of $x_1(t)$ and $x_2(t)$ can be expressed as

$$x(t) = \text{Re}[Ae^{i(\omega t + \alpha)}] \tag{E.8}$$

where A and α can be determined using Eqs. (1.43) and (1.44) as $A = 14.1477$ and $\alpha = 74.5963°$.

**1.10.3
Definitions**

The following definitions are useful in dealing with harmonic motion.

Cycle. The movement of a vibrating body from its undisturbed or equilibrium position to its extreme position in one direction, then to the equilibrium position, then to its extreme position in other direction, and back to equilibrium position is called a *cycle* of vibration. One revolution (i.e., angular displacement of 2π radians) of the pin P in Fig. 1.25 or one revolution of the vector $\overrightarrow{OP}$ in Fig. 1.26 constitutes a cycle.

Amplitude. The maximum displacement of a vibrating body from its equilibrium position is called the *amplitude* of vibration. In Figs. 1.25 and 1.26 the amplitude of vibration is equal to A.

Period of Oscillation. The time taken to complete one cycle of motion is known as the *period of oscillation* or *time period* and is denoted by τ. It is equal to the time required for the vector $\overrightarrow{OP}$ in Fig. 1.26 to rotate through an angle of 2π and hence

$$\tau = \frac{2\pi}{\omega} \tag{1.46}$$

where ω is called the circular frequency.

Frequency of Oscillation. The number of cycles per unit time is called the *frequency of oscillation* or simply the *frequency* and is denoted by f. Thus

$$f = \frac{1}{\tau} = \frac{\omega}{2\pi} \tag{1.47}$$

Here ω is called the circular frequency to distinguish it from the linear frequency $f = \omega/2\pi$. ω denotes the angular velocity of the cyclic motion; f is measured in cycles per second (Hertz) while ω is measured in radians per second.

Phase Angle. Consider two vibratory motions denoted by

$$x_1 = A_1 \sin \omega t \tag{1.48}$$
$$x_2 = A_2 \sin(\omega t + \phi) \tag{1.49}$$

The two harmonic motions given by Eqs. (1.48) and (1.49) are called *synchronous* because they have the same frequency or angular velocity ω. Two synchronous oscillations need not have the same amplitude, and they need not attain their maximum values at the same time. The motions given by Eqs. (1.48) and (1.49) can be represented graphically as shown in Fig. 1.31. In this figure, the second vector $\overrightarrow{OP_2}$ leads the first one $\overrightarrow{OP_1}$ by an angle ϕ, known as the *phase angle*. This means that the maximum of the second vector would occur ϕ radians earlier than that of the first vector. Note that instead of maxima, any other corresponding points can be taken for finding the phase angle. In Eqs. (1.48) and (1.49) or in Fig. 1.31, the two vectors are said to have a *phase difference* of ϕ.

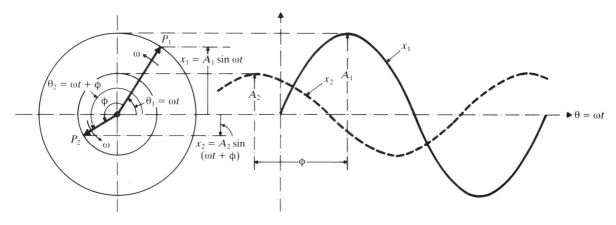

Figure 1.31 Phase difference between two vectors.

Natural Frequency. If a system, after an initial disturbance, is left to vibrate on its own, the frequency with which it oscillates without external forces is known as its *natural frequency*. As will be seen later, a vibratory system having n degrees of freedom will have, in general, n distinct natural frequencies of vibration.

1.11 HARMONIC ANALYSIS[†]

Although harmonic motion is simplest to handle, the motion of many vibratory systems is not harmonic. However, in many cases the vibrations are periodic—for example, the type shown in Fig. 1.32(a). Fortunately, any periodic function of time can be represented by Fourier series as an infinite sum of sine and cosine terms [1.31, 1.32].

**1.11.1
Fourier Series
Expansion**

If $x(t)$ is a periodic function with period τ, its Fourier series representation is given by

$$x(t) = \frac{a_0}{2} + a_1\cos \omega t + a_2\cos 2\omega t + \cdots + b_1\sin \omega t + b_2\sin 2\omega t + \cdots$$

$$= \frac{a_0}{2} + \sum_{n=1}^{\infty} (a_n\cos n\omega t + b_n\sin n\omega t) \tag{1.50}$$

where $\omega = 2\pi/\tau$ is the fundamental frequency and $a_0, a_1, a_2, \ldots, b_1, b_2, \ldots$ are constant coefficients. To determine the coefficients a_n and b_n, we multiply Eq. (1.50) by $\cos n\omega t$ and $\sin n\omega t$, respectively, and integrate over one period $\tau = 2\pi/\omega$, for example from 0 to $2\pi/\omega$. Then we notice that all terms except one on the

[†] The harmonic analysis forms a basis for Section 4.2.

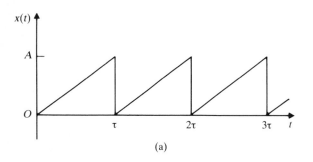

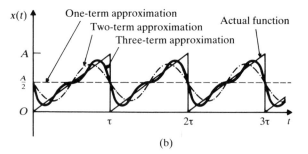

Figure 1.32   A periodic function.

right-hand side of the equation will be zero, and we obtain

$$a_0 = \frac{\omega}{\pi} \int_0^{2\pi/\omega} x(t)\, dt = \frac{2}{\tau} \int_0^\tau x(t)\, dt \tag{1.51}$$

$$a_n = \frac{\omega}{\pi} \int_0^{2\pi/\omega} x(t)\cos n\omega t\, dt = \frac{2}{\tau} \int_0^\tau x(t)\cos n\omega t\, dt \tag{1.52}$$

$$b_n = \frac{\omega}{\pi} \int_0^{2\pi/\omega} x(t)\sin n\omega t\, dt = \frac{2}{\tau} \int_0^\tau x(t)\sin n\omega t\, dt \tag{1.53}$$

The physical interpretation of Eq. (1.50) is that any periodic function can be represented as a sum of harmonic functions. Although the series in Eq. (1.50) is an infinite sum, we can approximate most periodic functions with the help of only a few harmonic functions. For example, the triangular wave of Fig. 1.32(a) can be represented closely by adding only three harmonic functions, as shown in Fig. 1.32(b).

Fourier series can also be represented by the sum of cosine terms only:

$$x(t) = c_0 + c_1\cos(\omega t - \phi_1) + c_2\cos(2\omega t - \phi_2) + \cdots \tag{1.54}$$

where

$$c_0 = a_0/2. \tag{1.55}$$

$$c_n = \left(a_n^2 + b_n^2 \right)^{1/2} \tag{1.56}$$

and

$$\phi_n = \tan^{-1}\left(\frac{b_n}{a_n} \right) \tag{1.57}$$

The Fourier series can also be represented in terms of complex numbers by writing Eq. (1.50) as

$$x(t) = \frac{a_0}{2} + \sum_{n=1}^{\infty} \left\{ a_n\left(\frac{e^{in\omega t} + e^{-in\omega t}}{2} \right) + b_n\left(\frac{e^{in\omega t} - e^{-in\omega t}}{2i} \right) \right\} \tag{1.58}$$

The harmonic functions $a_n\cos n\omega t$ or $b_n\sin n\omega t$ in Eq. (1.50) are called the *harmonics* of order n of the periodic function $x(t)$. The harmonic of order n has a

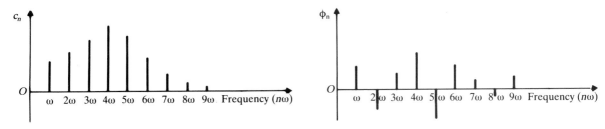

Figure 1.33 Frequency spectrum of a typical periodic function of time.

period τ/n. These harmonics can be plotted as vertical lines on a diagram of amplitude (a_n and b_n or c_n and ϕ_n) versus frequency ($n\omega$) called the *frequency spectrum* or *spectral diagram*. Figure 1.33 shows a typical frequency spectrum.

1.11.2 Even and Odd Functions

An even function satisfies the relation

$$x(-t) = x(t) \tag{1.59}$$

In this case, the Fourier series expansion of $x(t)$ contains only cosine terms:

$$x(t) = \frac{a_0}{2} + \sum_{n=1}^{\infty} a_n \cos n\omega t \tag{1.60}$$

where a_0 and a_n are given by Eqs. (1.51) and (1.52), respectively. An odd function satisfies the relation

$$x(-t) = -x(t) \tag{1.61}$$

In this case, the Fourier series expansion of $x(t)$ contains only sine terms:

$$x(t) = \sum_{n=1}^{\infty} b_n \sin n\omega t \tag{1.62}$$

where b_n are given by Eq. (1.53). In some cases, a given function may be considered as even or odd depending on the location of the coordinate axes. For example, the shifting of the vertical axis from (a) to (b) or (c) in Fig. 1.34(i) will make it an odd or even function. This means that we need to compute only the coefficients b_n or a_n. Similarly, a shift in the time axis from (d) to (e) amounts to adding a constant equal to the amount of shift. In the case of Fig. 1.34(ii), when the function is considered as an odd function, the Fourier series expansion becomes (see Problem 1.40)

$$x_1(t) = \frac{4A}{\pi} \sum_{n=1}^{\infty} \frac{1}{(2n-1)} \sin \frac{2\pi(2n-1)t}{\tau} \tag{1.63}$$

On the other hand, if the function is considered as an even function as shown in Fig. 1.34(iii), its Fourier series expansion becomes (see Problem 1.40)

$$x_2(t) = \frac{4A}{\pi} \sum_{n=1}^{\infty} \frac{(-1)^{n+1}}{(2n-1)} \cos \frac{2\pi(2n-1)t}{\tau} \tag{1.64}$$

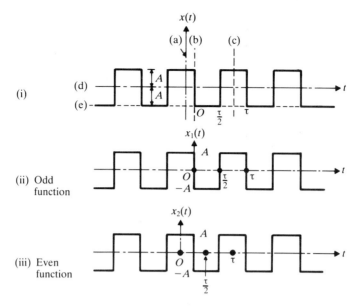

Figure 1.34 Even and odd functions.

Since the functions $x_1(t)$ and $x_2(t)$ represent the same wave, except for the location of the origin, there exists a relationship between their Fourier series expansions also. Noting that

$$x_1\left(t + \frac{\tau}{4}\right) = x_2(t) \tag{1.65}$$

we find from Eq. (1.63),

$$x_1\left(t + \frac{\tau}{4}\right) = \frac{4A}{\pi} \sum_{n=1}^{\infty} \frac{1}{(2n-1)} \sin \frac{2\pi(2n-1)}{\tau}\left(t + \frac{\tau}{4}\right)$$

$$= \frac{4A}{\pi} \sum_{n=1}^{\infty} \frac{1}{(2n-1)} \sin\left\{\frac{2\pi(2n-1)t}{\tau} + \frac{2\pi(2n-1)}{4}\right\} \tag{1.66}$$

Using the relation $\sin(A + B) = \sin A \cos B + \cos A \sin B$, Eq. (1.66) can be expressed as

$$x_1\left(t + \frac{\tau}{4}\right) = \frac{4A}{\pi} \sum_{n=1}^{\infty} \left\{ \frac{1}{(2n-1)} \sin \frac{2\pi(2n-1)t}{\tau} \cos \frac{2\pi(2n-1)}{4} \right.$$

$$\left. + \cos \frac{2\pi(2n-1)t}{\tau} \sin \frac{2\pi(2n-1)}{4} \right\} \tag{1.67}$$

Since $\cos[2\pi(2n-1)/4] = 0$ for $n = 1, 2, 3, \ldots$, and $\sin[2\pi(2n-1)/4] = (-1)^{n+1}$ for $n = 1, 2, 3, \ldots$, Eq. (1.67) reduces to

$$x_1\left(t + \frac{\tau}{4}\right) = \frac{4A}{\pi} \sum_{n=1}^{\infty} \frac{(-1)^{n+1}}{(2n-1)} \cos \frac{2\pi(2n-1)t}{\tau} \tag{1.68}$$

which can be identified to be the same as Eq. (1.64).

**1.11.3
Half Range
Expansions**

In some practical applications, the function $x(t)$ is defined only in the interval 0 to τ as shown in Fig. 1.35(a). In such a case, there is no condition of periodicity of the function since the function itself is not defined outside the interval 0 to τ. However, we can extend the function arbitrarily to include the interval $-\tau$ to 0 as shown in either Fig. 1.35(b) or Fig. 1.35(c). The extension of the function indicated in Fig. 1.35(b) results in an odd function $x_1(t)$, while the extension of the function shown in Fig. 1.35(c) results in an even function $x_2(t)$. Thus the Fourier series expansion of $x_1(t)$ yields only sine terms and that of $x_2(t)$ involves only cosine terms. These Fourier series expansions of $x_1(t)$ and $x_2(t)$ are known as half range expansions [1.40]. Any of these half range expansions can be used to find $x(t)$ in the interval 0 to τ.

**1.11.4
Numerical
Computation
of Coefficients**

For very simple forms of the function $x(t)$, the integrals of Eqs. (1.51) to (1.53) can be evaluated easily. However, the integration becomes involved if $x(t)$ does not have a simple form. In some practical applications, as in the case of experimental determination of the amplitude of vibration using a vibration transducer, the function $x(t)$ is not available in the form of a mathematical expression; only the values of $x(t)$ at a number of points $t_1, t_2, \ldots, t_N$ are available, as shown in Fig.

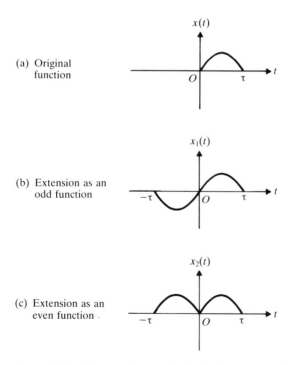

(a) Original function

(b) Extension as an odd function

(c) Extension as an even function

Figure 1.35 Extension of a function for half-range expansions.

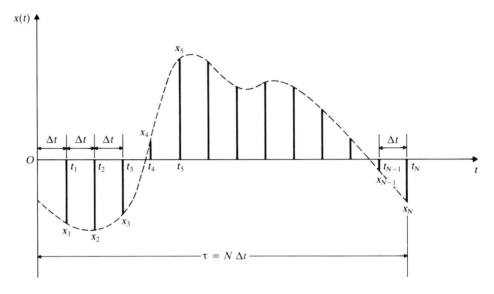

Figure 1.36   Values of the periodic function $x(t)$ at discrete points $t_1, t_2, \ldots, t_N$.

1.36. In these cases, the coefficients a_n and b_n of Eqs. (1.51) to (1.53) can be evaluated by using a numerical integration procedure like trapezoidal or Simpson's rule [1.33].

If $t_1, t_2, \ldots, t_N$ are assumed to be an even number of equidistant points over the period τ (N = even) with the corresponding values of $x(t)$ given by $x_1 = x(t_1)$, $x_2 = x(t_2), \ldots, x_N = x(t_N)$, respectively, the application of trapezoidal rule gives the coefficients a_n and b_n as[†] (by setting $\tau = N\Delta t$):

$$a_0 = \frac{2}{N} \sum_{i=1}^{N} x_i \tag{1.69}$$

$$a_n = \frac{2}{N} \sum_{i=1}^{N} x_i \cos \frac{2n\pi t_i}{\tau} \tag{1.70}$$

$$b_n = \frac{2}{N} \sum_{i=1}^{N} x_i \sin \frac{2n\pi t_i}{\tau} \tag{1.71}$$

EXAMPLE 1.6   **Fourier Series Expansion**

Find the Fourier series expansion of the function shown in Fig. 1.32(a).

Given: Saw-tooth type periodic function (Fig. 1.32a).

Find: Coefficients a_n and b_n in the Fourier series expansion of Eq. (1.50).

[†]N needs to be an even number for Simpson's rule, but not for the trapezoidal rule. Equations (1.69) to (1.71) assume that the periodicity condition, $x_0 = x_N$, holds true.

Approach: Fourier series expansion of a periodic function.

Solution. The function $x(t)$ can be represented within the first cycle as

$$x(t) = A\frac{t}{\tau}, 0 \leqslant t \leqslant \tau \tag{E.1}$$

where the period is given by $\tau = 2\pi/\omega$. To compute the Fourier coefficients a_n and b_n, we use Eqs. (1.51) to (1.53):

$$a_0 = \frac{\omega}{\pi} \int_0^{2\pi/\omega} x(t) \, dt = \frac{\omega}{\pi} \int_0^{2\pi/\omega} A\frac{t}{\tau} \, dt = \frac{\omega}{\pi} \frac{A}{\tau} \left(\frac{t^2}{2}\right)_0^{2\pi/\omega} = A \tag{E.2}$$

$$a_n = \frac{\omega}{\pi} \int_0^{2\pi/\omega} x(t) \cos n\omega t \cdot dt = \frac{\omega}{\pi} \int_0^{2\pi/\omega} A\frac{t}{\tau} \cos n\omega t \cdot dt$$

$$= \frac{A\omega}{\pi\tau} \int_0^{2\pi/\omega} t \cos n\omega t \cdot dt = \frac{A}{2\pi^2} \left[\frac{\cos n\omega t}{n^2} + \frac{\omega t \sin n\omega t}{n} \right]_0^{2\pi/\omega}$$

$$= 0, \, n = 1, 2, \ldots \tag{E.3}$$

$$b_n = \frac{\omega}{\pi} \int_0^{2\pi/\omega} x(t) \sin n\omega t \cdot dt = \frac{\omega}{\pi} \int_0^{2\pi/\omega} A\frac{t}{\tau} \sin n\omega t \cdot dt$$

$$= \frac{A\omega}{\pi\tau} \int_0^{2\pi/\omega} t \sin n\omega t \cdot dt = \frac{A}{2\pi^2} \left[\frac{\sin n\omega t}{n^2} - \frac{\omega t \cos n\omega t}{n} \right]_0^{2\pi/\omega}$$

$$= -\frac{A}{n\pi}, \, n = 1, 2, \ldots \tag{E.4}$$

Therefore the Fourier series expansion of $x(t)$ is

$$x(t) = \frac{A}{2} - \frac{A}{\pi} \sin \omega t - \frac{A}{2\pi} \sin 2\omega t - \cdots$$

$$= \frac{A}{\pi} \left[\frac{\pi}{2} - \left\{ \sin \omega t + \frac{1}{2} \sin 2\omega t + \frac{1}{3} \sin 3\omega t + \cdots \right\} \right] \tag{E.5}$$

The first three terms of the series are shown plotted in Fig. 1.32(b). It can be seen that the approximation reaches the sawtooth shape even with a small number of terms.

EXAMPLE 1.7   **Numerical Fourier Analysis**

The pressure fluctuations of water in a pipe, measured at 0.01 second intervals, are given in Table 1.1. These fluctuations are repetitive in nature. Make a harmonic analysis of the pressure fluctuations and determine the first three harmonics of the Fourier series expansion.

Given: Pressure fluctuations of water in a pipe at 0.01 second intervals.

Find: First three harmonics of the pressure fluctuation (i.e., $a_0, a_1, a_2, a_3, b_1, b_2, b_3$).

Approach: Fourier series expansion of a periodic function using numerical method [Eqs. (1.69) through (1.71)].

Solution. Since the given pressure fluctuations repeat every 0.12 sec, the period is $\tau = 0.12$ sec and the circular frequency of the first harmonic is 2π radians per 0.12 sec or $\omega = 2\pi/0.12$ = 52.36 rad/sec. As the number of observed values in each wave (N) is 12, we obtain from Eq. (1.69)

$$a_0 = \frac{2}{N} \sum_{i=1}^{N} p_i = \frac{1}{6} \sum_{i=1}^{12} p_i = 68166.7 \tag{E.1}$$

TABLE 1.1

Time Station, i	Time (sec), t_i	Pressure (kN / m^2), p_i
0	0	0
1	0.01	20
2	0.02	34
3	0.03	42
4	0.04	49
5	0.05	53
6	0.06	70
7	0.07	60
8	0.08	36
9	0.09	22
10	0.10	16
11	0.11	7
12	0.12	0

TABLE 1.2

i	t_i	p_i	$p_i\cos\dfrac{2\pi t_i}{0.12}$ ($n=1$)	$p_i\sin\dfrac{2\pi t_i}{0.12}$ ($n=1$)	$p_i\cos\dfrac{4\pi t_i}{0.12}$ ($n=2$)	$p_i\sin\dfrac{4\pi t_i}{0.12}$ ($n=2$)	$p_i\cos\dfrac{6\pi t_i}{0.12}$ ($n=3$)	$p_i\sin\dfrac{6\pi t_i}{0.12}$ ($n=3$)
1	0.01	20 000	17 320	10 000	10 000	17 320	0	20 000
2	0.02	34 000	17 000	29 444	− 17 000	29 444	− 34 000	0
3	0.03	42 000	0	42 000	− 42 000	0	0	− 42 000
4	0.04	49 000	− 24 500	42 434	− 24 500	− 42 434	49 000	0
5	0.05	53 000	− 45 898	26 500	26 500	− 45 898	0	53 000
6	0.06	70 000	− 70 000	0	70 000	0	− 70 000	0
7	0.07	60 000	− 51 960	− 30 000	30 000	51 960	0	− 60 000
8	0.08	36 000	− 18 000	− 31 176	− 18 000	31 176	36 000	0
9	0.09	22 000	0	− 22 000	− 22 000	0	0	22 000
10	0.10	16 000	8000	− 13 856	− 8000	− 13 856	− 16 000	0
11	0.11	7000	6062	− 3500	3500	− 6062	0	− 7000
12	0.12	0	0	0	0	0	0	0
$\sum\limits_{i=1}^{12}(\ )$		409 000	− 161 976	49 846	8500	21 650	− 35 000	− 14 000
$\dfrac{1}{6}\sum\limits_{i=1}^{12}(\ )$		68 166.7	− 26 996.0	8307.7	1416.7	3608.3	− 5833.3	− 2333.3

The coefficients a_n and b_n can be determined from Eqs. (1.70) and (1.71):

$$a_n = \frac{2}{N} \sum_{i=1}^{N} p_i \cos\frac{2n\pi t_i}{\tau} = \frac{1}{6} \sum_{i=1}^{12} p_i \cos\frac{2n\pi t_i}{0.12} \qquad \text{(E.2)}$$

$$b_n = \frac{2}{N} \sum_{i=1}^{N} p_i \sin\frac{2n\pi t_i}{\tau} = \frac{1}{6} \sum_{i=1}^{12} p_i \sin\frac{2n\pi t_i}{0.12} \qquad \text{(E.3)}$$

The computations involved in Eqs. (E.2) and (E.3) are shown in Table 1.2. From these calculations, the Fourier series expansion of the pressure fluctuations $p(t)$ can be obtained [see Eq. (1.50)]:

$$\begin{aligned} p(t) = {} & 34083.3 - 26996.0\cos 52.36t + 8307.7\sin 52.36t \\ & + 1416.7\cos 104.72t + 3608.3\sin 104.72t \\ & - 5833.3\cos 157.08t - 2333.3\sin 157.08t \\ & + \cdots \ N/m^2 \end{aligned} \qquad \text{(E.4)}$$

1.12 VIBRATION LITERATURE

The literature on vibrations is large and diverse. Several textbooks are available [1.34], and dozens of technical periodicals regularly publish papers relating to vibrations. This is primarily because vibration affects so many disciplines, from science of materials to machinery analysis to spacecraft structures. Researchers in many fields must be attentive to vibration research.

The most widely circulated journals that publish papers relating to vibrations are *Journal of Vibration, Acoustics, Stress, and Reliability in Design*; *Journal of Applied Mechanics*; *Journal of Sound and Vibration*; *AIAA Journal*; *ASCE Journal of Engineering Mechanics*; *Earthquake Engineering and Structural Dynamics*; *Bulletin of the Japan Society of Mechanical Engineers*; *International Journal of Solids and Structures*; *International Journal for Numerical Methods in Engineering*; *Journal of the Acoustical Society of America*; *Sound and Vibration*; *Vibrations, Mechanical Systems and Signal Processing, International Journal of Analytical and Experimental Modal Analysis*; and *Vehicle System Dynamics*. Many of these journals are cited in the chapter references.

In addition, *Shock and Vibration Digest* and *Applied Mechanics Reviews* are monthly abstract journals containing brief discussions of nearly every published vibration paper. Formulas and solutions in vibration engineering can be readily found in references [1.35–1.37].

1.13 COMPUTER PROGRAM

A FORTRAN computer program, in the form of subroutine FORIER, is given for the harmonic analysis of a function $x(t)$. The arguments of this subroutine are as follows:

N   =   Number of equidistant points at which the values of $x(t)$ are known. Input data.

M   =   Number of Fourier coefficients to be computed. Input data.

TIME   =   Time period of the function $x(t)$. Input data.

X   =   Array of dimension N, containing the known values of $x(t)$. $X(I) = x(t_i)$. Input data.

T   =   Array of dimension N, containing the known values of t. $T(I) = t_i$. Input data.

AZERO   = a_0 of Eq. (1.69). Output.

A   =   Array of dimension M, containing the computed values of a_n of Eq. (1.70). Output.

B   =   Array of dimension M, containing the computed values of b_n of Eq. (1.71). Output.

To illustrate the use of subroutine FORIER, consider Example 1.7 with M = 5 instead of M = 3. Thus we have N = 12, M = 5, and TIME = 0.12. The main program that calls subroutine FORIER, subroutine FORIER itself, and the output of the program are given below.

```
C ================================================================
C
C PROGRAM 1
C MAIN PROGRAM FOR CALLING THE SUBROUTINE FORIER
C
C ================================================================
C FOLLOWING 6 LINES CONTAIN PROBLEM-DEPENDENT DATA
      DIMENSION X(12),T(12),XSIN(12),XCOS(12),A(5),B(5)
      DATA N,M,TIME /12,5,0.12/
      DATA X /20000.0,34000.0,42000.0,49000.0,53000.0,70000.0,60000.0,
     2 36000.0,22000.0,16000.0,7000.0,0.0/
      DATA T /0.01,0.02,0.03,0.04,0.05,0.06,0.07,0.08,0.09,0.10,0.11,
     2 0.12/
C END OF PROBLEM-DEPENDENT DATA
      CALL FORIER (N,M,TIME,X,T,AZERO,A,B,XSIN,XCOS)
      PRINT 100
100   FORMAT (//,46H FOURIER SERIES EXPANSION OF THE FUNCTION X(T),//)
      PRINT 200, N,M,TIME
200   FORMAT (6H DATA:,//,37H NUMBER OF DATA POINTS IN ONE CYCLE =,I5,
     2 /,42H NUMBER OF FOURIER COEFFICIENTS REQUIRED =,I5,/,
     3 14H TIME PERIOD =,E15.8)
      PRINT 300, (T(I),I=1,N)
300   FORMAT (/,33H TIME AT VARIOUS STATIONS, T(I) =,/,(4E15.8,1X))
      PRINT 400, (X(I),I=1,N)
400   FORMAT (/,31H KNOWN VALUES OF X(I) AT T(I) =,/,(4E15.8,1X))
      PRINT 500
500   FORMAT (//,29H RESULTS OF FOURIER ANALYSIS:,/)
      PRINT 600, AZERO
600   FORMAT (8H AZERO =,2X,E15.8,//,31H VALUES OF I, A(I) AND B(I) ARE
     2 ,/)
      DO 700, I =1,M
700   PRINT 800, I,A(I),B(I)
800   FORMAT (I5,2X,E15.8,2X,E15.8)
      STOP
      END
```

```
C ================================================================
C
C SUBROUTINE FORIER
C
C ================================================================
      SUBROUTINE FORIER (N,M,TIME,X,T,AZERO,A,B,XSIN,XCOS)
      DIMENSION X(N),T(N),A(M),B(M),XSIN(N),XCOS(N)
      PI=3.1416
      SUMZ=0.0
      DO 100 I=1,N
  100 SUMZ=SUMZ+X(I)
      AZERO=2.0*SUMZ/REAL(N)
      DO 300 II=1,M
      SUMS=0.0
      SUMC=0.0
      DO 200 I=1,N
      THETA=2.0*PI*T(I)*REAL(II)/TIME
      XCOS(I)=X(I)*COS(THETA)
      XSIN(I)=X(I)*SIN(THETA)
      SUMS=SUMS+XSIN(I)
      SUMC=SUMC+XCOS(I)
  200 CONTINUE
      A(II)=2.0*SUMC/REAL(N)
      B(II)=2.0*SUMS/REAL(N)
  300 CONTINUE
      RETURN
      END

FOURIER SERIES EXPANSION OF THE FUNCTION X(T)

DATA:

NUMBER OF DATA POINTS IN ONE CYCLE =    12
NUMBER OF FOURIER COEFFICIENTS REQUIRED =     5
TIME PERIOD = 0.12000000E+00

TIME AT VARIOUS STATIONS, T(I) =
0.99999998E-02 0.20000000E-01 0.29999999E-01 0.39999999E-01
0.50000001E-01 0.59999999E-01 0.70000000E-01 0.79999998E-01
0.90000004E-01 0.10000000E+00 0.11000000E+00 0.12000000E+00

KNOWN VALUES OF X(I) AT T(I) =
0.20000000E+05 0.34000000E+05 0.42000000E+05 0.49000000E+05
0.53000000E+05 0.70000000E+05 0.60000000E+05 0.36000000E+05
0.22000000E+05 0.16000000E+05 0.70000000E+04 0.00000000E+00

RESULTS OF FOURIER ANALYSIS:

AZERO =   0.68166664E+05

VALUES OF I, A(I) AND B(I) ARE

    1   -0.26996299E+05    0.83075869E+04
    2    0.14166348E+04    0.36084932E+04
    3   -0.58332480E+04   -0.23334373E+04
    4   -0.58340521E+03    0.21650562E+04
    5   -0.21702822E+04   -0.64117188E+03
```

REFERENCES

1.1. S. P. Timoshenko, *History of Strength of Materials*, McGraw-Hill, New York, 1953.

1.2. J. T. Cannon and S. Dostrovsky, *The Evolution of Dynamics*: *Vibration Theory from 1687 to 1742*, Springer Verlag, New York, 1981.

1.3. L. L. Bucciarelli and N. Dworsky, *Sophie Germain*: *An Essay in the History of the Theory of Elasticity*, D. Reidel Publishing Co., Dordrecht, Holland, 1980.

1.4. J. W. Strutt (Baron Rayleigh), *The Theory of Sound*, Dover, New York, 1945.

1.5. R. B. Lindsay, *Lord Rayleigh: The Man and His Work*, Pergamon Press, Oxford, 1970.

1.6. N. Minorsky, *Nonlinear Oscillations*, D. Van Nostrand, Princeton, N.J., 1962.

1.7. J. J. Stoker, *Nonlinear Vibrations*, Interscience Publishers, New York, 1950.

1.8. S. H. Crandall and W. D. Mark, *Random Vibration in Mechanical Systems*, Academic Press, New York, 1963.

1.9. J. D. Robson, *Random Vibration*, Edinburgh University Press, Edinburgh, 1964.

1.10. T. G. Butler and D. Michel, "NASTRAN—A summary of the functions and capabilities of the NASA structural analysis computer system," NASA SP-260, 1971.

1.11. S. S. Rao, *The Finite Element Method in Engineering* (2nd Ed.), Pergamon Press, Oxford, 1989.

1.12. D. Radaj et al., "Finite element analysis, an automobile engineer's tool," *International Conference on Vehicle Structural Mechanics: Finite Element Application to Design*, Society of Automotive Engineers, Detroit, 1974.

1.13. R. E. D. Bishop, *Vibration* (2nd Ed.), Cambridge University Press, Cambridge, 1979.

1.14. M. H. Richardson and K. A. Ramsey, "Integration of dynamic testing into the product design cycle," *Sound and Vibration*, Vol. 15, No. 11, November 1981, pp. 14–27.

1.15. M. J. Griffin and E. M. Whitham, "The discomfort produced by impulsive whole-body vibration," *Journal of the Acoustical Society of America*, Vol. 65, No. 5, 1980, pp. 1277–1284.

1.16. J. E. Ruzicka, "Fundamental concepts of vibration control," *Sound and Vibration*, Vol. 5, No. 7, July 1971, pp. 16–23.

1.17. T. W. Black, "Vibratory finishing goes automatic" (Part 1: Types of machines; Part 2: Steps to automation), *Tool and Manufacturing Engineer*, July 1964, pp. 53–56; and August 1964, pp. 72–76.

1.18. S. Prakash and V. K. Puri, *Foundations for Machines: Analysis and Design*, John Wiley, New York, 1988.

1.19. E. O. Doebelin, *System Modeling and Response*, John Wiley, New York, 1980.

1.20. A. D. Dimarogonas, *Vibration Engineering*, West Publishing Co., St. Paul, 1976.

1.21. R. W. Fitzgerald, *Mechanics of Materials* (2nd Ed.), Addison-Wesley, Reading, Mass., 1982.

1.22. I. Cochin, *Analysis and Design of Dynamic Systems*, Harper & Row, New York, 1980.

1.23. F. Y. Chen, *Mechanics and Design of Cam Mechanisms*, Pergamon Press, New York, 1982.

1.24. W. T. Thomson, *Theory of Vibration with Applications* (2nd Ed.), Prentice-Hall, Englewood Cliffs, 1981.

1.25. S. H. Crandall, "The role of damping in vibration theory," *Journal of Sound and Vibration*, Vol. 11, No. 1, 1970, pp. 3–18.

1.26. C. W. Bert, "Material damping: An introductory review of mathematical models, measures, and experimental techniques," *Journal of Sound and Vibration*, Vol. 29, No. 2, 1973, pp. 129–153.

1.27. J. M. Gasiorek and W. G. Carter, *Mechanics of Fluids for Mechanical Engineers*, Hart Publishing Co., New York, 1968.

1.28. A. Mironer, *Engineering Fluid Mechanics*, McGraw-Hill, New York, 1979.

1.29. F. P. Beer and E. R. Johnston, *Vector Mechanics for Engineers* (3d Ed.), McGraw-Hill, New York, 1962.

1.30. A. Higdon and W. B. Stiles, *Engineering Mechanics* (2nd Ed.), Prentice-Hall, New York, 1955.

1.31. E. Kreyszig, *Advanced Engineering Mathematics* (4th Ed.), John Wiley, New York, 1979.

1.32. M. H. Richardson, "Fundamentals of the discrete Fourier transform," *Sound and Vibration*, Vol. 12, No. 3, March 1978, pp. 40–46.

1.33. C. F. Gerald and P. O. Wheatley, *Applied Numerical Analysis* (3d Ed.), Addison-Wesley, Reading, Mass., 1984.

1.34. N. F. Rieger, "The literature of vibration engineering," *Shock and Vibration Digest*, Vol. 14, No. 1, January 1982, pp. 5–13.

1.35. R. D. Blevins, *Formulas for Natural Frequency and Mode Shape*, Van Nostrand Reinhold, New York, 1979.

1.36. W. D. Pilkey and P. Y. Chang, *Modern Formulas for Statics and Dynamics*, McGraw-Hill, New York, 1978.

1.37. C. M. Harris (Ed.), *Shock and Vibration Handbook* (3rd Ed.), McGraw-Hill, New York, 1988.

1.38. N. O. Myklestad, *Fundamentals of Vibration Analysis*, McGraw-Hill, New York, 1956.

1.39. L. Meirovitch, *Analytical Methods in Vibrations*, Macmillan, New York, 1967.

1.40. M. C. Potter and J. Goldberg, *Mathematical Methods* (2nd Ed.), Prentice-Hall, Englewood Cliffs, 1987.

1.41. J. E. Shigley and C. R. Mischke, *Mechanical Engineering Design* (5th Ed.), McGraw-Hill, New York, 1989.

REVIEW QUESTIONS

1.1. What was Galileo's contribution to the development of vibration theory?

1.2. Give the names of two early investigators who derived the governing equation for the lateral vibration of prismatic bars.

1.3. Give two examples each of the bad and the good effects of vibration.

1.4. What are the three elementary parts of a vibrating system?

1.5. Define the degree of freedom of a vibrating system.

1.6. What is the difference between a discrete and a continuous system? Is it possible to solve any vibration problem as a discrete one?

1.7. What is the difference between free and forced vibration?

1.8. In vibration analysis, can we always disregard damping?

1.9. Can we identify a nonlinear vibration problem by looking at its governing differential equation?

1.10. What is the difference between deterministic and random vibration? Give two practical examples of each.

1.11. What methods are available for solving the governing equations of a vibration problem?

1.12. How do you connect several springs to increase the overall stiffness?

1.13. Define spring stiffness and damping constant.

1.14. What are the common types of damping?

1.15. What is the difference between harmonic motion and periodic motion?

1.16. State three different ways of expressing a periodic function in terms of its harmonics.

1.17. Define these terms: cycle, amplitude, phase angle, linear frequency, period, and natural frequency.

1.18. How are τ, ω, and f related to each other?

1.19. How can we obtain the frequency, phase, and amplitude of a harmonic motion from the corresponding rotating vector?

1.20. How do you add two harmonic motions having different frequencies?

1.21. Suggest two methods for finding the time derivative of a harmonic motion.

1.22. What is harmonic analysis?

1.23. Give the names of two technical journals and two abstract journals for vibration research.

1.24. What are half range expansions?

PROBLEMS

The problem assignments are organized as follows:

Problems	Section Covered	Topic Covered
1.1–1.3	1.6	Vibration analysis procedure
1.4–1.16, 1.21	1.7	Spring elements
1.10, 1.16–1.19	1.8	Mass elements
1.20, 1.22, 1.23	1.9	Damping elements
1.24–1.37	1.10	Harmonic motion
1.38–1.49	1.11	Harmonic analysis
1.50–1.53	1.13	Computer program

1.1.* A study of the response of a human body subjected to vibration/shock is important in many applications. In a standing posture, the masses of head, upper torso, hips, and

*The asterisk denotes a design type problem or a problem with no unique answer.

legs, and the elasticity/damping of neck, spinal column, abdomen, and legs influence the response characteristics. Develop a sequence of three improved approximations for modeling the human body.

1.2.* A reciprocating engine is mounted on a foundation as shown in Fig. 1.37. The unbalanced forces and moments developed in the engine are transmitted to the frame and the foundation. An elastic pad is placed between the engine and the foundation block to reduce the transmission of vibration. Develop two mathematical models of the system using a gradual refinement of the modeling process.

1.3.* An automobile moving over a rough road (Fig. 1.38) can be modeled considering (a) the weight of the car body, passengers, seats, front wheels, and rear wheels; (b) the elasticity of tires (suspension), main springs, and seats; and (c) damping of the seats, shock absorbers, and tires. Develop three mathematical models of the system using a gradual refinement in the modeling process.

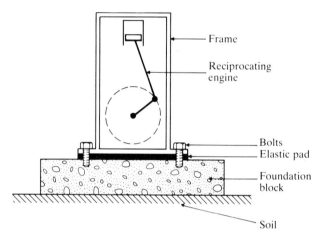

Figure 1.37 A reciprocating engine on foundation.   **Figure 1.38** An automobile moving on a rough road.

1.4. Determine the equivalent spring constant of the system shown in Fig. 1.39.

1.5. In Fig. 1.40, find the equivalent spring constant of the system in the direction of θ.

1.6. Find the equivalent torsional spring constant of the system shown in Fig. 1.41.

1.7. A machine of mass $m = 500$ kg is mounted on a simply supported steel beam of length $l = 2$ m having a rectangular cross-section (depth = 0.1, m, width = 1.2 m) and Young's modulus $E = 2.06 \times 10^{11}$ N/m^2. To reduce the vertical deflection of the beam, a spring of stiffness k is attached at the mid-span, as shown in Fig. 1.42. Determine the value of k needed to reduce the deflection of the beam to one-third of its original value. Assume that the mass of the beam is negligible.

1.8. Four identical rigid bars—each of length a—are connected to a spring of stiffness k to form a structure for carrying a vertical load P, as shown in Figs. 1.43(a) and (b). Find the equivalent spring constant of the system k_{eq}, for each case, disregarding the masses of the bars and the friction in the joints.

1.9. The tripod shown in Fig. 1.44 is used for mounting an electronic instrument that finds the distance between two points in space. The legs of the tripod are located symmetri-

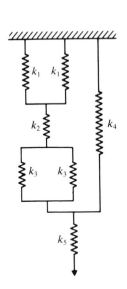

Figure 1.39

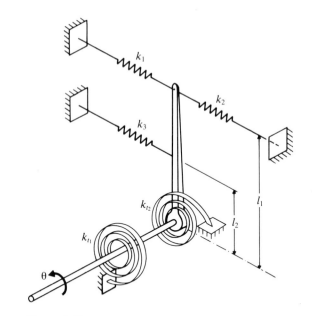

Figure 1.40

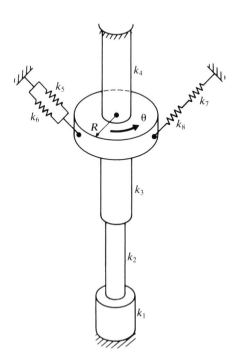

Figure 1.41

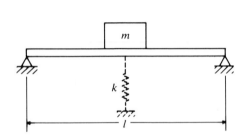

Figure 1.42

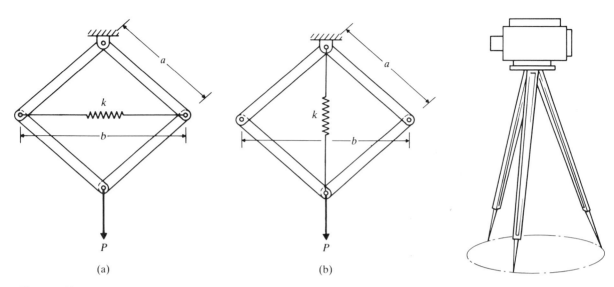

Figure 1.43

Figure 1.44

cally about the mid-vertical axis, each leg making an angle α with the vertical. If each leg has a length of l and axial stiffness of k, find the equivalent spring stiffness of the tripod in the vertical direction.

1.10. Find the equivalent spring constant and equivalent mass of the system shown in Fig. 1.45 with reference to θ.

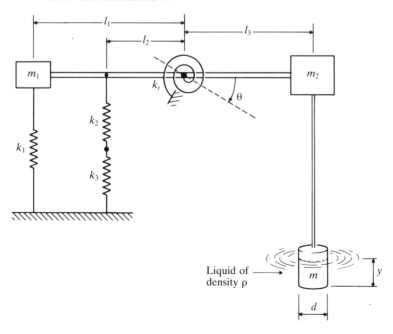

Figure 1.45

1.11. Find the length of the equivalent uniform hollow shaft of inner diameter d and thickness t that has the same axial spring constant as that of the solid conical shaft shown in Fig. 1.46.

1.12. The force-deflection characteristic of a spring is described by $F = 500x + 2x^3$ where the force (F) is in Newtons and the deflection (x) is in millimeters. Find (i) the linearized spring constant at $x = 10$ mm, and (ii) the spring forces at $x = 9$ mm and $x = 11$ mm using the linearized spring constant. Also find the error in the spring forces found in (ii).

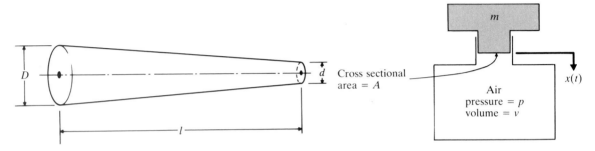

Figure 1.46   **Figure 1.47**

1.13. Figure 1.47 shows an air spring. This type of spring is generally used for obtaining very low natural frequencies while maintaining zero deflection under static loads. Find the spring constant of this air spring by assuming that the pressure p and volume v change adiabatically when the mass m moves.
Hint: $pv^\gamma = $ constant for an adiabatic process, where γ is the ratio of specific heats. For air, $\gamma = 1.4$.

1.14. Find the equivalent spring constant of the system shown in Fig. 1.48 in the direction of the load P.

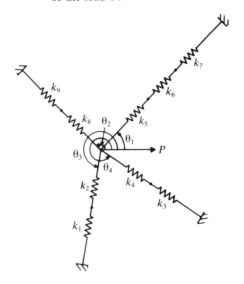

Figure 1.48

1.15.* Design a steel helical compression spring to satisfy the following requirements:

Spring stiffness $(k) \geqslant 8000$ N/mm

Fundamental natural frequency of vibration $(f_1) \geqslant 0.4$ Hz

Spring index $(D/d) \geqslant 6$

Number of active turns $(N) \geqslant 10$.

The stiffness and fundamental natural frequency of the spring are given by [1.41]:

$$k = \frac{Gd^4}{8D^3N} \qquad \text{and} \qquad f_1 = 1/2\sqrt{\frac{kg}{W}}$$

where G = shear modulus, d = wire diameter, D = coil diameter, W = weight of the spring, and g = acceleration due to gravity.

1.16. Two sector gears, located at the ends of links 1 and 2, are engaged together and rotate about O_1 and O_2, as shown in Fig. 1.49. If links 1 and 2 are connected to springs k_1 to k_4 and k_{t1} and k_{t2} as shown, find the equivalent torsional spring stiffness and equivalent mass moment of inertia of the system with reference to θ_1. Assume (a) the mass moment of inertia of link 1 (including the sector gear) about O_1 as J_1 and that of link 2 (including the sector gear) about O_2 as J_2, and (b) the angles θ_1 and θ_2 to be small.

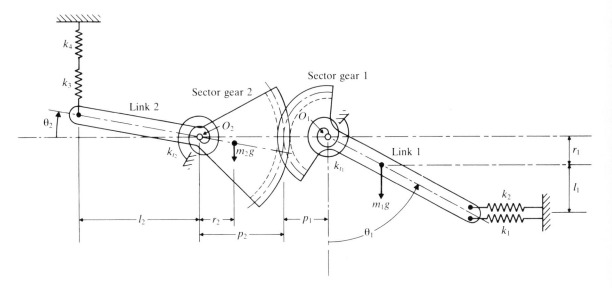

Figure 1.49

1.17. In Fig. 1.50, find the equivalent mass of the rocker arm assembly, referred to the x coordinate.

1.18. Find the equivalent mass moment of inertia of the gear train shown in Fig. 1.51, with reference to the driving shaft. In Fig. 1.51, J_i and n_i denote the mass moment of inertia and the number of teeth, respectively, of gear i, $i = 1, 2, \ldots, 2N$.

1.19. Two masses, having mass moments of inertia J_1 and J_2, are placed on rotating rigid shafts that are connected by gears, as shown in Fig. 1.52. If the number of teeth on

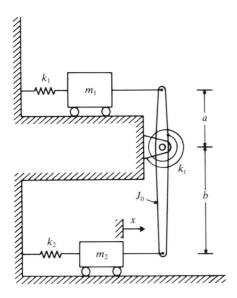

Figure 1.50

gears 1 and 2 are n_1 and n_2, respectively, find the equivalent mass moment of inertia corresponding to θ_1.

1.20. Find a single equivalent damping constant for the following cases:

   i. When three dampers are parallel.

   ii. When three dampers are in series.

   iii. When three dampers are connected to a rigid bar (Fig. 1.53), and the equivalent damper is at the site c_1.

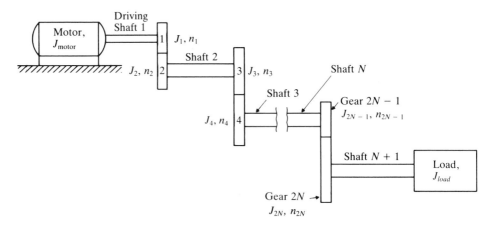

Figure 1.51

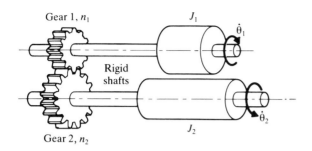

Figure 1.52 Rotational masses on geared shafts.

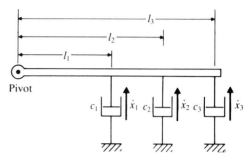

Figure 1.53 Dampers connected to a rigid bar.

iv. When three torsional dampers are located on geared shafts (Fig. 1.54), and the equivalent damper is at the location c_{t1}.

Hint: The energy dissipated by a viscous damper in a cycle during harmonic motion is given by $\pi c \omega X^2$, where c is the damping constant, ω is the frequency, and X is the amplitude of oscillation.

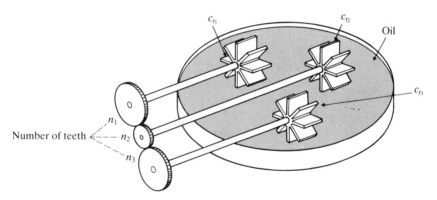

Figure 1.54 Dampers located on geared shafts.

1.21.* Design an air spring using a cylindrical container and a piston to achieve a spring constant of 75 lb/in. Assume that the maximum air pressure available is 200 psi.

1.22.* Design a shock absorber (piston-cylinder type dashpot) to obtain a damping constant of 10^5 lb-sec/in. using SAE 30 oil at 70° F. The diameter of the piston has to be less than 2.5 inches.

1.23. Develop an expression for the damping constant of the rotational damper shown in Fig. 1.55 in terms of D, d, l, h, ω, and μ, where ω denotes the constant angular velocity of the inner cylinder, and d and h represent the radial and axial clearances between the inner and outer cylinders.

1.24. Express the complex number $5 + 2i$ in the exponential form $Ae^{i\theta}$.

1.25. Add the two complex numbers $(1 + 2i)$ and $(3 - 4i)$ and express the result in the form $Ae^{i\theta}$.

1.26. Verify that the harmonic functions $x_1(t) = A_1\cos(\omega t + \phi)$ and $x_2(t) = A_2\sin(\omega t + \phi)$ satisfy the differential equation $d^2x/dt^2 + \omega^2 x = 0$.

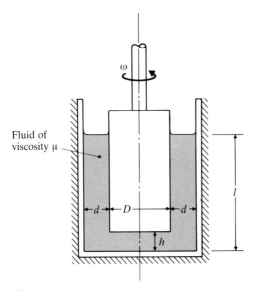

Figure 1.55

1.27. A machine is subjected to the motion $x(t) = A \cos(50t + \alpha)$ mm. The initial conditions are given by $x(0) = 3$ mm and $\dot{x}(0) = 1.0$ m/s.

  i. Find the constants A and α.

  ii. Express the motion in the form $x(t) = A_1 \cos \omega t + A_2 \sin \omega t$, and identify the constants A_1 and A_2.

1.28. Show that any linear combination of $\sin \omega t$ and $\cos \omega t$ such that $x(t) = A_1 \cos \omega t + A_2 \sin \omega t$ (A_1, A_2 = constants) represents a simple harmonic motion.

1.29. Find the sum of the two harmonic motions $x_1(t) = 5 \cos(3t + 1)$ and $x_2(t) = 10 \cos(3t + 2)$. Use:

  i. trigonometric relations

  ii. vector addition

  iii. complex number representation

1.30. If one of the components of the harmonic motion $x(t) = 10 \sin(\omega t + 60°)$ is $x_1(t) = 5 \sin(\omega t + 30°)$, find the other component.

1.31. Consider the two harmonic motions $x_1(t) = \frac{1}{2} \cos \frac{\pi}{2} t$ and $x_2(t) = \sin \pi t$. Is the sum $x_1(t) + x_2(t)$ a periodic motion? If so, what is its period?

1.32. Consider two harmonic motions of different frequencies: $x_1(t) = 2 \cos 2t$ and $x_2(t) = \cos 3t$. Is the sum $x_1(t) + x_2(t)$ a harmonic motion? If so, what is its period?

1.33. Consider the two harmonic motions $x_1(t) = \frac{1}{2} \cos \frac{\pi}{2} t$ and $x_2(t) = \cos \pi t$. Is the difference $x(t) = x_1(t) - x_2(t)$ a harmonic motion? If so, what is its period?

1.34. Whenever two harmonic motions $x_1(t)$ and $x_2(t)$ having slightly different frequencies are combined, the amplitude of the resulting motion $x(t)$ varies between a maximum and a minimum value. Every time the amplitude of $x(t)$ reaches a maximum, there is said to be a *beat*. What are the maximum and minimum amplitudes of the combined motion $x(t) = x_1(t) + x_2(t)$ when $x_1(t) = 3 \sin 30t$ and $x_2(t) = 3 \sin 29t$? Also find the frequency of beats corresponding to $x(t)$.

1.35. A harmonic motion has an amplitude of 0.05 m and a frequency of 10 Hz. Find its period, maximum velocity, and maximum acceleration.

1.36. An accelerometer mounted on a building frame indicates that the frame is vibrating harmonically at 15 cps, with a maximum acceleration of 0.5 g. Determine the amplitude and the maximum velocity of the building frame.

1.37. The maximum amplitude and the maximum acceleration of the foundation of a centrifugal pump were found to be $x_{max} = 0.25$ mm and $\ddot{x}_{max} = 0.4$ g. Find the operating speed of the pump.

1.38. Express the complex Fourier series expansion of Eq. (1.58) in the form

$$x(t) = \sum_{n=-\infty}^{\infty} c_n e^{in\omega t}$$

and identify the expression for c_n.

1.39. Prove that the sine Fourier components (b_n) are zero for even functions, that is, when $x(-t) = x(t)$. Also prove that the cosine Fourier components (a_0 and a_n) are zero for odd functions, that is, when $x(-t) = -x(t)$.

1.40. Find the Fourier series expansions of the functions shown in Figs. 1.34(ii) and (iii). Also, find their Fourier series expansions when the time axis is shifted down by a distance A.

1.41. The impact force created by a forging hammer can be modeled as shown in Fig. 1.56. Determine the Fourier series expansion of the impact force.

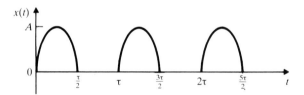

Figure 1.56

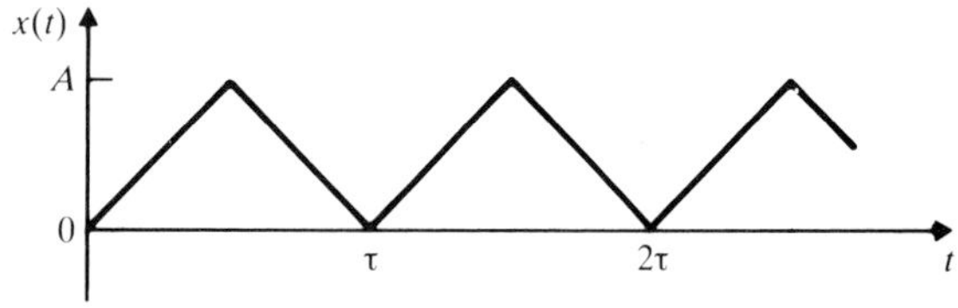

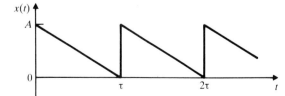

Figure 1.57

1.42–1.44. Find the Fourier series expansions of the periodic functions shown in Figs. 1.57 to 1.59. Also plot the corresponding frequency spectra.

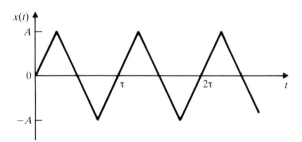

Figure 1.58

Figure 1.59

1.45. Conduct a harmonic analysis, including the first three harmonics, of the function given below:

t_i	0.02	0.04	0.06	0.08	0.10	0.12	0.14	0.16	0.18
x_i	9	13	17	29	43	59	63	57	49

t_i	0.20	0.22	0.24	0.26	0.28	0.30	0.32
x_i	35	35	41	47	41	13	7

1.46. In a centrifugal fan (Fig. 1.60a), the air at any point is subjected to an impulse each time a blade passes the point, as shown in Fig. 1.60(b). The frequency of these impulses is determined by the speed of rotation of the impeller n and the number of blades N in the impeller. For $n = 100$ rpm and $N = 4$, determine the first three harmonics of the pressure fluctuation shown in Fig. 1.60(b).

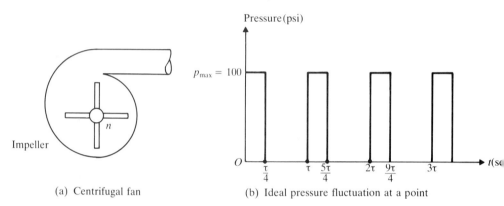

(a) Centrifugal fan   (b) Ideal pressure fluctuation at a point

Figure 1.60

1.47. The torque (M_t) variation with time, of an internal combustion engine, is given in Table 1.3. Make a harmonic analysis of the torque. Find the amplitudes of the first three harmonics.

TABLE 1.3

$t(s)$	$M_t(N-m)$	$t(s)$	$M_t(N-m)$	$t(s)$	$M_t(N-m)$
0.00050	770	0.00450	1890	0.00850	1050
0.00100	810	0.00500	1750	0.00900	990
0.00150	850	0.00550	1630	0.00950	930
0.00200	910	0.00600	1510	0.01000	890
0.00250	1010	0.00650	1390	0.01050	850
0.00300	1170	0.00700	1290	0.01100	810
0.00350	1370	0.00750	1190	0.01150	770
0.00400	1610	0.00800	1110	0.01200	750

1.48.* A slider crank mechanism is shown in Fig. 1.61. Derive an expression for the motion of the piston P in terms of the crank length r, connecting rod length l, and the constant angular velocity of the crank ω.

    i. Discuss the feasibility of using the mechanism for the generation of harmonic motion.

    ii. Find the value of l/r for which the amplitude of every higher harmonic is smaller than that of the first harmonic by a factor of at least 25.

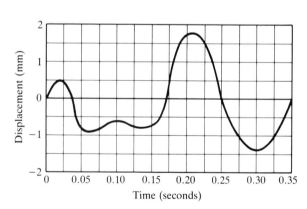

Figure 1.62

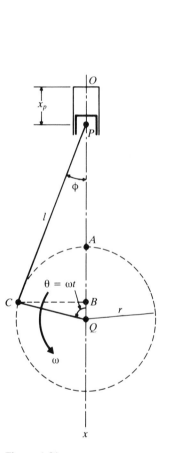

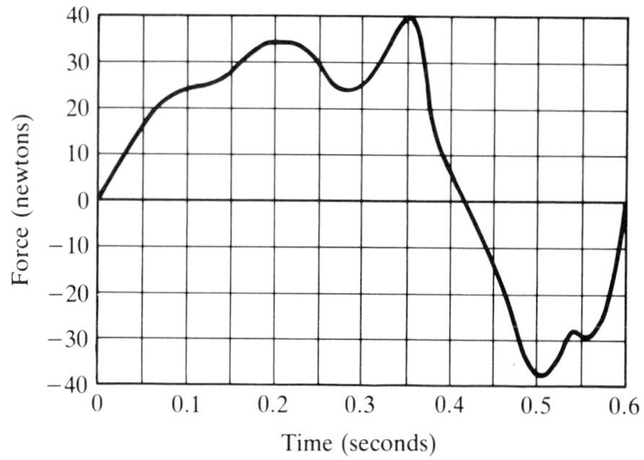

Figure 1.61

Figure 1.63

1.49. Make a harmonic analysis of the function shown in Fig. 1.62, including the first three harmonics.

1.50. Solve Problem 1.47 using subroutine FORIER.

1.51. Solve Problem 1.45 using subroutine FORIER.

1.52. Find the first six harmonics of the function shown in Fig. 1.62, using subroutine FORIER.

1.53. Use subroutine FORIER to conduct a harmonic analysis of the function shown in Fig. 1.63, including the first ten harmonics.

CHAPTER 2

Free Vibration of Single Degree of Freedom Systems

Sir Isaac Newton (1642–1727) was an English natural philosopher, a professor of mathematics at Cambridge University, and President of the Royal Society. His "Principia Mathematica" (1687), which deals with the laws and conditions of motion, is considered to be the greatest scientific work ever produced. The definitions of force, mass, and momentum, and his three laws of motion crop up continually in dynamics. Quite fittingly, the unit of force named "Newton" in SI units happens to be the approximate weight of an average apple, which inspired him to study the laws of gravity. (Courtesy of the Granger Collection)

2.1 INTRODUCTION

Figure 2.1(a) shows a spring-mass system that represents the simplest possible vibratory system. It is called a single degree of freedom system since one coordinate (x) is sufficient to specify the position of the mass at any time. There is no external force applied to the mass; hence the motion resulting from an initial disturbance will be a free vibration. Since there is no element that causes dissipation of energy during the motion of the mass, the amplitude of motion remains constant with time; it is an *undamped* system. In actual practice, except in a vacuum, the amplitude of free vibration diminishes gradually over time, due to the resistance offered by the surrounding medium (such as air). Such vibrations are said to be *damped*. The study of the free vibration of undamped and damped single degree of freedom systems is fundamental to the understanding of more advanced topics in vibrations.

Several mechanical and structural systems can be idealized as single degree of freedom systems. In many practical systems, the mass is distributed, but for a simple analysis, it can be approximated by a single point mass. Similarly, the elasticity of the system, which may be distributed throughout the system, can also be idealized by a single spring. For the cam-and-follower system shown in Fig. 1.21,

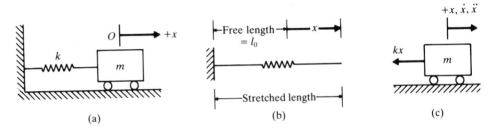

Figure 2.1. A spring-mass system in horizontal position.

for example, the various masses were replaced by an equivalent mass (m_{eq}) in Example 1.3. The elements of the follower system (pushrod, rocker arm, valve, and valve spring) are all elastic but can be reduced to a single equivalent spring of stiffness k_{eq}. For a simple analysis, the cam-and-follower system can thus be idealized as a single degree of freedom spring-mass system, as shown in Fig. 2.2. Similarly, the building frame shown in Fig. 2.3(a) can be idealized as a spring-mass system, as shown in Fig. 2.3(b). In this case, since the spring constant k is merely the ratio of force to deflection, it can be determined from the geometric and material properties of the columns. The mass of the idealized system is the same as that of the floor if we assume the mass of the columns to be negligible.

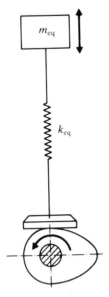

Figure 2.2. Equivalent spring-mass system for the cam-and-follower system of Fig. 1.21.

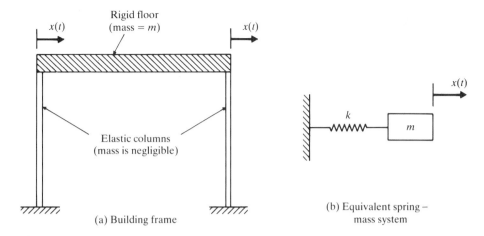

Figure 2.3.   Idealization of a building frame.

2.2 FREE VIBRATION OF AN UNDAMPED TRANSLATIONAL SYSTEM

**2.2.1
Equation of
Motion Using
Newton's
Second Law
of Motion**

Spring-mass System in Horizontal Position. Consider the undamped single degree of freedom system shown in Fig. 2.1(a). The mass is supported on frictionless rollers and can have translatory motion in the horizontal direction. The unstretched length of the spring is l_0. Let the mass be displaced a distance $+x$ from its rest position. This results in a spring force kx, as shown in Fig. 2.1(c). Newton's second law states that

$$\text{mass} \times \text{acceleration} = \text{resultant force on the mass} \qquad (2.1)$$

The application of Eq. (2.1) to the mass m yields the equation of motion

$$m\ddot{x} = -kx$$

or

$$m\ddot{x} + kx = 0 \qquad (2.2)$$

where $\ddot{x} = \dfrac{d^2x}{dt^2}$ is the acceleration of the mass.

Spring-mass System in Vertical Position. Consider the configuration of the spring-mass system shown in Fig. 2.4(a). The mass hangs at the lower end of a spring, which in turn is attached to a rigid support at its upper end. At rest the mass will hang in a position called the *static equilibrium position*, in which the upward spring force exactly balances the downward gravitational force on the mass. In this position the length of the spring is $l_0 + \delta_{st}$, where δ_{st} is the static deflection—the elongation due to the weight W of the mass m. From Fig. 2.4(a), we find that, for

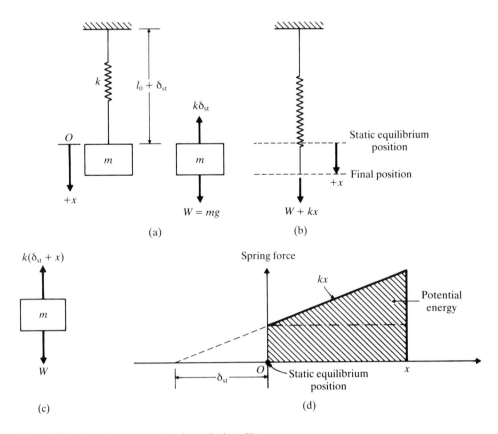

Figure 2.4. A spring-mass system in vertical position.

static equilibrium,

$$W = mg = k\delta_{st} \tag{2.3}$$

where g is the acceleration due to gravity. Let the mass be deflected a distance $+x$ from its static equilibrium position; then the spring force is $-k(x + \delta_{st})$, as shown in Fig. 2.4(c). The application of Newton's second law of motion to the mass m gives

$$m\ddot{x} = -k(x + \delta_{st}) + W$$

and since $k\delta_{st} = W$, we obtain

$$m\ddot{x} + kx = 0 \tag{2.4}$$

Notice that Eqs. (2.2) and (2.4) are identical. This indicates that when a mass moves in a vertical direction, we can ignore its weight, provided we measure x from its static equilibrium position.

Equation (2.2) can also be derived by using the conservation of energy principle. To apply this principle, first note that the system shown in Fig. 2.1(a) is conservative, since there is no energy dissipation due to damping. During vibration, the energy of the system is partly kinetic and partly potential. The kinetic energy T is stored in the mass by virtue of its velocity, and the potential energy U is stored in the spring by virtue of its elastic deformation. Due to the conservation of energy, we have

**2.2.2
Equation of
Motion Using
the Principle
of Conservation
of Energy**

$$T + U = \text{constant}$$

or

$$\frac{d}{dt}(T + U) = 0 \tag{2.5}$$

The kinetic and potential energies are given by

$$T = \tfrac{1}{2}m\dot{x}^2 \tag{2.6}$$

and*

$$U = \tfrac{1}{2}kx^2 \tag{2.7}$$

Substitution of Eqs. (2.6) and (2.7) into Eq. (2.5) yields the desired equation

$$m\ddot{x} + kx = 0 \tag{2.2}$$

**2.2.3
Solution**

The solution of Eq. (2.2) can be found by assuming

$$x(t) = Ce^{st} \tag{2.8}$$

where C and s are constants to be determined. Substitution of Eq. (2.8) into Eq. (2.2) gives

$$C(ms^2 + k) = 0$$

Since C cannot be zero, we have

$$ms^2 + k = 0 \tag{2.9}$$

and hence

$$s = \pm\left(-\frac{k}{m}\right)^{1/2} = \pm i\omega_n \tag{2.10}$$

where $i = (-1)^{1/2}$ and

$$\omega_n = \left(\frac{k}{m}\right)^{1/2} \tag{2.11}$$

Equation (2.9) is called the *auxiliary* or the *characteristic* equation corresponding to

---

* Equation (2.7) can also be derived by considering the weight of the mass (Fig. 2.4). Since the spring force is mg at $x = 0$, the potential energy of the spring under the deformation x will be $mgx + \tfrac{1}{2}kx^2$, as shown in Fig. 2.4(d). The potential energy of the system due to the change in elevation of the mass (note that $+x$ is downwards) is $-mgx$. Thus the net potential energy of the system about the static equilibrium position is given by

U = potential energy of the spring + change in potential energy due to change in elevation of the mass m

$$= mgx + \frac{1}{2}kx^2 - mgx = \frac{1}{2}kx^2$$

the differential Eq. (2.2). The two values of s given by Eq. (2.10) are the roots of the characteristic equation, also known as the *eigenvalues* or the *characteristic values* of the problem. Since both values of s satisfy Eq. (2.9), the general solution of Eq. (2.2) can be expressed as

$$x(t) = C_1 e^{i\omega_n t} + C_2 e^{-i\omega_n t} \tag{2.12}$$

where C_1 and C_2 are constants. By using the identities

$$e^{\pm iat} = \cos \alpha t \pm i \sin \alpha t$$

Eq. (2.12) can be rewritten as

$$x(t) = A_1 \cos \omega_n t + A_2 \sin \omega_n t \tag{2.13}$$

where A_1 and A_2 are new constants. The constants C_1 and C_2 or A_1 and A_2 can be determined from the initial conditions of the system. If the values of displacement $x(t)$ and velocity $\dot{x}(t) = (dx/dt)(t)$ are specified as x_0 and $\dot{x}_0$ at $t = 0$, we have, from Eq. (2.13),

$$x(t = 0) = A_1 = x_0$$
$$\dot{x}(t = 0) = \omega_n A_2 = \dot{x}_0 \tag{2.14}$$

Hence $A_1 = x_0$ and $A_2 = \dot{x}_0/\omega_n$. Thus the solution of Eq. (2.2) subject to the initial conditions of Eq. (2.14) is given by

$$x(t) = x_0 \cos \omega_n t + \frac{\dot{x}_0}{\omega_n} \sin \omega_n t \tag{2.15}$$

2.2.4
Harmonic
Motion

Equations (2.12), (2.13), and (2.15) are harmonic functions of time. The motion is symmetric about the equilibrium position of the mass m. The velocity is a maximum and the acceleration is zero each time the mass passes through this position. At the extreme displacements the velocity is zero and the acceleration is a maximum. Since this represents simple harmonic motion (see Sec. 1.10), the spring-mass system itself is called a *harmonic oscillator*. The quantity ω_n, given by Eq. (2.11), represents the natural frequency of vibration of the system.

Equation (2.13) can be expressed in a different form by introducing the notation

$$A_1 = A \cos \phi$$
$$A_2 = A \sin \phi \tag{2.16}$$

where A and ϕ are the new constants which can be expressed in terms of A_1 and A_2 as

$$A = \left(A_1^2 + A_2^2 \right)^{1/2} = \left[x_0^2 + \left(\frac{\dot{x}_0}{\omega_n} \right)^2 \right]^{1/2} = \text{amplitude}$$

$$\phi = \tan^{-1} \left(\frac{A_2}{A_1} \right) = \tan^{-1} \left(\frac{\dot{x}_0}{x_0 \omega_n} \right) = \text{phase angle} \tag{2.17}$$

Introducing Eq. (2.16) into Eq. (2.13), the solution can be written as

$$x(t) = A \cos(\omega_n t - \phi) \tag{2.18}$$

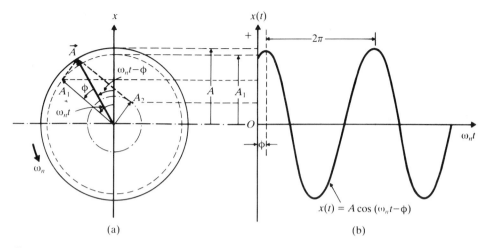

x(t) = A \cos(\omega_n t - \phi)

Figure 2.5. Graphical representation of the motion of a harmonic oscillator.

By using the relations

$$A_1 = A_0 \sin \phi_0$$

$$A_2 = A_0 \cos \phi_0 \tag{2.19}$$

Eq. (2.13) can also be expressed as

$$x(t) = A_0 \sin(\omega_n t + \phi_0) \tag{2.20}$$

where

$$A_0 = A = \left[x_0^2 + \left(\frac{\dot{x}_0}{\omega_n} \right)^2 \right]^{1/2} \tag{2.21}$$

and

$$\phi_0 = \tan^{-1} \left(\frac{x_0 \omega_n}{\dot{x}_0} \right) \tag{2.22}$$

The nature of harmonic oscillation can be represented graphically as in Fig. 2.5(a). If $\vec{A}$ denotes a vector of magnitude A which makes an angle $\omega_n t - \phi$ with respect to the vertical (x) axis, then the solution, Eq. (2.18), can be seen to be the projection of the vector $\vec{A}$ on the x-axis. The constants A_1 and A_2 of Eq. (2.13), given by Eq. (2.16), are merely the rectangular components of $\vec{A}$ along two orthogonal axes making angles ϕ and $-(\frac{\pi}{2} - \phi)$ with respect to the vector $\vec{A}$. Since the angle $\omega_n t - \phi$ is a linear function of time, it increases linearly with time; the entire diagram thus rotates anticlockwise at an angular velocity ω_n. As the diagram (Fig. 2.5a) rotates, the projection of $\vec{A}$ onto the x-axis varies harmonically so that the motion repeats itself every time the vector $\vec{A}$ sweeps an angle of 2π. The projection of $\vec{A}$, namely $x(t)$, is shown plotted in Fig. 2.5(b) as a function of time.

The phase angle ϕ can also be interpreted as the angle between the origin and the first peak.

Note the following aspects of the spring-mass system:

1. If the spring-mass system is in a vertical position, as shown in Fig. 2.4(a), the circular natural frequency can be expressed as

$$\omega_n = \left(\frac{k}{m}\right)^{1/2} \tag{2.23}$$

The spring constant k can be expressed in terms of the mass m from Eq. (2.3) as

$$k = \frac{W}{\delta_{st}} = \frac{mg}{\delta_{st}} \tag{2.24}$$

Substitution of Eq. (2.24) into Eq. (2.11) yields

$$\omega_n = \left(\frac{g}{\delta_{st}}\right)^{1/2} \tag{2.25}$$

Hence the natural frequency in cycles per second and the natural period are given by

$$f_n = \frac{1}{2\pi}\left(\frac{g}{\delta_{st}}\right)^{1/2} = \frac{1}{2\pi}\,\omega \tag{2.26}$$

$$\tau_n = \frac{1}{f_n} = 2\pi\left(\frac{\delta_{st}}{g}\right)^{1/2} = \frac{2\pi}{\omega} \tag{2.27}$$

Thus, when the mass vibrates in a vertical direction, we can compute the natural frequency and the period of vibration by simply measuring the static deflection δ_{st}. It is not necessary that we know the spring stiffness k and the mass m.

2. From Eq. (2.18), the velocity $\dot{x}(t)$ and the acceleration $\ddot{x}(t)$ of the mass m at time t can be obtained as

$$\dot{x}(t) = \frac{dx}{dt}(t) = -\omega_n A \sin(\omega_n t - \phi) = \omega_n A \cos\left(\omega_n t - \phi + \frac{\pi}{2}\right)$$

$$\ddot{x}(t) = \frac{d^2 x}{dt^2}(t) = -\omega_n^2 A \cos(\omega_n t - \phi) = \omega_n^2 A \cos(\omega_n t - \phi + \pi) \tag{2.28}$$

Equation (2.28) shows that the velocity leads the displacement by $\frac{\pi}{2}$ and the acceleration leads the displacement by π.

3. If the initial displacement (x_0) is zero, Eq. (2.18) becomes

$$x(t) = \frac{\dot{x}_0}{\omega_n}\cos\left(\omega_n t - \frac{\pi}{2}\right) = \frac{\dot{x}_0}{\omega_n}\sin \omega_n t \tag{2.29}$$

On the other hand, if the initial velocity $(\dot{x}_0)$ is zero, the solution becomes

$$x(t) = x_0 \cos \omega_n t \tag{2.30}$$

EXAMPLE 2.1   Natural Frequency of Hoisting System

Find the natural frequency of vibration in the vertical direction of the hoisting system shown in Fig. 1.16(a).

Given: Hoisting system of Fig. 1.16(a) with cantilever, rope, and weight.

Find: Natural frequency of vibration of the system in the vertical direction.

Approach: Single degree of freedom idealization.

Solution: The equivalent spring constant of the system (cantilever beam and rope) was derived in Example 1.1:

$$k_{eq} = \frac{k_b k_r}{k_b + k_r} = \frac{E}{4}\left(\frac{\pi a t^2 d^2}{\pi b^2 d^3 + lat^3} \right) \tag{E.1}$$

The cantilever, rope, and the weight being lifted can now be modeled as a single degree of freedom system, as shown in Fig. 1.16(c). This leads to the natural frequency of the system:

$$\omega_n = \left(\frac{k_{eq}}{m} \right)^{1/2} = \left(\frac{k_{eq} g}{W} \right)^{1/2} = \left[\frac{Eg}{4W}\left(\frac{\pi a t^2 d^2}{\pi b^2 d^3 + lat^3} \right) \right]^{1/2} \tag{E.2}$$

EXAMPLE 2.2   Natural Frequency of Pulley System

Determine the natural frequency of the system shown in Fig. 2.6. Assume the pulleys to be frictionless and of negligible mass.

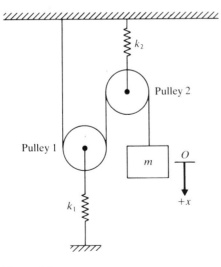

Figure 2.6

Given: System consisting of two pulleys and a mass as shown in Fig. 2.6.

Find: Natural frequency of vibration of the mass.

Approach: Single degree of freedom idealization.

Solution. Since the pulleys are frictionless and massless, the tension in the rope is constant and is equal to the weight W of the mass m. Thus the upward force acting on pulley 1 is $2W$, and the downward force acting on pulley 2 is $2W$. The center of pulley 1 moves up by a distance $2W/k_1$, and the center of pulley 2 moves down by $2W/k_2$. Thus the total movement of the mass m is

$$2\left(\frac{2W}{k_1} + \frac{2W}{k_2} \right)$$

as the rope on either side of the pulley is free to move the mass downward. If k_{eq} denotes the equivalent spring constant of the system,

$$\frac{\text{weight of the mass}}{\text{equivalent spring constant}} = \text{net displacement of the mass}$$

$$\frac{W}{k_{eq}} = 4W\left(\frac{1}{k_1} + \frac{1}{k_2} \right) = \frac{4W(k_1 + k_2)}{k_1 k_2}$$

$$k_{eq} = \frac{k_1 k_2}{4(k_1 + k_2)} \tag{E.1}$$

If the equation of motion of the mass is written as

$$m\ddot{x} + k_{eq}x = 0 \tag{E.2}$$

the natural frequency is given by

$$\omega_n = \left(\frac{k_{eq}}{m} \right)^{1/2} = \left[\frac{k_1 k_2}{4m(k_1 + k_2)} \right]^{1/2} \text{ rad/sec} \tag{E.3}$$

or

$$f_n = \frac{\omega_n}{2\pi} = \frac{1}{4\pi}\left[\frac{k_1 k_2}{m(k_1 + k_2)} \right]^{1/2} \text{ cycles/sec} \tag{E.4}$$

2.3 FREE VIBRATION OF AN UNDAMPED TORSIONAL SYSTEM

If a rigid body oscillates about a specific reference axis, the resulting motion is called *torsional vibration*. In this case, the displacement of the body is measured in terms of an angular coordinate. In a torsional vibration problem, the restoring moment may be due to the torsion of an elastic member or to the unbalanced moment of a force or couple.

Figure 2.7 shows a disc, which has a polar mass moment of inertia J_0, mounted at one end of a solid circular shaft, the other end of which is fixed. Let the angular rotation of the disc about the axis of the shaft be θ; θ also represents the angle of twist of the shaft. From the theory of torsion of circular shafts [2.1], we have the

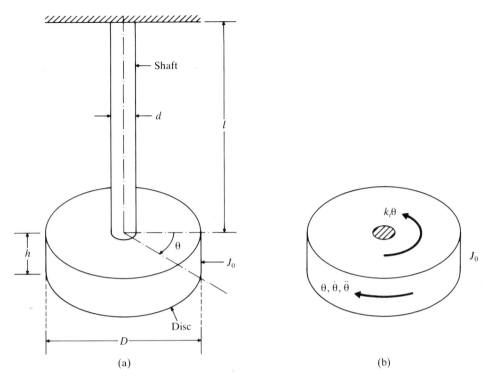

Figure 2.7. Torsional vibration of a disc.

relation

$$M_t = \frac{GJ\theta}{l} \tag{2.31}$$

where M_t is the torque that produces the twist θ, G is the shear modulus, l is the length of the shaft, J is the polar moment of inertia of the cross section of the shaft given by

$$J = \frac{\pi d^4}{32} \tag{2.32}$$

and d is the diameter of the shaft. If the disc is displaced by θ from its equilibrium position, the shaft provides a restoring torque of magnitude M_t. Thus the shaft acts as a torsional spring with a torsional spring constant

$$k_t = \frac{M_t}{\theta} = \frac{GJ}{l} = \frac{\pi Gd^4}{32l} \tag{2.33}$$

**2.3.1
Equation
of Motion**

The equation of the angular motion of the disc about its axis can be derived by using Newton's second law or the principle of conservation of energy. By considering the free body diagram of the disc (Fig. 2.7b), we can derive the equation of motion by applying Newton's second law of motion:

$$J_0\ddot{\theta} + k_t\theta = 0 \tag{2.34}$$

which can be seen to be identical to Eq. (2.2) if the polar mass moment of inertia J_0, the angular displacement θ, and the torsional spring constant k_t are replaced by the mass m, the displacement x, and the linear spring constant k, respectively. Thus the natural circular frequency of the torsional system is

$$\omega_n = \left(\frac{k_t}{J_0}\right)^{1/2} \tag{2.35}$$

and the period and frequency of vibration in cycles per second are

$$\tau_n = 2\pi\left(\frac{J_0}{k_t}\right)^{1/2} \tag{2.36}$$

$$f_n = \frac{1}{2\pi}\left(\frac{k_t}{J_0}\right)^{1/2} \tag{2.37}$$

Note the following aspects of this system:

1. If the cross section of the shaft supporting the disc is not circular, an appropriate torsional spring constant is to be used [2.4, 2.5].

2. The polar mass moment of inertia of a disc is given by

$$J_0 = \frac{\rho h\pi D^4}{32} = \frac{WD^2}{8g}$$

where ρ is the mass density, h is the thickness, D is the diameter, and W is the weight of the disc.

3. The torsional spring-inertia system shown in Fig. 2.7 is referred to as a *torsional pendulum*. One of the most important applications of a torsional pendulum is in a mechanical clock, where a ratchet and pawl convert the regular oscillation of a small torsional pendulum into the movements of the hands.

2.3.2 Solution

The general solution of Eq. (2.34) can be obtained, as in the case of Eq. (2.2):

$$\theta(t) = A_1\cos\omega_n t + A_2\sin\omega_n t \tag{2.38}$$

where ω_n is given by Eq. (2.35), and A_1 and A_2 can be determined from the initial conditions. If

$$\theta(t = 0) = \theta_0 \quad \text{and} \quad \dot{\theta}(t = 0) = \frac{d\theta}{dt}(t = 0) = \dot{\theta}_0 \tag{2.39}$$

the constants A_1 and A_2 can be found:

$$A_1 = \theta_0$$
$$A_2 = \dot{\theta}_0/\omega_n \tag{2.40}$$

Equation (2.38) can also be seen to represent a simple harmonic motion.

EXAMPLE 2.3   **Natural Frequency of Compound Pendulum**

Any rigid body pivoted at a point other than its center of mass will oscillate about the pivot point under its own gravitational force. Such a system is known as a compound pendulum (Fig. 2.8). Find the natural frequency of such a system.

Given: Rigid body oscillating about the pivot point O under gravity.

Find: Natural frequency of angular oscillations.

Approach: Idealize the system as a single degree of freedom torsional system.

Solution. Let O be the point of suspension and G be the center of mass of the compound pendulum, as shown in Fig. 2.8. Let the rigid body oscillate in the xy plane so that the coordinate θ can be used to describe its motion. Let d denote the distance between O and G, and J_0 the mass moment of inertia of the body about the z-axis (perpendicular to both x and y). For a displacement θ, the restoring torque (due to the weight of the body W) is $(Wd \sin \theta)$ and the equation of motion is

$$J_0 \ddot{\theta} + Wd \sin \theta = 0 \tag{E.1}$$

For small angles of oscillation, $\sin \theta \simeq \theta$. Hence Eq. (E.1) can be expressed as

$$J_0 \ddot{\theta} + Wd\theta = 0 \tag{E.2}$$

This gives the natural frequency of the compound pendulum:

$$\omega_n = \left(\frac{Wd}{J_0} \right)^{1/2} = \left(\frac{mgd}{J_0} \right)^{1/2} \tag{E.3}$$

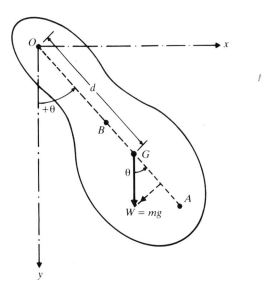

Figure 2.8

Comparing Eq. (E.3) with the natural frequency of a simple pendulum, $\omega_n = (g/l)^{1/2}$ (see Problem 2.27), we can find the length of the equivalent simple pendulum:

$$l = \frac{J_0}{md} \tag{E.4}$$

If J_0 is replaced by mk_0^2 where k_0 is the radius of gyration of the body about O, Eqs. (E.3) and (E.4) become

$$\omega_n = \left(\frac{gd}{k_0^2} \right)^{1/2} \tag{E.5}$$

$$l = \left(\frac{k_0^2}{d} \right) \tag{E.6}$$

If k_G denotes the radius of gyration of the body about G, we have

$$k_0^2 = k_G^2 + d^2 \tag{E.7}$$

and Eq. (E.6) becomes

$$l = \left(\frac{k_G^2}{d} + d \right) \tag{E.8}$$

If the line OG is extended to point A such that

$$GA = \frac{k_G^2}{d} \tag{E.9}$$

Eq. (E.8) becomes

$$l = GA + d = OA \tag{E.10}$$

Hence, from Eq. (E.5), ω_n is given by

$$\omega_n = \left\{ \frac{g}{(k_0^2/d)} \right\}^{1/2} = \left(\frac{g}{l} \right)^{1/2} = \left(\frac{g}{OA} \right)^{1/2} \tag{E.11}$$

This equation shows that, no matter whether the body is pivoted from O or A, its natural frequency is the same. The point A is called the *center of percussion.*

Center of Percussion. The concepts of compound pendulum and center of percussion can be used in many practical applications:

1. A hammer can be shaped to have the center of percussion at the hammer head while the center of rotation is at the handle. In this case, the impact force at the hammer head will not cause any normal reaction at the handle (Fig. 2.9a).

2. In a baseball bat, if the ball is made to strike at the center of percussion while the center of rotation is at the hands, no reaction perpendicular to the bat will be experienced by the batter (Fig. 2.9b). On the other hand, if the ball strikes the bat near the free end or near the hands, the batter will experience pain in the hands as a result of the reaction perpendicular to the bat.

3. In Izod (impact) testing of materials, the specimen is suitably notched and held in a vice fixed to the base of the machine (see Fig. 2.9c). A pendulum is released

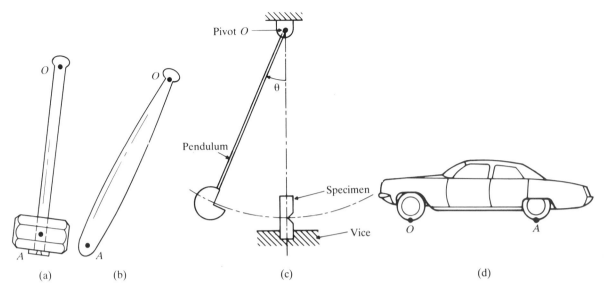

Figure 2.9

from a standard height, and the free end of the specimen is struck by the pendulum as it passes through its lowest position. The deformation and bending of the pendulum can be reduced if the center of percussion is located near the striking edge. In this case, the pivot will be free of any impulsive reaction.

4. In an automobile (shown in Fig. 2.9d), if the front wheels strike a bump, the passengers will not feel any reaction if the center of percussion of the vehicle is located near the rear axle. Similarly, if the rear wheels strike a bump at point A, no reaction will be felt at the front axle (point O) if the center of percussion is located near the front axle. It is desirable, therefore, to have the center of oscillation of the vehicle at one axle and the center of percussion at the other axle [2.2].

2.4 STABILITY CONDITIONS

Consider a uniform rigid bar that is pivoted at one end and connected symmetrically by two springs at the other end, as shown in Fig. 2.10(a). Assume that the mass of the bar is m and that the springs are unstretched when the bar is vertical. When the bar is displaced by an angle θ, the spring force in each spring is $kl \sin \theta$; the total spring force is $2kl \sin \theta$. The gravity force $W = mg$ acts vertically downward through the center of gravity, G. The moment about the point of rotation O due to the angular acceleration $\ddot{\theta}$ is $J_0\ddot{\theta} = (ml^2/3)\ddot{\theta}$. Thus the equation of motion of the

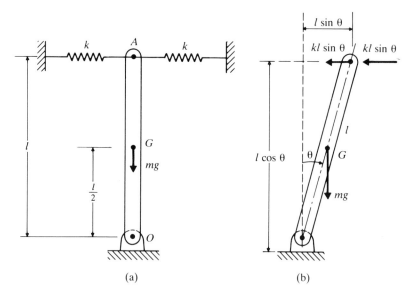

Figure 2.10

bar, for rotation about the point O, can be written as

$$\frac{ml^2}{3}\ddot{\theta} + (2kl\sin\theta)l\cos\theta - W\frac{l}{2}\sin\theta = 0 \qquad (2.41)$$

For small oscillations, Eq. (2.41) reduces to

$$\frac{ml^2}{3}\ddot{\theta} + 2kl^2\theta - \frac{Wl}{2}\theta = 0$$

or

$$\ddot{\theta} + \left(\frac{12kl^2 - 3Wl}{2ml^2}\right)\theta = 0 \qquad (2.42)$$

The solution of Eq. (2.42) depends on the sign of $(12kl^2 - 3Wl)/2ml^2$, as discussed below.

Case 1. When $(12kl^2 - 3Wl)/2ml^2 > 0$, the solution of Eq. (2.42) represents stable oscillations and can be expressed as

$$\theta(t) = A_1\cos\omega_n t + A_2\sin\omega_n t \qquad (2.43)$$

where A_1 and A_2 are constants and

$$\omega_n = \left(\frac{12kl^2 - 3Wl}{2ml^2}\right)^{1/2} \qquad (2.44)$$

Case 2. When $(12kl^2 - 3Wl)/2ml^2 = 0$, Eq. (2.42) reduces to $\ddot{\theta} = 0$ and the solution can be obtained directly by integrating twice as

$$\theta(t) = C_1 t + C_2 \qquad (2.45)$$

For the initial conditions $\theta(t = 0) = \theta_0$ and $\dot{\theta}(t = 0) = \dot{\theta}_0$, the solution becomes

$$\theta(t) = \dot{\theta}_0 t + \theta_0 \tag{2.46}$$

Equation (2.46) shows that the angular displacement increases linearly at a constant velocity $\dot{\theta}_0$. However, if $\dot{\theta}_0 = 0$, Eq. (2.46) denotes a static equilibrium position with $\theta = \theta_0$, that is, the pendulum remains in its original position, defined by $\theta = \theta_0$.

Case 3. When $(12kl^2 - 3Wl)/2ml^2 < 0$, we define

$$\alpha = \left(\frac{3Wl - 12kl^2}{2ml^2} \right)^{1/2}$$

and express the solution of Eq. (2.42) as

$$\theta(t) = B_1 e^{\alpha t} + B_2 e^{-\alpha t} \tag{2.47}$$

where B_1 and B_2 are constants. For the initial conditions $\theta(t = 0) = \theta_0$ and $\dot{\theta}(t = 0) = \dot{\theta}_0$, Eq. (2.47) becomes

$$\theta(t) = \frac{1}{2\alpha} \left[\left(\alpha\theta_0 + \dot{\theta}_0 \right) e^{\alpha t} + \left(\alpha\theta_0 - \dot{\theta}_0 \right) e^{-\alpha t} \right] \tag{2.48}$$

Equation (2.48) shows that $\theta(t)$ increases exponentially with time; hence the motion is unstable. The physical reason for this is that the restoring moment due to the spring $(2kl^2\theta)$, which tries to bring the system to equilibrium position, is less than the nonrestoring moment due to gravity $[-W(l/2)\theta]$, which tries to move the mass away from the equilibrium position. Although the stability conditions are illustrated with reference to Fig. 2.10 in this section, similar conditions need to be examined in the vibration analysis of many engineering systems.

2.5 ENERGY METHOD

For a single degree of freedom system, the equation of motion was derived using the energy method in Section 2.2.2. In this section, we shall use the energy method to find the natural frequencies of single degree of freedom systems. The principle of conservation of energy, in the context of an undamped vibrating system, can be restated as

$$T_1 + U_1 = T_2 + U_2 \tag{2.49}$$

where the subscripts 1 and 2 denote two different instants of time. Specifically, we use the subscript 1 to denote the time when the mass is passing through its static equilibrium position and choose $U_1 = 0$ as reference for the potential energy. If we let the subscript 2 indicate the time corresponding to the maximum displacement of the mass, we have $T_2 = 0$. Thus Eq. (2.49) becomes

$$T_1 + 0 = 0 + U_2 \tag{2.50}$$

If the system is undergoing harmonic motion, then T_1 and U_2 denote the maximum values of T and U, respectively, and Eq. (2.50) becomes

$$T_{\max} = U_{\max} \qquad (2.51)$$

The application of Eq. (2.51), which is also known as *Rayleigh's energy method*, gives the natural frequency of the system directly, as illustrated in the following examples.

EXAMPLE 2.4   **Manometer for Diesel Engine**

The exhaust from a single-cylinder four-stroke diesel engine is to be connected to a silencer, and the pressure therein is to be measured with a simple U-tube manometer (see Fig. 2.11). Calculate the minimum length of the manometer tube so that the natural frequency of oscillation of the mercury column will be 3.5 times slower than the frequency of the pressure fluctuations in the silencer at an engine speed of 600 revolutions per minute. The frequency of pressure fluctuations in the silencer is equal to

$$\frac{\text{number of cylinders} \times \text{speed of the engine}}{2}$$

Given: U-tube manometer, engine speed = 600 rpm, and natural frequency of oscillation = 3.5 times slower than the frequency of pressure fluctuations.

Find: Minimum length of the manometer tube.

Approach: Use energy method to find the natural frequency.

Solution

1. Natural frequency of oscillation of the liquid column: Let the datum in Fig. 2.11 be taken as the equilibrium position of the liquid. If the displacement of the liquid column from the

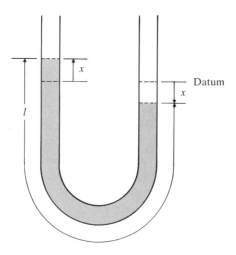

Figure 2.11

equilibrium position is denoted by x, the change in potential energy is given by

U = potential energy of raised liquid column + potential energy of depressed liquid column

= (weight of mercury raised $\times$ displacement of the C.G. of the segment) + (weight of mercury depressed $\times$ displacement of the C.G. of the segment)

$$= (Ax\gamma)\frac{x}{2} + (Ax\gamma)\frac{x}{2} = A\gamma x^2 \tag{E.1}$$

where A is the cross sectional area of the mercury column and γ is the specific weight of mercury. The change in kinetic energy is given by

$$T = \frac{1}{2}(\text{mass of mercury})(\text{velocity})^2$$

$$= \frac{1}{2}\frac{Al\gamma}{g}(\dot{x})^2 \tag{E.2}$$

where l is the length of the mercury column. By assuming harmonic motion, we can write

$$x(t) = X\cos\omega_n t \tag{E.3}$$

where X is the maximum displacement and ω_n is the natural frequency. By substituting Eq. (E.3) into Eqs. (E.1) and (E.2), we obtain

$$U = U_{max}\cos^2\omega_n t \tag{E.4}$$

$$T = T_{max}\sin^2\omega_n t \tag{E.5}$$

where

$$U_{max} = A\gamma X^2 \tag{E.6}$$

and

$$T_{max} = \frac{1}{2}\frac{A\gamma l\omega_n^2}{g}X^2 \tag{E.7}$$

By equating U_{max} to T_{max}, we obtain the natural frequency:

$$\omega_n = \left(\frac{2g}{l}\right)^{1/2} \tag{E.8}$$

2. *Length of the mercury column:* The frequency of pressure fluctuations in the silencer

$$= \frac{1 \times 600}{2}$$

$$= 300 \text{ rev/min}$$

$$= \frac{300 \times 2\pi}{60} = 10\pi \text{ rad/sec} \tag{E.9}$$

Thus the frequency of oscillations of the liquid column in the manometer is $10\pi/3.5 = 9.0$ rad/sec. By using Eq. (E.8), we obtain

$$\left(\frac{2g}{l}\right)^{1/2} = 9.0 \tag{E.10}$$

or

$$l = \frac{2.0 \times 9.81}{(9.0)^2} = 0.243 \text{ m} \tag{E.11}$$

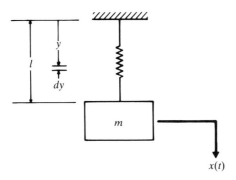

Figure 2.12

EXAMPLE 2.5   **Effect of Mass on ω_n of a Spring**

Determine the effect of the mass of the spring on the natural frequency of the spring-mass system shown in Fig. 2.12.

Given: Spring-mass system.

Find: Effect of mass of the spring on ω_n.

Approach: Add the kinetic energy of the spring to that of the attached mass and use the energy method to find the natural frequency.

Solution. Let l be the total length of the spring. If x denotes the displacement of the lower end of the spring (or mass m), the displacement at distance y from the support is given by $y(x/l)$. Similarly, if $\dot{x}$ denotes the velocity of the mass m, the velocity of a spring element located at distance y from the support is given by $y(\dot{x}/l)$. The kinetic energy of the spring element of length dy is

$$dT_s = \frac{1}{2}\left(\frac{m_s}{l}\, dy\right)\left(\frac{y\dot{x}}{l}\right)^2 \tag{E.1}$$

where m_s is the mass of the spring. The total kinetic energy of the system can be expressed as

$$T = \text{kinetic energy of mass }(T_m) + \text{kinetic energy of spring }(T_s)$$

$$= \frac{1}{2}m\dot{x}^2 + \int_{y=0}^{l}\frac{1}{2}\left(\frac{m_s}{l}\, dy\right)\left(\frac{y^2\dot{x}^2}{l^2}\right)$$

$$= \frac{1}{2}m\dot{x}^2 + \frac{1}{2}\frac{m_s}{3}\dot{x}^2 \tag{E.2}$$

The total potential energy of the system is given by

$$U = \tfrac{1}{2}kx^2 \tag{E.3}$$

By assuming a harmonic motion

$$x(t) = X\cos\omega_n t \tag{E.4}$$

where X is the maximum displacement of the mass and ω_n is the natural frequency, the maximum kinetic and potential energies can be expressed as

$$T_{max} = \frac{1}{2}\left(m + \frac{m_s}{3}\right)X^2\omega_n^2 \tag{E.5}$$

$$U_{max} = \frac{1}{2}kX^2 \tag{E.6}$$

By equating T_{max} and U_{max}, we obtain the expression for the natural frequency:

$$\omega_n = \left(\frac{k}{m + \frac{m_s}{3}}\right)^{1/2} \tag{E.7}$$

Thus the effect of the mass of the spring can be accounted by adding one third of its mass to the main mass [2.3].

2.6 FREE VIBRATION WITH VISCOUS DAMPING

**2.6.1
Equation
of Motion**

As stated in Section 1.9, the viscous damping force F is proportional to the velocity $\dot{x}$ or v and can be expressed as

$$F = -c\dot{x} \tag{2.52}$$

where c is the damping constant or coefficient of viscous damping and the negative sign indicates that the damping force is opposite to the direction of velocity. A single degree of freedom system with a viscous damper is shown in Fig. 2.13. If x is measured from the equilibrium position of the mass m, the application of Newton's law yields the equation of motion:

$$m\ddot{x} = -c\dot{x} - kx$$

or

$$m\ddot{x} + c\dot{x} + kx = 0 \tag{2.53}$$

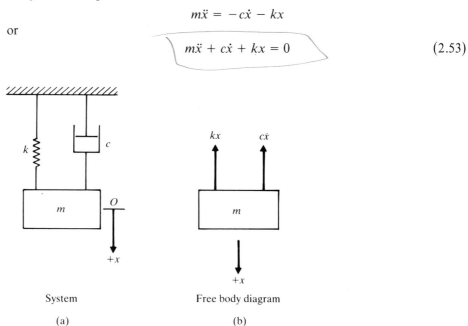

System

(a)

Free body diagram

(b)

Figure 2.13. Single degree of freedom system with viscous damper.

To solve Eq. (2.53), we assume a solution in the form

$$x(t) = Ce^{st} \tag{2.54}$$

where C and s are undetermined constants. Inserting this function into Eq. (2.53) leads to the characteristic equation

$$ms^2 + cs + k = 0 \tag{2.55}$$

the roots of which are

$$s_{1,2} = \frac{-c \pm \sqrt{c^2 - 4mk}}{2m} = -\frac{c}{2m} \pm \sqrt{\left(\frac{c}{2m}\right)^2 - \frac{k}{m}} \tag{2.56}$$

These roots give two solutions to Eq. (2.53):

$$x_1(t) = C_1 e^{s_1 t} \quad \text{and} \quad x_2(t) = C_2 e^{s_2 t} \tag{2.57}$$

Thus the general solution of Eq. (2.53) is given by a combination of the two solutions $x_1(t)$ and $x_2(t)$:

$$x(t) = C_1 e^{s_1 t} + C_2 e^{s_2 t}$$

$$= C_1 e^{\left\{-\frac{c}{2m} + \sqrt{\left(\frac{c}{2m}\right)^2 - \frac{k}{m}}\right\} t} + C_2 e^{\left\{-\frac{c}{2m} - \sqrt{\left(\frac{c}{2m}\right)^2 - \frac{k}{m}}\right\} t} \tag{2.58}$$

where C_1 and C_2 are arbitrary constants to be determined from the initial conditions of the system.

Critical Damping Constant and the Damping Ratio. The critical damping c_c is defined as the value of the damping constant c for which the radical in Eq. (2.56) becomes zero:

$$\left(\frac{c_c}{2m}\right)^2 - \frac{k}{m} = 0$$

or

$$c_c = 2m\sqrt{\frac{k}{m}} = 2\sqrt{km} = 2m\omega_n \tag{2.59}$$

For any damped system, the damping ratio ζ is defined as the ratio of the damping constant to the critical damping constant:

$$\zeta = c/c_c \tag{2.60}$$

Using Eqs. (2.60) and (2.59), we can write

$$\frac{c}{2m} = \frac{c}{c_c} \cdot \frac{c_c}{2m} = \zeta \omega_n \tag{2.61}$$

and hence

$$s_{1,2} = \left(-\zeta \pm \sqrt{\zeta^2 - 1}\right) \omega_n \tag{2.62}$$

Thus the solution, Eq. (2.58), can be written as

$$x(t) = C_1 e^{\left(-\zeta + \sqrt{\zeta^2 - 1}\right) \omega_n t} + C_2 e^{\left(-\zeta - \sqrt{\zeta^2 - 1}\right) \omega_n t} \tag{2.63}$$

The nature of the roots s_1 and s_2 and hence the behavior of the solution, Eq. (2.63), depends upon the magnitude of damping. It can be seen that the case $\zeta = 0$ leads to the undamped vibrations discussed in Section 2.2. Hence we assume that $\zeta \neq 0$ and consider the following three cases.

Case 1. Underdamped system ($\zeta < 1$ or $c < c_c$ or $c/2m < \sqrt{k/m}$). For this condition, ($\zeta^2 - 1$) is negative and the roots s_1 and s_2 can be expressed as

$$s_1 = \left(-\zeta + i\sqrt{1 - \zeta^2}\right)\omega_n$$
$$s_2 = \left(-\zeta - i\sqrt{1 - \zeta^2}\right)\omega_n$$

and the solution, Eq. (2.63), can be written in different forms:

$$
\begin{aligned}
x(t) &= C_1 e^{\left(-\zeta + i\sqrt{1-\zeta^2}\right)\omega_n t} + C_2 e^{\left(-\zeta - i\sqrt{1-\zeta^2}\right)\omega_n t} \\
&= e^{-\zeta\omega_n t}\left\{ C_1 e^{i\sqrt{1-\zeta^2}\,\omega_n t} + C_2 e^{-i\sqrt{1-\zeta^2}\,\omega_n t} \right\} \\
&= e^{-\zeta\omega_n t}\left\{ (C_1 + C_2)\cos\sqrt{1 - \zeta^2}\,\omega_n t + i(C_1 - C_2)\sin\sqrt{1 - \zeta^2}\,\omega_n t \right\} \\
&= e^{-\zeta\omega_n t}\left\{ C_1' \cos\sqrt{1 - \zeta^2}\,\omega_n t + C_2' \sin\sqrt{1 - \zeta^2}\,\omega_n t \right\} \\
&= X e^{-\zeta\omega_n t}\sin\left(\sqrt{1 - \zeta^2}\,\omega_n t + \phi\right) \\
&= X_0 e^{-\zeta\omega_n t}\cos\left(\sqrt{1 - \zeta^2}\,\omega_n t - \phi_0\right)
\end{aligned}
\tag{2.64}
$$

where (C_1', C_2'), (X, ϕ), and (X_0, ϕ_0) are arbitrary constants to be determined from the initial conditions.

For the initial conditions $x(t = 0) = x_0$ and $\dot{x}(t = 0) = \dot{x}_0$, C_1' and C_2' can be found:

$$C_1' = x_0 \quad\text{and}\quad C_2' = \frac{\dot{x}_0 + \zeta\omega_n x_0}{\sqrt{1 - \zeta^2}\,\omega_n} \tag{2.65}$$

and hence the solution becomes

$$x(t) = e^{-\zeta\omega_n t}\left\{ x_0\cos\sqrt{1 - \zeta^2}\,\omega_n t + \frac{\dot{x}_0 + \zeta\omega_n x_0}{\sqrt{1 - \zeta^2}\,\omega_n}\sin\sqrt{1 - \zeta^2}\,\omega_n t \right\} \tag{2.66}$$

The constants (X, ϕ) and (X_0, ϕ_0) can be expressed as

$$X = X_0 = \sqrt{(C_1')^2 + (C_2')^2} \tag{2.67}$$

$$\phi = \tan^{-1}(C_1'/C_2') \tag{2.68}$$

$$\phi_0 = \tan^{-1}(C_2'/C_1') \tag{2.69}$$

The motion described by Eq. (2.66) is a damped harmonic motion of angular frequency $\sqrt{1 - \zeta^2}\,\omega_n$, but because of the factor $e^{-\zeta\omega_n t}$, the amplitude decreases exponentially with time, as shown in Fig. 2.14. The quantity

$$\omega_d = \sqrt{1 - \zeta^2}\,\omega_n \tag{2.70}$$

is called the *frequency of damped vibration*. It can be seen that the frequency of

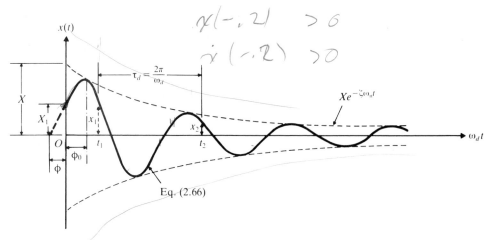

Figure 2.14. Underdamped solution.

damped vibration ω_d is always less than the undamped natural frequency ω_n. The decrease in the frequency of damped vibration with increasing amount of damping, given by Eq. (2.70), is shown graphically in Fig. 2.15. The underdamped case is very important in the study of mechanical vibrations, as it is the only case which leads to an oscillatory motion [2.10].

Case 2. Critically damped system ($\zeta = 1$ or $c = c_c$ or $c/2m = \sqrt{k/m}$). In this case the two roots s_1 and s_2 in Eq. (2.62) are equal:

$$s_1 = s_2 = -\frac{c_c}{2m} = -\omega_n \tag{2.71}$$

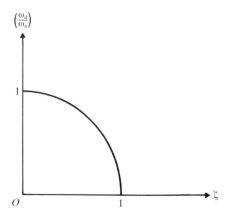

Figure 2.15. Variation of ω_d with damping.

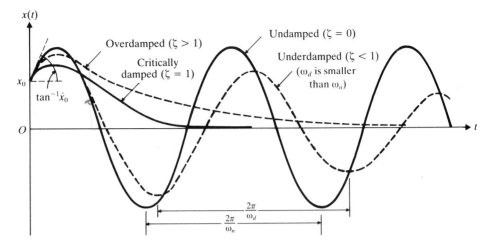

Figure 2.16. Comparison of motions with different types of damping.

Because of the repeated roots, the solution of Eq. (2.53) is given by [2.6]*

$$x(t) = (C_1 + C_2 t)e^{-\omega_n t} \tag{2.72}$$

The application of the initial conditions $x(t = 0) = x_0$ and $\dot{x}(t = 0) = \dot{x}_0$ for this case gives

$$C_1 = x_0$$
$$C_2 = \dot{x}_0 + \omega_n x_0 \tag{2.73}$$

and the solution becomes

$$x(t) = \left[x_0 + (\dot{x}_0 + \omega_n x_0)t \right] e^{-\omega_n t} \tag{2.74}$$

It can be seen that the motion represented by Eq. (2.74) is *aperiodic* (i.e., non-periodic). Since $e^{-\omega_n t} \to 0$ as $t \to \infty$, the motion will eventually diminish to zero, as indicated in Fig. 2.16.

Case 3. Overdamped system ($\zeta > 1$ or $c > c_c$ or $c/2m > \sqrt{k/m}$). As $\sqrt{\zeta^2 - 1} > 0$, Eq. (2.62) shows that the roots s_1 and s_2 are real and distinct and are given by

$$s_1 = \left(-\zeta + \sqrt{\zeta^2 - 1} \right)\omega_n < 0$$
$$s_2 = \left(-\zeta - \sqrt{\zeta^2 - 1} \right)\omega_n < 0$$

---

* Equation (2.72) can also be obtained by making ζ approach unity in the limit in Eq. (2.66). As $\zeta \to 1$, $\omega_d \to 0$; hence $\cos \omega_d t \to 1$ and $\sin \omega_d t \to \omega_d t$. Thus Eq. (2.66) yields

$$x(t) = e^{-\omega_n t}(C_1' + C_2'\omega_d t) = (C_1 + C_2 t)e^{-\omega_n t}$$

where $C_1 = C_1'$ and $C_2 = C_2'\omega_d$ are new constants.

with $s_2 \ll s_1$. In this case, the solution, Eq. (2.63), can be expressed as

$$x(t) = C_1 e^{(-\zeta + \sqrt{\zeta^2 - 1})\omega_n t} + C_2 e^{(-\zeta - \sqrt{\zeta^2 - 1})\omega_n t} \tag{2.75}$$

For the initial conditions $x(t = 0) = x_0$ and $\dot{x}(t = 0) = \dot{x}_0$, the constants C_1 and C_2 can be obtained:

$$C_1 = \frac{x_0 \omega_n \left(\zeta + \sqrt{\zeta^2 - 1} \right) + \dot{x}_0}{2 \omega_n \sqrt{\zeta^2 - 1}}$$

$$C_2 = \frac{-x_0 \omega_n \left(\zeta - \sqrt{\zeta^2 - 1} \right) - \dot{x}_0}{2 \omega_n \sqrt{\zeta^2 - 1}} \tag{2.76}$$

Equation (2.75) shows that the motion is aperiodic regardless of the initial conditions imposed on the system. Since roots s_1 and s_2 are both negative, the motion diminishes exponentially with time, as shown in Fig. 2.16.

Note the following two aspects of these systems:

1. The nature of the roots s_1 and s_2 with varying values of damping c or ζ can be shown in a complex plane. In Fig. 2.17, the horizontal and vertical axes are chosen as the real and imaginary axes. The semicircle represents the locus of the roots s_1 and s_2 for different values of ζ in the range $0 < \zeta < 1$. This figure permits us to see instantaneously the effect of the parameter ζ on the behavior of the system. We find that for $\zeta = 0$, we obtain the imaginary roots $s_1 = i\omega_n$

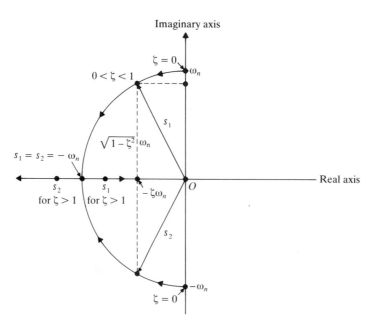

Figure 2.17. Locus of s_1 and s_2.

and $s_2 = -i\omega_n$, leading to the solution given in Eq. (2.12). For $0 < \zeta < 1$, the roots s_1 and s_2 are complex conjugate and are located symmetrically about the real axis. As the value of ζ approaches 1, both roots approach the point $-\omega_n$ on the real axis. If $\zeta > 1$, both roots lie on the real axis, one increasing and the other decreasing. In the limit when $\zeta \to \infty$, $s_1 \to 0$ and $s_2 \to -\infty$. The value $\zeta = 1$ can be seen to represent a transition stage, below which both roots are complex and above which both roots are real.

2. A critically damped system will have the smallest damping required for aperiodic motion; hence the mass returns to the position of rest in the shortest possible time without overshooting. The property of critical damping is used in many practical applications. For example, large guns have dashpots with critical damping value, so that they return to their original position after recoil in the minimum time without vibrating. If the damping provided were more than the critical value, some delay would be caused before the next firing.

**2.6.3
Logarithmic
Decrement**

The logarithmic decrement represents the rate at which the amplitude of a free damped vibration decreases. It is defined as the natural logarithm of the ratio of any two successive amplitudes. Let t_1 and t_2 denote the times corresponding to two consecutive amplitudes (displacements), measured one cycle apart for an underdamped system, as in Fig. 2.14. Using Eq. (2.64), we can form the ratio

$$\frac{x_1}{x_2} = \frac{X_0 e^{-\zeta\omega_n t_1}\cos(\omega_d t_1 - \phi_0)}{X_0 e^{-\zeta\omega_n t_2}\cos(\omega_d t_2 - \phi_0)} \tag{2.77}$$

But $t_2 = t_1 + \tau_d$ where $\tau_d = 2\pi/\omega_d$ is the period of damped vibration. Hence $\cos(\omega_d t_2 - \phi_0) = \cos(2\pi + \omega_d t_1 - \phi_0) = \cos(\omega_d t_1 - \phi_0)$, and Eq. (2.77) can be written as

$$\frac{x_1}{x_2} = \frac{e^{-\zeta\omega_n t_1}}{e^{-\zeta\omega_n(t_1+\tau_d)}} = e^{\zeta\omega_n\tau_d} \tag{2.78}$$

The logarithmic decrement δ can be obtained from Eq. (2.78):

$$\delta = \ln\frac{x_1}{x_2} = \zeta\omega_n\tau_d = \zeta\omega_n\frac{2\pi}{\sqrt{1-\zeta^2}\,\omega_n} = \frac{2\pi\zeta}{\sqrt{1-\zeta^2}} = \frac{2\pi}{\omega_d}\cdot\frac{c}{2m} \tag{2.79}$$

For small damping, Eq. (2.79) can be approximated:

$$\delta \simeq 2\pi\zeta \quad \text{if} \quad \zeta \ll 1 \tag{2.80}$$

Figure 2.18 shows the variation of the logarithmic decrement δ with ζ as given by Eqs. (2.79) and (2.80). It can be noticed that for values up to $\zeta = 0.3$, the two curves are difficult to distinguish.

The logarithmic decrement is dimensionless and is actually another form of the dimensionless damping ratio ζ. Once δ is known, ζ can be found by solving

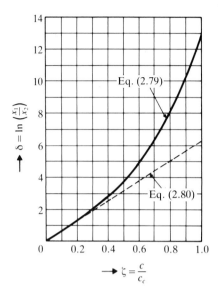

Figure 2.18. Variation of logarithmic decrement with damping.

Eq. (2.79):

$$\zeta = \frac{\delta}{\sqrt{(2\pi)^2 + \delta^2}} \tag{2.81}$$

If we use Eq. (2.80) instead of Eq. (2.79), we have

$$\zeta \simeq \frac{\delta}{2\pi} \tag{2.82}$$

If the damping in the given system is not known, we can determine it experimentally by measuring any two consecutive displacements x_1 and x_2. By taking the natural logarithm of the ratio of x_1 and x_2, we obtain δ. By using Eq. (2.81), we can compute the damping ratio ζ. In fact, the damping ratio ζ can also be found by measuring two displacements separated by any number of complete cycles. If x_1 and x_{m+1} denote the amplitudes corresponding to times t_1 and $t_{m+1} = t_1 + m\tau_d$ where m is an integer, we obtain

$$\frac{x_1}{x_{m+1}} = \frac{x_1}{x_2}\frac{x_2}{x_3}\frac{x_3}{x_4} \cdots \frac{x_m}{x_{m+1}} \tag{2.83}$$

Since any two successive displacements separated by one cycle satisfy the equation

$$\frac{x_j}{x_{j+1}} = e^{\zeta\omega_n\tau_d} \tag{2.84}$$

Eq. (2.83) becomes

$$\frac{x_1}{x_{m+1}} = \left(e^{\zeta\omega_n\tau_d}\right)^m = e^{m\zeta\omega_n\tau_d} \tag{2.85}$$

Equations (2.85) and (2.79) yield

$$\delta = \frac{1}{m} \ln\left(\frac{x_1}{x_{m+1}}\right) \tag{2.86}$$

which can be substituted into Eq. (2.81) or Eq. (2.82) to obtain the viscous damping ratio ζ.

2.6.4
Energy
Dissipated
in Viscous
Damping

In a viscously damped system, the rate of change of energy with time (dW/dt) is given by

$$\frac{dW}{dt} = \text{force} \times \text{velocity} = Fv = -cv^2 = -c\left(\frac{dx}{dt}\right)^2 \tag{2.87}$$

using Eq. (2.52). The negative sign in Eq. (2.87) denotes that energy dissipates with time. Assume a simple harmonic motion as $x(t) = X \sin \omega_d t$, where X is the amplitude of motion and the energy dissipated in a complete cycle is given by*

$$\Delta W = \int_{t=0}^{(2\pi/\omega_d)} c\left(\frac{dx}{dt}\right)^2 dt = \int_0^{2\pi} cX^2 \omega_d \cos^2 \omega_d t \cdot d(\omega_d t)$$
$$= \pi c \omega_d X^2 \tag{2.88}$$

This shows that the energy dissipated is proportional to the square of the amplitude of motion. It is to be noted that it is not a constant for given values of damping and amplitude, since ΔW is also a function of the frequency ω_d.

Equation (2.88) is valid even when there is a spring of stiffness k parallel to the viscous damper. To see this, consider the system shown in Fig. 2.19. The total force resisting motion can be expressed as

$$F = -kx - cv = -kx - c\dot{x} \tag{2.89}$$

If we assume simple harmonic motion

$$x(t) = X \sin \omega_d t \tag{2.90}$$

as before, Eq. (2.89) becomes

$$F = -kX \sin \omega_d t - c\omega_d X \cos \omega_d t \tag{2.91}$$

The energy dissipated in a complete cycle will be

$$\Delta W = \int_{t=0}^{2\pi/\omega_d} Fv \, dt$$
$$= \int_0^{2\pi/\omega_d} kX^2 \omega_d \sin \omega_d t \cdot \cos \omega_d t \cdot d(\omega_d t) + \int_0^{2\pi/\omega_d} c\omega_d X^2 \cos^2 \omega_d t \cdot d(\omega_d t)$$
$$= \pi c \omega_d X^2 \tag{2.92}$$

which can be seen to be identical with Eq. (2.88). This result is to be expected, since

---

* In the case of a damped system, simple harmonic motion $x(t) = X \cos \omega_d t$ is possible only when the steady-state response is considered under a harmonic force of frequency ω_d (see Section 3.4). The loss of energy due to the damper is supplied by the excitation under steady state forced vibration [2.7].

the spring force will not do any net work over a complete cycle or any integral number of cycles.

We can also compute the fraction of the total energy of the vibrating system that is dissipated in each cycle of motion ($\Delta W / W$), as follows. The total energy of the system W can be expressed either as the maximum potential energy ($\frac{1}{2}kX^2$) or as the maximum kinetic energy ($\frac{1}{2}mv_{max}^2 = \frac{1}{2}mX^2\omega_d^2$), the two being approximately equal for small values of damping. Thus

$$\frac{\Delta W}{W} = \frac{\pi c \omega_d X^2}{\frac{1}{2}m\omega_d^2 X^2} = 2\left(\frac{2\pi}{\omega_d}\right)\left(\frac{c}{2m}\right) = 2\delta \simeq 4\pi\zeta = \text{constant} \qquad (2.93)$$

using Eqs. (2.79) and (2.82). The quantity $\Delta W / W$ is called the *specific damping capacity* and is useful in comparing the damping capacity of engineering materials. Another quantity known as the *loss coefficient* is also used for comparing the damping capacity of engineering materials. The loss coefficient is defined as the ratio of the energy dissipated per radian and the total strain energy:

$$\text{loss coefficient} = \frac{(\Delta W / 2\pi)}{W} = \frac{\Delta W}{2\pi W} \qquad (2.94)$$

2.6.5
Torsional
Systems
with Viscous
Damping

The methods presented in Sections 2.6.1 through 2.6.4 for linear vibrations with viscous damping can be extended directly to viscously damped torsional (angular) vibrations. For this, consider a single degree of freedom torsional system with a viscous damper as shown in Fig. 2.20(a). The viscous damping torque is given by (Fig. 2.20b):

$$T = -c_t\dot{\theta} \qquad (2.95)$$

where c_t is the torsional viscous damping constant, $\dot{\theta} = d\theta / dt$ is the angular velocity of the disc, and the negative sign denotes that the damping torque is opposite the direction of angular velocity. The equation of motion can be derived as

$$J_0\ddot{\theta} + c_t\dot{\theta} + k_t\theta = 0 \qquad \{F \cdot L\} \qquad (2.96)$$

where J_0 = mass moment of inertia of the disc, k_t = spring constant of the system (restoring torque per unit angular displacement), and θ = angular displacement of the disc. The solution of Eq. (2.96) can be found exactly as in the case of linear vibrations. For example, in the underdamped case, the frequency of damped vibration is given by

$$\omega_d = \sqrt{1 - \zeta^2}\,\omega_n \qquad (2.97)$$

where

$$\omega_n = \sqrt{\frac{k_t}{J_0}} \qquad (2.98)$$

and

$$\zeta = \frac{c_t}{c_{tc}} = \frac{c_t}{2J_0\omega_n} = \frac{c_t}{2\sqrt{k_t J_0}} \qquad (2.99)$$

where c_{tc} is the critical torsional damping constant.

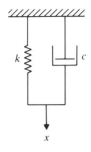

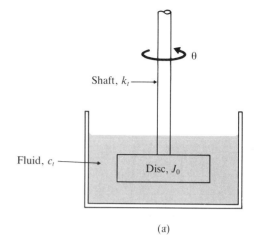

(a)

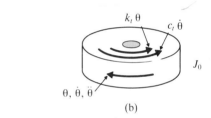

(b)

Figure 2.19

Figure 2.20

EXAMPLE 2.6   Shock Absorber for a Motorcycle

An underdamped shock absorber is to be designed for a motorcycle of mass 200 kg (Fig. 2.21a). When the shock absorber is subjected to an initial vertical velocity due to a road bump, the resulting displacement-time curve is to be as indicated in Fig. 2.21(b). Find the necessary stiffness and damping constants of the shock absorber if the damped period of vibration is to be 2 sec and the amplitude x_1 is to be reduced to one-fourth in one half cycle (i.e., $x_{1.5} = x_1/4$). Also find the minimum initial velocity that leads to a maximum displacement of 250 mm.

Given: Mass = 200 kg; displacement-time curve of the system (Fig. 2.21b); damped period of vibration = 2 sec, $x_{1.5} = x_1/4$; and maximum displacement = 250 mm.

Find: Stiffness (k), damping constant (c), and initial velocity $(\dot{x}_0)$, which results in a maximum displacement of 250 mm.

Approach: Equation for the logarithmic decrement in terms of the damping ratio, equation for the damped period of vibration, time corresponding to maximum displacement for an

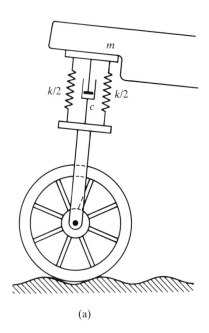

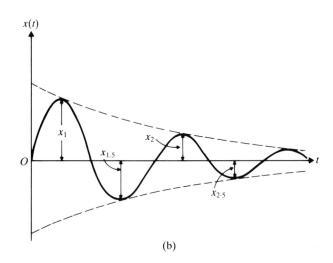

Figure 2.21

underdamped system, and envelope passing through the maximum points of an underdamped system.

Solution. Since $x_{1.5} = x_1/4$, $x_2 = x_{1.5}/4 = x_1/16$. Hence the logarithmic decrement becomes

$$\delta = \ln\left(\frac{x_1}{x_2}\right) = \ln(16) = 2.7726 = \frac{2\pi\zeta}{\sqrt{1-\zeta^2}} \qquad (\text{E.1})$$

from which the value of ζ can be found as $\zeta = 0.4037$. The damped period of vibration is given to be 2 sec. Hence

$$2 = \tau_d = \frac{2\pi}{\omega_d} = \frac{2\pi}{\omega_n\sqrt{1-\zeta^2}}$$

$$\omega_n = \frac{2\pi}{2\sqrt{1-(0.4037)^2}} = 3.4338 \text{ rad/sec}$$

The critical damping constant can be obtained:

$$c_c = 2m\omega_n = 2(200)(3.4338) = 1373.54 \text{ N-s/m}$$

Thus the damping constant is given by

$$c = \zeta c_c = (0.4037)(1373.54) = 554.4981 \text{ N-s/m}$$

and the stiffness by

$$k = m\omega_n^2 = (200)(3.4338)^2 = 2358.2652 \text{ N/m}$$

The displacement of the mass will attain its maximum value at time t_1, given by

$$\sin \omega_d t_1 = \sqrt{1 - \zeta^2}$$

(See Problem 2.45.) This gives

$$\sin \omega_d t_1 = \sin \pi t_1 = \sqrt{1 - (0.4037)^2} = 0.9149$$

or

$$t_1 = \frac{\sin^{-1}(0.9149)}{\pi} = 0.3678 \text{ sec}$$

The envelope passing through the maximum points (see Problem 2.45) is given by

$$x = \sqrt{1 - \zeta^2} \, Xe^{-\zeta\omega_n t} \qquad (E.2)$$

Since $x = 250$ mm, Eq. (E.2) gives at t_1

$$0.25 = \sqrt{1 - (0.4037)^2} \, Xe^{-(0.4037)(3.4338)(0.3678)}$$

or

$$X = 0.4550 \text{ m}.$$

The velocity of the mass can be obtained by differentiating the displacement

$$x(t) = Xe^{-\zeta\omega_n t} \sin \omega_d t$$

as

$$\dot{x}(t) = Xe^{-\zeta\omega_n t}(-\zeta\omega_n \sin \omega_d t + \omega_d \cos \omega_d t) \qquad (E.3)$$

When $t = 0$, Eq. (E.3) gives

$$\dot{x}(t = 0) = \dot{x}_0 = X\omega_d = X\omega_n\sqrt{1 - \zeta^2} = (0.4550)(3.4338)\left(\sqrt{1 - (0.4037)^2}\right)$$
$$= 1.4294 \text{ m/s}$$

EXAMPLE 2.7   **Analysis of a Cannon**

The schematic diagram of a large cannon is shown in Fig. 2.22 [2.8]. When the gun is fired, high-pressure gases accelerate the projectile inside the barrel to a very high velocity. The reaction force pushes the gun barrel in the opposite direction of the projectile. Since it is desirable to bring the gun barrel to rest in the shortest time without oscillation, it is made to translate backward against a critically damped spring-damper system called the *recoil mechanism*. In a particular case, the gun barrel and the recoil mechanism have a mass of 500 kg with a recoil spring of stiffness 10,000 N/m. The gun recoils 0.4 m upon firing. Find (1) the critical damping coefficient of the damper, (2) the initial recoil velocity of the gun, and (3) the time taken by the gun to return to a position 0.1 m from its initial position.

Given: Critically damped recoil mechanism with $m = 500$ kg, $k = 10,000$ N/m, and recoil distance $= 0.4$ m.

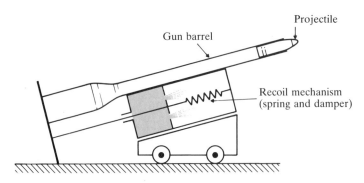

Figure 2.22

Find: Critical damping coefficient, recoil velocity, and time taken by the gun to return to a position 0.1 m from its initial position.

Approach: Use the response equation of a critically damped system.

Solution. 1. The undamped natural frequency of the system is

$$\omega_n = \sqrt{\frac{k}{m}} = \sqrt{\frac{10{,}000}{500}} = 4.4721 \text{ rad/sec}$$

and the critical damping coefficient (Eq. (2.59)) of the damper is

$$c_c = 2m\omega_n = 2(500)(4.4721) = 4472.1 \text{ N-s/m}$$

2. The response of a critically damped system is given by Eq. (2.72):

$$x(t) = (C_1 + C_2 t)e^{-\omega_n t} \qquad (E.1)$$

where $C_1 = x_0$ and $C_2 = \dot{x}_0 + \omega_n x_0$. The time t_1 at which $x(t)$ reaches a maximum value can be obtained by setting $\dot{x}(t) = 0$. The differentiation of Eq. (E.1) gives

$$x(t) = C_2 e^{-\omega_n t} - \omega_n (C_1 + C_2 t)e^{-\omega_n t}$$

Hence $\dot{x}(t) = 0$ yields

$$t_1 = \left(\frac{1}{\omega_n} - \frac{C_1}{C_2} \right) \qquad (E.2)$$

In this case, $x_0 = C_1 = 0$; hence Eq. (E.2) leads to $t_1 = 1/\omega_n$. Since the maximum value of $x(t)$ or the recoil distance is given to be $x_{max} = 0.4$ m, we have

$$x_{max} = x(t = t_1) = C_2 t_1 e^{-\omega_n t_1} = \frac{\dot{x}_0}{\omega_n} e^{-1} = \frac{\dot{x}_0}{e\omega_n}$$

or

$$\dot{x}_0 = x_{max}\omega_n e = (0.4)(4.4721)(2.7183) = 4.8626 \text{ m/s}$$

3. If t_2 denotes the time taken by the gun to return to a position 0.1 m from its initial position, we have

$$0.1 = C_2 t_2 e^{-\omega_n t_2} = 4.8626 t_2 e^{-4.4721 t_2} \qquad (E.3)$$

The solution of Eq. (E.3) gives $t_2 = 0.8258$ sec.

2.7 FREE VIBRATION WITH COULOMB DAMPING

In many mechanical systems, *Coulomb* or *dry-friction* dampers are used because of their mechanical simplicity and convenience [2.9]. Also in vibrating structures, whenever the components slide relative to each other, dry-friction damping appears internally. As stated in Section 1.9, Coulomb damping arises when bodies slide on dry surfaces. Coulomb's law of dry friction states that when two bodies are in contact, the force required to produce sliding is proportional to the normal force acting in the plane of contact. Thus the friction force F is given by

$$F = \mu N \tag{2.100}$$

where N is the normal force and μ is the coefficient of friction. The friction force acts in a direction opposite to the direction of velocity. Coulomb damping is sometimes called *constant damping*, since the damping force is independent of the displacement and velocity; it depends only on the normal force N between the sliding surfaces.

**2.7.1
Equation
of Motion**

Consider a single degree of freedom system with dry friction as shown in Fig. 2.23(a). Since the friction force varies with the direction of velocity, we need to consider two cases, as indicated in Figs. 2.23(b) and (c).

Case 1. When x is positive and dx/dt is positive, or when x is negative and dx/dt is positive (i.e., for the half cycle during which the mass moves from left to right), the equation of motion can be obtained using Newton's second law (see Fig. 2.23b)

$$m\ddot{x} = -kx - \mu N \quad \text{or} \quad m\ddot{x} + kx = -\mu N \tag{2.101}$$

This is a second order nonhomogeneous differential equation. The solution can be verified by substituting Eq. (2.102) into Eq. (2.101).

$$x(t) = A_1\cos \omega_n t + A_2\sin \omega_n t - \frac{\mu N}{k} \tag{2.102}$$

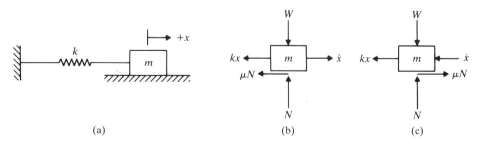

(a)   (b)   (c)

Figure 2.23. Spring-mass system with Coulomb damping.

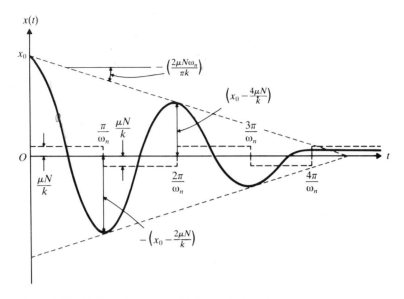

Figure 2.24. Motion of the mass with Coulomb damping.

where $\omega_n = \sqrt{k/m}$ is the frequency of vibration, and A_1 and A_2 are constants whose values depend on the initial conditions of this half cycle.

Case 2. When x is positive and dx/dt is negative, or when x is negative and dx/dt is negative (i.e., for the half cycle during which the mass moves from right to left), the equation of motion can be derived from Fig. 2.23(c) as

$$-kx + \mu N = m\ddot{x} \qquad \text{or} \qquad m\ddot{x} + kx = \mu N \qquad (2.103)$$

The solution of Eq. (2.103) is given by

$$x(t) = A_3\cos \omega_n t + A_4\sin \omega_n t + \frac{\mu N}{k} \qquad (2.104)$$

where A_3 and A_4 are constants to be found from the initial conditions of this half cycle. The term $\mu N/k$ appearing in Eqs. (2.102) and (2.104) is a constant representing the virtual displacement of the spring under the force μN, if it were applied as a static force. Equations (2.102) and (2.104) indicate that in each half cycle the motion is harmonic, with the equilibrium position changing from $\mu N/k$ to $-(\mu N/k)$ every half cycle as shown in Fig. 2.24.

**2.7.2
Solution**

To see the motion characteristics of the system more clearly, let us assume the initial conditions to be

$$x(t = 0) = x_0$$
$$\dot{x}(t = 0) = 0 \qquad (2.105)$$

That is, the system starts with zero velocity and displacement x_0 at $t = 0$. Since

$x = x_0$ at $t = 0$, the motion starts from right to left. Let $x_0, x_1, x_2, \ldots$ denote the amplitudes of motion at successive half cycles. Using Eqs. (2.104) and (2.105), we can evaluate the constants A_3 and A_4:

$$A_3 = x_0 - \frac{\mu N}{k}, \quad A_4 = 0$$

Thus Eq. (2.104) becomes

$$x(t) = \left(x_0 - \frac{\mu N}{k} \right) \cos \omega_n t + \frac{\mu N}{k} \qquad (2.106)$$

This solution is valid for half the cycle only, i.e., for $0 \leqslant t \leqslant \pi/\omega_n$. When $t = \pi/\omega_n$, the mass will be at its extreme left position and its displacement from equilibrium position can be found from Eq. (2.106):

$$-x_1 = x\left(t = \frac{\pi}{\omega_n} \right) = \left(x_0 - \frac{\mu N}{k} \right) \cos \pi + \frac{\mu N}{k} = -\left(x_0 - \frac{2\mu N}{k} \right) \quad (2.107)$$

Since the motion started with a displacement of $x = x_0$ and in a half cycle the value of x became $-(x_0 - (2\mu N/k))$, the reduction in magnitude of x in time π/ω_n is $2\mu N/k$.

In the second half cycle, the mass moves from left to right, so Eq. (2.102) is to be used. The initial conditions for this half cycle are

$$x(t = 0) = \text{value of } x \text{ at } t = \frac{\pi}{\omega_n} \text{ in Eq. (2.106)} = -\left(x_0 - \frac{2\mu N}{k} \right)$$

and

$$\dot{x}(t = 0) = \text{value of } \dot{x} \text{ at } t = \frac{\pi}{\omega_n} \text{ in Eq. (2.106)}$$

$$= \left\{ \text{value of } -\omega_n \left(x_0 - \frac{\mu N}{k} \right) \sin \omega_n t \text{ at } t = \frac{\pi}{\omega_n} \right\} = 0$$

Thus the constants in Eq. (2.102) become

$$-A_1 = -x_0 + \frac{3\mu N}{k}, \qquad A_2 = 0$$

so that Eq. (2.102) can be written as

$$x(t) = \left(x_0 - \frac{3\mu N}{k} \right) \cos \omega_n t - \frac{\mu N}{k} \qquad (2.108)$$

This equation is valid only for the second half cycle, that is, for $\pi/\omega_n \leqslant t \leqslant 2\pi/\omega_n$. At the end of this half cycle the value of $x(t)$ is

$$x_2 = x\left(t = \frac{\pi}{\omega_n} \right) \text{ in Eq. (2.108)} = x_0 - \frac{4\mu N}{k}$$

and

$$\dot{x}\left(t = \frac{\pi}{\omega_n} \right) \text{ in Eq. (2.108)} = 0$$

These become the initial conditions for the third half cycle, and the procedure can be continued until the motion stops. The motion stops when $x_n \leqslant \mu N/k$ since the restoring force exerted by the spring (kx) will then be less than the friction force μN. Thus the number of half cycles (r) that elapse before the motion ceases is given by

$$x_0 - r\frac{2\mu N}{k} \leqslant \frac{\mu N}{k}$$

that is,

$$r \geqslant \left\{\frac{x_0 - \dfrac{\mu N}{k}}{\dfrac{2\mu N}{k}}\right\} \tag{2.109}$$

Note the following characteristics of Coulomb damping:

1. In each successive cycle, the amplitude of motion is reduced by the amount $4\mu N/k$, so the amplitudes at the end of any two consecutive cycles are related:

$$X_m = X_{m-1} - \frac{4\mu N}{k} \tag{2.110}$$

2. As the amplitude is reduced by an amount $4\mu N/k$ in one cycle (i.e., in time $2\pi/\omega_n$), the slope of the enveloping straight lines (shown dotted) in Fig. 2.24 is

$$-\left(\frac{4\mu N}{k}\right)\bigg/\left(\frac{2\pi}{\omega_n}\right) = -\left(\frac{2\mu N\omega_n}{\pi k}\right)$$

The final position of the mass is usually displaced from equilibrium $(x = 0)$ position and represents a permanent displacement in which the friction force is locked in. Slight tapping will usually make the mass come to its equilibrium position.

3. The natural frequency of the system remains unaltered in Coulomb damping, in contrast to the viscous damping.

4. With viscous and hysteresis damping, the motion theoretically continues forever, perhaps with an infinitesimally small amplitude. In the case of Coulomb damping, however, the system comes to rest after some time.

**2.7.3
Torsional
Systems
with Coulomb
Damping**

If a constant frictional torque acts on a torsional system, the equation governing the angular oscillations of the system can be derived, similar to Eqs. (2.101) and (2.103), as

$$J_0\ddot{\theta} + k_t\theta = -T \tag{2.111}$$

and

$$J_0\ddot{\theta} + k_t\theta = T \tag{2.112}$$

where T denotes the constant damping torque (similar to μN for linear vibrations).

The solutions of Eqs. (2.111) and (2.112) are similar to those for linear vibrations. In particular, the frequency of vibration is given by

$$\omega_n = \sqrt{\frac{k_t}{J_0}}$$

(2.113)

and the amplitude of motion at the end of rth half cycle (θ_r) is given by

$$\theta_r = \theta_0 - r\frac{2T}{k_t}$$

(2.114)

where θ_0 is the initial angular displacement at $t = 0$ (with $\dot{\theta} = 0$ at $t = 0$). The motion ceases when

$$r \geqslant \left\{ \frac{\theta_0 - \dfrac{T}{k_t}}{\dfrac{2T}{k_t}} \right\}$$

(2.115)

EXAMPLE 2.8	Pulley Subjected to Coulomb Damping

A steel shaft of length 1 m and diameter 20 mm is fixed at one end and carries a pulley of mass moment of inertia 25 kg-m^2 at the other end. A band brake exerts a constant frictional torque of 400 N-m around the circumference of the pulley. If the pulley is displaced by 6° and released, determine (i) the number of cycles before the pulley comes to rest and (ii) the final settling position of the pulley.

Given: Steel shaft: length = 1 m, diameter = 20 mm, J_0 of pulley = 25 kg-m^2, frictional torque = T = 400 N-m, and θ_0 = 6°.

Find: (i) number of cycles before motion ceases and (ii) final settling position of pulley.

Approach: Torsional system with Coulomb damping.

Solution. (i) The number of half cycles that elapse before the angular motion of the pulley ceases is given by Eq. (2.115).

$$r \geqslant \left\{ \frac{\theta_0 - \dfrac{T}{k_t}}{\dfrac{2T}{k_t}} \right\}$$

(E.1)

where θ_0 = initial angular displacement = 6° = 0.10472 rad, k_t = torsional spring constant of the shaft given by

$$k_t = \frac{GJ}{l} = \frac{(8 \times 10^{10})\left\{ \frac{\pi}{32}(0.07)^4 \right\}}{1} = 62,832 \text{ N-m/rad}$$

and T = constant friction torque applied to the pulley = 400 N-m. Equation (E.1) gives

$$r \geqslant \frac{0.10472 - \left(\dfrac{400}{62832}\right)}{\left(\dfrac{800}{62832}\right)} = 7.72494$$

Thus the motion ceases after 8 half cycles.

(ii) The angular displacement after 8 half cycles is given by Eq. (2.114).

$$\theta = 0.10472 - 8 \times 2\left(\frac{400}{62832}\right) = 0.002861 \text{ rad} = 0.16393°$$

Thus the pulley stops at 0.16393° from the equilibrium position on the same side of the initial displacement.

2.8 FREE VIBRATION WITH HYSTERETIC DAMPING

Consider the spring-viscous damper arrangement shown in Fig. 2.25(a). For this system, the force (F) needed to cause a displacement $x(t)$ is given by

$$F = kx + c\dot{x} \tag{2.116}$$

For a harmonic motion of frequency ω and amplitude X,

$$x(t) = X \sin \omega t \tag{2.117}$$

Equations (2.116) and (2.117) yield

$$\begin{aligned} F(t) &= kX \sin \omega t + cX\omega \cos \omega t \\ &= kx \pm c\omega\sqrt{X^2 - (X \sin \omega t)^2} \\ &= kx \pm c\omega\sqrt{X^2 - x^2} \end{aligned} \tag{2.118}$$

When F versus x is plotted, Eq. (2.118) represents a closed loop as shown in Fig. 2.25(b). The area of the loop denotes the energy dissipated by the damper in a cycle of motion and is given by

$$\Delta W = \oint F \, dx = \int_0^{2\pi/\omega} (kX \sin \omega t + cX\omega \cos \omega t)(\omega X \cos \omega t) \, dt = \pi \omega c X^2 \tag{2.119}$$

Equation (2.119) has been derived in section 2.6.4 too [see Eq. (2.92)].

As stated in Section 1.9, the damping caused by the friction between the internal planes that slip or slide as the material deforms is called hysteresis (or solid or structural) damping. This causes a hysteresis loop to be formed in the stress-strain or force-displacement curve (see Fig. 2.26a). The energy loss in one loading and unloading cycle is equal to the area enclosed by the hysteresis loop [2.11–2.13]. The similarity between Figs. 2.25(b) and 2.26(a) can be used to define a hysteresis damping constant. It was found experimentally that the energy loss per cycle due to internal friction is independent of the frequency, but approximately proportional to the square of the amplitude. In order to achieve this observed behavior from Eq. (2.119), the damping coefficient c is assumed to be inversely proportional to the

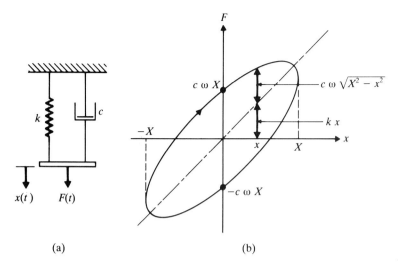

Figure 2.25

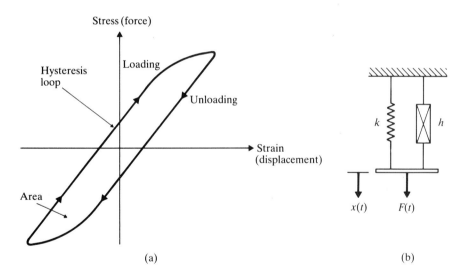

Figure 2.26

frequency as

$$c = \frac{h}{\omega} \tag{2.120}$$

where h is called the hysteresis damping constant. Equations (2.120) and (2.119) give

$$\Delta W = \pi h X^2 \tag{2.121}$$

Complex Stiffness. In Fig. 2.25(a), the spring and the damper are connected in parallel and for a general harmonic motion, $x = Xe^{i\omega t}$, the force is given by

$$F = kXe^{i\omega t} + c\omega i Xe^{i\omega t} = (k + i\omega c)x \qquad (2.122)$$

Similarly, if a spring and a hysteresis damper are connected in parallel as shown in Fig. 2.26(b), the force-displacement relation can be expressed as

$$F = (k + ih)x \qquad (2.123)$$

where

$$k + ih = k\left(1 + i\frac{h}{k}\right) = k(1 + i\beta) \qquad (2.124)$$

is called the complex stiffness of the system and $\beta = h/k$ is a constant indicating a dimensionless measure of damping.

Response of the System. In terms of β, the energy loss per cycle can be expressed as

$$\Delta W = \pi k \beta X^2 \qquad (2.125)$$

Under hysteresis damping, the motion can be considered to be nearly harmonic (since ΔW is small), and the decrease in amplitude per cycle can be determined using energy balance. For example, the energies at points P and Q (separated by half a cycle) in Fig. 2.27 are related as

$$\frac{kX_j^2}{2} - \frac{\pi k\beta X_j^2}{4} - \frac{\pi k\beta X_{j+0.5}^2}{4} = \frac{kX_{j+0.5}^2}{2}$$

$$\frac{X_j}{X_{j+0.5}} = \sqrt{\frac{2 + \pi\beta}{2 - \pi\beta}} \qquad (2.126)$$

Similarly, the energies at points Q and R give

$$\frac{X_{j+0.5}}{X_{j+1}} = \sqrt{\frac{2 + \pi\beta}{2 - \pi\beta}} \qquad (2.127)$$

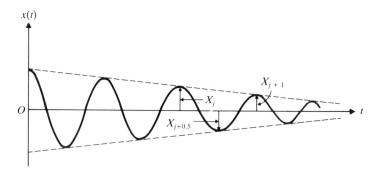

$x(t)$

X_{j+1}

X_j

$X_{j+0.5}$

Figure 2.27

Multiplication of Eqs. (2.126) and (2.127) gives

$$\frac{X_j}{X_{j+1}} = \frac{2 + \pi\beta}{2 - \pi\beta} = \frac{2 - \pi\beta + 2\pi\beta}{2 - \pi\beta} \simeq 1 + \pi\beta = \text{constant} \qquad (2.128)$$

The hysteresis logarithmic decrement can be defined as

$$\delta = \ln\left(\frac{X_j}{X_{j+1}}\right) \simeq \ln(1 + \pi\beta) \simeq \pi\beta \qquad (2.129)$$

Since the motion is assumed to be approximately harmonic, the corresponding frequency is defined by [2.10]

$$\omega = \sqrt{\frac{k}{m}} \qquad (2.130)$$

The equivalent viscous damping ratio ζ_{eq} can be found by equating the relations for the logarithmic decrement δ.

$$\delta \simeq 2\pi\zeta_{eq} \simeq \pi\beta = \frac{\pi h}{k}$$

$$\zeta_{eq} = \frac{\beta}{2} = \frac{h}{2k} \qquad (2.131)$$

Thus the equivalent damping constant c_{eq} is given by

$$c_{eq} = c_c \cdot \zeta_{eq} = 2\sqrt{mk} \cdot \frac{\beta}{2} = \beta\sqrt{mk} = \frac{\beta k}{\omega} = \frac{h}{\omega} \qquad (2.132)$$

Note that the method of finding an equivalent viscous damping coefficient for a structurally damped system is valid only for harmonic excitation. The above analysis assumes that the system responds approximately harmonically at the frequency ω.

EXAMPLE 2.9   **Estimation of Hysteretic Damping Constant**

The experimental measurements on a structure gave the force-deflection data shown in Fig. 2.28. From this data, estimate the hysteretic damping constant β and the logarithmic decrement δ.

Given: Experimental force-deflection curve.

Find: Hysteresis damping constant β and logarithmic decrement δ.

Approach: Equate the energy dissipated in a cycle (area enclosed by the hysteresis loop) to ΔW of Eq. (2.121).

Solution. The energy dissipated in each full load cycle is given by the area enclosed by the hysteresis curve. Each square in Fig. 2.28 denotes $100 \times 2 = 200$ N-mm. The area enclosed by the loop can be found as area ACB + area $ABDE$ + area $DFE \simeq \frac{1}{2}(AB)(CG) + (AB)(AE) + \frac{1}{2}(DE)(FH) = \frac{1}{2}(1.25)(1.8) + (1.25)(8) + \frac{1}{2}(1.25)(1.8) = 12.25$ square units.

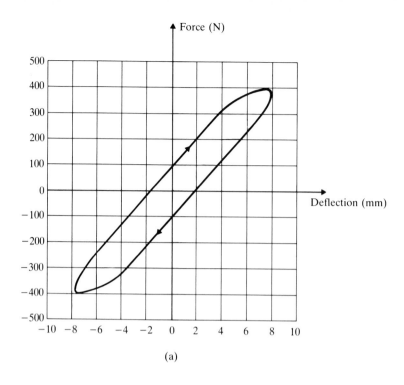

(a)

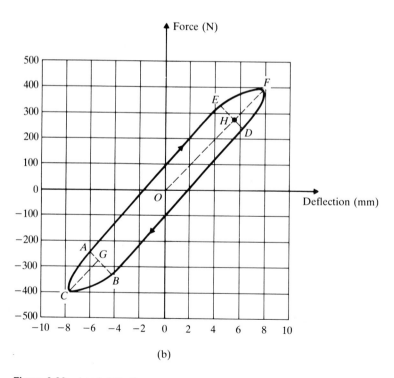

(b)

Figure 2.28. Load-deflection curve.

This area represents an energy of $12.25 \times 200/1000 = 2.5$ N-m. From Eq. (2.121), we have

$$\Delta W = \pi h X^2 = 2.5 \text{ N-m} \qquad (E.1)$$

Since the maximum deflection X is 0.008 m and the slope of the force-deflection curve (given approximately by the slope of the line OF) is $k = 400/8 = 50$ N/mm $= 50,000$ N/m, the hysteretic damping constant h is given by

$$h = \frac{\Delta W}{\pi X^2} = \frac{2.5}{\pi (0.008)^2} = 12433.95 \qquad (E.2)$$

and hence

$$\beta = \frac{h}{k} = \frac{12,433.95}{50,000} = 0.248679$$

The logarithmic decrement can be found

$$\delta \simeq \pi \beta = \pi (0.248679) = 0.78125 \qquad (E.3)$$

2.9 COMPUTER PROGRAM

A FORTRAN computer program, in the form of subroutine FREVIB, is given for the free vibration analysis of a viscously damped single degree of freedom system. The system may be underdamped, critically damped, or overdamped. The arguments of this subroutine are as follows:

M   =   Mass. Input data.

K   =   Spring stiffness. Input data.

C   =   Damping constant. Input data.

X0   =   Value of displacement of mass at time 0. Input data.

XD0   =   Value of velocity at time 0. Input data.

N   =   Number of time steps at which the value of $x(t)$ is to be printed. Input data.

DELT   =   Time interval between consecutive time steps (Δt). Input data.

X, XD, XDD =   Arrays of dimension N each, which contain computed values of displacement, velocity, and acceleration. X(I) = $x(t_i)$, XD(I) = $\dot{x}(t_i)$, XDD(I) = $\ddot{x}(t_i)$. Output.

T   =   Array of dimension N which contains the values of time. T(I) = t_i. Output.

To illustrate the use of the subroutine FREVIB, we consider an example with $m = 450$ kg, $k = 26519.2$ N/m, $c = 1000$ N-s/m, $x_0 = 0.539567$ m, $\dot{x}_0 = 1.0$ m/s, $\Delta t = 0.25$ s, and $N = 10$. The main program which calls FREVIB, the subroutine FREVIB, and the output of the program are given below.

```
C ======================================================================
C
C PROGRAM 2
C MAIN PROGRAM FOR CALLING FREVIB
C
C ======================================================================
      REAL M,K
C     THE FOLLOWING 3 LINES CONTAIN PROBLEM-DEPENDENT DATA
      DIMENSION X(10),XD(10),XDD(10),T(10)
      DATA M,K,C,XO,XDO,N,DELT/
     2 450.0,26519.2,1000.0,0.539567,1.0,10,0.25/
C     END OF PROBLEM-DEPENDENT DATA
      CALL FREVIB (M,K,C,XO,XDO,N,DELT,X,XD,XDD,T,II)
      PRINT 100
  100 FORMAT (/,24H FREE VIBRATION ANALYSIS,/,
     2 37H OF A SINGLE DEGREE OF FREEDOM SYSTEM,//,5H DATA)
      PRINT 200, M,K,C,XO,XDO,N,DELT
  200 FORMAT (/,7H M    =,E15.8,/,7H K    =,E15.8,/,7H C    =,E15.8,/,
     2 7H XO   =,E15.8,/,7H XDO  =,E15.8,/,7H N    =,I5,/,7H DELT =,
     3 E15.8)
      IF (II .EQ. 1) PRINT 500
      IF (II .EQ. 2) PRINT 600
      IF (II .EQ. 3) PRINT 700
      IF (II .EQ. 4) PRINT 800
      PRINT 900
      DO 300 I=1,N
      PRINT 400, I,T(I),X(I),XD(I),XDD(I)
  400 FORMAT (I5,4E15.6)
  300 CONTINUE
  500 FORMAT (//,19H SYSTEM IS UNDAMPED)
  600 FORMAT (//,23H SYSTEM IS UNDER DAMPED)
  700 FORMAT (//,28H SYSTEM IS CRITICALLY DAMPED)
  800 FORMAT (//,22H SYSTEM IS OVER DAMPED)
  900 FORMAT (//,9H RESULTS:,//,3X,2H I,3X,8H TIME(I),7X,5H X(I),10X,
     2 6H XD(I),9X,7H XDD(I),/)
      STOP
      END
C ======================================================================
C
C SUBROUTINE FREVIB
C
C ======================================================================
      SUBROUTINE FREVIB (M,K,C,XO,XDO,N,DELT,X,XD,XDD,T,II)
      DIMENSION X(N),XD(N),XDD(N),T(N)
      REAL M,K
      OMN=SQRT(K/M)
C     UNDAMPED SYSTEM
      IF (ABS(C) .GT. 1.0E-06) GO TO 100
      II=1
      OMN=SQRT(K/M)
      A=SQRT(XO**2+(XDO/OMN)**2)
      PHI=ATAN(XDO/(XO*OMN))
      DO 10 I=1,N
      IF (I .GT. 1) GO TO 20
```

```
         T(I)=DELT
         GO TO 30
  20     T(I)=T(I-1)+DELT
  30     TT=T(I)
         X(I)=A*COS(OMN*TT-PHI)
         XD(I)=A*OMN*COS(OMN*TT-PHI+1.5708)
         XDD(I)=-(C*XD(I)+K*X(I))/M
  10     CONTINUE
         GO TO 500
 100     CCRIT=2.0*SQRT(K*M)
         XAI=C/CCRIT
         IF (XAI - 1.0) 200,300,400
C        UNDERDAMPED SYSTEM
 200     II=2
         OMD=SQRT(1.0-(XAI**2))*OMN
         CP1=X0
         CP2=(XD0+XAI*OMN*X0)/OMD
         A=SQRT(CP1**2+CP2**2)
         PHI=ATAN(CP1/CP2)
         DO 110 I=1,N
         IF (I .GT. 1) GO TO 120
         T(I)=DELT
         GO TO 130
 120     T(I)=T(I-1)+DELT
 130     TT=T(I)
         X(I)=A*EXP(-XAI*OMN*TT)*SIN(OMD*TT+PHI)
         XD(I)=A*EXP(-XAI*OMN*TT)*(OMD*COS(OMD*TT+PHI)-XAI*OMN*SIN(OMD*
        2 TT+PHI))
         XDD(I)=-(C*XD(I)+K*X(I))/M
 110     CONTINUE
         GO TO 500
C        CRITICALLY DAMPED SYSTEM
 300     II=3
         DO 210 I=1,N
         IF (I .GT. 1) GO TO 220
         T(I)=DELT
         GO TO 230
 220     T(I)=T(I-1)+DELT
 230     TT=T(I)
         X(I)=(X0+(XD0+OMN*X0)*TT)*EXP(-OMN*TT)
         XD(I)=-(X0+(XD0+OMN*X0)*TT)*OMN*EXP(-OMN*TT)+(XD0+OMN*X0)*
        2 EXP(-OMN*TT)
         XDD(I)=-(C*XD(I)+K*X(I))/M
 210     CONTINUE
         GO TO 500
C        OVERDAMPED SYSTEM
 400     II=4
         X1=SQRT(XAI**2-1.0)
         C1=(X0*OMN*(XAI+X1)+XD0)/(2.0*OMN*X1)
         C2=(-X0*OMN*(XAI-X1)-XD0)/(2.0*OMN*X1)
         DO 310 I=1,N
         IF (I .GT. 1) GO TO 320
         T(I)=DELT
         GO TO 330
```

```
320   T(I)=T(I-1)+DELT
330   TT=T(I)
      X(I)=C1*EXP((-XAI+X1)*OMN*TT)+C2*EXP((-XAI-X1)*OMN*TT)
      XD(I)=C1*(-XAI+X1)*OMN*EXP((-XAI+X1)*OMN*TT)
     2 +C2*(-XAI-X1)*OMN*EXP((-XAI-X1)*OMN*TT)
      XDD(I)=-(C*XD(I)+K*X(I))/M
310   CONTINUE
500   RETURN
      END
```

FREE VIBRATION ANALYSIS
OF A SINGLE DEGREE OF FREEDOM SYSTEM

DATA

```
M     = 0.45000000E+03
K     = 0.26519199E+05
C     = 0.10000000E+04
X0    = 0.53956699E+00
XD0   = 0.10000000E+01
N     =    10
DELT  = 0.25000000E+00
```

SYSTEM IS UNDER DAMPED

RESULTS:

I	TIME(I)	X(I)	XD(I)	XDD(I)
1	0.250000E+00	0.192649E-01	-0.335069E+01	0.631066E+01
2	0.500000E+00	-0.318985E+00	0.106230E+01	0.164376E+02
3	0.750000E+00	0.144699E+00	0.140377E+01	-0.116468E+02
4	0.100000E+01	0.112366E+00	-0.129493E+01	-0.374428E+01
5	0.125000E+01	-0.137887E+00	-0.173140E+00	0.851065E+01
6	0.150000E+01	0.285635E-02	0.827508E+00	-0.200724E+01
7	0.175000E+01	0.777184E-01	-0.304712E+01	-0.390293E+01
8	0.200000E+01	-0.395868E-01	-0.326002E+00	0.305736E+01
9	0.225000E+01	-0.252620E-01	0.334008E+00	0.746487E+00
10	0.250000E+01	0.350478E-01	0.239573E-01	-0.211866E+01

REFERENCES

2.1. R. W. Fitzgerald, *Mechanics of Materials* (2nd ed.), Addison-Wesley, Reading, Mass., 1982.

2.2. R. F. Steidel, Jr., *An Introduction to Mechanical Vibrations* (2nd Ed.), John Wiley, New York, 1979.

2.3. W. Zambrano, "A brief note on the determination of the natural frequencies of a spring-mass system," *International Journal of Mechanical Engineering Education*, Vol. 9, October 1981, pp. 331–334; Vol. 10, July 1982, p. 216.

2.4. R. D. Blevins, *Formulas for Natural Frequency and Mode Shape*, Van Nostrand Reinhold, New York, 1979.

2.5. A. D. Dimarogonas, *Vibration Engineering*, West Publishing Co., St. Paul, 1976.

2.6. E. Kreyszig, *Advanced Engineering Mathematics* (4th Ed.), John Wiley, New York, 1979.

2.7. S. H. Crandall, "The role of damping in vibration theory," *Journal of Sound and Vibration*, Vol. 11, 1970, pp. 3–18.

2.8. I. Cochin, *Analysis and Design of Dynamic Systems*, Harper & Row, New York, 1980.

2.9. D. Sinclair, "Frictional vibrations," *Journal of Applied Mechanics*, Vol. 22, 1955, pp. 207–214.

2.10. T. K. Caughey and M. E. J. O'Kelly, "Effect of damping on the natural frequencies of linear dynamic systems," *Journal of the Acoustical Society of America*, Vol. 33, 1961, pp. 1458–1461.

2.11. E. E. Ungar, "The status of engineering knowledge concerning the damping of built-up structures," *Journal of Sound and Vibration*, Vol. 26, 1973, pp. 141–154.

2.12. W. Pinsker, "Structural damping," *Journal of the Aeronautical Sciences*, Vol. 16, 1949, p. 699.

2.13. R. H. Scanlan and A. Mendelson, "Structural damping," *AIAA Journal*, Vol. 1, 1963, pp. 938–939.

REVIEW QUESTIONS

2.1. Suggest a method for determining the damping constant of a highly damped vibrating system that uses viscous damping.

2.2. Can you apply the results of Section 2.2 to systems where the restoring force is not proportional to the displacement, that is, where k is not a constant?

2.3. State the quantities corresponding to m, c, k, and x for a torsional system.

2.4. What effect does a decrease in mass have on the frequency of a system?

2.5. What effect does a decrease in the stiffness of the system have on the natural period?

2.6. Why does the amplitude of free vibration gradually diminish in practical systems?

2.7. Why is it important to find the natural frequency of a vibrating system?

2.8. How many arbitrary constants must a general solution to a second order differential equation have? How are these constants determined?

2.9. Can the energy method be used to find the differential equation of motion of all single degree of freedom systems?

2.10. What assumptions are made in finding the natural frequency of a single degree of freedom system using the energy method?

2.11. Is the frequency of a damped free vibration smaller or greater than the natural frequency of the system?

2.12. What is the use of logarithmic decrement?

2.13. Is hysteresis damping a function of the maximum stress?

2.14. What is critical damping and what is its importance?

2.15. What happens to the energy dissipated by damping?

2.16. What is equivalent viscous damping? Is the equivalent viscous damping factor a constant?

2.17. What is the reason for studying the vibration of a single degree of freedom system?

2.18. How can you find the natural frequency of a system by measuring its static deflection?

2.19. Give two practical applications of a torsional pendulum.

2.20. Define these terms: damping ratio, logarithmic decrement, loss coefficient, and specific damping capacity.

2.21. In what ways is the response of a system with Coulomb damping different from that of systems with other types of damping?

2.22. What is complex stiffness?

2.23. Define the hysteresis damping constant.

2.24. Give three practical applications of the concept of center of percussion.

PROBLEMS

The problem assignments are organized as follows:

Problem	Section covered	Topic covered
2.1–2.25	2.2	Undamped translational systems
2.26–2.35	2.3	Undamped torsional systems
2.36–2.42	2.5	Energy method
2.43–2.54, 2.66	2.6	Systems with viscous damping
2.55–2.62	2.7	Systems with Coulomb damping
2.63–2.65	2.8	Systems with hysteretic damping
2.67–2.70	2.9	Computer program
2.71–2.73	—	Projects

2.1. An industrial press is mounted on a rubber pad to isolate it from its foundation. If the rubber pad is compressed 5 mm by the self-weight of the press, find the natural frequency of the system.

2.2. A spring-mass system has a natural period of 0.21 sec. What will be the new period if the spring constant is (i) increased by 50% and (ii) decreased by 50%?

2.3. A spring-mass system has a natural frequency of 10 Hz. When the spring constant is reduced by 800 N/m, the frequency is altered by 45%. Find the mass and spring constant of the original system.

2.4. A helical spring, when fixed at one end and loaded at the other, requires a force of 100 N to produce an elongation of 10 mm. The ends of the spring are now rigidly fixed, one end vertically above the other, and a mass of 10 kg is attached at the middle point of its length. Determine the time taken to complete one vibration cycle when the mass is set vibrating in the vertical direction.

2.5. The maximum velocity attained by the mass of a simple harmonic oscillator is 10 cm/sec, and the period of oscillation is 2 sec. If the mass is released with an initial displacement of 2 cm, find (a) the amplitude, (b) the initial velocity, (c) the maximum acceleration, and (d) the phase angle.

2.6. Three springs and a mass are attached to a rigid, weightless, bar PQ as shown in Fig. 2.29. Find the natural frequency of vibration of the system.

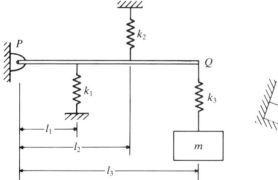

Figure 2.29   **Figure 2.30**

2.7. An automobile having a mass of 2000 kg deflects its suspension springs 0.02 m under static conditions. Determine the natural frequency of the automobile in the vertical direction by assuming damping to be negligible.

2.8. Find the natural frequency of vibration of a spring-mass system arranged on an inclined plane, as shown in Fig. 2.30.

2.9. Find the natural frequency of the system shown in Fig. 2.31 with and without the springs k_1 and k_2 in the middle of the elastic beam.

2.10. Find the natural frequency of the pulley system shown in Fig. 2.32 by neglecting the friction and the masses of the pulleys.

2.11. A rigid block of mass M is mounted on four elastic supports as shown in Fig. 2.33. A mass m drops from a height l and adheres to the rigid block without rebounding. If the spring constant of each elastic support is k, find the natural frequency of vibration of the system (a) without the mass m, and (b) with the mass m. Also find the resulting motion of the system in case (b).

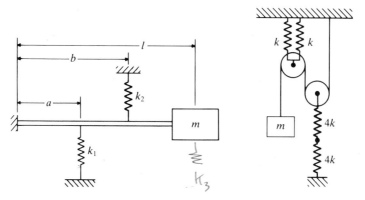

Figure 2.31   **Figure 2.32**

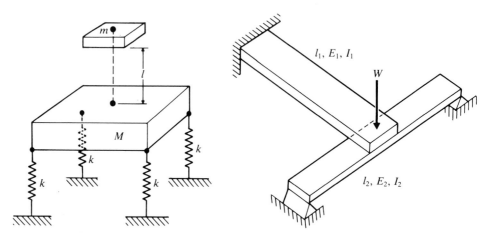

Figure 2.33   **Figure 2.34**

2.12. Derive the expression for the natural frequency of the system shown in Fig. 2.34. Note that the load W is applied at the tip of beam 1 and midpoint of beam 2.

2.13. A heavy machine weighing 9810 N is being lowered vertically down by a winch at a uniform velocity of 2 m/sec. The steel cable supporting the machine has a diameter of 0.01 m. The winch is suddenly stopped when the steel cable's length is 20 m. Find the period and amplitude of the ensuing vibration of the machine.

2.14. The natural frequency of a spring-mass system is found to be 2 Hz. When an additional mass of 1 kg is added to the original mass m, the natural frequency is reduced to 1 Hz. Find the spring constant k and the mass m.

2.15. Four weightless rigid links and a spring are arranged to support a weight W in two different ways as shown in Fig. 2.35. Determine the natural frequencies of vibration of the two arrangements.

2.16. Figure 2.36 shows a small mass m restrained by four linearly elastic springs, each of which has an unstretched length l, and an angle of orientation of 45° with respect to

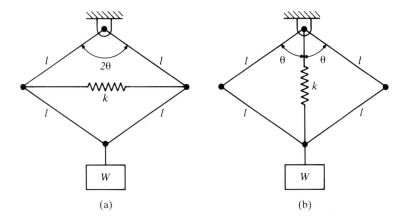

Figure 2.35

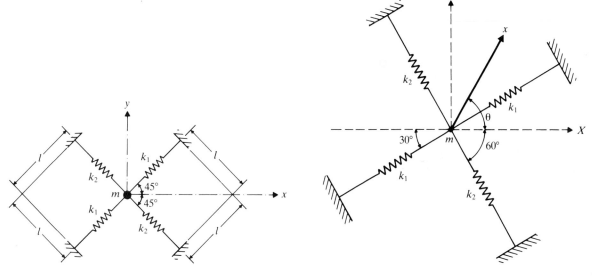

Figure 2.36

Figure 2.37

the x-axis. Determine the equation of motion for small displacements of the mass in the x direction.

2.17.* A mass m is supported by two sets of springs oriented at 30° and 120° with respect to the X axis, as shown in Fig. 2.37. A third pair of springs, with a stiffness of k_3 each, is to be designed so as to make the system have a constant natural frequency while vibrating in any direction x. Determine the necessary spring stiffness k_3 and the orientation of the springs with respect to the X axis.

2.18. A mass m is attached to a cord which is under a tension T, as shown in Fig. 2.38. Assuming that the tension T remains unchanged when the mass is displaced normal to

*The asterisk denotes a design problem or a problem with no unique answer.

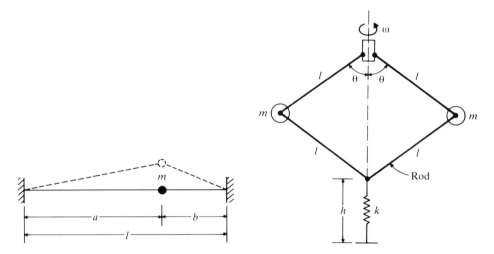

Figure 2.38   **Figure 2.39**

the cord, (a) write the differential equation of motion for small transverse vibrations and (b) find the natural frequency of vibration.

2.19. The schematic diagram of a centrifugal governor is shown in Fig. 2.39. The length of each rod is l, the mass of each ball is m and the free length of the spring is h. If the shaft speed is ω, determine the equilibrium position and the frequency for small oscillations about this position.

2.20. A square platform $PQRS$ and a car which it is supporting have a combined mass of M. The platform is suspended by four elastic wires from a fixed point O, as indicated in Fig. 2.40. The vertical distance between the point of suspension O and the horizontal equilibrium position of the platform is h. If the side of the platform is a and the stiffness of each wire is k, determine the period of vertical vibration of the platform.

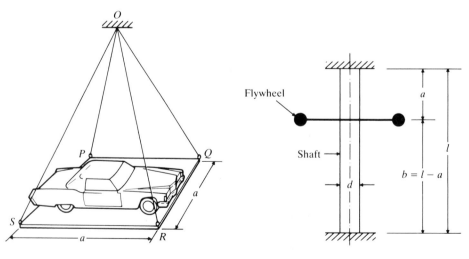

Figure 2.40   **Figure 2.41**

2.21. A flywheel is mounted on a vertical shaft, as shown in Fig. 2.41. The shaft has a diameter d and length l and is fixed at both ends. The flywheel has a weight of W and a radius of gyration of r. Find the natural frequency of the longitudinal, the transverse, and the torsional vibration of the system.

2.22. A building frame is modeled by four identical steel columns, of weight w each, and a rigid floor of weight W, as shown in Fig. 2.42. The columns are fixed at the ground and have a bending rigidity of EI each. Determine the natural frequency of horizontal vibration of the building frame by assuming the connection between the floor and the columns to be (a) pivoted as shown in Fig. 2.42(a), and (b) fixed against rotation as shown in Fig. 2.42(b). Include the effect of self weights of the columns.

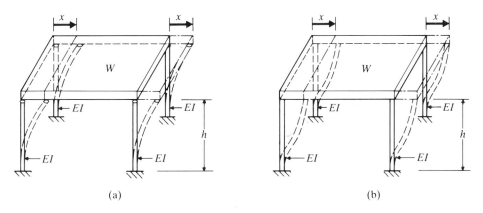

Figure 2.42

2.23. A helical spring of stiffness k is cut into two halves and a mass m is connected to the two halves as shown in Fig. 2.43(a). The natural time period of this system is found to be 0.5 sec. If an identical spring is cut so that one part is $\frac{1}{4}$ and the other part $\frac{3}{4}$ of the original length, and the mass m is connected to the two parts as shown in Fig. 2.43(b), what would be the natural period of the system?

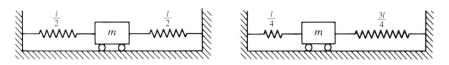

Figure 2.43

2.24. * Figure 2.44 shows a metal block supported on two identical cylindrical rollers rotating in opposite directions at the same angular speed. When the center of gravity of the block is initially displaced by a distance x, the block will be set into simple harmonic motion. If the frequency of motion of the block is found to be ω, determine the coefficient of friction between the block and the rollers.

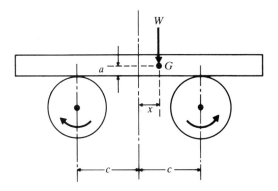

Figure 2.44

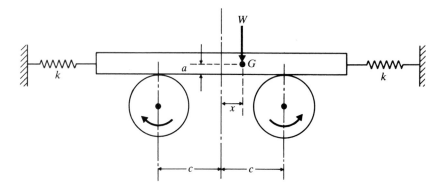

Figure 2.45

2.25.* If two identical springs of stiffness k each are attached to the metal block of Problem 2.24 as shown in Fig. 2.45, determine the coefficient of friction between the block and the rollers.

2.26. A pulley 250 mm in diameter drives a second pulley 1000 mm in diameter by means of a belt (see Fig. 2.46). The moment of inertia of the driven pulley is 0.2 kg-m². The belt connecting these pulleys is represented by two springs, each of stiffness k. For what value of k will the natural frequency be 6 Hz?

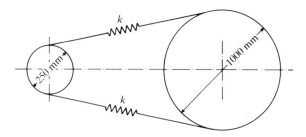

Figure 2.46

2.27. Derive an expression for the natural frequency of the simple pendulum shown in Fig. 1.5. Determine the period of oscillation of a simple pendulum having a mass $m = 5$ kg and a length $l = 0.5$ m.

2.28. A mass m is attached at the end of a bar of negligible mass and is made to vibrate in three different configurations, as indicated in Figs. 2.47(a) to (c). Find the configuration corresponding to the highest natural frequency.

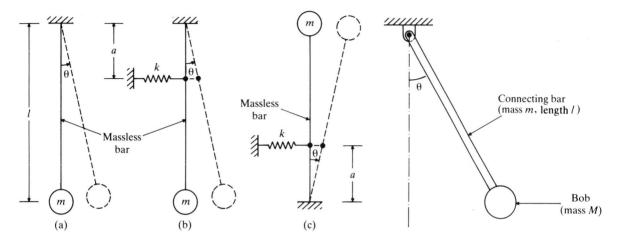

Figure 2.47

Figure 2.48

2.29. Find the natural frequency of the pendulum shown in Fig. 2.48 when the mass of the connecting bar is not negligible compared to the mass of the pendulum bob.

2.30. A steel shaft of 0.05 m diameter and 2 m length is fixed at one end and carries at the other end a steel disc of 1 m diameter and 0.1 m thickness, as shown in Fig. 2.7. Find the natural frequency of torsional vibration of the system.

2.31. A uniform slender rod of mass m and length l is hinged at point A and is attached to five springs as shown in Fig. 2.49. Find the natural frequency of the system if $k = 2000$ N/m, $k_t = 1000$ N-m/rad, $m = 10$ kg, and $l = 5$ m.

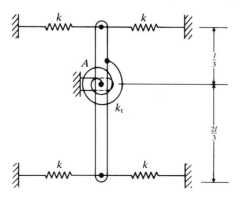

Figure 2.49

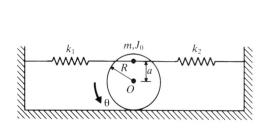

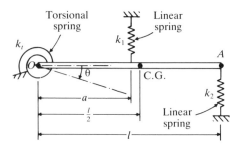

Figure 2.50   **Figure 2.51**

2.32. A cylinder of mass m and mass moment of inertia J_0 is free to roll without slipping but is restrained by two springs of stiffnesses k_1 and k_2 as shown in Fig. 2.50. Find its natural frequency of vibration. Also find the value of a that maximizes the natural frequency of vibration.

2.33. If the pendulum of Problem 2.27 is placed in a rocket moving vertically with an acceleration of 5 m/s^2, what will be its period of oscillation?

2.34. Find the equation of motion of the uniform rigid bar OA of length l and mass m shown in Fig. 2.51. Also find its natural frequency.

2.35. A uniform circular disc is pivoted at point O as shown in Fig. 2.52. Find the natural frequency of the system. Also find the maximum frequency of the system by varying the value of b.

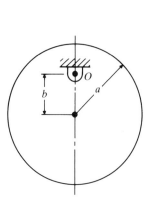

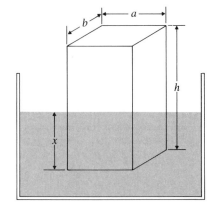

Figure 2.52   **Figure 2.53**

2.36. Solve problem 2.6 using Rayleigh's method.

2.37. Solve problem 2.10 using Rayleigh's method.

2.38. Find the natural frequency of the system shown in Fig. 2.36.

2.39. Solve problem 2.18 using Rayleigh's method.

2.40. Solve problem 2.31 using Rayleigh's method.

2.41. Solve problem 2.34 using Rayleigh's method.

2.42. A wooden rectangular prism of density ρ_w, height h and cross-section $a \times b$ is initially depressed in an oil tub and made to vibrate freely in the vertical direction (see Fig. 2.53). Find the natural frequency of vibration of the prism using Rayleigh's method.

Assume the density of oil as ρ_0. If the rectangular prism is replaced by a uniform circular cylinder of radius r, height h and density ρ_w, will there be any change in the natural frequency?

2.43. A simple pendulum is found to vibrate at a frequency of 0.5 Hz in vacuum and 0.45 Hz in a viscous fluid medium. Find the damping constant assuming the mass of the bob of the pendulum as 1 kg.

2.44. The ratio of successive amplitudes of a viscously damped single degree of freedom system is found to be $18:1$. Determine the ratio of successive amplitudes if the amount of damping is (a) doubled, and (b) halved.

2.45. Assuming that the phase angle is zero, show that the response $x(t)$ of an underdamped single degree of freedom system reaches a maximum value when

$$\sin \omega_d t = \sqrt{1 - \zeta^2}$$

and a minimum value when

$$\sin \omega_d t = -\sqrt{1 - \zeta^2}$$

Also show that the equations of the curves passing through the maximum and minimum values of $x(t)$ are given, respectively, by

$$x = \sqrt{1 - \zeta^2}\, X e^{-\zeta \omega_n t}$$

and

$$x = -\sqrt{1 - \zeta^2}\, X e^{-\zeta \omega_n t}$$

2.46. Derive an expression for the time at which the response of a critically damped system will attain its maximum value. Also find the expression for the maximum response.

2.47. A shock absorber is to be designed to limit its overshoot to 15% of its initial displacement when released. Find the damping ratio ζ_0 required. What will be the overshoot if ζ is made equal to (i) $\frac{3}{4}\zeta_0$, and (ii) $\frac{5}{4}\zeta_0$?

2.48. For a spring-mass-damper system, $m = 50$ kg and $k = 5000$ N/m. Find the following: (a) critical damping constant c_c, (b) damped natural frequency when $c = c_c/2$, and (c) logarithmic decrement.

2.49. A locomotive car of mass 2000 kg traveling at a velocity $v = 10$ m/sec is stopped at the end of tracks by a spring-damper system as shown in Fig. 2.54. If the stiffness of

Figure 2.54

the spring is $k = 40$ N/mm and the damping constant is $c = 20$ N-s/mm, determine (a) the maximum displacement of the car after engaging the springs and damper and (b) the time taken to reach the maximum displacement.

2.50. A torsional pendulum has a natural frequency of 200 cycles/min when vibrating in vacuum. The mass moment of inertia of the disc is 0.2 kg-m^2. It is then immersed in oil and its natural frequency is found to be 180 cycles/min. Determine the damping constant. If the disc, when placed in oil, is given an initial displacement of 2°, find its displacement at the end of the first cycle.

2.51. A body vibrating with viscous damping makes 5 complete oscillations per second, and in 50 cycles its amplitude diminishes to 10%. Determine the logarithmic decrement and the damping ratio. In what proportion will the period of vibration be decreased if damping is removed?

2.52.* The maximum permissible recoil distance of a gun is specified as 0.5 m. If the initial recoil velocity is to be between 8 m/sec and 10 m/sec, find the mass of the gun and the spring stiffness of the recoil mechanism. Assume that a critically damped dashpot is used in the recoil mechanism and the mass of the gun has to be at least 500 kg.

2.53. A viscously damped system has a stiffness of 5000 N/m, critical damping constant of 0.2 N-s/mm, and a logarithmic decrement of 2.0. If the system is given an initial velocity of 1 m/sec, determine the maximum displacement of the system.

2.54. Explain why an overdamped system never passes through the static equilibrium position when it is given (i) an initial displacement only and (ii) an initial velocity only.

2.55. A single degree of freedom system consists of a mass of 20 kg and a spring of stiffness 4000 N/m. The amplitudes of successive cycles are found to be $50, 45, 40, 35, \ldots$ mm. Determine the nature and magnitude of the damping force and the frequency of the damped vibration.

2.56. A mass of 20 kg slides back and forth on a dry surface due to the action of a spring having a stiffness of 10 N/mm. After four complete cycles, the amplitude has been found to be 100 mm. What is the average coefficient of friction between the two surfaces if the original amplitude was 150 mm? How much time has elapsed during the four cycles?

2.57. A 10-kg mass is connected to a spring of stiffness 3000 N/m and is released after giving an initial displacement of 100 mm. Assuming that the mass moves on a horizontal surface as shown in Fig. 2.23(a), determine the position at which the mass comes to rest. Assume the coefficient of friction between the mass and the surface to be 0.12.

2.58. A weight of 25 N is suspended from a spring that has a stiffness of 1000 N/m. The weight vibrates in the vertical direction under a constant damping force. When the weight is initially pulled downward a distance of 10 cm from its static equilibrium position and released, it comes to rest after exactly two complete cycles. Find the magnitude of the damping force.

2.59. A mass of 20 kg is suspended from a spring of stiffness 10,000 N/m. The vertical motion of the mass is subject to Coulomb friction of magnitude 50 N. If the spring is initially displaced downward by 5 cm from its static equilibrium position, determine (a) the number of half cycles elapsed before the mass comes to rest, (b) the time elapsed before the mass comes to rest, and (c) the final extension of the spring.

2.60. The Charpy impact test is a dynamic test in which a specimen is struck and broken by a pendulum (or hammer) and the energy absorbed in breaking the specimen is measured. The energy values serve as a useful guide for comparing the impact strengths of different materials. As shown in Fig. 2.55, the pendulum is suspended from a shaft, is released from a particular position, and is allowed to fall and break the specimen. If the pendulum is made to oscillate freely (with no specimen), find (a) an expression for the decrease in the angle of swing for each cycle caused by friction, (b) the solution for $\theta(t)$ if the pendulum is released from an angle θ_0, and (c) the number of cycles after which the motion ceases. Assume the mass of the pendulum as m and the coefficient of friction between the shaft and the bearing of the pendulum as μ.

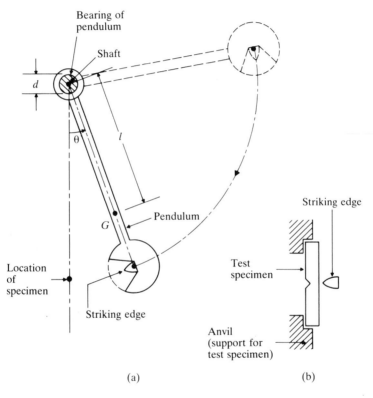

(a)   (b)

Figure 2.55

2.61. Find the equivalent viscous damping constant for Coulomb damping for sinusoidal vibration.

2.62. A single degree of freedom system consists of a mass, a spring, and a damper in which both dry friction and viscous damping act simultaneously. The free vibration amplitude is found to decrease by 1% per cycle when the amplitude is 20 mm and by 2% per cycle when the amplitude is 10 mm. Find the value of $(\mu N/k)$ for the dry friction component of the damping.

2.63. The experimentally observed force-deflection curve for a composite structure is shown in Fig. 2.56. Find the hysteresis damping constant, the logarithmic decrement and the equivalent viscous damping ratio corresponding to this curve.

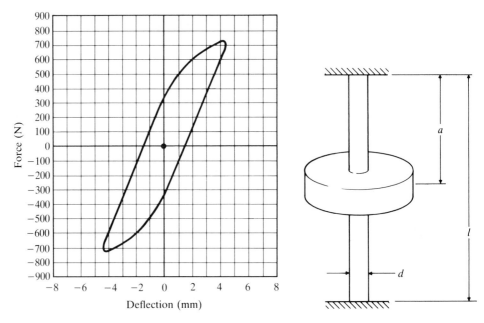

Figure 2.56   **Figure 2.57**

2.64. A panel made of fiber-reinforced composite material is observed to behave as a single degree of freedom system of mass 1 kg and stiffness 2 N/m. The ratio of successive amplitudes is found to be 1.1. Determine the value of the hysteresis damping constant β, the equivalent viscous damping constant c_{eq}, and the energy loss per cycle for an amplitude of 10 mm.

2.65. A built-up cantilever beam having a bending stiffness of 200 N/m supports a mass of 2 kg at its free end. The mass is displaced initially by 30 mm and released. If the amplitude is found to be 20 mm after 100 cycles of motion, estimate the hysteresis damping constant β of the beam.

2.66. The rotor of a dial indicator is connected to a torsional spring and a torsional viscous damper to form a single degree of freedom torsional system. The scale is graduated in equal divisions and the equilibrium position of the rotor corresponds to zero on the scale. When a torque of 2×10^{-3} N-m is applied, the angular displacement of the rotor is found to be 50° with the pointer showing 80 divisions on the scale. When the rotor is released from this position, the pointer swings first to -20 divisions in one second and then to 5 divisions in another second. Find (a) the mass moment of inertia of the rotor, (b) the undamped natural time period of the rotor, (c) the torsional damping constant, and (d) the torsional spring stiffness.

2.67–2.70.

Find the free vibration response of a viscously damped single degree of freedom system with $m = 4$ kg, $k = 2500$ N/m, $x_0 = 100$ mm, $\dot{x}_0 = -10$ m/s, $\Delta t = 0.01$ s, and N = 50 using the subroutine FREVIB for the following conditions:

(a) $c = 0$

(b) $c = 100$ N-s/m   *150*

(c) $c = 200$ N-s/m   *250*

(d) $c = 400$ N-s/m   *450*

Projects:

2.71. A water turbine of mass 1000 kg and mass moment of inertia 500 kg-m^2 is mounted on a steel shaft as shown in Fig. 2.57. The operational speed of the turbine is 2400 rpm. Assuming the ends of the shaft to be fixed, find the values of l, a, and d, such that the natural frequency of vibration of the turbine in each of the axial, transverse, and circumferential directions is greater than the operational speed of the turbine.

2.72. Design the columns for each of the building frames shown in Figs. 2.42(a) and (b) for minimum weight such that the natural frequency of vibration is greater than 50 Hz. The weight of the floor (W) is 4000 lb and the length of the columns (l) is 96 in. Assume that the columns are made of steel and have a tubular cross section with outer diameter d and wall thickness t.

2.73. One end of a uniform rigid bar of mass m is connected to a wall by a hinge joint O and the other end carries a concentrated mass M, as shown in Fig. 2.58. The bar rotates about the hinge point O against a torsional spring and a torsional damper. It is proposed to use this mechanism, in conjunction with a mechanical counter, to control entrance to an amusement park. Find the masses m and M, the stiffness of the torsional spring (k_t), and the damping force (F_d) necessary to satisfy the following specifications: (1) A viscous damper or a Coulomb damper can be used. (2) The bar has to return to within 5° of closing in less than 2 sec when released from an initial position of $\theta = 75°$.

Amusement park

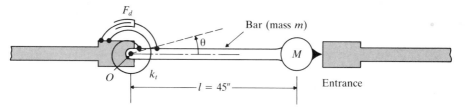

Figure 2.58

CHAPTER 3

Harmonically Excited Vibration

Charles Augustin de Coulomb (1736 – 1806) was a French military engineer and physicist. His early work on statics and mechanics was presented in his great memoir "The Theory of Simple Machines" in 1779, which describes the effect of resistance and the so-called "Coulomb's law of proportionality" between friction and normal pressure. In 1784, he obtained the correct solution to the problem of the small oscillations of a body subjected to torsion. He is well known for his laws of force for electrostatic and magnetic charges. His name is remembered through the unit of electric charge. (Courtesy of Brown Brothers)

3.1 INTRODUCTION

A dynamic system is often subjected to some type of external force or excitation, called the *forcing* or *exciting function*. This excitation is usually time-dependent. It may be harmonic, nonharmonic but periodic, nonperiodic, or random in nature. The response of a system to a harmonic excitation is called *harmonic response*. The nonperiodic excitation may have a long or short duration. The response of a dynamic system to suddenly applied nonperiodic excitations is called *transient response*.

In this chapter, we shall consider the dynamic response of a single degree of freedom system under harmonic excitations of the form $F(t) = F_0 e^{i(\omega t + \phi)}$ or $F(t) = F_0 \cos(\omega t + \phi)$ or $F(t) = F_0 \sin(\omega t + \phi)$, where F_0 is the amplitude, ω is the frequency, and ϕ is the phase angle of the harmonic excitation. The value of ϕ depends on the value of $F(t)$ at $t = 0$ and is usually taken to be zero. Under a harmonic excitation, the response of the system will also be harmonic. If the frequency of excitation coincides with the natural frequency of the system, the response of the system will be very large. This condition, known as resonance, is to be avoided to prevent failure of the system.

3.2 EQUATION OF MOTION

If a force $F(t)$ acts on a viscously damped spring-mass system as shown in Fig. 3.1, the equation of motion can be obtained using Newton's second law:

$$m\ddot{x} + c\dot{x} + kx = F(t) \tag{3.1}$$

Since this equation is nonhomogeneous, its general solution $x(t)$ is given by the sum of the homogeneous solution, $x_h(t)$, and the particular solution, $x_p(t)$. The homogeneous solution, which is the solution of the homogeneous equation

$$m\ddot{x} + c\dot{x} + kx = 0 \tag{3.2}$$

represents the free vibration of the system and was discussed in Chapter 2. As seen in Section 2.6.2, this free vibration dies out with time under each of the three possible conditions of damping (underdamping, critical damping, and overdamping) and under all possible initial conditions. Thus the general solution of Eq. (3.1) eventually reduces to the particular solution $x_p(t)$, which represents the steady-state vibration. The steady-state motion is present as long as the forcing function is present. The variations of homogeneous, particular, and general solutions with time for a typical case are shown in Fig. 3.2. It can be seen that $x_h(t)$ dies out and $x(t)$ becomes $x_p(t)$ after some time (τ in Fig. 3.2). The part of the motion that dies out due to damping (the free vibration part) is called *transient*. The rate at which the transient motion decays depends on the values of the system parameters k, c, and m. In this chapter, except in Section 3.3, we ignore the transient motion and derive only the particular solution of Eq. (3.1), which represents the steady-state response, under harmonic forcing functions.

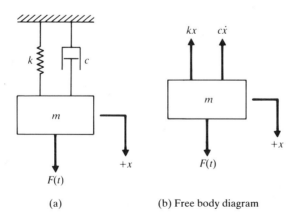

(a)   (b) Free body diagram

Figure 3.1   A spring-mass-damper system.

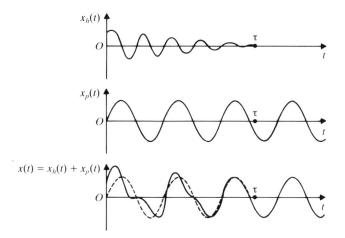

Figure 3.2 Homogeneous, particular, and general solutions of Eq. (3.1) for an underdamped case.

3.3 **RESPONSE OF AN UNDAMPED SYSTEM UNDER HARMONIC FORCE**

Before studying the response of a damped system, we consider an undamped system subjected to a harmonic force, for the sake of simplicity. If a force $F(t) = F_0 \cos \omega t$ acts on the mass m of an undamped system, the equation of motion, Eq. (3.1), reduces to

$$m\ddot{x} + kx = F_0 \cos \omega t \tag{3.3}$$

The homogeneous solution of this equation is given by

$$x_h(t) = C_1 \cos \omega_n t + C_2 \sin \omega_n t \tag{3.4}$$

where $\omega_n = (k/m)^{1/2}$ is the natural frequency of the system. Because the exciting force $F(t)$ is harmonic, the particular solution $x_p(t)$ is also harmonic and has the same frequency ω. Thus we assume a solution in the form

$$x_p(t) = X \cos \omega t \tag{3.5}$$

where X is a constant that denotes the maximum amplitude of $x_p(t)$. By substituting Eq. (3.5) into Eq. (3.3) and solving for X, we obtain

$$X = \frac{F_0}{k - m\omega^2} \tag{3.6}$$

Thus the total solution of Eq. (3.3) is

$$x(t) = C_1 \cos \omega_n t + C_2 \sin \omega_n t + \frac{F_0}{k - m\omega^2} \cos \omega t \tag{3.7}$$

Using the initial conditions $x(t = 0) = x_0$ and $\dot{x}(t = 0) = \dot{x}_0$, we find that

$$C_1 = x_0 - \frac{F_0}{k - m\omega^2}, \qquad C_2 = \frac{\dot{x}_0}{\omega_n} \tag{3.8}$$

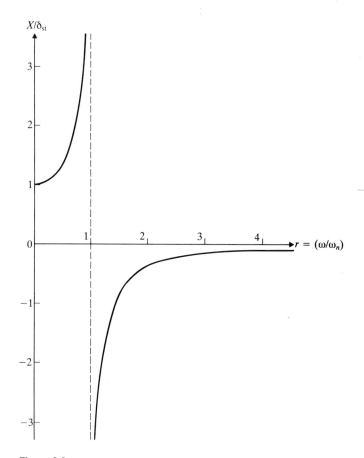

Figure 3.3

and hence

$$x(t) = \left(x_0 - \frac{F_0}{k - m\omega^2} \right) \cos \omega_n t + \left(\frac{\dot{x}_0}{\omega_n} \right) \sin \omega_n t + \left(\frac{F_0}{k - m\omega^2} \right) \cos \omega t \quad (3.9)$$

The maximum amplitude X in Eq. (3.6) can also be expressed as

$$\frac{X}{\delta_{st}} = \frac{1}{1 - \left(\dfrac{\omega}{\omega_n} \right)^2} \quad (3.10)$$

where $\delta_{st} = F_0/k$ denotes the deflection of the mass under a force F_0 and is sometimes called "static deflection" since F_0 is a constant (static) force. The quantity X/δ_{st} represents the ratio of the dynamic to the static amplitude of motion and is called the *magnification factor*, *amplification factor*, or *amplitude ratio*. The variation of the amplitude ratio, X/δ_{st}, with the frequency ratio $r = \omega/\omega_n$ [Eq.

$\delta_{ST} \equiv \dfrac{X_0}{Z_0} = \dfrac{F_0}{k}$

$X = X_0 \left(\dfrac{1}{1 - r^2} \right)$

$r \equiv \dfrac{\omega}{\omega_n} = $ freq ratio

$F(t) = F_0 \cos \omega t$

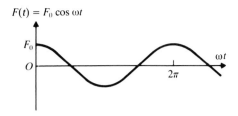

$x_p(t) = X \cos \omega t$

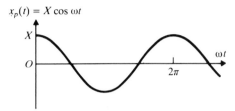

Figure 3.4

(3.10)] is shown in Fig. 3.3. From this figure, the response of the system can be identified to be of three types.

Case 1. When $0 < \omega/\omega_n < 1$, the denominator in Eq. (3.10) is positive and the response is given by Eq. (3.5) without change. The harmonic response of the system $x_p(t)$ is said to be in phase with the external force as shown in Fig. 3.4.

Case 2. When $\omega/\omega_n > 1$, the denominator in Eq. (3.10) is negative, and the steady-state solution can be expressed as

$$x_p(t) = -X \cos \omega t \qquad (3.11)$$

where the amplitude of motion X is redefined to be a positive quantity as

$$X = \frac{\delta_{st}}{\left(\dfrac{\omega}{\omega_n}\right)^2 - 1} \qquad (3.12)$$

The variations of $F(t)$ and $x_p(t)$ with time are shown in Fig. 3.5. Since $x_p(t)$ and $F(t)$ have opposite signs, the response is said to be 180° out of phase with the external force. Further, as $\omega/\omega_n \to \infty$, $X \to 0$. Thus the response of the system to a harmonic force of very high frequency is close to zero.

Case 3. When $\omega/\omega_n = 1$, the amplitude X given by Eq. (3.10) or (3.12) becomes infinite. This condition, for which the forcing frequency ω is equal to the natural frequency of the system ω_n, is called *resonance*. To find the response for this

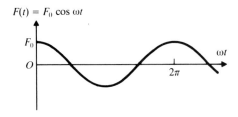

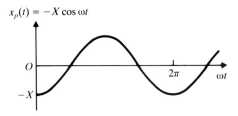

Figure 3.5

condition, we rewrite Eq. (3.9) as

$$x(t) = x_0 \cos \omega_n t + \frac{\dot{x}_0}{\omega_n} \sin \omega_n t + \delta_{st} \left[\frac{\cos \omega t - \cos \omega_n t}{1 - \left(\dfrac{\omega}{\omega_n} \right)^2} \right] \tag{3.13}$$

Since the last term of this equation takes an indefinite form for $\omega = \omega_n$, we apply L'Hospital's rule [3.1] to evaluate the limit of this term:

$$\lim_{\omega \to \omega_n} \left[\frac{\cos \omega t - \cos \omega_n t}{1 - \left(\dfrac{\omega}{\omega_n} \right)^2} \right] = \lim_{\omega \to \omega_n} \left[\frac{\dfrac{d}{d\omega}(\cos \omega t - \cos \omega_n t)}{\dfrac{d}{d\omega}\left(1 - \dfrac{\omega^2}{\omega_n^2} \right)} \right]$$

$$= \lim_{\omega \to \omega_n} \left[\frac{t \sin \omega t}{2 \dfrac{\omega}{\omega_n^2}} \right] = \frac{\omega_n t}{2} \sin \omega_n t. \tag{3.14}$$

Thus the response of the system at resonance becomes

$$x(t) = x_0 \cos \omega_n t + \frac{\dot{x}_0}{\omega_n} \sin \omega_n t + \frac{\delta_{st}\omega_n t}{2} \sin \omega_n t \tag{3.15}$$

It can be seen from Eq. (3.15) that at resonance, $x(t)$ increases indefinitely. The last term of Eq. (3.15) is shown in Fig. 3.6, from which the amplitude of the response can be seen to increase linearly with time.

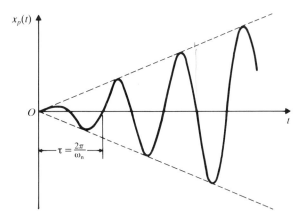

Figure 3.6

3.3.1
Total Response

The total response of the system, Eq. (3.7) or Eq. (3.9), can also be expressed as

$$x(t) = A\cos(\omega_n t - \phi) + \frac{\delta_{st}}{1 - \left(\dfrac{\omega}{\omega_n}\right)^2}\cos\omega t; \qquad \text{for } \frac{\omega}{\omega_n} < 1 \qquad (3.16)$$

$$x(t) = A\cos(\omega_n t - \phi) - \frac{\delta_{st}}{1 - \left(\dfrac{\omega}{\omega_n}\right)^2}\cos\omega t; \qquad \text{for } \frac{\omega}{\omega_n} > 1 \qquad (3.17)$$

where A and ϕ can be determined as in the case of Eq. (2.18). Thus the complete motion can be expressed as the sum of two cosine curves of different frequencies. In Eq. (3.16), the forcing frequency ω is smaller than the natural frequency, and the total response is shown in Fig. 3.7(a). In Eq. (3.17), the forcing frequency is greater than the natural frequency, and the total response appears as shown in Fig. 3.7(b).

3.3.2
Beating Phenomenon

If the forcing frequency is close to, but not exactly equal to, the natural frequency of the system, a phenomenon known as *beating* may occur. In this kind of vibration, the amplitude builds up and then diminishes in a regular pattern. The phenomenon of beating can be explained by considering the solution given by Eq. (3.9). If the initial conditions are taken as $x_0 = \dot{x}_0 = 0$, Eq. (3.9) reduces to

$$\begin{aligned} x(t) &= \frac{(F_0/m)}{\omega_n^2 - \omega^2}(\cos\omega t - \cos\omega_n t) \\ &= \frac{(F_0/m)}{\omega_n^2 - \omega^2}\left[2\sin\frac{\omega + \omega_n}{2}t \cdot \sin\frac{\omega_n - \omega}{2}t\right] \end{aligned} \qquad (3.18)$$

Let the forcing frequency ω be slightly less than the natural frequency:

$$\omega_n - \omega = 2\varepsilon \qquad (3.19)$$

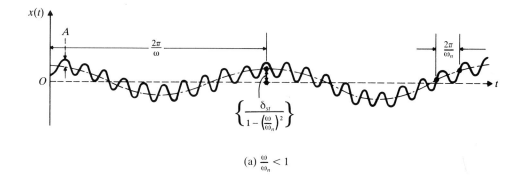

$$(a) \frac{\omega}{\omega_n} < 1$$

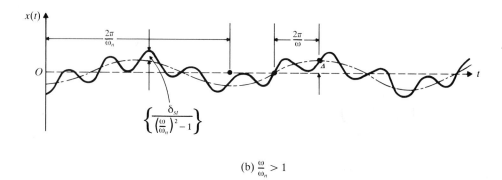

$$(b) \frac{\omega}{\omega_n} > 1$$

Figure 3.7

where ε is a small positive quantity. Then $\omega_n \approx \omega$ and

$$\omega + \omega_n \simeq 2\omega \tag{3.20}$$

Multiplication of Eqs. (3.19) and (3.20) gives

$$\omega_n^2 - \omega^2 = 4\varepsilon\omega \tag{3.21}$$

Use of Eqs. (3.19) to (3.21) in Eq. (3.18) gives

$$x(t) = \left(\frac{F_0/m}{2\varepsilon\omega} \sin \varepsilon t \right) \sin \omega t \tag{3.22}$$

Since ε is small, the function $\sin \varepsilon t$ varies slowly; its period, equal to $2\pi/\varepsilon$, is large. Thus Eq. (3.22) may be seen as representing vibration with period $2\pi/\omega$ and of variable amplitude equal to

$$\left(\frac{F_0/m}{2\varepsilon\omega} \right) \sin \varepsilon t$$

It can also be observed that the $\sin \omega t$ curve will go through several cycles, while the $\sin \varepsilon t$ wave goes through a single cycle, as shown in Fig. 3.8. Thus the amplitude builds up and dies down continuously. The time between the points of zero amplitude or the points of maximum amplitude is called the *period of beating* (τ_b)

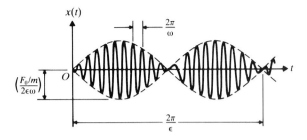

Figure 3.8

and is given by

$$\tau_b = \frac{2\pi}{2\varepsilon} = \frac{2\pi}{\omega_n - \omega} \tag{3.23}$$

with the frequency of beating defined as

$$\omega_b = 2\varepsilon = \omega_n - \omega$$

EXAMPLE 3.1   Plate Supporting a Pump

A reciprocating pump, weighing 150 lb, is mounted at the middle of a steel plate of thickness 0.5 in., width 20 in., and length 100 in., clamped along two edges as shown in Fig. 3.9. During operation of the pump, the plate is subjected to a harmonic force, $F(t) = 50 \cos 62.832\, t$ lb. Find the amplitude of vibration of the plate.

Given: Pump weight = 150 lb; plate dimensions: thickness (t) = 0.5 in., width (w) = 20 in., and length (l) = 100 in.; and harmonic force: $F(t) = 50 \cos 62.832\, t$ lb.

Find: Amplitude of vibration of the plate, X.

Approach: Find the stiffness of the plate by modeling it as a clamped beam. Use the equation for the response under harmonic excitation.

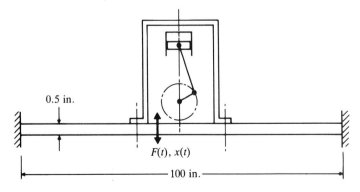

0.5 in.

$F(t), x(t)$

100 in.

Figure 3.9

Solution. The plate can be modeled as a fixed-fixed beam having Young's modulus (E) = 30 × 10^6 psi, length (l) = 100 in., and area moment of inertia (I) = $\frac{1}{12}(20)(0.5)^3$ = 0.2083 in^4. The bending stiffness of the beam is given by

$$k = \frac{192\,EI}{l^3} = \frac{192(30 \times 10^6)(0.2083)}{(100)^3} = 1200.0 \text{ lb/in.} \qquad \text{(E.1)}$$

The amplitude of harmonic response is given by Eq. (3.6) with F_0 = 50 lb, m = 150/386.4 lb-sec^2/in. (neglecting the weight of the steel plate), k = 1200.0 lb/in., and ω = 62.832 rad/sec. Thus Eq. (3.6) gives

$$X = \frac{F_0}{k - m\omega^2} = \frac{50}{1200.0 - (150/386.4)(62.832)^2} = -0.1504 \text{ in.} \qquad \text{(E.2)}$$

The negative sign indicates that the response $x(t)$ of the plate is out of phase with the excitation $F(t)$.

3.4 RESPONSE OF A DAMPED SYSTEM UNDER HARMONIC FORCE

If the forcing function is given by $F(t) = F_0 \cos \omega t$, the equation of motion becomes

$$m\ddot{x} + c\dot{x} + kx = F_0 \cos \omega t \quad \left(\overset{OR}{F_0 \sin \omega t}\right) \qquad (3.24)$$

The particular solution of Eq. (3.24) is also expected to be harmonic; we assume it in the form*

$$x_p(t) = X \cos(\omega t - \phi) \quad \left(X \sin(\omega t - \phi)\right) \qquad (3.25)$$

where X and ϕ are constants to be determined. X and ϕ denote the amplitude and phase angle of the response, respectively. By substituting Eq. (3.25) into Eq. (3.24), we arrive at

$$X\left[(k - m\omega^2)\cos(\omega t - \phi) - c\omega \sin(\omega t - \phi)\right] = F_0 \cos \omega t \qquad (3.26)$$

Using the trigonometric relations

$$\cos(\omega t - \phi) = \cos \omega t \cos \phi + \sin \omega t \sin \phi$$
$$\sin(\omega t - \phi) = \sin \omega t \cos \phi - \cos \omega t \sin \phi$$

in Eq. (3.26) and equating the coefficients of $\cos \omega t$ and $\sin \omega t$ on both sides of the resulting equation, we obtain

$$X\left[(k - m\omega^2)\cos \phi + c\omega \sin \phi\right] = F_0$$
$$X\left[(k - m\omega^2)\sin \phi - c\omega \cos \phi\right] = 0 \qquad (3.27)$$

Solution of Eqs. (3.27) gives

$$X = \frac{F_0}{\left[(k - m\omega^2)^2 + c^2\omega^2\right]^{1/2}} \qquad (3.28)$$

---

* Alternatively, we can assume $x_p(t)$ to be of the form $x_p(t) = C_1 \cos \omega t + C_2 \sin \omega t$, which also involves two constants C_1 and C_2. But the final result will be the same in both the cases.

and

$$\phi = \tan^{-1}\left(\frac{c\omega}{k - m\omega^2}\right) \tag{3.29}$$

By inserting the expressions of X and ϕ from Eqs. (3.28) and (3.29) into Eq. (3.25) we obtain the particular solution of Eq. (3.24). Figure 3.10 shows typical plots of the forcing function and (steady-state) response. Dividing both the numerator and denominator of Eq. (3.28) by k and making the following substitutions

$$\omega_n = \sqrt{\frac{k}{m}} = \text{undamped natural frequency,}$$

$$\zeta = \frac{c}{c_c} = \frac{c}{2m\omega_n} \; ; \; \frac{c}{m} = 2\zeta\omega_n,$$

$$X_0 \equiv \delta_{st} = \frac{F_0}{k} = \text{deflection under the static force } F_0, \text{ and}$$

MAG FACTOR
AMP FACTOR
AMP RATIO

$$r = \frac{\omega}{\omega_n} = \text{frequency ratio}$$

we obtain

$$\frac{X}{X_0} = \frac{X}{\delta_{st}} = \frac{1}{\left\{\left[1 - \left(\frac{\omega}{\omega_n}\right)^2\right]^2 + \left[2\zeta\frac{\omega}{\omega_n}\right]^2\right\}^{1/2}} = \frac{1}{\sqrt{(1 - r^2)^2 + (2\zeta r)^2}} \tag{3.30}$$

and

$$\phi = \tan^{-1}\left\{\frac{2\zeta\frac{\omega}{\omega_n}}{1 - \left(\frac{\omega}{\omega_n}\right)^2}\right\} = \tan^{-1}\left(\frac{2\zeta r}{1 - r^2}\right) \tag{3.31}$$

As stated in Section 3.3, the quantity X/δ_{st} is called the *magnification factor*, *amplification factor*, or *amplitude ratio*. The variations of X/δ_{st} and ϕ with the frequency ratio r and the damping ratio ζ are shown in Fig. 3.11. The following observations can be made from Eqs. (3.30) and (3.31) and from Fig. 3.11:

1. For an undamped system ($\zeta = 0$), Eq. (3.31) shows that the phase angle $\phi = 0$ (for $r < 1$) or $180°$ (for $r > 1$) and Eq. (3.30) reduces to Eq. (3.10).

2. The damping reduces the amplitude ratio for all values of the forcing frequency.

3. The reduction of the amplitude ratio in the presence of damping is very significant at or near resonance.

4. With damping, the maximum amplitude ratio (see Problem 3.11) occurs when

$$r = \sqrt{1 - 2\zeta^2} \quad \text{or} \quad \omega = \omega_n\sqrt{1 - 2\zeta^2} \tag{3.32}$$

which is lower than the undamped natural frequency ω_n and the damped natural frequency $\omega_d = \omega_n\sqrt{1 - \zeta^2}$.

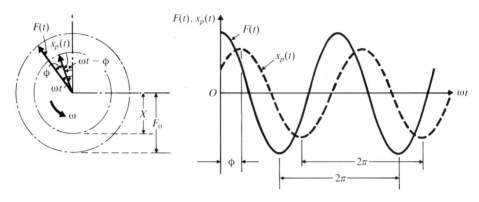

Figure 3.10 Graphical representation of forcing function and response.

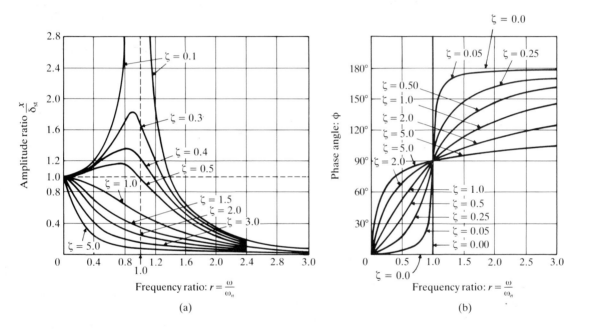

Figure 3.11 Variation of X and ϕ with frequency ratio r.

5. The maximum value of X (when $r = \sqrt{1 - 2\zeta^2}$) is given by

$$\left(\frac{X}{\delta_{st}}\right)_{max} = \frac{1}{2\zeta\sqrt{1 - \zeta^2}} \qquad (3.33)$$

and the value of X at $\omega = \omega_n$ by

$$\left(\frac{X}{\delta_{st}}\right)_{\omega = \omega_n} = \frac{1}{2\zeta} \qquad (3.34)$$

Equation (3.33) can be used for the experimental determination of the measure of damping present in the system. In a vibration test, if the maximum amplitude of the response $(X)_{max}$ is measured, the damping ratio of the system can be found using Eq. (3.33). Conversely, if the amount of damping is known, one can make an estimate of the maximum amplitude of vibration.

6. For $\zeta > 1/\sqrt{2}$, the graph of X has no peaks and for $\zeta = 0$, there is a discontinuity at $r = 1$.

7. The phase angle depends on the system parameters m, c, and k and the forcing frequency ω, but not on the amplitude F_0 of the forcing function.

8. The phase angle ϕ by which the response $x(t)$ or X lags the forcing function $F(t)$ or F_0 will be very small for small values of r. For very large values of r, the phase angle approaches 180° asymptotically. Thus the amplitude of vibration will be in phase with the exciting force for $r \ll 1$ and out of phase for $r \gg 1$. The phase angle at resonance will be 90° for all values of damping (ζ).

9. Below resonance ($\omega < \omega_n$), the phase angle increases with increase in damping. Above resonance ($\omega > \omega_n$), the phase angle decreases with increase in damping.

3.4.1 Total Response

The complete solution is given by $x(t) = x_h(t) + x_p(t)$ where $x_h(t)$ is given by Eq. (2.64). Thus

$$x(t) = X_0 e^{-\zeta \omega_n t} \cos(\omega_d t - \phi_0) + X \cos(\omega t - \phi) \tag{3.35}$$

where

$$\omega_d = \sqrt{1 - \zeta^2} \cdot \omega_n \tag{3.36}$$

$$r = \frac{\omega}{\omega_n} \tag{3.37}$$

X and ϕ are given by Eqs. (3.30) and (3.31), respectively, and X_0 and ϕ_0 can be determined from the initial conditions.

3.4.2 Quality Factor and Bandwidth

For small values of damping ($\zeta < 0.05$), we can take

$$\left(\frac{X}{\delta_{st}}\right)_{max} \simeq \left(\frac{X}{\delta_{st}}\right)_{\omega = \omega_n} = \frac{1}{2\zeta} = Q \tag{3.38}$$

The value of the amplitude ratio at resonance is also called Q factor or quality factor of the system, in analogy with some electrical-engineering applications, such as the tuning circuit of a radio, where the interest lies in an amplitude at resonance that is as large as possible [3.2]. The points R_1 and R_2, where the amplification factor falls to $Q/\sqrt{2}$, are called half power points because the power absorbed (ΔW) by the damper (or by the resistor in an electrical circuit), responding harmonically at a given frequency, is proportional to the square of the amplitude [see Eq. (2.88)]:

$$\Delta W = \pi c \omega X^2 \tag{3.39}$$

The difference between the frequencies associated with the half power points R_1 and R_2 is called the *bandwidth* of the system (see Fig. 3.12). To find the values of R_1 and R_2, we set $X/\delta_{st} = Q/\sqrt{2}$ in Eq. (3.30) so that

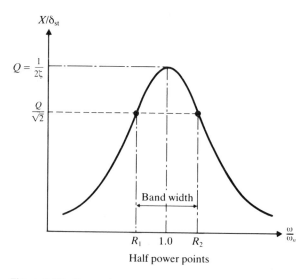

Figure 3.12 Harmonic response curve showing half power points and bandwidth.

$$\frac{1}{\sqrt{(1 - r^2)^2 + (2\zeta r)^2}} = \frac{Q}{\sqrt{2}} = \frac{1}{2\sqrt{2}\,\zeta}$$

or

$$r^4 - r^2(2 - 4\zeta^2) + (1 - 8\zeta^2) = 0 \qquad (3.40)$$

The solution of Eq. (3.40) gives

$$r_1^2 = 1 - 2\zeta^2 - 2\zeta\sqrt{1 + \zeta^2}, \qquad r_2^2 = 1 - 2\zeta^2 + 2\zeta\sqrt{1 + \zeta^2} \qquad (3.41)$$

For small values of ζ, Eq. (3.41) can be approximated as

$$r_1^2 = R_1^2 = \left(\frac{\omega_1}{\omega_n}\right)^2 \simeq 1 - 2\zeta, \qquad r_2^2 = R_2^2 = \left(\frac{\omega_2}{\omega_n}\right)^2 \simeq 1 + 2\zeta \qquad (3.42)$$

where $\omega_1 = \omega|_{R_1}$ and $\omega_2 = \omega|_{R_2}$. From Eq. (3.42),

$$\omega_2^2 - \omega_1^2 = (\omega_2 + \omega_1)(\omega_2 - \omega_1) = (R_2^2 - R_1^2)\omega_n^2 \simeq 4\zeta \qquad (3.43)$$

Using the relation

$$\omega_2 + \omega_1 = 2\omega_n \qquad (3.44)$$

in Eq. (3.43), we find that the bandwidth $\Delta\omega$ is given by

$$\Delta\omega = \omega_2 - \omega_1 \simeq 2\zeta/\omega_n \qquad (3.45)$$

Combining Eqs. (3.38) and (3.45), we obtain

$$Q \simeq \frac{1}{2\zeta} \simeq \frac{\omega_n}{\omega_2 - \omega_1} \qquad (3.46)$$

It can be seen that the quality factor Q can be used for estimating the equivalent viscous damping in a mechanical system.[†]

3.5 RESPONSE OF A DAMPED SYSTEM UNDER $F(t) = F_0 e^{i\omega t}$

Let the harmonic forcing function be represented in complex form as $F(t) = F_0 e^{i\omega t}$ so that the equation of motion becomes

$$m\ddot{x} + c\dot{x} + kx = F_0 e^{i\omega t} \tag{3.47}$$

Since the actual excitation is given only by the real part of $F(t)$, the response will also be given only by the real part of $x(t)$ where $x(t)$ is a complex quantity satisfying the differential equation (3.47). F_0 in Eq. (3.47) is, in general, a complex number. By assuming the particular solution $x_p(t)$

$$x_p(t) = X e^{i\omega t} \tag{3.48}$$

we obtain, by substituting Eq. (3.48) into Eq. (3.47),[*]

$$X = \frac{F_0}{(k - m\omega^2) + ic\omega} \tag{3.49}$$

Multiplying the numerator and denominator on the right side of Eq. (3.49) by $[(k - m\omega^2) - ic\omega]$ and separating the real and imaginary parts, we obtain

$$X = F_0 \left[\frac{k - m\omega^2}{(k - m\omega^2)^2 + c^2\omega^2} - i \frac{c\omega}{(k - m\omega^2)^2 + c^2\omega^2} \right] \tag{3.50}$$

Using the relation, $x + iy = A e^{i\phi}$ where $A = \sqrt{x^2 + y^2}$ and $\tan \phi = y/x$, Eq. (3.50) can be expressed as

$$X = \frac{F_0}{\left[(k - m\omega^2)^2 + c^2\omega^2 \right]^{1/2}} e^{-i\phi} \tag{3.51}$$

where

$$\phi = \tan^{-1} \left(\frac{c\omega}{k - m\omega^2} \right) \tag{3.52}$$

Thus the steady-state solution, Eq. (3.48), becomes

$$x_p(t) = \frac{F_0}{\left[(k - m\omega^2)^2 + (c\omega)^2 \right]^{1/2}} e^{i(\omega t - \phi)} \tag{3.53}$$

[†] The determination of the system parameters (m, c, and k) based on half-power points and other response characteristics of the system is considered in Section 10.8.

[*] Equation (3.49) can be written as $Z(i\omega) X = F_0$ where $Z(i\omega) = -m\omega^2 + i\omega c + k$ is called the *mechanical impedance* of the system [3.8].

Frequency Response. Equation (3.49) can be rewritten in the form

$$\frac{kX}{F_0} = \frac{1}{1 - r^2 + i2\zeta r} \equiv H(i\omega) \qquad (3.54)$$

where $H(i\omega)$ is known as the *complex frequency response* of the system. The absolute value of $H(i\omega)$ given by

$$|H(i\omega)| = \left|\frac{kX}{F_0}\right| = \frac{1}{\left[(1 - r^2)^2 + (2\zeta r)^2\right]^{1/2}} \qquad (3.55)$$

denotes the magnification factor defined in Eq. (3.30). Recalling that $e^{i\phi} = \cos\phi + i\sin\phi$, we can show that Eqs. (3.54) and (3.55) are related:

$$H(i\omega) = |H(i\omega)|e^{-i\phi} \qquad (3.56)$$

where ϕ is given by Eq. (3.52), which can also be expressed as

$$\phi = \tan^{-1}\left(\frac{2\zeta r}{1 - r^2}\right) \qquad (3.57)$$

Thus Eq. (3.53) can be expressed as

$$x_p(t) = \frac{F_0}{k}|H(i\omega)|e^{i(\omega t - \phi)} \qquad (3.58)$$

It can be seen that the complex frequency response function, $H(i\omega)$, contains both the magnitude and phase of the steady state response. The use of this function in the experimental determination of the system parameters (m, c, and k) is discussed in Section 10.8. If $F(t) = F_0\cos\omega t$, the corresponding steady-state solution is given by the real part of Eq. (3.53):

$$\begin{aligned} x_p(t) &= \frac{F_0}{\left[(k - m\omega^2)^2 + (c\omega)^2\right]^{1/2}}\cos(\omega t - \phi) \\ &= \mathrm{Re}\left[\frac{F_0}{k}H(i\omega)e^{i\omega t}\right] = \mathrm{Re}\left[\frac{F_0}{k}|H(i\omega)|e^{i(\omega t - \phi)}\right] \end{aligned} \qquad (3.59)$$

which can be seen to be the same as Eq. (3.25). Similarly, if $F(t) = F_0\sin\omega t$, the corresponding steady-state solution is given by the imaginary part of Eq. (3.53):

$$\begin{aligned} x_p(t) &= \frac{F_0}{\left[(k - m\omega^2)^2 + (c\omega)^2\right]^{1/2}}\sin(\omega t - \phi) \\ &= \mathrm{Im}\left[\frac{F_0}{k}|H(i\omega)|e^{i(\omega t - \phi)}\right] \end{aligned} \qquad (3.60)$$

Complex Vector Representation of Harmonic Motion. The harmonic excitation and the response of the damped system to that excitation can be represented graphically in the complex plane, and interesting interpretation can be given to the resulting

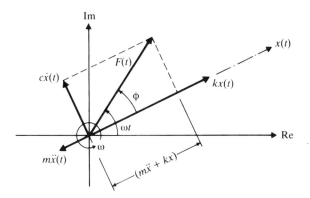

Figure 3.13   Representation of Eq. (3.47) in a complex plane.

diagram. We first differentiate Eq. (3.58) with respect to time and obtain

$$\text{velocity} = \dot{x}_p(t) = i\omega \frac{F_0}{k} |H(i\omega)| e^{i(\omega t - \phi)} = i\omega x_p(t)$$

$$\text{acceleration} = \ddot{x}_p(t) = (i\omega)^2 \frac{F_0}{k} |H(i\omega)| e^{i(\omega t - \phi)} = -\omega^2 x_p(t) \qquad (3.61)$$

Because i can be expressed as

$$i = \cos\frac{\pi}{2} + i\sin\frac{\pi}{2} = e^{i\pi} \qquad (3.62)$$

we can conclude that the velocity leads the displacement by the phase angle $\pi/2$ and that it is multiplied by ω. Similarly, -1 can be written as

$$-1 = \cos\pi + i\sin\pi = e^{\pi} \qquad (3.63)$$

Hence the acceleration leads the displacement by the phase angle π, and it is multiplied by ω^2.

Thus the various terms of the equation of motion (3.47) can be represented in the complex plane, as shown in Fig. 3.13. The interpretation of this figure is that the sum of the complex vectors $m\ddot{x}(t)$, $c\dot{x}(t)$, and $kx(t)$ balances $F(t)$, which is precisely what is required to satisfy Eq. (3.47). It is to be noted that the entire diagram rotates with angular velocity ω in the complex plane. If only the real part of the response is to be considered, then the entire diagram must be projected onto the real axis. Similarly, if only the imaginary part of the response is to be considered, then the diagram must be projected onto the imaginary axis. In Fig. 3.13, notice that the force $F(t) = F_0 e^{i\omega t}$ is represented as a vector located at an angle ωt to the real axis. This implies that F_0 is real. If F_0 is also complex, then the force vector $F(t)$ will be located at an angle of $(\omega t + \psi)$, where ψ is some phase angle introduced by F_0. In such a case, all the other vectors, namely, $m\ddot{x}$, $c\dot{x}$, and kx will be shifted by the same angle ψ. This is equivalent to multiplying both sides of Eq. (3.47) by $e^{i\psi}$.

3.6 RESPONSE OF A DAMPED SYSTEM UNDER THE HARMONIC MOTION OF THE BASE

Sometimes the base or support of a spring-mass-damper system undergoes harmonic motion, as shown in Fig. 3.14(a). Let $y(t)$ denote the displacement of the base and $x(t)$ the displacement of the mass from its static equilibrium position at time t. Then the net elongation of the spring is $x - y$ and the relative velocity between the two ends of the damper is $\dot{x} - \dot{y}$. From the free-body diagram shown in Fig. 3.14(b), we obtain the equation of motion:

$$m\ddot{x} + c(\dot{x} - \dot{y}) + k(x - y) = 0 \tag{3.64}$$

If $y(t) = Y \sin \omega t$, Eq. (3.64) becomes

$$m\ddot{x} + c\dot{x} + kx = A \sin \omega t + B \cos \omega t \tag{3.65}$$

where $A = kY$ and $B = c\omega Y$. This shows that giving excitation to the base is equivalent to applying harmonic force of magnitude $(kY \sin \omega t + c\omega Y \cos \omega t)$ to the mass. By using the solutions given in Eqs. (3.59) and (3.60), the steady-state response of the mass can be expressed as

$$x_p(t) = \frac{kY \sin(\omega t - \phi_1)}{\left[(k - m\omega^2)^2 + (c\omega)^2\right]^{1/2}} + \frac{\omega cY \cos(\omega t - \phi_1)}{\left[(k - m\omega^2)^2 + (c\omega)^2\right]^{1/2}} \tag{3.66}$$

The phase angle ϕ_1 will be the same for both the terms because it depends on the values of m, c, k, and ω, but not on the amplitude of the excitation. Equation (3.66) can be rewritten as

$$x_p(t) = X \cos(\omega t - \phi_1 - \phi_2)$$

$$\equiv Y \left[\frac{k^2 + (c\omega)^2}{(k - m\omega^2)^2 + (c\omega)^2}\right]^{1/2} \cos(\omega t - \phi_1 - \phi_2) \tag{3.67}$$

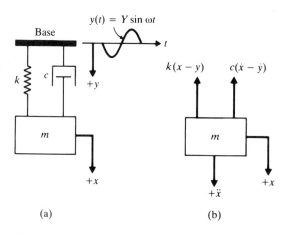

(a)

(b)

Figure 3.14   Base excitation.

$$F = kx$$

where the ratio of the amplitude of the response $x_p(t)$ to that of the base motion $y(t)$ is given by

$$\frac{X}{Y} = \left[\frac{k^2 + (c\omega)^2}{(k - m\omega^2)^2 + (c\omega)^2}\right]^{1/2} = \left[\frac{1 + (2\zeta r)^2}{(1 - r^2)^2 + (2\zeta r)^2}\right]^{1/2} \tag{3.68}$$

and ϕ_1 and ϕ_2 by

$$\phi_1 = \tan^{-1}\left(\frac{c\omega}{k - m\omega^2}\right) = \tan^{-1}\left(\frac{2\zeta r}{1 - r^2}\right)$$

$$\phi_2 = \tan^{-1}\left(\frac{k}{c\omega}\right) = \tan^{-1}\left(\frac{1}{2\zeta r}\right) \tag{3.69}$$

The ratio X/Y is called the displacement *transmissibility.*

Note that if the harmonic excitation of the base is expressed in complex form as $y(t) = \text{Re}(Ye^{i\omega t})$, the response of the system can be expressed as

$$x_p(t) = \text{Re}\left\{\left(\frac{1 + i2\zeta r}{1 - r^2 + i2\zeta r}\right)Ye^{i\omega t}\right\} \tag{3.70}$$

and the transmissibility as

$$\frac{X}{Y} = \left[1 + (2\zeta r)^2\right]^{1/2}|H(i\omega)| \tag{3.71}$$

where $|H(i\omega)|$ is given by Eq. (3.55).

**3.6.1
Force
Transmitted**

In Fig. 3.14(b), the force carried by the support F must be due to the spring and dashpot which are connected to it. It can be determined as follows:

$$F = k(x - y) + c(\dot{x} - \dot{y}) = -m\ddot{x} \tag{3.72}$$

From Eq. (3.67), Eq. (3.72) can be written as

$$F = m\omega^2 X \cos(\omega t - \phi_1 - \phi_2) = F_T \cos(\omega t - \phi_1 - \phi_2) \tag{3.73}$$

where F_T is the amplitude or maximum value of the transmitted force given by

$$\frac{F_T}{kY} = r^2\left[\frac{1 + (2\zeta r)^2}{(1 - r^2)^2 + (2\zeta r)^2}\right]^{1/2} \tag{3.74}$$

The ratio (F_T/kY) is known as the force transmissibility.[†] It can be noticed that the transmitted force is in phase with the motion of the mass $x(t)$. The variation of the

---

[†] The use of the concept of transmissibility in the design of vibration isolation systems is discussed in Chapter 9.

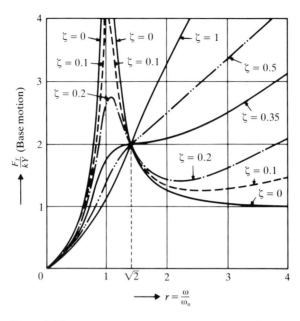

Figure 3.15

force transmitted to the base with the frequency ratio r is shown in Fig. 3.15 for different values of ζ.

3.6.2
Relative Motion

If $z = x - y$ denotes the motion of the mass relative to the base, the equation of motion, Eq. (3.64), can be rewritten as

$$m\ddot{z} + c\dot{z} + kz = -m\ddot{y} = m\omega^2 Y \sin \omega t \tag{3.75}$$

The steady-state solution of Eq. (3.75) is given by

$$z(t) = \frac{m\omega^2 Y \sin(\omega t - \phi_1)}{\left[(k - m\omega^2)^2 + (c\omega)^2\right]^{1/2}} = Z \sin(\omega t - \phi_1) \tag{3.76}$$

where Z, the amplitude of $z(t)$, can be expressed as

$$Z = \frac{m\omega^2 Y}{\sqrt{(k - m\omega^2)^2 + (c\omega)^2}} = Y\frac{r^2}{\sqrt{(1 - r^2)^2 + (2\zeta r)^2}} \tag{3.77}$$

and ϕ_1 by Eq. (3.69). The ratio Z/Y is shown graphically in Fig. 3.16.

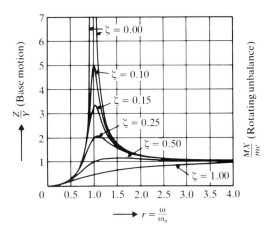

Figure 3.16 Variation of (Z/Y) or (MX/me) with frequency ratio $r = (\omega/\omega_n)$.

EXAMPLE 3.2   Vehicle Moving on a Rough Road

Figure 3.17(a) shows a simple model of a motor vehicle that can vibrate in the vertical direction while traveling over a rough road. The vehicle has a mass of 1200 kg. The suspension system has a spring constant of 400 kN/m and a damping ratio of $\zeta = 0.5$. If the vehicle speed is 100 km/hr, determine the displacement amplitude of the vehicle. The road surface varies sinusoidally with an amplitude of $Y = 0.05$ m and a wavelength of 6 m.

Given: Vehicle model: $m = 1200$ kg, $k = 400$ kN/m, $\zeta = 0.5$, and speed = 100 km/hr. Road surface: sinusoidal with $Y = 0.05$ m and period = 6 m.

Find: Displacement amplitude (X) of the vehicle.

Approach: Model the vehicle as a single degree of freedom system subjected to base motion as shown in Fig. 3.17(b).

Solution. The frequency ω of the base excitation can be found by dividing the vehicle speed by the length of one cycle of road roughness:

$$\omega = 2\pi f = 2\pi \left(\frac{100 \times 1000}{3600} \right) \frac{1}{6} = 29.0887 \text{ rad/sec}$$

The natural frequency of the vehicle is given by

$$\omega_n = \sqrt{\frac{k}{m}} = \left(\frac{400 \times 10^3}{1200} \right)^{1/2} = 18.2574 \text{ rad/sec}$$

and hence the frequency ratio r is

$$r = \frac{\omega}{\omega_n} = \frac{29.0887}{18.2574} = 1.5933$$

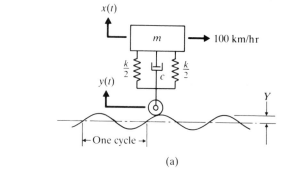

(a)

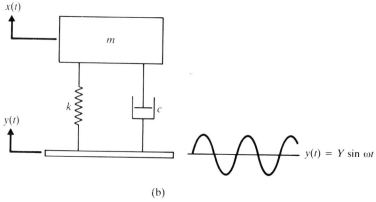

(b)

Figure 3.17 Vehicle moving over a rough road.

The amplitude ratio can be found from Eq. (3.63):

$$\frac{X}{Y} = \left\{ \frac{1 + (2\zeta r)^2}{(1 - r^2)^2 + (2\zeta r)^2} \right\}^{1/2} = \left\{ \frac{1 + (2 \times 0.5 \times 1.5933)^2}{(1 - 1.5933^2)^2 + (2 \times 0.5 \times 1.5933)^2} \right\}^{1/2}$$

$$= 0.8493$$

Thus the displacement amplitude of the vehicle is given by

$$X = 0.8493Y = 0.8493(0.05) = 0.0425 \text{ m}$$

EXAMPLE 3.3   Machine on Resilient Foundation

A heavy machine, weighing 3000 N, is supported on a resilient foundation. The static deflection of the foundation due to the weight of the machine is found to be 7.5 cm. It is observed that the machine vibrates with an amplitude of 1 cm when the base of the foundation is subjected to harmonic oscillation at the undamped natural frequency of the system with an amplitude of 0.25 cm. Find (1) the damping constant of the foundation, (2) the dynamic force amplitude on the base, and (3) the amplitude of the displacement of the machine relative to the base.

Given: Machine weight $(W) = 3000$ N, static deflection under $W = 7.5$ cm, and $X = 1$ cm, when $y(t) = 0.25 \sin \omega_n t$ cm.

Find: c, F_T, and Z.

Approach: Specialize the equations for X/Y, F_T, and Z for the case $\omega = \omega_r$.

Solution. (1) The stiffness of the foundation can be found from its static deflection:

$$k = \text{weight of machine}/\delta_{st} = 3000/0.075 = 40{,}000 \text{ N/m}$$

At resonance ($\omega = \omega_n$ or $r = 1$), Eq. (3.68) gives

$$\frac{X}{Y} = \frac{0.010}{0.0025} = 4 = \left[\frac{1 + (2\zeta)^2}{(2\zeta)^2} \right]^{1/2} \tag{E.1}$$

The solution of Eq. (E.1) gives $\zeta = 0.1291$. The damping constant is given by

$$c = \zeta \cdot c_c = \zeta 2\sqrt{km} = 0.1291 \times 2 \times \sqrt{40{,}000 \times (3000/9.81)} = 903.0512 \text{ N-s/m} \tag{E.2}$$

(2) The dynamic force amplitude on the base at $r = 1$ can be found from Eq. (3.74):

$$F_T = Yk \left[\frac{1 + 4\zeta^2}{4\zeta^2} \right]^{1/2} = kX = 40{,}000 \times 0.01 = 400 \text{ N} \tag{E.3}$$

(3) The amplitude of the relative displacement of the machine at $r = 1$ can be obtained from Eq. (3.77):

$$Z = \frac{Y}{2\zeta} = \frac{0.0025}{2 \times 0.1291} = 0.00968 \text{ m} \tag{E.4}$$

It can be noticed that $X = 0.01$ m, $Y = 0.0025$ m, and $Z = 0.00968$ m; therefore, $Z \neq X - Y$. This is due to the phase differences between x, y, and z.

3.7   RESPONSE OF A DAMPED SYSTEM UNDER ROTATING UNBALANCE

Unbalance in rotating machinery is one of the main causes of vibration. A simplified model of such a machine is shown in Fig. 3.18. The total mass of the machine is M, and there are two eccentric masses $m/2$ rotating in opposite directions with a constant angular velocity ω. The centrifugal force $(m e \omega^2)/2$ due to each mass will cause excitation of the mass M. We consider two equal masses $m/2$ rotating in opposite directions in order to have the horizontal components of excitation of the two masses cancel each other. However, the vertical components of excitation add together and act along the axis of symmetry $A - A$ in Fig. 3.18. If the angular position of the masses is measured from a horizontal position, the total vertical component of the excitation is always given by $F(t) = m e \omega^2 \sin \omega t$. The equation of motion can be derived by the usual procedure:

$$M\ddot{x} + c\dot{x} + kx = m e \omega^2 \sin \omega t \tag{3.78}$$

The solution of this equation will be identical to Eq. (3.60) if we replace m and F_0

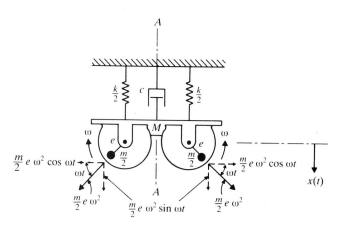

Figure 3.18 Rotating unbalanced masses.

by M and $me\omega^2$ respectively. This solution can also be expressed as

$$x_p(t) = X\sin(\omega t - \phi) = \text{Im}\left[\frac{me}{M}\left(\frac{\omega}{\omega_n}\right)^2 |H(i\omega)|e^{i(\omega t - \phi)}\right] \quad (3.79)$$

where $\omega_n = \sqrt{k/M}$ and X and ϕ denote the amplitude and the phase angle of vibration given by

$$X = \frac{me\omega^2}{\left[(k - M\omega^2)^2 + (c\omega)^2\right]^{1/2}} = \frac{me}{M}\left(\frac{\omega}{\omega_n}\right)^2 |H(i\omega)| \quad (3.80)$$

and

$$\phi = \tan^{-1}\left(\frac{c\omega}{k - M\omega^2}\right) \quad (3.81)$$

By defining $\zeta = c/c_c$ and $c_c = 2M\omega_n$, Eqs. (3.80) and (3.81) can be rewritten as

$$\frac{MX}{me} = \frac{r^2}{\left[(1 - r^2)^2 + (2\zeta r)^2\right]^{1/2}} = r^2|H(i\omega)| \quad (3.82)$$

and

$$\phi = \tan^{-1}\left(\frac{2\zeta r}{1 - r^2}\right) \quad (3.83)$$

The variation of MX/me with r for different values of ζ is shown in Fig. 3.16. On the other hand, the graph of ϕ versus r remains as in Fig. 3.11(b). The following observations can be made from Eq. (3.82) and Fig. 3.16:

1. All the curves begin at zero amplitude. The amplitude near resonance ($\omega = \omega_n$) is markedly affected by damping. Thus if the machine is to be run near resonance, damping should be introduced purposefully to avoid dangerous amplitudes.

2. At very high speeds (ω large), MX/me is almost unity, and the effect of damping is negligible.

3. The maximum of MX/me occurs when

$$\frac{d}{dr}\left(\frac{MX}{me}\right) = 0 \tag{3.84}$$

The solution of Eq. (3.84) gives

$$r = \frac{1}{\sqrt{1 - 2\zeta^2}} > 1$$

Accordingly, the peaks occur to the right of the resonance value of $r = 1$.

EXAMPLE 3.4   **Francis Water Turbine**

The schematic diagram of a Francis water turbine is shown in Fig. 3.19 in which water flows from A into the blades B and down into the tail race C. The rotor has a mass of 250 kg and an unbalance (me) of 5 kg-mm. The radial clearance between the rotor and the stator is 5 mm. The turbine operates in the speed range 600 to 6000 rpm. The steel shaft carrying the rotor can be assumed to be clamped at the bearings. Determine the diameter of the shaft so that the rotor is always clear of the stator at all the operating speeds of the turbine. Assume damping to be negligible.

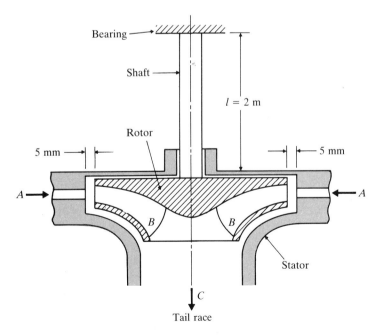

Figure 3.19

Given: Turbine: mass $(M) = 250$ kg, unbalance $(me) = 5$ kg-mm, and speed range = 600–6000 rpm. Shaft: length = 2 m and maximum radial deflection = 5 mm.

Find: Diameter of the shaft.

Approach: Equate the maximum amplitude (radial deflection) of rotor to 5 mm. Use the expression for the stiffness of a cantilever beam.

Solution. The maximum amplitude of the shaft (rotor) due to rotating unbalance can be obtained from Eq. (3.80) by setting $c = 0$ as

$$X = \frac{me\omega^2}{(k - M\omega^2)} = \frac{me\omega^2}{k(1 - r^2)} \tag{E.1}$$

where $me = 5$ kg-mm, $M = 250$ kg, and the limiting value of $X = 5$ mm. The value of ω ranges from

$$600 \text{ rpm} = 600 \times \frac{2\pi}{60} = 20\pi \text{ rad/sec}$$

to

$$6000 \text{ rpm} = 6000 \times \frac{2\pi}{60} = 200\pi \text{ rad/sec}$$

while the natural frequency of the system is given by

$$\omega_n = \sqrt{\frac{k}{M}} = \sqrt{\frac{k}{250}} = 0.0625\sqrt{k} \text{ rad/sec} \tag{E.2}$$

if k is in N/m. For $\omega = 20\pi$ rad/sec, Eq. (E.1) gives

$$0.005 = \frac{(5.0 \times 10^{-3}) \times (20\pi)^2}{k\left[1 - \dfrac{(20\pi)^2}{0.004\,k}\right]} = \frac{2\pi^2}{k - 10^5\pi^2}$$

$$k = 10.04 \times 10^4\pi^2 \text{ N/m} \tag{E.3}$$

For $\omega = 200\pi$ rad/sec, Eq. (E.1) gives

$$0.005 = \frac{(5.0 \times 10^{-3}) \times (200\pi)^2}{k\left[1 - \dfrac{(200\pi)^2}{0.004\,k}\right]} = \frac{200\pi^2}{k - 10^7\pi^2}$$

$$k = 10.04 \times 10^6\pi^2 \text{ N/m} \tag{E.4}$$

From Fig. 3.16, we find that the amplitude of vibration of the rotating shaft can be minimized by making $r = \omega/\omega_n$ very large. This means that ω_n must be made small compared to ω—that is, k must be made small. This can be achieved by selecting the value of k as $10.04 \times 10^4\pi^2$ N/m. Since the stiffness of a cantilever beam (shaft) supporting a load (rotor) at the end is given by

$$k = \frac{3EI}{l^3} = \frac{3E}{l^3}\left(\frac{\pi d^4}{64}\right) \tag{E.5}$$

the diameter of the beam (shaft) can be found:

$$d^4 = \frac{64kl^3}{3\pi E} = \frac{(64)(10.04 \times 10^4 \pi^2)(2^3)}{3\pi(2.07 \times 10^{11})} = 2.6005 \times 10^{-4} \text{ m}^4$$

or

$$d = 0.1270 \text{ m} = 127 \text{ mm} \tag{E.6}$$

3.8 FORCED VIBRATION WITH COULOMB DAMPING

For a single degree of freedom system with Coulomb or dry friction damping, subjected to a harmonic force $F(t) = F_0 \sin \omega t$ as in Fig. 3.20, the equation of motion is given by

$$m\ddot{x} + kx \pm \mu N = F(t) = F_0 \sin \omega t \tag{3.85}$$

where the sign of the friction force (μN) is positive (negative) when the mass moves from left to right (right to left). The exact solution of Eq. (3.85) is quite involved. However, we can expect that if the dry friction damping force is large, the motion of the mass will be discontinuous. On the other hand, if the dry friction force is small compared to the amplitude of the applied force F_0, the steady state solution is expected to be nearly harmonic. In this case, we can find an approximate solution of Eq. (3.85) by finding an equivalent viscous damping ratio. To find an equivalent viscous damping ratio, we equate the energy dissipated due to dry friction to the energy dissipated by an equivalent viscous damper during a full cycle of motion. If the amplitude of motion is denoted as X, the energy dissipated by the friction force μN in a quarter cycle is $\mu N X$. Hence in a full cycle, the energy dissipated by dry friction damping is given by

$$\Delta W = 4\mu N X \tag{3.86}$$

If the equivalent viscous damping constant is denoted as c_{eq}, the energy dissipated during a full cycle [see Eq. (2.92)] will be

$$\Delta W = \pi c_{eq} \omega X^2 \tag{3.87}$$

By equating Eqs. (3.86) and (3.87), we obtain

$$c_{eq} = \frac{4\mu N}{\pi \omega X} \tag{3.88}$$

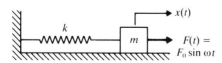

Figure 3.20

Thus the steady-state response is given by

$$x_p(t) = X \sin(\omega t - \phi) \tag{3.89}$$

where the amplitude X can be found from Eq. (3.60):

$$X = \frac{F_0}{\left[(k - m\omega^2)^2 + (c_{eq}\omega)^2\right]^{1/2}} = \frac{(F_0/k)}{\left[\left(1 - \frac{\omega^2}{\omega_n^2}\right)^2 + \left(2\zeta_{eq}\frac{\omega}{\omega_n}\right)^2\right]^{1/2}} \tag{3.90}$$

with

$$\zeta_{eq} = \frac{c_{eq}}{c_c} = \frac{c_{eq}}{2m\omega_n} = \frac{4\mu N}{2m\omega_n\pi\omega X} = \frac{2\mu N}{\pi m\omega\omega_n X} \tag{3.91}$$

Substitution of Eq. (3.91) into Eq. (3.90) gives

$$X = \frac{(F_0/k)}{\left[\left(1 - \frac{\omega^2}{\omega_n^2}\right)^2 + \left(\frac{4\mu N}{\pi k X}\right)^2\right]^{1/2}} \tag{3.92}$$

The solution of this equation gives the amplitude X as

only valid when positive

$$\frac{F_0}{k\,(1-r^2)}\sqrt{1 - \left(\frac{4\mu_N}{\pi F_0}\right)} \quad = \quad X = \frac{F_0}{k}\left[\frac{1 - \left(\frac{4\mu N}{\pi F_0}\right)^2}{\left(1 - \frac{\omega^2}{\omega_n^2}\right)^2}\right]^{1/2} \tag{3.93}$$

As stated earlier, Eq. (3.93) can be used only if the friction force is small compared to F_0. In fact, the limiting value of the friction force μN can be found from Eq. (3.93). To avoid imaginary values of X, we need to have

$\mu_N < .785\,F_0$

↑ *do first*

$\left(\frac{4\mu_N}{\pi F_0}\right) <$ good to calculate first

$$1 - \left(\frac{4\mu N}{\pi F_0}\right)^2 > 0 \quad \text{or} \quad \frac{F_0}{\mu N} > \frac{4}{\pi}.$$

If this condition is not satisfied, the exact analysis, given in Ref. [3.3], is to be used. The phase angle ϕ appearing in Eq. (3.89) can be found using Eq. (3.52):

$$\phi = \tan^{-1}\left(\frac{c_{eq}\omega}{k - m\omega^2}\right) = \tan^{-1}\left[\frac{2\zeta_{eq}\frac{\omega}{\omega_n}}{1 - \frac{\omega^2}{\omega_n^2}}\right] = \tan^{-1}\left\{\frac{\frac{4\mu N}{\pi k X}}{1 - \frac{\omega^2}{\omega_n^2}}\right\} \tag{3.94}$$

Substituting Eq. (3.93) into Eq. (3.94) for X, we obtain

$$\phi = \tan^{-1}\left[\frac{\dfrac{4\mu N}{\pi F_0}}{\left\{1 - \left(\dfrac{4\mu N}{\pi F_0}\right)^2\right\}^{1/2}}\right] \tag{3.95}$$

Equation (3.94) shows that $\tan\phi$ is a constant for a given value of $F_0/\mu N$. ϕ is discontinuous at $\omega/\omega_n = 1$ (resonance) since it takes a positive value for $\omega/\omega_n < 1$ and a negative value for $\omega/\omega_n > 1$. Thus Eq. (3.95) can also be expressed as

$$\phi = \tan^{-1}\left[\frac{\pm\dfrac{4\mu N}{\pi F_0}}{\left\{1 - \left(\dfrac{4\mu N}{\pi F_0}\right)^2\right\}^{1/2}}\right] \tag{3.96}$$

Equation (3.93) shows that friction serves to limit the amplitude of forced vibration for $\omega/\omega_n \neq 1$. However, at resonance ($\omega/\omega_n = 1$), the amplitude becomes infinite. This can be explained as follows. The energy directed into the system over one cycle when it is excited harmonically at resonance is

$$\Delta W' = \int_{\text{cycle}} F \cdot dx = \int_0^\tau F\frac{dx}{dt}\,dt$$

$$= \int_0^{\tau = 2\pi/\omega} F_0\sin\omega t \cdot \left[\omega X\cos(\omega t - \phi)\right]dt \tag{3.97}$$

Since Eq. (3.94) gives $\phi = 90°$ at resonance, Eq. (3.97) becomes

$$\Delta W' = F_0 X\omega\int_0^{2\pi/\omega}\sin^2\omega t\,dt = \pi F_0 X \tag{3.98}$$

The energy dissipated from the system is given by Eq. (3.86). Since $\pi F_0 X > 4\mu N X$ for X to be real-valued, $\Delta W' > \Delta W$ at resonance (see Fig. 3.21). Thus more energy

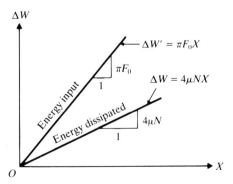

Figure 3.21

is directed into the system per cycle than is dissipated per cycle. This extra energy is used to build up the amplitude of vibration. For the nonresonant condition $(\omega/\omega_n \neq 1)$, the energy input can be found from Eq. (3.97):

$$\Delta W' = \omega F_0 X \int_0^{2\pi/\omega} \sin \omega t \cos(\omega t - \phi)\, dt = \pi F_0 X \sin \phi \qquad (3.99)$$

Due to the presence of $\sin \phi$ in Eq. (3.99), the input energy curve in Fig. 3.21 is made to coincide with the dissipated energy curve, so the amplitude is limited. Thus the phase of the motion ϕ can be seen to limit the amplitude of the motion.

The periodic response of a spring-mass system with Coulomb damping subjected to base excitation is given in Refs. [3.10, 3.11].

EXAMPLE 3.5   **Spring-Mass System with Coulomb Damping**

A spring-mass system, having a mass of 10 kg and a spring of stiffness of 4000 N/m, vibrates on a horizontal surface. The coefficient of friction is 0.12. When subjected to a harmonic force of frequency 2 Hz, the mass is found to vibrate with an amplitude of 40 mm. Find the amplitude of the harmonic force applied to the mass.

Given: Spring-mass system with Coulomb friction—$m = 10$ kg, $k = 4000$ N/m, $\mu = 0.12$, harmonic force with frequency = 2 Hz, vibration amplitude = 40 mm.

Find: Amplitude of the applied force.

Approach: Use Eq. (3.93).

Solution. The vertical force (weight) of the mass is $N = mg = 10 \times 9.81 = 98.1$ N. The natural frequency is

$$\omega_n = \sqrt{\frac{k}{m}} = \sqrt{\frac{4000}{10}} = 20 \text{ rad/sec}$$

and the frequency ratio is

$$\frac{\omega}{\omega_n} = \frac{2 \times 2\pi}{20} = 0.6283$$

The amplitude of vibration X is given by Eq. (3.93):

$$X = \frac{F_0}{k} \left[\frac{1 - \left(\dfrac{4\mu N}{\pi F_0} \right)^2}{\left\{ 1 - \left(\dfrac{\omega}{\omega_n} \right)^2 \right\}^2} \right]^{1/2}$$

$$0.04 = \frac{F_0}{4000} \left[\frac{1 - \left\{ \dfrac{4(0.12)(98.1)}{\pi F_0} \right\}^2}{\left(1 - 0.6283^2 \right)^2} \right]^{1/2}$$

The solution of this equation gives $F_0 = 97.9874$ N.

3.9 FORCED VIBRATION WITH HYSTERESIS DAMPING

Consider a single degree of freedom system with hysteresis damping and subjected to a harmonic force $F(t) = F_0 \sin \omega t$, as indicated in Fig. 3.22. The equation of motion of the mass can be derived, using Eq. (2.132), as

$$m\ddot{x} + \frac{\beta k}{\omega}\dot{x} + kx = F_0 \sin \omega t \tag{3.100}$$

where $(\beta k/\omega)\dot{x} = (h/\omega)\dot{x}$ denotes the damping force.* Although the solution of Eq. (3.100) is quite involved for a general forcing function $F(t)$, our interest is to find the response under a harmonic force.

The steady-state solution of Eq. (3.100) can be assumed:

$$x_p(t) = X \sin(\omega t - \phi) \tag{3.101}$$

By substituting Eq. (3.101) into Eq. (3.100), we obtain

$$X = \frac{F_0}{k\left[\left(1 - \dfrac{\omega^2}{\omega_n^2}\right)^2 + \beta^2\right]^{1/2}} \tag{3.102}$$

and

$$\phi = \tan^{-1}\left[\frac{\beta}{\left(1 - \dfrac{\omega^2}{\omega_n^2}\right)}\right] \tag{3.103}$$

Equations (3.102) and (3.103) are shown plotted in Fig. 3.23 for several values of β. A comparison of Fig. 3.23 with Fig. 3.11 for viscous damping reveals the following:

1. The amplitude ratio

$$\frac{X}{(F_0/k)}$$

attains its maximum value of $F_0/k\beta$ at the resonant frequency ($\omega = \omega_n$) in the case of hysteresis damping, while it occurs at a frequency below resonance ($\omega < \omega_n$) in the case of viscous damping.

2. The phase angle ϕ has a value of $\tan^{-1}(\beta)$ at $\omega = 0$ in the case of hysteresis damping, while it has a value of zero at $\omega = 0$ in the case of viscous damping. This indicates that the response can never be in phase with the forcing function in the case of hysteresis damping.

---

* In contrast to viscous damping, this damping force here can be seen to be a function of the forcing frequency ω (see Section 2.8).

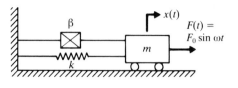

Figure 3.22

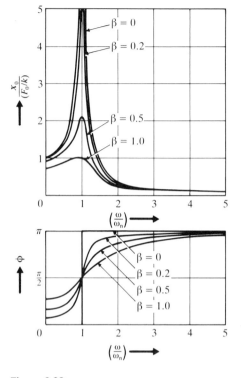

Figure 3.23

Note that if the harmonic excitation is assumed to be $F(t) = F_0 e^{i\omega t}$ in Fig. 3.22, the equation of motion becomes

$$m\ddot{x} + \frac{\beta k}{\omega}\dot{x} + kx = F_0 e^{i\omega t} \tag{3.104}$$

In this case, the response $x(t)$ is also a harmonic function involving the factor $e^{i\omega t}$. Hence $\dot{x}(t)$ is given by $i\omega x(t)$, and Eq. (3.104) becomes

$$m\ddot{x} + k(1 + i\beta)x = F_0 e^{i\omega t} \tag{3.105}$$

where the quantity $k(1 + i\beta)$ is called the *complex stiffness* or *complex damping* [3.7]. The steady-state solution of Eq. (3.105) is given by the real part of

$$x(t) = \frac{F_0 e^{i\omega t}}{k\left[1 - \left(\dfrac{\omega}{\omega_n}\right)^2 + i\beta\right]} \tag{3.106}$$

3.10 FORCED MOTION WITH OTHER TYPES OF DAMPING

Viscous damping is the simplest form of damping to use in practice, since it leads to linear equations of motion. In the cases of Coulomb and hysteretic damping, we defined equivalent viscous damping coefficients to simplify the analysis. Even for a more complex form of damping, we define an equivalent viscous damping coefficient, as illustrated in the following examples. The practical use of equivalent damping is discussed in Ref. [3.12].

EXAMPLE 3.6	Quadratic Damping

Quadratic or *velocity squared damping* is present whenever a body moves in a turbulent fluid flow.

Given: Velocity squared damping.

Find: Equivalent viscous damping coefficient and amplitude of steady state vibration of a single degree of freedom system having quadratic damping.

Approach: Equate energies dissipated per cycle during harmonic motion.

Solution. The damping force is assumed to be

$$F_d = \pm a(\dot{x})^2 \tag{E.1}$$

where a is a constant, $\dot{x}$ is the relative velocity across the damper, and the negative (positive) sign must be used in Eq. (E.1) when $\dot{x}$ is positive (negative). The energy dissipated per cycle during harmonic motion $x(t) = X \sin \omega t$ is given by

$$\Delta W = 2 \int_{-X}^{X} a(\dot{x})^2 \, dx = 2 X^3 \int_{-\pi/2}^{\pi/2} a\omega^2 \cos^3 \omega t \, d(\omega t) = \frac{8}{3} \omega^2 a X^3 \tag{E.2}$$

By equating this energy to the energy dissipated in an equivalent viscous damper [see Eq. (2.92)]:

$$\Delta W = \pi c_{eq} \omega X^2 \tag{E.3}$$

we obtain the equivalent viscous damping coefficient (c_{eq}):

$$c_{eq} = \frac{8}{3\pi} a \omega X \tag{E.4}$$

It can be noted that c_{eq} is not a constant but varies with ω and X. The amplitude of the steady-state response can be found from Eq. (3.30):

$$\frac{X}{\delta_{st}} = \frac{1}{\sqrt{\left(1 - r^2\right)^2 + \left(2\zeta_{eq} r\right)^2}} \tag{E.5}$$

where $r = \omega / \omega_n$ and

$$\zeta_{eq} = \frac{c_{eq}}{c_c} = \frac{c_{eq}}{2 m \omega_n} \tag{E.6}$$

Using Eqs. (E.4) and (E.6), Eq. (E.5) can be solved to obtain

$$X = \frac{3\pi m}{8ar^2}\left[-\frac{(1-r^2)^2}{2} + \sqrt{\frac{(1-r^2)^4}{4} + \left(\frac{8ar^2\delta_{st}}{3\pi m}\right)^2}\,\right]^{1/2} \tag{E.7}$$

3.11 SELF EXCITATION AND STABILITY ANALYSIS

The force acting on a vibrating system is usually external to the system and independent of the motion. However, there are systems for which the exciting force is a function of the motion parameters of the system, such as displacement, velocity, or acceleration. Such systems are called self-excited vibrating systems since the motion itself produces the exciting force (see Problem 3.46). The instability of rotating shafts, the flutter of turbine blades, the flow induced vibration of pipes, and the automobile wheel shimmy and aerodynamically induced motion of bridges are typical examples of self-excited vibrations.

A system is dynamically stable if the motion (or displacement) converges or remains steady with time. On the other hand, if the amplitude of displacement increases continuously (diverges) with time, it is said to be dynamically unstable. The motion diverges and the system becomes unstable if energy is fed into the system through self-excitation. To see the circumstances that lead to instability, we consider the equation of motion of a single degree of freedom system:

$$m\ddot{x} + c\dot{x} + kx = 0 \tag{3.107}$$

If a solution of the form $x(t) = Ce^{st}$, where C is a constant, is assumed, Eq. (3.107) leads to the characteristic equation

$$s^2 + \frac{c}{m}s + \frac{k}{m} = 0 \tag{3.108}$$

The roots of this equation are

$$s_{1,2} = -\frac{c}{2m} \pm \frac{1}{2}\left[\left(\frac{c}{m}\right)^2 - 4\left(\frac{k}{m}\right)\right]^{1/2} \tag{3.109}$$

Since the solution is assumed to be $x(t) = Ce^{st}$, the motion will be diverging and aperiodic if the roots s_1 and s_2 are real and positive. This situation can be avoided if c/m and k/m are positive. The motion will also diverge if the roots s_1 and s_2 are complex conjugates with positive real parts. To analyze the situation, let the roots s_1 and s_2 of Eq. (3.108) be expressed as

$$s_1 = p + iq, \qquad s_2 = p - iq \qquad (3.110)$$

where p and q are real numbers so that

$$(s - s_1)(s - s_2) = s^2 - (s_1 + s_2)s + s_1 s_2 = s^2 + \frac{c}{m}s + \frac{k}{m} = 0 \quad (3.111)$$

Equations (3.111) and (3.110) give

$$\frac{c}{m} = -(s_1 + s_2) = -2p, \qquad \frac{k}{m} = s_1 s_2 = p^2 + q^2 \qquad (3.112)$$

Equations (3.112) show that for negative p, c/m must be positive and for positive $p^2 + q^2$, k/m must be positive. Thus the system will be dynamically stable if c and k are positive (assuming that m is positive).

EXAMPLE 3.7   **Instability of a Vibrating System**

Find the value of free stream velocity u at which the airfoil section (single degree of freedom system) shown in Fig. 3.24 becomes unstable.[†]

Given: Single degree of freedom airfoil section in fluid flow.

Find: Velocity of the fluid which causes instability of the airfoil (or mass m).

Approach: Find the vertical force acting on the airfoil (or mass m) and obtain the condition that leads to zero damping.

Solution. The vertical force acting on the airfoil (or mass m) due to fluid flow can be expressed as [3.4]

$$F = \frac{1}{2}\rho u^2 D C_x \qquad (E.1)$$

where ρ = density of the fluid, u = free stream velocity, D = width of the cross section normal to the fluid flow direction, and C_x = vertical force coefficient, which can be expressed as

$$C_x = \frac{u_{rel}^2}{u^2}(C_L \cos\alpha + C_D \sin\alpha) \qquad (E.2)$$

where u_{rel} is the relative velocity of the fluid, C_L is the lift coefficient, C_D is the drag

---

[†]   The same analysis is valid for a vibrating structure such as a water tank (Fig. 3.25a) or galloping of an ice-coated power line (Fig. 3.25b) under wind loading [3.4–3.6].

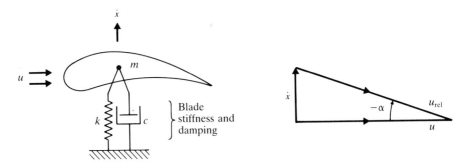

Figure 3.24

coefficient, and α is the angle of attack (see Fig. 3.24):

$$\alpha = -\tan^{-1}\left(\frac{\dot{x}}{u}\right) \tag{E.3}$$

For small angles of attack,

$$\alpha = -\frac{\dot{x}}{u} \tag{E.4}$$

and C_x can be approximated, using Taylor's series expansion about $\alpha = 0$, as

$$C_x \simeq C_x\Big|_{\alpha=0} + \frac{\partial C_x}{\partial \alpha}\Big|_{\alpha=0} \cdot \alpha \tag{E.5}$$

where, for small values of α, $u_{rel} \simeq u$ and Eq. (E.2) becomes

$$C_x = C_L \cos \alpha + C_D \sin \alpha \tag{E.6}$$

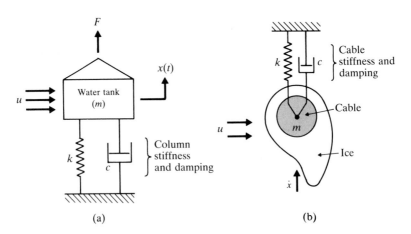

(a)   (b)

Figure 3.25

Equation (E.5) can be rewritten, using Eqs. (E.6) and (E.4), as

$$C_x = \left. (C_L \cos\alpha + C_D \sin\alpha)\right|_{\alpha=0} + \alpha \left[\frac{\partial C_L}{\partial\alpha}\cos\alpha - C_L\sin\alpha + \frac{\partial C_D}{\partial\alpha}\sin\alpha \right.$$

$$\left. + C_D \cos\alpha\right]\Big|_{\alpha=0}$$

$$= \left. C_L \right|_{\alpha=0} + \alpha \left.\frac{\partial C_x}{\partial\alpha}\right|_{\alpha=0}$$

$$= \left. C_L \right|_{\alpha=0} - \frac{\dot{x}}{u}\left\{ \left.\frac{\partial C_L}{\partial\alpha}\right|_{\alpha=0} + C_D|_{\alpha=0}\right\} \tag{E.7}$$

Substitution of Eq. (E.7) into Eq. (E.1) gives

$$F = \frac{1}{2}\rho u^2 D C_L\bigg|_{\alpha=0} - \frac{1}{2}\rho u D \frac{\partial C_x}{\partial\alpha}\bigg|_{\alpha=0}\dot{x} \tag{E.8}$$

The equation of motion of the airfoil (or mass m) is

$$m\ddot{x} + c\dot{x} + kx = F = \frac{1}{2}\rho u^2 D C_L\bigg|_{\alpha=0} - \frac{1}{2}\rho u D \frac{\partial C_x}{\partial\alpha}\bigg|_{\alpha=0}\dot{x} \tag{E.9}$$

The first term on the right-hand side of Eq. (E.9) produces a static displacement and hence only the second term can cause instability of the system. The equation of motion, considering only the second term on the right-hand side, is

$$m\ddot{x} + \underset{\sim}{c}\dot{x} + kx \equiv m\ddot{x} + \left[c + \frac{1}{2}\rho u D \frac{\partial C_x}{\partial\alpha}\bigg|_{\alpha=0}\right]\dot{x} + kx = 0 \tag{E.10}$$

Note that m includes the mass of the entrained fluid. We can see from Eq. (E.10) that the displacement of the airfoil (or mass m) will grow without bound (i.e., the system becomes unstable) if $\underset{\sim}{c}$ is negative. Hence the minimum velocity of the fluid for the onset of unstable oscillations is given by $\underset{\sim}{c} = 0$, or,

$$u = -\left\{ \frac{2c}{\rho D \dfrac{\partial C_x}{\partial\alpha}\bigg|_{\alpha=0}}\right\} \tag{E.11}$$

The value of $\dfrac{\partial C_x}{\partial\alpha}\bigg|_{\alpha=0} = -2.7$ for a square section in a steady flow [3.4].

3.12 COMPUTER PROGRAM

A FORTRAN computer program, in the form of subroutine HARESP, is given for finding the steady-state response of a viscously damped single degree of freedom system under the harmonic force $F_0 \cos\omega t$ or $F_0 \sin\omega t$. The arguments of the subroutine are as follows:

XM	=	Mass. Input data.
XC	=	Damping constant. Input data.
XK	=	Spring stiffness. Input data.

F0	=	Amplitude of the force, Input data.
OM	=	Forcing frequency, Input data.
IC	=	Integer for identifying the nature of the force. IC = 1 for cosine variation and IC = 0 for sine variation. Input data.
N	=	Number of time steps in a cycle at which the response is to be printed. Input data.
X, XD, XDD	=	Arrays of dimension N each, which contain the computed values of displacement, velocity, and acceleration. X(I) = $x(t_i)$, XD(I) = $\dot{x}(t_i)$, XDD(I) = $\ddot{x}(t_i)$. Output.
XAMP	=	Amplitude of the response (Eq. (3.51)). Output.
XPHI	=	Phase angle of the response (Eq. (3.52)). Output.

To illustrate the use of subroutine HARESP, an example is considered with $m = 5$ kg, $c = 20$ N-s/m, $k = 500$ N/s, $F_0 = 250$ N, $\omega = 40$ rad/s, $N = 20$, and $F(t) = F_0 \sin \omega t$. The main program, which calls HARESP, subroutine HARESP, and the output of the program are given below.

```
C   ================================================================
C
C PROGRAM 3
C MAIN PROGRAM WHICH CALLS HARESP
C
C   ================================================================
C FOLLOWING 2 LINES CONTAIN PROBLEM-DEPENDENT DATA
      DIMENSION X(20),XD(20),XDD(20)
      DATA XM,XC,XK,F0,OM,N,IC/5.0,20.0,500.0,250.0,40.0,20,0/
C END OF PROBLEM-DEPENDENT DATA
      CALL HARESP (XM,XC,XK,F0,OM,IC,N,X,XD,XDD,XAMP,XPHI)
      PRINT 100
100   FORMAT (//,40H STEADY STATE RESPONSE OF AN UNDERDAMPED,/,
     2 53H SINGLE DEGREE OF FREEDOM SYSTEM UNDER HARMONIC FORCE)
      PRINT 200, XM,XC,XK,F0,OM,IC,N
200   FORMAT (//,12H GIVEN DATA:,/,5H XM =,E15.8,/,5H XC =,E15.8,/,
     2 5H XK =,E15.8,/,5H F0 =,E15.8,/,5H OM =,E15.8,/,5H IC =,I2,/,
     3 5H N  =,I2)
      PRINT 300
300   FORMAT (//,10H RESPONSE:,//,5H    I ,3X,5H X(I),12X,6H XD(I),
     2 11X,7H XDD(I),/)
      DO 400 I=1,N
400   PRINT 500,I,X(I),XD(I),XDD(I)
500   FORMAT (I4,2X,E15.8,2X,E15.8,2X,E15.8)
      STOP
      END
```

```
C ====================================================================
C
C SUBROUTINE HARESP
C
C ====================================================================
      SUBROUTINE HARESP (XM,XC,XK,F0,OM,IC,N,X,XD,XDD,XAMP,XPHI)
      DIMENSION X(N),XD(N),XDD(N)
      OMN=SQRT(XK/XM)
      XAI=XC/(2.0*XM*OMN)
      DST=F0/XK
      R=OM/OMN
      XAMP=DST/SQRT((1.0-R**2)**2+(2.0*XAI*R)**2)
      XPHI=ATAN(2.0*XAI*R/(1.0-R**2))
      DELT=2.0*3.1416/(OM*REAL(N))
      IF (IC .EQ. 0) GO TO 20
      TIME=0.0
      DO 10 I=1,N
      TIME=TIME+DELT
      X(I)=XAMP*COS(OM*TIME-XPHI)
      XD(I)=-XAMP*OM*SIN(OM*TIME-XPHI)
10    XDD(I)=-XAMP*(OM**2)*COS(OM*TIME-XPHI)
      RETURN
20    TIME=0.0
      DO 30 I=1,N
      TIME=TIME+DELT
      X(I)=XAMP*SIN(OM*TIME-XPHI)
      XD(I)=XAMP*OM*COS(OM*TIME-XPHI)
30    XDD(I)=-XAMP*(OM**2)*SIN(OM*TIME-XPHI)

RETURN
END

STEADY STATE RESPONSE OF AN UNDERDAMPED
SINGLE DEGREE OF FREEDOM SYSTEM UNDER HARMONIC FORCE


GIVEN DATA:
XM = 0.50000000E+01
XC = 0.20000000E+02
XK = 0.50000000E+03
F0 = 0.25000000E+03
OM = 0.40000000E+02
IC = 0
N  =20


RESPONSE:

    I     X(I)              XD(I)              XDD(I)

    1   0.13528203E-01    0.12103548E+01    -0.21645124E+02
    2   0.22216609E-01    0.98389733E+00    -0.35546574E+02
    3   0.28730286E-01    0.66112888E+00    -0.45968460E+02
    4   0.32431632E-01    0.27364409E+00    -0.51890614E+02
    5   0.32958329E-01   -0.14062698E+00    -0.52733330E+02
```

6	0.30258821E-01	-0.54113245E+00	-0.48414116E+02
7	0.24597352E-01	-0.88866800E+00	-0.39355762E+02
8	0.16528117E-01	-0.11492138E+01	-0.26444986E+02
9	0.68409811E-02	-0.12972662E+01	-0.10945570E+02
10	-0.35157942E-02	-0.13183327E+01	0.56252708E+01
11	-0.13528424E-01	-0.12103508E+01	0.21645479E+02
12	-0.22216786E-01	-0.98389095E+00	0.35546856E+02
13	-0.28730409E-01	-0.66112041E+00	0.45968655E+02
14	-0.32431684E-01	-0.27363408E+00	0.51890697E+02
15	-0.32958303E-01	0.14063700E+00	0.52733284E+02
16	-0.30258721E-01	0.54114151E+00	0.48413952E+02
17	-0.24597190E-01	0.88867509E+00	0.39355507E+02
18	-0.16527895E-01	0.11492189E+01	0.26444633E+02
19	-0.68407385E-02	0.12972684E+01	0.10945182E+02
20	0.35160405E-02	0.13183316E+01	-0.56256652E+01

REFERENCES

3.1. G. B. Thomas and R. L. Finney, *Calculus and Analytic Geometry* (6th Ed.), Addison-Wesley, Reading, Mass., 1984.

3.2. J. W. Nilsson, *Electric Circuits*, Addison-Wesley, Reading, Mass., 1983.

3.3. J. P. Den Hartog, "Forced vibrations with combined Coulomb and viscous friction," *Journal of Applied Mechanics* (Transactions of ASME), Vol. 53, 1931, pp. APM 107–115.

3.4. R. D. Blevins, *Flow-Induced Vibration*, Van Nostrand Reinhold, New York, 1977.

3.5. J. C. R. Hunt and D. J. W. Richards, "Overhead line oscillations and the effect of aerodynamic dampers," *Proceedings of the Institute of Electrical Engineers*, London, Vol. 116, 1969, pp. 1869–1874.

3.6. K. P. Singh and A. I. Soler, *Mechanical Design of Heat Exchangers and Pressure Vessel Components*, Arcturus Publishers, Cherry Hill, New Jersey, 1984.

3.7. N. O. Myklestad, "The concept of complex damping," *Journal of Applied Mechanics*, Vol. 19, 1952, pp. 284–286.

3.8. R. Plunkett (Ed.), *Mechanical Impedance Methods for Mechanical Vibrations*, American Society of Mechanical Engineers, New York, 1958.

3.9. A. D. Dimarogonas, *Vibration Engineering*, West Publishing Co., St. Paul, 1976.

3.10. B. Westermo and F. Udwadia, "Periodic response of a sliding oscillator system to harmonic excitation," *Earthquake Engineering and Structural Dynamics*, Vol. 11, No. 1, 1983, pp. 135–146.

3.11. M. S. Hundal, "Response of a base excited system with Coulomb viscous friction," *Journal of Sound and Vibration*, Vol. 64, 1979, pp. 371–378.

3.12. J. P. Bandstra, "Comparison of equivalent viscous damping and nonlinear damping in discrete and continuous vibrating systems," *Journal of Vibration, Acoustics, Stress, and Reliability in Design*, Vol. 105, 1983, pp. 382–392.

REVIEW QUESTIONS

3.1. How are the amplitude, frequency, and phase of a steady-state vibration related to those of the applied harmonic force?

3.2. Explain why a constant force on the vibrating mass has no effect on the steady-state vibration.

3.3. Define the term *magnification factor*. How is the magnification factor related to the frequency ratio?

3.4. What will be the frequency of the applied force with respect to the natural frequency of the system if the magnification factor is less than unity?

3.5. What are the amplitude and the phase angle of the response of a viscously damped system in the neighborhood of resonance?

3.6. Is the phase angle corresponding to the peak amplitude of a viscously damped system ever larger than 90°?

3.7. Why is damping considered only in the neighborhood of resonance in most cases?

3.8. Show the various terms in the forced equation of motion of a viscously damped system in a vector diagram.

3.9. What happens to the response of an undamped system at resonance?

3.10. Define these terms: beating, quality factor, transmissibility, complex stiffness, quadratic damping.

3.11. Give a physical explanation of why the magnification factor is nearly equal to 1 for small values of r and is small for large values of r.

3.12. Will the force transmitted to the base of a spring-mounted machine decrease with the addition of damping?

3.13. How does the force transmitted to the base change as the speed of the machine increases?

3.14. If a vehicle vibrates badly while moving on a uniformly bumpy road, will a change in the speed improve the condition?

3.15. Is it possible to find the maximum amplitude of a damped forced vibration for any value of r by equating the energy dissipated by damping to the work done by the external force?

3.16. What assumptions are made about the motion of a forced vibration with nonviscous damping in finding the amplitude?

3.17. Is it possible to find the approximate value of the amplitude of a damped forced vibration without considering damping at all? If so, under what circumstances?

3.18. Is dry friction effective in limiting the reasonant amplitude?

3.19. How do you find the response of a viscously damped system under rotating unbalance?

3.20. What is the frequency of the response of a viscously damped system when the external force is $F_0 \sin \omega t$? Is this response harmonic?

3.21. What is the difference between the peak amplitude and the resonant amplitude?

3.22. Why is viscous damping used in most cases rather than other types of damping?

3.23. What is self-excited vibration?

PROBLEMS

The problem assignments are organized as follows:

Problems	Section Covered	Topic Covered
3.1–3.9	3.3	Undamped systems
3.10–3.16	3.4	Damped systems
3.17–3.22	3.6	Base excitation
3.23–3.32	3.7	Rotating unbalance
3.33–3.35	3.8	Response under Coulomb damping
3.36–3.3.37	3.9	Response under hysteresis damping
3.38–3.41	3.10	Response under other types of damping
3.46	3.11	Self excitation and stability
3.42–3.45	3.12	Computer program
3.47–3.48	—	Projects

3.1. A weight of 50 N is suspended from a spring of stiffness 4000 N/m and is subjected to a harmonic force of amplitude 60 N and frequency 6 Hz. Find (i) the extension of the spring due to the suspended weight, (ii) the static displacement of the spring due to the maximum applied force, and (iii) the amplitude of forced motion of the weight.

3.2. A spring-mass system is subjected to a harmonic force whose frequency is close to the natural frequency of the system. If the forcing frequency is 39.8 Hz and the natural frequency is 40.0 Hz, determine the period of beating.

3.3. A spring-mass system consists of a mass weighing 100 N and a spring with a stiffness of 2000 N/m. The mass is subjected to resonance by a harmonic force $F(t) = 25 \cos \omega t$ N. Find the amplitude of the forced motion at the end of (i) $\frac{1}{4}$ cycle, (ii) $2\frac{1}{2}$ cycles, and (iii) $5\frac{3}{4}$ cycles.

3.4. A mass m is suspended from a spring of stiffness 4000 N/m and is subjected to a harmonic force having an amplitude of 100 N and a frequency of 5 Hz. The amplitude of the forced motion of the mass is observed to be 20 mm. Find the value of m.

3.5. A spring-mass system with $m = 10$ kg and $k = 5000$ N/m is subjected to a harmonic force of amplitude 250 N and frequency ω. If the maximum amplitude of the mass is observed to be 100 mm, find the value of ω.

3.6. In Fig. 3.1(a), a periodic force $F(t) = F_0 \cos \omega t$ is applied at a point on the spring that is located at a distance of 25% of its length from the fixed support. Assuming that $c = 0$, find the steady state response of the mass m.

3.7.* Design a solid steel shaft supported in bearings which carries the rotor of a turbine at the middle. The rotor weighs 500 lb and delivers a power of 200 hp at 3000 rpm. In order to keep the stress due to the unbalance in the rotor small, the critical speed of the shaft is to be made one-fifth of the operating speed of the rotor. The length of the shaft is to be made equal to at least 30 times its diameter.

3.8. A hollow steel shaft, of length 100 in., outer diameter 4 in. and inner diameter 3.5 in., carries the rotor of a turbine, weighing 500 lb, at the middle and is supported at the ends in bearings. The clearance between the rotor and the stator is 0.5 in. The rotor has an eccentricity equivalent to a weight of 0.5 lb at a radius of 2 in. A limit switch is installed to stop the rotor whenever the rotor touches the stator. If the rotor operates at resonance, how long will it take to activate the limit switch? Assume the initial displacement and velocity of the rotor perpendicular to the shaft to be zero.

3.9. A steel cantilever beam, carrying a weight of 0.1 lb at the free end, is used as a frequency meter.[†] The beam has a length of 10 in., width of 0.2 in., and thickness of 0.05 in. The internal friction is equivalent to a damping ratio of 0.01. When the fixed end of the beam is subjected to a harmonic displacement $y(t) = 0.05 \cos \omega t$, the maximum tip displacement has been observed to be 2.5 in. Find the forcing frequency.

3.10. A spring-mass-damper system is subjected to a harmonic force. The amplitude is found to be 20 mm at resonance and 10 mm at a frequency 0.75 times the resonant frequency. Find the damping ratio of the system.

3.11. Find the frequency ratio $r = \omega/\omega_n$ at which the amplitude of a single degree of freedom damped system attains the maximum value. Also find the value of the maximum amplitude.

3.12. For the system shown in Fig. 3.26, x and y denote, respectively, the absolute displacements of the mass m and the end Q of the dashpot c_1. (i) Derive the equation of motion of the mass m, (ii) find the steady state displacement of the mass m, and (iii) find the force transmitted to the support at P, when the end Q is subjected to the harmonic motion $y(t) = Y \cos \omega t$.

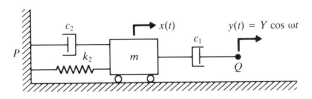

Figure 3.26

3.13. Show that, for small values of damping, the damping ratio ζ can be expressed as

$$\zeta = \frac{\omega_2 - \omega_1}{\omega_2 + \omega_1}$$

where ω_1 and ω_2 are the frequencies corresponding to the half power points.

3.14. A torsional system consists of a disc of mass moment of inertia $J_0 = 10$ kg-m^2, a torsional damper of damping constant $c_t = 300$ N-m-s/rad, and a steel shaft of diameter 4 cm and length 1 m (fixed at one end and attached to the disc at the other end). A steady angular oscillation of amplitude 2° is observed when a harmonic torque of magnitude 1000 N-m is applied to the disc. (i) Find the frequency of the applied torque. (ii) Find the maximum torque transmitted to the support.

3.15. For a vibrating system, $m = 10$ kg, $k = 2500$ N/m, and $c = 45$ N-s/m. A harmonic force of amplitude 180 N and frequency 3.5 Hz acts on the mass. If the initial displacement and velocity of the mass are 15 mm and 5 m/s, find the complete solution representing the motion of the mass.

[†] The use of cantilever beams as frequency meters is discussed in detail in Section 10.4.

3.16. The peak amplitude of a single degree of freedom system, under a harmonic excitation, is observed to be 0.2 in. If the undamped natural frequency of the system is 5 Hz, and the static deflection of the mass under the maximum force is 0.1 in., (i) estimate the damping ratio of the system, and (ii) find the frequencies corresponding to the amplitudes at half power.

3.17. A single story building frame is subjected to a harmonic ground acceleration as shown in Fig. 3.27. Find the steady state motion of the floor (mass m).

3.18. Find the horizontal displacement of the floor (mass m) of the building frame shown in Fig. 3.27 when the ground acceleration is given by $\ddot{x}_g = 100 \sin \omega t$ mm/sec². Assume $m = 2000$ kg, $k = 0.1$ MN/m, $\omega = 25$ rad/sec, and $x_g(t = 0) = \dot{x}_g(t = 0) = x(t = 0) = \dot{x}(t = 0) = 0$.

3.19. If the ground is subjected to a horizontal harmonic displacement with frequency $\omega = 200$ rad/sec and amplitude $X_g = 15$ mm in Fig. 3.27, find the amplitude of vibration of the floor (mass m). Assume the mass of the floor as 2000 kg and the stiffness of the columns as 0.5 MN/m.

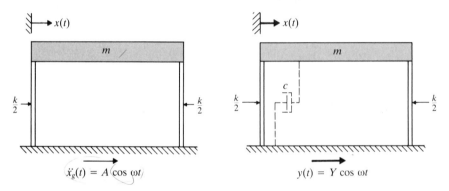

Figure 3.27   **Figure 3.28**

3.20. An automobile is modeled as a single degree of freedom system vibrating in the vertical direction. It is driven along a road whose elevation varies sinusoidally. The distance from peak to trough is 0.2 m and the distance along the road between the peaks is 35 m. If the natural frequency of the automobile is 2 Hz and the damping ratio of the shock absorbers is 0.15, determine the amplitude of vibration of the automobile at a speed of 60 km/hour. If the speed of the automobile is varied, find the most unfavorable speed for the passengers.

3.21. Derive Eq. (3.74).

3.22. A single story building frame is modeled by a rigid floor of mass m and columns of stiffness k as shown in Fig. 3.28. It is proposed to attach a damper as shown in Fig. 3.28 to absorb vibrations due to a horizontal ground motion $y(t) = Y \cos \omega t$. Derive an expression for the damping constant of the damper that absorbs maximum power.

3.23. One of the tail rotor blades of a helicopter has an unbalanced mass of $m = 0.5$ kg at a distance of $e = 0.15$ m from the axis of rotation, as shown in Fig. 3.29. The tail section has a length of 4 m, a mass of 240 kg, a flexural stiffness (EI) of 2.5 MN – m², and a damping ratio of 0.15. The mass of the tail rotor blades, including their drive system, is 20 kg. Determine the forced response of the tail section when the blades rotate at 1500 rpm.

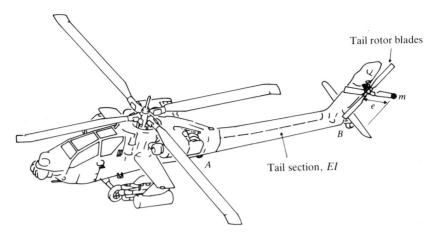

Figure 3.29

3.24. When an exhaust fan of mass 380 kg is supported on springs with negligible damping, the resulting static deflection is found to be 45 mm. If the fan has a rotating unbalance of 0.15 kg-m, find (i) the amplitude of vibration at 1750 rpm and (ii) the force transmitted to the ground at this speed.

3.25. A fixed-fixed steel beam, of length 5 m, width 0.5 m, and thickness 0.1 m, carries an electric motor of mass 75 kg and speed 1200 rpm at its mid-span, as shown in Fig. 3.30. A rotating force of magnitude $F_0 = 5000$ N is developed due to the unbalance in the rotor of the motor. Find the amplitude of steady-state vibrations by disregarding the mass of the beam. What will be the amplitude if the mass of the beam is considered?

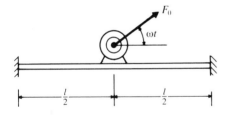

Figure 3.30

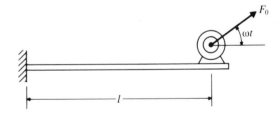

Figure 3.31

3.26.* If the electric motor of Problem 3.25 is to be mounted at the free end of a steel cantilever beam of length 5 m (Fig. 3.31), and the amplitude of vibration is to be limited to 0.5 cm, find the necessary cross-sectional dimensions of the beam. Include the weight of the beam in the computations.

3.27. A centrifugal pump, weighing 600 N and operating at 1000 rpm, is mounted on six springs of stiffness 6000 N/m each. Find the maximum permissible unbalance in order to limit the steady-state deflection to 5 mm peak-to-peak.

3.28.* An air compressor, weighing 1000 lb and operating at 1500 rpm, is to be mounted on a suitable isolator. A helical spring with a stiffness of 45,000 lb/in., another helical spring with a stiffness of 15,000 lb/in., and a shock absorber with a damping ratio of 0.15 are available for use. Select the best possible isolation system for the compressor.

3.29. A variable speed electric motor, having an unbalance, is mounted on an isolator. As the speed of the motor is increased from zero, the amplitudes of vibration of the motor have been observed to be 0.55 in. at resonance and 0.15 in. beyond resonance. Find the damping ratio of the isolator.

3.30. An electric motor weighing 750 lb and running at 1800 rpm is supported on four steel helical springs, each of which has 8 active coils with a wire diameter of 0.25 in. and a coil diameter of 3 in. The rotor has a weight of 100 lb with its center of mass located at a distance of 0.01 in. from the axis of rotation. Find the amplitude of vibration of the motor and the force transmitted through the springs to the base.

3.31. A small exhaust fan, rotating at 1500 rpm, is mounted on a 0.2 in. steel shaft. The rotor of the fan weighs 30 lb and has an eccentricity of 0.01 in. from the axis of rotation. (i) Find the maximum force transmitted to the bearings. (ii) Find the horse power needed to drive the shaft.

3.32. A rigid plate, weighing 100 lb, is hinged along an edge (P) and is supported on a dashpot with $c = 1$ lb-sec/in. at the opposite edge (Q) as shown in Fig. 3.32. A small fan weighing 50 lb and rotating at 750 rpm is mounted on the plate through a spring with $k = 200$ lb/in. If the center of gravity of the fan is located at 0.1 in. from its axis of rotation, find the steady state motion of the edge Q and the force transmitted to the point S.

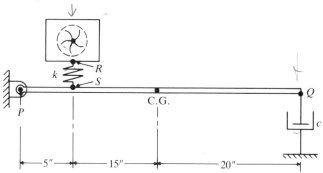

Figure 3.32

3.33. Derive Eq. (3.99).

3.34. Derive the equation of motion of the mass m shown in Fig. 3.33 when the pressure in

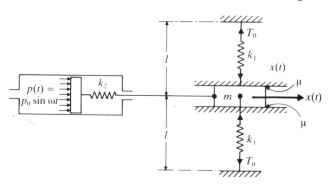

Figure 3.33

the cylinder fluctuates sinusoidally. The two springs with stiffnesses k_1 are initially under a tension of T_0, and the coefficient of friction between the mass and the contacting surfaces is μ.

3.35. A spring-mass system is subjected to Coulomb damping. When a harmonic force of amplitude 120 N and frequency 2.5173268 Hz is applied, the system is found to oscillate with an amplitude of 75 mm. Determine the coefficient of dry friction if $m = 2$ kg and $k = 2100$ N/m.

3.36. A load of 5000 N resulted in a static displacement of 0.05 m in a composite structure. A harmonic force of amplitude 1000 N is found to cause a resonant amplitude of 0.1 m. Find (i) the hysteresis damping constant of the structure, (ii) the energy dissipated per cycle at resonance, (iii) the steady state amplitude at one-quarter of the resonant frequency, and (iv) the steady state amplitude at thrice the resonant frequency.

3.37. The energy dissipated in hysteresis damping per cycle under harmonic excitation can be expressed in the general form

$$\Delta W = \pi \beta k X^\gamma \tag{E.1}$$

where γ is an exponent ($\gamma = 2$ was considered in Eq. (2.125)), and β is a coefficient of dimension $(\text{meter})^{2-\gamma}$. A spring-mass system having $k = 60$ kN/m vibrates under hysteresis damping. When excited harmonically at resonance, the steady-state amplitude is found to be 40 mm for an energy input of 3.8 N-m. When the resonant energy input is increased to 9.5 N-m, the amplitude is found to be 60 mm. Determine the values of β and γ in Eq. (E.1).

3.38. When a spring-mass-damper system is subjected to a harmonic force $F(t) = 5 \cos 3\pi t$ lb, the resulting displacement is given by $x(t) = 0.5 \cos(3\pi t - \pi/3)$ in. Find the work done (i) during the first 1 second, and (ii) during the first 4 seconds.

3.39. Find the equivalent viscous damping coefficient of a damper that offers a damping force of $F_d = c(\dot{x})^n$, where c and n are constants and $\dot{x}$ is the relative velocity across the damper. Also, find the amplitude of vibration.

3.40. Show that for a system with both viscous and Coulomb damping the approximate value of the steady-state amplitude is given by

$$X^2 \left[k^2 (1 - r^2)^2 + c^2 \omega^2 \right] + X \frac{8 \mu N c \omega}{\pi} + \left(\frac{16 \mu^2 N^2}{\pi^2} - F_0^2 \right) = 0$$

3.41. The equation of motion of a spring-mass-damper system is given by

$$m\ddot{x} \pm \mu N + c\dot{x}^3 + kx = F_0 \cos \omega t$$

Derive expressions for (i) the equivalent viscous damping constant, (ii) the steady-state amplitude, and (iii) the amplitude ratio at resonance.

3.42. Use subroutine HARESP to find the steady-state response of a torsional system with $J_0 = 6$ kg-m^2, $c_t = 210$ N-m-s/rad, $k_t = 14000$ N-m/rad, and $F(t) = 450 \sin 10t$ N-m.

3.43. Write a subroutine called TOTALR for finding the complete solution (homogeneous part plus particular integral) of a single degree of freedom system. Use this program to find the solution of Problem 3.15.

3.44. Find the steady-state solution of a single degree of freedom system with $m = 10$ kg, $c = 45$ N-s/m, $k = 2500$ N/m, $F(t) = 180 \cos 20t$ N, $x_0 = 0$, and $\dot{x}_0 = 10$ m/s, using subroutine HARESP.

3.45. Write a computer program for finding the total response of a spring-mass-viscous damper system subjected to base excitation. Use this program to find the solution of a problem with $m = 2$ kg, $c = 10$ N-s/m, $k = 100$ N/m, $y(t) = 0.1 \sin 25t$ m, $x_0 = 10$ mm, and $\dot{x}_0 = 5$ m/s.

3.46. Consider the equation of motion of a single degree of freedom system:

$$m\ddot{x} + c\dot{x} + kx = F$$

Derive the condition that leads to divergent oscillations in each of the following cases:

(a) When the forcing function is proportional to the displacement, $F(t) = F_0 x(t)$.

(b) When the forcing function is proportional to the velocity, $F(t) = F_0 \dot{x}(t)$.

(c) When the forcing function is proportional to the acceleration, $F(t) = F_0 \ddot{x}(t)$.

Projects:

3.47. The arrangement shown in Fig. 3.34 consists of two eccentric masses rotating in opposite directions at the same speed ω. It is to be used as a mechanical shaker over the frequency range 20 to 30 Hz. Find the values of ω, e, M, m, k, and c to satisfy the following requirements: (i) The mean power output of the shaker should be at least 1 hp over the specified frequency range. (ii) The amplitude of vibration of the masses should be between 0.1 and 0.2 in. (iii) The mass of the shaker (M) should be at least 50 times that of the eccentric mass (m).

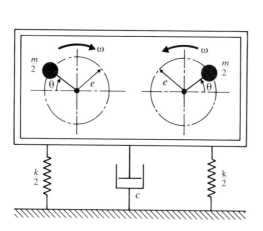

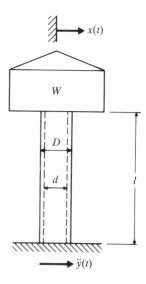

Figure 3.34

Figure 3.35

3.48. Design a minimum weight, hollow circular steel column for the water tank shown in Fig. 3.35. The weight of the tank (W) is 100,000 lb and the height is 50 ft. The stress induced in the column should not exceed the yield strength of the material, which is 30,000 psi, when subjected to a harmonic ground acceleration (due to an earthquake) of amplitude 0.5 g and frequency 15 Hz. In addition, the natural frequency of the water tank should be greater than 15 Hz. Assume a damping ratio of 0.15 for the column.

Vibration under General Forcing Conditions

Jean Baptiste Joseph Fourier (1768 – 1830) was a French mathematician and a professor at the Ecole Polytechnique in Paris. His works on heat flow, published in 1822, and on trigonometric series are well known. The expansion of a periodic function in terms of harmonic functions has been named after him as the "Fourier series." (Courtesy The Bettmann Archive, Inc.)

4.1 INTRODUCTION

This chapter deals with the vibration of a viscously damped single degree of freedom system under general forcing conditions. If the excitation is periodic but not harmonic, it can be replaced by a sum of harmonic functions using the harmonic analysis procedure discussed in Section 1.11. By the principle of superposition, the response of the system can then be determined by superposing the responses due to the individual harmonic forcing functions. On the other hand, if the system is subjected to a suddenly applied nonperiodic force, the response will be transient, since steady-state vibrations are not usually produced. The transient response of a system can be found using what is known as the *convolution integral*.

4.2 RESPONSE UNDER A GENERAL PERIODIC FORCE

When the external force $F(t)$ is periodic with period $\tau = 2\pi/\omega$, it can be expanded in a Fourier series (see Section 1.11):

$$F(t) = \frac{a_0}{2} + \sum_{j=1}^{\infty} a_j \cos j\omega t + \sum_{j=1}^{\infty} b_j \sin j\omega t \qquad (4.1)$$

where

$$a_j = \frac{2}{\tau} \int_0^\tau F(t)\cos j\omega t \, dt, \qquad j = 0, 1, 2, \dots \tag{4.2}$$

and

$$b_j = \frac{2}{\tau} \int_0^\tau F(t)\sin j\omega t \, dt, \qquad j = 1, 2, \dots \tag{4.3}$$

The equation of motion of the system can be expressed as

$$m\ddot{x} + c\dot{x} + kx = F(t) = \frac{a_0}{2} + \sum_{j=1}^\infty a_j \cos j\omega t + \sum_{j=1}^\infty b_j \sin j\omega t \tag{4.4}$$

The right-hand side of this equation is a constant plus a sum of harmonic functions. Using the principle of superposition, the steady-state solution of Eq. (4.4) is the sum of the steady-state solutions of the following equations:

$$m\ddot{x} + c\dot{x} + kx = \frac{a_0}{2} \tag{4.5}$$

$$m\ddot{x} + c\dot{x} + kx = a_j \cos j\omega t \tag{4.6}$$

$$m\ddot{x} + c\dot{x} + kx = b_j \sin j\omega t \tag{4.7}$$

Noting that the solution of Eq. (4.5) is given by

$$x_p(t) = \frac{a_0}{2k} \tag{4.8}$$

and using the results of Section 3.4, we can express the solutions of Eqs. (4.6) and (4.7), respectively, as

$$x_p(t) = \frac{(a_j/k)}{\sqrt{(1 - j^2 r^2)^2 + (2\zeta jr)^2}} \cos(j\omega t - \phi_j) \tag{4.9}$$

$$x_p(t) = \frac{(b_j/k)}{\sqrt{(1 - j^2 r^2)^2 + (2\zeta jr)^2}} \sin(j\omega t - \phi_j) \tag{4.10}$$

where

$$\phi_j = \tan^{-1}\left(\frac{2\zeta jr}{1 - j^2 r^2}\right) \tag{4.11}$$

and

$$r = \frac{\omega}{\omega_n} \tag{4.12}$$

Thus the complete steady-state solution of Eq. (4.4) is given by

$$x_p(t) = \frac{a_0}{2k} + \sum_{j=1}^\infty \frac{(a_j/k)}{\sqrt{(1 - j^2 r^2)^2 + (2\zeta jr)^2}} \cos(j\omega t - \phi_j)$$

$$+ \sum_{j=1}^\infty \frac{(b_j/k)}{\sqrt{(1 - j^2 r^2)^2 + (2\zeta jr)^2}} \sin(j\omega t - \phi_j) \tag{4.13}$$

It can be seen from the solution, Eq. (4.13), that the amplitude and phase shift corresponding to the jth term depend on j. If $j\omega = \omega_n$, for any j, the amplitude of the corresponding harmonic will be comparatively large. This will be particularly true for small values of j and ζ. Further, as j becomes larger, the amplitude becomes smaller and the corresponding terms tend to zero. Thus the first few terms are usually sufficient to obtain the response with reasonable accuracy.

The solution given by Eq. (4.13) denotes the steady-state response of the system. The transient part of the solution arising from the initial conditions can also be included to find the complete solution. To find the complete solution, we need to evaluate the arbitrary constants by setting the value of the complete solution and its derivative to the specified values of initial displacement $x(0)$ and the initial velocity $\dot{x}(0)$. This results in a complicated expression for the transient part of the total solution.

EXAMPLE 4.1 **Periodic Vibration of a Hydraulic Valve**

In the study of vibrations of valves used in hydraulic control systems, the valve and its elastic stem are modeled as a damped spring-mass system as shown in Fig. 4.1(a). In addition to the spring force and damping force, there is a fluid pressure force on the valve that changes with the amount of opening or closing of the valve. Find the steady-state response of the valve when the pressure in the chamber varies as indicated in Fig. 4.1(b). Assume $k = 2500$ N/m, $c = 10$ N-s/m, and $m = 0.25$ kg.

Given: Hydraulic control valve with $m = 0.25$ kg, $k = 2500$ N/m, and $c = 10$ N-s/m and pressure on the valve as given in Fig. 4.1(b).

Find: Steady-state response of the valve, $x_p(t)$.

Approach: Find the Fourier series expansion of the force acting on the valve. Add the responses due to individual harmonic force components.

Solution. The valve can be considered as a mass connected to a spring and a damper on one side and subjected to a forcing function $F(t)$ on the other side. The forcing function can be expressed as

$$F(t) = Ap(t) \tag{E.1}$$

where A is the cross sectional area of the chamber, given by

$$A = \frac{\pi(50)^2}{4} = 625\pi \text{ mm}^2 = 0.000625\pi \text{ m}^2 \tag{E.2}$$

and $p(t)$ is the pressure acting on the valve at any instant t. Since $p(t)$ is periodic with period $\tau = 2$ seconds and A is a constant, $F(t)$ is also a periodic function of period $\tau = 2$ seconds. The frequency of the forcing function is $\omega = (2\pi/\tau) = \pi$ rad/sec. $F(t)$ can be expressed in a Fourier series as:

$$F(t) = \frac{a_0}{2} + a_1\cos\omega t + a_2\cos 2\omega t + \cdots$$
$$+ b_1\sin\omega t + b_2\sin 2\omega t + \cdots \tag{E.3}$$

where a_j and b_j are given by Eqs. (4.2) and (4.3). Since the function $F(t)$ is given by

$$F(t) = \begin{cases} 50\,000At & \text{for } 0 \leqslant t \leqslant \dfrac{\tau}{2} \\ 50\,000A(2-t) & \text{for } \dfrac{\tau}{2} \leqslant t \leqslant \tau \end{cases} \qquad \text{(E.4)}$$

the Fourier coefficients a_j and b_j can be computed with the help of Eqs. (4.2) and (4.3):

$$a_0 = \frac{2}{2}\left[\int_0^1 50\,000At\,dt + \int_1^2 50\,000A(2-t)\,dt\right] = 50\,000A \qquad \text{(E.5)}$$

$$\begin{aligned} a_1 &= \frac{2}{2}\left[\int_0^1 50\,000At\cos\pi t\,dt + \int_1^2 50\,000A(2-t)\cos\pi t\,dt\right] \\ &= -\frac{2\times 10^5 A}{\pi^2} \end{aligned} \qquad \text{(E.6)}$$

$$b_1 = \frac{2}{2}\left[\int_0^1 50\,000At\sin\pi t\,dt + \int_1^2 50\,000A(2-t)\sin\pi t\,dt\right] = 0 \qquad \text{(E.7)}$$

$$a_2 = \frac{2}{2}\left[\int_0^1 50\,000At\cos 2\pi t\,dt + \int_1^2 50\,000A(2-t)\cos 2\pi t\,dt\right] = 0 \qquad \text{(E.8)}$$

$$b_2 = \frac{2}{2}\left[\int_0^1 50\,000At\sin 2\pi t\,dt + \int_1^2 50\,000A(2-t)\sin 2\pi t\,dt\right] = 0 \qquad \text{(E.9)}$$

$$\begin{aligned} a_3 &= \frac{2}{2}\left[\int_0^1 50\,000At\cos 3\pi t\,dt + \int_1^2 50\,000A(2-t)\cos 3\pi t\,dt\right] \\ &= -\frac{2\times 10^5 A}{9\pi^2} \end{aligned} \qquad \text{(E.10)}$$

$$b_3 = \frac{2}{2}\left[\int_0^1 50\,000At\sin 3\pi t\,dt + \int_1^2 50\,000A(2-t)\sin 3\pi t\,dt\right] = 0 \qquad \text{(E.11)}$$

Likewise, we can obtain $a_4 = a_6 = \cdots = b_4 = b_5 = b_6 = \cdots = 0$. By considering only the first three harmonics, the forcing function can be approximated:

$$F(t) \simeq 25\,000A - \frac{2\times 10^5 A}{\pi^2}\cos\omega t - \frac{2\times 10^5 A}{9\pi^2}\cos 3\omega t \qquad \text{(E.12)}$$

The steady-state response of the valve to the forcing function of Eq. (E.12) can be expressed as

$$\begin{aligned} x_p(t) &= \frac{25\,000A}{k} - \frac{\left(2\times 10^5 A/(k\pi^2)\right)}{\sqrt{(1-r^2)^2 + (2\zeta r)^2}}\cos(\omega t - \phi_1) \\ &\quad - \frac{\left(2\times 10^5 A/(9k\pi^2)\right)}{\sqrt{(1-9r^2)^2 + (6\zeta r)^2}}\cos(3\omega t - \phi_3) \end{aligned} \qquad \text{(E.13)}$$

The natural frequency of the valve is given by

$$\omega_n = \sqrt{\frac{k}{m}} = \sqrt{\frac{2500}{0.25}} = 100 \text{ rad/sec} \qquad \text{(E.14)}$$

and the forcing frequency ω by

$$\omega = \frac{2\pi}{\tau} = \frac{2\pi}{2} = \pi \text{ rad/sec} \qquad \text{(E.15)}$$

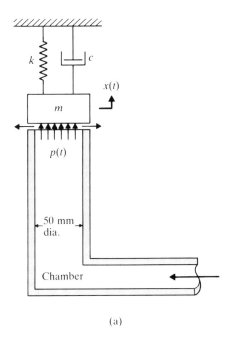

(a)

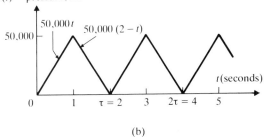

(b)

Figure 4.1

Thus the frequency ratio can be obtained:

$$r = \frac{\omega}{\omega_n} = \frac{\pi}{100} = 0.031416 \tag{E.16}$$

and the damping ratio:

$$\zeta = \frac{c}{c_c} = \frac{c}{2m\omega_n} = \frac{10.0}{2(0.25)(100)} = 0.2 \tag{E.17}$$

The phase angles ϕ_1 and ϕ_3 can be computed as follows:

$$\phi_1 = \tan^{-1}\left(\frac{2\zeta r}{1 - r^2}\right) = \tan^{-1}\left(\frac{2 \times 0.2 \times 0.031416}{1 - 0.031416^2}\right) = 0.0125664 \text{ rad} \tag{E.18}$$

and

$$\phi_3 = \tan^{-1}\left(\frac{6\zeta r}{1 - 9r^2}\right) = \tan^{-1}\left(\frac{6 \times 0.2 \times 0.031416}{1 - 9(0.031416)^2}\right) = 0.0380483 \text{ rad} \quad (\text{E.19})$$

In view of Eqs. (E.2) and (E.14) to (E.19), the solution can be written as

$$x_p(t) = 0.019635 - 0.015930 \cos(\pi t - 0.0125664)$$
$$- 0.0017828 \cos(3\pi t - 0.0380483) \; m \quad (\text{E.20})$$

4.3 RESPONSE UNDER A PERIODIC FORCE OF IRREGULAR FORM

In some cases, the force acting on a system may be quite irregular and may be determined only experimentally. Examples of such forces include wind-and earth-quake-induced forces. In such cases, the forces will be available in graphical form and no analytical expression can be found to describe $F(t)$. Sometimes, the value of $F(t)$ may be available only at a number of discrete points $t_1, t_2, \ldots, t_N$. In all these cases, it is possible to find the Fourier coefficients by using a numerical integration procedure, as described in Section 1.11. If $F_1, F_2, \ldots, F_N$ denote the values of $F(t)$ at $t_1, t_2, \ldots, t_N$, respectively, where N denotes an even number of equidistant points in one time period τ ($\tau = N\Delta t$), as shown in Fig. 4.2, the application of trapezoidal rule [4.1] gives

$$a_0 = \frac{2}{N} \sum_{i=1}^{N} F_i \quad (4.14)$$

$$a_j = \frac{2}{N} \sum_{i=1}^{N} F_i \cos \frac{2j\pi t_i}{\tau}, \qquad j = 1, 2, \ldots \quad (4.15)$$

$$b_j = \frac{2}{N} \sum_{i=1}^{N} F_i \sin \frac{2j\pi t_i}{\tau}, \qquad j = 1, 2, \ldots \quad (4.16)$$

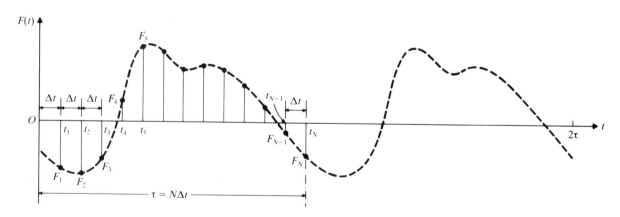

Figure 4.2

Once the Fourier coefficients a_0, a_j, and b_j are known, the steady-state response of the system can be found using Eq. (4.13) with

$$r = \left(\frac{2\pi}{\tau\omega_n}\right)$$

EXAMPLE 4.2   **Steady-State Vibration of a Hydraulic Valve**

Find the steady-state response of the valve in Example 4.1 if the pressure fluctuations in the chamber are found to be periodic. The values of pressure measured at 0.01 second intervals in one cycle are given below.

time, t_i (seconds)	0	0.01	0.02	0.03	0.04	0.05	0.06	0.07	0.08	0.09	0.10	0.11	0.12
$p_i = p(t_i)$ (kN/m²)	0	20	34	42	49	53	70	60	36	22	16	7	0

Given: Arbitrary pressure fluctuations on the valve, shown in Fig. 4.1(a).

Find: Steady state response of the valve.

Approach: Find Fourier series expansion of the pressure acting on the valve using numerical procedure. Add the responses due to individual harmonic force components.

Solution. The Fourier analysis of the pressure fluctuations (see Example 1.7) gives the result

$$p(t) = 34083.3 - 26996.0 \cos 52.36t + 8307.7 \sin 52.36t$$
$$+ 1416.7 \cos 104.72t + 3608.3 \sin 104.72t$$
$$- 5833.3 \cos 157.08t + 2333.3 \sin 157.08t + \cdots \text{ N/m}^2 \qquad \text{(E.1)}$$

Other quantities needed for the computation are

$$\omega = \frac{2\pi}{\tau} = \frac{2\pi}{0.12} = 52.36 \text{ rad/sec}$$

$$\omega_n = 100 \text{ rad/sec}$$

$$r = \frac{\omega}{\omega_n} = 0.5236$$

$$\zeta = 0.2$$

$$A = 0.000625\pi \text{ m}^2$$

$$\phi_1 = \tan^{-1}\left(\frac{2\zeta r}{1 - r^2}\right) = \tan^{-1}\left(\frac{2 \times 0.2 \times 0.5236}{1 - 0.5236^2}\right) = 16.1°$$

$$\phi_2 = \tan^{-1}\left(\frac{4\zeta r}{1 - 4r^2}\right) = \tan^{-1}\left(\frac{4 \times 0.2 \times 0.5236}{1 - 4 \times 0.5236^2}\right) = -77.01°$$

$$\phi_3 = \tan^{-1}\left(\frac{6\zeta r}{1 - 9r^2}\right) = \tan^{-1}\left(\frac{6 \times 0.2 \times 0.5236}{1 - 9 \times 0.5236^2}\right) = -23.18°$$

The steady-state response of the valve can be expressed, using Eq. (4.13), as

$$x_p(t) = \frac{34083.3A}{k} - \frac{(26996.0A/k)}{\sqrt{(1-r^2)^2 + (2\zeta r)^2}} \cos(52.36t - \phi_1)$$

$$+ \frac{(8309.7A/k)}{\sqrt{(1-r^2)^2 + (2\zeta r)^2}} \sin(52.36t - \phi_1)$$

$$+ \frac{(1416.7A/k)}{\sqrt{(1-4r^2)^2 + (4\zeta r)^2}} \cos(104.72t - \phi_2)$$

$$+ \frac{(3608.3A/k)}{\sqrt{(1-4r^2)^2 + (4\zeta r)^2}} \sin(104.72t - \phi_2)$$

$$- \frac{(5833.3A/k)}{\sqrt{(1-9r^2)^2 + (6\zeta r)^2}} \cos(157.08t - \phi_3)$$

$$+ \frac{(2333.3A/k)}{\sqrt{(1-9r^2)^2 + (6\zeta r)^2}} \sin(157.08t - \phi_3)$$

4.4 RESPONSE UNDER NONPERIODIC FORCE

We have seen that periodic forces of any general wave form can be represented by Fourier series as a superposition of harmonic components of various frequencies. The response of a linear system is then found by superposing the harmonic response to each of the exciting forces. When the exciting force $F(t)$ is nonperiodic, such as that due to the blast from an explosion, a different method of calculating the response is required. Various methods can be used to find the response of the system to an arbitrary excitation. Some of these methods are as follows:

1. by representing the excitation by a Fourier integral;
2. by using the method of convolution integral;
3. by using the method of Laplace transformation;
4. by first approximating $F(t)$ by a suitable interpolation model and then using a numerical procedure; and
5. by numerically integrating the equations of motion.

We shall discuss Methods 2, 3, and 4 in the following sections and Method 5 in Chapter 11.

4.5 CONVOLUTION INTEGRAL

A nonperiodic exciting force usually has a magnitude that varies with time; it acts for a specified period of time and then stops. The simplest form of such a force is the impulsive force. An impulsive force is one that has a large magnitude F and acts

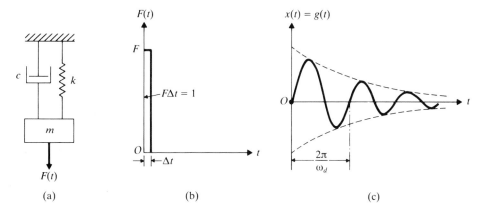

Figure 4.3

for a very short period of time Δt. From dynamics we know that impulse can be measured by finding the change in momentum of the system caused by it [4.2]. If $\dot{x}_1$ and $\dot{x}_2$ denote the velocities of the mass m before and after the application of the impulse, we have

$$\text{Impulse} = F\Delta t = m\dot{x}_2 - m\dot{x}_1 \tag{4.17}$$

By designating the magnitude of the impulse $F\Delta t$ by $\underset{\sim}{F}$, we can write, in general,

$$\underset{\sim}{F} = \int_t^{t+\Delta t} F\, dt \tag{4.18}$$

A unit impulse ($\underset{\sim}{f}$) is defined as

$$\underset{\sim}{f} = \lim_{\Delta t \to 0} \int_t^{t+\Delta t} F\, dt = F\, dt = 1 \tag{4.19}$$

It can be seen that in order for $F\, dt$ to have a finite value, F tends to infinity (since dt tends to zero). Although the unit impulse function has no physical meaning, it is a convenient tool in our present analysis.

**4.5.1
Response to
an Impulse**

We first consider the response of a single degree of freedom system to an impulse excitation; this case is important in studying the response under more general excitations. Consider a viscously damped spring-mass system subjected to a unit impulse at $t = 0$, as shown in Figs. 4.3(a) and (b). For an underdamped system, the solution of the equation of motion

$$m\ddot{x} + c\dot{x} + kx = 0 \tag{4.20}$$

is given by Eq. (2.66) as follows:

$$x(t) = e^{-\zeta\omega_n t}\left\{ x_0\cos\omega_d t + \frac{\dot{x}_0 + \zeta\omega_n x_0}{\omega_d}\sin\omega_d t \right\} \tag{4.21}$$

where

$$\zeta = \frac{c}{2m\omega_n} \tag{4.22}$$

$$\omega_d = \omega_n\sqrt{1 - \zeta^2} = \sqrt{\frac{k}{m} - \left(\frac{c}{2m}\right)^2} \tag{4.23}$$

$$\omega_n = \sqrt{\frac{k}{m}} \tag{4.24}$$

If the mass is at rest before the unit impulse is applied ($x = \dot{x} = 0$ for $t < 0$ or at $t = 0^-$), we obtain, from the impulse-momentum relation,

$$\text{Impulse} = \underset{\sim}{f} = 1 = m\dot{x}(t = 0) - m\dot{x}(t = 0^-) = m\dot{x}_0 \tag{4.25}$$

Thus the initial conditions are given by

$$x(t = 0) = x_0 = 0$$
$$\dot{x}(t = 0) = \dot{x}_0 = \frac{1}{m} \tag{4.26}$$

In view of Eq. (4.26), Eq. (4.21) reduces to

$$x(t) = g(t) = \frac{e^{-\zeta\omega_n t}}{m\omega_d}\sin\omega_d t \tag{4.27}$$

Equation (4.27) gives the response of a single degree of freedom system to a unit impulse, which is also known as the *impulse response function*, denoted by $g(t)$. The function $g(t)$, Eq. (4.27), is shown in Fig. 4.3(c).

If the magnitude of the impulse is $\underset{\sim}{F}$ instead of unity, the initial velocity $\dot{x}_0$ is $\underset{\sim}{F}/m$ and the response of the system becomes

$$x(t) = \frac{\underset{\sim}{F}e^{-\zeta\omega_n t}}{m\omega_d}\sin\omega_d t = \underset{\sim}{F}g(t) \tag{4.28}$$

If the impulse $\underset{\sim}{F}$ is applied at an arbitrary time $t = \tau$, as shown in Fig. 4.4(a), it will change the velocity at $t = \tau$ by an amount $\underset{\sim}{F}/m$. Assuming that $x = 0$ until the impulse is applied, the displacement x at any subsequent time t, caused by a change in the velocity at time τ, is given by Eq. (4.28) with t replaced by the time elapsed after the application of the impulse, that is, $t - \tau$. Thus we obtain

$$x(t) = \underset{\sim}{F}g(t - \tau) \tag{4.29}$$

This is shown in Fig. 4.4(b).

4.5.2
Response to
General Forcing
Condition

Now we consider the response of the system under an arbitrary external force $F(t)$, shown in Fig. 4.5. This force may be assumed to be made up of a series of impulses of varying magnitude. Assuming that at time τ, the force $F(\tau)$ acts on the system for a short period of time $\Delta\tau$, the impulse acting at $t = \tau$ is given by $F(\tau)\,\Delta\tau$. At any time t, the elapsed time since the impulse is $t - \tau$, so the response of the system at t due to this impulse alone is given by Eq. (4.29) with $\underset{\sim}{F} = F(\tau)\,\Delta\tau$:

$$\Delta x(t) = F(\tau)\,\Delta\tau g(t - \tau) \tag{4.30}$$

The total response at time t can be found by summing all the responses due to the

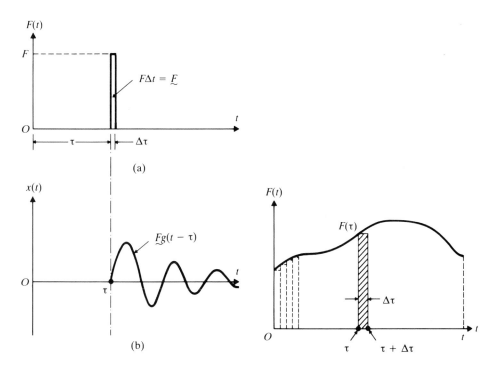

Figure 4.4

Figure 4.5 An arbitrary (nonperiodic) forcing function.

elementary impulses acting at all times τ:

$$x(t) \simeq \sum F(\tau) g(t - \tau) \, \Delta\tau \tag{4.31}$$

Letting $\Delta\tau \to 0$ and replacing the summation by integration, we obtain

$$x(t) = \int_0^t F(\tau) g(t - \tau) \, d\tau \tag{4.32}$$

By substituting Eq. (4.27) into Eq. (4.32), we obtain

$$x(t) = \frac{1}{m\omega_d} \int_0^t F(\tau) e^{-\zeta\omega_n(t-\tau)} \sin \omega_d(t - \tau) \, d\tau \tag{4.33}$$

which represents the response of an underdamped single degree of freedom system to the arbitrary excitation $F(t)$. Note that Eq. (4.33) does not consider the effect of initial conditions of the system. The integral in Eq. (4.32) or Eq. (4.33) is called the *convolution* or *Duhamel integral*. In many cases the function $F(t)$ has a form that permits an explicit integration of Eq. (4.33). In case such integration is not possible, it can be evaluated numerically without much difficulty, as illustrated in Section 4.8 and Chapter 11. An elementary discussion of the Duhamel integral in vibration analysis is given in Ref. [4.6].

**4.5.3
Response to
Base Excitation**

If a spring-mass-damper system is subjected to an arbitrary base excitation described by its displacement, velocity, or acceleration, the equation of motion can be expressed in terms of the relative displacement of the mass $z = x - y$ as follows (see Section 3.6.2)

$$m\ddot{z} + c\dot{z} + kz = -m\ddot{y} \tag{4.34}$$

This equation is similar to the equation

$$m\ddot{x} + c\dot{x} + kx = F \tag{4.35}$$

with the variable z replacing x and the term $-m\ddot{y}$ replacing the forcing function F. Hence all of the results derived for the force-excited system are applicable to the base-excited system also for z when the term F is replaced by $-m\ddot{y}$. For an underdamped system subjected to base excitation, the relative displacement can be found from Eq. (4.33):

$$z(t) = -\frac{1}{\omega_d}\int_0^t \ddot{y}(\tau)e^{-\zeta\omega_n(t-\tau)}\sin\omega_d(t-\tau)\,d\tau \tag{4.36}$$

EXAMPLE 4.3   **Step Force on a Compacting Machine**

A compacting machine, modeled as a single degree of freedom system, is shown in Fig. 4.6(a). The force acting on the mass m (m includes the masses of the piston, the platform, and the material being compacted) due to a sudden application of the pressure can be idealized as a step force, as shown in Fig. 4.6(b). Determine the response of the system.

Given: Compacting machine subjected to a step force.

Find: Response of the system.

Approach: Evaluate the Duhamel integral with $F(t) = F_0$.

Solution. Since the compacting machine is modeled as a mass-spring-damper system, the problem is to find the response of a damped single degree of freedom system subjected to a step force. By noting that $F(t) = F_0$, we can write Eq. (4.33) as

$$
\begin{aligned}
x(t) &= \frac{F_0}{m\omega_d}\int_0^t e^{-\zeta\omega_n(t-\tau)}\sin\omega_d(t-\tau)\,d\tau \\
&= \frac{F_0}{m\omega_d}\left[e^{-\zeta\omega_n(t-\tau)}\left\{\frac{\zeta\omega_n\sin\omega_d(t-\tau) + \omega_d\cos\omega_d(t-\tau)}{(\zeta\omega_n)^2 + (\omega_d)^2}\right\}\right]_{\tau=0}^{t} \\
&= \frac{F_0}{k}\left[1 - \frac{1}{\sqrt{1-\zeta^2}}\cdot e^{-\zeta\omega_n t}\cos(\omega_d t - \phi)\right]
\end{aligned}
\tag{E.1}
$$

where

$$\phi = \tan^{-1}\left(\frac{\zeta}{\sqrt{1-\zeta^2}}\right) \tag{E.2}$$

This response is shown in Fig. 4.6(c). If the system is undamped ($\zeta = 0$ and $\omega_d = \omega_n$), Eq.

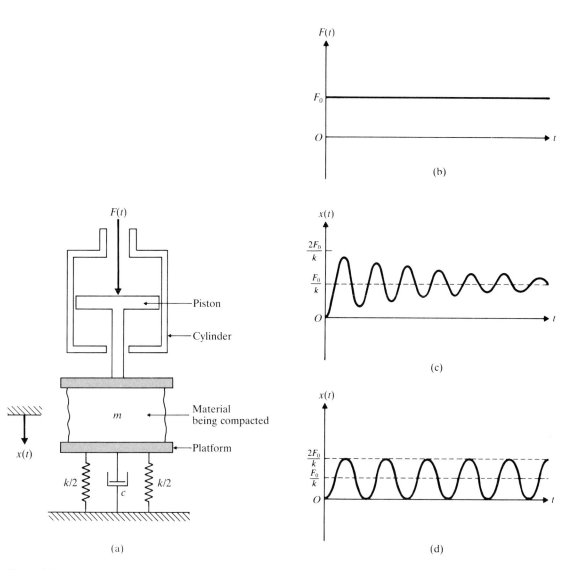

Figure 4.6

(E.1) reduces to

$$x(t) = \frac{F_0}{k}[1 - \cos \omega_n t] \qquad (E.3)$$

Equation (E.3) is shown graphically in Fig. 4.6(d). It can be seen that if the load is instantaneously applied to an undamped system, a maximum displacement of twice the static displacement will be attained, that is, $x_{\max} = 2F_0/k$.

EXAMPLE 4.4   **Time-Delayed Step Force**

Find the response of the compacting machine shown in Fig. 4.6(a) when it is subjected to the force shown in Fig. 4.7.

Given: Compacting machine, modeled as a damped single degree of freedom system, subjected to the force shown in Fig. 4.7.

Find: Response of the system.

Approach: Since the step force is time-delayed, substitute $t - t_0$ for t in the solution of Example 4.3.

Solution. Since the forcing function starts at $t = t_0$ instead of at $t = 0$, the response can be obtained from Eq. (E.1) of Example 4.3 by replacing t by $t - t_0$. This gives

$$x(t) = \frac{F_0}{k\sqrt{1 - \zeta^2}}\left[\sqrt{1 - \zeta^2} - e^{-\zeta\omega_n(t - t_0)}\cos\{\omega_d(t - t_0) - \phi\}\right] \qquad \text{(E.1)}$$

If the system is undamped, Eq. (E.1) reduces to

$$x(t) = \frac{F_0}{k}\left[1 - \cos\omega_n(t - t_0)\right] \qquad \text{(E.2)}$$

EXAMPLE 4.5   **Rectangular Pulse Load**

If the compacting machine shown in Fig. 4.6(a) is subjected to a constant force only during the time $0 \leqslant t \leqslant t_0$ (Fig. 4.8a), determine the response of the machine.

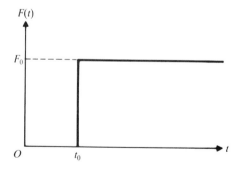

Figure 4.7

Given: Compacting machine, modeled as a spring-mass-damper system, subjected to a rectangular pulse type of load.

Find: Response of the machine.

Approach: Consider the rectangular pulse as a sum of a positive step function applied at $t = 0$ and a negative step function applied at $t = t_0$.

Solution. This forcing function can be considered as the sum of a step function of magnitude $+F_0$ beginning at $t = 0$ and a second step function of magnitude $-F_0$ starting at time $t = t_0$. Thus the response of the system can be obtained by subtracting Eq. (E.1) of Example 4.4 from Eq. (E.1) of Example 4.3. This gives

$$x(t) = \frac{F_0 e^{-\zeta \omega_n t}}{k\sqrt{1 - \zeta^2}} \left[-\cos(\omega_d t - \phi) + e^{\zeta \omega_n t_0} \cos\{\omega_d(t - t_0) - \phi\}\right] \qquad \text{(E.1)}$$

with

$$\phi = \tan^{-1}\left(\frac{\zeta}{\sqrt{1 - \zeta^2}}\right) \qquad \text{(E.2)}$$

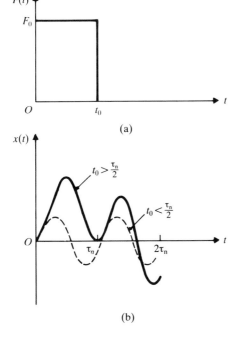

(a)

(b)

Figure 4.8

To see the vibration response graphically, we consider the system as undamped, so that Eq. (E.1) reduces to

$$x(t) = \frac{F_0}{k} \left[\cos \omega_n (t - t_0) - \cos \omega_n t \right] \qquad (E.3)$$

This response is shown in Fig. 4.8(b) for two cases: (1) $t_0 > \tau_n/2$, and (2) $t_0 < \tau_n/2$ where t_0 is the duration of the rectangular pulse and τ_n is the natural time period of the system. It can be seen that the peak occurs during the forced vibration era (that is, prior to t_0) for $t_0 > \tau_n/2$ while the peak occurs in the residual vibration era (that is, after t_0) if $t_0 < \tau_n/2$.

EXAMPLE 4.6   **Compacting Machine Under Linear Force**

Determine the response of the compacting machine shown in Fig. 4.9(a) when a linearly varying force (shown in Fig. 4.9b) is applied due to the motion of the cam.

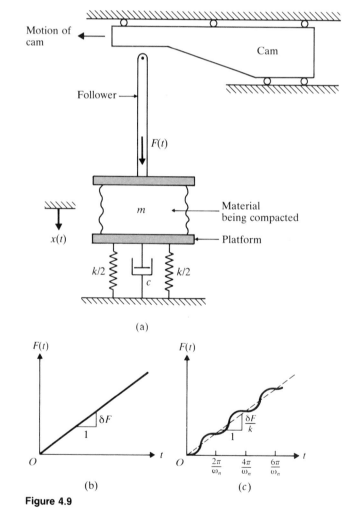

(a)

(b)

(c)

Figure 4.9

Given: Compacting machine, modeled as a single degree of freedom system, subjected to a ramp function.

Find: Response of the system.

Approach: Evaluate the Duhamel integral with $F(t) = \delta F \cdot t$.

Solution. The linearly varying force shown in Fig. 4.9(b) is known as the ramp function. This forcing function can be represented as $F(\tau) = \delta F \cdot \tau$, where δF denotes the rate of increase of the force F per unit time. By substituting this into Eq. (4.33), we obtain

$$
\begin{aligned}
x(t) &= \frac{\delta F}{m\omega_d} \int_0^t \tau e^{-\zeta\omega_n(t-\tau)} \sin \omega_d (t - \tau) \, d\tau \\
&= \frac{\delta F}{m\omega_d} \int_0^t (t - \tau) e^{-\zeta\omega_n(t-\tau)} \sin \omega_d (t - \tau)(-d\tau) \\
&\quad - \frac{\delta F \cdot t}{m\omega_d} \int_0^t e^{-\zeta\omega_n(t-\tau)} \sin \omega_d (t - \tau)(-d\tau)
\end{aligned}
$$

These integrals can be evaluated and the response expressed as follows. (See Problem 4.20.)

$$
x(t) = \frac{\delta F}{k} \left[t - \frac{2\zeta}{\omega_n} + e^{-\zeta\omega_n t} \left(\frac{2\zeta}{\omega_n} \cos \omega_d t - \left\{ \frac{\omega_d^2 - \zeta^2\omega_n^2}{\omega_n^2\omega_d} \right\} \sin \omega_d t \right) \right]
\tag{E.1}
$$

For an undamped system, Eq. (E.1) reduces to

$$
x(t) = \frac{\delta F}{\omega_n k} [\omega_n t - \sin \omega_n t]
\tag{E.2}
$$

Figure 4.9(c) shows the response given by Eq. (E.2).

EXAMPLE 4.7   Blast Load on a Building Frame

A building frame is modeled as an undamped single degree of freedom system (Fig. 4.10a). Find the response of the frame if it is subjected to a blast loading represented by the triangular pulse shown in Fig. 4.10(b).

Given: Building frame subjected to a triangular-pulse type blast loading.

Find: Displacement of the frame.

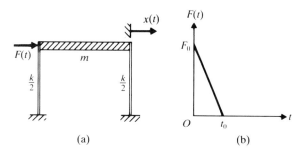

(a)   (b)

Figure 4.10

Approach: Model the building frame as an undamped single degree of freedom system and evaluate the Duhamel integral for the given load.

Solution. The forcing function is given by

$$F(\tau) = F_0\left(1 - \frac{\tau}{t_0}\right) \quad \text{for } 0 \leqslant \tau \leqslant t_0 \tag{E.1}$$

$$F(\tau) = 0 \quad \text{for } t_0 < \tau \tag{E.2}$$

Equation (4.33) gives, for an undamped system,

$$x(t) = \frac{1}{m\omega_n}\int_0^t F(\tau)\sin\omega_n(t - \tau)\, d\tau \tag{E.3}$$

Response during $0 \leqslant t \leqslant t_0$: Using Eq. (E.1) for $F(\tau)$ in Eq. (E.3) gives

$$
\begin{aligned}
x(t) &= \frac{F_0}{m\omega_n^2}\int_0^t\left(1 - \frac{\tau}{t_0}\right)[\sin\omega_n t\cos\omega_n\tau - \cos\omega_n t\sin\omega_n\tau]\, d(\omega_n\tau) \\
&= \frac{F_0}{k}\sin\omega_n t\int_0^t\left(1 - \frac{\tau}{t_0}\right)\cos\omega_n\tau \cdot d(\omega_n\tau) \\
&\quad - \frac{F_0}{k}\cos\omega_n t\int_0^t\left(1 - \frac{\tau}{t_0}\right)\sin\omega_n\tau \cdot d(\omega_n\tau)
\end{aligned}
\tag{E.4}
$$

By noting that integration by parts gives

$$\int \tau\cos\omega_n\tau \cdot d(\omega_n\tau) = \tau\sin\omega_n\tau + \frac{1}{\omega_n}\cos\omega_n\tau \tag{E.5}$$

and

$$\int \tau\sin\omega_n\tau \cdot d(\omega_n\tau) = -\tau\cos\omega_n\tau + \frac{1}{\omega_n}\sin\omega_n\tau \tag{E.6}$$

Eq. (E.4) can be written as

$$
\begin{aligned}
x(t) = \frac{F_0}{k}\Bigg\{ &\sin\omega_n t\left[\sin\omega_n t - \frac{t}{t_0}\sin\omega_n t - \frac{1}{\omega_n t_0}\cos\omega_n t + \frac{1}{\omega_n t_0}\right] \\
&- \cos\omega_n t\left[-\cos\omega_n t + 1 + \frac{t}{t_0}\cos\omega_n t - \frac{1}{\omega_n t_0}\sin\omega_n t\right]\Bigg\}
\end{aligned}
\tag{E.7}
$$

Simplifying this expression, we obtain

$$x(t) = \frac{F_0}{k}\left[1 - \frac{t}{t_0} - \cos\omega_n t + \frac{1}{\omega_n t_0}\sin\omega_n t\right] \tag{E.8}$$

Response during $t > t_0$: Here also we use Eq. (E.1) for $F(\tau)$, but the upper limit of integration in Eq. (E.3) will be t_0, since $F(\tau) = 0$ for $\tau > t_0$. Thus the response can be found from Eq. (E.7) by setting $t = t_0$ within the square brackets. This results in

$$x(t) = \frac{F_0}{k\omega_n t_0}\left[(1 - \cos\omega_n t_0)\sin\omega_n t - (\omega_n t_0 - \sin\omega_n t_0)\cos\omega_n t\right] \tag{E.9}$$

4.6 RESPONSE SPECTRUM

The graph showing the variation of the maximum response (maximum displacement, velocity, acceleration, or any other quantity) with the natural frequency (or natural period) of a single degree of freedom system to a specified forcing function is known as the *response spectrum*. Since the maximum response is plotted against the natural frequency (or natural period), the response spectrum gives the maximum response of all possible single degree of freedom systems. The response spectrum is widely used in earthquake engineering design [4.2, 4.5]. A review of recent literature on shock and seismic response spectra in engineering design is given in Ref. [4.7].

Once the response spectrum corresponding to a specified forcing function is available, we need to know just the natural frequency of the system to find its maximum response. Example 4.8 illustrates the construction of a response spectrum.

EXAMPLE 4.8   Response Spectrum of Sinusoidal Pulse

Find the undamped response spectrum for the sinusoidal pulse force shown in Fig. 4.11(a) using the initial conditions $x(0) = \dot{x}(0) = 0$.

Given: Single degree of freedom undamped system subjected to one-half period of a sinusoidal force.

Find: Response spectrum.

Approach: Find the response and express its maximum value in terms of its natural time period.

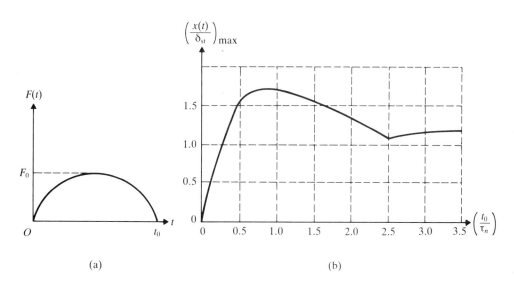

(a)

(b)

Figure 4.11

Solution. The equation of motion of an undamped system can be expressed as

$$m\ddot{x} + kx = F(t) = \begin{cases} F_0 \sin \omega t, & 0 \leqslant t \leqslant t_0 \\ 0, & t > t_0 \end{cases} \tag{E.1}$$

where

$$\omega = \frac{\pi}{t_0} \tag{E.2}$$

The solution of Eq. (E.1) can be obtained by superposing the homogeneous solution $x_c(t)$ and the particular solution $x_p(t)$ as

$$x(t) = x_c(t) + x_p(t) \tag{E.3}$$

that is,

$$x(t) = A \cos \omega_n t + B \sin \omega_n t + \left(\frac{F_0}{k - m\omega^2} \right) \sin \omega t \tag{E.4}$$

where A and B are constants and ω_n is the natural frequency of the system:

$$\omega_n = \frac{2\pi}{\tau_n} = \sqrt{\frac{k}{m}} \tag{E.5}$$

Using the initial conditions $x(0) = \dot{x}(0) = 0$ in Eq. (E.4), we can find the constants A and B as

$$A = 0, \qquad B = -\frac{F_0 \omega}{\omega_n (k - m\omega^2)} \tag{E.6}$$

Thus the solution becomes

$$x(t) = \frac{F_0/k}{1 - (\omega/\omega_n)^2} \left\{ \sin \omega t - \frac{\omega}{\omega_n} \sin \omega_n t \right\}, \qquad 0 \leqslant t \leqslant t_0 \tag{E.7}$$

which can be rewritten as

$$\frac{x(t)}{\delta_{st}} = \frac{1}{1 - \left(\frac{\tau_n}{2t_0} \right)^2} \left\{ \sin \frac{\pi t}{t_0} - \frac{\tau_n}{2t_0} \sin \frac{2\pi t}{\tau_n} \right\}, \qquad 0 \leqslant t \leqslant t_0 \tag{E.8}$$

where

$$\delta_{st} = \frac{F_0}{k} \tag{E.9}$$

The solution given by Eq. (E.8) is valid only during the period of force application, $0 \leqslant t \leqslant t_0$. Since there is no force applied for $t > t_0$, the solution can be expressed as a free vibration solution:

$$x(t) = A' \cos \omega_n t + B' \sin \omega_n t, \qquad t > t_0 \tag{E.10}$$

where the constants A' and B' can be found by using the values of $x(t = t_0)$ and $\dot{x}(t = t_0)$, given by Eq. (E.8), as initial conditions for the duration $t > t_0$. This gives

$$x(t = t_0) = \alpha \left[-\frac{\tau_n}{2t_0} \sin \frac{2\pi t_0}{\tau_n} \right] = A' \cos \omega_n t_0 + B' \sin \omega_n t_0 \tag{E.11}$$

$$\dot{x}(t = t_0) = \alpha \left\{ \frac{\pi}{t_0} - \frac{\pi}{t_0} \cos \frac{2\pi t_0}{\tau_n} \right\}$$

$$= -\omega_n A' \sin \omega_n t + \omega_n B' \cos \omega_n t \tag{E.12}$$

where

$$\alpha = \frac{\delta_{st}}{1 - \left(\dfrac{\tau_n}{2t_0}\right)^2} \tag{E.13}$$

Equations (E.11) and (E.12) can be solved to find A' and B' as

$$A' = \frac{\alpha\pi}{\omega_n t_0}\sin\omega_n t_0, \qquad B' = -\frac{\alpha\pi}{\omega_n t_0}[1 + \cos\omega_n t_0] \tag{E.14}$$

Equations (E.14) can be substituted into Eq. (E.10) to obtain

$$\frac{x(t)}{\delta_{st}} = \frac{(\tau_n/t_0)}{2\{1 - (\tau_n/2t_0)^2\}}\left[\sin 2\pi\left(\frac{t_0}{\tau_n} - \frac{t}{\tau_n}\right) - \sin 2\pi\frac{t}{\tau_n}\right], \qquad t \geq t_0 \tag{E.15}$$

Equations (E.8) and (E.15) give the response of the system in nondimensional form, that is, x/δ_{st} is expressed in terms of t/τ_n. Thus for any specified value of t_0/τ_n, the maximum value of x/δ_{st} can be found. This maximum value of x/δ_{st}, when plotted against t_0/τ_n, gives the response spectrum shown in Fig. 4.11(b). It can be observed that the maximum value of $(x/\delta_{st})_{max} \simeq 1.75$ occurs at a value of $t_0/\tau_n \simeq 0.75$.

In Example 4.8, the input force is simple and hence a closed form solution has been obtained for the response spectrum. However, if the input force is arbitrary, we can find the response spectrum only numerically. In such a case, Eq. (4.33) can be used to express the peak response of an undamped single degree of freedom system due to an arbitrary input force $F(t)$ as

$$x(t)\Big|_{max} = \frac{1}{m\omega_n}\int_0^t F(\tau)\sin\omega_n(t - \tau)\,d\tau\Big|_{max} \tag{4.37}$$

**4.6.1
Response
Spectrum
for Base
Excitation**

In the design of machinery and structures subjected to a ground shock, such as that caused by an earthquake, the response spectrum corresponding to the base excitation is useful. If the base of a damped single degree of freedom system is subjected to an acceleration $\ddot{y}(t)$, the equation of motion, in terms of the relative displacement $z = x - y$, is given by Eq. (4.34) and the response $z(t)$ by Eq. (4.36). In the case of a ground shock, the velocity response spectrum is generally used. The displacement and acceleration spectra are then expressed in terms of the velocity spectrum. For a harmonic oscillator (an undamped system under free vibration), we notice that

$$\ddot{x}|_{max} = -\omega_n^2 x|_{max} \quad \text{and} \quad \dot{x}|_{max} = \omega_n x|_{max}$$

Thus the acceleration and displacement spectra S_a and S_d can be obtained in terms of the velocity spectrum (S_v):

$$S_d = \frac{S_v}{\omega_n}, \qquad S_a = \omega_n S_v \tag{4.38}$$

To consider damping in the system, if we assume that the maximum relative displacement occurs after the shock pulse has passed, the subsequent motion must be harmonic. In such a case, we can use Eq. (4.38). The fictitious velocity associated with this apparent harmonic motion is called the *pseudo velocity* and its response spectrum, S_v, is called the *pseudo spectrum*. The velocity spectra of damped systems are used extensively in earthquake analysis.

To find the relative velocity spectrum, we differentiate Eq. (4.36) and obtain*

$$\dot{z}(t) = -\frac{1}{\omega_d} \int_0^t \ddot{y}(\tau) e^{-\zeta\omega_n(t-\tau)} \left[-\zeta\omega_n \sin \omega_d(t-\tau) \right.$$
$$\left. + \omega_d \cos \omega_d(t-\tau) \right] d\tau \qquad (4.39)$$

Equation (4.39) can be rewritten as

$$\dot{z}(t) = \frac{e^{-\zeta\omega_n t}}{\sqrt{1-\zeta^2}} \sqrt{P^2 + Q^2} \sin(\omega_d t - \phi) \qquad (4.40)$$

where

$$P = \int_0^t \ddot{y}(\tau) e^{\zeta\omega_n \tau} \cos \omega_d \tau \, d\tau \qquad (4.41)$$

$$Q = \int_0^t \ddot{y}(\tau) e^{\zeta\omega_n \tau} \sin \omega_d \tau \, d\tau \qquad (4.42)$$

and

$$\phi = \tan^{-1} \left\{ \frac{-\left(P\sqrt{1-\zeta^2} + Q\zeta \right)}{\left(P\zeta - Q\sqrt{1-\zeta^2} \right)} \right\} \qquad (4.43)$$

The velocity response spectrum, S_v, can be obtained from Eq. (4.40):

$$S_v = |\dot{z}(t)|_{max} = \left| \frac{e^{-\zeta\omega_n t}}{\sqrt{1-\zeta^2}} \sqrt{P^2 + Q^2} \right|_{max} \qquad (4.44)$$

Thus the pseudo response spectra are given by

$$S_d = |z|_{max} = \frac{S_v}{\omega_n}; \qquad S_v = |\dot{z}|_{max}; \qquad S_a = |\ddot{z}|_{max} = \omega_n S_v \qquad (4.45)$$

EXAMPLE 4.9   **Water Tank Subjected to Base Acceleration**

The water tank, shown in Fig. 4.12(a), is subjected to a linearly varying ground acceleration as shown in Fig. 4.12(b) due to an earthquake. The mass of the tank is m, the stiffness of the column is k, and damping is negligible. Find the response spectrum for the relative displacement, $z = x - y$, of the water tank.

---

*   The following relation is used in deriving Eq. (4.39) from Eq. (4.36):

$$\frac{d}{dt} \int_0^t f(t, \tau) \, d\tau = \int_0^t \frac{\partial f}{\partial t}(t, \tau) \, d\tau + f(t, \tau)|_{\tau = t}$$

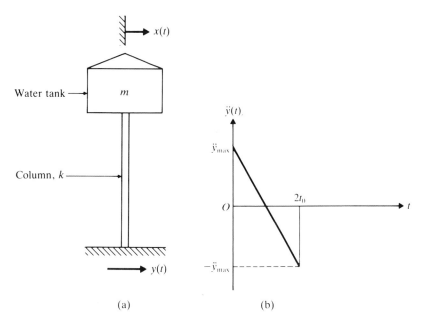

Figure 4.12

Given: Water tank subjected to the base acceleration shown in Fig. 4.12(b).

Find: Response spectrum of relative displacement of the tank.

Approach: Model the water tank as an undamped single degree of freedom system. Find the maximum relative displacement of the tank and express it as a function of ω_n.

Solution. The base acceleration can be expressed as

$$\ddot{y}(t) = \ddot{y}_{max}\left(1 - \frac{t}{t_0}\right) \quad \text{for } 0 \leqslant t \leqslant 2t_0 \tag{E.1}$$

$$\ddot{y}(t) = 0 \quad\quad\quad\quad \text{for } t > 2t_0 \tag{E.2}$$

Response during $0 \leqslant t \leqslant 2t_0$: By substituting Eq. (E.1) into Eq. (4.36), the response can be expressed, for an undamped system, as

$$z(t) = -\frac{1}{\omega_n}\ddot{y}_{max}\left[\int_0^t\left(1 - \frac{\tau}{t_0}\right)(\sin\omega_n t\cos\omega_n\tau - \cos\omega_n t\sin\omega_n\tau)\,d\tau\right] \tag{E.3}$$

This equation is the same as Eq. (E.4) of Example 4.7 except that $(-\ddot{y}_{max})$ appears in place of F_0/m. Hence $z(t)$ can be written, using Eq. (E.8) of Example 4.7, as

$$z(t) = -\frac{\ddot{y}_{max}}{\omega_n^2}\left[1 - \frac{t}{t_0} - \cos\omega_n t + \frac{1}{\omega_n t_0}\sin\omega_n t\right] \tag{E.4}$$

To find the maximum response z_{max} we set

$$\dot{z}(t) = -\frac{\ddot{y}_{max}}{t_0 \omega_n^2}[-1 + \omega_n t_0 \sin \omega_n t + \cos \omega_n t] = 0 \qquad (E.5)$$

This equation gives the time t_m at which z_{max} occurs:

$$t_m = \frac{2}{\omega_n} \tan^{-1}(\omega_n t_0) \qquad (E.6)$$

By substituting Eq. (E.6) into Eq. (E.4), the maximum response of the tank can be found:

$$z_{max} = -\frac{\ddot{y}_{max}}{\omega_n^2}\left[1 - \frac{t_m}{t_0} - \cos \omega_n t_m + \frac{1}{\omega_n t_0}\sin \omega_n t_m\right] \qquad (E.7)$$

Response during $t > 2t_0$: Since there is no excitation during this time, we can use the solution of the free vibration problem (Eq. (2.15)):

$$z(t) = z_0 \cos \omega_n t + \left(\frac{\dot{z}_0}{\omega_n}\right)\sin \omega_n t \qquad (E.8)$$

provided that we take the initial displacement and initial velocity as

$$z_0 = z(t = 2t_0) \quad \text{and} \quad \dot{z}_0 = \dot{z}(t = 2t_0) \qquad (E.9)$$

using Eq. (E.7). The maximum of $z(t)$ given by Eq. (E.8) can be identified as

$$z_{max} = \left[z_0^2 + \left(\frac{\dot{z}_0}{\omega_n}\right)^2\right]^{1/2} \qquad (E.10)$$

where z_0 and $\dot{z}_0$ are computed as indicated in Eq. (E.9).

4.7 LAPLACE TRANSFORMATION

The Laplace transform method can be used to find the response of a system under any type of excitation, including the harmonic and periodic types. This method can be used for the efficient solution of linear differential equations, particularly those with constant coefficients [4.3]. It permits the conversion of differential equations into algebraic ones, which are easier to manipulate. The major advantages of the method are that it can treat discontinuous functions without any particular difficulty and that it automatically takes into account the initial conditions.

The Laplace transform of a function $x(t)$, denoted symbolically as $\bar{x}(s) = \mathscr{L}x(t)$, is defined as

$$\bar{x}(s) = \mathscr{L}x(t) = \int_0^\infty e^{-st}x(t)\, dt \qquad (4.46)$$

where s is, in general, a complex quantity and is called the *subsidiary variable*. The function e^{-st} is called the *kernel* of the transformation. Since the integration is with respect to t, the transformation gives a function of s. In order to solve a vibration problem using the Laplace transform method, the following steps are necessary:

1. Write the equation of motion of the system.
2. Transform each term of the equation, using known initial conditions.
3. Solve for the transformed response of the system.
4. Obtain the desired solution (response) by using inverse Laplace transformation.

In order to solve the forced vibration equation

$$m\ddot{x} + c\dot{x} + kx = F(t) \tag{4.47}$$

by the Laplace transform method, it is necessary to find the transforms of the derivatives

$$\dot{x}(t) = \frac{dx}{dt}(t) \quad \text{and} \quad \ddot{x}(t) = \frac{d^2x}{dt^2}(t)$$

These can be found as follows:

$$\mathcal{L}\frac{dx}{dt}(t) = \int_0^\infty e^{-st}\frac{dx}{dt}(t)\,dt \tag{4.48}$$

This can be integrated by parts to obtain

$$\mathcal{L}\frac{dx}{dt}(t) = e^{-st}x(t)\bigg|_0^\infty + s\int_0^\infty e^{-st}x(t)\,dt = s\bar{x}(s) - x(0) \tag{4.49}$$

where $x(0) = x_0$ is the initial displacement of the mass m. Similarly, the Laplace transform of the second derivative of $x(t)$ can be obtained:

$$\mathcal{L}\frac{d^2x}{dt^2}(t) = \int_0^\infty e^{-st}\frac{d^2x}{dt^2}(t)\,dt = s^2\bar{x}(s) - sx(0) - \dot{x}(0) \tag{4.50}$$

where $\dot{x}(0) = \dot{x}_0$ is the initial velocity of the mass m. Since the Laplace transform of the force $F(t)$ is given by

$$\bar{F}(s) = \mathcal{L}F(t) = \int_0^\infty e^{-st}F(t)\,dt \tag{4.51}$$

we can transform both sides of Eq. (4.47) and obtain, using Eqs. (4.46) and (4.48) to (4.51),

$$m\mathcal{L}\ddot{x}(t) + c\mathcal{L}\dot{x}(t) + k\mathcal{L}x(t) = \mathcal{L}F(t)$$

or

$$(ms^2 + cs + k)\bar{x}(s) = \bar{F}(s) + m\dot{x}(0) + (ms + c)x(0) \tag{4.52}$$

where the right-hand side of Eq. (4.52) can be regarded as a generalized transformed excitation.

For the present, we take $\dot{x}(0)$ and $x(0)$ as zero, which is equivalent to ignoring the homogeneous solution of the differential equation (4.47). Then the ratio of the transformed excitation to the transformed response $\bar{Z}(s)$ can be expressed as

$$\bar{Z}(s) = \frac{\bar{F}(s)}{\bar{x}(s)} = ms^2 + cs + k \tag{4.53}$$

The function $\bar{Z}(s)$ is known as the *generalized impedance* of the system. The reciprocal of the function $\bar{Z}(s)$ is called the *admittance* or *transfer function* of the system and is denoted as $\bar{Y}(s)$:

$$\bar{Y}(s) = \frac{1}{\bar{Z}(s)} = \frac{\bar{x}(s)}{\bar{F}(s)} = \frac{1}{ms^2 + cs + k} = \frac{1}{m(s^2 + 2\zeta\omega_n s + \omega_n^2)} \quad (4.54)$$

It can be seen that by letting $s = i\omega$ in $\bar{Y}(s)$ and multiplying by k, we obtain the complex frequency response $H(i\omega)$ defined in Eq. (3.54). Equation (4.54) can also be expressed as

$$\bar{x}(s) = \bar{Y}(s)\bar{F}(s) \quad (4.55)$$

which indicates that the transfer function can be regarded as an algebraic operator that operates on the transformed force to yield the transformed response.

To find the desired response $x(t)$ from $\bar{x}(s)$, we have to take the inverse Laplace transform of $\bar{x}(s)$, which can be defined symbolically as

$$x(t) = \mathscr{L}^{-1}\bar{x}(s) = \mathscr{L}^{-1}\bar{Y}(s)\bar{F}(s) \quad (4.56)$$

In general, the operator $\mathscr{L}^{-1}$ involves a line integral in the complex domain, [4.9, 4.10]. Fortunately, we need not evaluate these integrals separately for each problem; such integrations have been carried out for various common forms of the function $F(t)$ and tabulated [4.4]. One such table is given in Appendix B. In order to find the solution using Eq. (4.56), we usually look for ways of decomposing $\bar{x}(s)$ into a combination of simple functions whose inverse transformations are available in Laplace transform tables. We can decompose $\bar{x}(s)$ conveniently by the method of partial fractions.

In the above discussion, we ignored the homogeneous solution by assuming $x(0)$ and $\dot{x}(0)$ as zero. We now consider the general solution by taking the initial conditions as $x(0) = x_0$ and $\dot{x}(0) = \dot{x}_0$. From Eq. (4.52), the transformed response $\bar{x}(s)$ can be obtained:

$$\bar{x}(s) = \frac{\bar{F}(s)}{m(s^2 + 2\zeta\omega_n s + \omega_n^2)} + \frac{s + 2\zeta\omega_n}{s^2 + 2\zeta\omega_n s + \omega_n^2}x_0 + \frac{1}{s^2 + 2\zeta\omega_n s + \omega_n^2}\dot{x}_0 \quad (4.57)$$

We can obtain the inverse transform of $\bar{x}(s)$ by considering each term on the right side of Eq. (4.57) separately. We also make use of the following relation [4.4]:

$$\mathscr{L}^{-1}\bar{f}_1(s)\bar{f}_2(s) = \int_0^t f_1(\tau)f_2(t - \tau)\, d\tau \quad (4.58)$$

By considering the first term on the right side of Eq. (4.57) as $\bar{f}_1(s)\bar{f}_2(s)$, where

$$\bar{f}_1(s) = \bar{F}(s) \quad \text{and} \quad \bar{f}_2(s) = \frac{1}{m(s^2 + 2\zeta\omega_n s + \omega_n^2)}$$

and by noting that $f_1(t) = \mathscr{L}^{-1}\bar{f}_1(s) = F(t)$, we obtain*

$$\mathscr{L}^{-1}\bar{f}_1(s)\bar{f}_2(s) = \frac{1}{m\omega_d}\int_0^t F(\tau)e^{-\zeta\omega_n(t-\tau)}\sin\omega_d(t-\tau)\,d\tau \qquad (4.59)$$

Considering the second term on the right side of Eq. (4.57), we find the inverse transform of the coefficient of x_0 from the table in Appendix B:

$$\mathscr{L}^{-1}\left(\frac{s + 2\zeta\omega_n}{s^2 + 2\zeta\omega_n s + \omega_n^2}\right) = \frac{1}{\sqrt{1-\zeta^2}}e^{-\zeta\omega_n t}\sin(\omega_d t + \phi_1) \qquad (4.60)$$

where

$$\phi_1 = \cos^{-1}(\zeta) \qquad (4.61)$$

Finally, the inverse transform of the coefficient of $\dot{x}_0$ in the third term on the right side of Eq. (4.57) can be obtained from the table in Appendix B:

$$\mathscr{L}^{-1}\left[\frac{1}{(s^2 + 2\zeta\omega_n s + \omega_n^2)}\right] = \frac{1}{\omega_d}e^{-\zeta\omega_n t}\sin\omega_d t \qquad (4.62)$$

Using Eqs. (4.57), (4.59), (4.60), and (4.62), the general solution of Eq. (4.47) can be expressed as

$$x(t) = \frac{x_0}{(1-\zeta^2)^{1/2}}e^{-\zeta\omega_n t}\sin(\omega_d t + \phi_1) + \frac{\dot{x}_0}{\omega_d}e^{-\zeta\omega_n t}\sin\omega_d t$$

$$+ \frac{1}{m\omega_d}\int_0^t F(\tau)e^{-\zeta\omega_n(t-\tau)}\sin\omega_d(t-\tau)\,d\tau \qquad (4.63)$$

EXAMPLE 4.10   **Response of a Compacting Machine**

Find the response of the compacting machine of Example 4.5 assuming the system to be underdamped (i.e., $\zeta < 1$).

Given: Compacting machine (spring-mass-damper system) subjected to a step force.

Find: Response of the system.

Approach: Use Laplace transformation technique.

Solution. The forcing function is given by

$$F(t) = \begin{cases} F_0 & \text{for } 0 \leqslant t \leqslant t_0 \\ 0 & \text{for } t > t_0 \end{cases} \qquad (\text{E.1})$$

By taking the Laplace transform of the governing differential equation, Eq. (4.47), we obtain Eq. (4.57), using Appendix B, with

$$\bar{F}(s) = \mathscr{L}F(t) = \frac{F_0(1 - e^{-t_0 s})}{s} \qquad (\text{E.1})$$

---

* The inverse transform of $\bar{f}_2(s)$ is obtained from the Laplace transform table in Appendix B.

Thus Eq. (4.57) can be written as

$$\bar{x}(s) = \frac{F_0(1 - e^{-t_0 s})}{ms(s^2 + 2\zeta\omega_n s + \omega_n^2)} + \frac{s + 2\zeta\omega_n}{s^2 + 2\zeta\omega_n s + \omega_n^2}x_0$$

$$+ \frac{1}{s^2 + 2\zeta\omega_n + \omega_n^2}\dot{x}_0$$

$$= \frac{F_0}{m\omega_n^2}\frac{1}{s\left(\dfrac{s^2}{\omega_n^2} + \dfrac{2\zeta s}{\omega_n} + 1\right)} - \frac{F_0}{m\omega_n^2}\frac{e^{-t_0 s}}{s\left(\dfrac{s^2}{\omega_n^2} + \dfrac{2\zeta s}{\omega_n} + 1\right)}$$

$$+ \frac{x_0}{\omega_n^2}\frac{s}{\left(\dfrac{s^2}{\omega_n^2} + \dfrac{2\zeta s}{\omega_n} + 1\right)} + \left(\frac{2\zeta x_0}{\omega_n} + \frac{\dot{x}_0}{\omega_n^2}\right)\frac{1}{\left(\dfrac{s^2}{\omega_n^2} + \dfrac{2\zeta s}{\omega_n} + 1\right)} \tag{E.2}$$

The inverse transform of Eq. (E.2) can be expressed by using the results in Appendix B as

$$x(t) = \frac{F_0}{m\omega_n^2}\left[1 - \frac{e^{-\zeta\omega_n t}}{\sqrt{1 - \zeta^2}}\sin\left\{\omega_n\sqrt{1 - \zeta^2}\,t + \phi_1\right\}\right]$$

$$- \frac{F_0}{m\omega_n^2}\left[1 - \frac{e^{-\zeta\omega_n(t - t_0)}}{\sqrt{1 - \zeta^2}}\sin\left\{\omega_n\sqrt{1 - \zeta^2}\,(t - t_0) + \phi_1\right\}\right]$$

$$- \frac{x_0}{\omega_n^2}\left[\frac{\omega_n^2 e^{-\zeta\omega_n t}}{\sqrt{1 - \zeta^2}}\sin\left\{\omega_n\sqrt{1 - \zeta^2}\,t - \phi_1\right\}\right]$$

$$+ \left(\frac{2\zeta x_0}{\omega_n} + \frac{\dot{x}_0}{\omega_n^2}\right)\left[\frac{\omega_n}{\sqrt{1 - \zeta^2}}e^{-\zeta\omega_n t}\sin\left(\omega_n\sqrt{1 - \zeta^2}\,t\right)\right] \tag{E.3}$$

where ϕ_1 is given by Eq. (4.61). Thus the response of the compacting machine can be expressed as

$$x(t) = \frac{F_0}{m\omega_n^2\sqrt{1 - \zeta^2}}\left[-e^{-\zeta\omega_n t}\sin\left(\omega_n\sqrt{1 - \zeta^2}\,t + \phi_1\right)\right.$$

$$\left. + e^{-\zeta\omega_n(t - t_0)}\sin\left\{\omega_n\sqrt{1 - \zeta^2}\,(t - t_0) + \phi_1\right\}\right]$$

$$- \frac{x_0}{\sqrt{1 - \zeta^2}}e^{-\zeta\omega_n t}\sin\left(\omega_n\sqrt{1 - \zeta^2}\,t - \phi_1\right)$$

$$+ \frac{(2\zeta\omega_n x_0 + \dot{x}_0)}{\omega_n\sqrt{1 - \zeta^2}}e^{-\zeta\omega_n t}\sin\left(\omega_n\sqrt{1 - \zeta^2}\,t\right) \tag{E.4}$$

Although the first part of Eq. (E.4) is expected to be the same as Eq. (E.1) of Example 4.5, it is difficult to see the equivalence in the present form of Eq. (E.4). However, for the undamped system, Eq. (E.4) reduces to

$$x(t) = \frac{F_0}{m\omega_n^2}\left[-\sin\left(\omega_n t + \frac{\pi}{2}\right) + \sin\left\{\omega_n(t - t_0) + \frac{\pi}{2}\right\}\right]$$

$$- x_0\sin\left(\omega_n t - \frac{\pi}{2}\right) + \frac{\dot{x}_0}{\omega_n}\sin\omega_n t$$

$$= \frac{F_0}{k}\left[\cos\omega_n(t - t_0) - \cos\omega_n t\right] + x_0\cos\omega_n t + \frac{\dot{x}_0}{\omega_n}\sin\omega_n t \tag{E.5}$$

The first or steady-state part of Eq. (E.5) can be seen to be identical to Eq. (E.3) of Example 4.5.

4.8 RESPONSE TO IRREGULAR FORCING CONDITIONS USING NUMERICAL METHODS

In the previous sections, it was assumed that the forcing functions $F(t)$ are available as functions of time in an explicit manner. In many practical problems, however, the forcing functions $F(t)$ are not available in the form of analytical expressions. When a forcing function is determined experimentally, $F(t)$ may be known as an irregular curve. Sometimes only the values of $F(t) = F_i$ at a series of points $t = t_i$ may be available, in the form of a diagram or a table. In such cases, we can fit polynomials or some such curves to the data and use them in the Duhamel integral, Eq. (4.33), to find the response of the system. Another, more common, method of finding the response involves dividing the time axis into a number of discrete points and using a simple variation of $F(t)$ during each time step. We shall present this numerical approach in this section, using several types of interpolation functions for $F(t)$ [4.8]. The direct numerical integration of the equations of motion is discussed in Chap. 11.

Method 1. Let the function $F(t)$ vary with time in an arbitrary manner, as indicated in Fig. 4.13. This forcing function can be approximated by a series of step functions having different magnitudes starting at different instants, as shown in Fig. 4.14. In this figure, the first step function starts at time $t = t_1 = 0$ and has a magnitude of ΔF_1, the second step function starts at time $t = t_2$ and has a magnitude of ΔF_2, and so forth. The response of the system in any time interval $t_{j-1} \leqslant t \leqslant t_j$ due to the step functions ΔF_i ($i = 1, 2, \ldots, j - 1$) can be found, using the results of Example 4.3:

$$
x(t) = \frac{1}{k} \sum_{i=1}^{j-1} \Delta F_i \left[1 - e^{-\zeta \omega_n (t - t_i)} \right.
$$
$$
\left. \times \left\{ \cos \omega_d (t - t_i) + \frac{\zeta \omega_n}{\omega_d} \sin \omega_d (t - t_i) \right\} \right] \tag{4.64}
$$

Thus the response of the system at $t = t_j$ becomes

$$
x_j = \frac{1}{k} \sum_{i=1}^{j-1} \Delta F_i \left[1 - e^{-\zeta \omega_n (t_j - t_i)} \right.
$$
$$
\left. \times \left\{ \cos \omega_d (t_j - t_i) + \frac{\zeta \omega_n}{\omega_d} \sin \omega_d (t_j - t_i) \right\} \right] \tag{4.65}
$$

Notice that the step function ΔF_i of step i is positive if the slope of the F-versus-t curve is positive, and it is negative if the slope of the F-versus-t curve is negative, as indicated in Fig. 4.14. For higher accuracy, the time steps taken should be small. In

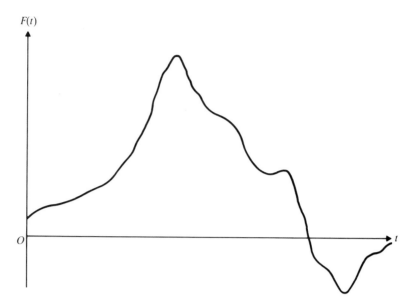

Figure 4.13

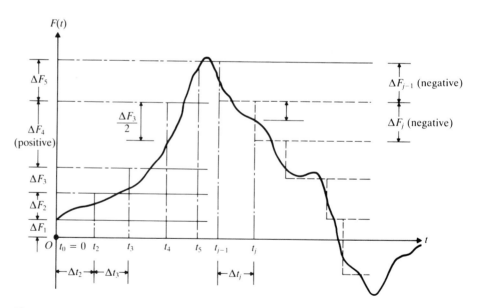

Figure 4.14

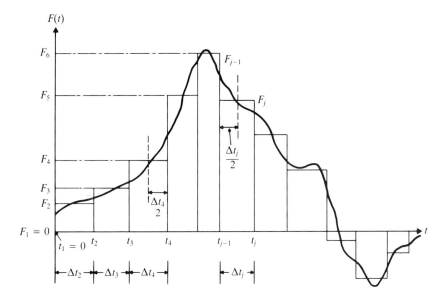

Figure 4.15

addition, it is desirable to make the force steps start, after the first one, at instants when the ordinates of the $F(t)$ curve are at the midheights of the steps, as shown in Fig. 4.14. In this case, the errors involved in approximating the $F(t)$ curve will be self-compensatory; that is, the areas lying above the $F(t)$ curve will be approximately equal to the areas lying below the $F(t)$ curve.

Method 2. Instead of approximating the $F(t)$ curve by a succession of step functions, we can approximate it by a series of rectangular impulses F_i, as shown in Fig. 4.15. These impulses F_i are positive or negative, depending on whether the curve $F(t)$ lies above or below the time (t) axis. As in the previous case, the magnitudes of F_i should be selected as the values of $F(t)$ at the midpoints of the time intervals Δt_i, as shown in Fig. 4.15, to make the errors self-compensating. The response of the system in any time interval $t_{j-1} \leqslant t \leqslant t_j$ can be found by adding the response due to F_j (applied in the interval Δt_j) to the response existing at $t = t_{j-1}$ (initial condition). This gives

$$x(t) = \frac{F_j}{k}\left[1 - e^{-\zeta\omega_n(t-t_{j-1})}\left\{\cos\omega_d\left(t - t_{j-1}\right) + \frac{\zeta\omega_n}{\omega_d}\sin\omega_d\left(t - t_{j-1}\right)\right\}\right]$$

$$+ e^{-\zeta\omega_n(t-t_{j-1})}\left\{x_{j-1}\cos\omega_d\left(t - t_{j-1}\right) + \frac{\dot{x}_{j-1} + \zeta\omega_n x_{j-1}}{\omega_d}\sin\omega_d\left(t - t_{j-1}\right)\right\}$$

$$(4.66)$$

By substituting $t = t_j$ in Eq. (4.66) the response of the system at the end of the interval Δt_j can be obtained:

$$x_j = \frac{F_j}{k}\left[1 - e^{-\zeta\omega_n\Delta t_j}\left\{\cos\omega_d \cdot \Delta t_j + \frac{\zeta\omega_n}{\omega_d}\sin\omega_d \cdot \Delta t_j\right\}\right]$$

$$+ e^{-\zeta\omega_n\Delta t_j}\left\{x_{j-1}\cos\omega_d \cdot \Delta t_j + \frac{\dot{x}_{j-1} + \zeta\omega_n x_{j-1}}{\omega_d}\sin\omega_d \cdot \Delta t_j\right\} \quad (4.67)$$

By differentiating Eq. (4.66) with respect to t and substituting $t = t_j$, we obtain the velocity $\dot{x}_j$ at the end of the interval Δt_j:

$$\dot{x}_j = \frac{F_j\omega_d}{k}e^{-\zeta\omega_n\Delta t_j}\left(1 + \frac{\zeta^2\omega_n^2}{\omega_d^2}\right)\sin\omega_d \cdot \Delta t_j + \omega_d e^{-\zeta\omega_n\cdot\Delta t_j}$$

$$\times\left\{-x_{j-1}\sin\omega_d \cdot \Delta t_j + \frac{\dot{x}_{j-1} + \zeta\omega_n x_{j-1}}{\omega_d}\cos\omega_d \cdot \Delta t_j\right.$$

$$\left. - \frac{\zeta\omega_n}{\omega_d}\left[x_{j-1}\cos\omega_d \cdot \Delta t_j + \frac{\dot{x}_{j-1} + \zeta\omega_n x_{j-1}}{\omega_d}\sin\omega_d \cdot \Delta t_j\right]\right\} \quad (4.68)$$

Equations (4.67) and (4.68) represent recurrence relations for computing the response at the end of jth time step. They also provide the initial conditions of x_j and $\dot{x}_j$ at the beginning of step $j + 1$. These equations may be sequentially applied to find the variations of displacement and velocity of the system with time.

Method 3. In the piecewise-constant types of approximations used in Methods 1 and 2, it is not always possible to make the areas above and below the $F(t)$ curve equal and make the errors self-compensating. Hence it is desirable to use a higher order interpolation, such as a piecewise linear or a piecewise quadratic approximation, for $F(t)$. In the piecewise linear interpolation, the variation of $F(t)$ in any time interval is assumed to be linear as shown in Fig. 4.16. In this case, the response of the system in the time interval $t_{j-1} \leqslant t \leqslant t_j$ can be found by adding the response due to the linear (ramp) function applied during the current interval to the response existing at $t = t_{j-1}$ (initial condition). This gives

$$x(t) = \frac{\Delta F_j}{k\Delta t_j}\left[t - t_{j-1} - \frac{2\zeta}{\omega_n} + e^{-\zeta\omega_n(t-t_{j-1})}\right.$$

$$\times\left\{\frac{2\zeta}{\omega_n}\cos\omega_d(t - t_{j-1})\right.$$

$$\left.\left. - \frac{\omega_d^2 - \zeta^2\omega_n^2}{\omega_n^2\omega_d}\sin\omega_d(t - t_{j-1})\right\}\right]$$

$$+ \frac{F_{j-1}}{k}\left[1 - e^{-\zeta\omega_n(t-t_{j-1})}\left\{\cos\omega_d(t - t_{j-1}) + \frac{\zeta\omega_n}{\omega_d}\sin\omega_d(t - t_{j-1})\right\}\right]$$

$$+ e^{-\zeta\omega_n(t-t_{j-1})}\left[x_{j-1}\cos\omega_d(t - t_{j-1}) + \frac{\dot{x}_{j-1} + \zeta\omega_n x_{j-1}}{\omega_d}\sin\omega_d(t - t_{j-1})\right]$$

$$(4.69)$$

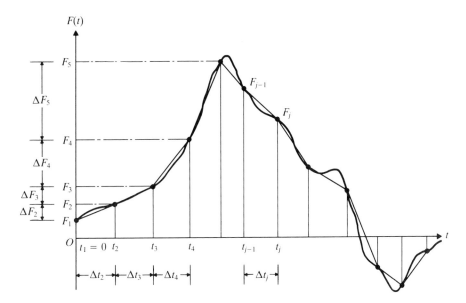

Figure 4.16

where $\Delta F_j = F_j - F_{j-1}$. By setting $t = t_j$ in Eq. (4.69), we obtain the response at the end of the interval Δt_j:

$$x_j = \frac{\Delta F_j}{k\,\Delta t_j}\left[\Delta t_j - \frac{2\zeta}{\omega_n} + e^{-\zeta\omega_n\Delta t_j}\left\{\frac{2\zeta}{\omega_n}\cos\omega_d\Delta t_j - \frac{\omega_d^2 - \zeta^2\omega_n^2}{\omega_n^2\omega_d}\sin\omega_d\Delta t_j\right\}\right]$$

$$+ \frac{F_{j-1}}{k}\left[1 - e^{-\zeta\omega_n\cdot\Delta t_j}\left\{\cos\omega_d\Delta t_j + \frac{\zeta\omega_n}{\omega_d}\sin\omega_d\Delta t_j\right\}\right]$$

$$+ e^{-\zeta\omega_n\Delta t_j}\left[x_{j-1}\cos\omega_d\Delta t_j + \frac{\dot{x}_{j-1} + \zeta\omega_n x_{j-1}}{\omega_d}\sin\omega_d\Delta t_j\right] \qquad (4.70)$$

By differentiating Eq. (4.69) with respect to t and substituting $t = t_j$, we obtain the velocity at the end of the interval:

$$\dot{x}_j = \frac{\Delta F_j}{k\,\Delta t_j}\left[1 - e^{-\zeta\omega_n\Delta t_j}\left\{\cos\omega_d\Delta t_j + \frac{\zeta\omega_n}{\omega_d}\sin\omega_d\Delta t_j\right\}\right]$$

$$+ \frac{F_{j-1}}{k}e^{-\zeta\omega_n\Delta t_j}\frac{\omega_n^2}{\omega_d}\cdot\sin\omega_d\cdot\Delta t_j + e^{-\zeta\omega_n\cdot\Delta t_j}$$

$$\times\left[\dot{x}_{j-1}\cos\omega_d\Delta t_j - \frac{\zeta\omega_n}{\omega_d}\left(\dot{x}_{j-1} + \frac{\omega_n}{\zeta}x_{j-1}\right)\sin\omega_d\Delta t_j\right] \qquad (4.71)$$

Equations (4.70) and (4.71) are the recurrence relations for finding the response of the system at the end of jth time step.

EXAMPLE 4.11   Damped Response Using Numerical Methods

Find the response of a spring-mass-damper system subjected to the forcing function

$$F(t) = F_0\left(1 - \sin\frac{\pi t}{2t_0}\right) \tag{E.1}$$

in the interval $0 \leqslant t \leqslant t_0$, using a numerical procedure. Assume $F_0 = 1$, $k = 1$, $m = 1$, $\zeta = 0.1$, and $t_0 = \tau_n/2$, where τ_n denotes the natural period of vibration given by

$$\tau_n = \frac{2\pi}{\omega_n} = \frac{2\pi}{(k/m)^{1/2}} = 2\pi \tag{E.2}$$

The values of x and $\dot{x}$ at $t = 0$ are zero.

Given: Spring-mass-damper system subjected to the force given by Eq. (E.1). $m = 1$, $k = 1$, $\zeta = 0.1$, $F_0 = 1$, $t_0 = \pi$, $x(0) = \dot{x}(0) = 0$.

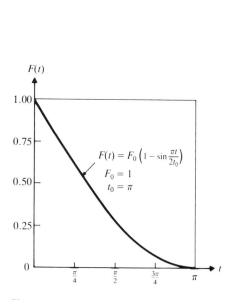

Figure 4.17

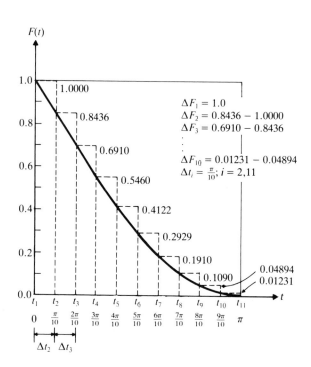

Figure 4.18

Find: Response of the system.

Approach: Use numerical methods.

Solution. Figure 4.17 shows the forcing function of Eq. (E.1). For the numerical computations, the time interval 0 to t_0 is divided into ten equal steps with

$$\Delta t_i = \frac{t_0}{10} = \frac{\pi}{10}; \qquad i = 2, 3, \ldots, 11 \qquad (E.3)$$

Four different methods are used to approximate the forcing function $F(t)$. In Fig. 4.18, $F(t)$ is approximated by a series of rectangular impulses, each starting at the beginning of the corresponding time step. A similar approximation, with the magnitude of the impulse at the end of the time step, is used in Fig. 4.19. The value of $F(t)$ at the middle of the time step is used as an impulse in Fig. 4.20. In Fig. 4.21, piecewise linear (trapezoidal) impulses are used to approximate the forcing function $F(t)$. The numerical results are given in Table 4.1. As can be expected from the idealizations, the results obtained by idealizations 1 and 2 (Figs. 4.18 and 4.19) overestimate and underestimate the true response, respectively. The results given by idealizations 3 and 4 are expected to lie between those given by idealizations 1 and 2. Further, the results obtained from idealization 4 will be the most accurate ones.

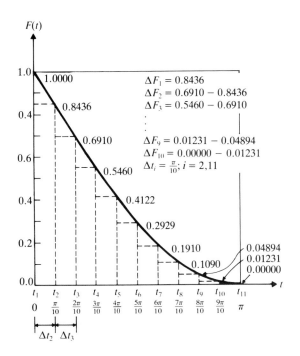

Figure 4.19

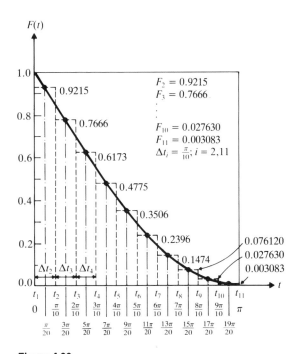

Figure 4.20

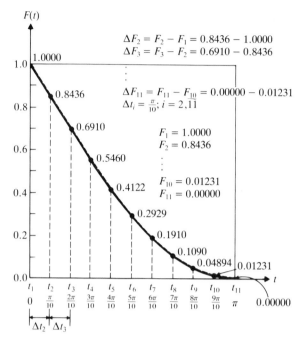

Figure 4.21

TABLE 4.1

		$x(t_i)$ Obtained According to			
i	t_i	**Fig. 4.18** **(Idealization 1)**	**Fig. 4.19** **(Idealization 2)**	**Fig. 4.20** **(Idealization 3)**	**Fig. 4.21** **(Idealization 4)**
1	0	0.00000	0.00000	0.00000	0.00000
2	0.1π	0.04794	0.04044	0.04417	0.04541
3	0.2π	0.17578	0.14729	0.16147	0.16377
4	0.3π	0.35188	0.29228	0.32190	0.32499
5	0.4π	0.54248	0.44609	0.49392	0.49746
6	0.5π	0.71540	0.58160	0.64790	0.65151
7	0.6π	0.84330	0.67659	0.75906	0.76238
8	0.7π	0.90630	0.71578	0.80986	0.81255
9	0.8π	0.89367	0.69214	0.79142	0.79323
10	0.9π	0.80449	0.60717	0.70403	0.70482
11	π	0.64730	0.47152	0.55672	0.55647

4.9 COMPUTER PROGRAMS

4.9.1
Response under an Arbitrary Periodic Forcing Function

A Fortran computer program, in the form of subroutine PERIOD, is given for finding the dynamic response of a damped oscillator excited by any periodic external force applied to the mass. The arguments of the subroutine are as follows:

XM	=	Mass of the system. Input data.
XK	=	Stiffness of the spring. Input data.
XAI	=	Damping ratio ζ. Input data.
N	=	Number of equidistant points at which the values of the force $F(t)$ are known. Input data.
M	=	Number of Fourier coefficients to be considered in the solution. Input data.
TIME	=	Time period of the function $F(t)$. Input data.
F	=	Array of dimension N which contains the known values of $F(t)$. $F(I) = F(t_i)$. Input data.
T	=	Array of dimension N which contains the known values of time t. $T(I) = t_i$. Input data.
FZERO	=	F_0. Output.
FC	=	Array of dimension M. $FC(J) = F_j$. Output.
X	=	Array of dimension N which contains the computed response at time t. $X(I) = x_i$. Output.

A sample problem and the listing of the program are given below.

```
C =============================================================
C
C PROGRAM 4
C MAIN PROGRAM WHICH CALLS PERIOD
C
C =============================================================
C FOLLOWING 10 LINES CONTAIN PROBLEM-DEPENDENT DATA
      DIMENSION F(24),T(24),XSIN(20),XCOS(20),PSI(20),PHI(20),FC(20),
     2 X(24),XPC(20),XPS(20)
      DATA XM,XK,XAI /100.0,100000.0,0.1/
      DATA N,M,TIME /24,20,0.12/
      DATA F/24000.0,48000.0,72000.0,96000.0,120000.0,96000.0,72000.0,
     2 48000.0,24000.0,0.0,0.0,0.0,0.0,0.0,0.0,0.0,0.0,0.0,0.0,
     3 0.0,0.0,0.0,0.0/
      DATA T/0.005,0.010,0.015,0.020,0.025,0.030,0.035,0.040,0.045,
     2 0.050,0.055,0.060,0.065,0.070,0.075,0.080,0.085,0.090,0.095,
     3 0.100,0.105,0.110,0.115,0.120/
C END OF PROBLEM-DEPENDENT DATA
      CALL PERIOD (XM,XK,XAI,N,M,TIME,F,T,XSIN,XCOS,PSI,PHI,FZERO,FC,
     2 X,XPC,XPS)
```

```
      PRINT 100, XM,XK,XAI,N,M,TIME
100   FORMAT (/,56H RESPONSE OF A SINGLE D.O.F. SYSTEM UNDER PERIODIC FO
     2RCE,//,6H XM  =,E15.6,/,6H XK  =,E15.6,/,6H XAI =,E15.6,/,
     36H N   =,I3,/,6H M   =,I3,/,6H TIME=,E15.6,/)
      PRINT 200
200   FORMAT (/,27H APPLIED FORCE AND RESPONSE,//,3H  I,3X,5H T(I),10X,
     2 5H F(I),10X,5H X(I),/)
      DO 400 I=1,N
400   PRINT 500, I,T(I),F(I),X(I)
500   FORMAT (I3,3E15.6)
      STOP
      END
C =================================================================
C
C SUBROUTINE PERIOD
C
C =================================================================
      SUBROUTINE PERIOD (XM,XK,XAI,N,M,TIME,F,T,XSIN,XCOS,PSI,PHI,
     2 FZERO,FC,X,XPC,XPS)
      DIMENSION F(N),T(N),XSIN(M),XCOS(M),PSI(M),PHI(M),FC(M),X(N)
     2 ,XPC(M),XPS(M)
      OMEG=2.0*3.1416/TIME
      OMEGN=SQRT(XK/XM)
      SUMZ=0.0
      DO 100 I=1,N
100   SUMZ=SUMZ+F(I)
      FZERO=2.0*SUMZ/REAL(N)
      DO 300 J=1,M
      SUMS=0.0
      SUMC=0.0
      DO 200 I=1,N
      THETA=REAL(J)*OMEG*T(I)
      FSIN=F(I)*SIN(THETA)
      FCOS=F(I)*COS(THETA)
      SUMS=SUMS+FSIN
      SUMC=SUMC+FCOS
200   CONTINUE
      AJ=2.0*SUMC/REAL(N)
      BJ=2.0*SUMS/REAL(N)
      R=OMEG/OMEGN
      PHI(J)=ATAN(2.0*XAI*REAL(J)*R/(1.0-(REAL(J)*R)**2))
      CON=SQRT((1.0-(REAL(J)*R)**2)**2+(2.0*XAI*REAL(J)*R)**2)
      XPC(J)=(AJ/XK)/CON
      XPS(J)=(BJ/XK)/CON
300   CONTINUE
      DO 400 I=1,N
      X(I)=FZERO/(2.0*XK)
      DO 500 J=1,M
500   X(I)=X(I)+XPC(J)*COS(REAL(J)*OMEG*T(I)-PHI(J))
     2 +XPS(J)*SIN(REAL(J)*OMEG*T(I)-PHI(J))
400   CONTINUE
      RETURN
      END
```

RESPONSE OF A SINGLE D.O.F. SYSTEM UNDER PERIODIC FORCE

```
XM   =   0.100000E+03
XK   =   0.100000E+06
XAI  =   0.100000E+00
N    =      24
M    =      20
TIME=   0.120000E+00
```

APPLIED FORCE AND RESPONSE

I	T(I)	F(I)	X(I)
1	0.500000E-02	0.240000E+05	0.393120E+00
2	0.100000E-01	0.480000E+05	0.451156E+00
3	0.150000E-01	0.720000E+05	0.496755E+00
4	0.200000E-01	0.960000E+05	0.523365E+00
5	0.250000E-01	0.120000E+06	0.525113E+00
6	0.300000E-01	0.960000E+05	0.497451E+00
7	0.350000E-01	0.720000E+05	0.447280E+00
8	0.400000E-01	0.480000E+05	0.382350E+00
9	0.450000E-01	0.240000E+05	0.310534E+00
10	0.500000E-01	0.000000E+00	0.239646E+00
11	0.550000E-01	0.000000E+00	0.176984E+00
12	0.600000E-01	0.000000E+00	0.124139E+00
13	0.650000E-01	0.000000E+00	0.821526E-01
14	0.700000E-01	0.000000E+00	0.517498E-01
15	0.750000E-01	0.000000E+00	0.333252E-01
16	0.800000E-01	0.000000E+00	0.269447E-01
17	0.850000E-01	0.000000E+00	0.323697E-01
18	0.900000E-01	0.000000E+00	0.490896E-01
19	0.950000E-01	0.000000E+00	0.763507E-01
20	0.100000E+00	0.000000E+00	0.113176E+00
21	0.105000E+00	0.000000E+00	0.158378E+00
22	0.110000E+00	0.000000E+00	0.210580E+00
23	0.115000E+00	0.000000E+00	0.268249E+00
24	0.120000E+00	0.000000E+00	0.329747E+00

**4.9.2
Response under
Arbitrary Forcing
Function
Using the
Methods
of Section 4.8**

A Fortran computer program is given for finding the response of a viscously damped single degree of freedom system under arbitrary forcing function using the methods outlined in Section 4.8. For illustration, the data of Example 4.11 is used. The following input data is required for this program.

F	=	Array containing the values of the forcing function at various time stations according to the idealization of Fig. 4.14 (Fig. 4.18 or 4.19 for Example 4.11).
FF	=	Array containing the values of the forcing function at various time stations according to the idealization of Fig. 4.16 (Fig. 4.20 or 4.21 for Example 4.11).
XAI	=	Damping factor.
OMN	=	Undamped natural frequency of the system.
DELT	=	Incremental time between consecutive time stations.
XK	=	Spring stiffness.

The program prints the values of $x(t_i)$ and $\dot{x}(t_i)$ given by four different methods at time stations $t_2, t_3, \ldots, t_{11}$. Although the program uses the data of Example 4.11 directly, it can be generalized to find the response under any arbitrary forcing function of any single degree of freedom system.

```
C =====================================================================
C
C PROGRAM 5
C RESPONSE OF A SINGLE D.O.F. SYSTEM UNDER ARBITRARY FORCING FUNCTION
C USING THE METHODS OF SECTION 4.8
C
C =====================================================================
C FOLLOWING 10 LINES CONTAIN PROBLEM-DEPENDENT DATA
      DIMENSION F(11),FF(11),DELF(11),T(11),X(11),XD(11),X1(11),
     2    XD1(11),X2(11),XD2(11),X3(11),XD3(11),X4(11),XD4(11)
      DATA F/0.0,1.0,0.84356554,0.69098301,0.54600950,0.41221475,
     2    0.29289322,0.19098301,0.10899348,0.04894348,0.01231166/
      DATA FF/1.0,0.92154090,0.76655464,0.61731657,0.47750144,
     2    0.35055195,0.23959404,0.14735984,0.07612047,0.02763008,
     3    0.00308267/
      DATA XAI,OMN,XK /0.1,1.0,1.0/
      DELT=3.14159265/10.0
      DATA NP,NP1,NP2 /11,10,9/
C NP = NUMBER OF POINTS AT WHICH VALUE OF F IS KNOWN, NP1=NP-1, NP2=NP-2
C END OF PROBLEM-DEPENDENT DATA
      XN=XAI*OMN
      PD=OMN*SQRT(1.0-XAI**2)
C SOLUTION ACCORDING TO METHOD 1 USING THE IDEALIZATION OF FIG. 4.18
      T(1)=0.0
      DO 10 I=2,NP
 10   T(I)=T(I-1)+DELT
      DO 20 I=1,NP1
 20   DELF(I)=F(I+1)-F(I)
      DO 40 J=2,NP
      X(J)=0.0
      XD(J)=0.0
      JM1=J-1
      DO 30 I=1,JM1
      X(J)=X(J)+(DELF(I)/XK)*(1.0-EXP(-XN*(T(J)-T(I)))*(COS(PD*(T(J)-
     2    T(I)))+(XN/PD)*SIN(PD*(T(J)-T(I)))))
C XD(J) OBTAINED BY DIFFERENTIATING EQ.(4.64)
      XD(J)=XD(J)+(DELF(I)/XK)*EXP(-XN*(T(J)-T(I)))*SIN(PD*(T(J)-
     2    T(I)))
 30   CONTINUE
 40   CONTINUE
      DO 50 I=2,NP
      X1(I)=X(I)
 50   XD1(I)=XD(I)
C SOLUTION ACCORDING TO METHOD 1 USING THE IDEALIZATION OF FIG. 4.19
      DO 60 K=2,NP2
 60   DELF(K)=DELF(K+1)
      DELF(1)=F(3)
```

```
      DELF(NP)=F(NP)
      DO 80 J=2,NP
      X(J)=0.0
      XD(J)=0.0
      JM1=J-1
      DO 70 I=1,JM1
      X(J)=X(J)+(DELF(I)/XK)*(1.0-EXP(-XN*(T(J)-T(I)))*(COS(PD*(T(J)-
     2 T(I)))+(XN/PD)*SIN(PD*(T(J)-T(I)))))
      XD(J)=XD(J)+(DELF(I)/XK)*EXP(-XN*(T(J)-T(I)))*SIN(PD*(T(J)-
     2 T(I)))
   70 CONTINUE
   80 CONTINUE
      DO 90 I=2,NP
      X2(I)=X(I)
   90 XD2(I)=XD(I)
C SOLUTION ACCORDING TO METHOD 2 USING THE IDEALIZATION OF FIG. 4.20
      X(1)=0.0
      XD(1)=0.0
      DO 100 J=2,NP
      DEL=DELT
      X(J)=(FF(J)/XK)*(1.0-EXP(-XN*DEL)*(COS(PD*DEL)+(XN/PD)*
     2 SIN(PD*DEL)))+EXP(-XN*DEL)*(X(J-1)*COS(PD*DEL)+((XD(J-1)
     3 +XN*X(J-1))/PD)*SIN(PD*DEL))
      XD(J)=(FF(J)*PD/XK)*EXP(-XN*DEL)*(1.0+XN**2/(PD**2))*SIN(PD*DEL)
     2 +PD*EXP(-XN*DEL)*(-X(J-1)*SIN(PD*DEL)+((XD(J-1)+XN*X(J-1))/PD)*
     3 COS(PD*DEL)-XN*(X(J-1)*COS(PD*DEL)+((XD(J-1)+XN*X(J-1))/PD)*
     4 SIN(PD*DEL))/PD)
  100 CONTINUE
      DO 110 I=2,NP
      X3(I)=X(I)
  110 XD3(I)=XD(I)
C SOLUTION ACCORDING TO METHOD 3 USING THE IDEALIZATION OF FIG. 4.21
      X(1)=0.0
      XD(1)=0.0
      DO 120 J=1,NP1
  120 F(J)=F(J+1)
      F(NP)=0.0
      DO 130 J=2,NP
      DELF(J)=F(J)-F(J-1)
      X(J)=(DELF(J)/(XK*DEL))*(DEL-(2.0*XAI/OMN)+EXP(-XN*DEL)*
     2 ((2.0*XAI/OMN)*COS(PD*DEL)-((PD**2-XN**2)/(OMN*OMN*PD))*
     3 SIN(PD*DEL)))+(F(J-1)/XK)*(1.0-EXP(-XN*DEL)*(COS(PD*DEL)
     4 +(XN/PD)*SIN(PD*DEL)))+EXP(-XN*DEL)*(X(J-1)*COS(PD*DEL)
     5 +((XD(J-1)+XN*X(J-1))/PD)*SIN(PD*DEL))
      XD(J)=(DELF(J)/(XK*DEL))*(1.0-EXP(-XN*DEL)*(((XN**2+PD**2)/
     2 (OMN**2))*COS(PD*DEL)+((XN**3+XN*PD*PD)/(PD*(OMN**2)))*
     3 SIN(PD*DEL)))+(F(J-1)/XK)*EXP(-XN*DEL)*((XN**2/PD)+PD)*
     4 SIN(PD*DEL)+EXP(-XN*DEL)*(XD(J-1)*COS(PD*DEL)-((XN*XD(J-1)
     5 +XN*XN*X(J-1)+PD*PD*X(J-1))/PD)*SIN(PD*DEL))
  130 CONTINUE
      DO 140 I=2,NP
      X4(I)=X(I)
  140 XD4(I)=XD(I)
      PRINT 150
```

```
150  FORMAT (//,6H VALUE,6X,10H METHOD #1,7X,10H METHOD #1,7X,
     2    10H METHOD #2,7X,10H METHOD #3)
     PRINT 160
160  FORMAT (3X,3H OF,5X,11H (FIG.4.18),6X,11H (FIG.4.19),6X,
     2    11H (FIG.4.20),6X,11H (FIG.4.21),/)
     PRINT 170
170  FORMAT (3X,2H I,7X,5H X(I),12X,5H X(I),12X,5H X(I),12X,5H X(I),
     2    /)
     DO 180 I=2,NP
180  PRINT 190,I,X1(I),X2(I),X3(I),X4(I)
190  FORMAT (I5,2X,E15.6,2X,E15.6,2X,E15.6,2X,E15.6)
     PRINT 200
200  FORMAT (//,3X,2H I,6X,6H XD(I),11X,6H XD(I),11X,6H XD(I),11X,
     2    6H XD(I),/)
     DO 210 I=2,NP
210  PRINT 190,I,XD1(I),XD2(I),XD3(I),XD4(I)
     STOP
     END
```

VALUE OF I	METHOD #1 (FIG.4.18) X(I)	METHOD #1 (FIG.4.19) X(I)	METHOD #2 (FIG.4.20) X(I)	METHOD #3 (FIG.4.21) X(I)
2	0.479360E-01	0.404372E-01	0.441750E-01	0.454151E-01
3	0.175781E+00	0.147294E+00	0.161471E+00	0.163773E+00
4	0.351883E+00	0.292277E+00	0.321877E+00	0.324989E+00
5	0.542483E+00	0.446091E+00	0.493842E+00	0.497464E+00
6	0.715396E+00	0.581603E+00	0.647699E+00	0.651514E+00
7	0.843296E+00	0.676586E+00	0.758676E+00	0.762379E+00
8	0.906301E+00	0.715783E+00	0.809225E+00	0.812552E+00
9	0.893674E+00	0.692145E+00	0.790486E+00	0.793231E+00
10	0.804490E+00	0.607167E+00	0.702788E+00	0.704820E+00
11	0.647299E+00	0.469170E+00	0.555198E+00	0.556465E+00

I	XD(I)	XD(I)	XD(I)	XD(I)
2	0.298008E+00	0.251389E+00	0.276010E+00	0.275640E+00
3	0.502976E+00	0.418148E+00	0.462605E+00	0.461687E+00
4	0.602270E+00	0.491876E+00	0.549249E+00	0.547683E+00
5	0.595174E+00	0.474576E+00	0.536630E+00	0.534405E+00
6	0.492171E+00	0.377744E+00	0.435845E+00	0.433036E+00
7	0.313187E+00	0.220601E+00	0.266613E+00	0.263378E+00
8	0.850188E-01	0.276649E-01	0.547668E-01	0.513272E-01
9	-0.161754E+00	-0.174042E+00	-0.170711E+00	-0.174093E+00
10	-0.396047E+00	-0.357870E+00	-0.380784E+00	-0.383830E+00
11	-0.589414E+00	-0.507466E+00	-0.549289E+00	-0.551730E+00

REFERENCES

4.1. M. J. Maron, *Applied Numerical Analysis*, Macmillan, New York, 1982.

4.2. M. Paz, *Structural Dynamics: Theory and Computation* (2nd Ed.), Van Nostrand Reinhold, New York, 1985.

4.3. E. Kreyszig, *Advanced Engineering Mathematics* (5th Ed.), John Wiley, New York, 1983.

4.4. F. Oberhettinger and L. Badii, *Tables of Laplace Transforms*, Springer Verlag, New York, 1973.

4.5. G. M. Hieber et al., "Understanding and measuring the shock response spectrum, Part I," *Sound and Vibration*, Vol. 8, March 1974, pp. 42–49.

4.6. R. E. D. Bishop, A. G. Parkinson, and J. W. Pendered, "Linear analysis of transient vibration," *Journal of Sound and Vibration*, Vol. 9, 1969, pp. 313–337.

4.7. Y. Matsuzaki and S. Kibe, "Shock and seismic response spectra in design problems," *Shock and Vibration Digest*, Vol. 15, October 1983, pp. 3–10.

4.8. S. Timoshenko, D. H. Young, and W. Weaver, Jr., *Vibration Problems in Engineering* (4th Ed.), Wiley, New York, 1974.

4.9. R. A. Spinelli, "Numerical inversion of a Laplace transform," *SIAM Journal of Numerical Analysis*, Vol. 3, 1966, pp. 636–649.

4.10. R. Bellman, R. E. Kalaba, and J. A. Lockett, *Numerical Inversion of the Laplace Transform*, American Elsevier, New York, 1966.

REVIEW QUESTIONS

4.1. What is the basis for expressing the response of a system under periodic excitation as a summation of several harmonic responses?

4.2. Indicate some methods for finding the response of a system under nonperiodic forces.

4.3. What is Duhamel integral? What is its use?

4.4. How are the initial conditions determined for a single degree of freedom system subjected to an impulse at $t = 0$?

4.5. Derive the equation of motion of a system subjected to base excitation.

4.6. What is a response spectrum?

4.7. What are the advantages of the Laplace transformation method?

4.8. What is the use of the pseudo spectrum?

4.9. How is the Laplace transform of a function $x(t)$ defined?

4.10. Define these terms: generalized impedance and admittance of a system.

4.11. State the interpolation models that can be used for approximating an arbitrary forcing function.

4.12. How many resonant conditions are there when the external force is not harmonic?

4.13. How do you compute the frequency of the first harmonic of a periodic force?

4.14. What is the relation between the frequencies of higher harmonics and the frequency of the first harmonic for a periodic excitation?

PROBLEMS

The problem assignments are organized as follows:

Problems	Section Covered	Topic Covered
4.1–4.5	4.2	Response under general periodic force
4.6–4.8	4.3	Periodic force of irregular form
4.9–4.21	4.5	Convolution integral
4.22–4.26	4.6	Response spectrum
4.27–4.29	4.7	Laplace transformation
4.30–4.33	4.8	Irregular forcing conditions using numerical methods
4.34–4.39	4.9	Computer program
4.40–4.41	—	Projects

4.1–4.4. Find the steady-state response of the hydraulic control valve shown in Fig. 4.1(a) to the forcing functions obtained by replacing $x(t)$ with $F(t)$ and A with F_0 in Figs. 1.56–1.59.

4.5. Find the steady-state response of a viscously damped system to the forcing function obtained by replacing $x(t)$ and A with $F(t)$ and F_0, respectively, in Fig. 1.32(a).

4.6. Find the response of a damped system with $m = 1$ kg, $k = 15$ kN/m, and $\zeta = 0.1$ under the action of a periodic forcing function, as shown in Fig. 1.62.

4.7. Find the response of a viscously damped system under the periodic force whose values are given in Problem 1.47. Assume that x_i denotes the value of the force in Newtons at time t_i seconds. Use $m = 0.5$ kg, $k = 8000$ N/m, and $\zeta = 0.06$.

4.8. Find the displacement of the water tank shown in Fig. 4.22(a) under the periodic force shown in Fig. 4.22(b) by treating it as an undamped single degree of freedom system. Use the numerical procedure described in Section 4.3.

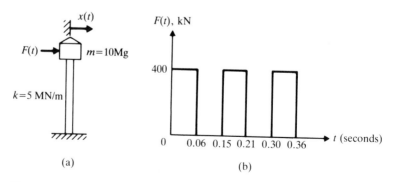

(a)   (b)

Figure 4.22

4.17. An automobile, having a mass of 1000 kg, runs over a road bump of the shape shown in Fig. 4.27. The speed of the automobile is 50 km/hr. If the undamped natural period of vibration in the vertical direction is 1.0 second, find the response of the car by assuming it as a single degree of freedom undamped system vibrating in the vertical direction.

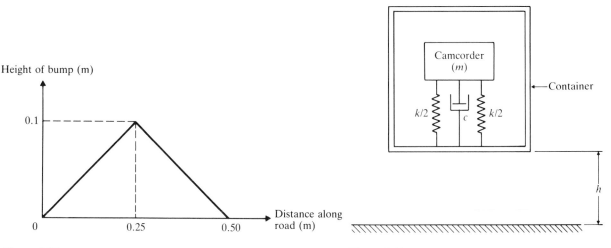

Height of bump (m)

0.1

0   0.25   0.50

Distance along road (m)

Figure 4.27

Camcorder (*m*)

Container

$k/2$ c $k/2$

h

Figure 4.28

4.18. A camcorder of mass m is packed in a container using a flexible packing material. The stiffness and damping constant of the packing material are given by k and c, respectively, and the mass of the container is negligible. If the container is dropped accidentally from a height of h onto a rigid floor (see Fig. 4.28), find the motion of the camcorder.

4.19. An airplane, taxiing on a runway, encounters a bump. As a result, the root of the wing is subjected to a displacement that can be expressed as

$$y(t) = \begin{cases} Y(t^2/t_0^2), & 0 \leq t \leq t_0 \\ 0, & t > t_0 \end{cases}$$

Find the response of the mass located at the tip of the wing if the stiffness of the wing is k (see Fig. 4.29).

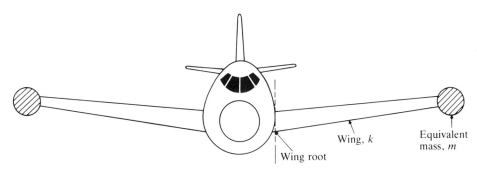

Wing, k Equivalent mass, m

Wing root

Figure 4.29

4.20. Derive Eq. (E.1) of Example 4.6.

4.21. In a static firing test of a rocket, the rocket is anchored to a rigid wall by a spring-damper system, as shown in Fig. 4.30(a). The thrust acting on the rocket reaches its maximum value F in a negligibly short time and remains constant until the burnout time t_0, as indicated in Fig. 4.30(b). The thrust acting on the rocket is given by $F = m_0 v$ where m_0 is the constant rate at which fuel is burnt and v is the velocity of the jet stream. The initial mass of the rocket is M, so that its mass at any time t is given by $m = M - m_0 t, 0 \le t \le t_0$. If the data are $k = 7.5 \times 10^6$ N/m, $c = 0.1 \times 10^6$ N-s/m, $m_0 = 10$ kg/s, $v = 2000$ m/s, $M = 2000$ kg, and $t_0 = 100$ s, (1) derive the equation of motion of the rocket, and (2) find the maximum steady-state displacement of the rocket by assuming an average (constant) mass of ($M - \frac{1}{2} m_0 t_0$).

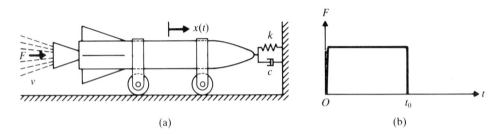

(a)   (b)

Figure 4.30

4.22. Derive the response spectrum of an undamped system for the rectangular pulse shown in Fig. 4.25(a). Plot $(x/\delta_{st})_{max}$ with respect to (t_0/τ_n).

4.23. Find the displacement response spectrum of an undamped system for the pulse shown in Fig. 4.25(c).

4.24. The base of an undamped spring-mass system is subjected to an acceleration excitation given by $a_0[1 - \sin(\pi t/2t_0)]$. Find the relative displacement of the mass z.

4.25. Find the response spectrum of the system considered in Example 4.7. Plot $\left(\dfrac{kx}{F_0}\right)_{max}$ versus $\omega_n t_0$ in the range $0 \le \omega_n t_0 \le 15$.

4.26.* A building frame is subjected to a blast load and the idealization of the frame and the load are shown in Fig. 4.10. If $m = 5000$ kg, $F_0 = 4$ MN, and $t_0 = 0.4$ s, find the minimum stiffness required if the displacement is to be limited to 10 mm.

4.27. Find the steady state response of an undamped single degree of freedom system subjected to the force $F(t) = F_0 e^{i\omega t}$ by using the method of Laplace transformation.

4.28. Find the response of a damped spring-mass system subjected to a step function of magnitude F_0 by using the method of Laplace transformation.

4.29. Find the response of an undamped system subjected to a square pulse $F(t) = F_0$ for $0 \le t \le t_0$ and 0 for $t > t_0$ by using the Laplace transformation method. Assume the initial conditions as zero.

4.30. Determine the expression for the velocity $\dot{x}_j$ for the damped response represented by Eq. (4.64).

4.31. Derive Eqs. (4.68) and (4.71).

4.32. Compare the values of $\dot{x}_j$ given by Eqs. (4.68) and (4.71) in the case of Example 4.11.

4.33. Derive the expressions for x_j and $\dot{x}_j$ according to the three interpolation functions considered in Section 4.8 for the undamped case. Using these expressions, find the solution of Example 4.11 by assuming the damping to be zero.

4.34. A damped single degree of freedom system has a mass $m = 2$, a spring of stiffness $k = 50$, and a damper with $c = 2$. A forcing function $F(t)$, whose magnitude is indicated in the following table, acts on the mass for one second. Find the response of the system by using the piecewise linear interpolation method described in Section 4.8.

Time (t_i)	$F(t_i)$
0.0	−8.0
0.1	−12.0
0.2	−15.0
0.3	−13.0
0.4	−11.0
0.5	−7.0
0.6	−4.0
0.7	3.0
0.8	10.0
0.9	15.0
1.0	18.0

Needs to be periodic

4.35. The equation of motion of an undamped system is given by $2\ddot{x} + 1500x = F(t)$ where the forcing function is defined by the curve shown in Fig. 4.31. Find the response of the system numerically for $0 \leqslant t \leqslant 0.5$. Assume the initial conditions as $x_0 = \dot{x}_0 = 0$ and the step size as $\Delta t = 0.01$.

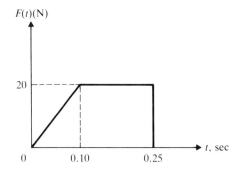

Figure 4.31

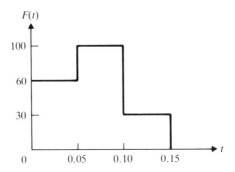

Figure 4.32

4.36. Solve Problem 4.35 if the system is viscously damped so that the equation of motion is $2\ddot{x} + 10\dot{x} + 1500x = F(t)$.

4.37. Write a computer program for finding the steady-state response of a single degree of freedom system subjected to an arbitrary force, by numerically evaluating the Duhamel integral. Using this program, solve Example 4.11.

4.38. Find the relative displacement of the water tank shown in Fig. 4.22(a) when its base is subjected to the earthquake acceleration record shown in Fig. 1.63, by assuming the ordinate to represent acceleration in g's. Use the program of Problem 4.37.

4.39. The differential equation of motion of an undamped system is given by $2\ddot{x} + 150x = F(t)$ with the initial conditions $x_0 = \dot{x}_0 = 0$. If $F(t)$ is as shown in Fig. 4.32, find the response of the problem using the computer program of Problem 4.37.

Projects:

4.40. Design a seismometer of the type shown in Fig. 4.33(a) (by specifying the values of a, m and k) to measure earthquakes. The seismometer should have a natural frequency of 10 Hz and the maximum relative displacement of the mass should be at least 2 cm when its base is subjected to the displacement shown in Fig. 4.33(b).

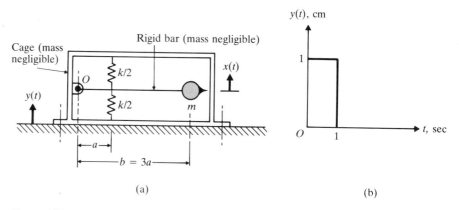

(a)

(b)

Figure 4.33

4.41. The cutting forces developed during two different machining operations are shown in Figs. 4.34(a) and (b). The inaccuracies (in the vertical direction) in the surface finish in the two cases were observed to be 0.1 mm and 0.05 mm, respectively. Find the equivalent mass and stiffness of the cutting head (Fig. 4.35) assuming it to be an undamped single degree of freedom system.

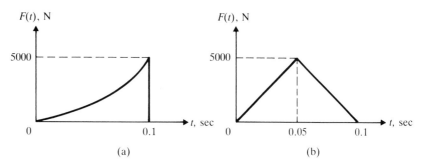

(a)   (b)

Figure 4.34

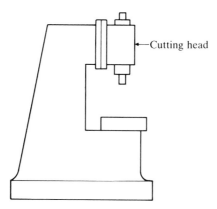

Figure 4.35

CHAPTER 5

Two Degree of Freedom Systems

Daniel Bernoulli (1700–1782) was a Swiss who became a professor of mathematics at St. Petersburg in 1725 after receiving his doctorate in medicine for his thesis on the action of lungs. He later became professor of anatomy and botany at Basel. He developed the theory of hydrostatics and hydrodynamics and ''Bernoulli's theorem'' is well known to engineers. He derived the equation of motion for the vibration of beams (the Euler-Bernoulli theory) and studied the problem of vibrating strings. Bernoulli was the first person to propose the principle of superposition of harmonics in free vibration. (Courtesy Culver Pictures)

5.1 INTRODUCTION

Systems that require two independent coordinates to describe their motion are called *two degree of freedom systems*. Some examples of systems having two degrees of freedom were shown in Fig. 1.7. We shall consider only two degree of freedom systems in this chapter, so as to provide a simple introduction to the behavior of systems with an arbitrarily large number of degrees of freedom, which is the subject of Chapter 6.

Consider the system shown in Fig. 5.1, in which a mass m is supported on two equal springs. Assuming that the mass is constrained to move in a vertical plane, we find that the position of the mass m at any time can be specified by a linear coordinate $x(t)$, indicating the vertical displacement of the center of gravity (C.G.) of the mass, and an angular coordinate $\theta(t)$, denoting the rotation of the mass m about its C.G. Instead of $x(t)$ and $\theta(t)$, we can also use $x_1(t)$ and $x_2(t)$ as independent coordinates to specify the motion of the system. Thus the system has two degrees of freedom. It is important to note that in this case the mass m is not treated as a point mass, but as a rigid body having two possible types of motion. (If it is a particle, there is no need to specify the rotation of the mass about its C.G.) The system shown in Fig. 5.2 does have one point mass m but is a two degree of

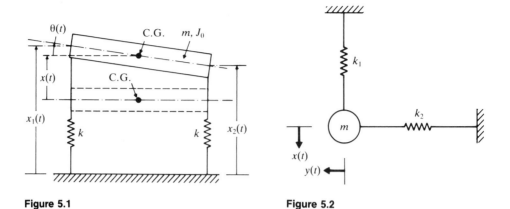

Figure 5.1   **Figure 5.2**

freedom system, because the mass has two possible types of motion (translations along the x and y directions). The general rule for the computation of the number of degrees of freedom can be stated as follows:

Number of degrees of freedom of the system $=$ Number of masses in the system $\times$ number of possible types of motion of each mass

There are two equations of motion for a two degree of freedom system, one for each mass (more precisely, for each degree of freedom). They are generally in the form of *coupled differential equations*—that is, each equation involves all the coordinates. If a harmonic solution is assumed for each coordinate, the equations of motion lead to a frequency equation that gives two natural frequencies for the system. If we give suitable initial excitation, the system vibrates at one of these natural frequencies. During free vibration at one of the natural frequencies, the amplitudes of the two degrees of freedom (coordinates) are related in a specific manner and the configuration is called a *normal mode*, *principal mode*, or *natural mode* of vibration. Thus a two degree of freedom system has two normal modes of vibration corresponding to the two natural frequencies.

If we give an arbitrary initial excitation to the system, the resulting free vibration will be a superposition of the two normal modes of vibration. However, if the system vibrates under the action of an external harmonic force, the resulting forced harmonic vibration takes place at the frequency of the applied force. Under harmonic excitation, resonance occurs (i.e., the amplitudes of the two coordinates will be maximum) when the forcing frequency is equal to one of the natural frequencies of the system.

As is evident from the systems shown in Figs. 5.1 and 5.2, the configuration of a system can be specified by a set of independent coordinates such as length, angle, or some other physical parameters. Any such set of coordinates is called *generalized coordinates*. Although the equations of motion of a two degree of freedom system are generally coupled so that each equation involves all the coordinates, it is always possible to find a particular set of coordinates such that each equation of motion contains only one coordinate. The equations of motion are then *uncoupled* and can be solved independently of each other. Such a set of coordinates, which lead to an uncoupled system of equations, is called *principal coordinates*.

5.2 EQUATIONS OF MOTION FOR FORCED VIBRATION

Consider a viscously damped two degree of freedom spring-mass system, shown in Fig. 5.3(a). The motion of the system is completely described by the coordinates $x_1(t)$ and $x_2(t)$, which define the positions of the masses m_1 and m_2 at any time t from the respective equilibrium positions. The external forces $F_1(t)$ and $F_2(t)$ act on the masses m_1 and m_2 respectively. The free-body diagrams of the masses m_1 and m_2 are shown in Fig. 5.3(b). The application of Newton's second law of motion to each of the masses gives the equations of motion:

$$m_1 \ddot{x}_1 + (c_1 + c_2)\dot{x}_1 - c_2\dot{x}_2 + (k_1 + k_2)x_1 - k_2 x_2 = F_1 \tag{5.1}$$
$$m_2 \ddot{x}_2 - c_2\dot{x}_1 + (c_2 + c_3)\dot{x}_2 - k_2 x_1 + (k_2 + k_3)x_2 = F_2 \tag{5.2}$$

It can be seen that Eq. (5.1) contains terms involving x_2 (namely, $-c_2\dot{x}_2$ and $-k_2 x_2$), whereas Eq. (5.2) contains terms involving x_1 (namely, $-c_2\dot{x}_1$ and $-k_2 x_1$). Hence they represent a system of two coupled differential equations. We can therefore expect that the motion of the mass m_1 will influence the motion of the mass m_2, and vice versa. Equations (5.1) and (5.2) can be written in matrix form as

$$[m]\ddot{\vec{x}}(t) + [c]\dot{\vec{x}}(t) + [k]\vec{x}(t) = \vec{F}(t) \tag{5.3}$$

where $[m]$, $[c]$, and $[k]$ are called the *mass*, *damping*, and *stiffness matrices*,

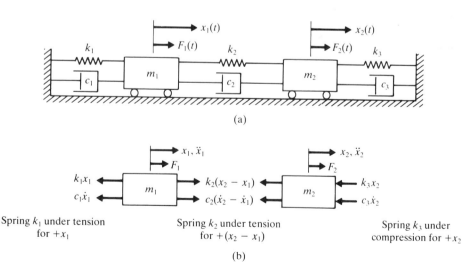

(a)

(b)

Figure 5.3 A two degree of freedom spring-mass-damper system.

respectively, and are given by

$$[m] = \begin{bmatrix} m_1 & 0 \\ 0 & m_2 \end{bmatrix}$$

$$[c] = \begin{bmatrix} c_1 + c_2 & -c_2 \\ -c_2 & c_2 + c_3 \end{bmatrix}$$

$$[k] = \begin{bmatrix} k_1 + k_2 & -k_2 \\ -k_2 & k_2 + k_3 \end{bmatrix}$$

and $\vec{x}(t)$ and $\vec{F}(t)$ are called the *displacement* and *force vectors*, respectively, and are given by

$$\vec{x}(t) = \left\{ \begin{array}{c} x_1(t) \\ x_2(t) \end{array} \right\}$$

and

$$\vec{F}(t) = \left\{ \begin{array}{c} F_1(t) \\ F_2(t) \end{array} \right\}$$

It can be seen that the matrices $[m]$, $[c]$, and $[k]$ are all 2×2 matrices whose elements are the known masses, damping coefficients, and stiffnesses of the system, respectively. Further, these matrices can be seen to be symmetric, so that

$$[m]^T = [m], \qquad [c]^T = [c], \qquad [k]^T = [k]$$

where the superscript T denotes the transpose of the matrix.

Notice that the equations of motion (5.1) and (5.2) become uncoupled (independent of one another) only when $c_2 = k_2 = 0$, which implies that the two masses m_1 and m_2 are not physically connected. In such a case, the matrices $[m]$, $[c]$, and $[k]$ become diagonal. The solution of the equations of motion (5.1) and (5.2) for any arbitrary forces $F_1(t)$ and $F_2(t)$ is difficult to obtain, mainly due to the coupling of the variables $x_1(t)$ and $x_2(t)$. We shall first consider the free vibration solution of Eqs. (5.1) and (5.2).

5.3 FREE VIBRATION ANALYSIS OF AN UNDAMPED SYSTEM

For the free vibration analysis of the system shown in Fig. 5.3(a), we set $F_1(t) = F_2(t) = 0$. Further, if damping is disregarded, $c_1 = c_2 = c_3 = 0$, and the equations of motion (5.1) and (5.2) reduce to

$$m_1 \ddot{x}_1(t) + (k_1 + k_2) x_1(t) - k_2 x_2(t) = 0 \qquad (5.4)$$

$$m_2 \ddot{x}_2(t) - k_2 x_1(t) + (k_2 + k_3) x_2(t) = 0 \qquad (5.5)$$

We are interested in knowing whether m_1 and m_2 can oscillate harmonically with the same frequency and phase angle but with different amplitudes. Assuming that it is possible to have harmonic motion of m_1 and m_2 at the same frequency ω and the same phase angle ϕ, we take the solutions of Eqs. (5.4) and (5.5) as

$$x_1(t) = X_1 \cos(\omega t + \phi)$$
$$x_2(t) = X_2 \cos(\omega t + \phi) \tag{5.6}$$

where X_1 and X_2 are constants that denote the maximum amplitudes of $x_1(t)$ and $x_2(t)$, and ϕ is the phase angle. Substituting Eq. (5.6) into Eqs. (5.4) and (5.5), we obtain

$$\left[\{-m_1\omega^2 + (k_1 + k_2)\}X_1 - k_2 X_2\right]\cos(\omega t + \phi) = 0$$
$$\left[-k_2 X_1 + \{-m_2\omega^2 + (k_2 + k_3)\}X_2\right]\cos(\omega t + \phi) = 0 \tag{5.7}$$

Since Eqs. (5.7) must be satisfied for all values of the time t, the terms between brackets must be zero. This yields

MATRIX FORM IN CLASS

$$\{-m_1\omega^2 + (k_1 + k_2)\}X_1 - k_2 X_2 = 0$$
$$- k_2 X_1 + \{-m_2\omega^2 + (k_2 + k_3)\}X_2 = 0 \tag{5.8}$$

which represent two simultaneous homogeneous algebraic equations in the unknowns X_1 and X_2. It can be seen that Eqs. (5.8) are satisfied by the trivial solution $X_1 = X_2 = 0$, which implies that there is no vibration. For a nontrivial solution of X_1 and X_2, the determinant of the coefficients of X_1 and X_2 must be zero:

$$\det\begin{bmatrix} \{-m_1\omega^2 + (k_1 + k_2)\} & -k_2 \\ -k_2 & \{-m_2\omega^2 + (k_2 + k_3)\} \end{bmatrix} = 0$$

CHARACTERISTIC or EQUATION

$$(m_1 m_2)\omega^4 - \{(k_1 + k_2)m_2 + (k_2 + k_3)m_1\}\omega^2$$
$$+ \{(k_1 + k_2)(k_2 + k_3) - k_2^2\} = 0 \tag{5.9}$$

Equation (5.9) is called the *frequency* or *characteristic equation* because solution of this equation yields the frequencies or the characteristic values of the system. The roots of Eq. (5.9) are given by

EIGEN VALUES $\rightarrow$ ω_{n1} ω_{n2}

NOTATION IN CLASS

$$\omega_1^2, \omega_2^2 = \frac{1}{2}\left\{\frac{(k_1 + k_2)m_2 + (k_2 + k_3)m_1}{m_1 m_2}\right\}$$
$$\mp \frac{1}{2}\left[\left\{\frac{(k_1 + k_2)m_2 + (k_2 + k_3)m_1}{m_1 m_2}\right\}^2\right.$$
$$\left. - 4\left\{\frac{(k_1 + k_2)(k_2 + k_3) - k_2^2}{m_1 m_2}\right\}\right]^{1/2} \tag{5.10}$$

This shows that it is possible for the system to have a nontrivial harmonic solution of the form of Eqs. (5.6) when ω is equal to ω_1 or ω_2 given by Eq. (5.10). We call ω_1 and ω_2 the *natural frequencies* of the system.

The values of X_1 and X_2 remain to be determined. These values depend on the natural frequencies ω_1 and ω_2. We shall denote the values of X_1 and X_2 corresponding to ω_1 as $X_1^{(1)}$ and $X_2^{(1)}$ and those corresponding to ω_2 as $X_1^{(2)}$ and $X_2^{(2)}$. Further, since the Eqs. (5.8) are homogeneous, only the ratios $r_1 = \{X_2^{(1)}/X_1^{(1)}\}$ and $r_2 = \{X_2^{(2)}/X_1^{(2)}\}$ can be found. For $\omega^2 = \omega_1^2$ and $\omega^2 = \omega_2^2$, Eqs. (5.8) give

$$r_1 = \frac{X_2^{(1)}}{X_1^{(1)}} = \frac{-m_1\omega_1^2 + (k_1 + k_2)}{k_2} = \frac{k_2}{-m_2\omega_1^2 + (k_2 + k_3)} \qquad (5.11)$$

AMPLITUDE

RATIOS

$$r_2 = \frac{X_2^{(2)}}{X_1^{(2)}} = \frac{-m_1\omega_2^2 + (k_1 + k_2)}{k_2} = \frac{k_2}{-m_2\omega_2^2 + (k_2 + k_3)} \qquad (5.12)$$

Notice that the two ratios given for each r_i ($i = 1, 2$) in Eqs. (5.11) and (5.12) are identical. The normal modes of vibration corresponding to ω_1^2 and ω_2^2 can be expressed, respectively, as

OR CALLED

→

NORMAL MODES

MODAL

$$\vec{X}^{(1)} = \left\{ \begin{array}{c} X_1^{(1)} \\ X_2^{(1)} \end{array} \right\} = \left\{ \begin{array}{c} X_1^{(1)} \\ r_1 X_1^{(1)} \end{array} \right\}$$

and   *VECTORS*

$$\vec{X}^{(2)} = \left\{ \begin{array}{c} X_1^{(2)} \\ X_2^{(2)} \end{array} \right\} = \left\{ \begin{array}{c} X_1^{(2)} \\ r_2 X_1^{(2)} \end{array} \right\}$$

The vectors $\vec{X}^{(1)}$ and $\vec{X}^{(2)}$, which denote the normal modes of vibration, are known as the *modal vectors* of the system.

The free vibration solution or the motion in time can be expressed as

$$\vec{x}^{(1)}(t) = \left\{ \begin{array}{c} x_1^{(1)}(t) \\ x_2^{(1)}(t) \end{array} \right\} = \left\{ \begin{array}{c} X_1^{(1)} \cos(\omega_1 t + \phi_1) \\ r_1 X_1^{(1)} \cos(\omega_1 t + \phi_1) \end{array} \right\} = \text{first mode} \qquad (5.13)$$

$$\vec{x}^{(2)}(t) = \left\{ \begin{array}{c} x_1^{(2)}(t) \\ x_2^{(2)}(t) \end{array} \right\} = \left\{ \begin{array}{c} X_1^{(2)} \cos(\omega_2 t + \phi_2) \\ r_2 X_1^{(2)} \cos(\omega_2 t + \phi_2) \end{array} \right\} = \text{second mode} \qquad (5.14)$$

where the constants $X_1^{(1)}$, $X_1^{(2)}$, ϕ_1, and ϕ_2 are determined by the initial conditions.

Initial Conditions. Since each of the two equations of motion, Eqs. (5.1) and (5.2), involves second-order time derivatives, we need to specify two initial conditions for each mass. As stated in Section 5.1, the system can be made to vibrate in its ith normal mode ($i = 1, 2$) by subjecting it to the specific initial conditions

leads to

$x_i^{(1)}$ $x_i^{(2)}$

ϕ_1 , ϕ_2

$$x_1(t = 0) = X_1^{(i)} = \text{some constant}, \qquad \dot{x}_1(t = 0) = 0,$$
$$x_2(t = 0) = r_1 X_1^{(i)}, \qquad \dot{x}_2(t = 0) = 0$$

However, for any other general initial conditions, both modes will be excited. The resulting motion, which is given by the general solution of Eqs. (5.4) and (5.5), can be obtained by superposing the two normal modes, Eqs. (5.13) and (5.14):

$$\vec{x}(t) = \vec{x}^{(1)}(t) + \vec{x}^{(2)}(t)$$

that is,

$$x_1(t) = x_1^{(1)}(t) + x_1^{(2)}(t) = X_1^{(1)} \cos(\omega_1 t + \phi_1) + X_1^{(2)} \cos(\omega_2 t + \phi_2)$$

$$x_2(t) = x_2^{(1)}(t) + x_2^{(2)}(t) = r_1 X_1^{(1)} \cos(\omega_1 t + \phi_1) + r_2 X_1^{(2)} \cos(\omega_2 t + \phi_2) \quad (5.15)$$

Thus if the initial conditions are given by

$$x_1(t = 0) = x_1(0), \quad \dot{x}_1(t = 0) = \dot{x}_1(0),$$

$$x_2(t = 0) = x_2(0), \quad \dot{x}_2(t = 0) = \dot{x}_2(0) \quad (5.16)$$

the constants $X_1^{(1)}$, $X_1^{(2)}$, ϕ_1, and ϕ_2 can be found by solving the following equations (obtained by substituting Eqs. (5.16) into Eqs. (5.15)):

$$x_1(0) = X_1^{(1)} \cos \phi_1 + X_1^{(2)} \cos \phi_2$$

$$\dot{x}_1(0) = -\omega_1 X_1^{(1)} \sin \phi_1 - \omega_2 X_1^{(2)} \sin \phi_2$$

$$x_2(0) = r_1 X_1^{(1)} \cos \phi_1 + r_2 X_1^{(2)} \cos \phi_2$$

$$\dot{x}_2(0) = -\omega_1 r_1 X_1^{(1)} \sin \phi_1 - \omega_2 r_2 X_1^{(2)} \sin \phi_2 \quad (5.17)$$

Equations (5.17) can be regarded as four algebraic equations in the unknowns $X_1^{(1)} \cos \phi_1$, $X_1^{(2)} \cos \phi_2$, $X_1^{(1)} \sin \phi_1$, and $X_1^{(2)} \sin \phi_2$. The solution of Eqs. (5.17) can be expressed as

$$X_1^{(1)} \cos \phi_1 = \left\{ \frac{r_2 x_1(0) - x_2(0)}{r_2 - r_1} \right\}, \qquad X_1^{(2)} \cos \phi_2 = \left\{ \frac{-r_1 x_1(0) + x_2(0)}{r_2 - r_1} \right\}$$

$$X_1^{(1)} \sin \phi_1 = \left\{ \frac{-r_2 \dot{x}_1(0) + \dot{x}_2(0)}{\omega_1(r_2 - r_1)} \right\}, \qquad X_1^{(2)} \sin \phi_2 = \left\{ \frac{r_1 \dot{x}_1(0) - \dot{x}_2(0)}{\omega_2(r_2 - r_1)} \right\}$$

from which we obtain the desired solution

$$X_1^{(1)} = \left[\left\{ X_1^{(1)} \cos \phi_1 \right\}^2 + \left\{ X_1^{(1)} \sin \phi_1 \right\}^2 \right]^{1/2}$$

$$= \frac{1}{(r_2 - r_1)} \left[\left\{ r_2 x_1(0) - x_2(0) \right\}^2 + \frac{\left\{ -r_2 \dot{x}_1(0) + \dot{x}_2(0) \right\}^2}{\omega_1^2} \right]^{1/2}$$

$$X_1^{(2)} = \left[\left\{ X_1^{(2)} \cos \phi_2 \right\}^2 + \left\{ X_1^{(2)} \sin \phi_2 \right\}^2 \right]^{1/2}$$

$$= \frac{1}{(r_2 - r_1)} \left[\left\{ -r_1 x_1(0) + x_2(0) \right\}^2 + \frac{\left\{ r_1 \dot{x}_1(0) - \dot{x}_2(0) \right\}^2}{\omega_2^2} \right]^{1/2}$$

$$\phi_1 = \tan^{-1} \left\{ \frac{X_1^{(1)} \sin \phi_1}{X_1^{(1)} \cos \phi_1} \right\} = \tan^{-1} \left\{ \frac{-r_2 \dot{x}_1(0) + \dot{x}_2(0)}{\omega_1 [r_2 x_1(0) - x_2(0)]} \right\}$$

$$\phi_2 = \tan^{-1} \left\{ \frac{X_1^{(2)} \sin \phi_2}{X_1^{(2)} \cos \phi_2} \right\} = \tan^{-1} \left\{ \frac{r_1 \dot{x}_1(0) - \dot{x}_2(0)}{\omega_2 [-r_1 x_1(0) + x_2(0)]} \right\} \quad (5.18)$$

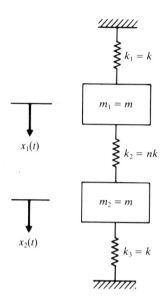

Figure 5.4

EXAMPLE 5.1   Frequencies of Spring-Mass System

Find the natural frequencies and mode shapes of a spring-mass system, shown in Fig. 5.4, which is constrained to move in the vertical direction only. Take $n = 1$.

Given: Two degree of freedom spring-mass system shown in Fig. 5.4.

Find: Natural frequencies and mode shapes.

Approach: Assume harmonic solution for free vibration and solve the resulting equations.

Solution. If we measure x_1 and x_2 from the static equilibrium positions of the masses m_1 and m_2, respectively, the equations of motion and the solution obtained for the system of Fig. 5.3(a) are also applicable to this case if we substitute $m_1 = m_2 = m$ and $k_1 = k_2 = k_3 = k$. Thus the equations of motion, Eqs. (5.4) and (5.5), are given by

$$m\ddot{x}_1 + 2kx_1 - kx_2 = 0$$
$$m\ddot{x}_2 - kx_1 + 2kx_2 = 0 \tag{E.1}$$

By assuming harmonic solution as

$$x_i(t) = X_i \cos(\omega t + \phi); \quad i = 1, 2 \tag{E.2}$$

the frequency equation can be obtained by substituting Eq. (E.2) into Eq. (E.1):

$$\begin{vmatrix} (-m\omega^2 + 2k) & (-k) \\ (-k) & (-m\omega^2 + 2k) \end{vmatrix} = 0$$

or

$$m^2\omega^4 - 4km\omega^2 + 3k^2 = 0 \tag{E.3}$$

The solution of Eq. (E.3) gives the natural frequencies

assign smalles ω_{n_1} then ω_{n_2}

$$\omega_1 = \left\{ \frac{4km - [16k^2m^2 - 12m^2k^2]^{1/2}}{2m^2} \right\}^{1/2} = \sqrt{\frac{k}{m}} \qquad (E.4)$$

$$\omega_2 = \left\{ \frac{4km + [16k^2m^2 - 12m^2k^2]^{1/2}}{2m^2} \right\}^{1/2} = \sqrt{\frac{3k}{m}} \qquad (E.5)$$

From Eqs. (5.11) and (5.12), the amplitude ratios are given by

$$r_1 = \frac{X_2^{(1)}}{X_1^{(1)}} = \frac{-m\omega_1^2 + 2k}{k} = \frac{k}{-m\omega_1^2 + 2k} = 1 \qquad (E.6)$$

$$r_2 = \frac{X_2^{(2)}}{X_1^{(2)}} = \frac{-m\omega_2^2 + 2k}{k} = \frac{k}{-m\omega_2^2 + 2k} = -1 \qquad (E.7)$$

The natural modes are given by Eqs. (5.13) and (5.14):

$$\text{First mode} = \vec{x}^{(1)}(t) = \left\{ \begin{array}{c} X_1^{(1)} \cos\left(\sqrt{\frac{k}{m}}\, t + \phi_1\right) \\ X_1^{(1)} \cos\left(\sqrt{\frac{k}{m}}\, t + \phi_1\right) \end{array} \right\} \qquad (E.8)$$

$$\text{Second mode} = \vec{x}^{(2)}(t) = \left\{ \begin{array}{c} X_1^{(2)} \cos\left(\sqrt{\frac{3k}{m}}\, t + \phi_2\right) \\ -X_1^{(2)} \cos\left(\sqrt{\frac{3k}{m}}\, t + \phi_2\right) \end{array} \right\} \qquad (E.9)$$

It can be seen from Eq. (E.8) that when the system vibrates in its first mode, the amplitudes of the two masses remain the same. This implies that the length of the middle spring remains constant. Thus the motions of m_1 and m_2 are in phase (see Fig. 5.5a). When the system vibrates in its second mode, Eq. (E.9) shows that the displacements of the two masses have the same magnitude with opposite signs. Thus the motions of m_1 and m_2 are 180° out of

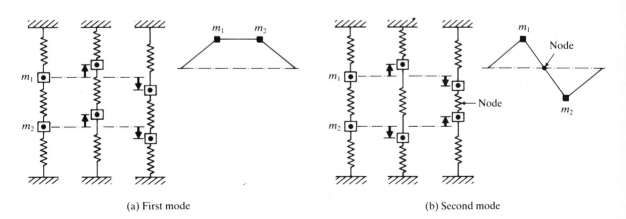

(a) First mode   (b) Second mode

Figure 5.5

phase (see Fig. 5.5b). In this case the midpoint of the middle spring remains stationary for all time t. Such a point is called a *node*. Using Eqs. (5.15), the motion (general solution) of the system can be expressed as

$$x_1(t) = X_1^{(1)} \cos\left(\sqrt{\frac{k}{m}}\, t + \phi_1\right) + X_1^{(2)} \cos\left(\sqrt{\frac{3k}{m}}\, t + \phi_2\right)$$

$$x_2(t) = X_1^{(1)} \cos\left(\sqrt{\frac{k}{m}}\, t + \phi_1\right) - X_1^{(2)} \cos\left(\sqrt{\frac{3k}{m}}\, t + \phi_2\right) \qquad \text{(E.10)}$$

EXAMPLE 5.2   **Initial Conditions to Excite Specific Mode**

Find the initial conditions that need to be applied to the system shown in Fig. 5.4 so as to make it vibrate in (1) the first mode, and (2) the second mode.

Given: Two degree of freedom spring mass system shown in Fig. 5.4.

Find: Initial conditions needed to make the system vibrate in one of the modes.

Approach: Specify the solution to be obtained for the first or second mode from the general solution for arbitrary initial conditions and solve the resulting equations.

Solution. For arbitrary initial conditions, the motion of the masses is described by Eqs. (5.15). In the present case, $r_1 = 1$ and $r_2 = -1$, so Eqs. (5.15) reduce to

$$x_1(t) = X_1^{(1)} \cos\left(\sqrt{\frac{k}{m}}\, t + \phi_1\right) + X_1^{(2)} \cos\left(\sqrt{\frac{3k}{m}}\, t + \phi_2\right)$$

$$x_2(t) = X_1^{(1)} \cos\left(\sqrt{\frac{k}{m}}\, t + \phi_1\right) - X_1^{(2)} \cos\left(\sqrt{\frac{3k}{m}}\, t + \phi_2\right) \qquad \text{(E.1)}$$

Assuming the initial conditions as in Eq. (5.16), the constants $X_1^{(1)}$, $X_1^{(2)}$, ϕ_1, and ϕ_2 can be obtained from Eqs. (5.18), using $r_1 = 1$ and $r_2 = -1$:

$$X_1^{(1)} = -\frac{1}{2}\left\{ [x_1(0) + x_2(0)]^2 + \frac{m}{k}[\dot{x}_1(0) + \dot{x}_2(0)]^2 \right\}^{1/2} \qquad \text{(E.2)}$$

$$X_1^{(2)} = -\frac{1}{2}\left\{ [-x_1(0) + x_2(0)]^2 + \frac{m}{3k}[\dot{x}_1(0) - \dot{x}_2(0)]^2 \right\}^{1/2} \qquad \text{(E.3)}$$

$$\phi_1 = \tan^{-1}\left\{ \frac{-\sqrt{m}\,[\dot{x}_1(0) + \dot{x}_2(0)]}{\sqrt{k}\,[x_1(0) + x_2(0)]} \right\} \qquad \text{(E.4)}$$

$$\phi_2 = \tan^{-1}\left\{ \frac{\sqrt{m}\,[\dot{x}_1(0) - \dot{x}_2(0)]}{\sqrt{3k}\,[-x_1(0) + x_2(0)]} \right\} \qquad \text{(E.5)}$$

(1) The first normal mode of the system is given by Eq. (E.8) of Example 5.1:

$$\vec{x}^{(1)}(t) = \left\{ \begin{array}{c} X_1^{(1)} \cos\left(\sqrt{\dfrac{k}{m}}\, t + \phi_1\right) \\[2mm] X_1^{(1)} \cos\left(\sqrt{\dfrac{k}{m}}\, t + \phi_1\right) \end{array} \right\} \qquad \text{(E.6)}$$

Comparison of Eqs. (E.1) and (E.6) shows that the motion of the system is identical with the first normal mode only if $X_1^{(2)} = 0$. This requires that (from Eq. (E.3))

$$x_1(0) = x_2(0) \quad \text{and} \quad \dot{x}_1(0) = \dot{x}_2(0) \tag{E.7}$$

(2) The second normal mode of the system is given by Eq. (E.9) of Example 5.1:

$$\vec{x}^{(2)}(t) = \left\{ \begin{array}{c} X_1^{(2)} \cos\left(\sqrt{\dfrac{3k}{m}}\, t + \phi_2\right) \\[3mm] -X_1^{(2)} \cos\left(\sqrt{\dfrac{3k}{m}}\, t + \phi_2\right) \end{array} \right\} \tag{E.8}$$

Comparison of Eqs. (E.1) and (E.8) shows that the motion of the system coincides with the second normal mode only if $X_1^{(1)} = 0$. This implies that (from Eq. (E.2))

$$x_1(0) = -x_2(0) \quad \text{and} \quad \dot{x}_1(0) = -\dot{x}_2(0) \tag{E.9}$$

5.4 TORSIONAL SYSTEM

Consider a torsional system consisting of two discs mounted on a shaft, as shown in Fig. 5.6. The three segments of the shaft have rotational spring constants k_{t1}, k_{t2}, and k_{t3}, as indicated in the figure. Also shown are the discs of mass moments of inertia J_1 and J_2, the applied torques M_{t1} and M_{t2}, and the rotational degrees of freedom θ_1 and θ_2. The differential equations of rotational motion for the discs J_1 and J_2 can be derived as follows:

$$\begin{aligned} J_1\ddot{\theta}_1 &= -k_{t1}\theta_1 + k_{t2}(\theta_2 - \theta_1) + M_{t1} \\ J_2\ddot{\theta}_2 &= -k_{t2}(\theta_2 - \theta_1) - k_{t3}\theta_2 + M_{t2} \end{aligned} \tag{5.19}$$

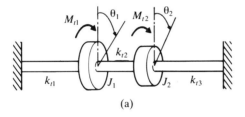

(a)

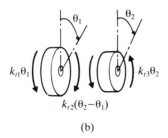

(b)

Figure 5.6

which upon rearrangement become

$$J_1\ddot{\theta}_1 + (k_{t1} + k_{t2})\theta_1 - k_{t2}\theta_2 = M_{t1}$$

$$J_2\ddot{\theta}_2 - k_{t2}\theta_1 + (k_{t2} + k_{t3})\theta_2 = M_{t2} \tag{5.20}$$

EXAMPLE 5.3   **Natural Frequencies of a Torsional System**

Find the natural frequencies and mode shapes for the torsional system shown in Fig. 5.7 for $J_1 = J_0$, $J_2 = 2J_0$, and $k_{t1} = k_{t2} = k_t$.

Given: Two degree of freedom torsional system shown in Fig. 5.7.

Find: Natural frequencies and mode shapes.

Approach: Assume harmonic solution for free vibration and solve the resulting equations.

Solution. The differential equations of motion, Eq. (5.20), reduce to (with $M_{t1} = M_{t2} = k_{t3} = 0$, $k_{t1} = k_{t2} = k_t$, $J_1 = J_0$, and $J_2 = 2J_0$):

$$J_0\ddot{\theta}_1 + 2k_t\theta_1 - k_t\theta_2 = 0$$

$$2J_0\ddot{\theta}_2 - k_t\theta_1 + k_t\theta_2 = 0 \tag{E.1}$$

Rearranging and substituting the harmonic solution

$$\theta_i(t) = \Theta_i \cos(\omega t + \phi); \qquad i = 1, 2 \tag{E.2}$$

gives the frequency equation:

$$2\omega^4 J_0^2 - 5\omega^2 J_0 k_t + k_t^2 = 0 \tag{E.3}$$

The solution of Eq. (E.3) gives the natural frequencies:

$$\omega_1 = \sqrt{\frac{k_t}{4J_0}(5 - \sqrt{17})} \qquad \text{and} \qquad \omega_2 = \sqrt{\frac{k_t}{4J_0}(5 + \sqrt{17})} \tag{E.4}$$

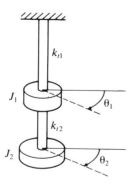

k_{t1}

J_1 θ_1

k_{t2}

J_2 θ_2

Figure 5.7

The amplitude ratios are given by

$$r_1 = \frac{\Theta_2^{(1)}}{\Theta_1^{(1)}} = 2 - \frac{(5 - \sqrt{17})}{4}$$

$$r_2 = \frac{\Theta_2^{(2)}}{\Theta_1^{(2)}} = 2 - \frac{(5 + \sqrt{17})}{4} \qquad (E.5)$$

5.5 COORDINATE COUPLING AND PRINCIPAL COORDINATES

As stated earlier, an n degree of freedom system requires n independent coordinates to describe its configuration. Usually, these coordinates are independent geometrical quantities measured from the equilibrium position of the vibrating body. However, it is possible to select some other set of n coordinates to describe the configuration of the system. The latter set may be, for example, different from the first set in that the coordinates may have their origin away from the equilibrium position of the body. There could be still other sets of coordinates to describe the configuration of the system. Each of these sets of n coordinates is called the *generalized coordinates*.

As an example, consider the lathe shown in Fig. 5.8. An accurate model of this machine tool would involve the consideration of the lathe bed as an elastic beam with lumped masses attached to it [5.1–5.3]. However, for simplified vibration analysis, the lathe bed can be considered as a rigid body having mass and inertia, and the headstock and tailstock can each be replaced by lumped masses. The bed can be assumed to be supported on springs at the ends. Thus the final model will be a rigid body of total mass m and mass moment of inertia J_0 about its C.G., resting on springs of stiffnesses k_1 and k_2, as shown in Fig. 5.9(a). For this two degree of freedom system, any of the following sets of coordinates may be used to describe

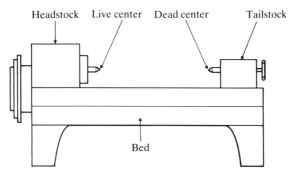

Headstock   Live center   Dead center   Tailstock

Bed

Figure 5.8

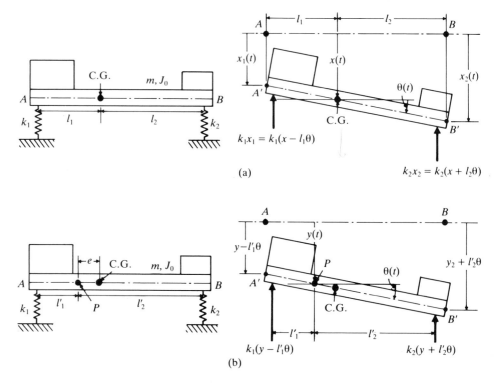

Figure 5.9

the motion:

1. Deflections $x_1(t)$ and $x_2(t)$ of the two ends of the lathe AB
2. Deflection $x(t)$ of the C.G. and rotation $\theta(t)$
3. Deflection $x_1(t)$ of the end A and rotation $\theta(t)$
4. Deflection $y(t)$ of point P located at a distance e to the left of the C.G. and rotation $\theta(t)$, as indicated in Fig. 5.9(b).

Thus any set of these coordinates—(x_1, x_2), (x, θ), (x_1, θ), and (y, θ)—represents the generalized coordinates of the system. Now we shall derive the equations of motion of the lathe using two different sets of coordinates to illustrate the concept of coordinate coupling.

Equations of Motion Using $x(t)$ and $\theta(t)$. From the free-body diagram shown in Fig. 5.9(a), with the positive values of the motion variables as indicated, the force equilibrium equation in the vertical direction can be written as

$$m\ddot{x} = -k_1(x - l_1\theta) - k_2(x + l_2\theta) \tag{5.21}$$

and the moment equation about the C.G. can be expressed as

$$J_0\ddot{\theta} = k_1(x - l_1\theta)l_1 - k_2(x + l_2\theta)l_2 \tag{5.22}$$

Equations (5.21) and (5.22) can be rearranged and written in matrix form as

$$\begin{bmatrix} m & 0 \\ 0 & J_0 \end{bmatrix} \begin{Bmatrix} \ddot{x} \\ \ddot{\theta} \end{Bmatrix} + \begin{bmatrix} (k_1 + k_2) & -(k_1l_1 - k_2l_2) \\ -(k_1l_1 - k_2l_2) & (k_1l_1^2 + k_2l_2^2) \end{bmatrix} \begin{Bmatrix} x \\ \theta \end{Bmatrix} = \begin{Bmatrix} 0 \\ 0 \end{Bmatrix} \tag{5.23}$$

It can be seen that each of the Eqs. (5.23) contain x and θ. They become independent of each other if the coupling term $(k_1l_1 - k_2l_2)$ is equal to zero—that is, if $k_1l_1 = k_2l_2$. If $k_1l_1 \neq k_2l_2$, the resultant motion of the lathe AB is both translational and rotational when either a displacement or torque is applied through the C.G. of the body as an initial condition. In other words, the lathe rotates in the vertical plane and has vertical motion as well, unless $k_1l_1 = k_2l_2$. This is known as *elastic* or *static coupling*.

Equations of Motion using $y(t)$ and $\theta(t)$. From Fig. 5.9(b), where $y(t)$ and $\theta(t)$ are used as the generalized coordinates of the system, the equations of motion for translation and rotation can be written as

$$m\ddot{y} = -k_1(y - l_1'\theta) - k_2(y + l_2'\theta) - me\ddot{\theta}$$
$$J_p\ddot{\theta} = k_1(y - l_1'\theta)l_1' - k_2(y + l_2'\theta)l_2' - me\ddot{y} \tag{5.24}$$

These equations can be rearranged and written in matrix form as

$$\begin{bmatrix} m & me \\ me & J_p \end{bmatrix} \begin{Bmatrix} \ddot{y} \\ \ddot{\theta} \end{Bmatrix} + \begin{bmatrix} (k_1 + k_2) & (k_2l_2' - k_1l_1') \\ (-k_1l_1' + k_2l_2') & (k_1l_1'^2 + k_2l_2'^2) \end{bmatrix} \begin{Bmatrix} y \\ \theta \end{Bmatrix} = \begin{Bmatrix} 0 \\ 0 \end{Bmatrix} \tag{5.25}$$

Both the equations of motion represented by Eq. (5.25) contain y and θ, so they are coupled equations. They contain static (or elastic) as well as dynamic (or mass) coupling terms. If $k_1l_1' = k_2l_2'$, the system will have *dynamic* or *inertia coupling* only. In this case, if the lathe moves up and down in the y direction, the inertia force $m\ddot{y}$, which acts through the center of gravity of the body, induces a motion in the θ direction, by virtue of the moment $m\ddot{y}e$. Similarly, a motion in the θ direction induces a motion of the lathe in the y direction due to the force $me\ddot{\theta}$.

Note the following characteristics of these systems:

1. In the most general case, a viscously damped two degree of freedom system has equations of motion in the following form: STATIC OR ELASTIC COUPLING

DYNAMIC
OR INERTIAL
coupling

$$\rightarrow \begin{bmatrix} m_{11} & m_{12} \\ m_{12} & m_{22} \end{bmatrix} \begin{Bmatrix} \ddot{x}_1 \\ \ddot{x}_2 \end{Bmatrix} + \begin{bmatrix} c_{11} & c_{12} \\ c_{12} & c_{22} \end{bmatrix} \begin{Bmatrix} \dot{x}_1 \\ \dot{x}_2 \end{Bmatrix} + \begin{bmatrix} k_{11} & k_{12} \\ k_{12} & k_{22} \end{bmatrix} \begin{Bmatrix} x_1 \\ x_2 \end{Bmatrix} = \begin{Bmatrix} 0 \\ 0 \end{Bmatrix} \tag{5.26} \quad C$$

This equation reveals the type of coupling present. If the stiffness matrix is not diagonal, the system has elastic or static coupling. If the damping matrix is not diagonal, the system has damping or velocity coupling. Finally, if the mass matrix is not diagonal, the system has mass or inertial coupling. Both velocity and mass coupling come under the heading of dynamic coupling.

2. The system vibrates in its own natural way regardless of the coordinates used. The choice of the coordinates is a mere convenience.

3. From Eqs. (5.23) and (5.25), it is clear that the nature of the coupling depends on the coordinates used and is not an inherent property of the system. It is possible to choose a system of coordinates $q_1(t)$ and $q_2(t)$ which give equations of motion that are uncoupled both statically and dynamically. Such coordinates are called *principal* or *natural coordinates*. The main advantage of using principal coordinates is that the resulting uncoupled equations of motion can be solved independently of one another.

The following example illustrates the method of finding the principal coordinates in terms of the geometrical coordinates.

EXAMPLE 5.4   Principal Coordinates of Spring-Mass System

Determine the principal coordinates for the system shown in Fig. 5.4.

Given: Two degree of freedom spring-mass system shown in Fig. 5.4.

Find: Principal coordinates.

Approach: Define two independent solutions as principal coordinates and express them in terms of the solutions $x_1(t)$ and $x_2(t)$.

Solution. The general motion of the system shown in Fig. 5.4 is given by Eqs. (E.10) of Example 5.1:

$$x_1(t) = B_1 \cos\left(\sqrt{\frac{k}{m}}\, t + \phi_1\right) + B_2 \cos\left(\sqrt{\frac{3k}{m}}\, t + \phi_2\right)$$

$$x_2(t) = B_1 \cos\left(\sqrt{\frac{k}{m}}\, t + \phi_1\right) - B_2 \cos\left(\sqrt{\frac{3k}{m}}\, t + \phi_2\right) \tag{E.1}$$

where $B_1 = X_1^{(1)}$, $B_2 = X_1^{(2)}$, ϕ_1 and ϕ_2 are constants. We define a new set of coordinates $q_1(t)$ and $q_2(t)$ such that

$$q_1(t) = B_1 \cos\left(\sqrt{\frac{k}{m}}\, t + \phi_1\right)$$

$$q_2(t) = B_2 \cos\left(\sqrt{\frac{3k}{m}}\, t + \phi_2\right) \tag{E.2}$$

Since $q_1(t)$ and $q_2(t)$ are harmonic functions, their corresponding equations of motion can be written as*

$$\ddot{q}_1 + \left(\frac{k}{m}\right) q_1 = 0$$

$$\ddot{q}_2 + \left(\frac{3k}{m}\right) q_2 = 0 \tag{E.3}$$

---

* Note that the equation of motion corresponding to the solution $q = B \cos(\omega t + \phi)$ is given by $\ddot{q} + \omega^2 q = 0$.

These equations represent a two degree of freedom system whose natural frequencies are $\omega_1 = \sqrt{k/m}$ and $\omega_2 = \sqrt{3k/m}$. Because there is neither static nor dynamic coupling in the equations of motion (E.3), $q_1(t)$ and $q_2(t)$ are principal coordinates. From Eqs. (E.1) and (E.2), we can write

$$x_1(t) = q_1(t) + q_2(t)$$
$$x_2(t) = q_1(t) - q_2(t) \tag{E.4}$$

The solution of Eqs. (E.4) gives the principal coordinates:

$$q_1(t) = \frac{1}{2}[x_1(t) + x_2(t)]$$
$$q_2(t) = \frac{1}{2}[x_1(t) - x_2(t)] \tag{E.5}$$

EXAMPLE 5.5   **Frequencies and Modes of an Automobile**

Determine the pitch (angular motion) and bounce (up and down linear motion) frequencies and the location of oscillation centers (nodes) of an automobile with the following data (see Fig. 5.10):

mass = m = 1000 kg

radius of gyration = r = 0.9 m

distance between front axle and C.G. = l_1 = 1.0 m

distance between rear axle and C.G. = l_2 = 1.5 m

front spring stiffness = k_f = 18 kN/m

rear spring stiffness = k_r = 22 kN/m

Given: Two degree of freedom automobile model, Fig. 5.10, with m = 1000 kg, r = 0.9 m, l_1 = 1.0 m, l_2 = 1.5 m, k_1 = 18 kN/m, and k_2 = 22 kN/m.

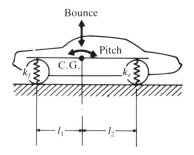

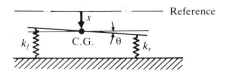

Figure 5.10

Find: Natural frequencies and mode shapes.

Approach: Assume harmonic solution for free vibration and solve the resulting equations.

Solution. If x and θ are used as independent coordinates, the equations of motion are given by Eq. (5.23) with $k_1 = k_f$, $k_2 = k_r$, and $J_0 = mr^2$. For free vibration, we assume a harmonic solution:

$$x(t) = X\cos(\omega t + \phi), \qquad \theta(t) = \Theta\cos(\omega t + \phi) \tag{E.1}$$

Using Eqs. (E.1) and (5.23), we obtain

$$\begin{bmatrix} (-m\omega^2 + k_1 + k_2) & (-k_1 l_1 + k_2 l_2) \\ (-k_1 l_1 + k_2 l_2) & (-J_0\omega^2 + k_1 l_1^2 + k_2 l_2^2) \end{bmatrix} \begin{Bmatrix} X \\ \Theta \end{Bmatrix} = \begin{Bmatrix} 0 \\ 0 \end{Bmatrix} \tag{E.2}$$

For the known data, Eq. (E.2) becomes

$$\begin{bmatrix} (-1000\omega^2 + 40,000) & 15,000 \\ 15,000 & (-810\omega^2 + 67,500) \end{bmatrix} \begin{Bmatrix} X \\ \Theta \end{Bmatrix} = \begin{Bmatrix} 0 \\ 0 \end{Bmatrix} \tag{E.3}$$

from which the frequency equation can be derived:

$$8.1\omega^4 - 999\omega^2 + 24,750 = 0 \tag{E.4}$$

The natural frequencies can be found from Eq. (E.4):

$$\omega_1 = 5.8593 \text{ rad/sec}, \qquad \omega_2 = 9.4341 \text{ rad/sec} \tag{E.5}$$

With these values, the ratio of amplitudes can be found from Eq. (E.3):

$$\frac{X^{(1)}}{\Theta^{(1)}} = -2.6461, \qquad \frac{X^{(2)}}{\Theta^{(2)}} = 0.3061 \tag{E.6}$$

The node locations can be obtained by noting that the tangent of a small angle is approximately equal to the angle itself. Thus, from Fig. 5.11, we find the distance between the

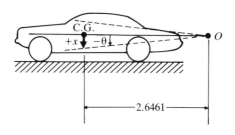

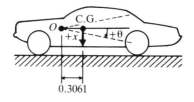

Figure 5.11

C.G. and the node as -2.6461 m for ω_1 and 0.3061 m for ω_2. The mode shapes are shown by dotted lines in Fig. 5.11.

5.6 FORCED VIBRATION ANALYSIS

The equations of motion of a general two degree of freedom system under external forces can be written as

$$
\begin{bmatrix} m_{11} & m_{12} \\ m_{12} & m_{22} \end{bmatrix} \begin{Bmatrix} \ddot{x}_1 \\ \ddot{x}_2 \end{Bmatrix} + \begin{bmatrix} c_{11} & c_{12} \\ c_{12} & c_{22} \end{bmatrix} \begin{Bmatrix} \dot{x}_1 \\ \dot{x}_2 \end{Bmatrix} + \begin{bmatrix} k_{11} & k_{12} \\ k_{12} & k_{22} \end{bmatrix} \begin{Bmatrix} x_1 \\ x_2 \end{Bmatrix} = \begin{Bmatrix} F_1 \\ F_2 \end{Bmatrix} \tag{5.27}
$$

Equations (5.1) and (5.2) can be seen to be special cases of Eq. (5.27), with $m_{11} = m_1$, $m_{22} = m_2$, and $m_{12} = 0$. We shall consider the external forces to be harmonic:

$$
F_j(t) = F_{j0} e^{i\omega t}, \qquad j = 1, 2 \tag{5.28}
$$

where ω is the forcing frequency. We can write the steady-state solutions as

$$
x_j(t) = X_j e^{i\omega t}, \qquad j = 1, 2 \tag{5.29}
$$

where X_1 and X_2 are, in general, complex quantities which depend on ω and the system parameters. Substitution of Eqs. (5.28) and (5.29) into Eq. (5.27) leads to

$$
\begin{bmatrix} \left(-\omega^2 m_{11} + i\omega c_{11} + k_{11}\right) & \left(-\omega^2 m_{12} + i\omega c_{12} + k_{12}\right) \\ \left(-\omega^2 m_{12} + i\omega c_{12} + k_{12}\right) & \left(-\omega^2 m_{22} + i\omega c_{22} + k_{22}\right) \end{bmatrix} \begin{Bmatrix} X_1 \\ X_2 \end{Bmatrix} = \begin{Bmatrix} F_{10} \\ F_{20} \end{Bmatrix} \tag{5.30}
$$

As in Section 3.5, we define the mechanical impedance $Z_{rs}(i\omega)$ as

$$
Z_{rs}(i\omega) = -\omega^2 m_{rs} + i\omega c_{rs} + k_{rs}, \qquad r, s = 1, 2 \tag{5.31}
$$

and write Eq. (5.30) as

$$
[Z(i\omega)] \vec{X} = \vec{F}_0 \tag{5.32}
$$

where

$$
[Z(i\omega)] = \begin{bmatrix} Z_{11}(i\omega) & Z_{12}(i\omega) \\ Z_{12}(i\omega) & Z_{22}(i\omega) \end{bmatrix} = \text{impedance matrix}
$$

$$
\vec{X} = \begin{Bmatrix} X_1 \\ X_2 \end{Bmatrix}
$$

and

$$
\vec{F}_0 = \begin{Bmatrix} F_{10} \\ F_{20} \end{Bmatrix}
$$

Equation (5.32) can be solved to obtain

$$
\vec{X} = [Z(i\omega)]^{-1} \vec{F}_0 \tag{5.33}
$$

where the inverse of the impedance matrix is given by

$$[Z(i\omega)]^{-1} = \frac{1}{Z_{11}(i\omega)Z_{22}(i\omega) - Z_{12}^2(i\omega)} \begin{bmatrix} Z_{22}(i\omega) & -Z_{12}(i\omega) \\ -Z_{12}(i\omega) & Z_{11}(i\omega) \end{bmatrix} \quad (5.34)$$

Equations (5.33) and (5.34) lead to the solution

$$X_1(i\omega) = \frac{Z_{22}(i\omega)F_{10} - Z_{12}(i\omega)F_{20}}{Z_{11}(i\omega)Z_{22}(i\omega) - Z_{12}^2(i\omega)}$$

$$X_2(i\omega) = \frac{-Z_{12}(i\omega)F_{10} + Z_{11}(i\omega)F_{20}}{Z_{11}(i\omega)Z_{22}(i\omega) - Z_{12}^2(i\omega)} \quad (5.35)$$

By substituting Eqs. (5.35) into Eqs. (5.29) we can find the complete solution, $x_1(t)$ and $x_2(t)$.

The analysis of a two degree of freedom system used as a vibration absorber is given in Section 9.10. Reference [5.4] deals with the impact response of a two degree of freedom system, while Ref. [5.5] considers the steady-state response under harmonic excitation.

EXAMPLE 5.6   Steady-State Response of Spring-Mass System

Find the steady-state response of the system shown in Fig. 5.12 when the mass m_1 is excited by the force $F_1(t) = F_{10} \cos \omega t$. Also, plot its frequency response curve.

Given: Two degree of freedom undamped spring-mass system subjected to the forcing function $F_1(t)$ as shown in Fig. 5.12.

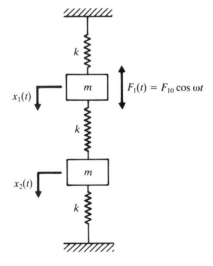

Figure 5.12

Find: Steady state response of the masses.

Approach: Assume harmonic solution and use the concept of mechanical impedance to find the response.

Solution. The equations of motion of the system can be expressed as

$$\begin{bmatrix} m & 0 \\ 0 & m \end{bmatrix} \begin{Bmatrix} \ddot{x}_1 \\ \ddot{x}_2 \end{Bmatrix} + \begin{bmatrix} 2k & -k \\ -k & 2k \end{bmatrix} \begin{Bmatrix} x_1 \\ x_2 \end{Bmatrix} = \begin{Bmatrix} F_{10} \cos \omega t \\ 0 \end{Bmatrix} \tag{E.1}$$

Comparison of Eq. (E.1) with Eq. (5.27) shows that

$$m_{11} = m_{22} = m, \quad m_{12} = 0, \quad c_{11} = c_{12} = c_{22} = 0,$$
$$k_{11} = k_{22} = 2k, \quad k_{12} = -k, \quad F_1 = F_{10} \cos \omega t, \quad F_2 = 0$$

We assume the solution to be:*

$$x_j(t) = X_j \cos \omega t; \quad j = 1, 2 \tag{E.2}$$

Equation (5.31) gives

$$Z_{11}(\omega) = Z_{22}(\omega) = -m\omega^2 + 2k, \quad Z_{12}(\omega) = -k \tag{E.3}$$

Hence X_1 and X_2 are given by Eqs. (5.35):

$$X_1(\omega) = \frac{(-\omega^2 m + 2k) F_{10}}{(-\omega^2 m + 2k)^2 - k^2} = \frac{(-\omega^2 m + 2k) F_{10}}{(-m\omega^2 + 3k)(-m\omega^2 + k)} \tag{E.4}$$

$$X_2(\omega) = \frac{kF_{10}}{(-m\omega^2 + 2k)^2 - k^2} = \frac{kF_{10}}{(-m\omega^2 + 3k)(-m\omega^2 + k)} \tag{E.5}$$

By defining $\omega_1^2 = k/m$ and $\omega_2^2 = 3k/m$, Eqs. (E.4) and (E.5) can be expressed as

$$X_1(\omega) = \frac{\left\{ 2 - \left(\dfrac{\omega}{\omega_1} \right)^2 \right\} F_{10}}{k \left[\left(\dfrac{\omega_2}{\omega_1} \right)^2 - \left(\dfrac{\omega}{\omega_1} \right)^2 \right] \left[1 - \left(\dfrac{\omega}{\omega_1} \right)^2 \right]} \tag{E.6}$$

$$X_2(\omega) = \frac{F_{10}}{k \left[\left(\dfrac{\omega_2}{\omega_1} \right)^2 - \left(\dfrac{\omega}{\omega_1} \right)^2 \right] \left[1 - \left(\dfrac{\omega}{\omega_1} \right)^2 \right]} \tag{E.7}$$

The responses X_1 and X_2 are shown in Fig. 5.13 in terms of the dimensionless parameter ω/ω_1. In the dimensionless parameter ω/ω_1, ω_1 was selected arbitrarily; ω_2 could have been selected just as easily. It can be seen that the amplitudes X_1 and X_2 become infinite when $\omega^2 = \omega_1^2$ or $\omega^2 = \omega_2^2$. Thus there are two resonance conditions for the system: one at ω_1 and another at ω_2. At all other values of ω, the amplitudes of vibration are finite. It can be noted

---

* Since $F_{10} \cos \omega t = \text{Real}(F_{10}e^{i\omega t})$, we shall assume the solution also to be $x_j = \text{Real}(X_j e^{i\omega t}) =$ $X_j \cos \omega t$, $j = 1, 2$. It can be verified that X_j are real for an undamped system.

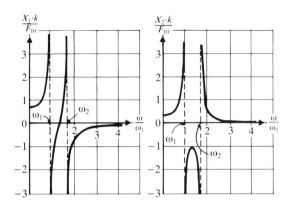

Figure 5.13.   Frequency response curves of Example 5.6.

from Fig. 5.13 that there is a particular value of the frequency ω at which the vibration of the first mass m_1, to which the force $F_1(t)$ is applied, is reduced to zero. This characteristic forms the basis of the dynamic vibration absorber discussed in Chapter 9.

5.7 SEMI-DEFINITE SYSTEMS

Semi-definite systems are also known as *unrestrained* or *degenerate systems*. An example of such a system is shown in Fig. 5.14. This arrangement may be considered to represent two railway cars of masses m_1 and m_2 with a coupling spring k. The equations of motion of the system can be written as

$$m_1\ddot{x}_1 + k(x_1 - x_2) = 0$$
$$m_2\ddot{x}_2 + k(x_2 - x_1) = 0 \tag{5.36}$$

For free vibration, we assume the motion to be harmonic:

$$x_j(t) = X_j \cos(\omega t + \phi_j), \qquad j = 1, 2 \tag{5.37}$$

Substitution of Eq. (5.37) into Eq. (5.36) gives

$$(-m_1\omega^2 + k)X_1 - kX_2 = 0$$
$$-kX_1 + (-m_2\omega^2 + k)X_2 = 0 \tag{5.38}$$

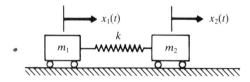

Figure 5.14

By equating the determinant of the coefficients of X_1 and X_2 to zero, we obtain the frequency equation as

$$\omega^2 \left[m_1 m_2 \omega^2 - k(m_1 + m_2) \right] = 0 \tag{5.39}$$

from which the natural frequencies can be obtained:

$$\omega_1 = 0 \quad \text{and} \quad \omega_2 = \sqrt{\frac{k(m_1 + m_2)}{m_1 m_2}} \tag{5.40}$$

It can be seen that one of the natural frequencies of the system is zero, which means that the system is not oscillating. In other words, the system moves as a whole without any relative motion between the two masses (rigid body translation). Such systems, which have one of the natural frequencies equal to zero, are called *semi-definite systems*. We can verify, by substituting ω_2 into Eq. (5.38), that $X_1^{(2)}$ and $X_2^{(2)}$ are opposite in phase. There would thus be a node at the middle of the spring.

5.8 SELF-EXCITATION AND STABILITY ANALYSIS

In Section 3.11, the stability conditions of a single degree of freedom system have been expressed in terms of the physical constants of the system. The procedure is extended to a two degree of freedom system in this section. When the system is subjected to self-exciting forces, the force terms can be combined with the damping/stiffness terms, and the resulting equations of motion can be expressed in matrix notation as

$$\begin{bmatrix} m_{11} & m_{12} \\ m_{21} & m_{22} \end{bmatrix} \begin{Bmatrix} \ddot{x}_1 \\ \ddot{x}_2 \end{Bmatrix} + \begin{bmatrix} c_{11} & c_{12} \\ c_{21} & c_{22} \end{bmatrix} \begin{Bmatrix} \dot{x}_1 \\ \dot{x}_2 \end{Bmatrix} + \begin{bmatrix} k_{11} & k_{12} \\ k_{21} & k_{22} \end{bmatrix} \begin{Bmatrix} x_1 \\ x_2 \end{Bmatrix} = \begin{Bmatrix} 0 \\ 0 \end{Bmatrix} \tag{5.41}$$

By substituting the solution

$$x_j(t) = X_j e^{st}, \quad j = 1, 2 \tag{5.42}$$

in Eq. (5.41) and setting the determinant of the coefficient matrix to zero, we obtain the characteristic equation of the form

$$a_0 s^4 + a_1 s^3 + a_2 s^2 + a_3 s + a_4 = 0 \tag{5.43}$$

The coefficients a_0, a_1, a_2, a_3, and a_4 are real numbers, since they are derived from the physical parameters of the system. If s_1, s_2, s_3, and s_4 denote the roots of Eq. (5.43), we have

$$(s - s_1)(s - s_2)(s - s_3)(s - s_4) = 0$$

or

$$s^4 - (s_1 + s_2 + s_3 + s_4)s^3 + (s_1 s_2 + s_1 s_3 + s_1 s_4 + s_2 s_3 + s_2 s_4 + s_3 s_4)s^2$$
$$- (s_1 s_2 s_3 + s_1 s_2 s_4 + s_1 s_3 s_4 + s_2 s_3 s_4)s + (s_1 s_2 s_3 s_4) = 0 \tag{5.44}$$

A comparison of Eqs. (5.43) and (5.44) yields

$$a_0 = 1$$
$$a_1 = -(s_1 + s_2 + s_3 + s_4)$$
$$a_2 = s_1 s_2 + s_1 s_3 + s_1 s_4 + s_2 s_3 + s_2 s_4 + s_3 s_4$$
$$a_3 = -(s_1 s_2 s_3 + s_1 s_2 s_4 + s_1 s_3 s_4 + s_2 s_3 s_4)$$
$$a_4 = s_1 s_2 s_3 s_4 \tag{5.45}$$

The criterion for stability is that the real parts of s_i ($i = 1, 2, 3, 4$) must be negative to avoid increasing exponentials in Eq. (5.42). Using the properties of a quartic equation, it can be derived that a necessary and sufficient condition for stability is that all the coefficients of the equation (a_0, a_1, a_2, a_3, and a_4) be positive and that the condition

$$a_1 a_2 a_3 > a_0 a_3^2 + a_4 a_1^2 \tag{5.46}$$

be fulfilled [5.8, 5.9]. A more general technique, which can be used to investigate the stability of an n degree of freedom system, is known as the Routh-Hurvitz criterion [5.10]. For the system under consideration, Eq. (5.43), the Routh-Hurvitz criterion states that the system will be stable if all the coefficients $a_0, a_1, \ldots, a_4$ are positive and the determinants defined below are positive:

$$T_1 = |a_1| > 0 \tag{5.47}$$

$$T_2 = \begin{vmatrix} a_1 & a_3 \\ a_0 & a_2 \end{vmatrix} = a_1 a_2 - a_0 a_3 > 0 \tag{5.48}$$

$$T_3 = \begin{vmatrix} a_1 & a_3 & 0 \\ a_0 & a_2 & a_4 \\ 0 & a_1 & a_3 \end{vmatrix} = a_1 a_2 a_3 - a_1^2 a_4 - a_0 a_3^2 > 0 \tag{5.49}$$

Equation (5.47) simply states that the coefficient a_1 must be positive, while the satisfaction of Eq. (5.49), coupled with the satisfaction of the conditions $a_3 > 0$ and $a_4 > 0$, implies the satisfaction of Eq. (5.48). Thus the necessary and sufficient condition for the stability of the system is that all the coefficients a_0, a_1, a_2, a_3, and a_4 be positive and that the inequality stated in (5.46) be satisfied.

5.9 COMPUTER PROGRAMS

The determination of the natural frequencies of a damped two degree of freedom system involves the solution of a fourth order polynomial equation. Similarly, a damped degenerate two degree of freedom system requires the determination of the roots of a cubic equation. An undamped system, on the other hand, requires the solution of a quadratic equation. This section presents three Fortran subroutines (QUADRA, CUBIC, and QUART) for the solution of quadratic, cubic, and quartic equations, respectively. The listing of these subroutines and typical main programs for calling them are given below. The input data required and the output of the programs are explained in the comment lines of the programs.

**5.9.1
Roots of
a Quadratic
Equation**

```
C =======================================================================
C
C PROGRAM 6
C MAIN PROGRAM WHICH CALLS QUADRA
C
C =======================================================================
C FOLLOWING 2 LINES CONTAIN PROBLEM-DEPENDENT DATA
C EXAMPLE X**2 - 2.0*X + 5.0 = 0.0
      DATA A1,A2,A3/1.0,-2.0,5.0/
C END OF PROBLEM-DEPENDENT DATA
      CALL QUADRA (A1,A2,A3,RR1,RR2,RI1,RI2)
      PRINT 10, A1,A2,A3
   10 FORMAT (/,2X,28H POLYNOMIAL COEFFICIENTS ARE,/,3E15.6,//,
     2    2X,10H ROOTS ARE,//,4X,5H REAL,14X,10H IMAGINARY)
      PRINT 20, RR1,RI1
      PRINT 20, RR2,RI2
   20 FORMAT (4X,E15.8,4X,E15.8)
      STOP
      END
C =======================================================================
C
C SUBROUTINE QUADRA
C
C =======================================================================
C     SOLUTION OF QUADRATIC EQUATION   A1*(X**2)+A2*(X)+A3 = 0
C     A1,A2,A3 ARE INPUT, (RR1,RI1) AND (RR2,RI2) ARE ROOTS (OUTPUT)
C     A1 MUST NOT BE EQUAL TO ZERO
      SUBROUTINE QUADRA (A1,A2,A3,RR1,RR2,RI1,RI2)
      RAD=A2**2-4.0*A1*A3
      IF (RAD) 20,10,10
   10 SRAD=SQRT(RAD)
      RR1=(-A2-SRAD)/(2.0*A1)
      RR2=(-A2+SRAD)/(2.0*A1)
      RI1=0.0
      RI2=0.0
      RETURN
   20 SRAD=SQRT(-RAD)
      RR1=-A2/(2.0*A1)
      RR2=RR1
      RI1=SRAD/(2.0*A1)
      RI2=-RI1
      RETURN
      END


POLYNOMIAL COEFFICIENTS ARE
0.100000E+01  -0.200000E+01   0.500000E+01

ROOTS ARE

      REAL              IMAGINARY
   0.10000000E+01     0.20000000E+01
   0.10000000E+01    -0.20000000E+01
```

5.9.2
Roots of
a Cubic Equation

```
C ================================================================
C
C PROGRAM 7
C MAIN PROGRAM FOR CALLING THE SUBROUTINE CUBIC
C
C ================================================================
        DIMENSION A(4),RR(3),RI(3)
        DATA A/1.0,0.0,6.0,20.0/
        PRINT 10
10      FORMAT (//,24H ROOTS OF CUBIC EQUATION,//,
     2    51H GIVEN POLYNOMIAL COEFFICIENTS A(1),A(2),A(3),A(4):,/)
        PRINT 20,(A(I),I=1,4)
20      FORMAT (4E15.6)
        CALL CUBIC (A,RR,RI)
        PRINT 30
30      FORMAT (//,38H ROOTS (REAL PART AND IMAGINARY PART):,/)
        DO 40 I=1,3
40      PRINT 50,RR(I),RI(I)
50      FORMAT (2E15.6)
        STOP
        END
C ================================================================
C
C SUBROUTINE CUBIC
C
C ================================================================
C     ROOTS OF CUBIC EQUATION   A(1)*(X**3)+A(2)*(X**2)+A(3)*X+A(4)=0
        SUBROUTINE CUBIC (A,RR,RI)
        DIMENSION A(4),RR(3),RI(3)
        DO 10 I=1,3
        RR(I)=0.0
10      RI(I)=0.0
        A0=A(1)
        A1=A(2)/3.0
        A2=A(3)/3.0
        A3=A(4)
        G=(A0**2)*A3-3.0*A0*A1*A2+2.0*(A1**3)
        H=A0*A2-A1**2
        Y1=G**2+4.0*(H**3)
        IF (Y1 .LT. 0.0) GO TO 100
        Y2=SQRT(Y1)
        Z1=(G+Y2)/2.0
        Z2=(G-Y2)/2.0
        IF(Z1 .LT. 0.0) GO TO 21
        Z3=Z1**(1.0/3.0)
        GO TO 22
21      Z3=(-Z1)**(1.0/3.0)
        Z3=-Z3
22      IF(Z2 .LT. 0.0) GO TO 23
        Z4=Z2**(1.0/3.0)
        GO TO 24
23      Z4=(-Z2)**(1.0/3.0)
        Z4=-Z4
```

```
   24    CONTINUE
         RR(1)=-(A1+Z3+Z4)/A0
         RR(2)=(-2.0*A1+Z3+Z4)/(2.0*A0)
         RI(2)=SQRT(3.0)*(Z4-Z3)/(2.0*A0)
         RR(3)=RR(2)
         RI(3)=-RI(2)
         GO TO 200
  100    SH=SQRT(-H)
         XK=2.0*SH
         THETA=ACOS(G/(2.0*H*SH))/3.0
         XY1=2.0*SH*COS(THETA)
         PI=3.1416
         XY2=2.0*SH*COS(THETA+(2.0*PI/3.0))
         XY3=2.0*SH*COS(THETA+(4.0*PI/3.0))
         RR(1)=(XY1-A1)/A0
         RR(2)=(XY2-A1)/A0
         RR(3)=(XY3-A1)/A0
  200    RETURN
         END
```

ROOTS OF CUBIC EQUATION

GIVEN POLYNOMIAL COEFFICIENTS A(1),A(2),A(3),A(4):

```
   0.100000E+01    0.000000E+00    0.600000E+01    0.200000E+02
```

ROOTS (REAL PART AND IMAGINARY PART):

```
  -0.200000E+01    0.000000E+00
   0.100000E+01   -0.300000E+01
   0.100000E+01    0.300000E+01
```

**5.9.3
Roots of
a Quartic
Equation**

```
C =========================================================================
C
C
C PROGRAM 8
C MAIN PROGRAM FOR CALLING THE SUBROUTINE QUART
C
C =========================================================================
C SOLUTION OF:  A(1)*(X**4)+A(2)*(X**3)+A(3)*(X**2)+A(4)*X+A(5)=0
         DIMENSION A(5),RR(4),RI(4)
C FOLLOWING LINE CONTAINS PROBLEM-DEPENDENT DATA
         DATA A/1.0,0.0,0.0,-8.0,12.0/
C END OF PROBLEM-DEPENDENT DATA
         PRINT 10,(A(I),I=1,5)
```

```
10    FORMAT (//,31H SOLUTION OF A QUARTIC EQUATION,//,6H DATA:,/,
   2    7H A(1) =,E15.6,/,7H A(2) =,E15.6,/,7H A(3) =,E15.6,/,
   3    7H A(4) =,E15.6,/,7H A(5) =,E15.6,/)
      CALL QUART (A,RR,RI)
      PRINT 20
20    FORMAT (/,7H ROOTS:,//,9H ROOT NO.,3X,10H REAL PART,5X,
   2    15H IMAGINARY PART,/)
      DO 30 I=1,4
30    PRINT 40,I,RR(I),RI(I)
40    FORMAT (I5,3X,E15.6,3X,E15.6)
      STOP
      END




C ==================================================================
C
C SUBROUTINE QUART
C
C ==================================================================
      SUBROUTINE QUART (A,RR,RI)
      DIMENSION A(5),RR(4),RI(4),B(4),RRC(3),RIC(3)
      DO 10 I=2,5
 10   A(I)=A(I)/A(1)
      B(1)=1.0
      B(2)=-A(3)
      B(3)=A(4)*A(2)-4.0*A(5)
      B(4)=A(5)*(4.0*A(3)-A(2)**2)-A(4)**2
      CALL CUBIC (B,RRC,RIC)
      IF (RIC(2) .NE. 0.0) GO TO 20
      X=AMAX1(RRC(1),RRC(2),RRC(3))
      RRC(1)=X
 20   X=RRC(1)/2.0
      IF ((X**2-A(5)) .GT. 0.0) GO TO 30
      Y=0.0
      Z=SQRT((A(2)/2.0)**2+2.0*X-A(3))
C ADD TO ABOVE EQUATION
      GO TO 40
 30   Y=SQRT(X**2-A(5))
      Z=-(A(4)-A(2)*X)/(2.0*Y)
 40   C1=1.0
      C2=A(2)/2.0+Z
      C3=X+Y
      CALL QUADRA (C1,C2,C3,QR1,QR2,QI1,QI2)
      RR(1)=QR1
      RR(2)=QR2
      RI(1)=QI1
      RI(2)=QI2
      C1=1.0
      C2=A(2)/2.0-Z
      C3=X-Y
      CALL QUADRA (C1,C2,C3,QR1,QR2,QI1,QI2)
      RR(3)=QR1
      RR(4)=QR2
      RI(3)=QI1
```

```
RI(4)=QI2
RETURN
END
```

SOLUTION OF A QUARTIC EQUATION

```
DATA:
A(1) =    0.100000E+01
A(2) =    0.000000E+00
A(3) =    0.000000E+00
A(4) =   -0.800000E+01
A(5) =    0.120000E+02
```

ROOTS:

ROOT NO.	REAL PART	IMAGINARY PART
1	-0.137091E+01	0.182709E+01
2	-0.137091E+01	-0.182709E+01
3	0.137091E+01	0.648457E+00
4	0.137091E+01	-0.648457E+00

REFERENCES

5.1. H. Sato, Y. Kuroda, and M. Sagara, "Development of the finite element method for vibration analysis of machine tool structure and its application," *Proceedings of the Fourteenth International Machine Tool Design and Research Conference*, Macmillan, London, 1974, pp. 545–552.

5.2. F. Koenigsberger and J. Tlusty, *Machine Tool Structures*, Pergamon Press, Oxford, 1970.

5.3. C. P. Reddy and S. S. Rao, "Automated optimum design of machine tool structures for static rigidity, natural frequencies and regenerative chatter stability," *Journal of Engineering for Industry*, Vol. 100, 1978, pp. 137–146.

5.4. M. S. Hundal, "Effect of damping on impact response of a two degree of freedom system," *Journal of Sound and Vibration*, Vol. 68, 1980, pp. 407–412.

5.5. J. A. Linnett, "The effect of rotation on the steady state response of a spring-mass system under harmonic excitation," *Journal of Sound and Vibration*, Vol. 35, 1974, pp. 1–11.

5.6. A. Hurwitz, "On the conditions under which an equation has only roots with negative real parts," in *Selected Papers on Mathematical Trends in Control Theory*, Dover Publications, New York, 1964, pp. 70–82.

5.7. R. C. Dorf, *Modern Control Systems* (3rd Ed.), Addison-Wesley, Reading, Mass., 1980.

5.8. J. P. Den Hartog, *Mechanical Vibrations* (4th Ed.), McGraw-Hill, New York, 1956.

5.9. R. H. Scanlan and R. Rosenbaum, *Introduction to the Study of Aircraft Vibration and Flutter*, Macmillan Co., New York, 1951.

5.10. L. A. Pipes and L. R. Harvill, *Applied Mathematics for Engineers and Physicists* (3rd Ed.), McGraw-Hill, New York, 1970.

REVIEW QUESTIONS

5.1. How do you determine the number of degrees of freedom of a lumped-mass system?

5.2. Define mass coupling, velocity coupling, and elastic coupling.

5.3. Is the nature of the coupling dependent on the coordinates used?

5.4. How many degrees of freedom does an airplane in flight have if it is treated as (a) a rigid body, and (b) an elastic body?

5.5. What are principal coordinates? What is their use?

5.6. Why are the mass, damping, and stiffness matrices symmetrical?

5.7. What is a node?

5.8. What is meant by static and dynamic coupling? How can you eliminate coupling of the equations of motion?

5.9. Define the impedance matrix.

5.10. How can we make a system vibrate in one of its natural modes?

5.11. What is a degenerate system? Give two examples of physical systems that are degenerate.

5.12. How many degenerate modes can a vibrating system have?

PROBLEMS

The problem assignments are organized as follows:

Problems	Section covered	Topic covered
5.1–5.19	5.3	Free vibration of undamped systems
5.20–5.23	5.4	Torsional systems
5.24–5.28	5.5	Coordinate coupling
5.29–5.41	5.6	Forced vibrations
5.42–5.46	5.7	Semi-definite systems
5.47	5.8	Stability analysis
5.48–5.50	5.9	Computer programs
5.51	—	Project

5.1. Find the natural frequencies of the system shown in Fig. 5.15, with $m_1 = m$, $m_2 = 2m$, $k_1 = k$, and $k_2 = 2k$. Determine the response of the system when $k = 1000$ N/m, $m = 20$ kg, and the initial values of the displacements of the masses m_1 and m_2 are 1 and -1, respectively.

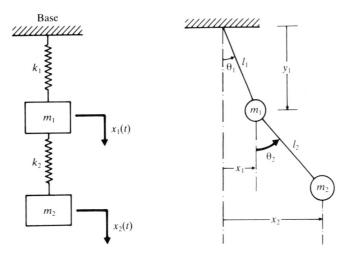

Figure 5.15   Figure 5.16

5.2. Set up the differential equations of motion for the double pendulum shown in Fig. 5.16, using the coordinates x_1 and x_2 and assuming small amplitudes. Find the natural frequencies, the ratios of amplitudes, and the locations of nodes for the two modes of vibration when $m_1 = m_2 = m$ and $l_1 = l_2 = l$.

5.3. Determine the natural modes of the system shown in Fig. 5.17 when $k_1 = k_2 = k_3 = k$.

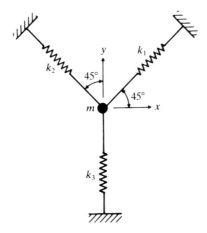

Figure 5.17

5.4. A machine tool, having a mass of $m = 1000$ kg and a mass moment of inertia of $J_0 = 300$ kg $-$ m^2, is supported on elastic supports as shown in Fig. 5.18. If the

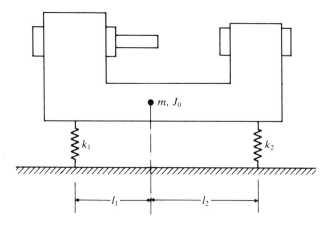

Figure 5.18

stiffnesses of the supports are given by $k_1 = 3000$ N/mm, and $k_2 = 2000$ N/mm, and the supports are located at $l_1 = 0.5$ m and $l_2 = 0.8$ m, find the natural frequencies and mode shapes of the machine tool.

5.5.   An overhead traveling crane can be modeled as shown in Fig. 5.19. The beam has an area moment of inertia (I) of 0.02 m⁴ and modulus of elasticity (E) of 2.06×10^{11} N/m², the truck has a mass (m_1) of 1000 kg, the load being lifted has a mass (m_2) of 5000 kg, and the cable through which the mass (m_2) is lifted has a stiffness (k) of 3.0×10^5 N/m. Determine the natural frequencies and mode shapes of the system. Assume the span of the beam as 40 m.

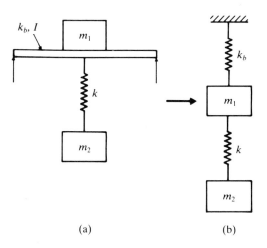

(a)   (b)

Figure 5.19

5.6.   One of the wheels and leaf springs of an automobile, traveling over a rough road, is shown in Fig. 5.20(a). For simplicity, all the wheels can be assumed to be identical and the system can be idealized as shown in Fig. 5.20(b). The automobile has a mass of $m_1 = 1000$ kg and the leaf springs have a total stiffness of $k_1 = 400$ kN/m. The wheels and axles have a mass of $m_2 = 300$ kg and the tires have a stiffness of

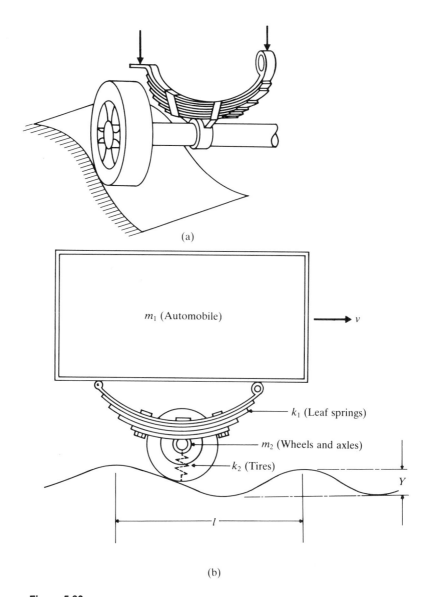

(a)

(b)

Figure 5.20

$k_2 = 500$ kN/m. If the road surface varies sinusoidally with an amplitude of $Y = 0.1$ m and a period of $l = 6$ m, find the critical velocities of the automobile.

5.7. Derive the equations of motion of the double pendulum shown in Fig. 5.16, using the coordinates θ_1 and θ_2. Also find the natural frequencies and mode shapes of the system for $m_1 = m_2 = m$ and $l_1 = l_2 = l$.

5.8. Find the natural frequencies and mode shapes of the system shown in Fig. 5.15 for $m_1 = m_2 = m$ and $k_1 = k_2 = k$.

5.9. The normal modes of a two degree of freedom system are orthogonal if $X^{(1)^T}[m]\vec{X}^{(2)} = 0$. Prove that the mode shapes of the system shown in Fig. 5.3(a) are orthogonal.

5.10. Find the natural frequencies of the system shown in Fig. 5.4 for $k_1 = 300$ N/m, $k_2 = 500$ N/m, $k_3 = 200$ N/m, $m_1 = 2$ kg, and $m_2 = 1$ kg.

5.11. Find the natural frequencies and mode shapes of the system shown in Fig. 5.15 for $m_1 = m_2 = 1$ kg, $k_1 = 2000$ N/m and $k_2 = 6000$ N/m.

5.12. Derive expressions for the displacements of the masses in Fig. 5.4 when $m_i = 25$ lb-sec^2/in, $i = 1, 2$ and $k_i = 50,000$ lb/in, $i = 1, 2, 3$.

5.13. For the system shown in Fig. 5.4, $m_1 = 1$ kg, $m_2 = 2$ kg, $k_1 = 2000$ N/m, $k_2 = 1000$ N/m, $k_3 = 3000$ N/m, and an initial velocity of 20 m/s is imparted to mass m_1. Find the resulting motion of the two masses.

5.14. For Problem 5.11, calculate $x_1(t)$ and $x_2(t)$ for the following initial conditions:
(a) $x_1(0) = 0.2$, $\dot{x}_1(0) = x_2(0) = \dot{x}_2(0) = 0$; and (b) $x_1(0) = 0.2$, $\dot{x}_1(0) = x_2(0) = 0$, $\dot{x}_2(0) = 5.0$.

5.15. A two-story building frame is modeled as shown in Fig. 5.21. The girders are assumed to be rigid, and the columns have flexural rigidities EI_1 and EI_2, with negligible masses. The stiffness of each column can be computed as

$$\frac{24EI_i}{h_i^3}, \quad i = 1, 2$$

For $m_1 = 2m$, $m_2 = m$, $h_1 = h_2 = h$, and $EI_1 = EI_2 = EI$, determine the natural frequencies and mode shapes of the frame.

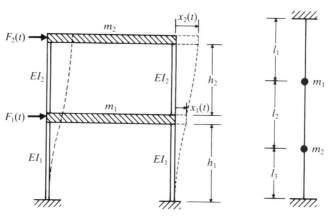

Figure 5.21

Figure 5.22

5.16. Figure 5.22 shows a system of two masses attached to a tightly stretched string, fixed at both ends. Determine the natural frequencies and mode shapes of the system for $m_1 = m_2 = m$ and $l_1 = l_2 = l_3 = l$.

5.17. Find the normal modes of the two-story building shown in Fig. 5.21 when $m_1 = 3m$, $m_2 = m$, $k_1 = 3k$, and $k_2 = k$, where k_1 and k_2 represent the total equivalent stiffnesses of the lower and upper columns, respectively.

5.18. A hoisting drum, having a weight W_1, is mounted at the end of a steel cantilever beam of thickness t, width a, and length b as shown in Fig. 5.23. The wire rope is made of steel and has a diameter of d and a suspended length of l. If the load hanging at the end of the rope is W_2, derive expressions for the natural frequencies of the system.

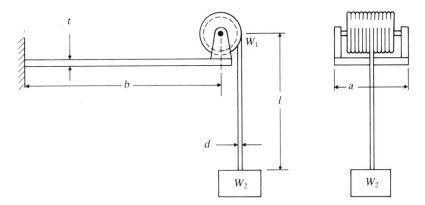

Figure 5.23

5.19.* Design the cantilever beam supporting the hoisting drum and the wire rope carrying the load in Problem 5.18 in order to have the natural frequencies of the system greater than 10 Hz when $W_1 = 1000$ lb and $W_2 = 500$ lb, $b = 30$ in., and $l = 60$ in.

5.20. Determine the natural frequencies and normal modes of the torsional system shown in Fig. 5.24 for $k_{t2} = 2k_{t1}$ and $J_2 = 2J_1$.

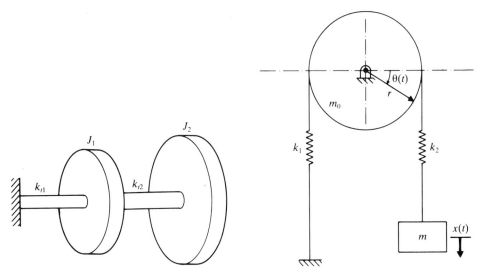

Figure 5.24   **Figure 5.25**

5.21. Determine the natural frequencies of the system shown in Fig. 5.25 by assuming that the rope passing over the cylinder does not slip.

5.22. Find the natural frequencies and mode shapes of the system shown in Fig. 5.6(a) by assuming that $J_1 = J_0$, $J_2 = 2J_0$, and $k_{t1} = k_{t2} = k_{t3} = k_t$.

5.23. Determine the normal modes of the torsional system shown in Fig. 5.7 when $k_{t1} = k_t$, $k_{t2} = 5k_t$, $J_1 = J_0$, and $J_2 = 5J_0$.

5.24. A simplified ride model of a military vehicle is shown in Fig. 5.26(b). This model can be used to obtain information about the bounce and pitch modes of the vehicle. If the total mass of the vehicle is m and the mass moment of inertia about its C.G. is J_0, derive the equations of motion of the vehicle using two different sets of coordinates, as indicated in Section 5.5.

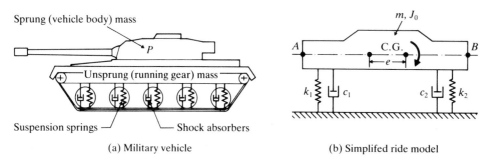

(a) Military vehicle   (b) Simplifed ride model

Figure 5.26

5.25. Find the natural frequencies and the amplitude ratios of the system shown in Fig. 5.27.

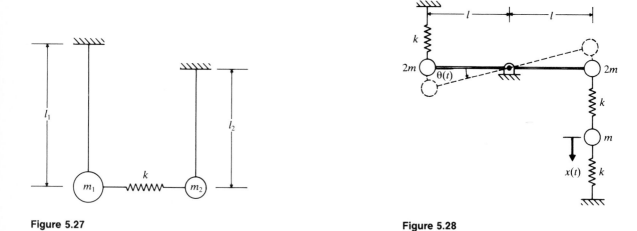

Figure 5.27   **Figure 5.28**

5.26. A rigid rod of negligible mass and length $2l$ is pivoted at the middle point and is constrained to move in the vertical plane by springs and masses, as shown in Fig. 5.28. Find the natural frequencies and mode shapes of the system.

5.27. An airfoil of mass m is suspended by a linear spring of stiffness k and a torsional spring of stiffness k_t in a wind tunnel, as shown in Fig. 5.29. The C.G. is located at a distance of e from point O. The mass moment of inertia of the airfoil about an axis passing through point O is J_0. Find the natural frequencies of the airfoil.

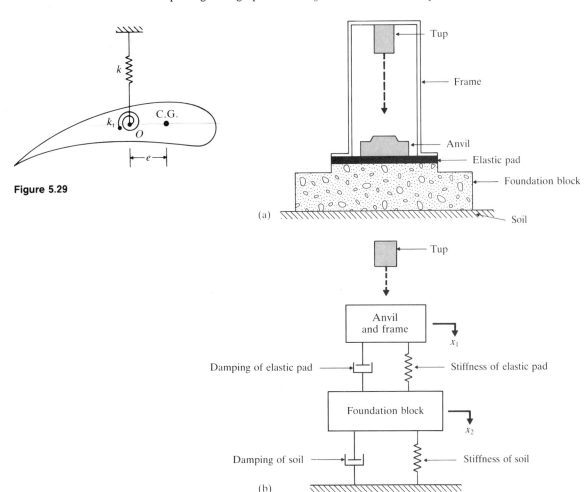

Figure 5.29

(a)

(b)

Figure 5.30

5.28. The expansion joints of a concrete highway, which are located at 15 m intervals, cause a series of impulses to affect cars running at a constant speed. Determine the speeds at which bounce motion and pitch motion are most likely to arise for the automobile of Example 5.5.

5.29. The weights of the tup, frame, anvil, and the foundation block in a forging hammer (Fig. 5.30) are 5000 lb, 40,000 lb, 60,000 lb and 140,000 lb, respectively. The stiffnesses of the elastic pad and the soil underneath the foundation block are 6×10^6 lb/in. and 3×10^6 lb/in., respectively. If the velocity of the tup before it strikes the anvil is 15 ft/sec, find (i) the natural frequencies of the system, and (ii) the magnitudes of displacement of the anvil and the foundation block. Assume the coefficient of restitution as 0.5 and damping to be negligible in the system.

5.30. Find (i) the natural frequencies of the system, and (ii) the responses of the anvil and the foundation block of the forging hammer shown in Fig. 5.30 when the time history of the force applied to the anvil is as shown in Fig. 5.31. Assume the following data:

$$\text{Mass of anvil and frame } (m_1) = 200 \text{ Mg}$$
$$\text{Mass of foundation block } (m_2) = 250 \text{ Mg}$$
$$\text{Stiffness of the elastic pad } (k_1) = 150 \text{ MN/m}$$
$$\text{Stiffness of the soil } (k_2) = 75 \text{ MN/m}$$
$$F_0 = 10^5 \text{ N and } T = 0.5 \text{ s}$$

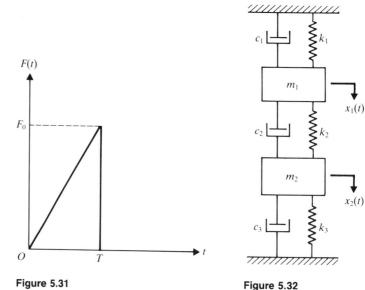

Figure 5.31   **Figure 5.32**

5.31. Derive the equations of motion for the free vibration of the system shown in Fig. 5.32. Assuming the solution as $x_i(t) = C_i e^{st}$, $i = 1, 2$, express the characteristic equation in the form

$$a_0 s^4 + a_1 s^3 + a_2 s^2 + a_3 s + a_4 = 0$$

Discuss the nature of possible solutions, $x_1(t)$ and $x_2(t)$.

5.32. Find the displacements $x_1(t)$ and $x_2(t)$ in Fig. 5.32 for $m_1 = 1$ kg, $m_2 = 2$ kg, $k_1 = k_2 = k_3 = 10,000$ N/m, and $c_1 = c_2 = c_3 = 2000$ N-s/m using the initial conditions $x_1(0) = 0.2$ m, $x_2(0) = 0.1$ m, and $\dot{x}_1(0) = \dot{x}_2(0) = 0$.

5.33. A centrifugal pump, having an unbalance of me, is supported on a rigid foundation of mass m_2 through isolator springs of stiffness k_1, as shown in Fig. 5.33. If the soil stiffness and damping are k_2 and c_2, find the displacements of the pump and the foundation for the following data: $mg = 0.5$ lb, $e = 6$ in., $m_1 g = 800$ lb, $k_1 = 2000$ lb/in., $m_2 g = 2000$ lb, $k_2 = 1000$ lb/in., $c_2 = 200$ lb-sec/in., and speed of pump = 1200 rpm.

5.34. A reciprocating engine of mass m_1 is mounted on a fixed-fixed beam of length l, width a, thickness t, and Young's modulus E as shown in Fig. 5.34. A spring-mass system

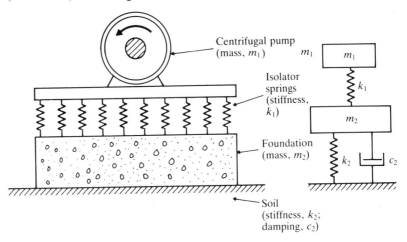

Figure 5.33

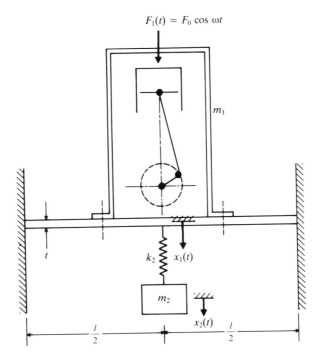

Figure 5.34

(k_2, m_2) is suspended from the beam as indicated in the figure. Find the relation between m_2 and k_2 that leads to no steady-state vibration of the beam when a harmonic force, $F_1(t) = F_0 \cos \omega t$, is developed in the engine during its operation.[†]

5.35. Find the steady-state response of the system shown in Fig. 5.15 by using the mechanical impedance method, when the mass m_1 is subjected to the force $F(t) = F_0 \sin \omega t$ in the direction of $x_1(t)$.

5.36. Find the steady-state response of the system shown in Fig. 5.15 when the base is subjected to a displacement $y(t) = Y_0 \cos \omega t$.

5.37. The mass m_1 of the two degree of freedom system shown in Fig. 5.15 is subjected to a force $F_0 \cos \omega t$. Assuming that the surrounding air damping is equivalent to $c = 200 \text{ N} \cdot \text{s/m}$, find the steady-state response of the two masses. Assume $m_1 = m_2 = 1$ kg, $k_1 = k_2 = 500 \text{ N/m}$, and $\omega = 1s^{-1}$.

5.38. Determine the steady-state vibration of the system shown in Fig. 5.3(a), assuming that $c_1 = c_2 = c_3 = 0$, $F_1(t) = F_{10} \cos \omega t$, and $F_2(t) = F_{20} \cos \omega t$.

5.39. In the system shown in Fig. 5.15, the mass m_1 is excited by a harmonic force having a maximum value of 50 N and a frequency of 2 Hz. Find the forced amplitude of each mass for $m_1 = 10$ kg, $m_2 = 5$ kg, $k_1 = 8000 \text{ N/m}$, and $k_2 = 2000 \text{ N/m}$.

5.40. Find the response of the two masses of the two-story frame shown in Fig. 5.21 under the ground displacement $y(t) = 0.2 \sin \pi t$ m. Assume the equivalent stiffness of the lower and upper columns to be 800 N/m and 600 N/m, respectively, and $m_1 = m_2 = 50$ kg.

5.41. Find the forced vibration response of the system shown in Fig. 5.12 when $F_1(t)$ is a step force of magnitude 5 N using the Laplace transformation method. Assume $x_1(0) = \dot{x}_1(0) = x_2(0) = \dot{x}_2(0) = 0$, $m = 1$ kg and $k = 100 \text{ N/m}$.

5.42. Determine the equations of motion and the natural frequencies of the system shown in Fig. 5.35.

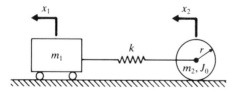

Figure 5.35

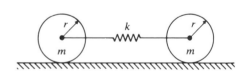

Figure 5.36

5.43. Two identical circular cylinders, of radius r and mass m each, are connected by a spring as shown in Fig. 5.36. Determine the natural frequencies of oscillation of the system.

5.44. The differential equations of motion for a two degree of freedom system are given by

$$a_1 \ddot{x}_1 + b_1 x_1 + c_1 x_2 = 0$$
$$a_2 \ddot{x}_2 + b_2 x_1 + c_2 x_2 = 0$$

Derive the condition to be satisfied for the system to be degenerate.

[†] The spring-mass system (k_2, m_2) added to make the amplitude of the first mass zero is known as a "vibration absorber." A detailed discussion of vibration absorbers is given in Section 9.10.

5.45. Find the angular displacements $\theta_1(t)$ and $\theta_2(t)$ of the system shown in Fig. 5.37 for the initial conditions $\theta_1(t = 0) = \theta_1(0)$, $\theta_2(t = 0) = \theta_2(0)$, and $\dot{\theta}_1(t = 0) = \dot{\theta}_2(t = 0) = 0$.

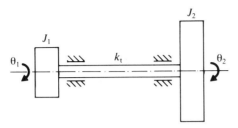

Figure 5.37

5.46. Determine the normal modes of the system shown in Fig. 5.7 with $k_{t1} = 0$. Show that the system with $k_{t1} = 0$ can be treated as a single degree of freedom system by using the coordinate $\alpha = \theta_1 - \theta_2$.

5.47. The transient vibrations of the drive line developed during the application of a cone (friction) clutch lead to unpleasant noise. To reduce the noise, a flywheel having a mass moment of inertia J_2 is attached to the drive line through a torsional spring k_{t2} and a viscous torsional damper c_{t2} as shown in Fig. 5.38. If the mass moment of inertia of the cone clutch is J_1 and the stiffness and damping constant of the drive line are given by k_{t1} and c_{t1}, respectively, derive the relations to be satisfied for the stable operation of the system.

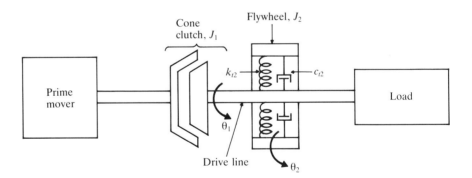

Figure 5.38

5.48. Find the response of the system shown in Fig. 5.3(a) using a numerical procedure when $k_1 = k$, $k_2 = 2k$, $k_3 = k$, $m_1 = 2m$, $m_2 = m$, $F_2(t) = 0$, and $F_1(t)$ is a rectangular pulse of magnitude 500 N and duration 0.5 sec. Assume $m = 10$ kg, $c_1 = c_2 = c_3 = 0$, and $k = 2000$ N/m.

5.49. (a) Find the roots of the frequency equation of the system shown in Fig. 5.3 using subroutine QUART with the following data: $m_1 = m_2 = 0.2$ lb-s²/in., $k_1 = k_2 =$

18 lb/in., $k_3 = 0$, $c_1 = c_2 = c_3 = 0$. (b) If the initial conditions are $x_1(0) = x_2(0) = 2$ in., $\dot{x}_1(0) = \dot{x}_2(0) = 0$, determine the displacements $x_1(t)$ and $x_2(t)$ of the masses.

5.50. Write a computer program for finding the steady-state response of a two degree of freedom system under the harmonic excitation $F_j(t) = F_{j0}e^{i\omega t}$; $j = 1,2$ using Eqs. (5.29) and (5.35). Use this program to find the response of a system with $m_{11} = m_{22} = 0.1$ lb-s^2/in., $m_{12} = 0$, $c_{11} = 1.0$ lb-s/in., $c_{12} = c_{22} = 0$, $k_{11} = 40$ lb/in., $k_{22} = 20$ lb/in., $k_{12} = -20$ lb/in., $F_{10} = 1$ lb, $F_{20} = 2$ lb, and $\omega = 5$ rad/s.

Project:

5.51. A step-cone pulley with a belt drive (Fig. 5.39) is used to change the cutting speeds in a lathe. The speed of the driving shaft is 350 rpm and the speeds of the output shaft are 150, 250, 450, and 750 rpm. The diameters of the driving and the driven pulleys, corresponding to 150 rpm output speed, are 250 mm and 1000 mm, respectively. The center distance between the shafts is 5 m. The mass moments of inertia of the driving and driven step cones are 0.1 and 0.2 kg-m^2, respectively. Find the cross sectional area of the belt to avoid resonance with any of the input/output speeds of the system. Assume the Young's modulus of the belt material as 10^{10} N/m^2.

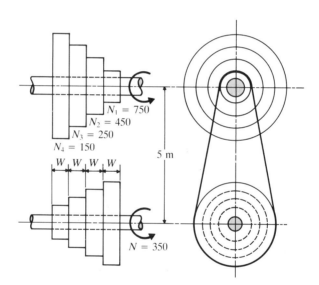

Figure 5.39

CHAPTER 6

Multidegree of Freedom Systems

Joseph Louis Lagrange (1736 – 1813) was an Italian-born mathematician famous for his work on theoretical mechanics. He was made professor of mathematics in 1755 at the Artillery School at Turin. Lagrange's masterpiece, his *Mèchanique*, contains what are now known as "Lagrange's equations," which are very useful in the study of vibrations. His work on elasticity and strength of materials, where he considered the strength and deflection of struts, is less well-known. (Courtesy Brown Brothers)

6.1 INTRODUCTION

All the concepts introduced in the preceding chapter can be directly extended to the case of multidegree of freedom systems. For example, there is one equation of motion for each degree of freedom; if generalized coordinates are used, there is one generalized coordinate for each degree of freedom. The equations of motion can be obtained from Newton's second law of motion or by using the influence coefficients defined in Section 6.3. However, it is often more convenient to derive the equations of motion of a multidegree of freedom system by using Lagrange's equations.

There are n natural frequencies, each associated with its own mode shape, for a system having n degrees of freedom. The method of determining the natural frequencies from the characteristic equation obtained by equating the determinant to zero also applies to these systems. However, as the number of degrees of freedom increases, the solution of the characteristic equation becomes more complex. The mode shapes exhibit a property known as *orthogonality*, which often enables us to simplify the analysis of multidegree of freedom systems.

6.2 MULTIDEGREE OF FREEDOM SPRING-MASS SYSTEM

Consider a simple n degree of freedom system, as shown in Fig. 6.1(a). With reference to the free-body diagram of a typical interior mass m_i, the equation of motion can be derived:

$$m_i \ddot{x}_i = -k_i(x_i - x_{i-1}) + k_{i+1}(x_{i+1} - x_i) + F_i; \qquad i = 2, 3, \ldots, n-1$$

or

$$m_i \ddot{x}_i - k_i x_{i-1} + (k_i + k_{i+1})x_i - k_{i+1}x_{i+1} = F_i;$$

$$i = 2, 3, \ldots, n-1 \qquad (6.1)$$

The equations of motion of the masses m_1 and m_n can be derived from Eq. (6.1) by setting $i = 1$ along with $x_0 = 0$ and $i = n$ along with $x_{n+1} = 0$, respectively:

$$m_1 \ddot{x}_1 + (k_1 + k_2)x_1 - k_2 x_2 = F_1 \qquad (6.2)$$

$$m_n \ddot{x}_n - k_n x_{n-1} + (k_n + k_{n+1})x_n = F_n \qquad (6.3)$$

Equations (6.1) to (6.3) can be expressed in matrix form as

$$[m]\ddot{\vec{x}} + [k]\vec{x} = \vec{F} \qquad (6.4)$$

where $[m]$ and $[k]$ are called the *mass matrix* and the *stiffness matrix*, respectively, and are given by

$$[m] = \begin{bmatrix} m_1 & 0 & 0 & \cdots & 0 & 0 \\ 0 & m_2 & 0 & \cdots & 0 & 0 \\ 0 & 0 & m_3 & \cdots & 0 & 0 \\ \vdots & & & & & \\ 0 & 0 & 0 & \cdots & 0 & m_n \end{bmatrix} \qquad (6.5)$$

$$[k] = \begin{bmatrix} (k_1 + k_2) & -k_2 & 0 & \cdots & 0 & 0 \\ -k_2 & (k_2 + k_3) & -k_3 & \cdots & 0 & 0 \\ 0 & -k_3 & (k_3 + k_4) & \cdots & 0 & 0 \\ \vdots & & & & & \\ 0 & 0 & 0 & \cdots & -k_n & (k_n + k_{n+1}) \end{bmatrix}$$
$$(6.6)$$

and $\vec{x}$, $\ddot{\vec{x}}$, and $\vec{F}$ are the displacement, acceleration, and force vectors, given by

$$\vec{x} = \begin{Bmatrix} x_1(t) \\ x_2(t) \\ \vdots \\ x_n(t) \end{Bmatrix}, \qquad \ddot{\vec{x}} = \begin{Bmatrix} \ddot{x}_1(t) \\ \ddot{x}_2(t) \\ \vdots \\ \ddot{x}_n(t) \end{Bmatrix}, \qquad \vec{F} = \begin{Bmatrix} F_1(t) \\ F_2(t) \\ \vdots \\ F_n(t) \end{Bmatrix} \qquad (6.7)$$

The spring-mass system considered above is a particular case of a general n degree of freedom spring-mass system. In their most general form, the mass and stiffness

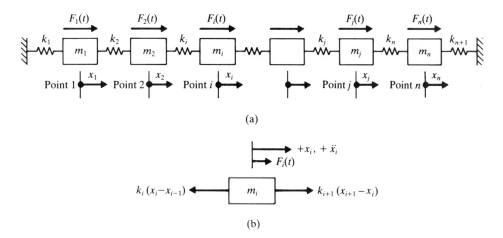

Figure 6.1

matrices are given by

$$[m] = \begin{bmatrix} m_{11} & m_{12} & m_{13} & \cdots & m_{1n} \\ m_{12} & m_{22} & m_{23} & \cdots & m_{2n} \\ \vdots & & & & \\ m_{1n} & m_{2n} & m_{3n} & \cdots & m_{nn} \end{bmatrix} \qquad (6.8)$$

and

$$[k] = \begin{bmatrix} k_{11} & k_{12} & k_{13} & \cdots & k_{1n} \\ k_{12} & k_{22} & k_{23} & \cdots & k_{2n} \\ \vdots & & & & \\ k_{1n} & k_{2n} & k_{3n} & \cdots & k_{nn} \end{bmatrix} \qquad (6.9)$$

6.3 INFLUENCE COEFFICIENTS

The equations of motion of a multidegree of freedom system can also be written in terms of influence coefficients, which are extensively used in structural engineering. For a linear spring, the force necessary to cause a unit elongation is called the *spring constant*. In more complex systems, we can express the relation between the displacement at a point and the forces acting at various other points of the system by means of influence coefficients. There are two types of influence coefficients: flexibility influence coefficients and stiffness influence coefficients. To illustrate the concept of an influence coefficient, consider the multidegree of freedom spring-mass system shown in Fig. 6.1.

Let the system be acted on by just one force F_j, and let the displacement at point i (i.e., mass m_i) due to F_j be x_{ij}. The flexibility influence coefficient, denoted by a_{ij}, is defined as the deflection at point i due to a unit load at point j. Since the deflection increases proportionately with the load for a linear system, we have

$$x_{ij} = a_{ij}F_j \tag{6.10}$$

If several forces F_j ($j = 1, 2, \ldots, n$) act at different points of the system, the total deflection at any point i can be found by summing up the contributions of all forces F_j:

$$x_i = \sum_{j=1}^{n} x_{ij} = \sum_{j=1}^{n} a_{ij}F_j, \qquad i = 1, 2, \ldots, n \tag{6.11}$$

Equation (6.11) can be expressed in matrix form as

$$\vec{x} = [a]\vec{F} \tag{6.12}$$

where $\vec{x}$ and $\vec{F}$ are the displacement and force vectors defined in Eq. (6.7) and $[a]$ is the flexibility matrix given by

$$[a] = \begin{bmatrix} a_{11} & a_{12} & \cdots & a_{1n} \\ a_{21} & a_{22} & \cdots & a_{2n} \\ \vdots & & & \\ a_{n1} & a_{n2} & \cdots & a_{nn} \end{bmatrix} \tag{6.13}$$

The stiffness influence coefficient, denoted by k_{ij}, is defined as the force at point i due to a unit displacement at point j when all the points other than the point j are fixed. The total force at point i, F_i, can be obtained by summing up the forces due to all displacements x_j ($j = 1, 2, \ldots, n$):

$$F_i = \sum_{j=1}^{n} k_{ij}x_j, \qquad i = 1, 2, \ldots, n \tag{6.14}$$

Equation (6.14) can be stated in matrix form as

$$\vec{F} = [k]\vec{x} \tag{6.15}$$

where $[k]$ is the stiffness matrix given by

$$[k] = \begin{bmatrix} k_{11} & k_{12} & \cdots & k_{1n} \\ k_{21} & k_{22} & \cdots & k_{2n} \\ \vdots & & & \\ k_{n1} & k_{n2} & \cdots & k_{nn} \end{bmatrix} \tag{6.16}$$

An examination of Eqs. (6.12) and (6.15) indicates that the flexibility and stiffness matrices are related. If we substitute Eq. (6.15) into Eq. (6.12), we obtain

$$\vec{x} = [a]\vec{F} = [a][k]\vec{x} \tag{6.17}$$

from which we can obtain the relation

$$[a][k] = [I] \tag{6.18}$$

where $[I]$ denotes the unit matrix. Equation (6.18) is equivalent to

$$[k] = [a]^{-1}, \qquad [a] = [k]^{-1} \tag{6.19}$$

That is, the stiffness and flexibility matrices are the inverse of one another. The use of dynamic stiffness influence coefficients in the vibration of nonuniform beams is discussed in Ref. [6.10].

Note the following aspects of influence coefficients:

1. Since the deflection at point i due to a unit load at point j is the same as the deflection at point j due to a unit load at point i for a linear system (Maxwell's reciprocity theorem [6.1]), we have $a_{ij} = a_{ji}$. By a similar reasoning, we have $k_{ij} = k_{ji}$.

2. The flexibility and stiffness influence coefficients can be calculated from the principles of solid mechanics.

3. The influence coefficients for torsional systems can be defined in terms of unit torque and the angular deflection it causes. For example, in a multirotor torsional system, a_{ij} can be defined as the angular displacement of point i (rotor i) due to a unit torque at point j.

EXAMPLE 6.1   **Flexibility Influence Coefficients**

Find the flexibility influence coefficients of the system shown in Fig. 6.2(a).

Given: Three degree of freedom spring-mass system, Fig. 6.2(a).

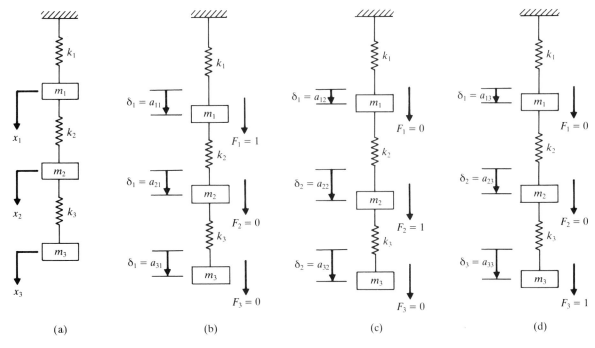

Figure 6.2

Find: Flexibility influence coefficients, a_{ij}.

Approach: Use the definition of a_{ij}.

Solution. Let x_1, x_2, and x_3 denote the displacements of the masses m_1, m_2, and m_3, respectively. The flexibility influence coefficients a_{ij} of the system can be determined in terms of the spring stiffnesses k_1, k_2, and k_3 as follows. If we apply a unit force at mass m_1 and no force at the other masses ($F_1 = 1$, $F_2 = F_3 = 0$), as shown in Fig. 6.2(b), the deflection of the mass m_1 is equal to $\delta_1 = 1/k_1 = a_{11}$. Since the other two masses m_2 and m_3 move (undergo rigid body translation) by the same amount of deflection δ_1, we have, by definition,

$$a_{21} = a_{31} = \delta_1 = \frac{1}{k_1}$$

Next, we apply a unit force at mass m_2 and no force at masses m_1 and m_3, as shown in Fig. 6.2(c). Since the two springs k_1 and k_2 offer resistance, the deflection of mass m_2 is given by

$$\delta_2 = \frac{1}{k_{eq}} = \frac{1}{k_1} + \frac{1}{k_2} = \frac{k_1 + k_2}{k_1 k_2} = a_{22}$$

The mass m_3 undergoes the same displacement δ_2 (rigid body translation) while the mass m_1 moves through a smaller distance given by $\delta_1 = 1/k_1$. Hence

$$a_{32} = \delta_2 = \frac{k_1 + k_2}{k_1 k_2} \qquad \text{and} \qquad a_{12} = \delta_1 = \frac{1}{k_1}$$

Finally, when we apply a unit force to mass m_3 and no force to masses m_1 and m_2, as shown in Fig. 6.2(d), the displacement of mass m_3 is given by

$$\delta_3 = \frac{1}{k_{eq}} = \frac{1}{k_1} + \frac{1}{k_2} + \frac{1}{k_3} = \frac{k_1 k_2 + k_2 k_3 + k_3 k_1}{k_1 k_2 k_3} = a_{33}$$

while the displacements of masses m_2 and m_1 are given by

$$\delta_2 = \frac{1}{k_1} + \frac{1}{k_2} = \frac{k_1 + k_2}{k_1 k_2} = a_{23}$$

and

$$\delta_1 = \frac{1}{k_1} = a_{13}$$

According to Maxwell's reciprocity theorem, we have

$$a_{ij} = a_{ji}$$

Thus the flexibility matrix of the system is given by

$$[a] = \begin{bmatrix} \dfrac{1}{k_1} & \dfrac{1}{k_1} & \dfrac{1}{k_1} \\[2mm] \dfrac{1}{k_1} & \left(\dfrac{1}{k_1} + \dfrac{1}{k_2}\right) & \left(\dfrac{1}{k_1} + \dfrac{1}{k_2}\right) \\[2mm] \dfrac{1}{k_1} & \left(\dfrac{1}{k_1} + \dfrac{1}{k_2}\right) & \left(\dfrac{1}{k_1} + \dfrac{1}{k_2} + \dfrac{1}{k_3}\right) \end{bmatrix} \qquad (E.1)$$

The stiffness matrix of the system can be found from the relation $[k] = [a]^{-1}$ or can be derived by using the definition of k_{ij} (see Problem 6.8):

$$[k] = \begin{bmatrix} (k_1 + k_2) & -k_2 & 0 \\ -k_2 & (k_2 + k_3) & -k_3 \\ 0 & -k_3 & k_3 \end{bmatrix} \tag{E.2}$$

EXAMPLE 6.2   **Flexibility Matrix of a Beam**

Derive the flexibility matrix of the weightless beam shown in Fig. 6.3(a). The beam is simply supported at both ends, and the three masses are placed at equal intervals. Assume the beam to be uniform with stiffness EI.

Given: Beam carrying three masses, Fig. 6.3(a).

Find: Flexibility matrix, $[a]$.

Approach: Use the definition of a_{ij} along with beam deflection formula.

Solution. Let x_1, x_2, and x_3 denote the total transverse deflection of the masses m_1, m_2, and m_3, respectively. From the known formula for the deflection of a pinned-pinned beam [6.2], the influence coefficients a_{1j} ($j = 1, 2, 3$) can be found by applying a unit load at the location of m_1 (see Fig. 6.3b):

$$a_{11} = \frac{9}{768} \frac{l^3}{EI}, \qquad a_{12} = \frac{11}{768} \frac{l^3}{EI}, \qquad a_{13} = \frac{7}{768} \frac{l^3}{EI} \tag{E.1}$$

Similarly, by applying a unit load at the locations of m_2 and m_3 separately, we obtain

$$a_{21} = a_{12} = \frac{11}{768} \frac{l^3}{EI}, \qquad a_{22} = \frac{1}{48} \frac{l^3}{EI}, \qquad a_{23} = \frac{11}{768} \frac{l^3}{EI} \tag{E.2}$$

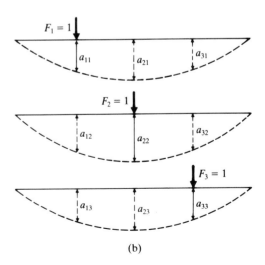

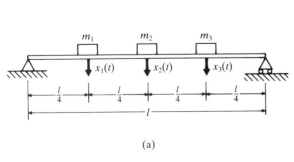

(a)   (b)

Figure 6.3

and

$$a_{31} = a_{13} = \frac{7}{768} \frac{l^3}{EI}, \qquad a_{32} = a_{23} = \frac{11}{768} \frac{l^3}{EI}, \qquad a_{33} = \frac{9}{768} \frac{l^3}{EI} \qquad (E.3)$$

Thus the flexibility matrix of the system is given by

$$[a] = \frac{l^3}{768EI} \begin{bmatrix} 9 & 11 & 7 \\ 11 & 16 & 11 \\ 7 & 11 & 9 \end{bmatrix} \qquad (E.4)$$

6.4 POTENTIAL AND KINETIC ENERGY EXPRESSIONS IN MATRIX FORM

Let x_i denote the displacement of mass m_i and F_i the force applied in the direction of x_i at mass m_i in an n degree of freedom system similar to the one shown in Fig. 6.1. The elastic potential energy (also known as *strain energy* or *energy of deformation*) of the ith spring is given by

$$V_i = \frac{1}{2} F_i x_i \qquad (6.20)$$

The total potential energy can be expressed as

$$V = \sum_{i=1}^{n} V_i = \frac{1}{2} \sum_{i=1}^{n} F_i x_i \qquad (6.21)$$

Since

$$F_i = \sum_{j=1}^{n} k_{ij} x_j \qquad (6.22)$$

Eq. (6.21) becomes

$$V = \frac{1}{2} \sum_{i=1}^{n} \left(\sum_{j=1}^{n} k_{ij} x_j \right) x_i = \frac{1}{2} \sum_{i=1}^{n} \sum_{j=1}^{n} k_{ij} x_i x_j \qquad (6.23)$$

Equation (6.23) can also be written in matrix form as*

$$V = \frac{1}{2} \vec{x}^T [k] \vec{x} \qquad (6.24)$$

where the displacement vector is given by Eq. (6.7) and the stiffness matrix is given by

$$[k] = \begin{bmatrix} k_{11} & k_{12} & \cdots & k_{1n} \\ k_{21} & k_{22} & \cdots & k_{2n} \\ \vdots & & & \\ k_{n1} & k_{n2} & \cdots & k_{nn} \end{bmatrix} \qquad (6.25)$$

---

*   Since the indices i and j can be interchanged in Eq. (6.23), we have the relation $k_{ij} = k_{ji}$.

The kinetic energy associated with mass m_i is, by definition, equal to

$$T_i = \frac{1}{2} m_i \dot{x}_i^2 \tag{6.26}$$

The total kinetic energy of the system can be expressed as

$$T = \sum_{i=1}^{n} T_i = \frac{1}{2} \sum_{i=1}^{n} m_i \dot{x}_i^2 \tag{6.27}$$

which can be written in matrix form as

$$T = \frac{1}{2} \dot{\vec{x}}^T [m] \dot{\vec{x}} \tag{6.28}$$

where the velocity vector $\dot{\vec{x}}$ is given by

$$\dot{\vec{x}} = \begin{Bmatrix} \dot{x}_1 \\ \dot{x}_2 \\ \vdots \\ \dot{x}_n \end{Bmatrix}$$

and the mass matrix $[m]$ is a diagonal matrix given by

$$[m] = \begin{bmatrix} m_1 & & & 0 \\ & m_2 & & \\ & & \cdot\cdot\cdot & \\ 0 & & & m_n \end{bmatrix} \tag{6.29}$$

If generalized coordinates (q_i), discussed in Section 6.5, are used instead of the physical displacements (x_i), the kinetic energy can be expressed as

$$T = \frac{1}{2} \dot{\vec{q}}^T [m] \dot{\vec{q}} \tag{6.30}$$

where $\dot{\vec{q}}$ is the vector of generalized velocities, given by

$$\dot{\vec{q}} = \begin{Bmatrix} \dot{q}_1 \\ \dot{q}_2 \\ \vdots \\ \dot{q}_n \end{Bmatrix} \tag{6.31}$$

and $[m]$ is called the *generalized mass matrix*, given by

$$[m] = \begin{bmatrix} m_{11} & m_{12} & \cdots & m_{1n} \\ m_{21} & m_{22} & \cdots & m_{2n} \\ \vdots & & & \\ m_{n1} & m_{n2} & \cdots & m_{nn} \end{bmatrix} \tag{6.32}$$

with $m_{ij} = m_{ji}$. The generalized mass matrix given by Eq. (6.32) is full, as opposed to the diagonal mass matrix of Eq. (6.29).

It can be seen that the potential energy is a quadratic function of the displacements, and the kinetic energy is a quadratic function of the velocities. Hence they are said to be in quadratic form. Since kinetic energy, by definition, cannot be negative and vanishes only when all the velocities vanish, Eqs. (6.28) and (6.30) are called *positive definite quadratic forms* and the mass matrix $[m]$ is called a *positive definite matrix*. On the other hand, the potential energy expression, Eq. (6.24), is a positive definite quadratic form, but the matrix $[k]$ is positive definite only if the system is a stable one. There are systems for which the potential energy is zero without the displacements or coordinates $x_1, x_2, \ldots, x_n$ being zero. In these cases the potential energy will be a positive quadratic function rather than positive definite; correspondingly, the matirx $[k]$ is said to be positive. A system for which $[k]$ is positive and $[m]$ is positive definite is called a semi-definite system (see Section 6.11).

6.5 GENERALIZED COORDINATES AND GENERALIZED FORCES

The equations of motion of a vibrating system can be formulated in a number of different coordinate systems. As stated earlier, n independent coordinates are necessary to describe the motion of a system having n degrees of freedom. Any set of n independent coordinates is called generalized coordinates, usually designated by $q_1, q_2, \ldots, q_n$. The generalized coordinates may be lengths, angles, or any other set of numbers that define the configuration of the system at any time uniquely. They are also independent of the conditions of constraint.

To illustrate the concept of generalized coordinates, consider the triple pendulum shown in Fig. 6.4. The configuration of the system can be specified by the six coordinates (x_j, y_j), $j = 1, 2, 3$. However, these coordinates are not independent but are constrained by the relations

$$x_1^2 + y_1^2 = l_1^2$$
$$(x_2 - x_1)^2 + (y_2 - y_1)^2 = l_2^2$$
$$(x_3 - x_2)^2 + (y_3 - y_2)^2 = l_3^2 \tag{6.33}$$

Since the coordinates (x_j, y_j), $j = 1, 2, 3$ are not independent, they cannot be called generalized coordinates. Without the constraints of Eq. (6.33), each of the masses m_1, m_2, and m_3 will be free to occupy any position in the x, y plane. The constraints eliminate three degrees of freedom from the six coordinates (two for each mass) and the system, thus, has only three degrees of freedom. If the angular displacements $\theta_j(j = 1, 2, 3)$ are used to specify the locations of the masses $m_j(j = 1, 2, 3)$ at any time, there will be no constraints on θ_j. Thus they form a set of generalized coordinates and are denoted as $q_j = \theta_j$, $j = 1, 2, 3$.

When external forces act on the system, the configuration of the system changes. The new configuration of the system can be obtained by changing the generalized coordinates q_j by δq_j, $j = 1, 2, \ldots, n$, where n denotes the number of generalized coordinates (or degrees of freedom) of the system. If U_j denotes the

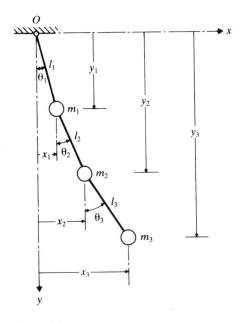

Figure 6.4

work done in changing the generalized coordinate q_j by the amount δq_j, the corresponding generalized force Q_j can be defined as

$$Q_j = \frac{U_j}{\delta q_j}, \qquad j = 1, 2, \ldots, n \qquad (6.34)$$

where Q_j will be a force (moment) when q_j is a linear (angular) displacement.

6.6 LAGRANGE'S EQUATIONS

The equations of motion of a vibrating system can often be derived in a simple manner in terms of generalized coordinates by the use of Lagrange's equations [6.3]. Lagrange's equations can be stated, for an n degree of freedom system, as

$$\frac{d}{dt}\left(\frac{\partial T}{\partial \dot{q}_j}\right) - \frac{\partial T}{\partial q_j} + \frac{\partial V}{\partial q_j} = Q_j^{(n)}, \qquad j = 1, 2, \ldots, n \qquad (6.35)$$

where $\dot{q}_j = \partial q_j / \partial t$ is the generalized velocity and $Q_j^{(n)}$ is the nonconservative generalized force corresponding to the generalized coordinate q_j. The forces represented by $Q_j^{(n)}$ may be dissipative (damping) forces or other external forces that are not derivable from a potential function. For example, if F_{xk}, F_{yk}, and F_{zk} represent the external forces acting on the kth mass of the system in the x, y, and z

directions, respectively, then the generalized force $Q_j^{(n)}$ can be computed as follows:

$$Q_j^{(n)} = \sum_k \left(F_{xk} \frac{\partial x_k}{\partial q_j} + F_{yk} \frac{\partial y_k}{\partial q_j} + F_{zk} \frac{\partial z_k}{\partial q_j} \right) \tag{6.36}$$

where x_k, y_k, and z_k are the displacements of the kth mass in the x, y, and z directions, respectively. For a conservative system, $Q_j^{(n)} = 0$, so Eqs. (6.35) take the form

$$\frac{d}{dt}\left(\frac{\partial T}{\partial \dot{q}_j} \right) - \frac{\partial T}{\partial q_j} + \frac{\partial V}{\partial q_j} = 0, \qquad j = 1, 2, \ldots, n \tag{6.37}$$

Equations (6.35) or (6.37) represent a system of n differential equations, one corresponding to each of the n generalized coordinates. Thus the equations of motion of the vibrating system can be derived, provided the energy expressions are available.

EXAMPLE 6.3   **Equations of Motion of a Torsional System**

The arrangement of the compressor, turbine, and generator in a thermal power plant is shown in Fig. 6.5. This arrangement can be considered as a torsional system where J_i denote the mass moments of inertia of the three components (compressor, turbine, and generator), M_{ti} indicate the external moments acting on the components, and k_{ti} represent the torsional spring constants of the shaft between the components as indicated in Fig. 6.5. Derive the equations of motion of the system using Lagrange's equations by treating the angular displacements of the components θ_i as generalized coordinates.

Given: Compressor-turbine-generator arrangement with known mass moments of inertia (J_i), external moments (M_{ti}), and stiffnesses (k_{ti}).

Find: Equations of motion.

Approach: Use Lagrange's equations.

Solution. Here $q_1 = \theta_1$, $q_2 = \theta_2$, and $q_3 = \theta_3$, and the kinetic energy of the system is given by

$$T = \frac{1}{2} J_1 \dot{\theta}_1^2 + \frac{1}{2} J_2 \dot{\theta}_2^2 + \frac{1}{2} J_3 \dot{\theta}_3^2 \tag{E.1}$$

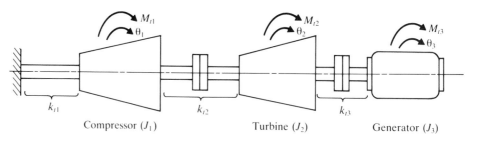

Figure 6.5

For the shaft, the potential energy is equal to the work done by the shaft as it returns from the dynamic configuration to the reference equilibrium position. Thus if θ denotes the angular displacement, for a shaft having a torsional spring constant k_t, the potential energy is equal to the work done in causing an angular displacement θ of the shaft:

$$V = \int_0^\theta (k_t \theta) \, d\theta = \frac{1}{2} k_t \theta^2 \tag{E.2}$$

Thus the total potential energy of the system can be expressed as

$$V = \frac{1}{2} k_{t1} \theta_1^2 + \frac{1}{2} k_{t2} (\theta_2 - \theta_1)^2 + \frac{1}{2} k_{t3} (\theta_3 - \theta_2)^2 \tag{E.3}$$

There are external moments applied to the components, so Eq. (6.36) gives

$$Q_j^{(n)} = \sum_{k=1}^{3} M_{tk} \frac{\partial \theta_k}{\partial q_j} = \sum_{k=1}^{3} M_{tk} \frac{\partial \theta_k}{\partial \theta_j} \tag{E.4}$$

from which we can obtain

$$Q_1^{(n)} = M_{t1} \frac{\partial \theta_1}{\partial \theta_1} + M_{t2} \frac{\partial \theta_2}{\partial \theta_1} + M_{t3} \frac{\partial \theta_3}{\partial \theta_1} = M_{t1}$$

$$Q_2^{(n)} = M_{t1} \frac{\partial \theta_1}{\partial \theta_2} + M_{t2} \frac{\partial \theta_2}{\partial \theta_2} + M_{t3} \frac{\partial \theta_3}{\partial \theta_2} = M_{t2}$$

$$Q_3^{(n)} = M_{t1} \frac{\partial \theta_1}{\partial \theta_3} + M_{t2} \frac{\partial \theta_2}{\partial \theta_3} + M_{t3} \frac{\partial \theta_3}{\partial \theta_3} = M_{t3} \tag{E.5}$$

Substituting Eqs. (E.1), (E.3), and (E.5) in Lagrange's equations, Eq. (6.35), we obtain for $j = 1, 2, 3$ the equations of motion:

$$J_1 \ddot{\theta}_1 + (k_{t1} + k_{t2}) \theta_1 - k_{t2} \theta_2 = M_{t1}$$
$$J_2 \ddot{\theta}_2 + (k_{t2} + k_{t3}) \theta_2 - k_{t2} \theta_1 - k_{t3} \theta_3 = M_{t2}$$
$$J_3 \ddot{\theta}_3 + k_{t3} \theta_3 - k_{t3} \theta_2 = M_{t3} \tag{E.6}$$

which can be expressed in matrix form as

$$\begin{bmatrix} J_1 & 0 & 0 \\ 0 & J_2 & 0 \\ 0 & 0 & J_3 \end{bmatrix} \begin{Bmatrix} \ddot{\theta}_1 \\ \ddot{\theta}_2 \\ \ddot{\theta}_3 \end{Bmatrix} + \begin{bmatrix} (k_{t1} + k_{t2}) & -k_{t2} & 0 \\ -k_{t2} & (k_{t2} + k_{t3}) & -k_{t3} \\ 0 & -k_{t3} & k_{t3} \end{bmatrix} \begin{Bmatrix} \theta_1 \\ \theta_2 \\ \theta_3 \end{Bmatrix} = \begin{Bmatrix} M_{t1} \\ M_{t2} \\ M_{t3} \end{Bmatrix} \tag{E.7}$$

6.7 GENERAL EQUATIONS OF MOTION IN MATRIX FORM

We can derive the equations of motion of a multidegree of freedom system in matrix form from Lagrange's equations,*

$$\frac{d}{dt} \left(\frac{\partial T}{\partial \dot{x}_i} \right) - \frac{\partial T}{\partial x_i} + \frac{\partial V}{\partial x_i} = F_i, \qquad i = 1, 2, \ldots, n \tag{6.38}$$

---

* The generalized coordinates are denoted as x_i instead of q_i and the generalized forces as F_i instead of $Q_i^{(n)}$ in Eq. (6.38).

where F_i is the nonconservative generalized force corresponding to the ith generalized coordinate x_i and $\dot{x}_i$ is the time derivative of x_i (generalized velocity). The kinetic and potential energies of a multidegree of freedom system can be expressed in matrix form as indicated in Section 6.4:

$$T = \frac{1}{2}\dot{\vec{x}}^T[m]\dot{\vec{x}} \tag{6.39}$$

$$V = \frac{1}{2}\vec{x}^T[k]\vec{x} \tag{6.40}$$

where $\vec{x}$ is the column vector of the generalized coordinates

$$\vec{x} = \begin{Bmatrix} x_1 \\ x_2 \\ \vdots \\ \vdots \\ x_n \end{Bmatrix} \tag{6.41}$$

From the theory of matrices, we obtain, by taking note of the symmetry of $[m]$,

$$\frac{\partial T}{\partial \dot{x}_i} = \frac{1}{2}\vec{\delta}^T[m]\dot{\vec{x}} + \frac{1}{2}\dot{\vec{x}}^T[m]\vec{\delta} = \vec{\delta}^T[m]\dot{\vec{x}}$$

$$= \vec{m}_i^T\dot{\vec{x}}, \qquad i = 1, 2, \ldots, n \tag{6.42}$$

where δ_{ji} is the Kronecker delta ($\delta_{ji} = 1$ if $j = i$ and $= 0$ if $j \neq i$), $\vec{\delta}$ is the column vector of Kronecker deltas whose elements in the rows for which $j \neq i$ are equal to zero and whose element in the row $i = j$ is equal to 1, and $\vec{m}_i^T$ is a row vector which is identical to the ith row of the matrix $[m]$. All the relations represented by Eq. (6.42) can be expressed as

$$\frac{\partial T}{\partial \dot{x}_i} = \vec{m}_i^T\dot{\vec{x}} \tag{6.43}$$

Differentiation of Eq. (6.43) with respect to time gives

$$\frac{d}{dt}\left(\frac{\partial T}{\partial \dot{x}_i}\right) = \vec{m}_i^T\ddot{\vec{x}}, \qquad i = 1, 2, \ldots, n \tag{6.44}$$

since the mass matrix is not a function of time. Further, the kinetic energy is a function of only the velocities $\dot{x}_i$, and so

$$\frac{\partial T}{\partial x_i} = 0, \qquad i = 1, 2, \ldots, n \tag{6.45}$$

Similarly, we can differentiate Eq. (6.40), taking note of the symmetry of $[k]$:

$$\frac{\partial V}{\partial x_i} = \frac{1}{2}\vec{\delta}^T[k]\vec{x} + \frac{1}{2}\vec{x}^T[k]\vec{\delta} = \vec{\delta}^T[k]\vec{x}$$

$$= \vec{k}_i^T\vec{x}, \qquad i = 1, 2, \ldots, n \tag{6.46}$$

where $\vec{k}_i^T$ is a row vector identical to the ith row of the matrix $[k]$. By substituting Eqs. (6.44) to (6.46) into Eq. (6.38), we obtain the desired equations of motion in matrix form:

$$[m]\ddot{\vec{x}} + [k]\vec{x} = \vec{F} \tag{6.47}$$

where

$$\vec{F} = \begin{Bmatrix} F_1 \\ F_2 \\ \vdots \\ F_n \end{Bmatrix} \tag{6.48}$$

Note that if the system is conservative, there are no nonconservative forces F_i, so the equations of motion become

$$[m]\ddot{\vec{x}} + [k]\vec{x} = \vec{0} \tag{6.49}$$

Note also that if the generalized coordinates x_i are same as the actual (physical) displacements, the mass matrix $[m]$ is a diagonal matrix.

6.8 EIGENVALUE PROBLEM

The solution of Eq. (6.49) corresponds to the undamped free vibration of the system. In this case, if the system is given some energy in the form of initial displacements or initial velocities or both, it vibrates indefinitely because there is no dissipation of energy. We can find the solution of Eq. (6.49) by assuming a solution of the form

$$x_i(t) = X_i T(t), \qquad i = 1, 2, \ldots, n \tag{6.50}$$

where X_i is a constant and T is a function of time t. Equation (6.50) shows that the amplitude ratio of two coordinates

$$\left\{ \frac{x_i(t)}{x_j(t)} \right\}$$

is independent of time. Physically, this means that all coordinates have synchronous motions. The configuration of the system does not change its shape during motion, but its amplitude does. The configuration of the system, given by the vector

$$\vec{X} = \begin{Bmatrix} X_1 \\ X_2 \\ \vdots \\ X_n \end{Bmatrix}$$

is known as the *mode shape* of the system. Substituting Eq. (6.50) into Eq. (6.49), we obtain

$$[m]\vec{X}\ddot{T}(t) + [k]\vec{X}T(t) = \vec{0} \tag{6.51}$$

Equation (6.51) can be written in scalar form as n separate equations:

$$\left(\sum_{j=1}^{n} m_{ij} X_j \right) \ddot{T}(t) + \left(\sum_{j=1}^{n} k_{ij} X_j \right) T(t) = 0, \qquad i = 1, 2, \ldots, n \tag{6.52}$$

from which we can obtain the relations

$$-\frac{\ddot{T}(t)}{T(t)} = \frac{\left(\sum\limits_{j=1}^{n} k_{ij} X_j\right)}{\left(\sum\limits_{j=1}^{n} m_{ij} X_j\right)}, \qquad i = 1, 2, \ldots, n \tag{6.53}$$

Since the left-hand side of Eq. (6.53) is independent of the index i, and the right-hand side is independent of t, both sides must be equal to a constant. By assuming this constant* as ω^2, we can write Eqs. (6.53) as

$$\ddot{T}(t) + \omega^2 T(t) = 0 \tag{6.54}$$

$$\sum\limits_{j=1}^{n} \left(k_{ij} - \omega^2 m_{ij}\right) X_j = 0, \qquad i = 1, 2, \ldots, n$$

or

$$\left[[k] - \omega^2 [m]\right] \vec{X} = \vec{0} \tag{6.55}$$

The solution of Eq. (6.54) can be expressed as

$$T(t) = C_1 \cos(\omega t + \phi) \tag{6.56}$$

where C_1 and ϕ are constants, known as the *amplitude* and the *phase angle*, respectively. Equation (6.56) shows that all the coordinates can perform a harmonic motion with the same frequency ω and the same phase angle ϕ. However, the frequency ω cannot take any arbitrary value; it has to satisfy Eq. (6.55). Since Eq. (6.55) represents a set of n linear homogeneous equations in the unknowns X_i ($i = 1, 2, \ldots, n$), the trivial solution is $X_1 = X_2 = \cdots = X_n = 0$. For a nontrivial solution of Eq. (6.55), the determinant Δ of the coefficient matrix must be zero. That is,

$$\Delta = |k_{ij} - \omega^2 m_{ij}| = |[k] - \omega^2 [m]| = 0 \tag{6.57}$$

Equation (6.55) represents what is known as the *eigenvalue* or *characteristic value* problem, Eq. (6.57) is called the *characteristic equation*, ω^2 is known as the *eigenvalue* or the *characteristic value*, and ω is called the *natural frequency* of the system.

The expansion of Eq. (6.57) leads to an nth order polynomial equation in ω^2. The solution (roots) of this polynomial or characteristic equation gives n values of ω^2. It can be shown that all the n roots are real and positive when the matrices $[k]$ and $[m]$ are symmetric and positive definite [6.4], as in the present case. If $\omega_1^2, \omega_2^2, \ldots, \omega_n^2$ denote the n roots in ascending order of magnitude, their positive square roots give the n natural frequencies of the system $\omega_1 \leqslant \omega_2 \leqslant \cdots \leqslant \omega_n$. The lowest value ($\omega_1$) is called the *fundamental or first natural frequency*. In general, all

---

* The constant is assumed to be a positive number, ω^2, so as to obtain a harmonic solution to the resulting Eq. (6.54). Otherwise, the solution of $T(t)$ and hence that of $x(t)$ become exponential, which violates the physical limitations of finite total energy.

the natural frequencies ω_i are distinct, although in some cases two natural frequencies might possess the same value.

6.9 SOLUTION OF THE EIGENVALUE PROBLEM

Several methods are available to solve an eigenvalue problem. We shall consider an elementary method in this section.

6.9.1
Solution of the
Characteristic
(Polynomial)
Equation

Equation (6.55) can also be expressed as

$$[\lambda[k] - [m]] \vec{X} = \vec{0} \tag{6.58}$$

where

$$\lambda = \frac{1}{\omega^2} \tag{6.59}$$

By premultiplying Eq. (6.58) by $[k]^{-1}$, we obtain

$$[\lambda[I] - [D]] \vec{X} = \vec{0}$$

or

$$\lambda[I]\vec{X} = [D]\vec{X} \tag{6.60}$$

where $[I]$ is the identity matrix and

$$[D] = [k]^{-1}[m] \tag{6.61}$$

is called the *dynamical matrix*. The eigenvalue problem of Eq. (6.60) is known as the *standard eigenvalue problem*. For a nontrivial solution of $\vec{X}$, the characteristic determinant must be zero—that is,

$$\Delta = |\lambda[I] - [D]| = 0 \tag{6.62}$$

On expansion, Eq. (6.62) gives an nth degree polynomial in λ, known as the *characteristic* or *frequency equation*. If the degree of freedom of the system (n) is large, the solution of this polynomial equation becomes quite tedious. We must use some numerical method, several of which are available to find the roots of a polynomial equation [6.5].

EXAMPLE 6.4   **Natural Frequencies of a Three Degree-of-Freedom System**

Find the natural frequencies and mode shapes of the system shown in Fig. 6.2 for $k_1 = k_2 = k_3 = k$ and $m_1 = m_2 = m_3 = m$.

Given: Three degree of freedom spring-mass system with equal masses and stiffnesses.

Find: Natural frequencies and mode shapes.

Approach: Find the eigenvalues and eigenvectors of the dynamical matrix.

Solution. The dynamical matrix is given by

$$[D] = [k]^{-1}[m] \equiv [a][m] \tag{E.1}$$

where the flexibility and mass matrices can be obtained from Example 6.1:

$$[a] = \frac{1}{k} \begin{bmatrix} 1 & 1 & 1 \\ 1 & 2 & 2 \\ 1 & 2 & 3 \end{bmatrix} \tag{E.2}$$

and

$$[m] = m \begin{bmatrix} 1 & 0 & 0 \\ 0 & 1 & 0 \\ 0 & 0 & 1 \end{bmatrix} \tag{E.3}$$

Thus

$$[D] = \frac{m}{k} \begin{bmatrix} 1 & 1 & 1 \\ 1 & 2 & 2 \\ 1 & 2 & 3 \end{bmatrix} \tag{E.4}$$

By setting the characteristic determinant equal to zero, we obtain the frequency equation:

$$\Delta = |\lambda[I] - [D]| = \left| \begin{bmatrix} \lambda & 0 & 0 \\ 0 & \lambda & 0 \\ 0 & 0 & \lambda \end{bmatrix} - \frac{m}{k} \begin{bmatrix} 1 & 1 & 1 \\ 1 & 2 & 2 \\ 1 & 2 & 3 \end{bmatrix} \right| = 0 \tag{E.5}$$

where

$$\lambda = \frac{1}{\omega^2} \tag{E.6}$$

By dividing throughout by λ, Eq. (E.5) gives

$$\begin{vmatrix} 1 - \alpha & -\alpha & -\alpha \\ -\alpha & 1 - 2\alpha & -2\alpha \\ -\alpha & -2\alpha & 1 - 3\alpha \end{vmatrix} = \alpha^3 - 5\alpha^2 + 6\alpha - 1 = 0 \tag{E.7}$$

where

$$\alpha = \frac{m}{k\lambda} = \frac{m\omega^2}{k} \tag{E.8}$$

The roots of the cubic equation (E.7) are given by

$$\alpha_1 = \frac{m\omega_1^2}{k} = 0.19806, \qquad \omega_1 = 0.44504 \sqrt{\frac{k}{m}} \tag{E.9}$$

$$\alpha_2 = \frac{m\omega_2^2}{k} = 1.5553, \qquad \omega_2 = 1.2471 \sqrt{\frac{k}{m}} \tag{E.10}$$

$$\alpha_3 = \frac{m\omega_3^2}{k} = 3.2490, \qquad \omega_3 = 1.8025 \sqrt{\frac{k}{m}} \tag{E.11}$$

Once the natural frequencies are known, the mode shapes or *eigenvectors* can be calculated, using Eq. (6.60):

$$[\lambda_i[I] - [D]] \vec{X}^{(i)} = \vec{0}, \qquad i = 1, 2, 3 \tag{E.12}$$

where

$$\vec{X}^{(i)} = \begin{Bmatrix} X_1^{(i)} \\ X_2^{(i)} \\ X_3^{(i)} \end{Bmatrix}$$

denotes the ith mode shape. The procedure is outlined below.

First mode: By substituting the value of ω_1 $\left(\text{i.e., } \lambda_1 = 5.0489\dfrac{m}{k}\right)$ in Eq. (E.12), we obtain

$$\left[5.0489\frac{m}{k}\begin{bmatrix} 1 & 0 & 0 \\ 0 & 1 & 0 \\ 0 & 0 & 1 \end{bmatrix} - \frac{m}{k}\begin{bmatrix} 1 & 1 & 1 \\ 1 & 2 & 2 \\ 1 & 2 & 3 \end{bmatrix} \right] \begin{Bmatrix} X_1^{(1)} \\ X_2^{(1)} \\ X_3^{(1)} \end{Bmatrix} = \begin{Bmatrix} 0 \\ 0 \\ 0 \end{Bmatrix}$$

That is,

$$\begin{bmatrix} 4.0489 & -1.0 & -1.0 \\ -1.0 & 3.0489 & -2.0 \\ -1.0 & -2.0 & 2.0489 \end{bmatrix} \begin{Bmatrix} X_1^{(1)} \\ X_2^{(1)} \\ X_3^{(1)} \end{Bmatrix} = \begin{Bmatrix} 0 \\ 0 \\ 0 \end{Bmatrix} \tag{E.13}$$

Equation (E.13) denotes a system of three homogeneous linear equations in the three unknowns $X_1^{(1)}$, $X_2^{(1)}$, and $X_3^{(1)}$. Any two of these unknowns can be expressed in terms of the remaining one. If we choose, arbitrarily, to express $X_2^{(1)}$ and $X_3^{(1)}$ in terms of $X_1^{(1)}$, we obtain from the first two rows of Eq. (E.13).

$$X_2^{(1)} + X_3^{(1)} = 4.0489X_1^{(1)}$$

$$3.0489X_2^{(1)} - 2.0X_3^{(1)} = X_1^{(1)} \tag{E.14}$$

Once Eqs. (E.14) are satisfied, the third row of Eq. (E.13) is satisfied automatically. The solution of Eqs. (E.14) can be obtained:

$$X_2^{(1)} = 1.8019X_1^{(1)} \qquad \text{and} \qquad X_3^{(1)} = 2.2470X_1^{(1)} \tag{E.15}$$

Thus the first mode shape is given by

$$\vec{X}^{(1)} = X_1^{(1)} \begin{Bmatrix} 1.0 \\ 1.8019 \\ 2.2470 \end{Bmatrix} \tag{E.16}$$

where the value of $X_1^{(1)}$ can be chosen arbitrarily.

Second mode: The substitution of the value of ω_2 $\left(\text{i.e., } \lambda_2 = 0.6430\dfrac{m}{k}\right)$ in Eq. (E.12) leads to

$$\left[0.6430\frac{m}{k}\begin{bmatrix} 1 & 0 & 0 \\ 0 & 1 & 0 \\ 0 & 0 & 1 \end{bmatrix} - \frac{m}{k}\begin{bmatrix} 1 & 1 & 1 \\ 1 & 2 & 2 \\ 1 & 2 & 3 \end{bmatrix} \right] \begin{Bmatrix} X_1^{(2)} \\ X_2^{(2)} \\ X_3^{(2)} \end{Bmatrix} = \begin{Bmatrix} 0 \\ 0 \\ 0 \end{Bmatrix}$$

that is,

$$\begin{bmatrix} -0.3570 & -1.0 & -1.0 \\ -1.0 & -1.3570 & -2.0 \\ -1.0 & -2.0 & -2.3570 \end{bmatrix} \begin{Bmatrix} X_1^{(2)} \\ X_2^{(2)} \\ X_3^{(2)} \end{Bmatrix} = \begin{Bmatrix} 0 \\ 0 \\ 0 \end{Bmatrix} \tag{E.17}$$

As before, the first two rows of Eq. (E.17) can be used to obtain

$$-X_2^{(2)} - X_3^{(2)} = 0.3570 X_1^{(2)}$$
$$-1.3570 X_2^{(2)} - 2.0 X_3^{(2)} = X_1^{(2)} \tag{E.18}$$

The solution of Eqs. (E.18) leads to

$$X_2^{(2)} = 0.4450 X_1^{(2)} \qquad \text{and} \qquad X_3^{(2)} = -0.8020 X_1^{(2)} \tag{E.19}$$

Thus the second mode shape can be expressed as

$$\vec{X}^{(2)} = X_1^{(2)} \begin{Bmatrix} 1.0 \\ 0.4450 \\ -0.8020 \end{Bmatrix} \tag{E.20}$$

where the value of $X_1^{(2)}$ can be chosen arbitrarily.

Third mode. To find the third mode, we substitute the value of ω_3 $\left(\text{i.e., } \lambda_3 = 0.3078 \dfrac{m}{k} \right)$ in Eq. (E.12) and obtain

$$\left[0.3078 \frac{m}{k} \begin{bmatrix} 1 & 0 & 0 \\ 0 & 1 & 0 \\ 0 & 0 & 1 \end{bmatrix} - \frac{m}{k} \begin{bmatrix} 1 & 1 & 1 \\ 1 & 2 & 2 \\ 1 & 2 & 3 \end{bmatrix} \right] \begin{Bmatrix} X_1^{(3)} \\ X_2^{(3)} \\ X_3^{(3)} \end{Bmatrix} = \begin{Bmatrix} 0 \\ 0 \\ 0 \end{Bmatrix}$$

that is,

$$\begin{bmatrix} -0.6922 & -1.0 & -1.0 \\ -1.0 & -1.6922 & -2.0 \\ -1.0 & -2.0 & -2.6922 \end{bmatrix} \begin{Bmatrix} X_1^{(3)} \\ X_2^{(3)} \\ X_3^{(3)} \end{Bmatrix} = \begin{Bmatrix} 0 \\ 0 \\ 0 \end{Bmatrix} \tag{E.21}$$

The first two rows of Eq. (E.21) can be written as

$$-X_2^{(3)} - X_3^{(3)} = 0.6922 X_1^{(3)}$$
$$-1.6922 X_2^{(3)} - 2.0 X_3^{(3)} = X_1^{(3)} \tag{E.22}$$

Equations (E.22) give

$$X_2^{(3)} = -1.2468 X_1^{(3)} \qquad \text{and} \qquad X_3^{(3)} = 0.5544 X_1^{(3)} \tag{E.23}$$

Hence the third mode shape can be written as

$$\vec{X}^{(3)} = X_1^{(3)} \begin{Bmatrix} 1.0 \\ -1.2468 \\ 0.5544 \end{Bmatrix} \tag{E.24}$$

where the value of $X_1^{(3)}$ is arbitrary. The values of $X_1^{(1)}$, $X_1^{(2)}$, and $X_1^{(3)}$ are usually taken as 1, and the mode shapes are shown in Fig. 6.6.

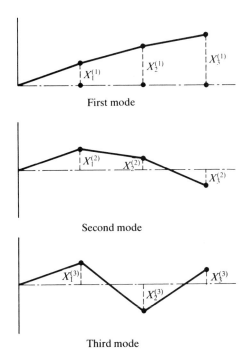

First mode

Second mode

Third mode

Figure 6.6

6.9.2
Orthogonality of
Normal Modes

In the previous section we considered a method of finding the n natural frequencies ω_i and the corresponding normal modes or modal vectors $\vec{X}^{(i)}$. We shall now see an important property of the normal modes—*orthogonality*. The natural frequency ω_i and the corresponding modal vector $\vec{X}^{(i)}$ satisfy Eq. (6.55) so that

$$\omega_i^2 [m] \vec{X}^{(i)} = [k] \vec{X}^{(i)} \tag{6.63}$$

If we consider another natural frequency ω_j and the corresponding modal vector $\vec{X}^{(j)}$, they also satisfy Eq. (6.55) so that

$$\omega_j^2 [m] \vec{X}^{(j)} = [k] \vec{X}^{(j)} \tag{6.64}$$

By premultiplying Eqs. (6.63) and (6.64) by $\vec{X}^{(j)^T}$ and $\vec{X}^{(i)^T}$ respectively, we obtain, by considering the symmetry of the matrices $[k]$ and $[m]$,

$$\omega_i^2 \vec{X}^{(j)^T}[m] \vec{X}^{(i)} = \vec{X}^{(j)^T}[k] \vec{X}^{(i)} \equiv \vec{X}^{(i)^T}[k] \vec{X}^{(j)} \tag{6.65}$$

$$\omega_j^2 \vec{X}^{(i)^T}[m] \vec{X}^{(j)} \equiv \omega_j^2 \vec{X}^{(j)^T}[m] \vec{X}^{(i)} = \vec{X}^{(i)^T}[k] \vec{X}^{(j)} \tag{6.66}$$

By subtracting Eq. (6.66) from Eq. (6.65), we obtain

$$\left(\omega_i^2 - \omega_j^2 \right) \vec{X}^{(j)^T} [m] \vec{X}^{(i)} = 0 \tag{6.67}$$

In general, $\omega_i^2 \neq \omega_j^2$, so Eq. (6.67) leads to*

$$\vec{X}^{(j)^T} [m] \vec{X}^{(i)} = 0, \qquad i \neq j \tag{6.68}$$

From Eqs. (6.65) and (6.66), we obtain, in view of Eq. (6.68),

$$\vec{X}^{(j)^T} [k] \vec{X}^{(i)} = 0, \qquad i \neq j \tag{6.69}$$

Equations (6.68) and (6.69) indicate that the modal vectors $\vec{X}^{(i)}$ and $\vec{X}^{(j)}$ are orthogonal with respect to both mass and stiffness matrices.

When $i = j$, the left-hand sides of Eqs. (6.68) and (6.69) are not equal to zero, but yield the generalized mass and stiffness coefficients of the ith mode:

$$M_{ii} = \vec{X}^{(i)^T} [m] \vec{X}^{(i)}, \qquad i = 1, 2, \ldots, n \tag{6.70}$$

$$K_{ii} = \vec{X}^{(i)^T} [k] \vec{X}^{(i)}, \qquad i = 1, 2, \ldots, n \tag{6.71}$$

Equations (6.70) and (6.71) can be written in matrix form as

$$[M] = \begin{bmatrix} M_{11} & & & 0 \\ & M_{22} & & \\ & & \ddots & \\ 0 & & & M_{nn} \end{bmatrix} = [X]^T [m][X] \tag{6.72}$$

$$[K] = \begin{bmatrix} K_{11} & & & 0 \\ & K_{22} & & \\ & & \ddots & \\ 0 & & & K_{nn} \end{bmatrix} = [X]^T [k][X] \tag{6.73}$$

where $[X]$ is called the *modal matrix*, in which the ith column corresponds to the ith modal vector:

$$[X] = \begin{bmatrix} \vec{X}^{(1)} & \vec{X}^{(2)} & \cdots & \vec{X}^{(n)} \end{bmatrix} \tag{6.74}$$

In many cases, we normalize the modal vectors $\vec{X}^{(i)}$ such that $[M] = [I]$—that is,

$$\vec{X}^{(i)^T} [m] \vec{X}^{(i)} = 1, \qquad i = 1, 2, \ldots, n \tag{6.75}$$

---

* In the case of repeated eigenvalues, $\omega_i = \omega_j$, the associated modal vectors are orthogonal to all the remaining modal vectors, but are not usually orthogonal to each other.

In this case the matrix $\lceil K \rfloor$ reduces to

$$[K] = \lceil \omega_i^2 \rfloor = \begin{bmatrix} \omega_1^2 & & & 0 \\ & \omega_2^2 & & \\ & & \ddots & \\ 0 & & & \omega_n^2 \end{bmatrix} \tag{6.76}$$

EXAMPLE 6.5   **Orthonormalization of Eigenvectors**

Orthonormalize the eigenvectors of Example 6.4 with respect to the mass matrix.

Given: Eigenvectors (of Example 6.4).

Find: Eigenvectors, orthonormalized with respect to the matrix $[m]$.

Approach: Multiply each eigenvector by a constant and find its value from the relation $\vec{X}^{(i)^T}[m]\vec{X}^{(i)} = 1$, $i = 1, 2, 3$.

Solution. The eigenvectors of Example 6.4 are given by

$$\vec{X}^{(1)} = X_1^{(1)} \begin{Bmatrix} 1.0 \\ 1.8019 \\ 2.2470 \end{Bmatrix}$$

$$\vec{X}^{(2)} = X_1^{(2)} \begin{Bmatrix} 1.0 \\ 0.4450 \\ -0.8020 \end{Bmatrix}$$

$$\vec{X}^{(3)} = X_1^{(3)} \begin{Bmatrix} 1.0 \\ -1.2468 \\ 0.5544 \end{Bmatrix}$$

The mass matrix is given by

$$[m] = m \begin{bmatrix} 1 & 0 & 0 \\ 0 & 1 & 0 \\ 0 & 0 & 1 \end{bmatrix}$$

The eigenvector $\vec{X}^{(i)}$ is said to be $[m]$-orthonormal if the following condition is satisfied:

$$\vec{X}^{(i)^T}[m]\vec{X}^{(i)} = 1 \tag{E.1}$$

Thus for $i = 1$, Eq. (E.1) leads to

$$m\left(X_1^{(1)}\right)^2 (1.0^2 + 1.8019^2 + 2.2470^2) = 1$$

or

$$X_1^{(1)} = \frac{1}{\sqrt{m(9.2959)}} = \frac{0.3280}{\sqrt{m}}$$

Similarly, for $i = 2$ and $i = 3$, Eq. (E.1) gives

$$m\left(X_1^{(2)}\right)^2 (1.0^2 + 0.4450^2 + \{-0.8020\}^2) = 1 \qquad \text{or} \qquad X_1^{(2)} = \frac{0.7370}{\sqrt{m}}$$

and

$$m\left(X_1^{(3)}\right)^2\left(1.0^2 + \{-1.2468\}^2 + 0.5544^2\right) = 1 \quad \text{or} \quad X_1^{(3)} = \frac{0.5911}{\sqrt{m}}$$

6.9.3
Repeated
Eigenvalues

When the characteristic equation possesses repeated roots, the corresponding mode shapes are not unique. To see this, let $\vec{X}^{(1)}$ and $\vec{X}^{(2)}$ be the mode shapes corresponding to the repeated eigenvalue $\lambda_1 = \lambda_2 = \lambda$ and $\vec{X}^{(3)}$ be the mode shape corresponding to a different eigenvalue λ_3. Equation (6.60) can be written as

$$[D]\vec{X}^{(1)} = \lambda\vec{X}^{(1)} \tag{6.77}$$

$$[D]\vec{X}^{(2)} = \lambda\vec{X}^{(2)} \tag{6.78}$$

$$[D]\vec{X}^{(3)} = \lambda_3\vec{X}^{(3)} \tag{6.79}$$

By multiplying Eq. (6.77) by a constant p and adding to Eq. (6.78), we obtain

$$[D]\left(p\vec{X}^{(1)} + \vec{X}^{(2)}\right) = \lambda\left(p\vec{X}^{(1)} + \vec{X}^{(2)}\right) \tag{6.80}$$

This shows that the new mode shape, $(p\vec{X}^{(1)} + \vec{X}^{(2)})$, which is a linear combination of the first two, also satisfies Eq. (6.60), so the mode shape corresponding to λ is not unique. Any $\vec{X}$ corresponding to λ must be orthogonal to $\vec{X}^{(3)}$ if it is to be a normal mode. If all three modes are orthogonal, they will be linearly independent and can be used to describe the free vibration resulting from any initial conditions.

The response of a multidegree of freedom system with repeated natural frequencies to force and displacement excitation was presented by Mahalingam and Bishop [6.16].

EXAMPLE 6.6   **Repeated Eigenvalues**

Determine the eigenvalues and eigenvectors of a vibrating system for which

$$[m] = \begin{bmatrix} 1 & 0 & 0 \\ 0 & 2 & 0 \\ 0 & 0 & 1 \end{bmatrix} \quad \text{and} \quad [k] = \begin{bmatrix} 1 & -2 & 1 \\ -2 & 4 & -2 \\ 1 & -2 & 1 \end{bmatrix}$$

Given: Mass and stiffness matrices.

Find: Eigenvalues and eigenvectors.

Approach: Use Eq. (6.80) for repeated eigenvalues.

Solution. The eigenvalue equation $[[k] - \lambda[m]]\vec{X} = \vec{0}$ can be written in the form

$$\begin{bmatrix} (1-\lambda) & -2 & 1 \\ -2 & 2(2-\lambda) & -2 \\ 1 & -2 & (1-\lambda) \end{bmatrix}\begin{Bmatrix} X_1 \\ X_2 \\ X_3 \end{Bmatrix} = \begin{Bmatrix} 0 \\ 0 \\ 0 \end{Bmatrix} \tag{E.1}$$

where $\lambda = \omega^2$. The characteristic equation gives

$$|[k] - \lambda[m]| = \lambda^2(\lambda - 4) = 0$$

so

$$\lambda_1 = 0, \lambda_2 = 0, \lambda_3 = 4 \tag{E.2}$$

Eigenvector for $\lambda_3 = 4$: Using $\lambda_3 = 4$, Eq. (E.1) gives

$$-3X_1^{(3)} - 2X_2^{(3)} + X_3^{(3)} = 0$$
$$-2X_1^{(3)} - 4X_2^{(3)} - 2X_3^{(3)} = 0$$
$$X_1^{(3)} - 2X_2^{(3)} - 3X_3^{(3)} = 0 \tag{E.3}$$

If $X_1^{(3)}$ is set equal to 1, Eqs. (E.3) give the eigenvector $\vec{X}^{(3)}$:

$$\vec{X}^{(3)} = \begin{Bmatrix} 1 \\ -1 \\ 1 \end{Bmatrix} \tag{E.4}$$

Eigenvector for $\lambda_1 = \lambda_2 = 0$: The value $\lambda_1 = 0$ or $\lambda_2 = 0$ indicates that the system is degenerate (see Section 6.11). Using $\lambda_1 = 0$ in Eq. (E.1), we obtain

$$X_1^{(1)} - 2X_2^{(1)} + X_3^{(1)} = 0$$
$$-2X_1^{(1)} + 4X_2^{(1)} - 2X_3^{(1)} = 0$$
$$X_1^{(1)} - 2X_2^{(1)} + X_3^{(1)} = 0 \tag{E.5}$$

All these equations are of the form

$$X_1^{(1)} = 2X_2^{(1)} - X_3^{(1)}$$

Thus the eigenvector corresponding to $\lambda_1 = \lambda_2 = 0$ can be written as

$$\vec{X}^{(1)} = \begin{Bmatrix} 2X_2^{(1)} - X_3^{(1)} \\ X_2^{(1)} \\ X_3^{(1)} \end{Bmatrix} \tag{E.6}$$

If we choose $X_2^{(1)} = 1$ and $X_3^{(1)} = 1$, we obtain

$$\vec{X}^{(1)} = \begin{Bmatrix} 1 \\ 1 \\ 1 \end{Bmatrix} \tag{E.7}$$

If we select $X_2^{(1)} = 1$ and $X_3^{(1)} = -1$, Eq. (E.6) gives

$$\vec{X}^{(1)} = \begin{Bmatrix} 3 \\ 1 \\ -1 \end{Bmatrix} \tag{E.8}$$

As shown earlier in Eq. (6.80), $\vec{X}^{(1)}$ and $\vec{X}^{(2)}$ are not unique; any linear combination of $\vec{X}^{(1)}$ and $\vec{X}^{(2)}$ will also satisfy the original Eq. (E.1). Note that $\vec{X}^{(1)}$ given by Eq. (E.6) is orthogonal to $\vec{X}^{(3)}$ of Eq. (E.4) for all values of $X_2^{(1)}$ and $X_3^{(1)}$, since

$$\vec{X}^{(3)^T} [m] \vec{X}^{(1)} = \begin{pmatrix} 1 & -1 & 1 \end{pmatrix} \begin{bmatrix} 1 & 0 & 0 \\ 0 & 2 & 0 \\ 0 & 0 & 1 \end{bmatrix} \begin{Bmatrix} 2X_2^{(1)} - X_3^{(1)} \\ X_2^{(1)} \\ X_3^{(1)} \end{Bmatrix} = 0$$

6.10 EXPANSION THEOREM

The eigenvectors, due to their property of orthogonality, are linearly independent.* Hence they form a basis in the n-dimensional space.[†] This means that any vector in the n-dimensional space can be expressed by a linear combination of the n linearly independent vectors. If $\vec{x}$ is an arbitrary vector in n-dimensional space, it can be expressed as

$$\vec{x} = \sum_{i=1}^{n} c_i \vec{X}^{(i)} \qquad (6.81)$$

where c_i are constants. By premultiplying Eq. (6.81) throughout by $\vec{X}^{(i)^T}[m]$, the value of the constant c_i can be determined as

$$c_i = \frac{\vec{X}^{(i)^T}[m]\vec{x}}{\vec{X}^{(i)^T}[m]\vec{X}^{(i)}} = \frac{\vec{X}^{(i)^T}[m]\vec{x}}{M_{ii}}, \qquad i = 1, 2, \ldots, n \qquad (6.82)$$

where M_{ii} is the generalized mass in the ith normal mode. If the modal vectors $\vec{X}^{(i)}$ are normalized according to Eq. (6.75), c_i is given by

$$c_i = \vec{X}^{(i)^T}[m]\vec{x}, \qquad i = 1, 2, \ldots, n \qquad (6.83)$$

Equation (6.83) represents what is known as the *expansion theorem* [6.6]. It is very useful in finding the response of multidegree of freedom systems subjected to arbitrary forcing conditions according to a procedure called *modal analysis*.

6.11 UNRESTRAINED SYSTEMS

As stated in Section 5.7, an unrestrained system is one that has no restraints or supports and that can move as a rigid body. It is not uncommon to see, in practice, systems that are not attached to any stationary frame. A common example is the motion of two railway cars with masses m_1 and m_2 and a coupling spring k. Such systems are capable of moving as rigid bodies, which can be considered as modes of oscillation with zero frequency. For a conservative system, the kinetic and potential energies are given by Eqs. (6.27) and (6.24), respectively. By definition, the kinetic energy is always positive, so the mass matrix $[m]$ is a positive definite matrix. However, the stiffness matrix $[k]$ is a semi-definite matrix; V is zero without the displacement vector $\vec{x}$ being zero for unrestrained systems. To see this, consider the equation of motion for free vibration in normal coordinates:

$$\ddot{q}(t) + \omega^2 q(t) = 0 \qquad (6.84)$$

For $\omega = 0$, the solution of Eq. (6.84) can be expressed as

$$q(t) = \alpha + \beta t \qquad (6.85)$$

---

* A set of vectors is called linearly independent if no vector in the set can be obtained by a linear combination of the remaining ones.

[†] Any set of n linearly independent vectors in an n-dimensional space is called a *basis* in that space.

where α and β are constants. Equation (6.85) represents a rigid body translation. Let the modal vector of a multidegree of freedom system corresponding to the rigid body mode be denoted by $\vec{X}^{(0)}$. The eigenvalue problem, Eq. (6.58), can be expressed as

$$\omega^2 [m] \vec{X}^{(0)} = [k] \vec{X}^{(0)} \tag{6.86}$$

With $\omega = 0$, Eq. (6.86) gives

$$[k] \vec{X}^{(0)} = \vec{0}$$

That is,

$$k_{11} X_1^{(0)} + k_{12} X_2^{(0)} + \cdots + k_{1n} X_n^{(0)} = 0$$
$$k_{21} X_1^{(0)} + k_{22} X_2^{(0)} + \cdots + k_{2n} X_n^{(0)} = 0$$
$$\vdots$$
$$k_{n1} X_1^{(0)} + k_{n2} X_2^{(0)} + \cdots + k_{nn} X_n^{(0)} = 0 \tag{6.87}$$

If the system undergoes rigid body translation, not all the components $X_i^{(0)}$, $i = 1, 2, \ldots, n$ are zero—that is, the vector $\vec{X}^{(0)}$ is not equal to $\vec{0}$. Hence, in order to satisfy Eq. (6.87), the determinant of $[k]$ must be zero. Thus the stiffness matrix of an unrestrained system (having zero natural frequency) is singular. If $[k]$ is singular, the potential energy is given by

$$V = \frac{1}{2} \vec{X}^{(0)^T} [k] \vec{X}^{(0)} \tag{6.88}$$

by virtue of Eq. (6.87). The mode $\vec{X}^{(0)}$ is called a *zero mode* or *rigid body mode*. If we substitute any vector $\vec{X}$ other than $\vec{X}^{(0)}$ and $\vec{0}$ for $\vec{x}$ in Eq. (6.24), the potential energy V becomes a positive quantity. The matrix $[k]$ is then a positive semi-definite matrix. This is why an unrestrained system is also called a *semi-definite system*.

Note that a multidegree of freedom system can have at most six rigid body modes with the corresponding frequencies equal to zero. There can be three modes for rigid body translation, one for translation along each of the three Cartesian coordinates, and three modes for rigid body rotation, one for rotation about each of the three Cartesian coordinates. We can determine the mode shapes and natural frequencies of a semi-definite system by the procedures outlined in Section 6.9.

EXAMPLE 6.7   **Natural Frequencies of a Free System**

Find the natural frequencies and mode shapes of the system shown in Fig. 6.7 for $m_1 = m_2 = m_3 = m$ and $k_1 = k_2 = k$.

Given: Unrestrained spring-mass system, Fig. 6.7.

Find: Natural frequencies and mode shapes.

Approach: Find a rigid body mode corresponding to zero frequency.

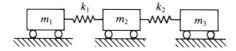

Figure 6.7

Solution. The kinetic energy of the system can be written as

$$T = \frac{1}{2}\left(m_1 \dot{x}_1^2 + m_2 \dot{x}_2^2 + m_3 \dot{x}_3^2 \right) = \frac{1}{2}\dot{\vec{x}}^T[m]\dot{\vec{x}} \tag{E.1}$$

where

$$\vec{x} = \begin{Bmatrix} x_1 \\ x_2 \\ x_3 \end{Bmatrix}, \qquad \dot{\vec{x}} = \begin{Bmatrix} \dot{x}_1 \\ \dot{x}_2 \\ \dot{x}_3 \end{Bmatrix}$$

and

$$[m] = \begin{bmatrix} m_1 & 0 & 0 \\ 0 & m_2 & 0 \\ 0 & 0 & m_3 \end{bmatrix} \tag{E.2}$$

The elongations of the springs k_1 and k_2 are $(x_2 - x_1)$ and $(x_3 - x_2)$, respectively, so the potential energy of the system is given by

$$V = \frac{1}{2}\left\{ k_1(x_2 - x_1)^2 + k_2(x_3 - x_2)^2 \right\} = \frac{1}{2}\vec{x}^T[k]\vec{x} \tag{E.3}$$

where

$$[k] = \begin{bmatrix} k_1 & -k_1 & 0 \\ -k_1 & k_1 + k_2 & -k_2 \\ 0 & -k_2 & k_2 \end{bmatrix} \tag{E.4}$$

It can be verified that the stiffness matrix $[k]$ is singular. Furthermore, if we take all the displacement components to be the same as $x_1 = x_2 = x_3 = c$ (rigid body motion), the potential energy V can be seen to be equal to zero.

To find the natural frequencies and the mode shapes of the system, we express the eigenvalue problem as

$$\left[[k] - \omega^2[m] \right] \vec{X} = \vec{0} \tag{E.5}$$

Since $[k]$ is singular, we cannot find its inverse $[k]^{-1}$, so the dynamical matrix $[D] = [m]^{-1}[k]$. Hence we set the determinant of the coefficient matrix of $\vec{X}$ in Eq. (E.5) equal to zero. For $k_1 = k_2 = k$ and $m_1 = m_2 = m_3 = m$, this yields

$$\begin{vmatrix} (k - \omega^2 m) & -k & 0 \\ -k & (2k - \omega^2 m) & -k \\ 0 & -k & (k - \omega^2 m) \end{vmatrix} = 0 \tag{E.6}$$

The expansion of the determinant in Eq. (E.6) leads to

$$m^3\omega^6 - 4m^2k\omega^4 + 3mk^2\omega^2 = 0 \tag{E.7}$$

By setting

$$\underset{\sim}{\lambda} = \omega^2 \tag{E.8}$$

Eq. (E.7) can be rewritten as

$$m\underset{\sim}{\lambda}\left(\underset{\sim}{\lambda} - \frac{k}{m}\right)\left(\underset{\sim}{\lambda} - \frac{3k}{m}\right) = 0 \tag{E.9}$$

As $m \neq 0$, the roots of Eq. (E.9) are

$$\underset{\sim}{\lambda}_1 = \omega_1^2 = 0$$

$$\underset{\sim}{\lambda}_2 = \omega_2^2 = \frac{k}{m}$$

$$\underset{\sim}{\lambda}_3 = \omega_3^2 = \frac{3k}{m} \tag{E.10}$$

The first natural frequency ω_1 can be observed to be zero in Eq. (E.10). To find the mode shapes, we substitute the values of ω_1, ω_2, and ω_3 into Eq. (E.5) and solve for $\vec{X}^{(1)}$, $\vec{X}^{(2)}$, and $\vec{X}^{(3)}$, respectively. For $\omega_1 = 0$, Eq. (E.5) gives

$$kX_1^{(1)} - kX_2^{(1)} = 0$$
$$-kX_1^{(1)} + 2kX_2^{(1)} - kX_3^{(1)} = 0$$
$$-kX_2^{(1)} + kX_3^{(1)} = 0 \tag{E.11}$$

By fixing the value of one component of $\vec{X}^{(1)}$—say, $X_1^{(1)}$ as 1—Eqs. (E.11) can be solved to obtain

$$X_2^{(1)} = X_1^{(1)} = 1 \qquad \text{and} \qquad X_3^{(1)} = X_2^{(1)} = 1$$

Thus the first (rigid body) mode $\vec{X}^{(1)}$ corresponding to $\omega_1 = 0$ is given by

$$\vec{X}^{(1)} = \begin{Bmatrix} 1 \\ 1 \\ 1 \end{Bmatrix} \tag{E.12}$$

For $\omega_2 = (k/m)^{1/2}$, Eq. (E.5) yields

$$-kX_2^{(2)} = 0$$
$$-kX_1^{(2)} + kX_2^{(2)} - kX_3^{(2)} = 0$$
$$-kX_2^{(2)} = 0 \tag{E.13}$$

By fixing the value of one component of $\vec{X}^{(2)}$—say, $X_1^{(2)}$ as 1—Eqs. (E.13) can be solved to obtain

$$X_2^{(2)} = 0 \qquad \text{and} \qquad X_3^{(2)} = -X_1^{(2)} = -1$$

Thus the second mode $\vec{X}^{(2)}$ corresponding to $\omega_2 = (k/m)^{1/2}$ is given by

$$\vec{X}^{(2)} = \begin{Bmatrix} 1 \\ 0 \\ -1 \end{Bmatrix} \tag{E.14}$$

For $\omega_3 = (3k/m)^{1/2}$, Eq. (E.5) gives

$$-2kX_1^{(3)} - kX_2^{(3)} = 0$$
$$-kX_1^{(3)} - kX_2^{(3)} - kX_3^{(3)} = 0$$
$$-kX_2^{(3)} - 2kX_3^{(3)} = 0 \tag{E.15}$$

By fixing the value of one component of $\vec{X}^{(3)}$—say, $X_1^{(3)}$ as 1—Eqs. (E.15) can be solved to obtain

$$X_2^{(3)} = -2\,X_1^{(3)} = -2 \quad \text{and} \quad X_3^{(3)} = -\frac{1}{2}\,X_2^{(3)} = 1$$

Thus the third mode $\vec{X}^{(3)}$ corresponding to $\omega_3 = (3k/m)^{1/2}$ is given by

$$\vec{X}^{(3)} = \left\{ \begin{array}{c} 1 \\ -2 \\ 1 \end{array} \right\} \tag{E.16}$$

6.12 FORCED VIBRATION

When external forces act on a multidegree of freedom system, the system undergoes forced vibration. For a system with n coordinates or degrees of freedom, the governing equations of motion are a set of n coupled ordinary differential equations of second order. The solution of these equations becomes more complex when the degree of freedom of the system (n) is large and/or when the forcing functions are nonperiodic.* In such cases, a more convenient method known as *modal analysis* can be used to solve the problem. In this method, the expansion theorem is used, and the displacements of the masses are expressed as a linear combination of the normal modes of the system. This linear transformation uncouples the equations of motion so that we obtain a set of n uncoupled differential equations of second order. The solution of these equations, which is equivalent to the solution of the equations of n single degree of freedom systems, can be readily obtained. We shall now consider the procedure of modal analysis.

Modal Analysis. The equations of motion of a multidegree of freedom system under external forces are given by

$$[m]\ddot{\vec{x}} + [k]\vec{x} = \vec{F} \tag{6.89}$$

where $\vec{F}$ is the vector of arbitrary external forces. To solve Eq. (6.89) by modal analysis, it is necessary first to solve the eigenvalue problem,

$$\omega^2[m]\vec{X} = [k]\vec{X} \tag{6.90}$$

and find the natural frequencies $\omega_1, \omega_2, \ldots, \omega_n$ and the corresponding normal modes $\vec{X}^{(1)}, \vec{X}^{(2)}, \ldots, \vec{X}^{(n)}$. According to the expansion theorem, the solution vector of Eq. (6.89) can be expressed by a linear combinaton of the normal modes:

$$\vec{x}(t) = q_1(t)\,\vec{X}^{(1)} + q_2(t)\,\vec{X}^{(2)} + \cdots + q_n(t)\,\vec{X}^{(n)} \tag{6.91}$$

where $q_1(t), q_2(t), \ldots, q_n(t)$ are time-dependent generalized coordinates, also known as the *modal participation coefficients*. By defining a modal matrix $[X]$ in which the jth column is the vector $\vec{X}^{(j)}$—that is,

$$[X] = \left[\vec{X}^{(1)} \quad \vec{X}^{(2)} \quad \ldots \quad \vec{X}^{(n)} \right] \tag{6.92}$$

---

* The dynamic response of multidegree of freedom systems with statistical properties was considered in Ref. [6.15].

Eq. (6.91) can be rewritten as

$$\vec{x}(t) = [X]\vec{q}(t) \tag{6.93}$$

where

$$\vec{q}(t) = \begin{Bmatrix} q_1(t) \\ q_2(t) \\ \vdots \\ q_n(t) \end{Bmatrix} \tag{6.94}$$

Since $[X]$ is not a function of time, we obtain from Eq. (6.93)

$$\ddot{\vec{x}}(t) = [X]\ddot{\vec{q}}(t) \tag{6.95}$$

Using Eqs. (6.93) and (6.95), Eq. (6.89) can be written as

$$[m][X]\ddot{\vec{q}} + [k][X]\vec{q} = \vec{F} \tag{6.96}$$

Premultiplying Eq. (6.96) throughout by $[X]^T$, we obtain

$$[X]^T[m][X]\ddot{\vec{q}} + [X]^T[k][X]\vec{q} = [X]^T\vec{F} \tag{6.97}$$

If the normal modes are normalized according to Eqs. (6.68) and (6.69), we have

$$[X]^T[m][X] = [I] \tag{6.98}$$

$$[X]^T[k][X] = \lceil \omega^2 \rfloor \tag{6.99}$$

By defining the vector of generalized forces $\vec{Q}(t)$ associated with the generalized coordinates $\vec{q}(t)$ as

$$\vec{Q}(t) = [X]^T\vec{F}(t) \tag{6.100}$$

Eq. (6.97) can be expressed, using Eqs. (6.98) and (6.99), as

$$\ddot{\vec{q}}(t) + \lceil \omega^2 \rfloor \vec{q}(t) = \vec{Q}(t) \tag{6.101}$$

Equation (6.101) denotes a set of n uncoupled differential equations of second order*

$$\ddot{q}_i(t) + \omega_i^2 q_i(t) = Q_i(t), \qquad i = 1, 2, \ldots, n \tag{6.102}$$

---

* It is possible to approximate the solution vector $\vec{x}(t)$ by only the first $r(r < n)$ modal vectors (instead of n vectors as in Eq. (6.91)):

$$\underset{n \times 1}{\vec{x}(t)} = \underset{n \times r}{[X]} \, \underset{r \times 1}{\vec{q}(t)}$$

where

$$[X] = \begin{bmatrix} \vec{X}^{(1)} & \vec{X}^{(2)} & \cdots & \vec{X}^{(r)} \end{bmatrix} \quad \text{and} \quad \vec{q}(t) = \begin{Bmatrix} q_1(t) \\ q_2(t) \\ \vdots \\ q_r(t) \end{Bmatrix}$$

This leads to only r uncoupled differential equations

$$\ddot{q}_i(t) + \omega_i^2 q_i(t) = Q_i(t), \qquad i = 1, 2, \ldots, r$$

instead of n equations. The resulting solution $\vec{x}(t)$ will be an approximate solution. This procedure is called the mode displacement method. An alternate procedure, known as the mode acceleration method, for finding an approximate solution is indicated in Problem 6.40.

It can be seen that Eqs. (6.102) have precisely the form of the differential equation describing the motion of an undamped single degree of freedom system. The solution of Eqs. (6.102) can be expressed (see Eq. (4.33)) as:

$$q_i(t) = q_i(0) \cos \omega_i t + \left(\frac{\dot{q}_i(0)}{\omega_i} \right) \sin \omega_i t + \frac{1}{\omega_i} \int_0^t Q_i(\tau) \sin \omega_i(t - \tau) \, d\tau,$$

$$i = 1, 2, \ldots, n \qquad (6.103)$$

The initial generalized displacements $q_i(0)$ and the initial generalized velocities $\dot{q}_i(0)$ can be obtained from the initial values of the physical displacements $x_i(0)$ and physical velocities $\dot{x}_i(0)$:

$$\vec{q}(0) = [X]^T [m] \vec{x}(0) \qquad (6.104)$$

$$\dot{\vec{q}}(0) = [X]^T [m] \dot{\vec{x}}(0) \qquad (6.105)$$

where

$$\vec{q}(0) = \begin{Bmatrix} q_1(0) \\ q_2(0) \\ \vdots \\ q_n(0) \end{Bmatrix}, \quad \dot{\vec{q}}(0) = \begin{Bmatrix} \dot{q}_1(0) \\ \dot{q}_2(0) \\ \vdots \\ \dot{q}_n(0) \end{Bmatrix}, \quad \vec{x}(0) = \begin{Bmatrix} x_1(0) \\ x_2(0) \\ \vdots \\ x_n(0) \end{Bmatrix}, \quad \dot{\vec{x}}(0) = \begin{Bmatrix} \dot{x}_1(0) \\ \dot{x}_2(0) \\ \vdots \\ \dot{x}_n(0) \end{Bmatrix}$$

Once the generalized displacements $q_i(t)$ are found, using Eqs. (6.103) to (6.105), the physical displacements $x_i(t)$ can be found with the help of Eq. (6.93).

6.13 VISCOUSLY DAMPED SYSTEMS

Modal analysis, as presented in Section 6.12, applies only to undamped systems. In many cases, the influence of damping upon the response of a vibratory system is minor and can be disregarded. However, the effect of damping must be considered if the response of the system is required for a relatively long period of time compared to the natural periods of the system. Further, if the frequency of excitation (in the case of a periodic force) is at or near one of the natural frequencies of the system, damping is of primary importance and must be taken into account. In general, since the effects are not known in advance, damping must be considered in the vibration analysis of any system. In this section, we shall consider the equations of motion of a damped multidegree of freedom system and their solution using Lagrange's equations. If the system has viscous damping, its motion will be resisted by a force whose magnitude is proportional to that of the velocity but in the opposite direction. It is convenient to introduce a function R, known as Rayleigh's dissipation function, in deriving the equations of motion by means of Lagrange's equations [6.7]. This function is defined as

$$R = \frac{1}{2}\dot{\vec{x}}^T[c]\dot{\vec{x}} \tag{6.106}$$

where the matrix $[c]$ is called the *damping matrix* and is positive definite, like the mass and stiffness matrices. Lagrange's equations, in this case [6.8], can be written as

$$\frac{d}{dt}\left(\frac{\partial T}{\partial \dot{x}_i}\right) - \frac{\partial T}{\partial x_i} + \frac{\partial R}{\partial \dot{x}_i} + \frac{\partial V}{\partial x_i} = F_i, \qquad i = 1, 2, \ldots, n \tag{6.107}$$

where F_i is the force applied to mass m_i. By substituting Eqs. (6.24), (6.28), and (6.106) into Eq. (6.107), we obtain the equations of motion of a damped multidegree of freedom system in matrix form:

$$[m]\ddot{\vec{x}} + [c]\dot{\vec{x}} + [k]\vec{x} = \vec{F} \tag{6.108}$$

For simplicity, we shall consider a special system for which the damping matrix can be expressed as a linear combination of the mass and stiffness matrices:

$$[c] = \alpha[m] + \beta[k] \tag{6.109}$$

where α and β are constants. This type of damping is known as *proportional damping* because $[c]$ is proportional to a linear combination of $[m]$ and $[k]$. By substituting Eq. (6.109) into Eq. (6.108) we obtain

$$[m]\ddot{\vec{x}} + [\alpha[m] + \beta[k]]\dot{\vec{x}} + [k]\vec{x} = \vec{F} \tag{6.110}$$

By expressing the solution vector $\vec{x}$ as a linear combination of the natural modes of the undamped system, as in the case of Eq. (6.93),

$$\vec{x}(t) = [X]\vec{q}(t) \tag{6.111}$$

Eq. (6.110) can be rewritten as

$$[m][X]\ddot{\vec{q}}(t) + [\alpha[m] + \beta[k]][X]\dot{\vec{q}}(t) \\ + [k][X]\vec{q}(t) = \vec{F}(t) \tag{6.112}$$

Premultiplication of Eq. (6.112) by $[X]^T$ leads to

$$[X]^T[m][X]\ddot{\vec{q}} + \left[\alpha[X]^T[m][X] + \beta[X]^T[k][X]\right]\dot{\vec{q}} \\ + [X]^T[k][X]\vec{q} = [X]^T\vec{F} \tag{6.113}$$

If the eigenvectors $\vec{X}^{(j)}$ are normalized according to Eqs. (6.68) and (6.69), Eq. (6.113) reduces to

$$[I]\ddot{\vec{q}}(t) + [\alpha[I] + \beta[\ulcorner\omega^2\lrcorner]]\dot{\vec{q}}(t) + [\ulcorner\omega^2\lrcorner]\vec{q}(t) = \vec{Q}(t)$$

that is,

$$\ddot{q}_i(t) + \left(\alpha + \omega_i^2\beta\right)\dot{q}_i(t) + \omega_i^2 q_i(t) = Q_i(t),$$
$$i = 1, 2, \ldots, n \qquad (6.114)$$

where ω_i is the ith natural frequency of the undamped system and

$$\vec{Q}(t) = [X]^T\vec{F}(t) \qquad (6.115)$$

By writing

$$\alpha + \omega_i^2\beta = 2\zeta_i\omega_i \qquad (6.116)$$

where ζ_i is called the *modal damping ratio* for the ith normal mode, Eqs. (6.114) can be rewritten as

$$\ddot{q}_i(t) + 2\zeta_i\omega_i\dot{q}_i(t) + \omega_i^2 q_i(t) = Q_i(t), \qquad i = 1, 2, \ldots, n \qquad (6.117)$$

It can be seen that each of the n equations represented by this expression is uncoupled from all of the others. Hence we can find the response of the ith mode in the same manner as that of a viscously damped single degree of freedom system. The solution of Eqs. (6.117), when $\zeta_i < 1$, can be expressed as

$$q_i(t) = e^{-\zeta_i\omega_i t}\left\{\cos\omega_{di}t + \frac{\zeta_i}{\sqrt{1 - \zeta_i^2}}\sin\omega_{di}t\right\}q_i(0)$$

$$+ \left\{\frac{1}{\omega_{di}}e^{-\zeta_i\omega_i t}\sin\omega_{di}t\right\}\dot{q}_i(0) + \frac{1}{\omega_{di}}\int_0^t Q_i(\tau)e^{-\zeta_i\omega_i(t-\tau)}\sin\omega_{di}(t - \tau)\,d\tau,$$

$$i = 1, 2, \ldots, n \qquad (6.118)$$

where

$$\omega_{di} = \omega_i\sqrt{1 - \zeta_i^2} \qquad (6.119)$$

Note the following aspects of these systems:

1. It has been shown by Caughey [6.9] that the condition given by Eq. (6.109) is sufficient but not necessary for the existence of normal modes in damped systems. The necessary condition is that the transformation that diagonalizes the damping matrix also uncouples the coupled equations of motion. This condition is less restrictive than Eq. (6.109) and covers more possibilities.

2. In the general case of damping, the damping matrix cannot be diagonalized simultaneously with the mass and stiffness matrices. In this case, the eigenvalues of the system are either real and negative or complex with negative real parts. The complex eigenvalues exist as conjugate pairs; the associated eigenvectors also consist of complex conjugate pairs. A common procedure for finding the solution of the eigenvalue problem of a damped system involves the transformation of the n coupled second order equations of motion into $2n$ uncoupled first order equations [6.6].

3. The error bounds and numerical methods in the modal analysis of dynamic systems were discussed in Refs. [6.11, 6.12].

EXAMPLE 6.8	**Equations of Motion of a Dynamic System**

Derive the equations of motion of the system shown in Fig. 6.8.

Given: Three degree of freedom spring-mass-damper system, Fig. 6.8.

Find: Equations of motion.

Approach: Use Lagrange's equations in conjunction with Rayleigh's dissipation function.

Solution. The kinetic energy of the system is

$$T = \tfrac{1}{2}\left(m_1 \dot{x}_1^2 + m_2 \dot{x}_2^2 + m_3 \dot{x}_3^2 \right) \tag{E.1}$$

The potential energy has the form

$$V = \tfrac{1}{2}\left[k_1 x_1^2 + k_2 (x_2 - x_1)^2 + k_3 (x_3 - x_2)^2 \right] \tag{E.2}$$

and Rayleigh's dissipation function is

$$R = \tfrac{1}{2}\left[c_1 \dot{x}_1^2 + c_2 (\dot{x}_2 - \dot{x}_1)^2 + c_3 (\dot{x}_3 - \dot{x}_2)^2 + c_4 \dot{x}_2^2 + c_5 (\dot{x}_3 - \dot{x}_1)^2 \right] \tag{E.3}$$

Lagrange's equations can be written as

$$\frac{d}{dt}\left(\frac{\partial T}{\partial \dot{x}_i} \right) - \frac{\partial T}{\partial x_i} + \frac{\partial R}{\partial \dot{x}_i} + \frac{\partial V}{\partial x_i} = F_i, \qquad i = 1, 2, 3 \tag{E.4}$$

By substituting Eqs. (E.1) to (E.3) into Eq. (E.4), we obtain the differential equations of

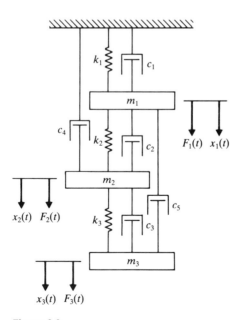

Figure 6.8

motion:

$$[m]\ddot{\vec{x}} + [c]\dot{\vec{x}} + [k]\vec{x} = \vec{F} \tag{E.5}$$

where

$$[m] = \begin{bmatrix} m_1 & 0 & 0 \\ 0 & m_2 & 0 \\ 0 & 0 & m_3 \end{bmatrix} \tag{E.6}$$

$$[c] = \begin{bmatrix} c_1 + c_2 + c_5 & -c_2 & -c_5 \\ -c_2 & c_2 + c_3 + c_4 & -c_3 \\ -c_5 & -c_3 & c_3 + c_5 \end{bmatrix} \tag{E.7}$$

$$[k] = \begin{bmatrix} k_1 + k_2 & -k_2 & 0 \\ -k_2 & k_2 + k_3 & -k_3 \\ 0 & -k_3 & k_3 \end{bmatrix} \tag{E.8}$$

$$\vec{x} = \begin{Bmatrix} x_1(t) \\ x_2(t) \\ x_3(t) \end{Bmatrix} \quad \text{and} \quad \vec{F} = \begin{Bmatrix} F_1(t) \\ F_2(t) \\ F_3(t) \end{Bmatrix} \tag{E.9}$$

EXAMPLE 6.9   **Steady State Response of a Forced System**

Find the steady-state response of the system shown in Fig. 6.8 when the masses are subjected to the simple harmonic forces $F_1 = F_2 = F_3 = F_0 \cos \omega t$, where $\omega = 1.75\sqrt{k/m}$. Assume that $m_1 = m_2 = m_3 = m$, $k_1 = k_2 = k_3 = k$, $c_4 = c_5 = 0$, and the damping ratio in each normal mode is given by $\zeta_i = 0.01$, $i = 1, 2, 3$.

Given: Three degree of freedom spring-mass-damper system, Fig. 6.8.

Find: Steady-state response under harmonic forces.

Approach: Uncouple the equations of motion and find the response.

Solution. The (undamped) natural frequencies of the system (see Example 6.4) are given by

$$\omega_1 = 0.44504\sqrt{\frac{k}{m}}$$

$$\omega_2 = 1.2471\sqrt{\frac{k}{m}}$$

$$\omega_3 = 1.8025\sqrt{\frac{k}{m}} \tag{E.1}$$

and the corresponding $[m]$-orthonormal mode shapes (see Example 6.5) are given by

$$\vec{X}^{(1)} = \frac{0.3280}{\sqrt{m}} \begin{Bmatrix} 1.0 \\ 1.8019 \\ 2.2470 \end{Bmatrix}, \qquad \vec{X}^{(2)} = \frac{0.7370}{\sqrt{m}} \begin{Bmatrix} 1.0 \\ 1.4450 \\ -0.8020 \end{Bmatrix},$$

$$\vec{X}^{(3)} = \frac{0.5911}{\sqrt{m}} \begin{Bmatrix} 1.0 \\ -1.2468 \\ 0.5544 \end{Bmatrix} \tag{E.2}$$

Thus the modal vector can be expressed as

$$[X] = [\vec{X}^{(1)} \vec{X}^{(2)} \vec{X}^{(3)}] = \frac{1}{\sqrt{m}} \begin{bmatrix} 0.3280 & 0.7370 & 0.5911 \\ 0.5911 & 0.3280 & -0.7370 \\ 0.7370 & -0.5911 & 0.3280 \end{bmatrix} \qquad (E.3)$$

The generalized force vector $\vec{Q}(t)$ can be obtained:

$$\vec{Q}(t) = [X]^T \vec{F}(t) = \frac{1}{\sqrt{m}} \begin{bmatrix} 0.3280 & 0.5911 & 0.7370 \\ 0.7370 & 0.3280 & -0.5911 \\ 0.5911 & -0.7370 & 0.3280 \end{bmatrix} \begin{Bmatrix} F_0 \cos \omega t \\ F_0 \cos \omega t \\ F_0 \cos \omega t \end{Bmatrix}$$

$$= \begin{Bmatrix} Q_{10} \\ Q_{20} \\ Q_{30} \end{Bmatrix} \cos \omega t \qquad (E.4)$$

where

$$Q_{10} = 1.6561 \frac{F_0}{\sqrt{m}}, \qquad Q_{20} = 0.4739 \frac{F_0}{\sqrt{m}}, \qquad Q_{30} = 0.1821 \frac{F_0}{\sqrt{m}} \qquad (E.5)$$

If the generalized coordinates or the modal participation factors for the three principal modes are denoted as $q_1(t)$, $q_2(t)$, and $q_3(t)$, the equations of motion can be expressed as

$$\ddot{q}_i(t) + 2\zeta_i \omega_i \dot{q}_i(t) + \omega_i^2 q_i(t) = Q_i(t), \qquad i = 1, 2, 3 \qquad (E.6)$$

The steady-state solution of Eqs. (E.6) can be written as

$$q_i(t) = q_{i0} \cos(\omega t - \phi), \qquad i = 1, 2, 3 \qquad (E.7)$$

where

$$q_{i0} = \frac{Q_{i0}}{\omega_i^2} \frac{1}{\left[\left\{ 1 - \left(\frac{\omega}{\omega_i} \right)^2 \right\}^2 + \left(2\zeta_i \frac{\omega}{\omega_i} \right)^2 \right]^{1/2}} \qquad (E.8)$$

and

$$\phi_i = \tan^{-1} \left\{ \frac{2\zeta_i \frac{\omega}{\omega_i}}{1 - \left(\frac{\omega}{\omega_i} \right)^2} \right\} \qquad (E.9)$$

By substituting the values given in Eqs. (E.5) and (E.1) into Eqs. (E.8) and (E.9), we obtain

$$q_{10} = 0.57815 \frac{F_0 \sqrt{m}}{k}, \qquad \phi_1 = \tan^{-1}(-0.00544)$$

$$q_{20} = 0.31429 \frac{F_0 \sqrt{m}}{k}, \qquad \phi_2 = \tan^{-1}(-0.02988)$$

$$q_{30} = 0.92493 \frac{F_0 \sqrt{m}}{k}, \qquad \phi_3 = \tan^{-1}(0.33827) \qquad (E.10)$$

Finally the steady-state response can be found using Eq. (6.111).

6.14 SELF-EXCITATION AND STABILITY ANALYSIS

In a number of damped vibratory systems, friction leads to negative damping instead of positive damping. This leads to the instability (or self-excited vibration) of the system. In general, for an n degree of freedom system shown in Fig. 6.9, the equations of motion will be a set of second order linear differential equations [as given by Eqs. (6.108) or (6.117)]:

$$[m]\ddot{\vec{x}} + [c]\dot{\vec{x}} + [k]\vec{x} = \vec{F} \tag{6.120}$$

The method presented in Section 5.8 can be extended to study the stability of the system governed by Eq. (6.120). Accordingly, we assume a solution of the form

$$x_j(t) = C_j e^{st}, \qquad j = 1, 2, \ldots, n$$

or

$$\vec{x}(t) = \vec{C} e^{st} \tag{6.121}$$

where s is a complex number to be determined, C_j is the amplitude of x_j and

$$\vec{C} = \begin{Bmatrix} C_1 \\ C_2 \\ \vdots \\ C_n \end{Bmatrix}$$

The real part of s determines the damping and its imaginary part gives the natural frequency of the system. The substitution of Eq. (6.121) into the free vibration equations (obtained by setting $\vec{F} = \vec{0}$ in Eq. (6.120)) leads to

$$([m]s^2 + [c]s + [k])\vec{C} e^{st} = \vec{0} \tag{6.122}$$

For a nontrivial solution of C_j, the determinant of the coefficients of C_j is set equal to zero, which leads to the "characteristic equation," similar to Eq. (6.57):

$$D(s) = \left| [m]s^2 + [c]s + [k] \right| = 0 \tag{6.123}$$

The expansion of Eq. (6.123) leads to a polynomial in s of order $m = 2n$, which can be expressed in the form:

$$D(s) = a_0 s^m + a_1 s^{m-1} + a_2 s^{m-2} + \cdots + a_{m-1} s + a_m = 0 \tag{6.124}$$

The stability or instability of the system depends on the roots of the polynomial equation, $D(s) = 0$. Let the roots of Eq. (6.124) be denoted as

$$s_j = b_j + i\omega_j, \qquad j = 1, 2, \ldots, m \tag{6.125}$$

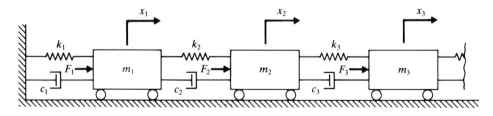

Figure 6.9

If the real parts of all the roots b_j are negative numbers, there will be decaying time functions, $e^{b_j t}$, in Eq. (6.121), and hence the solution (system) will be stable. On the other hand, if one or more roots s_j have a positive real part, then the solution of Eq. (6.120) will contain one or more exponentially increasing time functions $e^{b_j t}$, and hence the solution (system) will be unstable. If there is a purely imaginary root of the form $s_j = i\omega_j$, it will lead to an oscillatory solution $e^{i\omega_j t}$, which represents a borderline case between stability and instability. If s_j is a multiple root, the above conclusion still holds unless it is a pure imaginary number as $s_j = i\omega_j$. In this case, the solution contains functions of the type $e^{i\omega_j t}$, $t e^{i\omega_j t}$, $t^2 e^{i\omega_j t}$, ..., which increase with time. Thus the multiple roots with purely imaginary values indicate the instability of the system. Thus, for a linear system governed by Eq. (6.120) to be stable, it is necessary and sufficient that the roots of Eq. (6.124) should have nonpositive real parts, and that, if any purely imaginary root exists, it should not appear as a multiple root.

Since finding the roots of the polynomial equation (6.124) is a lengthy procedure, a simplified procedure, known as Routh-Hurwitz stability criterion [6.13, 6.14], can be used to investigate the stability of the system. In order to apply this procedure, the following mth order determinant T_m is defined in terms of the coefficients of the polynomial equation (6.124) as:

$$T_m = \begin{vmatrix} a_1 & a_3 & a_5 & a_7 & \cdots & a_{2m-1} \\ a_0 & a_2 & a_4 & a_6 & \cdots & a_{2m-2} \\ 0 & a_1 & a_3 & a_5 & \cdots & a_{2m-3} \\ 0 & a_0 & a_2 & a_4 & \cdots & a_{2m-4} \\ 0 & 0 & a_1 & a_3 & \cdots & a_{2m-5} \\ \vdots & & & & & \\ \vdots & & \cdot & \cdot & \cdots & a_m \end{vmatrix} \qquad (6.126)$$

Then the following subdeterminants, indicated by the dashed lines in Eq. (6.126), are defined:

$$T_1 = a_1 \qquad (6.127)$$

$$T_2 = \begin{vmatrix} a_1 & a_3 \\ a_0 & a_2 \end{vmatrix} \qquad (6.128)$$

$$T_3 = \begin{vmatrix} a_1 & a_3 & a_5 \\ a_0 & a_2 & a_4 \\ 0 & a_1 & a_3 \end{vmatrix} \qquad (6.129)$$

$$\vdots$$

In constructing these subdeterminants, all the coefficients a_i with $i > m$ or $i < 0$ are to be replaced by zeros. According to the Routh-Hurwitz criterion, a necessary and sufficient condition for the stability of the system is that all the coefficients a_0, $a_1, \ldots, a_m$ must be positive and also that all the determinants $T_1, T_2, \ldots, T_m$ must be positive.

6.15 COMPUTER PROGRAMS

**6.15.1
Generating the
Characteristic
Polynomial from
the Matrix**

A Fortran program, in the form of subroutine POLCOF, is given for expanding a determinantal equation to a polynomial form. Thus the equation

$$
|[A] - x[I]| = \left\| \begin{bmatrix} a_{11} & a_{12} & \cdots & a_{1n} \\ a_{21} & a_{22} & \cdots & a_{2n} \\ \vdots & & & \\ a_{n1} & a_{n2} & \cdots & a_{nn} \end{bmatrix} - x \begin{bmatrix} 1 & 0 & \cdots & 0 \\ 0 & 1 & \cdots & 0 \\ \vdots & & & \\ 0 & 0 & \cdots & 1 \end{bmatrix} \right\| = 0
$$

is expanded:

$$
c_{n+1}x^n + c_n x^{n-1} + \cdots + c_2 x + c_1 = 0
$$

The arguments of the subroutine are as follows:

A $\quad=\quad$ Array of dimension $N \times N$. Contains the matrix $[A]$. Input data.

B, C $\quad=\quad$ Arrays of dimension $N \times N$ each.

P, S $\quad=\quad$ Arrays of dimension N each.

PCF $\quad=\quad$ Array of dimension NP. Contains the polynomial coefficients in the order $c_1, c_2, \ldots, c_{n+1}$. Thus the coefficient of the highest order term is stored as the last number. Output.

N $\quad=\quad$ Order of the matrix $[A]$. Input data.

NP $\quad=\quad$ Number of polynomial coefficients $= N + 1$. Input data.

A main program for calling subroutine POLCOF is written with the data:

$$
N = 3, \ NP = 4, \ [A] = \begin{bmatrix} 2 & -1 & 0 \\ -1 & 2 & -1 \\ 0 & -1 & 2 \end{bmatrix}
$$

The program listing and the output are given below.

```
C ============================================================
C
C PROGRAM 9
C MAIN PROGRAM WHICH CALLS POLCOF
C
C ============================================================
C FOLLOWING 4 LINES CONTAIN PROBLEM-DEPENDENT DATA
      DIMENSION A(3,3),B(3,3),C(3,3),P(3),S(3),PCF(4)
      N=3
      NP=4
      DATA A/2.0,-1.0,0.0,-1.0,2.0,-1.0,0.0,-1.0,2.0/
```

```
C END OF PROBLEM-DEPENDENT DATA
      CALL POLCOF (A,B,C,P,S,PCF,N,NP)
      PRINT 10
   10 FORMAT (//,49H POLYNOMIAL EXPANSION OF A DETERMINANTAL EQUATION)
      PRINT 20
   20 FORMAT (//,21H DATA: DETERMINANT A:,/)
      DO 30 I=1,N
   30 PRINT 40,(A(I,J),J=1,N)
   40 FORMAT (4E15.6)
      PRINT 50
   50 FORMAT (/,35H RESULT: POLYNOMIAL COEFFICIENTS IN,/,
     2    53H PCF(NP)*(X**N)+PCF(N)*(X**N-1)+...+PCF(2)*X+PCF(1)=0,/)
      PRINT 60,(PCF(I),I=1,NP)
   60 FORMAT (4E15.6)
      STOP
      END
C ===============================================================
C
C SUBROUTINE POLCOF
C
C ===============================================================
      SUBROUTINE POLCOF (A,B,C,P,S,PCF,N,NP)
      DIMENSION A(N,N),B(N,N),C(N,N),P(N),S(N),PCF(NP)
      DO 10 I=1,N
      P(I)=0.0
      DO 10 J=1,N
   10 B(I,J)=0.0
      DO 20 J=1,N
   20 B(J,J)=1.0
      DO 60 K=1,N
      S(K)=0.0
      DO 30 I=1,N
      DO 30 J=1,N
   30 C(I,J)=B(I,J)
      CALL MATMUL (B,C,A,N,N,N)
      DO 40 J=1,N
   40 S(K)=S(K)+B(J,J)
      P(1)=-S(1)
      IF (K .EQ. 1) GO TO 60
      KM=K-1
      DO 50 I=1,KM
   50 P(K)=P(K)-(1.0/FLOAT(K))*(S(I)*P(K-I))
      P(K)=P(K)-S(K)/FLOAT(K)
   60 CONTINUE
      DO 70 I=1,N
   70 PCF(I)=P(NP-I)
      PCF(NP)=1.0
      RETURN
      END
C ===============================================================
C
C SUBROUTINE MATMUL
C
C ===============================================================
```

```
C MATRIX MULTIPLICATION SUBROUTINE:   A = B * C
C B(L,M) AND C(M,N) ARE INPUT MATRICES, A(L,N) IS OUTPUT MATRIX
      SUBROUTINE MATMUL (A,B,C,L,M,N)
      DIMENSION A(L,N),B(L,M),C(M,N)
      DO 10 I=1,L
      DO 10 J=1,N
      A(I,J)=0.0
      DO 10 K=1,M
   10 A(I,J)=A(I,J)+B(I,K)*C(K,J)
      RETURN
      END
```

POLYNOMIAL EXPANSION OF A DETERMINANTAL EQUATION

DATA: DETERMINANT A:

```
  0.200000E+01   -0.100000E+01    0.000000E+00
 -0.100000E+01    0.200000E+01   -0.100000E+01
  0.000000E+00   -0.100000E+01    0.200000E+01
```

RESULT: POLYNOMIAL COEFFICIENTS IN
PCF(NP)*(X**N)+PCF(N)*(X**N-1)+...+PCF(2)*X+PCF(1)=0

```
 -0.400000E+01    0.100000E+02   -0.600000E+01    0.100000E+01
```


6.15.2
Roots of an *n*th
Order Polynomial
Equation with
Complex
Coefficients

A Fortran program, in the form of subroutine CROOTS, is given to find the roots of the polynomial equation

$$a_{n+1}x^n + a_n x^{n-1} + \cdots + a_2 x + a_1 = 0$$

The program takes the polynomial coefficients $a_1, a_2, \ldots, a_n, a_{n+1}$ in complex form as input data and gives the roots in the complex form. If the coefficients of the polynomial are real numbers, they can always be given in complex form by assuming the imaginary parts to be zero. The arguments of subroutine CROOTS are as follows:

A	=	Complex array of dimension *NPS* containing the complex polynomial coefficients in the order $a_1, a_2, \ldots, a_{n+1}$. Input data.
N	=	Order of the polynomial *n*. Input data.
NP1	=	Number of polynomial coefficients $= N + 1$. Input data.
EPS	=	Convergence requirement. A small value such as 10^{-6} is to be used. Input data.
IMAX	=	Maximum number of iterations to be used in finding the roots. A number in the range 50 to 100 is to be used. Input data.

X   =   Complex array of dimension N containing the computed complex roots. Output.

J   =   Number of roots actually computed. Output.

ID   =   A value of ID equal to zero indicates that the program could not find all the roots. Output.

This program is used to compute the roots of the equation

$$x^3 - 6x^2 + 11x - 6 = 0$$

The main program for this problem, subroutine CROOTS, and the output of the program are given below.

```
C ================================================================
C
C PROGRAM 10
C MAIN PROGRAM FOR CALLING CROOTS
C
C ================================================================
C FOLLOWING 6 LINES CONTAIN PROBLEM-DEPENDENT DATA
C A(NP1) DENOTES THE COEFFICIENT OF (X**N) IN THE POLYNOMIAL
      COMPLEX A(4),X(3)
      DATA N,NP1,IMAX,EPS/3,4,50,1.0E-06/
      A(4)=(1.0,0.0)
      A(3)=(-6.0,0.0)
      A(2)=(11.0,0.0)
      A(1)=(-6.0,0.0)
C END OF PROBLEM-DEPENDENT DATA
      CALL CROOTS (A,N,NP1,EPS,IMAX,X,J,ID)
      PRINT 10
 10   FORMAT (//,28H COMPLEX ROOTS OF POLYNOMIAL,/)
      DO 20 I=1,N
 20   PRINT 30,I,X(I)
 30   FORMAT ((I5,2E15.6))
      STOP
      END
C ================================================================
C
C SUBROUTINE CROOTS
C
C ================================================================
      SUBROUTINE CROOTS (A,N,NP1,EPS,IMAX,X,J,ID)
      COMPLEX A(NP1),X(N)
      COMPLEX XOLD,XNEW,ZX0,ZX1,ZX2,ALN,ALO,HN,HO,DELTA,ZG,ZR,DEN
     2    ,AZ1,AZ2
      DO 10 J=1,N
 10   X(J)=(0.0,0.0)
      J=0
      IF (N .EQ. 1) GO TO 100
 20   J=J+1
      XOLD=(0.0,0.0)
      ZX0=A(J)-A(J+1)+A(J+2)
      ZX1=A(J)+A(J+1)+A(J+2)
      ZX2=A(J)
      ALN=(-0.5,0.0)
```

```
               HN=(-1.0,0.0)
               DO 70 K=1,IMAX
               ALO=ALN
               HO=HN
               DELTA=(1.0,0.0)+ALN
               ZG=ZX0*(ALO**2)-ZX1*(DELTA**2)+ZX2*(ALO+DELTA)
               ZR=CSQRT(ZG**2-4.0*ZX2*DELTA*ALO*(ZX0*ALO-ZX1*DELTA+ZX2))
               AZ3=REAL(ZR)
               AZ4=AIMAG(ZR)
               AZ1=CMPLX(AZ3,-AZ4)
               AZ2=ZG*AZ1
               ZZZSS=REAL (AZ2)
               IF (ZZZSS .LT. 0.0) GO TO 30
               DEN=ZG+ZR
               GO TO 40
      30       DEN=ZG-ZR
      40       ALN=-2.0*ZX2*DELTA
               ALN=ALN/DEN
               IF (REAL(ALN) .GT. 1.0E25 .OR. AIMAG(ALN) .GT. 1.0E25) ALN=
              2    (1.0,0.0)
      50       HN=ALN*HO
               XNEW=XOLD+HN
               IF (CABS((XNEW-XOLD)/XNEW) .LT. EPS) GO TO 80
               ZX0=ZX1
               ZX1=ZX2
               ZS=CABS(ZX2)
               ZX2=A(NP1)
               NJ1=N-J+1
               DO 60 II=1,NJ1
               I=N-II+1
      60       ZX2=ZX2*XNEW+A(I)
               IF (CABS(ZX2/ZS) .LT. 10.0) GO TO 70
               ALN=0.5*ALN
               GO TO 50
      70       XOLD=XNEW
               X(J)=XNEW
               ID=0
               RETURN
      80       X(J)=XNEW
               JP1=J+1
               NJ1=N-JP1+1
               DO 90 II=1,NJ1
               L=N-II+1
      90       A(L)=A(L+1)*X(J)+A(L)
               IF (JP1 .LT. N) GO TO 20
      100      X(N)=-A(N)/A(NP1)
               J=N
               RETURN
               END
```

COMPLEX ROOTS OF POLYNOMIAL

```
    1    0.100000E+01     0.157652E-13
    2    0.200000E+01    -0.315303E-13
    3    0.300000E+01     0.157652E-13
```

**6.15.3
Modal Analysis of
a Multidegree of
Freedom System**

Subroutine MODAL is given for the modal analysis of a multidegree of freedom system. The arguments of this subroutine are as follows:

XM	=	Array of size $N \times N$, in which the mass matrix $[m]$ is stored. Input data.
OM	=	Array of size $NVEC$, in which the natural frequencies are stored. Input data.
T	=	Array of size $NSTEP$, in which the times $t_1, t_2, \ldots, t_{NSTEP}$ are stored.
Z	=	Array of size $NVEC$, in which the modal damping ratios of various modes are stored. Input data.
X0	=	Array of size N, in which the initial values $x_1(0), x_2(0), \ldots, x_n(0)$ are stored. Input data.
XD0	=	Array of size N, in which the initial values $\dot{x}_1(0), \dot{x}_2(0), \ldots, \dot{x}_n(0)$ are stored. Input data.
Y0, YD0	=	Arrays of size $NVEC$.
Q	=	Array of size $NVEC \times NSTEP$.
F	=	Array of size $N \times NSTEP$, in which the magnitudes of the forces applied to the different masses at times $t_1, t_2, \ldots, t_{NSTEP}$ are stored. Input data.
DELT	=	Time step used for calculation, $\Delta t = t_{i+1} - t_i$. Input data.
EV	=	Array of size $N \times NVEC$, in which the normal modes are stored columnwise. Input data.
EVT	=	Array of size $NVEC \times N$ = transpose of the matrix EV.
XMX	=	Array of size $N \times NVEC$.
XTMX	=	Array of size $NVEC \times NVEC$.
X	=	Array of size $N \times NSTEP$, in which the displacements of the masses $m_1, m_2, \ldots, m_n$ at various time stations $t_1, t_2, \ldots, t_{NSTEP}$ are stored. Output.
U, V	=	Arrays of size $NVEC \times NSTEP$.
NSTEP	=	Number of time stations or integration points $t_1, t_2, \ldots, t_{NSTEP}$. Input data.
N	=	Number of degrees of freedom of the system. Input data.
NVEC	=	Number of modes used in the analysis. Input data.

To illustrate the use of the program, the solution of Example 6.9 is considered:

$$[m]\ddot{\vec{x}} + [c]\dot{\vec{x}} + [k]\vec{x} = \vec{f} \quad \text{with} \quad \vec{x}(0) = \vec{x}_0 \quad \text{and} \quad \dot{\vec{x}}(0) = \dot{\vec{x}}_0$$

Data:

$$[m] = \begin{bmatrix} 1 & 0 & 0 \\ 0 & 1 & 0 \\ 0 & 0 & 1 \end{bmatrix}, \qquad [EV] = \begin{bmatrix} 1.0000 & 1.0000 & 1.0000 \\ 1.8019 & 0.4450 & -1.2468 \\ 2.2470 & -0.8020 & 0.5544 \end{bmatrix}$$

$$n = 3, \qquad NVEC = 3, \qquad \zeta_i = 0.01 \quad \text{for} \quad i = 1, 2, 3$$

$$\omega_1 = 0.89008, \qquad \omega_2 = 1.4942, \qquad \omega_3 = 3.6050$$

$$X0 = \begin{Bmatrix} 0 \\ 0 \\ 0 \end{Bmatrix}, \qquad XD0 = \begin{Bmatrix} 0 \\ 0 \\ 0 \end{Bmatrix}, \qquad \vec{f} = \begin{Bmatrix} F_0 \\ F_0 \\ F_0 \end{Bmatrix} \cos \omega t, \qquad \omega = 3.5, \qquad F_0 = 2.0$$

$$NSTEP = 20, \qquad DELT = 0.1$$

The force array F is generated in the main program that calls subroutine MODAL. The program listing and the output are given below.

```
C ============================================================
C
C PROGRAM 11
C MAIN PROGRAM FOR CALLING THE SUBROUTINE MODAL
C
C ============================================================
      DIMENSION XM(3,3),OM(3),Z(3),X0(3),XD0(3),Y0(3),YD0(3),EV(3,3),
     2    EVT(3,3),XMX(3,3),XTMX(3,3),T(20),F(3,20),X(3,20),U(3,20),
     3    V(3,20),Q(3,20)
      DATA N,NVEC,NSTEP,DELT/3,3,20,0.1/
      DATA XM/1.0,0.0,0.0,0.0,1.0,0.0,0.0,0.0,1.0/
      OMF=3.5
      DATA OM/0.89008,1.4942,3.6050/
      DATA Z/0.01,0.01,0.01/
      DATA X0/0.0,0.0,0.0/
      DATA XD0/0.0,0.0,0.0/
      DATA (EV(I,1),I=1,3)/1.0,1.8019,2.2470/
      DATA (EV(I,2),I=1,3)/1.0,0.4450,-0.8020/
      DATA (EV(I,3),I=1,3)/1.0,-1.2468,0.5544/
      DO 5 I=1,NSTEP
      TIME=REAL(I)*DELT
5     F(1,I)=2.0*COS(3.5*TIME)
      DO 10 I=1,20
      F(2,I)=F(1,I)
10    F(3,I)=F(1,I)
      DO 20 I=1,NVEC
      DO 20 J=1,N
20    EVT(I,J)=EV(J,I)
      CALL MODAL (XM,OM,OMF,T,Z,X0,XD0,Y0,YD0,Q,F,DELT,EV,EVT,XMX,
     2    XTMX,X,U,V,NSTEP,N,NVEC)
      PRINT 30
30    FORMAT (//,40H RESPONSE OF SYSTEM USING MODAL ANALYSIS,/)
      DO 40 I=1,N
40    PRINT 50,I,(X(I,J),J=1,NSTEP)
50    FORMAT (/,11H COORDINATE,I5,/,(1X,5E14.6))
      STOP
      END
```

```
C ================================================================
C
C SUBROUTINE MODAL
C
C ================================================================
      SUBROUTINE MODAL (XM,OM,OMF,T,Z,X0,XD0,Y0,YD0,Q,F,DELT,EV,EVT,
     2  XMX,XTMX,X,U,V,NSTEP,N,NVEC)
      DIMENSION XM(N,N),OM(NVEC),T(NSTEP),Z(NVEC),X0(N),XD0(N),
     2  Y0(NVEC),YD0(NVEC),Q(NVEC,NSTEP),F(N,NSTEP),EV(N,NVEC),
     3  EVT(NVEC,N),XMX(N,NVEC),XTMX(NVEC,NVEC),X(N,NSTEP),U(NVEC,
     4  NSTEP),V(NVEC,NSTEP)
      T(1)=DELT
      DO 10 I=2,NSTEP
  10  T(I)=T(I-1)+DELT
C NORMALIZATION OF MODAL MATRIX WITH RESPECT TO THE MASS MATRIX
      CALL MATMUL (XMX,XM,EV,N,N,NVEC)
      CALL MATMUL (XTMX,EVT,XMX,NVEC,N,NVEC)
      DO 30 I=1,NVEC
      DO 20 J=1,N
  20  EV(J,I)=EV(J,I)/SQRT(XTMX(I,I))
  30  CONTINUE
C CONVERTION OF INFORMATION TO NORMAL COORDINATES
      DO 40 I=1,NVEC
      Y0(I)=0.0
  40  YD0(I)=0.0
      DO 60 I=1,NVEC
      DO 50 J=1,N
      Y0(I)=Y0(I)+EV(J,I)*X0(J)
  50  YD0(I)=YD0(I)+EV(J,I)*XD0(J)
  60  CONTINUE
      DO 70 I=1,NVEC
      DO 70 J=1,N
  70  EVT(I,J)=EV(J,I)
      CALL MATMUL (Q,EVT,F,NVEC,N,NSTEP)
      DO 100 I=1,NVEC
      R=OMF/OM(I)
      PP=Y0(I)
      QQ=YD0(I)
      ZI=Z(I)
      OMEG=OM(I)
      OMD=OMEG*SQRT(1.0-ZI**2)
      DO 90 J=1,NSTEP
      IF (J .EQ. 1) GO TO 80
      PP=U(I,J-1)
      QQ=V(I,J-1)
  80  C1=EXP(-ZI*OMEG*DELT)
      C2=COS(OMD*DELT)
      C3=SIN(OMD*DELT)
      C4=(QQ+OMEG*ZI*PP)/OMD
      C5=OMEG*ZI/OMD
      C6=Q(I,J)/(OMEG**2)
      U(I,J)=C1*(PP*C2+C3*C4)+C6*(1.0-C1*(C2+C3*C5))
      V(I,J)=OMD*C1*(-PP*C3+C2*C4-C5*(PP*C2+C3*C4))+C6*OMD*C1*C3*
     2  (1.0+C5**2)
```

```
   90   CONTINUE
   100  CONTINUE
C FINDING THE SOLUTION IN THE ORIGINAL COORDINATES
        CALL MATMUL (X,EV,U,N,NVEC,NSTEP)
        RETURN
        END
```

RESPONSE OF SYSTEM USING MODAL ANALYSIS

COORDINATE   1
```
  0.936322E-02   0.354378E-01   0.731183E-01   0.115538E+00   0.155010E+00
  0.184085E+00   0.196572E+00   0.188401E+00   0.158198E+00   0.107508E+00
  0.406290E-01  -0.359245E-01  -0.114247E+00  -0.186060E+00  -0.243782E+00
 -0.281509E+00  -0.295776E+00  -0.286009E+00  -0.254587E+00  -0.206525E+00
```

COORDINATE   2
```
  0.939528E-02   0.358739E-01   0.751277E-01   0.121321E+00   0.167711E+00
  0.207370E+00   0.233936E+00   0.242311E+00   0.229239E+00   0.193704E+00
  0.137094E+00   0.631125E-01  -0.225565E-01  -0.112803E+00  -0.199814E+00
 -0.275895E+00  -0.334293E+00  -0.369942E+00  -0.380052E+00  -0.364457E+00
```

COORDINATE   3
```
  0.937892E-02   0.356481E-01   0.740876E-01   0.118353E+00   0.161294E+00
  0.195874E+00   0.216076E+00   0.217673E+00   0.198772E+00   0.160053E+00
  0.104682E+00   0.379048E-01  -0.336272E-01  -0.102705E+00  -0.162484E+00
 -0.207353E+00  -0.233644E+00  -0.240091E+00  -0.227975E+00  -0.200947E+00
```

**6.15.4
Solution of
Simultaneous
Linear Equations**

Subroutine SIMUL is given for solving a system of N linear equations of the form $[A]\vec{X} = \vec{B}$. The following arguments are used:

A = Array of size $N \times N$. It is used to store the matrix $[A]$ in the beginning (Input) and contains the inverse of the matrix $[A]$ upon return from the subroutine SIMUL (Output).

B = Array of size N. It is used to store the vector $\vec{B}$ in the beginning (Input) and contains the solution vector $\vec{X}$ upon return from subroutine SIMUL (Output).

N = Number of equations to be solved. Input data.

IND = Zero if only the inverse $[A]^{-1}$ is required and = any non-zero integer if $\vec{X}$ is needed. Input data.

LA, S = Arrays of size N each.

LB = Array of size $N \times 2$.

The program listing and typical results are given below.

```
C ================================================================
C
C PROGRAM 12
C MAIN PROGRAM WHICH CALLS SIMUL
C
C ================================================================
C FOLLOWING 4 LINES CONTAIN PROBLEM-DEPEMDENT DATA
      DIMENSION A(3,3),B(3),LA(3),LB(3,2),S(3)
      DATA A/1.0,2.0,3.0,10.0,0.0,3.0,1.0,1.0,2.0/
      DATA B/7.0,0.0,14.0/
      DATA N,IND/3,1/
C END OF PROBLEM-DEPENDENT DATA
      PRINT 100
  100 FORMAT (//,42H SOLUTION OF SIMULTANEOUS LINEAR EQUATIONS)
      PRINT 200, ((A(I,J),J=1,N),I=1,N)
  200 FORMAT (//,2X,28H ORIGINAL COEFFICIENT MATRIX,//,3(E12.4,1X))
      PRINT 300, (B(I),I=1,N)
  300 FORMAT (//,2X,23H RIGHT HAND SIDE VECTOR,//,3(E12.4,1X))
      CALL SIMUL (A,B,N,IND,LA,LB,S)
      PRINT 400, ((A(I,J),J=1,N),I=1,N)
  400 FORMAT`(//,2X,30H INVERSE OF COEFFICIENT MATRIX,//,3(E12.4,1X))
      PRINT 500, (B(I),I=1,N)
  500 FORMAT (//,2X,16H SOLUTION VECTOR,//,3(E12.4,1X))
      STOP
      END
C ================================================================
C
C SUBROUTINE SIMUL
C
C ================================================================
      SUBROUTINE SIMUL (A,B,N,IND,LA,LB,S)
      DIMENSION A(N,N),B(N),LA(N),LB(N,2),S(N)
      DO 100 I=1,N
  100 LA(I)=0
      DO 250 K=1,N
      Z=0.0
      DO 150 I=1,N
      IF (LA(I) .EQ. 1) GO TO 150
      DO 140 J=1,N
      IF (LA(J)-1) 130,140,300
  130 IF (ABS(Z) .GE. ABS(A(I,J))) GO TO 140
      IA=I
      IB=J
      Z=A(I,J)
  140 CONTINUE
  150 CONTINUE
      LA(IB)=LA(IB)+1
      IF (IA .EQ. IB) GO TO 190
      DO 160 I=1,N
      Z=A(IA,I)
      A(IA,I)=A(IB,I)
  160 A(IB,I)=Z
      IF (IND .EQ. 0) GO TO 190
      Z=B(IA)
```

```
                  B(IA)=B(IB)
                  B(IB)=Z
          190     LB(K,1)=IA
                  LB(K,2)=IB
                  S(K)=A(IB,IB)
                  A(IB,IB)=1.0
                  DO 200 I=1,N
          200     A(IB,I)=A(IB,I)/S(K)
                  IF (IND .EQ. 0) GO TO 220
                  B(IB)=B(IB)/S(K)
          220     DO 250 I=1,N
                  IF(I .EQ. IB) GO TO 250
                  Z=A(I,IB)
                  A(I,IB)=0.0
                  DO 230 J=1,N
          230     A(I,J)=A(I,J)-A(IB,J)*Z
                  IF (IND .EQ. 0) GO TO 250
                  B(I)=B(I)-B(IB)*Z
          250     CONTINUE
                  DO 270 I=1,N
                  J=N-I+1
                  IF (LB(J,1) .EQ. LB(J,2)) GO TO 270
                  IA=LB(J,1)
                  IB=LB(J,2)
                  DO 260 K=1,N
                  Z=A(K,IA)
                  A(K,IA)=A(K,IB)
                  A(K,IB)=Z
          260     CONTINUE
          270     CONTINUE
          300     RETURN
                  END


       SOLUTION OF SIMULTANEOUS LINEAR EQUATIONS

          ORIGINAL COEFFICIENT MATRIX

          0.1000E+01     0.1000E+02     0.1000E+01
          0.2000E+01     0.0000E+00     0.1000E+01
          0.3000E+01     0.3000E+01     0.2000E+01

          RIGHT HAND SIDE VECTOR

          0.7000E+01     0.0000E+00     0.1400E+02

          INVERSE OF COEFFICIENT MATRIX

          0.4286E+00     0.2429E+01     -0.1429E+01
          0.1429E+00     0.1429E+00     -0.1429E+00
         -0.8571E+00    -0.3857E+01      0.2857E+01

          SOLUTION VECTOR

         -0.1700E+02    -0.1000E+01      0.3400E+02
```

REFERENCES

6.1. F. W. Beaufait, *Basic Concepts of Structural Analysis*, Prentice-Hall, Englewood Cliffs, N.J., 1977.

6.2. R. J. Roark and W. C. Young, *Formulas for Stress and Strain* (5th Ed.), McGraw-Hill, New York, 1975.

6.3. D. A. Wells, *Theory and Problems of Lagrangian Dynamics*, Schaum's Outline Series, McGraw-Hill, New York, 1967.

6.4. J. H. Wilkinson, *The Algebraic Eigenvalue Problem*, Clarendon Press, Oxford, 1965.

6.5. A. Ralston, *A First Course in Numerical Analysis*, McGraw-Hill, New York, 1965.

6.6. L. Meirovitch, *Analytical Methods in Vibrations*, Macmillan, New York, 1967.

6.7. J. W. Strutt, Lord Rayleigh, *The Theory of Sound*, Macmillan, London, 1877 (reprinted by Dover Publications, New York in 1945).

6.8. W. C. Hurty and M. F. Rubinstein, *Dynamics of Structures*, Prentice-Hall, Englewood Cliffs, N.J., 1964.

6.9. T. K. Caughey, "Classical normal modes in damped linear dynamic systems," *Journal of Applied Mechanics*, Vol. 27, 1960, pp. 269–271.

6.10. A. Avakian and D. E. Beskos, "Use of dynamic stiffness influence coefficients in vibrations of non-uniform beams," letter to the editor, *Journal of Sound and Vibration*, Vol. 47, 1976, pp. 292–295.

6.11. I. Gladwell and P. M. Hanson, "Some error bounds and numerical experiments in modal methods for dynamics of systems," *Earthquake Engineering and Structural Dynamics*, Vol. 12, 1984, pp. 9–36.

6.12. R. Bajan, A. R. Kukreti, and C. C. Feng, "Method for improving incomplete modal coupling," *Journal of Engineering Mechanics*, Vol. 109, 1983, pp. 937–949.

6.13. E. J. Routh, *Advanced Rigid Dynamics*, Macmillan, New York, 1905.

6.14. D. W. Nicholson and D. J. Inman, "Stable response of damped linear systems," *Shock and Vibration Digest*, Vol. 15, November 1983, pp. 19–25.

6.15. P. C. Chen and W. W. Soroka, "Multidegree dynamic response of a system with statistical properties," *Journal of Sound and Vibration*, Vol. 37, 1974, pp. 547–556.

6.16. S. Mahalingam and R. E. D. Bishop, "The response of a system with repeated natural frequencies to force and displacement excitation," *Journal of Sound and Vibration*, Vol. 36, 1974, pp. 285–295.

REVIEW QUESTIONS

6.1. Define the flexibility and stiffness influence coefficients. What is the relation between them?

6.2. Write the equations of motion of a multidegree of freedom system in matrix form using (1) the flexibility matrix and (2) the stiffness matrix.

6.3. Express the potential and kinetic energy of an n degree of freedom system, using matrix notation.

6.4. What is a generalized mass matrix?

6.5. Why is the mass matrix $[m]$ always positive definite?

6.6. Is the stiffness matrix $[k]$ always positive definite? Why?

6.7. What is the difference between generalized coordinates and Cartesian coordinates?

6.8. State Lagrange's equations.

6.9. What is an eigenvalue problem?

6.10. What is a mode shape? How is it computed?

6.11. How many distinct natural frequencies can exist for an n degree of freedom system?

6.12. What is a dynamical matrix? What is its use?

6.13. How is the frequency equation derived for a multidegree of freedom system?

6.14. What is meant by the orthogonality of normal modes? What are orthonormal modal vectors?

6.15. What is a basis in n-dimensional space?

6.16. What is the expansion theorem? What is its importance?

6.17. Explain the modal analysis procedure.

6.18. What is a rigid body mode? How is it determined?

6.19. What is a degenerate system?

6.20. How can we find the response of a multidegree of freedom system using the first few modes only?

6.21. Define Rayleigh's dissipation function.

6.22. Define these terms: proportional damping, modal damping ratio, and modal participation factor.

6.23. When do we get complex eigenvalues?

6.24. What is the reason for the occurrence of an irregular mode instead of a principal mode of vibration?

PROBLEMS

The problem assignments are organized as follows:

Problems	Section covered	Topic covered
6.1–6.9	6.3	Influence coefficients
6.10	6.4	Potential and kinetic energies
6.11	6.5	Generalized coordinates
6.12–6.16	6.6	Lagrange's equations
6.17–6.18	6.8	Eigenvalue problem
6.19–6.34	6.9	Solution of the eigenvalue problem
6.35–6.36	6.11	Unrestrained systems
6.37–6.41	6.12	Forced vibration
6.42–6.43	6.13	Viscously damped systems
6.44–6.50	6.15	Computer programs
6.51	—	Project

6.1. Find the flexibility and stiffness influence coefficients of the torsional system shown in Fig. 6.10. Also write the equations of motion of the system.

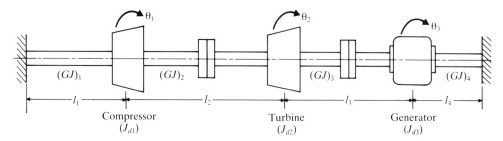

Figure 6.10

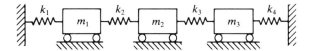

Figure 6.11

6.2. Find the flexibility and stiffness influence coefficients of the system shown in Fig. 6.11. Also, derive the equations of motion of the system.

6.3. An airplane wing, Fig. 6.12(a), is modeled as a three degree of freedom lumped mass system as shown in Fig. 6.12(b). Derive the flexibility matrix and the equations of

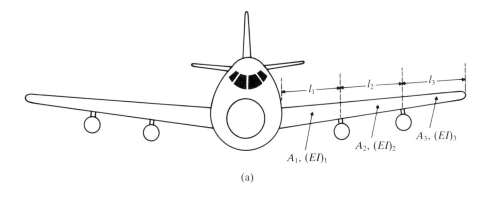

(a)

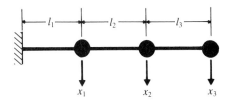

(b)

Figure 6.12

motion of the wing by assuming that all $A_i = A$, $(EI)_i = EI$, $l_i = l$, and the root is fixed.

6.4. Determine the flexibility matrix of the uniform beam shown in Fig. 6.13. Disregard the mass of the beam compared to the concentrated masses placed on the beam and assume all $l_i = l$.

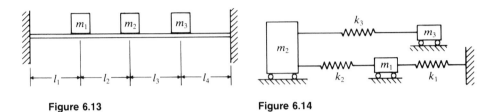

Figure 6.13   **Figure 6.14**

6.5. Derive the flexibility and stiffness matrices of the spring-mass system shown in Fig. 6.14, assuming that all the contacting surfaces are frictionless.

6.6. Drive the equations of motion for the tightly stretched string carrying three masses, as shown in Fig. 6.15. Assume the ends of the string to be fixed.

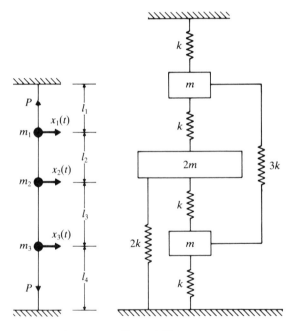

Figure 6.15   **Figure 6.16**

6.7. Derive the equations of motion of the system shown in Fig. 6.16.

6.8. Find the stiffness influence coefficients for the spring-mass system shown in Fig. 6.2(a).

6.9. Four identical springs, each having a stiffness k, are arranged symmetrically at 90° from each other, as shown in Fig. 2.33. Find the influence coefficient of the junction point in an arbitrary direction.

6.10. Show that the stiffness matrix of the spring-mass system shown in Fig. 6.1 is a band matrix along the diagonal.

6.11. For the four-story shear building shown in Fig. 6.17, there is no rotation of the horizontal section at the level of floors. Assuming that the floors are rigid and the total mass is concentrated at the levels of the floors, derive the equations of motion of the building using (a) Newton's second law of motion and (b) Lagrange's equations.

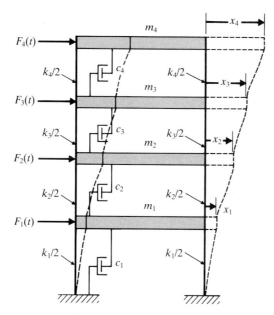

Figure 6.17   **Figure 6.18**

6.12. Derive the equations of motion of the system shown in Fig. 6.18 by using Lagrange's equations with x and θ as generalized coordinates.

6.13. Derive the equations of motion of the system shown in Fig. 5.9(a), using Lagrange's equations with (1) x_1 and x_2 as generalized coordinates and (2) x and θ as generalized coordinates.

6.14. Derive the equations of motion of the system shown in Fig. 6.11, using Lagrange's equations.

6.15. Derive the equations of motion of the triple pendulum shown in Fig. 6.4, using Lagrange's equations.

6.16.* When an airplane undergoes symmetric vibrations, the fuselage can be idealized as a concentrated central mass M_0 and the wings can be modeled as rigid bars carrying end masses M, as shown in Fig. 6.19(b). The flexibility between the wings and the fuselage can be represented by two torsional springs of stiffness k_t each. (i) Derive the equations of motion of the airplane, using Lagrange's equations with x and θ as generalized coordinates. (ii) Find the natural frequencies and mode shapes of the airplane. (iii) Find the torsional spring constant in order to have the natural frequency of vibration, in torsional mode, greater than 2 Hz when $M_0 = 1000$ kg, $M = 500$ kg, and $l = 6$ m.

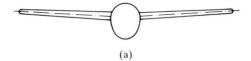

(a)

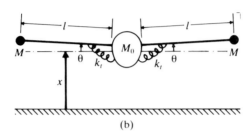

(b)

Figure 6.19

6.17. Set up the eigenvalue problem of Example 6.4 in terms of the coordinates $q_1 = x_1$, $q_2 = x_2 - x_1$ and $q_3 = x_3 - x_2$, and solve the resulting problem. Compare the results obtained with those of Example 6.4 and draw conclusions.

6.18. Derive the frequency equation of the system shown in Fig. 6.11.

6.19. Find the natural frequencies and mode shapes of the system shown in Fig. 6.2 when $k_1 = k$, $k_2 = 2k$, $k_3 = 3k$, $m_1 = m$, $m_2 = 2m$, and $m_3 = 3m$. Plot the mode shapes.

6.20. Set up the matrix equation of motion and determine the three principal modes of vibration for the system shown in Fig. 6.2 with $k_1 = 3k$, $k_2 = k_3 = k$, $m_1 = 3m$, and $m_2 = m_3 = m$. Check the orthogonality of the modes found.

6.21. Find the natural frequencies of the system shown in Fig. 6.4 with $l_1 = 20$ cm, $l_2 = 30$ cm, $l_3 = 40$ cm, $m_1 = 1$ kg, $m_2 = 2$ kg, and $m_3 = 3$ kg.

6.22.* (a) Find the natural frequencies of the system shown in Fig. 6.13 with $m_1 = m_2 = m_3 = m$ and $l_1 = l_2 = l_3 = l_4 = l/4$. (b) Find the natural frequencies of the beam when $m = 10$ kg, $l = 0.5$ m, cross section is solid circular section with diameter 2.5 cm, and the material is steel. (c) Consider using hollow circular, solid rectangular, or hollow rectangular cross section for the beam to achieve the same natural frequencies as in (b). Identify the cross section corresponding to the least weight of the beam.

6.23. The frequency equation of a three degree of freedom system is given by

$$\begin{vmatrix} \lambda - 5 & -3 & -2 \\ -3 & \lambda - 6 & -4 \\ -1 & -2 & \lambda - 6 \end{vmatrix} = 0$$

Find the roots of this equation.

6.24. Determine the eigenvalues and eigenvectors of the system shown in Fig. 6.12, taking $k_1 = k_2 = k_3 = k_4 = k$ and $m_1 = m_2 = m_3 = m$.

6.25. Find the natural frequencies and mode shapes of the system shown in Fig. 6.12 for $k_1 = k_2 = k_3 = k_4 = k$, $m_1 = 2m$, $m_2 = 3m$, and $m_3 = 2m$.

6.26. Find the natural frequencies and principal modes of the triple pendulum shown in Fig. 6.4, assuming that $l_1 = l_2 = l_3 = l$ and $m_1 = m_2 = m_3 = m$.

6.27. Find the natural frequencies and mode shapes of the system considered in Problem 6.5 with $m_1 = m$, $m_2 = 2m$, $m_3 = m$, $k_1 = k_2 = k$, and $k_3 = 2k$.

6.28. Show that the natural frequencies of the system shown in Fig. 6.2(a), with $k_1 = 3k$, $k_2 = k_3 = k$, $m_1 = 4m$, $m_2 = 2m$, and $m_3 = m$, are given by $\omega_1 = 0.46\sqrt{k/m}$, $\omega_2 = \sqrt{k/m}$, and $\omega_3 = 1.34\sqrt{k/m}$. Find the eigenvectors of the system.

6.29. Find the natural frequencies of the system considered in Problem 6.6 with $m_1 = 2m$, $m_2 = m$, $m_3 = 3m$, and $l_1 = l_2 = l_3 = l_4 = l$.

6.30. Find the natural frequencies and principal modes of the torsional system shown in Fig. 6.11 for $(GJ)_i = GJ$, $i = 1, 2, 3, 4$, $J_{d1} = J_{d2} = J_{d3} = J_0$, and $l_1 = l_2 = l_3 = l_4 = l$.

6.31. The mass matrix $[m]$ and the stiffness matrix $[k]$ of a uniform bar are

$$[m] = \frac{\rho A l}{4} \begin{bmatrix} 1 & 0 & 0 \\ 0 & 2 & 0 \\ 0 & 0 & 1 \end{bmatrix} \quad \text{and} \quad [k] = \frac{2AE}{l} \begin{bmatrix} 1 & -1 & 0 \\ -1 & 2 & -1 \\ 0 & -1 & 1 \end{bmatrix}$$

where ρ is the density, A is the cross-sectional area, E is Young's modulus, and l is the length of the bar. Find the natural frequencies of the system by finding the roots of the characteristic equation. Also find the principal modes.

6.32. The mass matrix of a vibrating system is given by

$$[m] = \begin{bmatrix} 1 & 0 & 0 \\ 0 & 2 & 0 \\ 0 & 0 & 1 \end{bmatrix}$$

and the eigenvectors by

$$\begin{Bmatrix} 1 \\ -1 \\ 1 \end{Bmatrix}, \quad \begin{Bmatrix} 1 \\ 1 \\ 1 \end{Bmatrix} \quad \text{and} \quad \begin{Bmatrix} 0 \\ 1 \\ 2 \end{Bmatrix}$$

Find the $[m]$-orthonormal modal matrix of the system.

6.33. For the system shown in Fig. 6.20, (a) determine the characteristic polynomial $\Delta(\omega^2) = \det |[k] - \omega^2[m]|$, (b) plot $\Delta(\omega^2)$ from $\omega^2 = 0$ to $\omega^2 = 4.0$ (using increments $\Delta\omega^2 = 0.2$), and (c) find ω_1^2, ω_2^2, and ω_3^2.

6.34. (a) Two of the eigenvectors of a vibrating system are known to be

$$\begin{Bmatrix} 0.2754946 \\ 0.3994672 \\ 0.4490562 \end{Bmatrix} \quad \text{and} \quad \begin{Bmatrix} 0.6916979 \\ 0.2974301 \\ -0.3389320 \end{Bmatrix}$$

Prove that these are orthogonal with respect to the mass matrix

$$[m] = \begin{bmatrix} 1 & 0 & 0 \\ 0 & 2 & 0 \\ 0 & 0 & 3 \end{bmatrix}$$

Find the remaining $[m]$-orthogonal eigenvector. (b) If the stiffness matrix of the

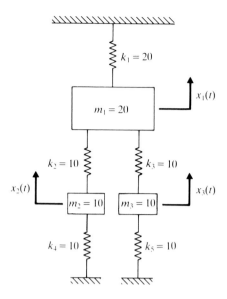

Figure 6.20

system is given by

$$\begin{bmatrix} 6 & -4 & 0 \\ -4 & 10 & 0 \\ 0 & 0 & 6 \end{bmatrix}$$

determine all the natural frequencies of the system, using the eigenvectors of part (a).

6.35. Find the natural frequencies and mode shapes of the system shown in Fig. 6.7 with $m_1 = m$, $m_2 = 2m$, $m_3 = 3m$, and $k_1 = k_2 = k$.

6.36. Find the modal matrix for the semi-definite system shown in Fig. 6.21 for $J_1 = J_2 = J_3 = J_0$, $k_{t1} = k_t$, and $k_{t2} = 2k_t$.

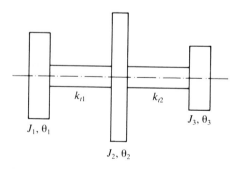

Figure 6.21

6.37. Determine the amplitudes of motion of the three masses in Fig. 6.22 when a harmonic force $F(t) = F_0 \sin \omega t$ is applied to the lower left mass with $m = 1$ kg, $k = 1000$ N/m, $F_0 = 5$ N and $\omega = 10$ rad/sec using the mode superposition method.

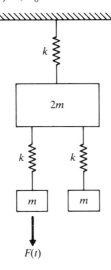

Figure 6.22

6.38. (a) Determine the natural frequencies and mode shapes of the torsional system shown in Fig. 6.5 for $k_{t1} = k_{t2} = k_{t3} = k_t$ and $J_1 = J_2 = J_3 = J_0$. (b) If a torque $M_{t3}(t) = M_{t0} \cos \omega t$, with $M_{t0} = 500$ N-m and $\omega = 100$ rad/sec, acts on the generator (J_3), find the amplitude of each component. Assume $M_{t1} = M_{t2} = 0$, $k_t = 100$ N-m/rad, and $J_0 = 1$ kg-m^2.

6.39. Using the results of Problems 6.2 and 6.24, determine the modal matrix $[X]$ of the system shown in Fig. 6.11 and derive the uncoupled equations of motion.

6.40. An approximate solution of a multidegree of freedom system can be obtained using the mode acceleration method. According to this method, the equations of motion of an undamped system, for example, are expressed as

$$\vec{x} = [k]^{-1}\left(\vec{F} - [m]\ddot{\vec{x}}\right) \tag{E.1}$$

and $\ddot{\vec{x}}$ is approximated using the first r modes ($r < n$) as

$$\underset{n \times 1}{\ddot{\vec{x}}} = \underset{n \times r}{[X]} \underset{r \times 1}{\ddot{\vec{q}}} \tag{E.2}$$

Since $([k] - \omega_i^2[m])\vec{X}^{(i)} = \vec{0}$, Eq. (E.1) can be written as

$$\vec{x}(t) = [k]^{-1}\vec{F}(t) - \sum_{i=1}^{r} \frac{1}{\omega_i^2}\vec{X}^{(i)}\ddot{q}_i(t) \tag{E.3}$$

Find the approximate response of the system described in Example 6.9 (without damping), using the mode acceleration method with $r = 1$.

6.41. Determine the response of the system in Problem 6.19 to the initial conditions $x_1(0) = 1$, $\dot{x}_1(0) = 0$, $x_2(0) = 2$, $\dot{x}_2(0) = 1$, $x_3(0) = 1$, and $\dot{x}_3(0) = -1$. Assume $k/m = 1$.

6.42. Find the steady state response of the system shown in Fig. 6.9 with $k_1 = k_2 = k_3 = k_4 = 100$ N/m, $c_1 = c_2 = c_3 = c_4 = 1$ N-s/m, $m_1 = m_2 = m_3 = 1$ kg, $F_1(t) = F_0 \cos \omega t$, $F_0 = 10$ N and $\omega = 1$ rad/sec. Assume that the spring k_4 and the damper c_4 are connected to a rigid wall at the right end. Use the mechanical impedance method described in Section 5.6 for solution.

6.43. An airplane wing, Fig. 6.23(a), is modeled as a twelve degree of freedom lumped mass system as shown in Fig. 6.23(b). The first three natural frequencies and mode shapes, obtained experimentally, are given below.

Mode shape	Degree of Freedom												
	0	**1**	**2**	**3**	**4**	**5**	**6**	**7**	**8**	**9**	**10**	**11**	**12**
$\vec{X}^{(1)}$	0.0	0.126	0.249	0.369	0.483	0.589	0.686	0.772	0.846	0.907	0.953	0.984	1.000
$\vec{X}^{(2)}$	0.0	−0.375	−0.697	−0.922	−1.017	−0.969	−0.785	−0.491	−0.127	0.254	0.599	0.860	1.000
$\vec{X}^{(3)}$	0.0	0.618	1.000	1.000	0.618	0.000	−0.618	−1.000	−1.000	−0.618	0.000	0.618	1.000

$\omega_1 = 225$ rad/sec, $\omega_2 = 660$ rad/sec, and $\omega_3 = 1100$ rad/sec. If the fuselage of the airplane is subjected to a known vertical motion $x_0(t)$, derive the uncoupled equations for determining the dynamic response of the wing by approximating it as a linear

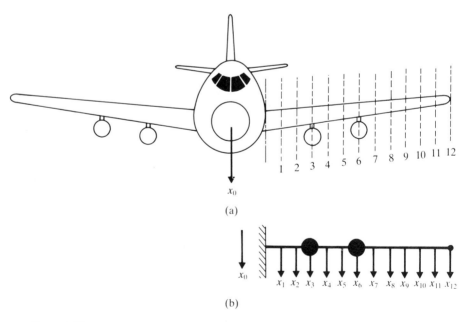

(a)

(b)

Figure 6.23

combination of the first three normal modes. *Hint:* The equation of motion of the airplane wing can be written, similar to Eq. (3.64), as

$$[m]\ddot{\vec{x}} + [c]\left(\dot{\vec{x}} - \dot{x}_0\vec{u}_1\right) + [k]\left(\vec{x} - x_0\vec{u}_1\right) = \vec{0}$$

or

$$[m]\ddot{\vec{x}} + [c]\dot{\vec{x}} + [k]\vec{x} = -x_0[m]\vec{u}_1$$

where $\vec{u}_1 = \{1, 0, 0, \ldots, 0\}^T$ is a unit vector.

6.44. Write a computer program for finding the eigenvectors using the known eigenvalues in Eq. (6.55). Find the mode shapes of Problem 6.25 using this program.

6.45. Write a computer program for generating the $[m]$-orthonormal modal matrix $[X]$. The program should accept the number of degrees of freedom, the normal modes, and the mass matrix as input. Solve Problem 6.32 using this program.

6.46. Generate the characteristic polynomial of Problem 6.23 using subroutine POLCOF.

6.47. Find the characteristic values of Problem 6.46 using subroutine CROOTS.

6.48. Write a computer program for finding the natural frequencies and mode shapes of a multidegree of freedom system when the mass and stiffness matrices are known, using the following steps:

  1. Find the dynamical matrix using subroutines DECOMP and MATMUL.

  2. Find the characteristic polynomial using subroutine POLCOF.

  3. Find the natural frequencies using subroutine CROOTS.

  4. Find the mode shapes using subroutine SIMUL.

  Solve Problem 6.21 using this program.

6.49. The equations of motion of an undamped system in SI units are given by

$$\begin{bmatrix} 2 & 0 & 0 \\ 0 & 2 & 0 \\ 0 & 0 & 2 \end{bmatrix}\ddot{\vec{x}} + \begin{bmatrix} 16 & -8 & 0 \\ -8 & 16 & -8 \\ 0 & -8 & 16 \end{bmatrix}\vec{x} = \begin{Bmatrix} 10\sin\omega t \\ 0 \\ 0 \end{Bmatrix}$$

Find the steady-state response of the system when $\omega = 5$ rad/s, using subroutine MODAL.

6.50. Find the response of the system in Problem 6.49 by varying ω between 1 and 10 rad/sec in increments of 1 rad/sec. Plot the graphs showing the variations of magnitudes of the first peaks of $x_i(t)$, $i = 1, 2, 3$ with respect to ω.

Project:

6.51. A heavy machine tool mounted on the first floor of a building, Fig. 6.24(a), has been modeled as a three degree of freedom system as indicated in Fig. 6.24(b). (i) For $k_1 = 5000$ lb/in., $k_2 = 500$ lb/in., $k_3 = 2000$ lb/in., $c_1 = c_2 = c_3 = 10$ lb-sec/in., $m_f = 50$ lb-sec^2/in., $m_b = 10$ lb-sec^2/in., $m_h = 2$ lb-sec^2/in., and $F(t) = 1000\cos 60t$ lb, find the steady state vibration of the system using the mechanical

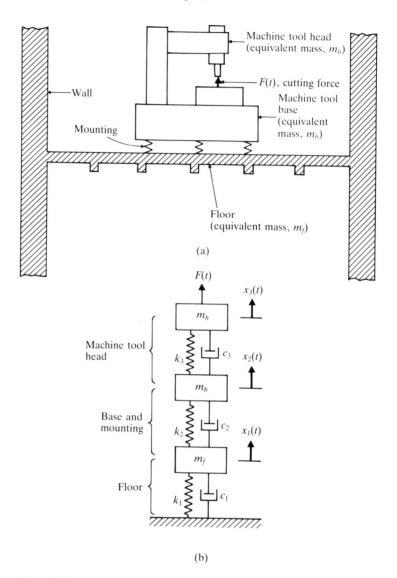

Figure 6.24

impedance method described in Section 5.6. (ii) If the maximum response of the machine tool head (x_3) has to be reduced by 25%, how should the stiffness of the mounting (k_2) be changed? (iii) Is there any better way of achieving the goal stated in (ii)? Give the details.

CHAPTER 7

Determination of Natural Frequencies and Mode Shapes

John William Strutt, Lord Rayleigh (1842–1919), was an English physicist who held the positions of professor of experimental physics at Cambridge University, professor of natural philosophy at the Royal Institution in London, president of the Royal Society, and chancellor of Cambridge University. His works in optics and acoustics are well known, with *Theory of Sound* (1877) considered as a standard reference even today. The method of computing approximate natural frequencies of vibrating bodies using an energy approach has become known as "Rayleigh's method." (Courtesy Brown Brothers)

7.1 INTRODUCTION

In the preceding chapter, the natural frequencies (eigenvalues) and the natural modes (eigenvectors) of a multidegree of freedom system were found by setting the characteristic determinant equal to zero. Although this is an exact method, the expansion of the characteristic determinant and the solution of the resulting nth degree polynomial equation to obtain the natural frequencies can become quite tedious for large values of n. Several analytical and numerical methods have been developed to compute the natural frequencies and mode shapes of multidegree of freedom systems. In this chapter, we shall consider Dunkerley's formula, Rayleigh's method, Holzer's method, the matrix iteration method, and Jacobi's method. Dunkerley's formula and Rayleigh's method are useful for estimating the fundamental natural frequency only. Holzer's method is essentially a tabular method that can be used to find partial or full solutions to eigenvalue problems. The matrix iteration method finds one natural frequency at a time, usually starting from the lowest value. The method can thus be terminated after finding the required number of natural frequencies and mode shapes. When all the natural frequencies and mode shapes are required, Jacobi's method can be used; it finds all the eigenvalues and eigenvectors simultaneously.

7.2 DUNKERLEY'S FORMULA

Dunkerley's formula gives the approximate value of the fundamental frequency of a composite system in terms of the natural frequencies of its component parts. It is derived by making use of the fact that the higher natural frequencies of most vibratory systems are large compared to their fundamental frequencies [7.1–7.3]. To derive Dunkerley's formula, consider a general n degree of freedom system whose eigenvalues can be determined by solving the frequency equation, Eq. (6.57):

$$\left| -[k] + \omega^2[m] \right| = 0$$

or

$$\left| -\frac{1}{\omega^2}[I] + [a][m] \right| = 0 \tag{7.1}$$

For a lumped mass system with a diagonal mass matrix, Eq. (7.1) becomes

$$\left| -\frac{1}{\omega^2}\begin{bmatrix} 1 & 0 & \cdots & 0 \\ 0 & 1 & \cdots & 0 \\ \vdots & & & \\ 0 & 0 & \cdots & 1 \end{bmatrix} + \begin{bmatrix} a_{11} & a_{12} & \cdots & a_{1n} \\ a_{21} & a_{22} & \cdots & a_{2n} \\ \vdots & & & \\ a_{n1} & a_{n2} & \cdots & a_{nn} \end{bmatrix} \begin{bmatrix} m_1 & 0 & \cdots & 0 \\ 0 & m_2 & \cdots & 0 \\ \vdots & & & \\ 0 & 0 & \cdots & m_n \end{bmatrix} \right| = 0$$

that is,

$$\left| \begin{array}{cccc} \left(-\dfrac{1}{\omega^2} + a_{11}m_1\right) & a_{12}m_2 & \cdots & a_{1n}m_n \\ a_{21}m_1 & \left(-\dfrac{1}{\omega^2} + a_{22}m_2\right) & \cdots & a_{2n}m_n \\ \vdots & \vdots & & \vdots \\ a_{n1}m_1 & a_{n2}m_2 & \cdots & \left(-\dfrac{1}{\omega^2} + a_{nn}m_n\right) \end{array} \right| = 0 \tag{7.2}$$

The expansion of Eq. (7.2) leads to

$$\left(\frac{1}{\omega^2}\right)^n - \left(a_{11}m_1 + a_{22}m_2 + \cdots + a_{nn}m_n\right)\left(\frac{1}{\omega^2}\right)^{n-1}$$
$$+ \left(a_{11}a_{22}m_1m_2 + a_{11}a_{33}m_1m_3 + \cdots + a_{n-1,n-1}a_{nn}m_{n-1}m_n\right.$$
$$\left. - a_{12}a_{21}m_1m_2 - \cdots - a_{n-1,n}a_{n,n-1}m_{n-1}m_n\right)\left(\frac{1}{\omega^2}\right)^{n-2} - \cdots = 0 \tag{7.3}$$

This is a polynomial equation of nth degree in $(1/\omega^2)$. Let the roots of Eq. (7.3) be denoted as $1/\omega_1^2, 1/\omega_2^2, \ldots, 1/\omega_n^2$. Thus

$$\left(\frac{1}{\omega^2} - \frac{1}{\omega_1^2}\right)\left(\frac{1}{\omega^2} - \frac{1}{\omega_2^2}\right) \cdots \left(\frac{1}{\omega^2} - \frac{1}{\omega_n^2}\right)$$
$$= \left(\frac{1}{\omega^2}\right)^n - \left(\frac{1}{\omega_1^2} + \frac{1}{\omega_2^2} + \cdots + \frac{1}{\omega_n^2}\right)\left(\frac{1}{\omega^2}\right)^{n-1} - \cdots = 0 \tag{7.4}$$

Equating the coefficient of $(1/\omega^2)^{n-1}$ in Eqs. (7.4) and (7.3) gives

$$\frac{1}{\omega_1^2} + \frac{1}{\omega_2^2} + \cdots + \frac{1}{\omega_n^2} = a_{11}m_1 + a_{22}m_2 + \cdots + a_{nn}m_n \qquad (7.5)$$

In most cases, the higher frequencies $\omega_2, \omega_3, \ldots, \omega_n$ are considerably larger than the fundamental frequency ω_1, and so

$$\frac{1}{\omega_i^2} \ll \frac{1}{\omega_1^2}, \qquad i = 2, 3, \ldots, n$$

Thus, Eq. (7.5) can be approximately written as

$$\frac{1}{\omega_1^2} \simeq a_{11}m_1 + a_{22}m_2 + \cdots + a_{nn}m_n \qquad (7.6)$$

This equation is known as *Dunkerley's formula*. The fundamental frequency given by Eq. (7.6) will always be smaller than the exact value. In some cases, it will be more convenient to rewrite Eq. (7.6) as

$$\frac{1}{\omega_1^2} \simeq \frac{1}{\omega_{1n}^2} + \frac{1}{\omega_{2n}^2} + \cdots + \frac{1}{\omega_{nn}^2} \qquad (7.7)$$

where $\omega_{in} = (1/a_{ii}m_i)^{1/2} = (k_{ii}/m_i)^{1/2}$ denotes the natural frequency of a single degree of freedom system consisting of mass m_i and spring of stiffness k_{ii}, $i = 1, 2, \ldots, n$. The use of Dunkerley's formula for finding the lowest frequency of elastic systems is presented in Refs. [7.4, 7.5].

EXAMPLE 7.1   **Fundamental Frequency of a Beam**

Estimate the fundamental natural frequency of a simply supported beam carrying three identical equally spaced masses, as shown in Fig. 7.1.

Given: Simply supported beam with equal masses as shown in Fig. 7.1.

Find: Fundamental natural frequency.

Approach: Dunkerley's formula.

Solution. The flexibility influence coefficients (see Example 6.2) required for the application of Dunkerley's formula are given by

$$a_{11} = a_{33} = \frac{3}{256} \frac{l^3}{EI}, \qquad a_{22} = \frac{1}{48} \frac{l^3}{EI} \qquad (E.1)$$

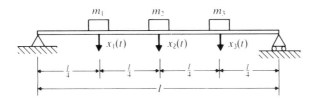

Figure 7.1

Equation (7.6) thus gives, using $m_1 = m_2 = m_3 = m$,

$$\frac{1}{\omega_1^2} \simeq \left(\frac{3}{256} + \frac{1}{48} + \frac{3}{256} \right) \frac{ml^3}{EI} = 0.04427 \frac{ml^3}{EI}$$

$$\omega_1 \simeq 4.75375 \sqrt{\frac{EI}{ml^3}}$$

This value can be compared with the exact value of the fundamental frequency

$$4.934 \sqrt{\frac{EI}{ml^3}}$$

7.3 RAYLEIGH'S METHOD

Rayleigh's method was presented in Section 2.5 to find the natural frequencies of single degree of freedom systems. The method can be extended to find the approximate value of the fundamental natural frequency of a discrete system.* The method is based on *Rayleigh's principle*, which can be stated as follows [7.6]:

> *The frequency of vibration of a conservative system vibrating about an equilibrium position has a stationary value in the neighborhood of a natural mode. This stationary value, in fact, is a minimum value in the neighborhood of the fundamental natural mode.*

We shall now derive an expression for the approximate value of the first natural frequency of a multidegree of freedom system according to Rayleigh's method.

The kinetic and potential energies of an n degree of freedom discrete system can be expressed as

$$T = \frac{1}{2} \dot{\vec{x}}^T [m] \dot{\vec{x}} \tag{7.8}$$

$$V = \frac{1}{2} \vec{x}^T [k] \vec{x} \tag{7.9}$$

To find the natural frequencies, we assume harmonic motion to be

$$\vec{x} = \vec{X} \cos \omega t \tag{7.10}$$

where $\vec{X}$ denotes the vector of amplitudes (mode shape) and ω represents the natural frequency of vibration. If the system is conservative, the maximum kinetic energy is equal to the maximum potential energy:

$$T_{\max} = V_{\max} \tag{7.11}$$

---

* Rayleigh's method for continuous systems is presented in Section 8.7.

By substituting Eq. (7.10) into Eqs. (7.8) and (7.9), we find

$$T_{max} = \frac{1}{2} \vec{X}^T [m] \vec{X} \omega^2 \tag{7.12}$$

$$V_{max} = \frac{1}{2} \vec{X}^T [k] \vec{X} \tag{7.13}$$

By equating T_{max} and V_{max}, we obtain*

$$\omega^2 = \frac{\vec{X}^T [k] \vec{X}}{\vec{X}^T [m] \vec{X}} \tag{7.14}$$

The right-hand side of Eq. (7.14) is known as *Rayleigh's quotient* and is denoted as $R(\vec{X})$.

7.3.1 Properties of Rayleigh's Quotient

As stated earlier, $R(\vec{X})$ has a stationary value when the arbitrary vector $\vec{X}$ is in the neighborhood of any eigenvector $\vec{X}^{(r)}$. To prove this, we express the arbitrary vector $\vec{X}$ in terms of the normal modes of the system, $\vec{X}^{(i)}$, as

$$\vec{X} = c_1 \vec{X}^{(1)} + c_2 \vec{X}^{(2)} + c_3 \vec{X}^{(3)} + \cdots \tag{7.15}$$

Then

$$\vec{X}^T [k] \vec{X} = c_1^2 \vec{X}^{(1)^T} [k] \vec{X}^{(1)} + c_2^2 \vec{X}^{(2)^T} [k] \vec{X}^{(2)} + c_3^2 \vec{X}^{(3)^T} [k] \vec{X}^{(3)} + \cdots \tag{7.16}$$

and

$$\vec{X}^T [m] \vec{X} = c_1^2 \vec{X}^{(1)^T} [m] \vec{X}^{(1)} + c_2^2 \vec{X}^{(2)^T} [m] \vec{X}^{(2)} + c_3^2 \vec{X}^{(3)^T} [m] \vec{X}^{(3)} + \cdots \tag{7.17}$$

as the cross terms of the form $c_i c_j \vec{X}^{(i)^T} [k] \vec{X}^{(j)}$ and $c_i c_j \vec{X}^{(i)^T} [m] \vec{X}^{(j)}$, $i \neq j$, are zero by the orthogonality property. Using Eqs. (7.16) and (7.17) and the relation

$$\vec{X}^{(i)^T} [k] \vec{X}^{(i)} = \omega_i^2 \vec{X}^{(i)^T} [m] \vec{X}^{(i)} \tag{7.18}$$

the Rayleigh's quotient of Eq. (7.14) can be expressed as

$$\omega^2 = R(\vec{X}) = \frac{c_1^2 \omega_1^2 \vec{X}^{(1)^T} [m] \vec{X}^{(1)} + c_2^2 \omega_2^2 \vec{X}^{(2)^T} [m] \vec{X}^{(2)} + \cdots}{c_1^2 \vec{X}^{(1)^T} [m] \vec{X}^{(1)} + c_2^2 \vec{X}^{(2)^T} [m] \vec{X}^{(2)} + \cdots} \tag{7.19}$$

If the normal modes are normalized, this equation becomes

$$\omega^2 = R(\vec{X}) = \frac{c_1^2 \omega_1^2 + c_2^2 \omega_2^2 + \cdots}{c_1^2 + c_2^2 + \cdots} \tag{7.20}$$

If $\vec{X}$ differs little from the eigenvector $\vec{X}^{(r)}$, the coefficient c_r will be much larger

---

* Equation (7.14) can also be obtained from the relation $[k]\vec{X} = \omega^2 [m]\vec{X}$. Premultiplying this equation by $\vec{X}^T$ and solving the resulting equation gives Eq. (7.14).

than the remaining coefficients c_i $(i \neq r)$ and Eq. (7.20) can be written as

$$
R(\vec{X}) = \frac{c_r^2 \omega_r^2 + c_r^2 \displaystyle\sum_{\substack{i=1,2,\dots \\ i \neq r}} \left(\frac{c_i}{c_r}\right)^2 \omega_i^2}{c_r^2 + c_r^2 \displaystyle\sum_{\substack{i=1,2,\dots \\ i \neq r}} \left(\frac{c_i}{c_r}\right)^2}
\tag{7.21}
$$

Since $|c_i/c_r| = \varepsilon_i \ll 1$ where ε_i is a small number for all $i \neq r$, Eq. (7.21) gives

$$
R(\vec{X}) = \omega_r^2 \{1 + 0(\varepsilon^2)\}
\tag{7.22}
$$

where $0(\varepsilon^2)$ represents an expression in ε of the second order or higher. Equation (7.22) indicates that if the arbitrary vector $\vec{X}$ differs from the eigenvector $\vec{X}^{(r)}$ by a small quantity of the first order, $R(\vec{X})$ differs from the eigenvalue ω_r^2 by a small quantity of the second order. This means that Rayleigh's quotient has a stationary value in the neighborhood of an eigenvector.

The stationary value is actually a minimum value in the neighborhood of the fundamental mode, $\vec{X}^{(1)}$. To see this, let $r = 1$ in Eq. (7.21) and write

$$
\begin{aligned}
R(\vec{X}) &= \frac{\omega_1^2 + \displaystyle\sum_{i=2,3,\dots} \left(\frac{c_i}{c_1}\right)^2 \omega_i^2}{\left\{1 + \displaystyle\sum_{i=2,3,\dots} \left(\frac{c_i}{c_1}\right)^2\right\}} \\
&\simeq \omega_1^2 + \sum_{i=2,3,\dots} \varepsilon_i^2 \omega_i^2 - \omega_1^2 \sum_{i=2,3,\dots} \varepsilon_i^2 \\
&\simeq \omega_1^2 + \sum_{i=2,3,\dots} \left(\omega_i^2 - \omega_1^2\right) \varepsilon_i^2
\end{aligned}
\tag{7.23}
$$

Since, in general, $\omega_i^2 > \omega_1^2$ for $i = 2, 3, \dots$, Eq. (7.23) leads to

$$
R(\vec{X}) \geq \omega_1^2
\tag{7.24}
$$

which shows that Rayleigh's quotient is never lower than the first eigenvalue. By proceeding in a similar manner, we can show that

$$
R(\vec{X}) \leq \omega_n^2
\tag{7.25}
$$

which means that Rayleigh's quotient is never higher than the highest eigenvalue. Thus Rayleigh's quotient provides an upper bound for ω_1^2 and a lower bound for ω_n^2.

7.3.2
Computation of
the Fundamental
Natural
Frequency

Equation (7.14) can be used to find an approximate value of the first natural frequency (ω_1) of the system. For this, we select a trial vector $\vec{X}$ to represent the first natural mode $\vec{X}^{(1)}$ and substitute it on the right-hand side of Eq. (7.14). This yields the approximate value of ω_1^2. Because Rayleigh's quotient is stationary, remarkably good estimates of ω_1^2 can be obtained even if the trial vector $\vec{X}$ deviates greatly from the true natural mode $\vec{X}^{(1)}$. Obviously, the estimated value of the fundamental frequency ω_1 is more accurate if the trial vector ($\vec{X}$) chosen resembles the true natural mode $\vec{X}^{(1)}$ closely. Rayleigh's method is compared with Dunkerley's and other methods in Refs. [7.7–7.9].

EXAMPLE 7.2   **Fundamental Frequency of a 3 Degree-of-Freedom System**

Estimate the fundamental frequency of vibration of the system shown in Fig. 7.2. Assume that $m_1 = m_2 = m_3 = m$, $k_1 = k_2 = k_3 = k$, and the mode shape is

$$\vec{X} = \begin{Bmatrix} 1 \\ 2 \\ 3 \end{Bmatrix}$$

Given: A three degree of freedom system given in Fig. 7.2 with $m_1 = m_2 = m_3 = m$, $k_1 = k_2 = k_3 = k$, and mode shape $\vec{X} = \begin{Bmatrix} 1 \\ 2 \\ 3 \end{Bmatrix}$.

Find: Fundamental frequency of vibration.

Approach: Rayleigh's quotient.

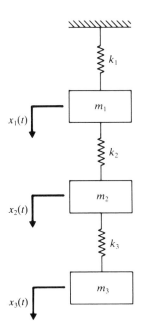

Figure 7.2

Solution. The stiffness and mass matrices of the system are:

$$[k] = k \begin{bmatrix} 2 & -1 & 0 \\ -1 & 2 & -1 \\ 0 & -1 & 1 \end{bmatrix} \tag{E.1}$$

$$[m] = m \begin{bmatrix} 1 & 0 & 0 \\ 0 & 1 & 0 \\ 0 & 0 & 1 \end{bmatrix} \tag{E.2}$$

By substituting the assumed mode shape in the expression for Rayleigh's quotient, we obtain

$$R(\vec{X}) = \omega^2 = \frac{\vec{X}^T[k]\vec{X}}{\vec{X}^T[m]\vec{X}} = \frac{(1 \quad 2 \quad 3)k \begin{bmatrix} 2 & -1 & 0 \\ -1 & 2 & -1 \\ 0 & -1 & 1 \end{bmatrix} \begin{Bmatrix} 1 \\ 2 \\ 3 \end{Bmatrix}}{(1 \quad 2 \quad 3)m \begin{bmatrix} 1 & 0 & 0 \\ 0 & 1 & 0 \\ 0 & 0 & 1 \end{bmatrix} \begin{Bmatrix} 1 \\ 2 \\ 3 \end{Bmatrix}}$$

$$= 0.2143 \frac{k}{m} \tag{E.3}$$

$$\omega_1 = 0.4629 \sqrt{\frac{k}{m}} \tag{E.4}$$

This value is 4.0225% larger than the exact value of $0.4450\sqrt{k/m}$. The exact fundamental mode shape (see Example 6.4) in this case is

$$\vec{X}^{(1)} = \begin{Bmatrix} 1.0000 \\ 1.8019 \\ 2.2470 \end{Bmatrix} \tag{E.5}$$

**7.3.3
Fundamental
Frequency of
Beams and Shafts**

Although the procedure outlined above is applicable to all discrete systems, a simpler equation can be derived for the fundamental frequency of the lateral vibration of a beam or a shaft carrying several masses such as pulleys, gears, or flywheels. In these cases, the static deflection curve is used as an approximation of the dynamic deflection curve.

Consider a shaft carrying several masses, as shown in Fig. 7.3. The shaft is assumed to have negligible mass. The potential energy of the system is the strain energy of the deflected shaft, which is equal to the work done by the static loads. Thus

$$V_{\text{max}} = \frac{1}{2}(m_1 g w_1 + m_2 g w_2 + \cdots) \tag{7.26}$$

where $m_i g$ is the static load due to the mass m_i, and w_i is the total static deflection of mass m_i due to all the masses. For harmonic oscillation (free vibration), the maximum kinetic energy due to the masses is

$$T_{\text{max}} = \frac{\omega^2}{2}(m_1 w_1^2 + m_2 w_2^2 + \cdots) \tag{7.27}$$

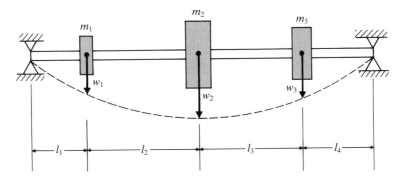

Figure 7.3

where ω is the frequency of oscillation. Equating V_{max} and T_{max}, we obtain

$$\omega = \left\{ \frac{g(m_1 w_1 + m_2 w_2 + \cdots)}{(m_1 w_1^2 + m_2 w_2^2 + \cdots)} \right\}^{1/2} \tag{7.28}$$

EXAMPLE 7.3   Fundamental Frequency of a Shaft with Rotors

Estimate the fundamental frequency of lateral vibration of a shaft carrying three rotors (masses) as shown in Fig. 7.3 with $m_1 = 20$ kg, $m_2 = 50$ kg, $m_3 = 40$ kg, $l_1 = 1$ m, $l_2 = 3$ m, $l_3 = 4$ m, and $l_4 = 2$ m. The shaft is made of steel with solid circular cross section of diameter 10 cm.

Given: Simply supported shaft carrying three masses as shown in Fig. 7.3. $m_1 = 20$ kg, $m_2 = 50$ kg, $m_3 = 40$ kg, $l_1 = 1$ m, $l_2 = 3$ m, $l_3 = 4$ m, $l_4 = 2$ m, diameter of shaft $(d) = 10$ cm, and $E = 207 \times 10^9$ N/m^2.

Find: Fundamental frequency of lateral vibration.

Approach: Eq. (7.28) along with the static deflection curve for beams.

Solution. From strength of materials, the deflection of the beam shown in Fig. 7.4 due to a static load P [7.10] is given by

$$w(x) = \begin{cases} \dfrac{Pbx}{6EIl}(l^2 - b^2 - x^2); & 0 \leqslant x \leqslant a \tag{E.1} \\[2mm] -\dfrac{Pa(l-x)}{6EIl}[a^2 + x^2 - 2lx]; & a \leqslant x \leqslant l \tag{E.2} \end{cases}$$

Deflection due to the weight of m_1: At the location of mass m_1 (with $x = 1$ m, $b = 9$ m, and

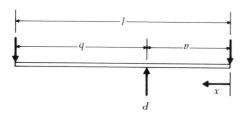

Figure 7.4

$l = 10$ m in Eq. (E.1)):

$$w_1' = \frac{(20 \times 9.81)(9)(1)}{6EI(10)}(100 - 81 - 1) = \frac{529.74}{EI} \tag{E.3}$$

At the location of m_2 (with $a = 1$ m, $x = 4$ m, and $l = 10$ m in Eq. (E.2)):

$$w_2' = -\frac{(20 \times 9.81)(1)(6)}{6EI(10)}[1 + 16 - 2(10)(4)] = \frac{1236.06}{EI} \tag{E.4}$$

At the location of m_3 (with $a = 1$ m, $x = 8$ m, and $l = 10$ m in Eq. (E.2)):

$$w_3' = -\frac{(20 \times 9.81)(1)(2)}{6EI(10)}[1 + 64 - 2(10)(8)] = \frac{621.3}{EI} \tag{E.5}$$

Deflection due to the weight of m_2: At the location of m_1 (with $x = 1$ m, $b = 6$ m, and $l = 10$ m in Eq. (E.1)):

$$w_1'' = \frac{(50 \times 9.81)(6)(1)}{6EI(10)}(100 - 36 - 1) = \frac{3090.15}{EI} \tag{E.6}$$

At the location of m_2 (with $x = 4$ m, $b = 6$ m, and $l = 10$ m in Eq. (E.1)):

$$w_2'' = \frac{(50 \times 9.81)(6)(4)}{6EI(10)}(100 - 36 - 16) = \frac{9417.6}{EI} \tag{E.7}$$

At the location of m_3 (with $a = 4$ m, $x = 8$ m, and $l = 10$ m in Eq. (E.2)):

$$w_3'' = -\frac{(50 \times 9.81)(4)(2)}{6EI(10)}[16 + 64 - 2(10)(8)] = \frac{5232.0}{EI} \tag{E.8}$$

Deflection due to the weight of m_3: At the location of m_1 (with $x = 1$ m, $b = 2$ m, and $l = 10$ m in Eq. (E.1)):

$$w_1''' = \frac{(40 \times 9.81)(2)(1)}{6EI(10)}(100 - 4 - 1) = \frac{1242.6}{EI} \tag{E.9}$$

At the location of m_2 (with $x = 4$ m, $b = 2$ m, and $l = 10$ m in Eq. (E.1)):

$$w_2''' = \frac{(40 \times 9.81)(2)(4)}{6EI(10)}(100 - 4 - 16) = \frac{4185.6}{EI} \tag{E.10}$$

At the location of m_3 (with $x = 8$ m, $b = 2$ m, and $l = 10$ m in Eq. (E.1)):

$$w_3''' = \frac{(40 \times 9.81)(2)(8)}{6EI(10)}(100 - 4 - 64) = \frac{3348.48}{EI} \tag{E.11}$$

The total deflections of the masses m_1, m_2, and m_3 are

$$w_1 = w_1' + w_1'' + w_1''' = \frac{4862.49}{EI}$$

$$w_2 = w_2' + w_2'' + w_2''' = \frac{14839.26}{EI}$$

$$w_3 = w_3' + w_3'' + w_3''' = \frac{9201.78}{EI}$$

Substituting into Eq. (7.28), we find the fundamental natural frequency:

$$\omega = \left\{ \frac{9.81(20 \times 4862.49 + 50 \times 14839.26 + 40 \times 9201.78)\, EI}{20 \times (4862.49)^2 + 50 \times (14839.26)^2 + 40 \times (9201.78)^2} \right\}^{1/2} = 6.5591 \times 10^{-4} \sqrt{EI}$$

$$(E.12)$$

For the shaft, $E = 2.07 \times 10^{11}$ N/m^2 and $I = \pi(0.1)^4/64 = 4.90875 \times 10^{-6}$ m^4 and hence Eq. (E.12) gives

$$\omega = 0.66117 \text{ rad/sec}$$

7.4 HOLZER'S METHOD

Holzer's method is essentially a trial-and-error scheme to find the natural frequencies of undamped, damped, semi-definite, fixed, or branched vibrating systems involving linear and angular displacements [7.11, 7.12]. The method can also be programmed for computer applications. A trial frequency of the system is first assumed, and a solution is found when the assumed frequency satisfies the constraints of the system. This generally requires several trials. Depending on the trial frequency used, the fundamental as well as the higher frequencies of the system can be determined. The method also gives the mode shapes.

7.4.1 Torsional Systems

Consider the undamped torsional semi-definite system shown in Fig. 7.5. The equations of motion of the discs can be derived as follows:

$$J_1\ddot{\theta}_1 + k_{t1}(\theta_1 - \theta_2) = 0 \qquad (7.29)$$

$$J_2\ddot{\theta}_2 + k_{t1}(\theta_2 - \theta_1) + k_{t2}(\theta_2 - \theta_3) = 0 \qquad (7.30)$$

$$J_3\ddot{\theta}_3 + k_{t2}(\theta_3 - \theta_2) = 0 \qquad (7.31)$$

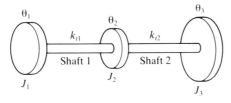

Figure 7.5

Since the motion is harmonic in a natural mode of vibration, we assume that $\theta_i = \Theta_i \cos(\omega t + \phi)$ in Eqs. (7.29) to (7.31) and obtain

$$\omega^2 J_1 \Theta_1 = k_{t1}(\Theta_1 - \Theta_2) \tag{7.32}$$

$$\omega^2 J_2 \Theta_2 = k_{t1}(\Theta_2 - \Theta_1) + k_{t2}(\Theta_2 - \Theta_3) \tag{7.33}$$

$$\omega^2 J_3 \Theta_3 = k_{t2}(\Theta_3 - \Theta_2) \tag{7.34}$$

Summing these equations gives

$$\sum_{i=1}^{3} \omega^2 J_i \Theta_i = 0 \tag{7.35}$$

Equation (7.35) states that the sum of the inertia torques of the semi-definite system must be zero. This equation can be treated as another form of the frequency equation, and the trial frequency must satisfy this requirement.

In Holzer's method, a trial frequency ω is assumed, and Θ_1 is arbitrarily chosen as unity. Next, Θ_2 is computed from Eq. (7.32), and then Θ_3 is found from Eq. (7.33). Thus we obtain

$$\Theta_1 = 1 \tag{7.36}$$

$$\Theta_2 = \Theta_1 - \frac{\omega^2 J_1 \Theta_1}{k_{t1}} \tag{7.37}$$

$$\Theta_3 = \Theta_2 - \frac{\omega^2}{k_{t2}}(J_1 \Theta_1 + J_2 \Theta_2) \tag{7.38}$$

These values are substituted in Eq. (7.35) to verify whether the constraint is satisfied. If Eq. (7.35) is not satisfied, a new trial value of ω is assumed and the process repeated. Equations (7.35), (7.37), and (7.38) can be generalized for an n disc system as follows:

$$\sum_{i=1}^{n} \omega^2 J_i \Theta_i = 0 \tag{7.39}$$

$$\Theta_i = \Theta_{i-1} - \frac{\omega^2}{k_{t_{i-1}}} \left(\sum_{k=1}^{i-1} J_k \Theta_k \right), \qquad i = 2, 3, \ldots, n \tag{7.40}$$

Thus the method uses Eqs. (7.39) and (7.40) repeatedly for different trial frequencies. If the assumed trial frequency is not a natural frequency of the system, Eq. (7.39) is not satisfied. The resultant torque in Eq. (7.39) represents a torque applied at the last disc. This torque M_t is then plotted for the chosen ω. When the calculation is repeated with other values of ω, the resulting graph appears as shown in Fig. 7.6. From this graph, the natural frequencies of the system can be identified as the values of ω at which $M_t = 0$. The amplitudes Θ_i ($i = 1, 2, \ldots, n$) corresponding to the natural frequencies are the mode shapes of the system.

Holzer's method can also be applied to systems with fixed ends. At a fixed end, the amplitude of vibration must be zero. In this case, the natural frequencies can be found by plotting the resulting amplitude (instead of the resultant torque) against the assumed frequencies. For a system with one end free and the other end fixed,

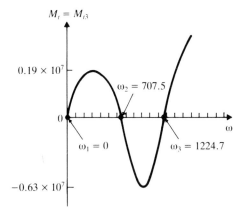

Figure 7.6

Eq. (7.40) can be used for checking the amplitude at the fixed end. An improvement of Holzer's method was presented in Refs. [7.13, 7.14].

EXAMPLE 7.4 **Natural Frequencies of a Torsional System**

The arrangement of the compressor, turbine, and generator in a thermal power plant is shown in Fig. 7.7. Find the natural frequencies and mode shapes of the system.

Given: Compressor-turbine-generator arrangement shown in Fig. 7.7.

Find: Natural frequencies and mode shapes.

Approach: Holzer's method.

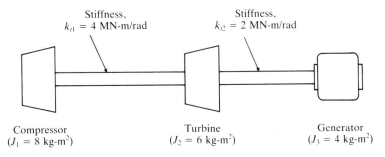

Stiffness,
$k_{t1} = 4$ MN-m/rad

Stiffness,
$k_{t2} = 2$ MN-m/rad

Compressor
($J_1 = 8$ kg-m^2)

Turbine
($J_2 = 6$ kg-m^2)

Generator
($J_3 = 4$ kg-m^2)

Figure 7.7

TABLE 7.1

Parameters of the System	Quantity	Trial					
		1	2	3	...	71	72
	ω	0	10	20		700	710
	ω^2	0	100	400		490000	504100
Station 1:							
$J_1 = 8$	Θ_1	1.0	1.0	1.0		1.0	1.0
$k_{t1} = 4 \times 10^6$	$M_{t1} = \omega^2 J_1 \Theta_1$	0	800	3200		0.392E7	0.403E7
Station 2:							
$J_2 = 6$	$\Theta_2 = 1 - \dfrac{M_{t1}}{k_{t1}}$	1.0	0.9998	0.9992		0.0200	−0.0082
$k_{t2} = 2 \times 10^6$	$M_{t2} = M_{t1} + \omega^2 J_2 \Theta_2$	0	1400	5598		0.398E7	0.401E7
Station 3:							
$J_3 = 4$	$\Theta_3 = \Theta_2 - \dfrac{M_{t2}}{k_{t2}}$	1.0	0.9991	0.9964		−1.9690	−2.0120
$k_{t3} = 0$	$M_{t3} = M_{t2} + \omega^2 J_3 \Theta_3$	0	1800	7192		0.119E6	−0.494E5

Solution. This system represents an unrestrained or free-free torsional system. Table 7.1 shows its parameters and the sequence of computations. The calculations for the trial frequencies $\omega = 0, 10, 20, 700$, and 710 are shown in this table. The quantity M_{t3} denotes the torque to the right of Station 3 (generator) which must be zero at the natural frequencies. Figure 7.5 shows the graph of M_{t3} versus ω. Closely spaced trial values of ω are used in the vicinity of $M_{t3} = 0$ to obtain accurate values of the first two flexible mode shapes, shown in Fig. 7.8. Note that the value $\omega = 0$ corresponds to the rigid body rotation.

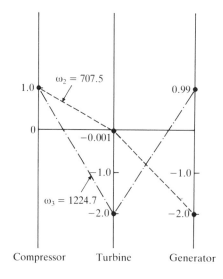

Figure 7.8

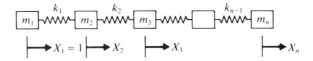

Figure 7.9

**7.4.2
Spring-Mass
Systems**

Although Holzer's method has been extensively applied to torsional systems, the procedure is equally applicable to the vibration analysis of spring-mass systems. The equations of motion of a spring-mass system (see Fig. 7.9) can be expressed as

$$m_1\ddot{x}_1 + k_1(x_1 - x_2) = 0 \tag{7.41}$$
$$m_2\ddot{x}_2 + k_1(x_2 - x_1) + k_2(x_2 - x_3) = 0 \tag{7.42}$$
$$\cdots$$

For harmonic motion, $x_i(t) = X_i \cos \omega t$, where X_i is the amplitude of mass m_i, and Eqs. (7.41) and (7.42) can be rewritten as

$$\omega^2 m_1 X_1 = k_1(X_1 - X_2) \tag{7.43}$$
$$\omega^2 m_2 X_2 = k_1(X_2 - X_1) + k_2(X_2 - X_3) = -\omega^2 m_1 X_1 + k_2(X_2 - X_3) \tag{7.44}$$
$$\cdots$$

The procedure of Holzer's method starts with a trial frequency ω and the amplitude of mass m_1 as $X_1 = 1$. Equations (7.43) and (7.44) can then be used to obtain the amplitudes of the masses $m_2, m_3, \ldots, m_i$:

$$X_2 = X_1 - \frac{\omega^2 m_1 X_1}{k_1} \tag{7.45}$$

$$X_3 = X_2 - \frac{\omega^2}{k_2}(m_1 X_1 + m_2 X_2) \tag{7.46}$$

$$\cdots$$

$$X_i = X_{i-1} - \frac{\omega^2}{k_{i-1}}\left(\sum_{k=1}^{i-1} m_k X_k\right), \qquad i = 2, 3, \ldots, n \tag{7.47}$$

As in the case of torsional systems, the resultant force applied to the last (nth) mass can be computed as follows:

$$F = \sum_{i=1}^{n} \omega^2 m_i X_i \tag{7.48}$$

The calculations are repeated with several other trial frequencies ω. The natural frequencies are identified as those values of ω that give $F = 0$ for a free-free system. For this, it is convenient to plot a graph between F and ω, using the same procedure for spring-mass systems as for torsional systems.

7.5 MATRIX ITERATION METHOD

The matrix iteration method assumes that the natural frequencies are distinct and well separated such that $\omega_1 < \omega_2 < \cdots < \omega_n$. The iteration is started by selecting a trial vector $\vec{X}_1$, which is then premultiplied by the dynamical matrix $[D]$. The resulting column vector is then normalized, usually by making one of its components to unity. The normalized column vector is premultiplied by $[D]$ to obtain a third column vector, which is normalized in the same way as before and becomes still another trial column vector. The process is repeated until the successive normalized column vectors converge to a common vector: the fundamental eigenvector. The normalizing factor gives the largest value of $\lambda = 1/\omega^2$—that is, the smallest or the fundamental natural frequency [7.15]. The convergence of the process can be explained as follows.

According to the expansion theorem, any arbitrary n-dimensional vector $\vec{X}_1$ can be expressed as a linear combination of the n orthogonal eigenvectors of the system $\vec{X}^{(i)}$, $i = 1, 2, \ldots, n$:

$$\vec{X}_1 = c_1 \vec{X}^{(1)} + c_2 \vec{X}^{(2)} + \cdots + c_n \vec{X}^{(n)} \tag{7.49}$$

where $c_1, c_2, \ldots, c_n$ are constants. In the iteration method, the trial vector $\vec{X}_1$ is selected arbitrarily and is therefore a known vector. The modal vectors $\vec{X}^{(i)}$, although unknown, are constant vectors because they depend upon the properties of the system. The constants c_i are unknown numbers to be determined. According to the iteration method, we premultiply $\vec{X}_1$ by the matrix $[D]$. In view of Eq. (7.49), this gives

$$[D]\vec{X}_1 = c_1[D]\vec{X}^{(1)} + c_2[D]\vec{X}^{(2)} + \cdots + c_n[D]\vec{X}^{(n)} \tag{7.50}$$

Now, according to Eq. (6.60), we have

$$[D]\vec{X}^{(i)} = \lambda_i[I]\vec{X}^{(i)} = \frac{1}{\omega_i^2}\vec{X}^{(i)}; \qquad i = 1, 2, \ldots, n \tag{7.51}$$

Substitution of Eq. (7.51) into Eq. (7.50) yields

$$[D]\vec{X}_1 = \vec{X}_2$$
$$= \frac{c_1}{\omega_1^2}\vec{X}^{(1)} + \frac{c_2}{\omega_2^2}\vec{X}^{(2)} + \cdots + \frac{c_n}{\omega_n^2}\vec{X}^{(n)} \tag{7.52}$$

where $\vec{X}_2$ is the second trial vector. We now repeat the process and premultiply $\vec{X}_2$ by $[D]$ to obtain, by Eqs. (7.49) and (6.60),

$$[D]\vec{X}_2 = \vec{X}_3$$
$$= \frac{c_1}{\omega_1^4}\vec{X}^{(1)} + \frac{c_2}{\omega_2^4}\vec{X}^{(2)} + \cdots + \frac{c_n}{\omega_n^4}\vec{X}^{(n)} \tag{7.53}$$

By repeating the process we obtain, after the rth iteration,

$$[D]\vec{X}_r = \vec{X}_{r+1}$$
$$= \frac{c_1}{\omega_1^{2r}}\vec{X}^{(1)} + \frac{c_2}{\omega_2^{2r}}\vec{X}^{(2)} + \cdots + \frac{c_n}{\omega_n^{2r}}\vec{X}^{(n)} \tag{7.54}$$

Since the natural frequencies are assumed to be $\omega_1 < \omega_2 < \cdots < \omega_n$, a sufficiently large value of r yields

$$\frac{1}{\omega_1^{2r}} \gg \frac{1}{\omega_2^{2r}} \gg \cdots \gg \frac{1}{\omega_n^{2r}} \qquad (7.55)$$

Thus the first term on the right-hand side of Eq. (7.54) becomes the only significant one. Hence we have

$$\vec{X}_{r+1} = \frac{c_1}{\omega_1^{2r}} \vec{X}^{(1)} \qquad (7.56)$$

which means that the $(r + 1)$th trial vector becomes identical to the fundamental modal vector to within a multiplicative constant. Since

$$\vec{X}_r = \frac{c_1}{\omega_1^{2(r-1)}} \vec{X}^{(1)} \qquad (7.57)$$

the fundamental natural frequency ω_1 can be found by taking the ratio of any two corresponding components in the vectors $\vec{X}_r$ and $\vec{X}_{r+1}$:

$$\omega_1^2 \simeq \frac{X_{i,r}}{X_{i,r+1}}, \qquad \text{for any } i = 1, 2, \ldots, n \qquad (7.58)$$

where $X_{i,r}$ and $X_{i,r+1}$ are the ith elements of the vectors $\vec{X}_r$ and $\vec{X}_{r+1}$, respectively.

Discussion

1. In the above proof, nothing has been said about the normalization of the successive trial vectors $\vec{X}_i$. Actually, it is not necessary to establish the proof of convergence of the method. The normalization amounts to a readjustment of the constants $c_1, c_2, \ldots, c_n$ in each iteration.

2. Although it is theoretically necessary to have $r \to \infty$ for the convergence of the method, in practice only a finite number of iterations suffices to obtain a reasonably good estimate of ω_1.

3. The actual number of iterations necessary to find the value of ω_1 to within a desired degree of accuracy depends on how closely the arbitrary trial vector $\vec{X}_1$ resembles the fundamental mode $\vec{X}^{(1)}$ and on how well ω_1 and ω_2 are separated. The required number of iterations is less if ω_2 is very large compared to ω_1.

4. The method has a distinct advantage in that any computational errors made do not yield incorrect results. Any error made in premultiplying $\vec{X}_i$ by $[D]$ results in a vector other than the desired one, $\vec{X}_{i+1}$. But this wrong vector can be considered as a new trial vector. This may delay the convergence but does not produce wrong results.

5. One can take any set of n numbers for the first trial vector $\vec{X}_1$ and still achieve convergence to the fundamental modal vector. Only in the unusual case in which the trial vector $\vec{X}_1$ is exactly proportional to one of the modes $\vec{X}^{(i)}$ $(i \ne 1)$ does the method fail to converge to the first mode. In such a case, the premultiplication of $\vec{X}^{(i)}$ by $[D]$ results in a vector proportional to $\vec{X}^{(i)}$ itself.

7.5.1
Convergence to the Highest Natural Frequency

To obtain the highest natural frequency ω_n and the corresponding mode shape or eigenvector $\vec{X}^{(n)}$ by the matrix iteration method, we first rewrite Eq. (6.60) as

$$[D]^{-1}\vec{X} = \omega^2[I]\vec{X} = \omega^2\vec{X} \qquad (7.59)$$

where $[D]^{-1}$ is the inverse of the dynamical matrix $[D]$ given by

$$[D]^{-1} = [m]^{-1}[k] \qquad (7.60)$$

Now we select any arbitrary trial vector $\vec{X}_1$ and premultiply it by $[D]^{-1}$ to obtain an improved trial vector $\vec{X}_2$. The sequence of trial vectors $\vec{X}_{i+1}$ ($i = 1, 2, \ldots$) obtained by premultiplying by $[D]^{-1}$ converges to the highest normal mode $\vec{X}^{(n)}$. It can be seen that the procedure is similar to the one already described. The constant of proportionality in this case is ω^2 instead of $1/\omega^2$.

7.5.2
Computation of Intermediate Natural Frequencies

Once the first natural frequency ω_1 (or the largest eigenvalue $\lambda_1 = 1/\omega_1^2$) and the corresponding eigenvector $\vec{X}^{(1)}$ are determined, we can proceed to find the higher natural frequencies and the corresponding mode shapes by the matrix iteration method. Before we proceed, it should be remembered that any arbitrary trial vector premultiplied by $[D]$ would lead again to the largest eigenvalue. It is thus necessary to remove the largest eigenvalue from the matrix $[D]$. The succeeding eigenvalues and eigenvectors can be obtained by eliminating the root λ_1 from the characteristic or frequency equation

$$\left|[D] - \lambda[I]\right| = 0 \qquad (7.61)$$

A procedure known as *matrix deflation* can be used for this purpose [7.16]. To find the eigenvector $\vec{X}^{(i)}$ by this procedure, the previous eigenvector $\vec{X}^{(i-1)}$ is normalized with respect to the mass matrix such that

$$\vec{X}^{(i-1)^T}[m]\vec{X}^{(i-1)} = 1 \qquad (7.62)$$

Then the deflated matrix $[D_i]$ is constructed as

$$[D_i] = [D_{i-1}] - \lambda_{i-1}\vec{X}^{(i-1)}\vec{X}^{(i-1)^T}[m], \qquad i = 2, 3, \ldots, n \qquad (7.63)$$

where $[D_1] = [D]$. Once $[D_i]$ is constructed, the iterative scheme

$$\vec{X}_{r+1} = [D_i]\vec{X}_r \qquad (7.64)$$

is used, where $\vec{X}_1$ is an arbitrary trial eigenvector.

EXAMPLE 7.5   **Natural Frequencies of a 3-Degree-of-Freedom System**

Find the natural frequencies and mode shapes of the system shown in Fig. 7.2 for $k_1 = k_2 = k_3 = k$ and $m_1 = m_2 = m_3 = m$ by the matrix iteration method.

Given: A three degree of freedom spring-mass system, shown in Fig. 7.2, with $m_1 = m_2 = m_3 = m$ and $k_1 = k_2 = k_3 = k$.

Find: Natural frequencies and mode shapes.

Approach: Matrix iteration method.

Solution. The mass and stiffness matrices of the system are given in Example 7.2. The flexibility matrix is

$$[a] = [k]^{-1} = \frac{1}{k}\begin{bmatrix} 1 & 1 & 1 \\ 1 & 2 & 2 \\ 1 & 2 & 3 \end{bmatrix} \tag{E.1}$$

and so the dynamical matrix is

$$[k]^{-1}[m] = \frac{m}{k}\begin{bmatrix} 1 & 1 & 1 \\ 1 & 2 & 2 \\ 1 & 2 & 3 \end{bmatrix} \tag{E.2}$$

The eigenvalue problem can be stated as

$$[D]\vec{X} = \lambda \vec{X} \tag{E.3}$$

where

$$[D] = \begin{bmatrix} 1 & 1 & 1 \\ 1 & 2 & 2 \\ 1 & 2 & 3 \end{bmatrix} \tag{E.4}$$

and

$$\lambda = \frac{k}{m} \cdot \frac{1}{\omega^2} \tag{E.5}$$

First natural frequency: By assuming the first trial eigenvector or mode shape to be

$$\vec{X}_1 = \begin{Bmatrix} 1 \\ 1 \\ 1 \end{Bmatrix} \tag{E.6}$$

the second trial eigenvector can be obtained:

$$\vec{X}_2 = [D]\vec{X}_1 = \begin{Bmatrix} 3 \\ 5 \\ 6 \end{Bmatrix} \tag{E.7}$$

By making the first element equal to unity, we obtain

$$\vec{X}_2 = 3.0 \begin{Bmatrix} 1.0000 \\ 1.6667 \\ 2.0000 \end{Bmatrix} \tag{E.8}$$

and the corresponding eigenvalue is given by

$$\lambda_1 \simeq 3.0 \quad \text{or} \quad \omega_1 \simeq 0.5773 \sqrt{\frac{k}{m}} \tag{E.9}$$

The subsequent trial eigenvector can be obtained from the relation

$$\vec{X}_{i+1} = [D]\vec{X}_i \tag{E.10}$$

and the corresponding eigenvalues are given by

$$\lambda_1 \simeq X_{1, i+1} \tag{E.11}$$

where $X_{1, i+1}$ is the first component of the vector $\vec{X}_{i+1}$ before normalization. The various trial eigenvectors and eigenvalues obtained by using Eqs. (E.10) and (E.11) are shown below:

i	$\vec{X}_i$ with $X_{1,i} = 1$	$\vec{X}_{i+1} = [D]\vec{X}_i$	$\lambda_1 \simeq X_{1,i+1}$	ω_1
1	$\begin{Bmatrix} 1 \\ 1 \\ 1 \end{Bmatrix}$	$\begin{Bmatrix} 3 \\ 5 \\ 6 \end{Bmatrix}$	3.0	$0.5773\sqrt{\dfrac{k}{m}}$
2	$\begin{Bmatrix} 1.00000 \\ 1.66667 \\ 2.00000 \end{Bmatrix}$	$\begin{Bmatrix} 4.66667 \\ 8.33333 \\ 10.33333 \end{Bmatrix}$	4.66667	$0.4629\sqrt{\dfrac{k}{m}}$
3	$\begin{Bmatrix} 1.0000 \\ 1.7857 \\ 2.2143 \end{Bmatrix}$	$\begin{Bmatrix} 5.00000 \\ 9.00000 \\ 11.2143 \end{Bmatrix}$	5.00000	$0.4472\sqrt{\dfrac{k}{m}}$
$\vdots$				
7	$\begin{Bmatrix} 1.00000 \\ 1.80193 \\ 2.24697 \end{Bmatrix}$	$\begin{Bmatrix} 5.04891 \\ 9.09781 \\ 11.34478 \end{Bmatrix}$	5.04891	$0.44504\sqrt{\dfrac{k}{m}}$
8	$\begin{Bmatrix} 1.00000 \\ 1.80194 \\ 2.24698 \end{Bmatrix}$	$\begin{Bmatrix} 5.04892 \\ 9.09783 \\ 11.34481 \end{Bmatrix}$	5.04892	$0.44504\sqrt{\dfrac{k}{m}}$

It can be seen that the mode shape and the natural frequency converged (to the fourth decimal place) in eight iterations. Thus the first eigenvalue and the corresponding natural frequency and mode shape are given by

$$\lambda_1 = 5.04892, \qquad \omega_1 = 0.44504\sqrt{\frac{k}{m}}$$

$$\vec{X}^{(1)} = \begin{Bmatrix} 1.00000 \\ 1.80194 \\ 2.24698 \end{Bmatrix} \tag{E.12}$$

Second natural frequency: To compute the second eigenvalue and the eigenvector, we must first produce a deflated matrix:

$$[D_2] = [D_1] - \lambda_1 \vec{X}^{(1)} \vec{X}^{(1)^T}[m] \tag{E.13}$$

This equation, however, calls for a normalized vector $\vec{X}^{(1)}$ satisfying $\vec{X}^{(1)^T}[m]\vec{X}^{(1)} = 1$. Let the normalized vector be denoted as

$$\vec{X}^{(1)} = \alpha \begin{Bmatrix} 1.00000 \\ 1.80194 \\ 2.24698 \end{Bmatrix}$$

where α is a constant whose value must be such that

$$\vec{X}^{(1)^T}[m]\vec{X}^{(1)} = \alpha^2 m \begin{Bmatrix} 1.00000 \\ 1.80194 \\ 2.24698 \end{Bmatrix}^T \begin{bmatrix} 1 & 0 & 0 \\ 0 & 1 & 0 \\ 0 & 0 & 1 \end{bmatrix} \begin{Bmatrix} 1.00000 \\ 1.80194 \\ 2.24698 \end{Bmatrix}$$

$$= \alpha^2 m (9.29591) = 1 \tag{E.14}$$

from which we obtain $\alpha = 0.32799m^{-1/2}$. Hence the first normalized eigenvector is

$$\vec{X}^{(1)} = m^{-1/2} \begin{Bmatrix} 0.32799 \\ 0.59102 \\ 0.73699 \end{Bmatrix} \tag{E.15}$$

Next we use Eq. (E.13) and form the first deflated matrix

$$[D_2] = \begin{bmatrix} 1 & 1 & 1 \\ 1 & 2 & 2 \\ 1 & 2 & 3 \end{bmatrix} - 5.04892 \begin{Bmatrix} 0.32799 \\ 0.59102 \\ 0.73699 \end{Bmatrix} \begin{Bmatrix} 0.32799 \\ 0.59102 \\ 0.73699 \end{Bmatrix}^{T} \begin{bmatrix} 1 & 0 & 0 \\ 0 & 1 & 0 \\ 0 & 0 & 1 \end{bmatrix}$$

$$= \begin{bmatrix} 0.45684 & 0.02127 & -0.22048 \\ 0.02127 & 0.23641 & -0.19921 \\ -0.22048 & -0.19921 & 0.25768 \end{bmatrix} \tag{E.16}$$

Since the trial vector can be chosen arbitrarily, we again take

$$\vec{X}_1 = \begin{Bmatrix} 1 \\ 1 \\ 1 \end{Bmatrix} \tag{E.17}$$

By using the iterative scheme

$$\vec{X}_{i+1} = [D_2] \vec{X}_i \tag{E.18}$$

we obtain $\vec{X}_2$:

$$\vec{X}_2 = \begin{Bmatrix} 0.25763 \\ 0.05847 \\ -0.16201 \end{Bmatrix} = 0.25763 \begin{Bmatrix} 1.00000 \\ 0.22695 \\ -0.62885 \end{Bmatrix} \tag{E.19}$$

Hence λ_2 can be found from the general relation

$$\lambda_2 \simeq X_{1,i+1} \tag{E.20}$$

as 0.25763. Continuation of this procedure gives the following results:

i	$\vec{X}_i$ with $X_{1,i} = 1$	$\vec{X}_{i+1} = [D_2]\vec{X}_i$	$\lambda_2 \simeq X_{1,i+1}$	ω_2
1	$\begin{Bmatrix} 1 \\ 1 \\ 1 \end{Bmatrix}$	$\begin{Bmatrix} 0.25763 \\ 0.05847 \\ -0.16201 \end{Bmatrix}$	0.25763	$1.97016\sqrt{\dfrac{k}{m}}$
2	$\begin{Bmatrix} 1.00000 \\ 0.22695 \\ -0.62885 \end{Bmatrix}$	$\begin{Bmatrix} 0.60032 \\ 0.20020 \\ -0.42773 \end{Bmatrix}$	0.60032	$1.29065\sqrt{\dfrac{k}{m}}$
$\vdots$				
10	$\begin{Bmatrix} 1.00000 \\ 0.44443 \\ -0.80149 \end{Bmatrix}$	$\begin{Bmatrix} 0.64300 \\ 0.28600 \\ -0.51554 \end{Bmatrix}$	0.64300	$1.24708\sqrt{\dfrac{k}{m}}$
11	$\begin{Bmatrix} 1.00000 \\ 0.44479 \\ -0.80177 \end{Bmatrix}$	$\begin{Bmatrix} 0.64307 \\ 0.28614 \\ -0.51569 \end{Bmatrix}$	0.64307	$1.24701\sqrt{\dfrac{k}{m}}$

Thus the converged second eigenvalue and the eigenvector are

$$\lambda_2 = 0.64307, \qquad \omega_2 = 1.24701\sqrt{\frac{k}{m}}$$

$$\vec{X}^{(2)} = \left\{ \begin{array}{c} 1.00000 \\ 0.44496 \\ -0.80192 \end{array} \right\} \tag{E.21}$$

Third natural frequency: For the third eigenvalue and the eigenvector, we use a similar procedure. The detailed calculations are left as an exercise to the reader. Note that before computing the deflated matrix $[D_3]$, we need to normalize $\vec{X}^{(2)}$ by using Eq. (7.62), which gives

$$\vec{X}^{(2)} = m^{-1/2} \left\{ \begin{array}{c} 0.73700 \\ 0.32794 \\ -0.59102 \end{array} \right\} \tag{E.22}$$

7.6 JACOBI'S METHOD

The matrix iteration method described in the preceding section produces the eigenvalues and eigenvectors of a matrix $[D]$ one at a time. Jacobi's method is also an iterative method but produces all the eigenvalues and eigenvectors of $[D]$ simultaneously, where $[D] = [d_{ij}]$ is a real symmetric matrix of order $n \times n$. The method is based on a theorem in linear algebra that states that a real symmetric matrix $[D]$ has only real eigenvalues and that there exists a real orthogonal matrix $[R]$ such that $[R]^T[D][R]$ is diagonal [7.17]. The diagonal elements are the eigenvalues, and the columns of the matrix $[R]$ are the eigenvectors. In Jacobi's method, the matrix $[R]$ is generated as a product of several rotation matrices [7.18] of the form

$$\begin{matrix} & \overset{i\text{th column} \qquad j\text{th column}}{} \\ [R_1] = \underset{n \times n}{} & \begin{bmatrix} 1 & 0 & & & & & \\ 0 & 1 & & & & & \\ & & \ddots & & & & \\ & & & \cos\theta & & -\sin\theta & \\ & & & & \ddots & & \\ & & & \sin\theta & & \cos\theta & \\ & & & & & & \ddots \\ & & & & & & & 1 \end{bmatrix} \begin{matrix} \\ \\ \\ i\text{th row} \\ \\ j\text{th row} \\ \\ \end{matrix} \end{matrix} \tag{7.65}$$

where all elements other than those appearing in columns and rows i and j are identical with those of the identity matrix $[I]$. If the sine and cosine entries appear in positions (i, i), (i, j), (j, i), and (j, j), then the corresponding elements of $[R_1]^T[D][R_1]$ can be computed as follows:

$$\underline{d}_{ii} = d_{ii}\cos^2\theta + 2d_{ij}\sin\theta\cos\theta + d_{jj}\sin^2\theta \tag{7.66}$$

$$\underline{d}_{ij} = \underline{d}_{ji} = (d_{jj} - d_{ii})\sin\theta\cos\theta + d_{ij}(\cos^2\theta - \sin^2\theta) \tag{7.67}$$

$$\underline{d}_{jj} = d_{ii}\sin^2\theta - 2d_{ij}\sin\theta\cos\theta + d_{jj}\cos^2\theta \tag{7.68}$$

If θ is chosen to be

$$\tan 2\theta = \left(\frac{2d_{ij}}{d_{ii} - d_{jj}} \right) \tag{7.69}$$

then it makes $\underline{d}_{ij} = \underline{d}_{ji} = 0$. Thus each step of Jacobi's method reduces a pair of off-diagonal elements to zero. Unfortunately, in the next step, while the method reduces a new pair of zeros, it introduces nonzero contributions to formerly zero positons. However, successive matrices of the form

$$[R_2]^T [R_1]^T [D][R_1][R_2], \qquad [R_3]^T [R_2]^T [R_1]^T [D][R_1][R_2][R_3], \ldots$$

converge to the required diagonal form; the final matrix $[R]$, whose columns give the eigenvectors, then becomes

$$[R] = [R_1][R_2][R_3]\ldots \tag{7.70}$$

EXAMPLE 7.6 Eigenvalue Solution Using Jacobi Method

Find the eigenvalues and eigenvectors of the matrix

$$[D] = \begin{bmatrix} 1 & 1 & 1 \\ 1 & 2 & 2 \\ 1 & 2 & 3 \end{bmatrix}$$

using Jacobi's method.

Given: Dynamical matrix $[D]$.

Find: Eigenvalues and eigenvectors.

Approach: Jacobi's method.

Solution. We start with the largest off-diagonal term $d_{23} = 2$ in the matrix $[D]$ and try to reduce it to zero. From Eq. (7.69),

$$\theta_1 = \frac{1}{2} \tan^{-1} \left(\frac{2d_{23}}{d_{22} - d_{33}} \right) = \frac{1}{2} \tan^{-1} \left(\frac{4}{2-3} \right) = -37.981878°$$

$$[R_1] = \begin{bmatrix} 1.0 & 0.0 & 0.0 \\ 0.0 & 0.7882054 & 0.6154122 \\ 0.0 & -0.6154122 & 0.7882054 \end{bmatrix}$$

$$[D'] = [R_1]^T [D][R_1] = \begin{bmatrix} 1.0 & 0.1727932 & 1.4036176 \\ 0.1727932 & 0.4384472 & 0.0 \\ 1.4036176 & 0.0 & 4.5615525 \end{bmatrix}$$

Next we try to reduce the largest off-diagonal term of $[D']$, namely, $d'_{13} = 1.4036176$ to zero. Equation (7.69) gives

$$\theta_2 = \frac{1}{2} \tan^{-1} \left(\frac{2d'_{13}}{d'_{11} - d'_{33}} \right) = \frac{1}{2} \tan^{-1} \left(\frac{2.8072352}{1.0 - 4.5615525} \right) = -19.122686°$$

$$[R_2] = \begin{bmatrix} 0.9448193 & 0.0 & 0.3275920 \\ 0.0 & 1.0 & 0.0 \\ -0.3275920 & 0.0 & 0.9448193 \end{bmatrix}$$

$$[D''] = [R_2]^T [D'][R_2] = \begin{bmatrix} 0.5133313 & 0.1632584 & 0.0 \\ 0.1632584 & 0.4384472 & 0.0566057 \\ 0.0 & 0.0566057 & 5.0482211 \end{bmatrix}$$

The largest off-diagonal element in $[D'']$ is $d_{12}'' = 0.1632584$. θ_3 can be obtained from Eq. (7.69) as

$$\theta_3 = \frac{1}{2} \tan^{-1}\left(\frac{2d_{12}''}{d_{11}'' - d_{22}''}\right) = \frac{1}{2} \tan^{-1}\left(\frac{0.3265167}{0.5133313 - 0.4384472}\right) = 38.541515°$$

$$[R_3] = \begin{bmatrix} 0.7821569 & -0.6230815 & 0.0 \\ 0.6230815 & 0.7821569 & 0.0 \\ 0.0 & 0.0 & 1.0 \end{bmatrix}$$

$$[D'''] = [R_3]^T[D''][R_3] = \begin{bmatrix} 0.6433861 & 0.0 & 0.0352699 \\ 0.0 & 0.3083924 & 0.0442745 \\ 0.0352699 & 0.0442745 & 5.0482211 \end{bmatrix}$$

Assuming that all the off-diagonal terms in $[D''']$ are close to zero, we can stop the process here. The diagonal elements of $[D''']$ give the eigenvalues (values of $1/\omega^2$) as 0.6433861, 0.3083924, and 5.0482211. The corresponding eigenvectors are given by the columns of the matrix $[R]$ where

$$[R] = [R_1][R_2][R_3] = \begin{bmatrix} 0.7389969 & -0.5886994 & 0.3275920 \\ 0.3334301 & 0.7421160 & 0.5814533 \\ -0.5854125 & -0.3204631 & 0.7447116 \end{bmatrix}$$

The iterative process can be continued for obtaining a more accurate solution. The present eigenvalues can be compared with the exact values: 0.6431041, 0.3079786, and 5.0489173.

7.7 STANDARD EIGENVALUE PROBLEM

In the preceding chapter, the eigenvalue problem was stated as

$$[k]\vec{X} = \omega^2[m]\vec{X} \tag{7.71}$$

which can be rewritten in the form of a standard eigenvalue problem [7.19] as

$$[D]\vec{X} = \lambda\vec{X} \tag{7.72}$$

where

$$[D] = [k]^{-1}[m] \tag{7.73}$$

and

$$\lambda = \frac{1}{\omega^2} \tag{7.74}$$

In general, the matrix $[D]$ is nonsymmetric, although the matrices $[k]$ and $[m]$ are both symmetric. Since Jacobi's method (described in Section 7.6) is applicable only to symmetric matrices $[D]$, we can adopt the following procedure [7.18] to derive a standard eigenvalue problem with a symmetric matrix $[D]$.

Assuming that the matrix $[k]$ is symmetric and positive definite, we can use Choleski decomposition (see Section 7.7.1) and express $[k]$ as

$$[k] = [U]^T[U] \tag{7.75}$$

where $[U]$ is an upper triangular matrix. Using this relation, the eigenvalue problem of Eq. (7.71) can be stated as

$$\lambda [U]^T [U] \vec{X} = [m] \vec{X} \qquad (7.76)$$

Premultiplying this equation by $([U]^T)^{-1}$, we obtain

$$\lambda [U] \vec{X} = ([U]^T)^{-1} [m] \vec{X} = ([U]^T)^{-1} [m][U]^{-1}[U] \vec{X} \qquad (7.77)$$

By defining a new vector $\vec{Y}$ as

$$\vec{Y} = [U] \vec{X} \qquad (7.78)$$

Eq. (7.77) can be written as a standard eigenvalue problem:

$$[D]\vec{Y} = \lambda \vec{Y} \qquad (7.79)$$

where

$$[D] = ([U]^T)^{-1} [m][U]^{-1} \qquad (7.80)$$

Thus, to formulate $[D]$ according to Eq. (7.80), we first decompose the symmetric matrix $[k]$ as shown in Eq. (7.75), find $[U]^{-1}$ and $([U]^T)^{-1} = ([U]^{-1})^T$ as outlined in the next section, and then carry out the matrix multiplication as stated in Eq. (7.80). The solution of the eigenvalue problem stated in Eq. (7.79) yields λ_i and $\vec{Y}^{(i)}$. Then we apply inverse transformation and find the desired eigenvectors:

$$\vec{X}^{(i)} = [U]^{-1} \vec{Y}^{(i)} \qquad (7.81)$$

**7.7.1
Choleski
Decomposition**

Any symmetric and positive definite matrix $[A]$ of order $n \times n$ can be decomposed uniquely [7.20]:

$$[A] = [U]^T [U] \qquad (7.82)$$

where $[U]$ is an upper triangular matrix given by

$$[U] = \begin{bmatrix} u_{11} & u_{12} & u_{13} & \cdots & u_{1n} \\ 0 & u_{22} & u_{23} & \cdots & u_{2n} \\ 0 & 0 & u_{33} & \cdots & u_{3n} \\ \vdots & & & & \\ 0 & 0 & 0 & \cdots & u_{nn} \end{bmatrix} \qquad (7.83)$$

with

$$u_{11} = (a_{11})^{1/2}$$

$$u_{1j} = \frac{a_{1j}}{u_{11}}, \qquad j = 2, 3, \ldots, n$$

$$u_{ii} = \left(a_{ii} - \sum_{k=1}^{i-1} u_{ki}^2 \right)^{1/2}, \qquad i = 2, 3, \ldots, n$$

$$u_{ij} = \frac{1}{u_{ii}} \left(a_{ij} - \sum_{k=1}^{i-1} u_{ki} u_{kj} \right), \qquad i = 2, 3, \ldots, n \text{ and } j = i + 1, i + 2, \ldots, n$$

$$u_{ij} = 0, \qquad i > j \qquad (7.84)$$

Inverse of the Matrix $[U]$. If the inverse of the upper triangular matrix $[U]$ is denoted as $[\alpha_{ij}]$, the elements α_{ij} can be determined from the relation

$$[U][U]^{-1} = [I] \tag{7.85}$$

which gives

$$\alpha_{ii} = \frac{1}{u_{ii}}$$

$$\alpha_{ij} = \frac{-1}{u_{ii}} \left(\sum_{k=i+1}^{j} u_{ik} \alpha_{kj} \right), \qquad i < j$$

$$\alpha_{ij} = 0, \qquad i > j \tag{7.86}$$

Thus the inverse of $[U]$ is also an upper triangular matrix.

EXAMPLE 7.7   **Decomposition of a Symmetric Matrix**

Decompose the matrix

$$[A] = \begin{bmatrix} 5 & 1 & 0 \\ 1 & 3 & 2 \\ 0 & 2 & 8 \end{bmatrix}$$

into the form of Eq. (7.82).

Given: Symmetric and positive definite matrix $[A]$.

Find: Upper triangular matrix $[U]$ such that $[A] = [U]^T[U]$.

Approach: Choleski decomposition.

Solution. Equations (7.84) give

$$u_{11} = \sqrt{a_{11}} = \sqrt{5} = 2.2360680$$
$$u_{12} = a_{12}/u_{11} = 1/2.236068 = 0.4472136$$
$$u_{13} = a_{13}/u_{11} = 0$$
$$u_{22} = \left[a_{22} - u_{12}^2 \right]^{1/2} = \left(3 - 0.4472136^2 \right)^{1/2} = 1.6733201$$
$$u_{33} = \left[a_{33} - u_{13}^2 - u_{23}^2 \right]^{1/2}$$

where

$$u_{23} = (a_{23} - u_{12}u_{13})/u_{22} = (2 - 0.4472136 \times 0)/1.6733201 = 1.1952286$$
$$u_{33} = (8 - 0^2 - 1.1952286^2)^{1/2} = 2.5634799$$

Since $u_{ij} = 0$ for $i > j$, we have

$$[U] = \begin{bmatrix} 2.2360680 & 0.4472136 & 0.0 \\ 0.0 & 1.6733201 & 1.1952286 \\ 0.0 & 0.0 & 2.5634799 \end{bmatrix}$$

7.7.2
Other Solution Methods

Several other methods have been developed for finding the numerical solution of an eigenvalue problem [7.18, 7.21]. Bathe and Wilson [7.22] have done a comparative study of some of these methods. Recent emphasis has been on the economical solution of large eigenproblems [7.23, 7.24]. The estimation of natural frequencies by the use of Sturm sequences is presented in Refs. [7.25] and [7.26]. An alternative way to solve a class of lumped mechanical vibration problems using topological methods is presented in Ref. [7.27].

7.8 COMPUTER PROGRAMS

7.8.1
Jacobi's Method

A Fortran program, in the form of subroutine JACOBI, is given for finding the eigenvalues and eigenvectors of a real symmetric matrix $[D]$ according to Jacobi's method. The arguments of this subroutine are as follows:

D   =   Array of size $N \times N$, containing the elements of the matrix $[D]$. Input data. The diagonal elements of the array contain the eigenvalues upon return to the main program; $D(I, I) = 1/\omega_i^2$. Output.

N   =   Order of the matrix $[D]$. Input data.

E   =   Array of size $N \times N$ in which the eigenvectors are stored columnwise. Output.

EPS   =   Convergence specification. A small quantity on the order of 10^{-5} is to be used. Input data.

ITMAX   =   Maximum number of iterations or rotations permitted. Input data.

By way of illustration, Example 7.6 is solved using subroutine JACOBI. The main program, subroutine JACOBI, and the output are given below.

```
C ================================================================
C
C PROGRAM 13
C MAIN PROGRAM FOR CALLING JACOBI
C
C ================================================================
C FOLLOWING 3 LINES CONTAIN PROBLEM-DEPENDENT DATA
      DIMENSION D(3,3),E(3,3)
      DATA N,ITMAX,EPS/3,200,1.0E-05/
      DATA D/1.0,1.0,1.0,1.0,2.0,2.0,1.0,2.0,3.0/
C END OF PROBLEM-DEPENDENT DATA
      PRINT 50
   50 FORMAT (//,37H EIGENVALUE SOLUTION BY JACOBI METHOD)
      PRINT 40
   40 FORMAT (/,13H GIVEN MATRIX)
      DO 30 I=1,N
   30 PRINT 20,(D(I,J),J=1,N)
   20 FORMAT (3E15.6)
      CALL JACOBI (D,N,E,EPS,ITMAX)
      PRINT 10, (D(I,I),I=1,N), ((E(I,J),J=1,N),I=1,N)
   10 FORMAT (/,1X,17H EIGEN VALUES ARE,/,3E15.6,//,1X,
     2 14H EIGEN VECTORS,/,6X,5HFIRST,10X,6HSECOND,9X,5HTHIRD,/,
     3 (1X,3E15.6))
      STOP
      END
```

```
C ================================================================
C
C SUBROUTINE JACOBI
C
C ================================================================
      SUBROUTINE JACOBI (D,N,E,EPS,ITMAX)
      DIMENSION D(N,N),E(N,N)
      ITER=0
      DO 110 I=1,N
      DO 110 J=1,N
      E(I,J)=0.0
110   E(I,I)=1.0
120   ZZ=0.0
      NM1=N-1
      DO 130 I=1,NM1
      IP1=I+1
      DO 130 J=IP1,N
      IF (ABS(D(I,J)) .LE. ZZ) GO TO 130
      ZZ=ABS(D(I,J))
      IR=I
      IC=J
130   CONTINUE
      IF (ITER .EQ. 0) YY=ZZ*EPS
      IF (ZZ .LE. YY) GO TO 210
      DIF=D(IR,IR) - D(IC,IC)
      TANZ=(-DIF+SQRT(DIF**2+4.0*ZZ**2))/(2.0*D(IR,IC))
      COSZ=1.0/SQRT(1.0+TANZ**2)
      SINZ=COSZ*TANZ
      DO 140 I=1,N
      ZZZ=E(I,IR)
      E(I,IR)=COSZ*ZZZ+SINZ*E(I,IC)
140   E(I,IC)=COSZ*E(I,IC)-SINZ*ZZZ
      I=1
150   IF (I .EQ. IR) GO TO 160
      YYY=D(I,IR)
      D(I,IR)=COSZ*YYY+SINZ*D(I,IC)
      D(I,IC)=COSZ*D(I,IC)-SINZ*YYY
      I=I+1
      GO TO 150
160   I=IR+1
170   IF (I .EQ. IC) GO TO 180
      YYY=D(IR,I)
      D(IR,I)=COSZ*YYY+SINZ*D(I,IC)
      D(I,IC)=COSZ*D(I,IC)-SINZ*YYY
      I=I+1
      GO TO 170
180   I=IC+1
190   IF (I .GT. N) GO TO 200
      ZZZ=D(IR,I)
      D(IR,I)=COSZ*ZZZ+SINZ*D(IC,I)
      D(IC,I)=COSZ*D(IC,I)-SINZ*ZZZ
      I=I+1
      GO TO 190
200   YYY=D(IR,IR)
      D(IR,IR)=YYY*COSZ**2+D(IR,IC)*2.0*COSZ*SINZ+D(IC,IC)
```

```
      2    *SINZ**2
           D(IC,IC)=D(IC,IC)*COSZ**2+YYY*SINZ**2-D(IR,IC)*2.0*COSZ*SINZ
           D(IR,IC)=0.0
           ITER=ITER+1
           IF (ITER .LT. ITMAX) GO TO 120
   210     RETURN
           END
```

EIGENVALUE SOLUTION BY JACOBI METHOD

GIVEN MATRIX
```
      0.100000E+01    0.100000E+01    0.100000E+01
      0.100000E+01    0.200000E+01    0.200000E+01
      0.100000E+01    0.200000E+01    0.300000E+01
```

EIGEN VALUES ARE
```
      0.504892E+01    0.643104E+00    0.307979E+00
```

EIGEN VECTORS
```
         FIRST            SECOND            THIRD
      0.327985E+00    -0.736984E+00     0.590999E+00
      0.591007E+00    -0.327977E+00    -0.736981E+00
      0.736978E+00     0.591004E+00     0.327991E+00
```

7.8.2
Matrix Iteration
Method

Subroutine MITER is given for implementing the matrix iteration method. This subroutine uses the following arguments:

D   =   Array of size $N \times N$, containing the matrix $[D]$. Input data.

X, XX   =   Arrays of size N each.

XS   =   Array of size N. Contains the initial guess vector such as $\left\{ \begin{matrix} 1 \\ 1 \\ \vdots \\ 1 \end{matrix} \right\}$. Input data.

N   =   Order of the matrix $[D]$. Input data.

NVEC   =   Number of eigenvalues and eigenvectors to be found. Input data.

B, C   =   Arrays of size $N \times N$ each.

XM   =   Array of size $N \times N$, containing the mass matrix $[m]$. Input data.

EPS   =   Convergence requirement. A small number on the order of 10^{-5} is to be used. Input data.

FREQ   =   Array of size $NVEC$, containing the computed natural frequencies. Output.

EIG   =   Array of size $N \times NVEC$, containing the computed eigenvectors columnwise. Output.

Example 7.5 is solved using subroutine MITER. The main program that calls subroutine MITER, subroutine MITER, and the output of the program are given below.

```
C ==============================================================
C
C PROGRAM 14
C MAIN PROGRAM FOR CALLING THE SUBROUTINE MITER
C
C ==============================================================
C FOLLOWING 8 LINES CONTAIN PROBLEM-DEPENDENT DATA
      DIMENSION D(3,3),X(3),XS(3),B(3,3),C(3,3),XX(3),XM(3,3),FREQ(3),
     2 EIG(3,3)
      N=3
      NVEC=3
      DATA D/1.0,1.0,1.0,1.0,2.0,2.0,1.0,2.0,3.0/
      DATA XM/1.0,0.0,0.0,0.0,1.0,0.0,0.0,0.0,1.0/
      EPS=0.00001
      DATA XS/1.0,1.0,1.0/
C END OF PROBLEM-DEPENDENT DATA
      CALL MITER (D,X,XS,N,NVEC,B,C,XX,XM,EPS,FREQ,EIG)
      PRINT 10
   10 FORMAT (//,34H SOLUTION OF EIGENVALUE PROBLEM BY,/,
     2    24H MATRIX ITERATION METHOD)
      PRINT 20,(FREQ(I),I=1,NVEC)
   20 FORMAT (//,20H NATURAL FREQUENCIES,//,3(E15.8,1X))
      PRINT 30
   30 FORMAT (//,26H MODE SHAPES (COLUMNWISE):,/)
      DO 40 I=1,NVEC
   40 PRINT 50,(EIG(I,J),J=1,N)
   50 FORMAT (3(E15.8,1X),/)
      STOP
      END
C==============================================================
C
C SUBROUTINE MITER
C
C==============================================================
      SUBROUTINE MITER (D,X,XS,N,NVEC,B,C,XX,XM,EPS,FREQ,EIG)
      DIMENSION D(N,N),X(N),XS(N),B(N,N),C(N,N),XX(N),XM(N,N),
     2    FREQ(NVEC),EIG(N,NVEC)
      CON=XS(1)
      DO 10 I=1,N
   10 X(I)=XS(I)/CON
   50 ICON=0
      CALL MULT(D,X,N,XX)
      ALAM=XX(1)
      DO 20 I=1,N
   20 XX(I)=XX(I)/ALAM
      DO 30 I=1,N
   30 IF(ABS((XX(I)-X(I))/X(I)) .GT. EPS) ICON=1
      DO 40 I=1,N
   40 X(I)=XX(I)
      IF (ICON .EQ. 0) GO TO 60
      GO TO 50
   60 ICON=0
      FREQ(1)=SQRT(1.0/ALAM)
      DO 70 I=1,N
   70 EIG(I,1)=X(I)
```

```
        II=1
100     II=II+1
        SUM=0.0
        DO 110 I=1,N
110     SUM=SUM+X(I)**2
        ALP=SQRT(1.0/SUM)
        DO 120 I=1,N
120     X(I)=X(I)*ALP
        DO 130 I=1,N
        DO 130 J=1,N
130     C(I,J)=X(I)*X(J)
        CALL MATMUL (B,C,XM,N,N,N)
        DO 140 I=1,N
        DO 140 J=1,N
140     D(I,J)=D(I,J)-ALAM*B(I,J)
        CON=XS(1)
        DO 150 I=1,N
150     X(I)=XS(I)/CON
250     ICON=0
        CALL MULT (D,X,N,XX)
        ALAM=XX(1)
        DO 220 I=1,N
220     XX(I)=XX(I)/ALAM
        DO 230 I=1,N
230     IF (ABS((XX(I)-X(I))/X(I)) .GT. EPS) ICON=1
        DO 240 I=1,N
240     X(I)=XX(I)
        IF (ICON .EQ. 0) GO TO 260
        GO TO 250
260     ICON=0
        FREQ(II)=SQRT(1.0/ALAM)
        DO 270 I=1,N
270     EIG(I,II)=X(I)
        IF (II .EQ. NVEC) GO TO 300
        GO TO 100
300     RETURN
        END
C =================================================================
C
C SUBROUTINE MULT
C
C =================================================================
        SUBROUTINE MULT (D,X,N,XX)
        DIMENSION D(N,N),X(N),XX(N)
        DO 20 I=1,N
        XX(I)=0.0
        DO 10 J=1,N
10      XX(I)=XX(I)+D(I,J)*X(J)
20      CONTINUE
        RETURN
        END
```

```
SOLUTION OF EIGENVALUE PROBLEM BY
MATRIX ITERATION METHOD

NATURAL FREQUENCIES

0.44504240E+00   0.12469811E+01   0.18019377E+01

MODE SHAPES (COLUMNWISE):

0.10000000E+01   0.10000000E+01   0.10000000E+01

0.18019372E+01   0.44503731E+00  -0.12469926E+01

0.22469788E+01  -0.80193609E+00   0.55496848E+00
```

7.8.3
Choleski
Decomposition

Subroutine DECOMP is given for decomposing a matrix $[A]$ of order N using Choleski decomposition. The program gives the upper triangular matrix $[U]$ as output. The listing of DECOMP is given below.

```
C ================================================================
C
C PROGRAM 15
C MAIN PROGRAM WHICH CALLS DECOMP
C
C ================================================================
      DIMENSION A(3,3),U(3,3)
      DATA A/5.0,1.0,0.0,1.0,3.0,2.0,0.0,2.0,8.0/
      N=3
      CALL DECOMP (A,U,N)
      PRINT 10
10    FORMAT (/,25H UPPER TRIANGULAR MATRIX:,/)
      DO 30 I=1,N
      PRINT 20, (U(I,J),J=1,N)
20    FORMAT (3E15.8)
30    CONTINUE
      STOP
      END
C ================================================================
C
C SUBROUTINE DECOMP
C
C ================================================================
      SUBROUTINE DECOMP (A,U,N)
      DIMENSION A(N,N),U(N,N)
      DO 10 I=1,N
```

```
        DO 10 J=1,N
10      U(I,J)=0.0
        U(1,1)=SQRT(A(1,1))
        DO 90 J=2,N
90      U(1,J)=A(1,J)/U(1,1)
        DO 40 I=2,N
        IM=I-1
        SUM=0.0
        DO 30 K=1,IM
30      SUM=SUM+U(K,I)**2
        U(I,I)=SQRT(A(I,I)-SUM)
        J=I+1
        SUM=0.0
        DO 50 K=1,IM
50      SUM=SUM+U(K,I)*U(K,J)
        U(I,J)=(A(I,J)-SUM)/U(I,I)
40      CONTINUE
        RETURN
        END
```

UPPER TRIANGULAR MATRIX:

```
0.22360680E+01  0.44721359E+00  0.00000000E+00
0.00000000E+00  0.16733201E+01  0.11952286E+01
0.00000000E+00  0.00000000E+00  0.25634799E+01
```

7.8.4 Eigenvalue Solution Using Choleski Decomposition

A Fortran program is written for solving the general eigenvalue problem

$$[k]\vec{X} = \omega^2[m]\vec{X}$$

The problem is converted into a special eigenvalue problem

$$[D]\vec{Y} = \frac{1}{\omega^2}[I]\vec{Y}$$

by generating the matrix $[D]$ using the relation

$$[D] = \left([U]^T\right)^{-1}[m][U]^{-1}$$

The following data is needed for this program:

BK = Array of size $N \times N$, containing the matrix $[k]$.

BM = Array of size $N \times N$, containing the matrix $[m]$.

ND = Order of the matrices $[k]$ and $[m]$.

The following quantities need to be defined and dimensioned in the program:

U, UI, UTI = Arrays of size $N \times N$, indicating the matrices $[U]$, $[U]^{-1}$, and $([U]^{-1})^T$ respectively.

BMU, UMU, EV = Arrays of size $N \times N$ each.

XF = Array of size $N \times N$. The computed eigenvectors $\vec{X}^{(i)}$ are stored columnwise in XF.

To illustrate the use of the program, the matrices of Example 7.2 are used with

$$[k] = \begin{bmatrix} 2 & -1 & 0 \\ -1 & 2 & -1 \\ 0 & -1 & 1 \end{bmatrix} \quad \text{and} \quad [m] = \begin{bmatrix} 1 & 0 & 0 \\ 0 & 1 & 0 \\ 0 & 0 & 1 \end{bmatrix}$$

The listing of the program and the output are given below.

```
C =======================================================================
C
C PROGRAM 16
C DETERMINATION OF EIGENVALUES AND EIGENVECTORS BY FINDING THE MATRIX
C [D] ACCORDING TO THE RELATION [D] = [UTI][M][UI]
C
C =======================================================================
C FOLLOWING 5 LINES CONTAIN PROBLEM-DEPENDENT DATA
      DIMENSION BK(3,3),BM(3,3),U(3,3),UI(3,3),UTI(3,3),BMU(3,3),
     2   UMU(3,3),XF(3,3),EV(3,3)
      DATA BK/2.0,-1.0,0.0,-1.0,2.0,-1.0,0.0,-1.0,1.0/
      DATA BM/1.0,0.0,0.0,0.0,1.0,0.0,0.0,0.0,1.0/
      ND=3
C END OF PROBLEM-DEPENDENT DATA
C DECOMPOSING THE MATRIX [BK] INTO TRIANGULAR MATRICES
      CALL DECOMP (BK,U,ND)
      PRINT 100
  100 FORMAT (//,29H UPPER TRIANGULAR MATRIX [U]:,/)
      DO 110 I=1,ND
  110 PRINT 120,(U(I,J),J=1,ND)
  120 FORMAT (4(1X,E15.6))
      DO 130 I=1,ND
      DO 130 J=1,ND
  130 UI(I,J)=0.0
      DO 140 I=1,ND
  140 UI(I,I)=1.0/U(I,I)
      DO 160 J=1,ND
      DO 160 II=1,ND
      I=ND-II+1
      IF (I .GE. J) GO TO 160
      IP=I+1
      SUM=0.0
      DO 150 K=IP,J
  150 SUM =SUM+U(I,K)*UI(K,J)
      UI(I,J)=-SUM/U(I,I)
  160 CONTINUE
      PRINT 170
  170 FORMAT (/,46H INVERSE OF THE UPPER TRIANGULAR MATRIX, [UI],/)
      DO 180 I=1,ND
  180 PRINT 120,(UI(I,J),J=1,ND)
      DO 190 I=1,ND
      DO 190 J=1,ND
  190 UTI(I,J)=UI(J,I)
      CALL MATMUL (BMU,BM,UI,ND,ND,ND)
      CALL MATMUL (UMU,UTI,BMU,ND,ND,ND)
      PRINT 200
  200 FORMAT (/,29H MATRIX [UMU] = [UTI][M][UI]:,/)
```

```
      DO 210 I=1,ND
 210  PRINT 120,(UMU(I,J),J=1,ND)
      CALL JACOBI (UMU,ND,EV,1.0E-05,200)
      PRINT 230
 230  FORMAT (/,13H EIGENVALUES:,/)
      PRINT 120,(UMU(I,I),I=1,ND)
      PRINT 240
 240  FORMAT (/,27H EIGENVECTORS (COLUMNWISE):,/)
      CALL MATMUL (XF,UI,EV,ND,ND,ND)
      DO 250 I=1,ND
 250  PRINT 120,(XF(I,J),J=1,ND)
      STOP
      END
```

UPPER TRIANGULAR MATRIX [U]:

```
   0.141421E+01   -0.707107E+00    0.000000E+00
   0.000000E+00    0.122474E+01   -0.816497E+00
   0.000000E+00    0.000000E+00    0.577350E+00
```

INVERSE OF THE UPPER TRIANGULAR MATRIX, [UI],

```
   0.707107E+00    0.408248E+00    0.577350E+00
   0.000000E+00    0.816497E+00    0.115470E+01
   0.000000E+00    0.000000E+00    0.173205E+01
```

MATRIX [UMU] = [UTI][M][UI]:

```
   0.500000E+00    0.288675E+00    0.408248E+00
   0.288675E+00    0.833333E+00    0.117851E+01
   0.408248E+00    0.117851E+01    0.466667E+01
```

EIGENVALUES:

```
   0.504892E+01    0.643104E+00    0.307979E+00
```

EIGENVECTORS (COLUMNWISE):

```
   0.736973E+00   -0.590976E+00    0.328051E+00
   0.132799E+01   -0.263064E+00   -0.408952E+00
   0.165597E+01    0.473971E+00    0.181988E+00
```

REFERENCES

7.1. S. Dunkerley, "On the whirling and vibration of shafts," *Philosophical Transactions of the Royal Society of London*, 1894, Series A, Vol. 185, Part I, pp. 279–360.

7.2. B. Atzori, "Dunkerley's formula for finding the lowest frequency of vibration of elastic systems," letter to the editor, *Journal of Sound and Vibration*, 1974, Vol. 36, pp. 563–564.

7.3. H. H. Jeffcott, "The periods of lateral vibration of loaded shafts—The rational derivation of Dunkerley's empirical rule for determining whirling speeds," *Proceedings of the Royal Society of London*, 1919, Series A, Vol. 95, No. A666, pp. 106–115.

7.4. M. Endo and O. Taniguchi, "An extension of the Southwell–Dunkerley methods for synthesizing frequencies," *Journal of Sound and Vibration*, 1976, "Part I: Principles," Vol. 49, pp. 501–516, and "Part II: Applications," Vol. 49, pp. 517–533.

7.5. A. Rutenberg, "A lower bound for Dunkerley's formula in continuous elastic systems," *Journal of Sound and Vibration*, 1976, Vol. 45, pp. 249–252.

7.6. G. Temple and W. G. Bickley, *Rayleigh's Principle and Its Applications to Engineering*, Dover, New York, 1956.

7.7. N. G. Stephen, "Rayleigh's, Dunkerley's and Southwell's methods," *International Journal of Mechanical Engineering Education*, January 1983, Vol. 11, pp. 45–51.

7.8. A. Rutenberg, "Dunkerley's formula and alternative approximations," letter to the editor, *Journal of Sound and Vibration*, 1975, Vol. 39, pp. 530–531.

7.9. R. Jones, "Approximate expressions for the fundamental frequency of vibration of several dynamic systems," *Journal of Sound and Vibration*, 1976, Vol. 44, pp. 475–478.

7.10. R. W. Fitzgerald, *Mechanics of Materials* (2nd Ed.), Addison-Wesley, Reading, Mass., 1982.

7.11. H. Holzer, *Die Berechnung der Drehschwin gungen*, Julius Springer, Berlin, 1921.

7.12. H. E. Fettis, "A modification of the Holzer method for computing uncoupled torsion and bending modes," *Journal of the Aeronautical Sciences*, October 1949, pp. 625–634; May 1954, pp. 359–360.

7.13. S. H. Crandall and W. G. Strang, "An improvement of the Holzer Table based on a suggestion of Rayleigh's," *Journal of Applied Mechanics*, 1957, Vol. 24, p. 228.

7.14. S. Mahalingam, "An improvement of the Holzer method," *Journal of Applied Mechanics*, 1958, Vol. 25, p. 618.

7.15. S. Mahalingam, "Iterative procedures for torsional vibration analysis and their relationships," *Journal of Sound and Vibration*, 1980, Vol. 68, pp. 465–467.

7.16. L. Meirovitch, *Computational Methods in Structural Dynamics*, Sijthoff and Noordhoff, The Netherlands, 1980.

7.17. J. H. Wilkinson and G. Reinsch, *Linear Algebra*, Springer Verlag, New York, 1971.

7.18. J. W. Wilkinson, *The Algebraic Eigenvalue Problem*, Oxford University Press, London, 1965.

7.19. R. S. Martin and J. H. Wilkinson, "Reduction of a symmetric eigenproblem $Ax = \lambda Bx$ and related problems to standard form," *Numerical Mathematics*, 1968, Vol. 11, pp. 99–110.

7.20. G. B. Haggerty, *Elementary Numerical Analysis with Programming*, Allyn & Bacon, Boston, 1972.

7.21. A. Jennings, "Eigenvalue methods for vibration analysis," *Shock and Vibration Digest*, Part I, February 1980, Vol. 12, pp. 3–16; Part II, January 1984, Vol. 16, pp. 25–33.

7.22. K. Bathe and E. L. Wilson, "'Solution methods for eigenvalue problems in structural mechanics," *International Journal for Numerical Methods in Engineering*, 1973, Vol. 6, pp. 213–226.

7.23. E. Cohen and H. McCallion, "Economical methods for finding eigenvalues and eigenvectors," *Journal of Sound and Vibration*, 1967, Vol. 5, pp. 397–406.

7.24. A. J. Fricker, "A method for solving high-order real symmetric eigenvalue problems," *International Journal for Numerical Methods in Engineering*, 1983, Vol. 19, pp. 1131–1138.

7.25. G. Longbottom and K. F. Gill, "The estimation of natural frequencies by use of Sturm sequences," *International Journal of Mechanical Engineering Education*, 1976, Vol. 4, pp. 319–329.

7.26. K. K. Gupta, "Solution of eigenvalue problems by Sturm sequence method," *International Journal for Numerical Methods in Engineering*, 1972, Vol. 4, pp. 379–404.

7.27. W. K. Chen and F. Y. Chen, "Topological analysis of a class of lumped vibrational systems," *Journal of Sound and Vibration*, 1969, Vol. 10, pp. 198–207.

REVIEW QUESTIONS

7.1. Name a few methods for finding the fundamental natural frequency of a multidegree of freedom system.

7.2. What is the basic assumption made in deriving Dunkerley's formula?

7.3. What is Rayleigh's principle?

7.4. State whether we get a lower bound or an upper bound to the fundamental natural frequency if we use (a) Dunkerley's formula and (b) Rayleigh's method.

7.5. What is Rayleigh's quotient?

7.6. What is the basic principle used in Holzer's method?

7.7. What is the matrix iteration method?

7.8. Can we use any trial vector $\vec{X}_1$ in the matrix iteration method to find the largest natural frequency?

7.9. How do you find the intermediate natural frequencies using the matrix iteration method?

7.10. What is the difference between the matrix iteration method and Jacobi's method?

7.11. What is a rotation matrix? What is its purpose in Jacobi's method?

7.12. What is a standard eigenvalue problem?

7.13. What is the role of Choleski decomposition in deriving a standard eigenvalue problem?

7.14. How do you find the inverse of an upper triangular matrix?

PROBLEMS

The problem assignments are organized as follows:

Problems	Section covered	Topic covered
7.1–7.6, 7.32	7.2	Dunkerley's formula
7.7–7.12	7.3	Rayleigh's method
7.13–7.17	7.4	Holzer's method
7.18–7.23	7.5	Matrix iteration method
7.24–7.25, 7.29	7.6	Jacobi's method
7.26–7.28, 7.30–7.31	7.7	Standard eigenvalue problem
7.33–7.39	7.8	Computer programs
7.40, 7.41	—	Projects

7.1. Estimate the fundamental frequency of the beam shown in Fig. 6.3 using Dunkerley's formula for the following data: (a) $m_1 = m_3 = 5m$, $m_2 = m$; and (b) $m_1 = m_3 = m$, $m_2 = 5m$.

7.2. Find the fundamental frequency of the torsional system shown in Fig. 6.5, using Dunkerley's formula for the following data: (a) $J_1 = J_2 = J_3 = J_0$; $k_{t1} = k_{t2} = k_{t3} = K_t$; and (b) $J_1 = J_0$, $J_2 = 2J_0$, $J_3 = 3J_0$; $k_{t1} = k_t$, $k_{t2} = 2k_t$, $k_{t3} = 3k_t$.

7.3. Estimate the fundamental frequency of the shaft shown in Fig. 7.3, using Dunkerley's formula for the following data: $m_1 = m$, $m_2 = 2m$, $m_3 = 3m$, $l_1 = l_2 = l_3 = l_4 = l/4$.

7.4. The natural frequency of vibration, in bending, of the wing of a military aircraft is found to be 20 Hz. Find the new frequency of bending vibration of the wing when a weapon, weighing 2000 lb, is attached at the tip of the wing (Fig. 7.10). The stiffness of the wing tip, in bending, is known to be 50,000 lb/ft.

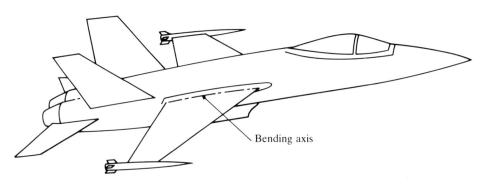

Bending axis

Figure 7.10

7.5. In an overhead crane (see Fig. 7.11) the trolley weighs ten times the weight of the girder. Estimate the fundamental frequency of the system using Dunkerley's formula.

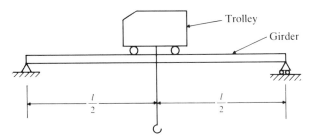

Trolley

Girder

$\frac{l}{2}$

$\frac{l}{2}$

Figure 7.11

7.6. Determine the fundamental natural frequency of the stretched string system shown in Fig. 5.22 with $m_1 = m_2 = m$ and $l_1 = l_2 = l_3 = l$, using Dunkerley's formula.

7.7. Determine the first natural frequency of vibration of the system shown in Fig. 7.2, using Rayleigh's method. Assume $k_1 = k$, $k_2 = 2k$, $k_3 = 3k$, and $m_1 = m$, $m_2 = 2m$, $m_3 = 3m$.

7.8. Find the fundamental natural frequency of the torsional system shown in Fig. 6.5 using Rayleigh's method. Assume that $J_1 = J_0$, $J_2 = 2J_0$, $J_3 = 3J_0$, and $k_{t1} = k_{t2} = k_{t3} = k_t$.

7.9. Solve Problem 7.6 using Rayleigh's method.

7.10. Determine the fundamental natural frequency of the system shown in Fig. 5.22 when $m_1 = m$, $m_2 = 5m$, $l_1 = l_2 = l_3 = l$, using Rayleigh's method.

7.11. A two-story shear building is shown in Fig. 7.12 in which the floors are assumed to be rigid. Compute the first natural frequency of the building, using Rayleigh's method, for $m_1 = 2m$, $m_2 = m$, $h_1 = h_2 = h$, and $k_1 = k_2 = 3EI/h^3$. Assume the first mode configuration to be the same as the static equilibrium shape due to loads proportional to the floor weights.

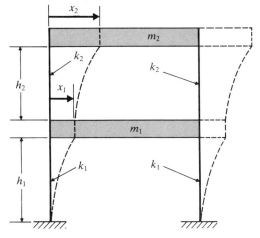

Figure 7.12

7.12. Prove that Rayleigh's quotient is never higher than the highest eigenvalue.

7.13. Find the natural frequencies and mode shapes of the system shown in Fig. 6.7, with $m_1 = 100$ kg, $m_2 = 20$ kg, $m_3 = 200$ kg, $k_1 = 8000$ N/m, and $k_2 = 4000$ N/m, using Holzer's method.

7.14. The stiffness and mass matrices of a vibrating system are given by

$$[k] = k \begin{bmatrix} 2 & -1 & 0 \\ -1 & 2 & -1 \\ 0 & -1 & 3 \end{bmatrix}, \quad [m] = m \begin{bmatrix} 1 & 0 & 0 \\ 0 & 1 & 0 \\ 0 & 0 & 2 \end{bmatrix}$$

Determine all the principal modes and the natural frequencies, using Holzer's method.

7.15. For the torsional system shown in Fig. 6.5, determine a principal mode and the corresponding frequency by Holzer's method. Assume $k_{t1} = k_{t2} = k_{t3} = k_t$ and $J_1 = J_2 = J_3 = J_0$.

7.16. Find the natural frequencies and mode shapes of the shear building shown in Fig. 7.12 using Holzer's method. Assume that $m_1 = 2m$, $m_2 = m$, $h_1 = h_2 = h$, $k_1 = 2k$, $k_2 = k$, and $k = 3EI/h^3$.

7.17. Find the natural frequencies and mode shapes of the system shown in Fig. 6.21 with $J_1 = 10$ kg-m^2, $J_2 = 5$ kg-m^2, $J_3 = 1$ kg-m^2, and $k_{t1} = k_{t2} = 1 \times 10^6$ N-m/rad, using Holzer's method.

7.18. The largest eigenvalue of the matrix

$$[D] = \begin{bmatrix} 2.5 & -1 & 0 \\ -1 & 5 & -\sqrt{2} \\ 0 & -\sqrt{2} & 10 \end{bmatrix}$$

is given by $\lambda_1 = 10.38068$. Find the other eigenvalues and all the eigenvectors of the matrix, using the matrix iteration method. Assume $[m] = [I]$.

7.19. The mass and stiffness matrices of a spring-mass system are known to be

$$[m] = m \begin{bmatrix} 1 & 0 & 0 \\ 0 & 1 & 0 \\ 0 & 0 & 2 \end{bmatrix} \quad \text{and} \quad [k] = k \begin{bmatrix} 2 & -1 & 0 \\ -1 & 3 & -2 \\ 0 & -2 & 2 \end{bmatrix}$$

Find the natural frequencies and mode shapes of the system, using the matrix iteration method.

7.20. Find the natural frequencies and mode shapes of the system shown in Fig. 6.2 with $k_1 = k$, $k_2 = 2k$, $k_3 = 3k$, and $m_1 = m_2 = m_3 = m$, using the matrix iteration method.

7.21. Find the natural frequencies of the system shown in Fig. 6.10, using the matrix iteration method. Assume that $J_{d1} = J_{d2} = J_{d3} = J_0$, $l_i = l$, and $(GJ)_i = GJ$ for $i = 1$ to 4.

7.22. Solve Problem 7.6 using the matrix iteration method.

7.23. The stiffness and mass matrices of a vibrating system are given by

$$[k] = k \begin{bmatrix} 4 & -2 & 0 & 0 \\ -2 & 3 & -1 & 0 \\ 0 & -1 & 2 & -1 \\ 0 & 0 & -1 & 1 \end{bmatrix} \quad \text{and} \quad [m] = m \begin{bmatrix} 3 & 0 & 0 & 0 \\ 0 & 2 & 0 & 0 \\ 0 & 0 & 1 & 0 \\ 0 & 0 & 0 & 1 \end{bmatrix}$$

Find the fundamental frequency and the mode shape of the system, using the matrix iteration method.

7.24. Find the eigenvalues and eigenvectors of the matrix

$$[D] = \begin{bmatrix} 3 & -2 & 0 \\ -2 & 5 & -3 \\ 0 & -3 & 3 \end{bmatrix}$$

using Jacobi's method.

7.25. Find the eigenvalues and eigenvectors of the matrix

$$[D] = \begin{bmatrix} 3 & 2 & 1 \\ 2 & 2 & 1 \\ 1 & 1 & 1 \end{bmatrix}$$

using Jacobi's method.

7.26. Decompose the matrix

$$[A] = \begin{bmatrix} 4 & -2 & 6 & 4 \\ -2 & 2 & -1 & 3 \\ 6 & -1 & 22 & 13 \\ 4 & 3 & 13 & 46 \end{bmatrix}$$

using the Choleski decomposition technique.

7.27. Find the inverse of the following matrix, using the decomposition $[A] = [U]^T[U]$:

$$[A] = \begin{bmatrix} 5 & -1 & 1 \\ -1 & 6 & -4 \\ 1 & -4 & 3 \end{bmatrix}$$

7.28. Find the inverse of the following matrix, using Choleski decomposition:

$$[A] = \begin{bmatrix} 2 & 5 & 8 \\ 5 & 16 & 28 \\ 8 & 28 & 54 \end{bmatrix}$$

7.29. Find the eigenvalues of the matrix $[A]$ given in Problem 7.26, using Jacobi's method.

7.30. Convert Problem 7.23 to a standard eigenvalue problem with a symmetric matrix.

7.31. Using the Choleski decomposition technique, express the following matrix as the product of two triangular matrices:

$$[A] = \begin{bmatrix} 16 & -20 & -24 \\ -20 & 89 & -50 \\ -24 & -50 & 280 \end{bmatrix}$$

7.32.* Design a minimum weight tubular section for the shaft shown in Fig. 7.3 to achieve a fundamental frequency of vibration of 0.5 Hz. Assume $m_1 = 20$ kg, $m_2 = 50$ kg, $m_3 = 40$ kg, $l_1 = 1$ m, $l_2 = 3$ m, $l_3 = 4$ m, $l_4 = 2$ m, and $E = 2.07 \times 10^{11}$ N/m^2.

7.33. Find the eigenvalues and eigenvectors of the matrix $[D]$ given in Problem 7.18, using subroutine JACOBI.

7.34. Solve the general eigenvalue problem stated in Problem 7.19, using the computer program of Section 7.8.4.

7.35. Solve the general eigenvalue problem stated in Problem 7.23, using the computer program of Section 7.8.4.

7.36. Find the eigenvalues and eigenvectors of Problem 7.14, using the computer program of Section 7.8.4.

7.37. Solve Problem 7.25, using subroutine MITER.

7.38. Find the eigensolution of the matrix $[A]$ of Problem 7.27, using subroutine JACOBI.

7.39. Find the eigensolution of the matrix $[A]$ given in Problem 7.31, using subroutine JACOBI.

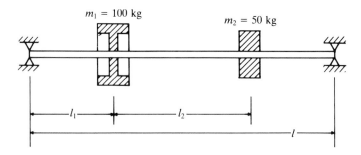

Figure 7.13

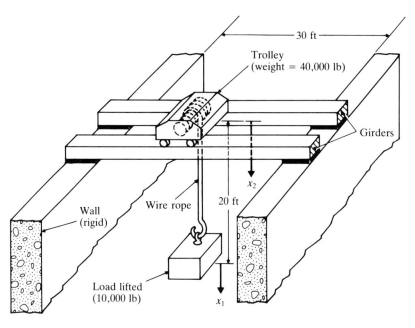

Figure 7.14

Projects:

7.40. A flywheel of mass $m_1 = 100$ kg and a pulley of mass $m_2 = 50$ kg are to be mounted on a shaft of length $l = 2$ m as shown in Fig. 7.13. Determine their locations l_1 and l_2 to maximize the fundamental frequency of vibration of the system.

7.41. A simplified diagram of an overhead traveling crane is shown in Fig. 7.14. The girder, with square cross section, and the wire rope, with circular cross section, are made up of steel. Design the girders and the wire rope such that the natural frequencies of the system are greater than the operating speed, 1500 rpm, of an electric motor located in the trolley.

CHAPTER 8

Continuous Systems

Stephen Prokf'yevich Timoshenko (1878–1972), a Russian-born engineer who emigrated to the U.S.A., was one of the most widely known authors of books in the field of elasticity, strength of materials, and vibrations. He held the chair of mechanics at the University of Michigan and later at Stanford University, and is regarded as the father of engineering mechanics in the U.S.A. The improved theory he presented in 1921 for the vibration of beams has become known as the Timoshenko beam theory. (Courtesy Stanford University School of Engineering)

8.1 INTRODUCTION

We have so far dealt with discrete systems where mass, damping, and elasticity were assumed to be present only at certain discrete points in the system. There are many cases, known as *distributed* or *continuous systems*, in which it is not possible to identify discrete masses, dampers, or springs. We must then consider the continuous distribution of the mass, damping, and elasticity and assume that each of the infinite number of points of the system can vibrate. This is why a continuous system is also called a *system of infinite degrees of freedom*.

If a system is modeled as a discrete one, the governing equations are ordinary differential equations, which are relatively easy to solve. On the other hand, if the system is modeled as a continuous one, the governing equations are partial differential equations, which are more difficult. However, the information obtained from a discrete model of a system may not be as accurate as that obtained from a continuous model. The choice between the two models must be made carefully, with due consideration of factors such as the purpose of the analysis, the influence of the analysis on design, and the computational time available.

In this chapter, we shall consider the vibration of simple continuous systems—strings, bars, shafts, beams, and membranes. A more specialized treatment of the vibration of continuous structural elements is given in Refs. [8.1–8.3]. In general, the frequency equation of a continuous system is a transcendental equation that yields an infinite number of natural frequencies and normal modes. This is in contrast to the behavior of discrete systems, which yield a finite number of such frequencies and modes. We need to apply boundary conditions to find the natural frequencies of a continuous system. The question of boundary conditons does not arise in the case of discrete systems except in an indirect way, because the influence coefficients depend on the manner in which the system is supported.

8.2 TRANSVERSE VIBRATION OF A STRING OR CABLE

**8.2.1
Equation of
Motion**

Consider a tightly stretched elastic string or cable of length l subjected to a transverse force $f(x, t)$ per unit length, as shown in Fig. 8.1(a). The transverse displacement of the string, $w(x, t)$, is assumed to be small. Equilibrium of the forces in the z direction gives (see Fig. 8.1b):

$$\text{net force acting on an element} = \text{inertia force acting on the element}$$

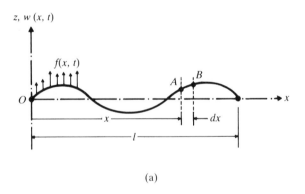

(a)

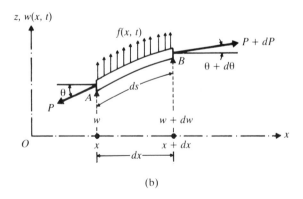

(b)

Figure 8.1 A vibrating string.

or

$$(P + dP)\sin(\theta + d\theta) + f\,dx - P\sin\theta = \rho\,dx\frac{\partial^2 w}{\partial t^2} \tag{8.1}$$

where P is the tension, ρ is the mass per unit length, and θ is the angle the deflected string makes with the x axis. For an elemental length dx,

$$dP = \frac{\partial P}{\partial x}\,dx \tag{8.2}$$

$$\sin\theta \simeq \tan\theta = \frac{\partial w}{\partial x} \tag{8.3}$$

and

$$\sin(\theta + d\theta) \simeq \tan(\theta + d\theta) = \frac{\partial w}{\partial x} + \frac{\partial^2 w}{\partial x^2}\,dx \tag{8.4}$$

Hence the forced vibration equation of the nonuniform string, Eq. (8.1), can be simplified to

$$\frac{\partial}{\partial x}\left[P\frac{\partial w(x, t)}{\partial x}\right] + f(x, t) = \rho(x)\frac{\partial^2 w(x, t)}{\partial t^2} \tag{8.5}$$

If the string is uniform and the tension is constant, Eq. (8.5) reduces to

$$P\frac{\partial^2 w(x, t)}{\partial x^2} + f(x, t) = \rho\frac{\partial^2 w(x, t)}{\partial t^2} \tag{8.6}$$

If $f(x, t) = 0$, we obtain the free vibration equation

$$P\frac{\partial^2 w(x, t)}{\partial x^2} = \rho\frac{\partial^2 w(x, t)}{\partial t^2} \tag{8.7}$$

or

$$c^2\frac{\partial^2 w}{\partial x^2} = \frac{\partial^2 w}{\partial t^2} \tag{8.8}$$

where

$$c = \left(\frac{P}{\rho}\right)^{1/2} \tag{8.9}$$

Equation (8.8) is also known as the *wave equation*.

**8.2.2
Initial and
Boundary
Conditions**

The equation of motion, Eq. (8.5) or its special forms (8.6) and (8.7), is a partial differential equation of the second order. Since the order of the highest derivative of w with respect to x and t in this equation is two, we need to specify two boundary and two initial conditions in finding the solution $w(x, t)$. If the string has a known deflection $w_0(x)$ and velocity $\dot{w}_0(x)$ at time $t = 0$, the initial conditions are specified as

$$w(x, t = 0) = w_0(x)$$
$$\frac{\partial w}{\partial t}(x, t = 0) = \dot{w}_0(x) \tag{8.10}$$

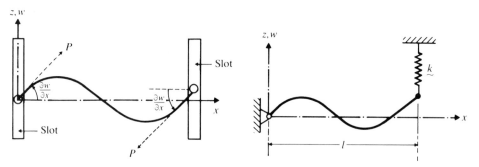

Figure 8.2   **Figure 8.3**

If the string is fixed at an end, say $x = 0$, the displacement w must always be zero, and so the boundary condition is

$$w(x = 0, t) = 0, \qquad t \geqslant 0 \tag{8.11}$$

If the string or cable is connected to a pin that can move in a perpendicular direction as shown in Fig. 8.2, the end cannot support a transverse force. Hence the boundary condition becomes

$$P(x)\frac{\partial w(x, t)}{\partial x} = 0 \tag{8.12}$$

If the end $x = 0$ is free and P is a constant, then Eq. (8.12) becomes

$$\frac{\partial w(0, t)}{\partial x} = 0, \qquad t \geqslant 0 \tag{8.13}$$

If the end $x = l$ is constrained elastically as shown in Fig. 8.3, the boundary condition becomes

$$P(x)\frac{\partial w(x, t)}{\partial x}\bigg|_{x=l} = -\underset{\sim}{k}w(x, t)\big|_{x=l}, \qquad t \geqslant 0 \tag{8.14}$$

where $\underset{\sim}{k}$ is the spring constant.

8.2.3
Free Vibration of
a Uniform String

The free vibration equation, Eq. (8.8), can be solved by the method of separation of variables. In this method, the solution is written as the product of a function $W(x)$ (which depends only on x) and a function $T(t)$ (which depends only on t) [8.4]:

$$w(x, t) = W(x)T(t) \tag{8.15}$$

Substitution of Eq. (8.15) into Eq. (8.8) leads to

$$\frac{c^2}{W}\frac{d^2W}{dx^2} = \frac{1}{T}\frac{d^2T}{dt^2} \tag{8.16}$$

Since the left-hand side of this equation depends only on x and the right-hand side depends only on t, their common value must be a constant—say, a—so that

$$\frac{c^2}{W}\frac{d^2W}{dx^2} = \frac{1}{T}\frac{d^2T}{dt^2} = a \tag{8.17}$$

The equations implied in Eq. (8.17) can be written as

$$\frac{d^2W}{dx^2} - \frac{a}{c^2}W = 0 \tag{8.18}$$

$$\frac{d^2T}{dt^2} - aT = 0 \tag{8.19}$$

Since the constant a is generally negative (see Problem 8.9), we can set $a = -\omega^2$ and write Eqs. (8.18) and (8.19) as

$$\frac{d^2W}{dx^2} + \frac{\omega^2}{c^2}W = 0 \tag{8.20}$$

$$\frac{d^2T}{dt^2} + \omega^2 T = 0 \tag{8.21}$$

The solutions of these equations are given by

$$W(x) = A \cos\frac{\omega x}{c} + B \sin\frac{\omega x}{c} \tag{8.22}$$

$$T(t) = C \cos \omega t + D \sin \omega t \tag{8.23}$$

where ω is the frequency of vibration and the constants A, B, C, and D can be evaluated from the boundary and initial conditions.

8.2.4
Free Vibration of a String with Both Ends Fixed

If the string is fixed at both ends, the boundary conditions are $w(0, t) = w(l, t) = 0$ for all time $t \geqslant 0$. Hence, from Eq. (8.15), we obtain

$$W(0) = 0 \tag{8.24}$$
$$W(l) = 0 \tag{8.25}$$

In order to satisfy Eq. (8.24), A must be zero in Eq. (8.22). Equation (8.25) requires that

$$B \sin\frac{\omega l}{c} = 0 \tag{8.26}$$

Since B cannot be zero for a nontrivial solution, we have

$$\sin\frac{\omega l}{c} = 0 \tag{8.27}$$

Equation (8.27) is called the *frequency* or *characteristic equation* and is satisfied by several values of ω. The values of ω are called the *eigenvalues* (or *natural frequencies* or *characteristic values*) of the problem. The nth natural frequency is given by

$$\frac{\omega_n l}{c} = n\pi, \quad n = 1, 2, \ldots$$

or

$$\omega_n = \frac{nc\pi}{l}, \quad n = 1, 2, \ldots \tag{8.28}$$

The solution $w_n(x, t)$ corresponding to ω_n can be expressed as

$$w_n(x, t) = W_n(x)T_n(t) = \sin\frac{n\pi x}{l}\left[C_n\cos\frac{nc\pi t}{l} + D_n\sin\frac{nc\pi t}{l}\right] \tag{8.29}$$

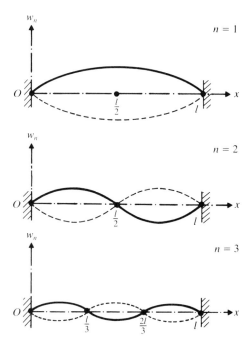

Figure 8.4 Mode shapes of a string.

where C_n and D_n are arbitrary constants. The solution $w_n(x, t)$ is called the *nth mode of vibration* or *nth harmonic* or *nth normal mode* of the string. In this mode, each point of the string vibrates with an amplitude proportional to the value of W_n at that point, with the circular frequency $\omega_n = (nc\pi)/l$. The function $W_n(x)$ is called the nth normal or characteristic function. The first three modes of vibration are shown in Fig. 8.4. The mode corresponding to $n = 1$ is called the *fundamental mode*, and ω_1 is called the *fundamental frequency*. The fundamental period is

$$\tau_1 = \frac{2\pi}{\omega_1} = \frac{2l}{c}$$

The points at which $w_n = 0$ for all times are called *nodes*. Thus the fundamental mode has two nodes, at $x = 0$ and $x = l$; the second mode has three nodes, at $x = 0$, $x = l/2$, and $x = l$; etc.

The general solution of Eq. (8.8), which satisfies the boundary conditions of Eqs. (8.24) and (8.25), is given by the superposition of all $w_n(x, t)$:

$$w(x, t) = \sum_{n=1}^{\infty} w_n(x, t)$$

$$= \sum_{n=1}^{\infty} \sin\frac{n\pi x}{l}\left[C_n\cos\frac{nc\pi t}{l} + D_n\sin\frac{nc\pi t}{l}\right] \tag{8.30}$$

This equation gives all possible vibrations of the string; the particular vibration that occurs is uniquely determined by the specified initial conditions. The initial conditions give unique values of the constants C_n and D_n. If the initial conditions are specified as in Eq. (8.10), we obtain

$$\sum_{n=1}^{\infty} C_n \sin \frac{n\pi x}{l} = w_0(x) \tag{8.31}$$

$$\sum_{n=1}^{\infty} \frac{nc\pi}{l} D_n \sin \frac{n\pi x}{l} = \dot{w}_0(x) \tag{8.32}$$

which can be seen to be Fourier sine series expansions of $w_0(x)$ and $\dot{w}_0(x)$ in the interval $0 \leqslant x \leqslant l$. The values of C_n and D_n can be determined by multiplying Eqs. (8.31) and (8.32) by $\sin(n\pi x/l)$ and integrating with respect to x from 0 to l:

$$C_n = \frac{2}{l} \int_0^l w_0(x) \sin \frac{n\pi x}{l} \, dx \tag{8.33}$$

$$D_n = \frac{2}{nc\pi} \int_0^l \dot{w}_0(x) \sin \frac{n\pi x}{l} \, dx \tag{8.34}$$

EXAMPLE 8.1   **Dynamic Response of a Plucked String**

If a string of length l, fixed at both ends, is plucked at its midpoint as shown in Fig. 8.5 and then released, determine its subsequent motion.

Given: String: length $= l$, fixed at both ends, initial deflection $=$ as shown in Fig. 8.5.

Find: $w(x, t)$ for $t > 0$.

Approach: General solution of Eq. (8.30).

Solution. The solution is given by Eq. (8.30) with C_n and D_n given by Eqs. (8.33) and (8.34), respectively. Since there is no initial velocity, $\dot{w}_0(x) = 0$, and so $D_n = 0$. Thus the solution of Eq. (8.30) reduces to

$$w(x, t) = \sum_{n=1}^{\infty} C_n \sin \frac{n\pi x}{l} \cos \frac{nc\pi t}{l} \tag{E.1}$$

where

$$C_n = \frac{2}{l} \int_0^l w_0(x) \sin \frac{n\pi x}{l} \, dx \tag{E.2}$$

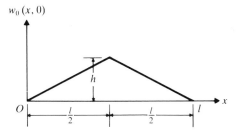

Figure 8.5   Initial deflection of the string.

The initial deflection $w_0(x)$ is given by

$$w_0(x) = \begin{cases} \dfrac{2hx}{l} & \text{for } 0 \leqslant x \leqslant \dfrac{l}{2} \\ \dfrac{2h(l-x)}{l} & \text{for } \dfrac{l}{2} \leqslant x \leqslant l \end{cases} \qquad \text{(E.3)}$$

By substituting Eq. (E.3) into Eq. (E.2), C_n can be evaluated:

$$C_n = \frac{2}{l} \left\{ \int_0^{l/2} \frac{2hx}{l} \sin \frac{n\pi x}{l} \, dx + \int_{l/2}^l \frac{2h}{l}(l-x) \sin \frac{n\pi x}{l} \, dx \right\}$$

$$= \begin{cases} \dfrac{8h}{\pi^2 n^2} \sin \dfrac{n\pi}{2} & \text{for } n = 1,3,5,\ldots \\ 0 & \text{for } n = 2,4,6,\ldots \end{cases} \qquad \text{(E.4)}$$

By using the relation

$$\sin \frac{n\pi}{2} = (-1)^{(n-1)/2}, \qquad n = 1,3,5,\ldots \qquad \text{(E.5)}$$

the desired solution can be expressed as

$$w(x,t) = \frac{8h}{\pi^2} \left\{ \sin \frac{\pi x}{l} \cos \frac{\pi ct}{l} - \frac{1}{9} \sin \frac{3\pi x}{l} \cos \frac{3\pi ct}{l} + \cdots \right\} \qquad \text{(E.6)}$$

In this case, no even harmonics are excited.

8.2.5 Traveling-Wave Solution

The solution of the wave equation, Eq. (8.8), for a string of infinite length can be expressed as [8.5]

$$w(x,t) = w_1(x - ct) + w_2(x + ct) \qquad \text{(8.35)}$$

where w_1 and w_2 are arbitrary functions of $(x - ct)$ and $(x + ct)$, respectively. To show that Eq. (8.35) is the correct solution of Eq. (8.8), we first differentiate Eq. (8.35):

$$\frac{\partial^2 w(x,t)}{\partial x^2} = w_1''(x - ct) + w_2''(x + ct) \qquad \text{(8.36)}$$

$$\frac{\partial^2 w(x,t)}{\partial t^2} = c^2 w_1''(x - ct) + c^2 w_2''(x + ct) \qquad \text{(8.37)}$$

Substitution of these equations into Eq. (8.8) reveals that the wave equation is satisfied. In Eq. (8.35), $w_1(x - ct)$ and $w_2(x + ct)$ represent waves that propagate in the positive and negative directions of the x axis, respectively, with a velocity c.

For a given problem, the arbitrary functions w_1 and w_2 are determined from the initial conditions, Eq. (8.10). Substitution of Eq. (8.35) into Eq. (8.10) gives, at $t = 0$,

$$w_1(x) + w_2(x) = w_0(x) \qquad \text{(8.38)}$$
$$-cw_1'(x) + cw_2'(x) = \dot{w}_0(x) \qquad \text{(8.39)}$$

where the prime indicates differentiation with respect to the respective argument at $t = 0$ (that is, with respect to x). Integration of Eq. (8.39) yields

$$-w_1(x) + w_2(x) = \frac{1}{c} \int_{x_0}^x \dot{w}_0(x') \, dx' \qquad \text{(8.40)}$$

where x_0 is a constant. Solution of Eqs. (8.38) and (8.40) gives w_1 and w_2:

$$w_1(x) = \frac{1}{2}\left[w_0(x) - \frac{1}{c}\int_{x_0}^{x}\dot{w}_0(x')\,dx'\right] \tag{8.41}$$

$$w_2(x) = \frac{1}{2}\left[w_0(x) + \frac{1}{c}\int_{x_0}^{x}\dot{w}_0(x')\,dx'\right] \tag{8.42}$$

By replacing x by $(x - ct)$ and $(x + ct)$, respectively, in Eqs. (8.41) and (8.42), we obtain the total solution:

$$w(x, t) = w_1(x - ct) + w_2(x + ct)$$

$$= \frac{1}{2}\left[w_0(x - ct) + w_0(x + ct)\right] + \frac{1}{2c}\int_{x-ct}^{x+ct}\dot{w}_0(x')\,dx' \tag{8.43}$$

Note:

1. As can be seen from Eq. (8.43), there is no need to apply boundary conditions to the problem.

2. The solution given by Eq. (8.43) can be expressed as

$$w(x, t) = w_D(x, t) + w_V(x, t) \tag{8.44}$$

where $w_D(x, t)$ denotes the waves propagating due to the known initial displacement $w_0(x)$ with zero initial velocity, and $w_V(x, t)$ represents waves traveling due only to the known initial velocity $\dot{w}_0(x)$ with zero initial displacement.

The transverse vibration of a string fixed at both ends excited by the transverse impact of an elastic load at an intermediate point was considered in [8.6]. A review of the literature on the dynamics of cables and chains was given by Triantafyllou [8.7].

8.3 LONGITUDINAL VIBRATION OF A BAR OR ROD

**8.3.1
Equation of
Motion and
Solution**

Consider an elastic bar of length l with varying cross-sectional area $A(x)$, as shown in Fig. 8.6. The forces acting on the cross sections of a small element of the bar are given by P and $P + dP$ with

$$P = \sigma A = EA\frac{\partial u}{\partial x} \tag{8.45}$$

where σ is the axial stress, E is Young's modulus, u is the axial displacement, and $\partial u/\partial x$ is the axial strain. If $f(x, t)$ denotes the external force per unit length, the summation of the forces in the x direction gives the equation of motion:

$$(P + dP) + f\,dx - P = \rho A\,dx\frac{\partial^2 u}{\partial t^2} \tag{8.46}$$

where ρ is the mass density of the bar. By using the relation $dP = (\partial P/\partial x)\,dx$ and Eq. (8.45), the equation of motion for the forced longitudinal vibration of a

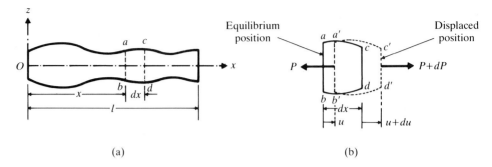

Figure 8.6 Longitudinal vibration of a bar.

nonuniform bar, Eq. (8.46), can be expressed as

$$\frac{\partial}{\partial x}\left[EA(x)\frac{\partial u(x, t)}{\partial x}\right] + f(x, t) = \rho(x)A(x)\frac{\partial^2 u}{\partial t^2}(x, t) \tag{8.47}$$

For a uniform bar, Eq. (8.47) reduces to

$$EA\frac{\partial^2 u}{\partial x^2}(x, t) + f(x, t) = \rho A\frac{\partial^2 u}{\partial t^2}(x, t) \tag{8.48}$$

The free vibration equation can be obtained from Eq. (8.48), by setting $f = 0$, as

$$c^2\frac{\partial^2 u}{\partial x^2}(x, t) = \frac{\partial^2 u}{\partial t^2}(x, t) \tag{8.49}$$

where

$$c = \sqrt{\frac{E}{\rho}} \tag{8.50}$$

Note that Eqs. (8.47) to (8.50) can be seen to be similar to Eqs. (8.5), (8.6), (8.8), and (8.9), respectively. The solution of Eq. (8.49), which can be obtained as in the case of Eq. (8.8), can thus be written as

$$u(x, t) = U(x)T(t) \equiv \left(\underset{\sim}{A}\cos\frac{\omega x}{c} + \underset{\sim}{B}\sin\frac{\omega x}{c}\right)(C\cos\omega t + D\sin\omega t)* \tag{8.51}$$

where the function $U(x)$ depends only on x and the function $T(t)$ depends only on t. If the bar has known initial axial displacement $u_0(x)$ and initial velocity $\dot{u}_0(x)$, the initial conditions can be stated as

$$u(x, t = 0) = u_0(x)$$

$$\frac{\partial u}{\partial t}(x, t = 0) = \dot{u}_0(x) \tag{8.52}$$

---

* We use $\underset{\sim}{A}$ and $\underset{\sim}{B}$ in this section; A is used to denote the cross-sectional area of the bar.

End conditions of bar	Bounty conditions	Frequency equation	Mode shape (normal function)	Natural frequencies
Fixed-free 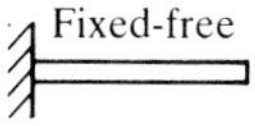	$u(0, t) = 0$ $\dfrac{\partial u}{\partial x}(l, t) = 0$	$\cos \dfrac{\omega l}{c} = 0$	$U_n(x) = C_n \sin \dfrac{(2n + 1)\,\pi x}{2l}$	$\omega_n = \dfrac{(2n + 1)\,\pi c}{2l}$; $n = 0, 1, 2, \ldots$
Free-free 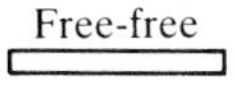	$\dfrac{\partial u}{\partial x}(0, t) = 0$ $\dfrac{\partial u}{\partial x}(l, t) = 0$	$\sin \dfrac{\omega l}{c} = 0$	$U_n(x) = C_n \cos \dfrac{n\pi x}{l}$	$\omega_n = \dfrac{n\pi c}{l}$; $n = 0, 1, 2, \ldots$
Fixed-fixed 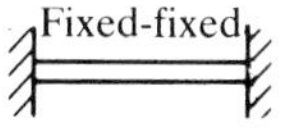	$u(0, t) = 0$ $u(l, t) = 0$	$\sin \dfrac{\omega l}{c} = 0$	$U_n(x) = C_n \cos \dfrac{n\pi x}{l}$	$\omega_n = \dfrac{n\pi c}{l}$; $n = 1, 2, 3, \ldots$

Figure 8.7 Common boundary conditions for a bar in longitudinal vibration.

The common boundary conditions and the corresponding frequency equations for the longitudinal vibration of uniform bars are shown in Fig. 8.7.

EXAMPLE 8.2 **Boundary Conditions for a Bar**

A uniform bar of cross sectional area A, length l, and Young's modulus E is connected at both ends by springs, dampers, and masses as shown in Fig. 8.8(a). State the boundary conditions.

Given: Bar with attached spring, damper, and mass at each end.

Find: Boundary conditions.

Approach: Equilibrium of forces.

Solution. The free body diagrams of the masses m_1 and m_2 are shown in Fig. 8.8(b). From this, we find that at the left end ($x = 0$), the force developed in the bar due to positive u and $\partial u/\partial x$ must be equal to the sum of spring, damper and inertia forces:

$$AE\frac{\partial u}{\partial x}(0, t) = k_1 u(0, t) + c_1 \frac{\partial u}{\partial t}(0, t) + m_1 \frac{\partial^2 u}{\partial t^2}(0, t) \qquad (E.1)$$

Similarly at the right end ($x = l$), the force developed in the bar due to positive u and $\partial u/\partial x$ must be equal to the negative sum of spring, damper, and inertia forces:

$$AE\frac{\partial u}{\partial x}(l, t) = -k_2 u(l, t) - c_2 \frac{\partial u}{\partial t}(l, t) - m_2 \frac{\partial^2 u}{\partial t^2}(l, t) \qquad (E.2)$$

(a)

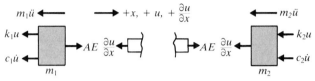

Free body diagram of mass m_1 Free body diagram of mass m_2

(b)

Figure 8.8

8.3.2
Orthogonality of
Normal Functions

The normal functions for the longitudinal vibration of bars satisfy the orthogonality relation

$$\int_0^l U_i(x)U_j(x)\,dx = 0 \tag{8.53}$$

where $U_i(x)$ and $U_j(x)$ denote the normal functions corresponding to the ith and jth natural frequencies ω_i and ω_j, respectively. When $u(x,t) = U_i(x)T(t)$ and $u(x,t) = U_j(x)T(t)$ are assumed as solutions, Eq. (8.49) gives

$$c^2\frac{d^2U_i(x)}{dx^2} + \omega_i^2 U_i(x) = 0 \quad \text{or} \quad c^2 U_i''(x) + \omega_i^2 U_i(x) = 0 \tag{8.54}$$

and

$$c^2\frac{d^2U_j(x)}{dx^2} + \omega_j^2 U_j(x) = 0 \quad \text{or} \quad c^2 U_j''(x) + \omega_j^2 U_j(x) = 0 \tag{8.55}$$

where $U_i'' = \dfrac{d^2U_i}{dx^2}$ and $U_j'' = \dfrac{d^2U_j}{dx^2}$. Multiplication of Eq. (8.54) by U_j and Eq. (8.55) by U_i gives:

$$c^2 U_i'' U_j + \omega_i^2 U_i U_j = 0 \tag{8.56}$$
$$c^2 U_j'' U_i + \omega_j^2 U_j U_i = 0 \tag{8.57}$$

Subtraction of Eq. (8.57) from Eq. (8.56) and integration from 0 to l results in

$$\int_0^l U_i U_j\,dx = -\frac{c^2}{\omega_i^2 - \omega_j^2}\int_0^l (U_i'' U_j - U_j'' U_i)\,dx$$

$$= -\frac{c^2}{\omega_i^2 - \omega_j^2}\big[U_i' U_j - U_j' U_i\big]\Big|_0^l \tag{8.58}$$

The right-hand side of Eq. (8.58) can be proved to be zero for any combination of boundary conditions. For example, if the bar is fixed at $x = 0$ and free at $x = l$,

$$u(0, t) = 0, \qquad t \geqslant 0 \qquad \text{or} \qquad U(0) = 0 \tag{8.59}$$

$$\frac{\partial u}{\partial x}(l, t) = 0, \qquad t \geqslant 0 \qquad \text{or} \qquad U'(l) = 0 \tag{8.60}$$

Thus $(U_i' U_j - U_j' U_i)|_{x=l} = 0$ due to U' being zero (Eq. (8.60)) and $(U_i' U_j - U_j' U_i|_{x=0} = 0$ due to U being zero (Eq. (8.59)). Equation (8.58) thus reduces to Eq. (8.53), which is also known as the *orthogonality principle for the normal functions*.

EXAMPLE 8.3 **Free Vibrations of a Fixed-Free Bar**

Find the natural frequencies and the free vibration solution of a bar fixed at one end and free at the other.

Given: Fixed-free bar in longitudinal vibration.

Find: Natural frequencies and free vibration solution.

Approach: Use proper boundary conditions in the general solution, Eq. (8.51).

Solution. Let the bar be fixed at $x = 0$ and free at $x = l$, so that the boundary conditions can be expressed as

$$u(0, t) = 0, \qquad t \geqslant 0 \tag{E.1}$$

$$\frac{\partial u}{\partial x}(l, t) = 0, \qquad t \geqslant 0 \tag{E.2}$$

The use of Eq. (E.1) in Eq. (8.51) gives $A = 0$, while the use of Eq. (E.2) gives the frequency equation

$$B \frac{\omega}{c} \cos \frac{\omega l}{c} = 0 \qquad \text{or} \qquad \cos \frac{\omega l}{c} = 0 \tag{E.3}$$

The eigenvalues or natural frequencies are given by

$$\frac{\omega_n l}{c} = (2n + 1)\frac{\pi}{2}, \qquad n = 0, 1, 2, \ldots$$

or

$$\omega_n = \frac{(2n + 1)\pi c}{2l}, \qquad n = 0, 1, 2, \ldots \tag{E.4}$$

Thus the total (free vibration) solution of Eq. (8.49) can be written as

$$u(x, t) = \sum_{n=0}^{\infty} u_n(x, t) = \sum_{n=0}^{\infty} \sin\frac{(2n + 1)\pi x}{2l}\left[C_n \cos\frac{(2n + 1)\pi ct}{2l}\right.$$
$$\left. + D_n \sin\frac{(2n + 1)\pi ct}{2l}\right] \tag{E.5}$$

where the values of the constants C_n and D_n can be determined from the initial conditions, as in Eqs. (8.33) and (8.34):

$$C_n = \frac{2}{l} \int_0^l u_0(x) \sin\frac{(2n + 1)\pi x}{2l} \, dx \tag{E.6}$$

$$D_n = \frac{4}{(2n + 1)\pi c} \int_0^l \dot{u}_0(x) \sin\frac{(2n + 1)\pi x}{2l} \, dx \tag{E.7}$$

EXAMPLE 8.4   **Natural Frequencies of a Bar Carrying a Mass**

Find the natural frequencies of a bar with one end fixed and a mass attached at the other end, as in Fig. 8.9.

Given: Bar with one end fixed and a mass at the other end.

Find: Natural frequencies.

Approach: Apply proper boundary conditions to the general solution of Eq. (8.51).

Solution. The equation governing the axial vibration of the bar is given by Eq. (8.49) and the solution by Eq. (8.51). The boundary condition at the fixed end ($x = 0$)

$$u(0, t) = 0 \tag{E.1}$$

leads to $\underset{\sim}{A} = 0$ in Eq. (8.51). At the end $x = l$, the tensile force in the bar must be equal to the inertia force of the vibrating mass M, and so

$$AE\frac{\partial u}{\partial x}(l, t) = -M\frac{\partial^2 u}{\partial t^2}(l, t) \tag{E.2}$$

With the help of Eq. (8.51), this equation can be expressed as

$$AE\frac{\omega}{c}\cos\frac{\omega l}{c}(C\cos\omega t + D\sin\omega t) = M\omega^2\sin\frac{\omega l}{c}(C\cos\omega t + D\sin\omega t)$$

That is,

$$\frac{AE\omega}{c}\cos\frac{\omega l}{c} = M\omega^2\sin\frac{\omega l}{c}$$

or

$$\alpha\tan\alpha = \beta \tag{E.3}$$

where

$$\alpha = \frac{\omega l}{c} \tag{E.4}$$

and

$$\beta = \frac{AEl}{c^2 M} = \frac{A\rho l}{M} = \frac{m}{M} \tag{E.5}$$

where m is the mass of the bar. Equation (E.3) is the frequency equation (in the form of a transcendental equation) whose solution gives the natural frequencies of the system. The first two natural frequencies are given in Table 8.1 for different values of the parameter β.

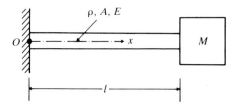

Figure 8.9

TABLE 8.1					
	Value of the mass ratio β				
	0.01	**0.1**	**1.0**	**10.0**	**100.0**
Value of α_1 $(\omega_1 = \dfrac{\alpha_1 c}{l})$	0.1000	0.3113	0.8602	1.4291	1.5549
Value of α_2 $(\omega_2 = \dfrac{\alpha_2 c}{l})$	3.1448	3.1736	3.4267	4.3063	4.6658

Note: If the mass of the bar is negligible compared to the mass attached, $m \simeq 0$,

$$c = \left(\frac{E}{\rho}\right)^{1/2} = \left(\frac{EAl}{m}\right)^{1/2} \to \infty \qquad \text{and} \qquad \alpha = \frac{\omega l}{c} \to 0$$

In this case

$$\tan\frac{\omega l}{c} \simeq \frac{\omega l}{c}$$

and the frequency equation (E.3) can be taken as

$$\left(\frac{\omega l}{c}\right)^2 = \beta$$

This gives the approximate value of the fundamental frequency:

$$\omega_1 = \frac{c}{l}\beta^{1/2} = \frac{c}{l}\left(\frac{\rho Al}{M}\right)^{1/2} = \left(\frac{EA}{lM}\right)^{1/2} = \left(\frac{g}{\delta_s}\right)^{1/2}$$

where

$$\delta_s = \frac{Mgl}{EA}$$

represents the static elongation of the bar under the action of the load Mg.

EXAMPLE 8.5   Vibrations of a Bar Subjected to Initial Force

A bar of uniform cross-sectional area A, density ρ, modulus of elasticity E, and length l is fixed at one end and free at the other end. It is subjected to an axial force F_0 at its free end, as shown in Fig. 8.10(a). Study the resulting vibrations if the force F_0 is suddenly removed.

Given: Fixed-free bar subjected to an axial force at the free end.

Find: Resulting vibrations when the force is suddenly removed.

Approach: Use proper initial conditions and boundary conditions.

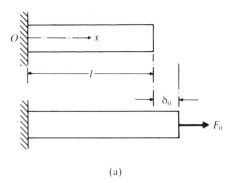

(a)

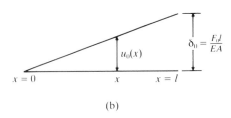

(b)

Figure 8.10

Solution. The tensile strain induced in the bar due to F_0 is

$$\varepsilon = \frac{F_0}{EA}$$

Thus the displacement of the bar just before the force F_0 is removed (initial displacement) is given by (see Fig. 8.10b)

$$u_0 = u(x,0) = \varepsilon x = \frac{F_0 x}{EA}, \qquad 0 \leqslant x \leqslant l \tag{E.1}$$

Since the initial velocity is zero, we have

$$\dot{u}_0 = \frac{\partial u}{\partial t}(x,0) = 0, \qquad 0 \leqslant x \leqslant l \tag{E.2}$$

The general solution of a bar fixed at one end and free at the other end is given by Eq. (E.5) of Example 8.3:

$$u(x,t) = \sum_{n=0}^{\infty} u_n(x,t) = \sum_{n=0}^{\infty} \sin \frac{(2n+1)\pi x}{2l} \left[C_n \cos \frac{(2n+1)\pi ct}{2l} \right.$$
$$\left. + D_n \sin \frac{(2n+1)\pi ct}{2l} \right] \tag{E.3}$$

where C_n and D_n are given by Eqs. (E.6) and (E.7) of Example 8.3. Since $\dot{u}_0 = 0$, we obtain $D_n = 0$. By using the initial displacement of Eq. (E.1) in Eq. (E.6) of Example 8.3, we obtain

$$C_n = \frac{2}{l} \int_0^l \frac{F_0 x}{EA} \cdot \sin \frac{(2n+1)\pi x}{2l} \, dx = \frac{8F_0 l}{EA\pi^2} \frac{(-1)^n}{(2n+1)^2} \tag{E.4}$$

Thus the solution becomes

$$u(x, t) = \frac{8F_0 l}{EA\pi^2} \sum_{n=0}^{\infty} \frac{(-1)^n}{(2n+1)^2} \sin\frac{(2n+1)\pi x}{2l} \cos\frac{(2n+1)\pi ct}{2l} \qquad (E.5)$$

Equations (E.3) and (E.5) indicate that the motion of a typical point at $x = x_0$ on the bar is composed of the amplitudes

$$C_n \sin\frac{(2n+1)\pi x_0}{2l}$$

corresponding to the circular frequencies

$$\frac{(2n+1)\pi c}{2l}$$

8.4 TORSIONAL VIBRATION OF A SHAFT OR ROD

Figure 8.11 represents a nonuniform shaft subjected to an external torque $f(x, t)$ per unit length. If $\theta(x, t)$ denotes the angle of twist of the cross section, the relation between the torsional deflection and the twisting moment $M_t(x, t)$ is given by [8.8]

$$M_t(x, t) = GJ(x)\frac{\partial\theta}{\partial x}(x, t) \qquad (8.61)$$

where G is the shear modulus and $GJ(x)$ is the torsional stiffness, with $J(x)$

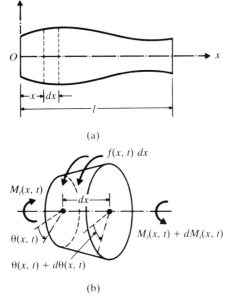

(a)

(b)

Figure 8.11 Torsional vibration of a shaft.

denoting the polar moment of inertia of the cross section in the case of a circular section. If the mass polar moment of inertia of the shaft per unit length is J_0, the inertia torque acting on an element of length dx becomes

$$J_0 \, dx \frac{\partial^2 \theta}{\partial t^2}$$

If an external torque $f(x, t)$ acts on the shaft per unit length, the application of Newton's second law yields the equation of motion:

$$(M_t + dM_t) + f \, dx - M_t = J_0 \, dx \frac{\partial^2 \theta}{\partial t^2} \tag{8.62}$$

By expressing dM_t as

$$\frac{\partial M_t}{\partial x} \, dx$$

and using Eq. (8.61), the forced torsional vibration equation for a nonuniform shaft can be obtained:

$$\frac{\partial}{\partial x}\left[GJ(x)\frac{\partial \theta}{\partial x}(x, t)\right] + f(x, t) = J_0(x)\frac{\partial^2 \theta}{\partial t^2}(x, t) \tag{8.63}$$

For a uniform shaft, Eq. (8.63) takes the form

$$GJ\frac{\partial^2 \theta}{\partial x^2}(x, t) + f(x, t) = J_0\frac{\partial^2 \theta}{\partial t^2}(x, t) \tag{8.64}$$

which, in the case of free vibration, reduces to

$$c^2 \frac{\partial^2 \theta}{\partial x^2}(x, t) = \frac{\partial^2 \theta}{\partial t^2}(x, t) \tag{8.65}$$

where

$$c = \sqrt{\frac{GJ}{J_0}} \tag{8.66}$$

Notice that Eqs. (8.63) to (8.66) are similar to the equations derived in the cases of transverse vibration of a string and longitudinal vibration of a bar. If the shaft has a uniform cross section, $J_0 = \rho J$. Hence Eq. (8.66) becomes

$$c = \sqrt{\frac{G}{\rho}} \tag{8.67}$$

If the shaft is given an angular displacement $\theta_0(x)$ and an angular velocity $\dot{\theta}_0(x)$ at $t = 0$, the initial conditions can be stated as

$$\theta(x, t = 0) = \theta_0(x)$$
$$\frac{\partial \theta}{\partial t}(x, t = 0) = \dot{\theta}_0(x) \tag{8.68}$$

End conditions of shaft	Bounty conditions	Frequency equation	Mode shape (normal function)	Natural frequencies
Fixed-free	$\theta(0, t) = 0$ $\dfrac{\partial \theta}{\partial x}(l, t) = 0$	$\cos \dfrac{\omega l}{c} = 0$	$\Theta(x) = C_n \sin \dfrac{(2n + 1)\,\pi x}{2l}$	$\omega_n = \dfrac{(2n + 1)\,\pi c}{2l}$; $n = 0, 1, 2, \ldots$
Free-free	$\dfrac{\partial \theta}{\partial x}(0, t) = 0$ $\dfrac{\partial \theta}{\partial x}(l, t) = 0$	$\sin \dfrac{\omega l}{c} = 0$	$\Theta(x) = C_n \cos \dfrac{n\pi x}{l}$	$\omega_n = \dfrac{n\pi c}{l}$; $n = 0, 1, 2, \ldots$
Fixed-fixed	$\theta(0, t) = 0$ $\theta(l, t) = 0$	$\sin \dfrac{\omega l}{c} = 0$	$\Theta(x) = C_n \cos \dfrac{n\pi x}{l}$	$\omega_n = \dfrac{n\pi c}{l}$; $n = 1, 2, 3, \ldots$

Figure 8.12 Boundary conditions for a shaft (rod) subjected to torsional vibration.

The general solution of Eq. (8.65) can be expressed as

$$\theta(x, t) = \left(A \cos \frac{\omega x}{c} + B \sin \frac{\omega x}{c}\right)(C \cos \omega t + D \sin \omega t) \qquad (8.69)$$

The common boundary conditions for the torsional vibration of uniform shafts are indicated in Fig. 8.12 along with the corresponding frequency equations and the normal functions.

EXAMPLE 8.6   **Natural Frequencies of a Milling Cutter**

Find the natural frequencies of the plane milling cutter shown in Fig. 8.13 when the free end of the shank is fixed. Assume the torsional rigidity of the shank as GJ and the mass moment of inertia of the cutter as J_0.

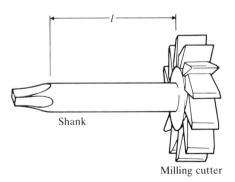

Shank

Milling cutter

Figure 8.13

Given: Milling cutter of mass moment of inertia J_0, shank of length l, and torsional rigidity GJ.

Find: Natural frequencies of torsional vibration.

Approach: Model the system as a uniform shaft with one end fixed and other end carrying a cutter with J_0.

Solution. The general solution is given by Eq. (8.61). By using the fixed boundary condition $\theta(0, t) = 0$, we obtain from Eq. (8.61), $A = 0$. The boundary condition at $x = l$ can be stated as

$$GJ\frac{\partial \theta}{\partial x}(l, t) = -J_0\frac{\partial^2 \theta}{\partial t^2}(l, t) \tag{E.1}$$

That is,

$$BGJ\frac{\omega}{c}\cos\frac{\omega l}{c} = BJ_0\omega^2\sin\frac{\omega l}{c}$$

or

$$\frac{\omega l}{c}\tan\frac{\omega l}{c} = \frac{J\rho l}{J_0} = \frac{J_{\text{rod}}}{J_0} \tag{E.2}$$

Equation (E.2) can be expressed as

$$\alpha \tan \alpha = \beta \quad \text{where } \alpha = \frac{\omega l}{c} \quad \text{and} \quad \beta = \frac{J_{\text{rod}}}{J_0} \tag{E.3}$$

The solution of Eq. (E.3), and thus the natural frequencies of the system, can be obtained as in the case of Example 8.4.

8.5 LATERAL VIBRATION OF BEAMS

**8.5.1
Equation of
Motion**

Consider the free body diagram of an element of a beam shown in Fig. 8.14 where $M(x, t)$ is the bending moment, $V(x, t)$ is the shear force, and $f(x, t)$ is the external force per unit length of the beam. Since the inertia force acting on the

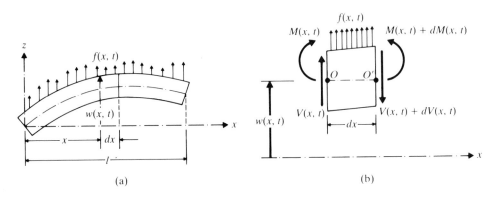

Figure 8.14 A beam in bending.

element of the beam is

$$\rho A(x)\, dx \frac{\partial^2 w}{\partial t^2}(x, t)$$

the force equation of motion in the z direction gives

$$-(V + dV) + f(x, t)\, dx + V = \rho A(x)\, dx \frac{\partial^2 w}{\partial t^2}(x, t) \qquad (8.70)$$

where ρ is the mass density and $A(x)$ is the cross-sectional area of the beam. The moment equation of motion about the y axis passing through the point 0 in Fig. 8.14 leads to

$$(M + dM) - (V + dV)\, dx + f(x, t)\, dx \frac{dx}{2} - M = 0 \qquad (8.71)$$

By writing

$$dV = \frac{\partial V}{\partial x}\, dx \qquad \text{and} \qquad dM = \frac{\partial M}{\partial x}\, dx$$

and disregarding terms involving second powers in dx, Eqs. (8.70) and (8.71) can be written as

$$-\frac{\partial V}{\partial x}(x, t) + f(x, t) = \rho A(x)\frac{\partial^2 w}{\partial t^2}(x, t) \qquad (8.72)$$

$$\frac{\partial M}{\partial x}(x, t) - V(x, t) = 0 \qquad (8.73)$$

By using the relation $V = \partial M/\partial x$ from Eq. (8.73), Eq. (8.72) becomes

$$-\frac{\partial^2 M}{\partial x^2}(x, t) + f(x, t) = \rho A(x)\frac{\partial^2 w}{\partial t^2}(x, t) \qquad (8.74)$$

From the elementary theory of bending of beams (also known as the *Euler-Bernoulli* or *thin beam theory*), the relationship between bending moment and deflection can be expressed as [8.8]

$$M(x, t) = EI(x)\frac{\partial^2 w}{\partial x^2}(x, t) \qquad (8.75)$$

where E is Young's modulus and $I(x)$ is the moment of inertia of the beam cross section about the y axis. Inserting Eq. (8.75) into Eq. (8.74), we obtain the equation of motion for the forced lateral vibration of a nonuniform beam:

$$\frac{\partial^2}{\partial x^2}\left[EI(x)\frac{\partial^2 w}{\partial x^2}(x, t) \right] + \rho A(x)\frac{\partial^2 w}{\partial t^2}(x, t) = f(x, t) \qquad (8.76)$$

For a uniform beam, Eq. (8.76) reduces to

$$EI\frac{\partial^4 w}{\partial x^4}(x, t) + \rho A\frac{\partial^2 w}{\partial t^2}(x, t) = f(x, t) \qquad (8.77)$$

For free vibration, $f(x, t) = 0$, and so the equation of motion becomes

$$c^2 \frac{\partial^4 w}{\partial x^4}(x, t) + \frac{\partial^2 w}{\partial t^2}(x, t) = 0 \tag{8.78}$$

where

$$c = \sqrt{\frac{EI}{\rho A}} \tag{8.79}$$

8.5.2
Initial Conditions

Since the equation of motion involves a second order derivative with respect to time and a fourth order derivative with respect to x, two initial conditions and four boundary conditions are needed for finding a unique solution for $w(x, t)$. Usually, the values of lateral displacement and velocity are specified as $w_0(x)$ and $\dot{w}_0(x)$ at $t = 0$, so that the initial conditions become

$$w(x, t = 0) = w_0(x)$$
$$\frac{\partial w}{\partial t}(x, t = 0) = \dot{w}_0(x) \tag{8.80}$$

8.5.3
Free Vibration

The free vibration solution can be found using the method of separation of variables as

$$w(x, t) = W(x)T(t) \tag{8.81}$$

Substituting Eq. (8.81) into Eq. (8.78) and rearranging leads to

$$\frac{c^2}{W(x)} \frac{d^4 W(x)}{dx^4} = -\frac{1}{T(t)} \frac{d^2 T(t)}{dt^2} = a = \omega^2 \tag{8.82}$$

where $a = \omega^2$ is a positive constant (see Problem 8.42). Equation (8.82) can be written as two equations:

$$\frac{d^4 W(x)}{dx^4} - \beta^4 W(x) = 0 \tag{8.83}$$

$$\frac{d^2 T(t)}{dt^2} + \omega^2 T(t) = 0 \tag{8.84}$$

where

$$\beta^4 = \frac{\omega^2}{c^2} = \frac{\rho A \omega^2}{EI} \tag{8.85}$$

The solution of Eq. (8.84) can be expressed as

$$T(t) = A \cos \omega t + B \sin \omega t \tag{8.86}$$

where A and B are constants that can be found from the initial conditions. For the solution of Eq. (8.83), we assume

$$W(x) = Ce^{sx} \tag{8.87}$$

where C and s are constants, and derive the auxiliary equation as

$$s^4 - \beta^4 = 0 \tag{8.88}$$

The roots of this equation are

$$s_{1,2} = \pm\beta, \qquad s_{3,4} = \pm i\beta \tag{8.89}$$

Hence the solution of Eq. (8.83) becomes

$$W(x) = C_1 e^{\beta x} + C_2 e^{-\beta x} + C_3 e^{i\beta x} + C_4 e^{-i\beta x} \tag{8.90}$$

where C_1, C_2, C_3, and C_4 are constants. Equation (8.90) can also be expressed as

$$W(x) = C_1 \cos\beta x + C_2 \sin\beta x + C_3 \cosh\beta x + C_4 \sinh\beta x \tag{8.91}$$

or

$$\begin{aligned} W(x) = & \, C_1(\cos\beta x + \cosh\beta x) + C_2(\cos\beta x - \cosh\beta x) \\ & + C_3(\sin\beta x + \sinh\beta x) + C_4(\sin\beta x - \sinh\beta x) \end{aligned} \tag{8.92}$$

where C_1, C_2, C_3, and C_4, in each case, are different constants. The constants C_1, C_2, C_3, and C_4 can be found from the boundary conditions. The natural frequencies of the beam are computed from Eq. (8.85) as

$$\omega = \beta^2 \sqrt{\frac{EI}{\rho A}} = (\beta l)^2 \sqrt{\frac{EI}{\rho A l^4}} \tag{8.93}$$

8.5.4 Boundary Conditions

The common boundary conditions are as follows:

(i) Free end:

$$\text{Bending moment} = EI\frac{\partial^2 w}{\partial x^2} = 0,$$

$$\text{Shear force} = \frac{\partial}{\partial x}\left(EI\frac{\partial^2 w}{\partial x^2}\right) = 0 \tag{8.94}$$

(ii) Simply supported (pinned) end:

$$\text{Deflection} = w = 0, \qquad \text{Bending moment} = EI\frac{\partial^2 w}{\partial x^2} = 0 \tag{8.95}$$

(iii) Fixed (clamped) end:

$$\text{Deflection} = 0, \qquad \text{Slope} = \frac{\partial w}{\partial x} = 0 \tag{8.96}$$

The frequency equations, the mode shapes (normal functions) and the natural frequencies for beams with common boundary conditions are given in Fig. 8.15 [8.13, 8.17]. We shall now consider some other possible boundary conditions for a beam.

(iv) When an end is connected to a linear spring, damper and mass (Fig. 8.16):

When the end of a beam undergoes a transverse displacement w and slope $\partial w/\partial x$, with velocity $\partial w/\partial t$ and acceleration $\partial^2 w/\partial t^2$, the resisting forces due to the spring, damper, and mass are proportional to w, $\partial w/\partial t$, and $\partial^2 w/\partial t^2$,

End conditions of beam	Frequency equation	Mode shape (normal function)	Value of $\beta_n l$
Pinned-pinned 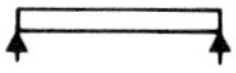	$\sin \beta_n l = 0$	$W_n(x) = C_n[\sin \beta_n x]$	$\beta_1 l = \pi$ $\beta_2 l = 2\pi$ $\beta_3 l = 3\pi$ $\beta_4 l = 4\pi$
Free-free 	$\cos \beta_n l \cdot \cosh \beta_n l = 1$	$W_n(x) = C_n[\sin \beta_n x + \sinh \beta_n x$ $+ \alpha_n (\cos \beta_n x + \cosh \beta_n x)]$ where $\alpha_n = \left(\dfrac{\sin \beta_n l - \sinh \beta_n l}{\cosh \beta_n l - \cos \beta_n l} \right)$	$\beta_1 l = 4.730041$ $\beta_2 l = 7.853205$ $\beta_3 l = 10.995608$ $\beta_4 l = 14.137165$ ($\beta l = 0$ for rigid body mode)
Fixed-fixed 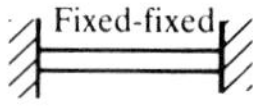	$\cos \beta_n l \cdot \cosh \beta_n l = 1$	$W_n(x) = C_n[\sinh \beta_n x - \sin \beta_n x$ $+ \alpha_n (\cosh \beta_n x - \cos \beta_n x)]$ where $\alpha_n = \left(\dfrac{\sinh \beta_n l - \sin \beta_n l}{\cos \beta_n l - \cosh \beta_n l} \right)$	$\beta_1 l = 4.730041$ $\beta_2 l = 7.853205$ $\beta_3 l = 10.995608$ $\beta_4 l = 14.137165$
Fixed-free 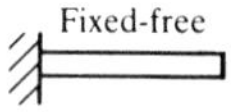	$\cos \beta_n l \cdot \cosh \beta_n l = -1$	$W_n(x) = C_n[\sin \beta_n x - \sinh \beta_n x$ $- \alpha_n (\cos \beta_n x - \cosh \beta_n x)]$ where $\alpha_n = \left(\dfrac{\sin \beta_n l + \sinh \beta_n l}{\cos \beta_n l + \cosh \beta_n l} \right)$	$\beta_1 l = 1.875104$ $\beta_2 l = 4.694091$ $\beta_3 l = 7.854757$ $\beta_4 l = 10.995541$
Fixed-pinned 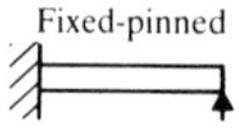	$\tan \beta_n l - \tanh \beta_n l = 0$	$W_n(x) = C_n[\sin \beta_n x + \sinh \beta_n x$ $+ \alpha_n (\cosh \beta_n x - \cos \beta_n x)]$ where $\alpha_n = \left(\dfrac{\sin \beta_n l - \sinh \beta_n l}{\cos \beta_n l - \cosh \beta_n l} \right)$	$\beta_1 l = 3.926602$ $\beta_2 l = 7.068583$ $\beta_3 l = 10.210176$ $\beta_4 l = 13.351768$
Pinned-free 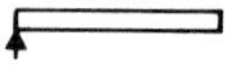	$\tan \beta_n l - \tanh \beta_n l = 0$	$W_n(x) = C_n[\sin \beta_n x \, \alpha_n \sinh \beta_n x]$ where $\alpha_n = \left(\dfrac{\sin \beta_n l}{\sinh \beta_n l} \right)$	$\beta_1 l = 3.926602$ $\beta_2 l = 7.068583$ $\beta_3 l = 10.210176$ $\beta_4 l = 13.351768$ ($\beta l = 0$ for rigid body mode)

Figure 8.15 Common boundary conditions for the transverse vibration of a beam.

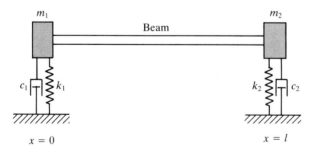

Figure 8.16

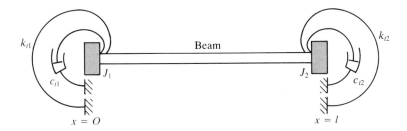

Figure 8.17

respectively. This resisting force is balanced by the shear force at the end. Thus

$$\frac{\partial}{\partial x}\left(EI\frac{\partial^2 w}{\partial x^2}\right) = a\left[kw + c\frac{\partial w}{\partial t} + m\frac{\partial^2 w}{\partial t^2}\right] \tag{8.97}$$

where $a = -1$ for left end and $+1$ for right end of the beam. In addition, the bending moment must be zero; hence

$$EI\frac{\partial^2 w}{\partial x^2} = 0 \tag{8.98}$$

(v) When an end is connected to a torsional spring, torsional damper, and rotational inertia (Fig. 8.17):

In this case, the boundary conditions are

$$EI\frac{\partial^2 w}{\partial x^2} = a\left[k_t\frac{\partial w}{\partial x} + c_t\frac{\partial^2 w}{\partial x\partial t} + J_0\frac{\partial^3 w}{\partial x\partial t^2}\right] \tag{8.99}$$

where $a = -1$ for left end and $+1$ for the right end of the beam, and

$$\frac{\partial}{\partial x}\left[EI\frac{\partial^2 w}{\partial x^2}\right] = 0 \tag{8.100}$$

**8.5.5
Orthogonality of
Normal Functions**

The normal functions $W(x)$ satisfy Eq. (8.83):

$$c^2\frac{d^4 W}{dx^4}(x) - \omega^2 W(x) = 0 \tag{8.101}$$

Let $W_i(x)$ and $W_j(x)$ be the normal functions corresponding to the natural frequencies ω_i and $\omega_j (i \neq j)$ so that

$$c^2\frac{d^4 W_i}{dx^4} - \omega_i^2 W_i = 0 \tag{8.102}$$

and

$$c^2\frac{d^4 W_j}{dx^4} - \omega_j^2 W_j = 0 \tag{8.103}$$

Multiplying Eq. (8.102) by W_j and Eq. (8.103) by W_i, subtracting the resulting equations one from the other, and integrating from 0 to l gives

$$\int_0^l \left[c^2 \frac{d^4 W_i}{dx^4} W_j - \omega_i^2 W_i W_j \right] dx - \int_0^l \left[c^2 \frac{d^4 W_j}{dx^4} W_i - \omega_j^2 W_j W_i \right] dx = 0$$

or

$$\int_0^l W_i W_j \, dx = -\frac{c^2}{\omega_i^2 - \omega_j^2} \int_0^l \left(W_i'''' W_j - W_i W_j'''' \right) dx \qquad (8.104)$$

where a prime indicates differentiation with respect to x. The right-hand side of Eq. (8.104) can be evaluated using integration by parts to obtain:

$$\int_0^l W_i W_j \, dx = -\frac{c^2}{\omega_i^2 - \omega_j^2} \left[W_i W_j''' - W_j W_i''' + W_j' W_i'' - W_i' W_j'' \right] \Big|_0^l \qquad (8.105)$$

The right-hand side of Eq. (8.105) can be shown to be zero for any combination of free, fixed, or simply supported end conditions. At a free end, the bending moment and shear force are equal to zero so that

$$W'' = 0, \qquad W''' = 0 \qquad (8.106)$$

For a fixed end, the deflection and slope are zero:

$$W = 0, \qquad W' = 0 \qquad (8.107)$$

At a simply supported end, the bending moment and deflection are zero:

$$W'' = 0, \qquad W = 0 \qquad (8.108)$$

Since each term on the right-hand side of Eq. (8.105) is zero at $x = 0$ or $x = l$ for any combination of the boundary conditions in Eqs. (8.106) to (8.108), Eq. (8.105) reduces to

$$\int_0^l W_i W_j \, dx = 0 \qquad (8.109)$$

which proves the orthogonality of normal functions for the transverse vibration of beams.

EXAMPLE 8.7   **Natural Frequencies of a Fixed-Pinned Beam**

Determine the natural frequencies of vibration of a uniform beam fixed at $x = 0$ and simply supported at $x = l$.

Given: Uniform beam fixed at $x = 0$ and simply supported at $x = l$.

Find: Natural frequencies.

Approach: Use proper boundary conditions in the free vibration solution, Eq. (8.91).

Solution. The boundary conditions can be stated as

$$W(0) = 0 \tag{E.1}$$

$$\frac{dW}{dx}(0) = 0 \tag{E.2}$$

$$W(l) = 0 \tag{E.3}$$

$$EI\frac{d^2W}{dx^2}(l) = 0 \quad \text{or} \quad \frac{d^2W}{dx^2}(l) = 0 \tag{E.4}$$

Condition (E.1) leads to

$$C_1 + C_3 = 0 \tag{E.5}$$

in Eq. (8.91), while Eqs. (E.2) and (8.91) give

$$\frac{dW}{dx}\bigg|_{x=0} = \beta\left[-C_1\sin\beta x + C_2\cos\beta x + C_3\sinh\beta x + C_4\cosh\beta x\right]_{x=0} = 0$$

or

$$\beta[C_2 + C_4] = 0 \tag{E.6}$$

Thus the solution, Eq. (8.91), becomes

$$W(x) = C_1(\cos\beta x - \cosh\beta x) + C_2(\sin\beta x - \sinh\beta x) \tag{E.7}$$

Applying conditions (E.3) and (E.4) to Eq. (E.7) yields

$$C_1(\cos\beta l - \cosh\beta l) + C_2(\sin\beta l - \sinh\beta l) = 0 \tag{E.8}$$

$$- C_1(\cos\beta l + \cosh\beta l) - C_2(\sin\beta l + \sinh\beta l) = 0 \tag{E.9}$$

For a nontrivial solution of C_1 and C_2, the determinant of their coefficients must be zero—that is,

$$\begin{vmatrix} (\cos\beta l - \cosh\beta l) & (\sin\beta l - \sinh\beta l) \\ -(\cos\beta l + \cosh\beta l) & -(\sin\beta l + \sinh\beta l) \end{vmatrix} = 0 \tag{E.10}$$

Expanding the determinant gives the frequency equation

$$\cos\beta l \sinh\beta l - \sin\beta l \cosh\beta l = 0$$

or

$$\tan\beta l = \tanh\beta l \tag{E.11}$$

The roots of this equation, $\beta_n l$, give the natural frequencies of vibration:

$$\omega_n = (\beta_n l)^2\left(\frac{EI}{\rho A l^4}\right)^{1/2}, \quad n = 1, 2, \ldots \tag{E.12}$$

where the values of $\beta_n l$, $n = 1, 2, \ldots$ satisfying Eq. (E.11) are given in Fig. 8.15. If the value of C_2 corresponding to β_n is denoted as C_{2n}, it can be expressed in terms of C_{1n} from Eq. (E.8) as

$$C_{2n} = -C_{1n}\left(\frac{\cos\beta_n l - \cosh\beta_n l}{\sin\beta_n l - \sinh\beta_n l}\right) \tag{E.13}$$

Hence Eq. (E.7) can be written as

$$W_n(x) = C_{1n}\left[(\cos\beta_n x - \cosh\beta_n x) - \left(\frac{\cos\beta_n l - \cosh\beta_n l}{\sin\beta_n l - \sinh\beta_n l}\right)\right.$$
$$\left. \times(\sin\beta_n x - \sinh\beta_n x)\right] \tag{E.14}$$

The normal modes of vibration can be obtained by the use of Eq. (8.81):

$$w_n(x, t) = W_n(x)(A_n \cos \omega_n t + B_n \sin \omega_n t) \qquad \text{(E.15)}$$

with $W_n(x)$ given by Eq. (E.14). The general or total solution of the fixed-simply supported beam can be expressed by the sum of the normal modes:

$$w(x, t) = \sum_{n=1}^{\infty} w_n(x, t) \qquad \text{(E.16)}$$

8.5.6 Effect of Axial Force

The problem of vibrations of a beam under the action of axial force finds application in the study of vibrations of cables and guy wires. For example, although the vibrations of a cable can be found by treating it as an equivalent string, many cables have failed due to fatigue caused by alternating flexure. The alternating flexure is produced by the regular shedding of vortices from the cable in a light wind. We must therefore consider the effects of axial force and bending stiffness on lateral vibrations in the study of fatigue failure of cables.

To find the effect of an axial force $P(x, t)$ on the bending vibrations of a beam, consider the equation of motion of an element of the beam, as shown in Fig. 8.18. For the vertical motion, we have

$$-(V + dV) + f dx + V + (P + dP)\sin(\theta + d\theta) - P \sin \theta = \rho A \, dx \frac{\partial^2 w}{\partial t^2} \quad \text{(8.110)}$$

and for the rotational motion about 0,

$$(M + dM) - (V + dV) \, dx + f dx \frac{dx}{2} - M = 0 \qquad \text{(8.111)}$$

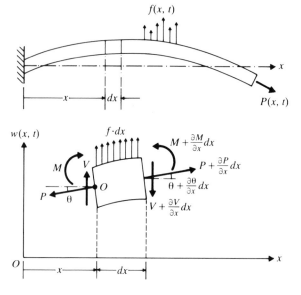

Figure 8.18   An element of a beam under axial load.

For small deflections,

$$\sin(\theta + d\theta) \simeq \theta + d\theta = \theta + \frac{\partial \theta}{\partial x}\, dx = \frac{\partial w}{\partial x} + \frac{\partial^2 w}{\partial x^2}\, dx$$

With this, Eqs. (8.110), (8.111), and (8.75) can be combined to obtain a single differential equation of motion:

$$\frac{\partial^2}{\partial x^2}\left[EI \frac{\partial^2 w}{\partial x^2}\right] + \rho A \frac{\partial^2 w}{\partial t^2} - P\frac{\partial^2 w}{\partial x^2} = f \qquad (8.112)$$

For the free vibration of a uniform beam, Eq. (8.112) reduces to

$$EI \frac{\partial^4 w}{\partial x^4} + \rho A \frac{\partial^2 w}{\partial t^2} - P\frac{\partial^2 w}{\partial x^2} = 0 \qquad (8.113)$$

The solution of Eq. (8.113) can be obtained using the method of separation of variables as

$$w(x, t) = W(x)(A \cos \omega t + B \sin \omega t) \qquad (8.114)$$

Substitution of Eq. (8.114) into Eq. (8.113) gives

$$EI \frac{d^4 W}{dx^4} - P\frac{d^2 W}{dx^2} - \rho A \omega^2 W = 0 \qquad (8.115)$$

By assuming the solution $W(x)$ to be

$$W(x) = Ce^{sx} \qquad (8.116)$$

in Eq. (8.115), the auxiliary equation can be obtained:

$$s^4 - \frac{P}{EI} s^2 - \frac{\rho A \omega^2}{EI} = 0 \qquad (8.117)$$

The roots of Eq. (8.117) are

$$s_1^2, s_2^2 = \frac{P}{2EI} \pm \left(\frac{P^2}{4E^2 I^2} + \frac{\rho A \omega^2}{EI}\right)^{1/2} \qquad (8.118)$$

and so the solution can be expressed as (with absolute value of s_2)

$$W(x) = C_1 \cosh s_1 x + C_2 \sinh s_1 x + C_3 \cos s_2 x + C_4 \sin s_2 x \qquad (8.119)$$

where the constants C_1 to C_4 are to be determined from the boundary conditions.

EXAMPLE 8.8   **Beam Subjected to an Axial Compressive Force**

Find the natural frequencies of a simply supported beam subjected to an axial compressive force.

Given: Simply supported beam subjected to axial compressive force.

Find: Natural frequencies.

Approach: Use proper boundary conditions in the free vibration solution, Eq. (8.119).

Solution. The boundary conditions are

$$W(0) = 0 \tag{E.1}$$

$$\frac{d^2W}{dx^2}(0) = 0 \tag{E.2}$$

$$W(l) = 0 \tag{E.3}$$

$$\frac{d^2W}{dx^2}(l) = 0 \tag{E.4}$$

Equations (E.1) and (E.2) require that $C_1 = C_3 = 0$ in Eq. (8.119), and so

$$W(x) = C_2 \sinh s_1 x + C_4 \sin s_2 x \tag{E.5}$$

The application of Eqs. (E.3) and (E.4) to Eq. (E.5) leads to

$$\sinh s_1 l \cdot \sin s_2 l = 0 \tag{E.6}$$

Since $\sinh s_1 l > 0$ for all values of $s_1 l \neq 0$, the only roots to this equation are

$$s_2 l = n\pi, \qquad n = 0, 1, 2, \dots \tag{E.7}$$

Thus Eqs. (E.7) and (8.118) give the natural frequencies of vibration:

$$\omega_n = \frac{\pi^2}{l^2} \sqrt{\frac{EI}{\rho A}} \left(n^4 + \frac{n^2 P l^2}{\pi^2 EI} \right)^{1/2} \tag{E.8}$$

Since the axial force P is compressive, P is negative. Further, from strength of materials, the smallest Euler buckling load for a simply supported beam is given by [8.9]

$$P_{\text{cri}} = \frac{\pi^2 EI}{l^2} \tag{E.9}$$

Thus Eq. (E.8) can be written as

$$\omega_n = \frac{\pi^2}{l^2} \left(\frac{EI}{\rho A} \right)^{1/2} \left(n^4 - n^2 \frac{P}{P_{\text{cri}}} \right)^{1/2} \tag{E.10}$$

The following observations can be made from the present example:

1. If $P = 0$, the natural frequency will be same as that of a simply supported beam given in Fig. 8.15.
2. If $EI = 0$, the natural frequency (see Eq. (E.8)) reduces to that of a taut string.
3. If $P > 0$, the natural frequency increases as the tensile force stiffens the beam.
4. As $P \rightarrow P_{\text{cri}}$, the natural frequency approaches zero for $n = 1$.

8.5.7
Effects of Rotary
Inertia and Shear
Deformation

If the cross-sectional dimensions are not small compared to the length of the beam, we need to consider the effects of rotary inertia and shear deformation. The procedure, presented by Timoshenko [8.10], is known as the *thick beam theory* or *Timoshenko beam theory*. Consider the element of the beam shown in Fig. 8.19. If the effect of shear deformation is disregarded, the tangent to the deflected center line $O'T$ coincides with the normal to the face $Q'R'$ (since cross sections normal to the center line remain normal even after deformation). Due to shear deformation, the tangent to the deformed center line $O'T$ will not be perpendicular to the face

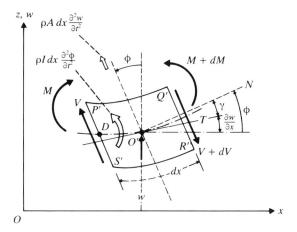

Figure 8.19

$Q'R'$. The angle γ between the tangent to the deformed center line ($O'T$) and the normal to the face ($O'N$) denotes the shear deformation of the element. Since positive shear on the right face $Q'R'$ acts downwards, we have, from Fig. 8.19,

$$\gamma = \phi - \frac{\partial w}{\partial x} \tag{8.120}$$

where ϕ denotes the slope of the deflection curve due to bending deformation alone. Note that because of shear alone, the element undergoes distortion but no rotation.

The bending moment M and the shear force V are related to ϕ and w by the formulas*

$$M = EI \frac{\partial \phi}{\partial x} \tag{8.121}$$

and

$$V = kAG\gamma = kAG\left(\phi - \frac{\partial w}{\partial x}\right) \tag{8.122}$$

where G denotes the modulus of rigidity of the material of the beam and k is a constant, also known as *Timoshenko's shear coefficient*, which depends on the shape of the cross section. For a rectangular section the value of k is $5/6$; for a circular section it is $9/10$ [8.11].

---

* Equation (8.121) is similar to Eq. (8.75). Eq. (8.122) can be obtained as follows:

shear force = shear stress × area = shear strain × shear modulus × area

or

$$V = \gamma GA$$

This equation is modified as $V = kAG\gamma$ by introducing a factor k on the right-hand side to take care of the shape of the cross section.

The equations of motion for the element shown in Fig. 8.19 can be derived as follows:

For translation in the z direction:

$$-[V(x, t) + dV(x, t)] + f(x, t)\, dx + V(x, t)$$

$$= \rho A(x)\, dx \frac{\partial^2 w}{\partial t^2}(x, t)$$

$$\equiv \text{translational inertia of the element} \qquad (8.123)$$

For rotation about a line passing through point D and parallel to the y axis:

$$[M(x, t) + dM(x, t)] + [V(x, t) + dV(x, t)]\, dx + f(x, t)\, dx \frac{dx}{2} - M(x, t)$$

$$= \rho I(x)\, dx \frac{\partial^2 \phi}{\partial t^2} \equiv \text{rotary inertia of the element} \qquad (8.124)$$

Using the relations

$$dV = \frac{\partial V}{\partial x}\, dx \qquad \text{and} \qquad dM = \frac{\partial M}{\partial x}\, dx$$

along with Eqs. (8.121) and (8.122) and disregarding terms involving second powers in dx, Eqs. (8.123) and (8.124) can be expressed as

$$-kAG\left(\frac{\partial \phi}{\partial x} - \frac{\partial^2 w}{\partial x^2}\right) + f(x, t) = \rho A \frac{\partial^2 w}{\partial t^2} \qquad (8.125)$$

$$EI \frac{\partial^2 \phi}{\partial x^2} - kAG\left(\phi - \frac{\partial w}{\partial x}\right) = \rho I \frac{\partial^2 \phi}{\partial t^2} \qquad (8.126)$$

By solving Eq. (8.125) for $\partial \phi / \partial x$ and substituting the result in Eq. (8.126), we obtain the desired equation of motion for the forced vibration of a uniform beam:

$$EI \frac{\partial^4 w}{\partial x^4} + \rho A \frac{\partial^2 w}{\partial t^2} - \rho I\left(1 + \frac{E}{kG}\right) \frac{\partial^4 w}{\partial x^2 \partial t^2} + \frac{\rho^2 I}{kG} \frac{\partial^4 w}{\partial t^4}$$

$$+ \frac{EI}{kAG} \frac{\partial^2 f}{\partial x^2} - \frac{\rho I}{kAG} \frac{\partial^2 f}{\partial t^2} - f = 0 \qquad (8.127)$$

For free vibration, $f = 0$, and Eq. (8.127) reduces to

$$EI \frac{\partial^4 w}{\partial x^4} + \rho A \frac{\partial^2 w}{\partial t^2} - \rho I\left(1 + \frac{E}{kG}\right) \frac{\partial^4 w}{\partial x^2 \partial t^2} + \frac{\rho^2 I}{kG} \frac{\partial^4 w}{\partial t^4} = 0 \qquad (8.128)$$

The following boundary conditions are to be applied in the solution of Eq. (8.127) or (8.128).

1. Fixed end:

$$\phi = w = 0$$

2. Simply supported end:

$$EI\frac{\partial \phi}{\partial x} = w = 0$$

3. Free end:

$$kAG\left(\frac{\partial w}{\partial x} - \phi\right) = EI\frac{\partial \phi}{\partial x} = 0$$

EXAMPLE 8.9   **Natural Frequencies of a Simply Supported Beam**

Determine the effects of rotary inertia and shear deformation on the natural frequencies of a simply supported uniform beam.

Given: Simply supported uniform beam.

Find: Natural frequencies, including the effects of rotary inertia and shear deformation.

Approach: Find the solution that satisfies the boundary conditions and study the influence of rotary inertia and shear deformation.

Solution. Be defining

$$\alpha^2 = \frac{EI}{\rho A} \quad \text{and} \quad r^2 = \frac{I}{A} \tag{E.1}$$

Eq. (8.128) can be written as

$$\alpha^2 \frac{\partial^4 w}{\partial x^4} + \frac{\partial^2 w}{\partial t^2} - r^2\left(1 + \frac{E}{kG}\right)\frac{\partial^4 w}{\partial x^2 \partial t^2} + \frac{\rho r^2}{kG}\frac{\partial^4 w}{\partial t^4} = 0 \tag{E.2}$$

We can express the solution of Eq. (E.2) as

$$w(x, t) = C \sin\frac{n\pi x}{l} \cos \omega_n t \tag{E.3}$$

which satisfies the necessary boundary conditions at $x = 0$ and $x = l$. Here, C is a constant and ω_n is the nth natural frequency. By substituting Eq. (E.3) into Eq. (E.2), we obtain the frequency equation:

$$\omega_n^4\left(\frac{\rho r^2}{kG}\right) - \omega_n^2\left(1 + \frac{n^2\pi^2 r^2}{l^2} + \frac{n^2\pi^2 r^2}{l^2}\frac{E}{kG}\right) + \left(\frac{\alpha^2 n^4 \pi^4}{l^4}\right) = 0 \tag{E.4}$$

It can be seen that Eq. (E.4) is a quadratic equation in ω_n^2 for any given n; there are two values of ω_n that satisfy Eq. (E.4). The smaller value corresponds to the bending deformation mode, while the larger one corresponds to the shear deformation mode.

The values of the ratio of ω_n given by Eq. (E.4) to the natural frequency given by the classical theory (in Fig. 8.15)* are plotted for three values of E/kG in Fig. 8.20 [8.22].

---

* The theory used for the derivation of the equation of motion (8.76), which disregards the effects of rotary inertia and shear deformation, is called the *classical* or *Euler-Bernoulli theory*.

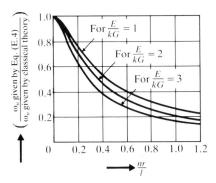

Figure 8.20

Note the following aspects of rotary inertia and shear deformation:

1. If the effect of rotary inertia alone is considered, the resulting equation of motion does not contain any term involving the shear coefficient k. Hence we obtain (from Eq. (8.128)):

$$EI\frac{\partial^4 w}{\partial x^4} + \rho A\frac{\partial^2 w}{\partial t^2} - \rho I\frac{\partial^4 w}{\partial x^2\,\partial t^2} = 0 \tag{E.5}$$

In this case the frequency equation (E.4) reduces to

$$\omega_n^2 = \frac{\alpha^2 n^4 \pi^4}{l^4\left(1 + \dfrac{n^2\pi^2 r^2}{l^2}\right)} \tag{E.6}$$

2. If the effect of shear deformation alone is considered, the resulting equation of motion does not contain the terms originating from $\rho I(\partial^2\phi/\partial t^2)$ in Eq. (8.126). Thus we obtain the equation of motion:

$$EI\frac{\partial^4 w}{\partial x^4} + \rho A\frac{\partial^2 w}{\partial t^2} - \frac{EI\rho}{kG}\frac{\partial^4 w}{\partial x^2\,\partial t^2} = 0 \tag{E.7}$$

and the corresponding frequency equation:

$$\omega_n^2 = \frac{\alpha^2 n^4 \pi^4}{l^4\left(1 + \dfrac{n^2\pi^2 r^2}{l^2}\dfrac{E}{kG}\right)} \tag{E.8}$$

3. If both the effects of rotary inertia and shear deformation are disregarded, Eq. (8.128) reduces to the classical equation of motion, Eq. (8.78):

$$EI\frac{\partial^4 w}{\partial x^4} + \rho A\frac{\partial^2 w}{\partial t^2} = 0 \tag{E.9}$$

and Eq. (E.4) to

$$\omega_n^2 = \frac{\alpha^2 n^4 \pi^4}{l^4} \tag{E.10}$$

8.5.8
Other Effects

The transverse vibration of tapered beams is presented in Refs. [8.12, 8.14]. The natural frequencies of continuous beams are discussed by Wang [8.15]. The dynamic response of beams resting on elastic foundation is considered in Ref. [8.16].

The effect of support flexibility on the natural frequencies of beams is presented in [8.18, 8.19]. A treatment of the problem of natural vibrations of a system of elastically connected Timoshenko beams is given in Ref. [8.20]. A comparison of the exact and approximate solutions of vibrating beams is made by Hutchinson [8.30]. The steady-state vibration of damped beams is considered in Ref. [8.21].

8.6 VIBRATION OF MEMBRANES

A membrane is a plate that is subjected to tension and has negligible bending resistance. Thus a membrane bears the same relationship to a plate as a string bears to a beam. A drumhead is an example of a membrane.

8.6.1 Equation of Motion

To derive the equation of motion of a membrane, consider the membrane to be bounded by a plane curve S in the xy plane, as shown in Fig. 8.21. Let $f(x, y, t)$ denote the pressure loading acting in the z direction and P the intensity of tension

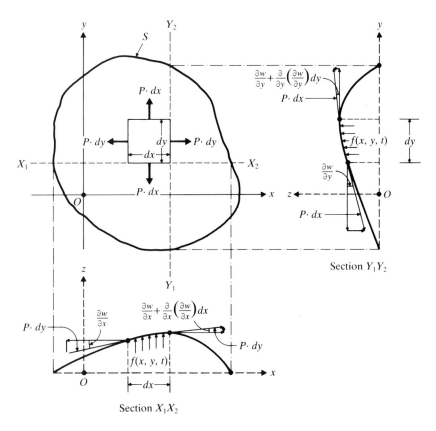

Figure 8.21. A membrane under uniform tension.

at a point that is equal to the product of the tensile stress and the thickness of the membrane. The magnitude of P is usually constant throughout the membrane, as in a drumhead. If we consider an elemental area $dx\,dy$, forces of magnitude $P\,dx$ and $P\,dy$ act on the sides parallel to the y and x axes, respectively, as shown in Fig. 8.21. The net forces acting along the z direction due to these forces are

$$\left(P\frac{\partial^2 w}{\partial y^2}\,dx\,dy\right) \quad \text{and} \quad \left(P\frac{\partial^2 w}{\partial x^2}\,dx\,dy\right)$$

The pressure force along the z direction is $f(x, y, t)\,dx\,dy$, and the inertia force is

$$\rho(x, y)\frac{\partial^2 w}{\partial t^2}\,dx\,dy$$

where $\rho(x, y)$ is the mass per unit area. The equation of motion for the forced transverse vibration of the membrane can be obtained:

$$P\left(\frac{\partial^2 w}{\partial x^2} + \frac{\partial^2 w}{\partial y^2}\right) + f = \rho\frac{\partial^2 w}{\partial t^2} \tag{8.129}$$

If the external force $f(x, y, t) = 0$, Eq. (8.129) gives the free vibration equation

$$c^2\left(\frac{\partial^2 w}{\partial x^2} + \frac{\partial^2 w}{\partial y^2}\right) = \frac{\partial^2 w}{\partial t^2} \tag{8.130}$$

where

$$c = \left(\frac{P}{\rho}\right)^{1/2} \tag{8.131}$$

Equations (8.129) and (8.130) can be expressed as

$$P\nabla^2 w + f = \rho\frac{\partial^2 w}{\partial t^2} \tag{8.132}$$

and

$$c^2\nabla^2 w = \frac{\partial^2 w}{\partial t^2} \tag{8.133}$$

where

$$\nabla^2 = \frac{\partial^2}{\partial x^2} + \frac{\partial^2}{\partial y^2} \tag{8.134}$$

is the Laplacian operator.

**8.6.2
Initial and
Boundary
Conditions**

Since the equation of motion, Eq. (8.129) or (8.130), involves second order partial derivatives with respect to each of t, x, and y, we need to specify two initial conditions and four boundary conditions to find a unique solution of the problem. Usually, the displacement and velocity of the membrane at $t = 0$ are specified as

$w_0(x, y)$ and $\dot{w}_0(x, y)$. Hence the initial conditions are given by

$$w(x, y, 0) = w_0(x, y)$$

$$\frac{\partial w}{\partial t}(x, y, 0) = \dot{w}_0(x, y) \qquad (8.135)$$

The boundary conditions are of the following types:

1. If the membrane is fixed at any point (x_1, y_1) on a segment of the boundary, we have

$$w(x_1, y_1, t) = 0, \qquad t \geqslant 0 \qquad (8.136)$$

2. If the membrane is free to deflect transversely (in the z direction) at a different point (x_2, y_2) of the boundary, then the force component in the z direction must be zero. Thus

$$P \frac{\partial w}{\partial n}(x_2, y_2, t) = 0, \qquad t \geqslant 0 \qquad (8.137)$$

where $\partial w / \partial n$ represents the derivative of w with respect to a direction n normal to the boundary at the point (x_2, y_2).

The solution of the equation of motion of the vibrating membrane was presented in Refs. [8.23–8.25].

EXAMPLE 8.10   **Free Vibrations of a Rectangular Membrane**

Find the free vibration solution of a rectangular membrane of sides a and b along the x and y axes, respectively.

Given: Rectangular membrane of sides a and b.

Find: Free vibration solution.

Approach: Use method of separation of variables.

Solution. By using the method of separation of variables, $w(x, y, t)$ can be assumed to be

$$w(x, y, t) = W(x, y)T(t) = X(x)Y(y)T(t) \qquad (E.1)$$

By using Eqs. (E.1) and (8.130), we obtain

$$\frac{d^2 X(x)}{dx^2} + \alpha^2 X(x) = 0 \qquad (E.2)$$

$$\frac{d^2 Y(y)}{dy^2} + \beta^2 Y(y) = 0 \qquad (E.3)$$

$$\frac{d^2 T(t)}{dt^2} + \omega^2 T(t) = 0 \qquad (E.4)$$

where α^2 and β^2 are constants related to ω^2 as follows:

$$\beta^2 = \frac{\omega^2}{c^2} - \alpha^2 \tag{E.5}$$

The solutions of Eqs. (E.2) to (E.4) are given by

$$X(x) = C_1 \cos \alpha x + C_2 \sin \alpha x \tag{E.6}$$

$$Y(y) = C_3 \cos \beta y + C_4 \sin \beta y \tag{E.7}$$

$$T(t) = A \cos \omega t + B \sin \omega t \tag{E.8}$$

where the constants C_1 to C_4, A, and B can be determined from the boundary and initial conditions.

8.7 RAYLEIGH'S METHOD

Rayleigh's method can be applied to find the fundamental natural frequency of continuous systems. This method is much simpler than exact analysis for systems with varying distributions of mass and stiffness. Although the method is applicable to all continuous systems, we shall apply it only to beams in this section.* Consider the beam shown in Fig. 8.14. In order to apply Rayleigh's method, we need to derive expressions for the maximum kinetic and potential energies and Rayleigh's quotient. The kinetic energy of the beam can be expressed as

$$T = \frac{1}{2} \int_0^l \dot{w}^2 dm = \frac{1}{2} \int_0^l \dot{w}^2 \rho A(x)\, dx \tag{8.138}$$

The maximum kinetic energy can be found by assuming a harmonic variation $w(x, t) = W(x) \cos \omega t$:

$$T_{max} = \frac{\omega^2}{2} \int_0^l \rho A(x) W^2(x)\, dx \tag{8.139}$$

The potential energy of the beam V is the same as the work done in deforming the beam. By disregarding the work done by the shear forces, we have

$$V = \frac{1}{2} \int_0^l M\, d\theta \tag{8.140}$$

where M is the bending moment given by Eq. (8.75) and θ is the slope of the deformed beam given by $\theta = \dfrac{\partial w}{\partial x}$. Thus Eq. (8.140) can be rewritten as

$$V = \frac{1}{2} \int_0^l \left(EI \frac{\partial^2 w}{\partial x^2} \right) \frac{\partial^2 w}{\partial x^2}\, dx = \frac{1}{2} \int_0^l EI \left(\frac{\partial^2 w}{\partial x^2} \right)^2 dx \tag{8.141}$$

---

* An integral equation approach for the determination of the fundamental frequency of vibrating beams is presented by Penny and Reed [8.26].

Since the maximum value of $w(x, t)$ is $W(x)$, the maximum value of V is given by

$$V_{\max} = \frac{1}{2} \int_0^l EI(x) \left(\frac{d^2 W(x)}{dx^2} \right)^2 dx \tag{8.142}$$

By equating $T_{\max}$ to $V_{\max}$, we obtain Rayleigh's quotient:

$$R(\omega) = \omega^2 = \frac{\int_0^l EI \left(\frac{d^2 W(x)}{dx^2} \right)^2 dx}{\int_0^l \rho A (W(x))^2 dx} \tag{8.143}$$

Thus the natural frequency of the beam can be found once the deflection $W(x)$ is known. In general, $W(x)$ is not known and must therefore be assumed. Generally, the static equilibrium shape is assumed for $W(x)$ to obtain the fundamental frequency. It is to be noted that the assumed shape $W(x)$ unintentionally introduces a constraint on the system (which amounts to adding additional stiffness to the system), and so the frequency given by Eq. (8.143) is higher than the exact value [8.27].

For a stepped beam, Eq. (8.143) can be more conveniently written as

$$R(\omega) = \omega^2 = \frac{E_1 I_1 \int_0^{l_1} \left(\frac{d^2 W}{dx^2} \right)^2 dx + E_2 I_2 \int_{l_1}^{l_2} \left(\frac{d^2 W}{dx^2} \right)^2 dx + \cdots}{\rho A_1 \int_0^{l_1} W^2 dx + \rho A_2 \int_{l_1}^{l_2} W^2 dx + \cdots} \tag{8.144}$$

where E_i, I_i, A_i, and l_i correspond to the ith step $(i = 1, 2, \dots)$.

EXAMPLE 8.11   Fundamental Frequency of a Tapered Beam

Find the fundamental frequency of transverse vibration of the nonuniform cantilever beam shown in Fig. 8.22, using the deflection shape $W(x) = (1 - x/l)^2$.

Given: Uniformly tapered cantilever beam (Fig. 8.22) and a deflection function.

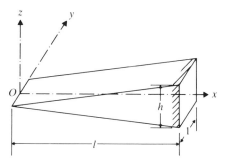

Figure 8.22

Find: Fundamental frequency of vibration.

Approach: Use the given deflection shape in Rayleigh's quotient.

Solution. The given deflection shape can be verified to satisfy the boundary conditions of the beam. The cross-sectional area A and the moment of inertia I of the beam can be expressed as

$$A(x) = \frac{hx}{l}, \qquad I(x) = \frac{1}{12}\left(\frac{hx}{l}\right)^3 \tag{E.1}$$

Thus Rayleigh's quotient gives

$$\omega^2 = \frac{\int_0^l E\left(\frac{h^3 x^3}{12 l^3}\right)\left(\frac{2}{l^2}\right)^2 dx}{\int_0^l \rho\left(\frac{hx}{l}\right)\left(1 - \frac{x}{l}\right)^4 dx} = 2.5\frac{Eh^2}{\rho l^4}$$

or

$$\omega = 1.5811\left(\frac{Eh^2}{\rho l^4}\right)^{1/2} \tag{E.2}$$

The exact value of the frequency, for this case [8.2], is known to be

$$\omega_1 = 1.5343\left(\frac{Eh^2}{\rho l^4}\right)^{1/2} \tag{E.3}$$

Thus the value of ω_1 given by Rayleigh's method can be seen to be 3.0503 percent higher than the exact value.

8.8 THE RAYLEIGH-RITZ METHOD

The Rayleigh-Ritz method can be considered an extension of Rayleigh's method. It is based on the premise that a closer approximation to the exact natural modes can be obtained by superposing a number of assumed functions than by using a single assumed function, as in Rayleigh's method. If the assumed functions are suitably chosen, this method provides not only the approximate value of the fundamental frequency but also the approximate values of the higher natural frequencies and the mode shapes. An arbitrary number of functions can be used, and the number of frequencies that can be obtained is equal to the number of functions used. A large number of functions, although it involves more computational work, leads to more accurate results.

In the case of transverse vibration of beams, if n functions are chosen for approximating the deflection $W(x)$, we can write

$$W(x) = c_1 w_1(x) + c_2 w_2(x) + \cdots + c_n w_n(x) \tag{8.145}$$

where $w_1(x), w_2(x), \ldots, w_n(x)$ are known linearly independent functions of the spatial coordinate x, which satisfy all the boundary conditions of the problem, and $c_1, c_2, \ldots, c_n$ are coefficients to be found. The coefficients c_i are to be determined so that the assumed functions $w_i(x)$ provide the best possible approximation to the natural modes. To obtain such approximations, the coefficients c_i are adjusted and the natural frequency is made stationary at the natural modes. For this we substitute Eq. (8.145) in Rayleigh's quotient, Eq. (8.143), and the resulting expression is partially differentiated with respect to each of the coefficients c_i. To make the natural frequency stationary, we set each of the partial derivatives equal to zero and obtain

$$\frac{\partial(\omega^2)}{\partial c_i} = 0, \qquad i = 1, 2, \ldots, n \tag{8.146}$$

Equation (8.146) denotes a set of n linear algebraic equations in the coefficients $c_1, c_2, \ldots, c_n$ and also contains the undetermined quantity ω^2. This defines an algebraic eigenvalue problem similar to the ones that arise in multidegree of freedom systems. The solution of this eigenvalue problem generally gives n natural frequencies ω_i^2, $i = 1, 2, \ldots, n$ and n eigenvectors, each containing a set of numbers for $c_1, c_2, \ldots, c_n$. For example, the ith eigenvector corresponding to ω_i may be expressed as

$$\vec{C}^{(i)} = \begin{Bmatrix} c_1^{(i)} \\ c_2^{(i)} \\ \vdots \\ c_n^{(i)} \end{Bmatrix} \tag{8.147}$$

When this eigenvector—the values of $c_1^{(i)}, c_2^{(i)}, \ldots, c_n^{(i)}$—is substituted into Eq. (8.145), we obtain the best possible approximation to the ith mode of the beam. A method of reducing the size of the eigenproblem in the Rayleigh-Ritz method is presented in Ref. [8.28]. A new approach, which combines the advantages of the Rayleigh-Ritz analysis and the finite element method is given in Ref. [8.29]. The basic Rayleigh-Ritz procedure is illustrated with the help of the following example.

EXAMPLE 8.12   First Two Frequencies of a Tapered Beam

Find the natural frequencies of the tapered cantilever beam of Example 8.11 by using the Rayleigh-Ritz method.

Given: Uniformly tapered cantilever beam (Fig. 8.22).

Find: First two natural frequencies.

Approach: Use Rayleigh-Ritz method with a two-term solution.

Solution. We assume the deflection functions $w_i(x)$ to be

$$w_1(x) = \left(1 - \frac{x}{l}\right)^2 \tag{E.1}$$

$$w_2(x) = \frac{x}{l}\left(1 - \frac{x}{l}\right)^2 \tag{E.2}$$

$$w_3(x) = \frac{x^2}{l^2}\left(1 - \frac{x}{l}\right)^2 \tag{E.3}$$

$$\vdots$$

If we use a one-term approximation,

$$W(x) = c_1\left(1 - \frac{x}{l}\right)^2 \tag{E.4}$$

the fundamental frequency will be the same as the one found in Example 8.11. Now we use a two-term approximation,

$$W(x) = c_1\left(1 - \frac{x}{l}\right)^2 + c_2\frac{x}{l}\left(1 - \frac{x}{l}\right)^2 \tag{E.5}$$

Rayleigh's quotient is given by

$$R[W(x)] = \omega^2 = \frac{X}{Y} \tag{E.6}$$

where

$$X = \int_0^l EI(x)\left(\frac{d^2W(x)}{dx^2}\right)^2 dx \tag{E.7}$$

and

$$Y = \int_0^l \rho A(x)[W(x)]^2\, dx \tag{E.8}$$

If Eq. (E.5) is substituted, Eq. (E.6) becomes a function of c_1 and c_2. The conditions that make ω^2 or $R[W(x)]$ stationary are

$$\frac{\partial(\omega^2)}{\partial c_1} = \frac{Y\frac{\partial X}{\partial c_1} - X\frac{\partial Y}{\partial c_1}}{Y^2} = 0 \tag{E.9}$$

$$\frac{\partial(\omega^2)}{\partial c_2} = \frac{Y\frac{\partial X}{\partial c_2} - X\frac{\partial Y}{\partial c_2}}{Y^2} = 0 \tag{E.10}$$

These equations can be rewritten as

$$\frac{\partial X}{\partial c_1} - \frac{X}{Y}\frac{\partial Y}{\partial c_1} = \frac{\partial X}{\partial c_1} - \omega^2\frac{\partial Y}{\partial c_1} = 0 \tag{E.11}$$

$$\frac{\partial X}{\partial c_2} - \frac{X}{Y}\frac{\partial Y}{\partial c_2} = \frac{\partial X}{\partial c_2} - \omega^2\frac{\partial Y}{\partial c_2} = 0 \tag{E.12}$$

By substituting Eq. (E.5) into Eqs. (E.7) and (E.8), we obtain

$$X = \frac{Eh^3}{3l^3}\left(\frac{c_1^2}{4} + \frac{c_2^2}{10} + \frac{c_1 c_2}{5}\right) \tag{E.13}$$

$$Y = \rho hl\left(\frac{c_1^2}{30} + \frac{c_2^2}{280} + \frac{2c_1 c_2}{105}\right) \tag{E.14}$$

With the help of Eqs. (E.13) and (E.14), Eqs. (E.11) and (E.12) can be expressed as

$$\begin{bmatrix} \left(\frac{1}{2} - \omega^2 \cdot \frac{1}{15}\right) & \left(\frac{1}{5} - \omega^2 \cdot \frac{2}{105}\right) \\ \left(\frac{1}{5} - \omega^2 \cdot \frac{2}{105}\right) & \left(\frac{1}{5} - \omega^2 \cdot \frac{1}{140}\right) \end{bmatrix} \begin{Bmatrix} c_1 \\ c_2 \end{Bmatrix} = \begin{Bmatrix} 0 \\ 0 \end{Bmatrix} \tag{E.15}$$

where

$$\underset{\sim}{\omega}^2 = \frac{3\omega^2 \rho l^4}{Eh^2} \qquad (E.16)$$

By setting the determinant of the matrix in Eq. (E.15) equal to zero, we obtain the frequency equation:

$$\frac{1}{8820} \underset{\sim}{\omega}^4 - \frac{13}{1400} \underset{\sim}{\omega}^2 + \frac{3}{50} = 0 \qquad (E.17)$$

The roots of Eq. (E.17) are given by $\underset{\sim}{\omega}_1 = 2.6599$ and $\underset{\sim}{\omega}_2 = 8.6492$. Thus the natural frequencies of the tapered beam are

$$\omega_1 \simeq 1.5367 \left(\frac{Eh^2}{\rho l^4} \right)^{1/2} \qquad (E.18)$$

and

$$\omega_2 \simeq 4.9936 \left(\frac{Eh^2}{\rho l^4} \right)^{1/2} \qquad (E.19)$$

8.9 COMPUTER PROGRAM

A Fortran computer program, in the form of subroutine NONEQN, is given for finding the roots of a nonlinear equation of the form $F(Y) = 0$. The following arguments are used in this subroutine:

N   =   Number of roots to be determined. Input data.

X   =   Array of dimension N. Contains the computed values of the roots. Output.

XS   =   Initial guess for the first root. Input data.

XINC =   Initial increment to be used in searching for the root. Input data.

NINT =   Maximum number of subintervals to be used. A value on the order of 50 is to be used. Input data.

ITER =   Maximum number of iterations permitted in finding a root. A value on the order of 100 is to be used. Input data.

EPS   =   Convergence requirement. A small value on the order of 10^{-6} is to be used. Input data.

An external function routine, "Function $F(Y)$" is to be provided by the user. In it, the nonlinear function F must be defined in terms of the independent parameter Y.

For illustration, the roots of the frequency equation of a fixed-pinned beam are determined using subroutine NONEQN. Since the frequency equation (see Fig. 8.15) is given by

$$\tan \beta_n l - \tanh \beta_n l = 0$$

the external function is defined as

$$F = TAN(Y) - TANH(Y)$$

Five roots are found (N = 5) by using an initial guess value of XS = 2.0 with an increment of XINC = 0.1. The main program, subroutine NONEQN, and the output of the program are given below.

```
C==============================================================================
C
C PROGRAM 17
C MAIN PROGRAM FOR CALLING THE SUBROUTINE NONEQN
C
C==============================================================================
C FOLLOWING 3 LINES CONTAIN PROBLEM-DEPENDENT DATA
      DIMENSION X(5)
      DATA N,XS,XINC,NINT,ITER,EPS
     2   /5,2.0,0.1,50,100,0.000001/
C END OF PROBLEM-DEPENDENT DATA
      PRINT 10,N,XS,XINC,NINT,ITER,EPS
  10  FORMAT (//,28H ROOTS OF NONLINEAR EQUATION,//,6H DATA:,/,
     2    2H N,4X,2H =,I4,/,3H XS,3X,2H =,E15.6,/,8H XINC  =,E15.6,/,
     3    8H NINT  =,I4,/,8H ITER  =,I4,/,8H EPS   =,E15.6)
      CALL NONEQN (N,X,XS,XINC,NINT,ITER,EPS)
      PRINT 20,(X(I),I=1,N)
  20  FORMAT (//,7H ROOTS:,/,(E15.6))
      STOP
      END


C==========================================================================
C
C SUBROUTINE NONEQN
C
C =========================================================================
      SUBROUTINE NONEQN (N,X,XS,XINC,NINT,ITER,EPS)
      DATA NP,II,CC,DELX /1,0,0.0,0.0/
      DIMENSION X(N)
      A=F(XS)
      IF (A) 13,11,13
  11  IF (XS) 13,12,13
  12  II=II+1
      X(II)=0.0
      XS=XS+XINC
      A=F(XS)
  13  AA=A
      A=F(XS)
      IF (CC .LT. XS) CC=XS
      IF (NP .GT. NINT) GO TO 18
      XS=XS+XINC
      NP=NP+1
      IF (A/AA) 14,16,13
  14  XS=XS-XINC
      XINCN=XINC/5.0
      I=1
```

```
15        DELX=F(XS)*XINCN/(F(XS+XINCN)-F(XS))
          XS=XS-DELX
          ITER1=I
          IF (ABS(DELX) .LE. EPS) GO TO 16
          IF (CC .LT. XS) CC=XS
          I=I+1
          IF (I-ITER) 15,15,16
16        II=II+1
          X(II)=XS
          XS=CC+XINC
          IF (II-N) 17,18,18
17        A=F(XS)
          XS=XS+XINC
          IF (A .EQ. 0.0) A=F(XS)
          GO TO 13
18        CONTINUE
          RETURN
          END
C======================================================================
C
C FUNCTION F(Y)
C THIS FUNCTION IS PROBLEM-DEPENDENT
C
C ======================================================================
          FUNCTION F(Y)
          F = TAN(Y)-TANH(Y)
          RETURN
          END
```

ROOTS OF NONLINEAR EQUATION

DATA:
```
N     =    5
XS    =    0.200000E+01
XINC  =    0.100000E+00
NINT  =   50
ITER  =  100
EPS   =    0.100000E-05
```

ROOTS:
```
   0.392660E+01
   0.706858E+01
   0.102102E+02
   0.133518E+02
   0.164934E+02
```

REFERENCES

8.1. S. K. Clark, *Dynamics of Continuous Elements*, Prentice-Hall, Englewood Cliffs, N.J., 1972.

8.2. S. Timoshenko, D. H. Young, and W. Weaver, Jr., *Vibration Problems in Engineering* (4th Ed.), John Wiley, New York, 1974.

8.3. A. Leissa, *Vibration of Plates*, NASA SP-160, Washington, D.C., 1969.

8.4. I. S. Habib, *Engineering Analysis Methods*, Lexington Books, Lexington, Mass., 1975.

8.5. J. D. Achenbach, *Wave Propagation in Elastic Solids*, North-Holland Publishing Company, Amsterdam, 1973.

8.6. K. K. Deb, "Dynamics of a string and an elastic hammer," *Journal of Sound and Vibration*, Vol. 40, 1975, pp. 243–248.

8.7. M. S. Triantafyllou, "Linear dynamics of cables and chains," *Shock and Vibration Digest*, Vol. 16, March 1984, pp. 9–17.

8.8. R. W. Fitzgerald, *Mechanics of Materials* (2nd Ed.), Addison-Wesley, Reading, Mass., 1982.

8.9. S. P. Timoshenko and J. Gere, *Theory of Elastic Stability* (2nd Ed.), McGraw-Hill, New York, 1961.

8.10. S. P. Timoshenko, "On the correction for shear of the differential equation for transverse vibration of prismatic bars," *Philosophical Magazine*, Series 6, Vol. 41, 1921, pp. 744–746.

8.11. G. R. Cowper, "The shear coefficient in Timoshenko's beam theory," *Journal of Applied Mechanics*, Vol. 33, 1966, pp. 335–340.

8.12. G. W. Housner and W. O. Keightley, "Vibrations of linearly tapered beams," Part I, *Transactions of ASCE*, Vol. 128, 1963, pp. 1020–1048.

8.13. C. M. Harris (Ed.), *Shock and Vibration Handbook* (3rd ed.), McGraw-Hill, New York, 1988.

8.14. J. H. Gaines and E. Volterra, "Transverse vibrations of cantilever bars of variable cross section," *Journal of the Acoustical Society of America*, Vol. 39, 1966, pp. 674–679.

8.15. T. M. Wang, "Natural frequencies of continuous Timoshenko beams," *Journal of Sound and Vibration*, Vol. 13, 1970, pp. 409–414.

8.16. S. L. Grassie, R. W. Gregory, D. Harrison, and K. L. Johnson, "The dynamic response of railway track to high frequency vertical excitation," *Journal of Mechanical Engineering Science*, Vol. 24, June 1982, pp. 77–90.

8.17. A. Dimarogonas, *Vibration Engineering*, West Publishing Co., St. Paul, 1976.

8.18. T. Justine and A. Krishnan, "Effect of support flexibility on fundamental frequency of beams," *Journal of Sound and Vibration*, Vol. 68, 1980, pp. 310–312.

8.19. K. A. R. Perkins, "The effect of support flexibility on the natural frequencies of a uniform cantilever," *Journal of Sound and Vibration*, Vol. 4, 1966, pp. 1–8.

8.20. S. S. Rao, "Natural frequencies of systems of elastically connected Timoshenko beams," *Journal of the Acoustical Society of America*, Vol. 55, 1974, pp. 1232–1237.

8.21. A. M. Ebner and D. P. Billington, "Steady state vibration of damped Timoshenko beams," *Journal of Structural Division* (ASCE), Vol. 3, 1968, p. 737.

8.22. M. Levinson and D. W. Cooke, "On the frequency spectra of Timoshenko beams," *Journal of Sound and Vibration*, Vol. 84, 1982, pp. 319–326.

8.23. N. Y. Olcer, "General solution to the equation of the vibrating membrane," *Journal of Sound and Vibration*, Vol. 6, 1967, pp. 365–374.

8.24. G. R. Sharp, "Finite transform solution of the vibrating annular membrane," *Journal of Sound and Vibration*, Vol. 6, 1967, pp. 117–128.

8.25. J. Mazumdar, "A review of approximate methods for determining the vibrational modes of membranes," *Shock and Vibration Digest*, Vol. 14, February 1982, pp. 11–17.

8.26. J. E. Penny and J. R. Reed, "An integral equation approach to the fundamental frequency of vibrating beams," *Journal of Sound and Vibration*, Vol. 19, 1971, pp. 393–400.

8.27. G. Temple and W. G. Bickley, *Rayleigh's Principle and Its Application to Engineering*, Dover, New York, 1956.

8.28. W. L. Craver, Jr. and D. M. Egle, "A method for selection of significant terms in the assumed solution in a Rayleigh-Ritz analysis," *Journal of Sound and Vibration*, Vol. 22, 1972, pp. 133–142.

8.29. L. Klein, "Transverse vibrations of non-uniform beams," *Journal of Sound and Vibration*, Vol. 37, 1974, pp. 491–505.

8.30. J. R. Hutchinson, "Transverse vibrations of beams: Exact versus approximate solutions," *Journal of Applied Mechanics*, Vol. 48, 1981, pp. 923–928.

REVIEW QUESTIONS

8.1. How does a continuous system differ from a discrete system in the nature of its equation of motion?

8.2. How many natural frequencies does a continuous system have?

8.3. Are the boundary conditions important in a discrete system? Why?

8.4. What is a wave equation? What is a traveling-wave solution?

8.5. What is the significance of wave velocity?

8.6. State the boundary conditions to be specified at a simply supported end of a beam if (a) thin beam theory is used and (b) Timoshenko beam theory is used.

8.7. State the possible boundary conditions at the ends of a string.

8.8. What is the main difference in the nature of the frequency equations of a discrete system and a continuous system?

8.9. What is the effect of a tensile force on the natural frequencies of a beam?

8.10. Under what circumstances does the frequency of vibration of a beam subjected to an axial load become zero?

8.11. Why does the natural frequency of a beam become lower if the effects of shear deformation and rotary inertia are considered?

8.12. Give two practical examples of the vibration of membranes.

8.13. What is the basic principle used in Rayleigh's method?

8.14. Why is the natural frequency given by Rayleigh's method always larger than the true value of ω_1?

8.15. What is the difference between Rayleigh's method and the Rayleigh-Ritz method?

8.16. What is Rayleigh's quotient?

PROBLEMS

The problem assignments are organized as follows:

Problems	Section covered	Topic covered
8.1–8.13	8.2	Transverse vibration of strings
8.14–8.18	8.3	Longitudinal vibration of bars
8.19–8.26	8.4	Torsional vibration of shafts
8.27–8.42, 8.50	8.5	Beams
8.43–8.49	8.6	Membranes
8.51–8.59	8.7	Rayleigh's method
8.60–8.64	8.8	The Rayleigh-Ritz method
8.65–8.67	8.9	Computer program
8.68	—	Project

8.1. Determine the velocity of wave propagation in a cable of mass $\rho = 5$ kg/m when stretched by a tension $P = 4000$ N.

8.2. A steel wire of 2 mm diameter is fixed between two points located 2 m apart. The tensile force in the wire is 250 N. Determine (a) the fundamental frequency of vibration and (b) the velocity of wave propagation in the wire.

8.3. A stretched cable of length 2 m has a fundamental frequency of 3000 Hz. Find the frequency of the third mode. How are the fundamental and third mode frequencies changed if the tension is increased by 20%?

8.4. Find the time it takes for a transverse wave to travel along a transmission line from one tower to another one 300 m away. Assume the horizontal component of the cable tension as 30,000 N and the mass of the cable as 2 kg/m of length.

8.5. A cable of length l and mass ρ per unit length is stretched under a tension P. One end of the cable is connected to a mass m, which can move in a frictionless slot and the other end is fastened to a spring of stiffness k as shown in Fig. 8.23. Derive the frequency equation for the transverse vibration of the cable.

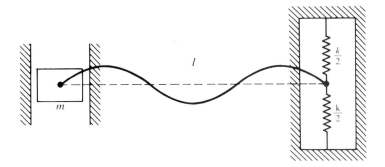

Figure 8.23

8.6. The cord of a musical instrument is fixed at both ends and has a length 2 m, diameter 0.5 mm, and density 7800 kg/m³. Find the tension required in order to have a fundamental frequency of (a) 1 Hz and (b) 5 Hz.

8.7. A cable of length l and mass ρ per unit length is stretched under a tension P. One end of the cable is fixed and the other end is connected to a pin, which can move in a frictionless slot. Find the natural frequencies of vibration of the cable.

8.8. Find the free vibration solution of a cord fixed at both ends when its initial conditions are given by

$$w(x,0) = 0, \qquad \frac{\partial w}{\partial t}(x,0) = \frac{2ax}{l} \qquad \text{for} \ \ 0 \leqslant x \leqslant \frac{l}{2}$$

and

$$\frac{\partial w}{\partial t}(x,0) = 2a\left(1 - \frac{x}{l}\right) \qquad \text{for} \ \frac{l}{2} \leqslant x \leqslant l$$

8.9. Prove that the constant a in Eqs. (8.18) and (8.19) is negative for common boundary conditions. *Hint*: Multiply Eq. (8.18) by $W(x)$ and integrate with respect to x from 0 to l.

8.10.* The cable between two electric transmission towers has a length of 2000 m. It is clamped at its ends under a tension P (Fig. 8.24). The density of the cable material is 8890 kg/m³. If the first four natural frequencies are required to lie between 0 and 20 Hz, determine the necessary cross sectional area of the cable and the initial tension.

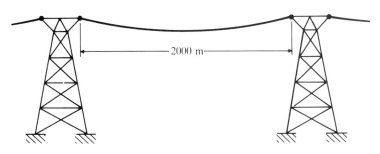

Figure 8.24

8.11. If a string of length l, fixed at both ends, is given an initial transverse displacement of h at $x = l/3$ and then released, determine its subsequent motion. Compare the deflection shapes of the string at times $t = 0$, $l/(4c)$, $l/(3c)$, $l/(2c)$, and l/c by considering the first four terms of the series solution.

8.12. A cord of length l is made to vibrate in a viscous medium. Derive the equation of motion considering the viscous damping force.

8.13. Determine the free vibration solution of a string fixed at both ends under the initial conditions $w(x,0) = w_0 \sin\dfrac{\pi x}{l}$ and $\dfrac{\partial w}{\partial t}(x,0) = 0$.

8.14. Derive an equation for the principal modes of longitudinal vibration of a uniform bar having both ends free.

8.15. Derive the frequency equation for the longitudinal vibration of the systems shown in Fig. 8.25.

8.16.* A thin bar of length l and mass m is clamped at one end and free at the other. What mass M must be attached to the free end in order to decrease the fundamental frequency of longitudinal vibration by 50% from its fixed-free value?

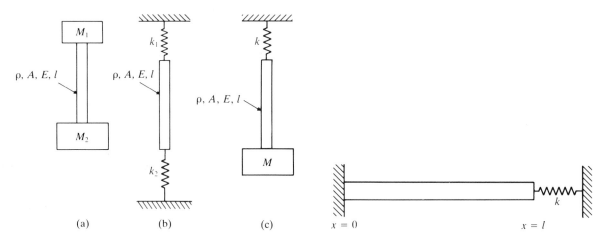

Figure 8.25   **Figure 8.26**

8.17. Show that the normal functions corresponding to the longitudinal vibration of the bar shown in Fig. 8.26 are orthogonal.

8.18. Derive the frequency equation for the longitudinal vibration of a stepped bar having two different cross-sectional areas A_1 and A_2 over lengths l_1 and l_2, respectively. Assume fixed-free end conditions.

8.19. A torsional system consists of a shaft with a disc of mass moment of inertia J_0 mounted at its center. If both ends of the shaft are fixed, find the response of the system in free torsional vibration of the shaft. Assume that the disc is given a zero initial angular displacement and an initial velocity of $\dot{\theta}_0$.

8.20. Find the natural frequencies for torsional vibration of a fixed-fixed shaft.

8.21. A uniform shaft of length l and torsional stiffness GJ is connected at both ends by torsional springs, torsional dampers, and discs with inertias as shown in Fig. 8.27. State the boundary conditions.

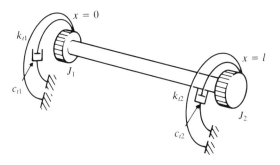

Figure 8.27

8.22. Solve Problem 8.20 if one end of the shaft is fixed and the other free.

8.23. Derive the frequency equation for the torsional vibration of a uniform shaft carrying rotors of mass moment of inertia J_1 and J_2, one at each end.

8.24. An external torque $M_t(t) = M_{t0}\cos \omega t$ is applied at the free end of a fixed-free uniform shaft. Find the steady-state vibration of the shaft.

8.25. Find the fundamental frequency for torsional vibration of a shaft of length 2 m and diameter 50 mm when both the ends are fixed. The density of the material is 7800 kg/m^3 and the modulus of rigidity is 0.8×10^{11} N/m^2.

8.26. A uniform shaft, supported at $x = 0$ and rotating at an angular velocity ω, is suddenly stopped at the end $x = 0$. If the end $x = l$ is free, determine the subsequent angular displacement response of the shaft.

8.27. Compute the first three natural frequencies and the corresponding mode shapes of the transverse vibrations of a uniform beam of rectangular cross-section (100 mm × 300 mm) with $l = 2$ m, $E = 20.5 \times 10^{10}$ N/m^2, and $\rho = 7.83 \times 10^3$ kg/m^3 for the following cases:

    a. When both ends are simply supported

    b. When both ends are built-in (clamped)

    c. When one end is fixed and the other end is free

    d. When both ends are free.

    Plot the mode shapes.

8.28. Derive an expression for the natural frequencies for the lateral vibration of a uniform fixed-free beam.

8.29. Prove that the normal functions of a uniform beam, whose ends are connected by springs as shown in Fig. 8.28, are orthogonal.

8.30. Derive an expression for the natural frequencies for the transverse vibration of a uniform beam with both ends simply supported.

8.31. Derive the expression for the natural frequencies for the lateral vibration of a uniform beam suspended as a pendulum, neglecting the effect of dead weight.

8.32. Find the cross sectional area (A), and the area moment of inertia (I) of a simply supported steel beam of length 1 m for which the first three natural frequencies lie in the range 1500 Hz–5000 Hz.

8.33. A uniform beam, simply supported at both ends, is found to vibrate in its first mode with an amplitude of 10 mm at its center. If $A = 120$ mm^2, $I = 1000$ mm^4, $E = 20.5 \times 10^{10}$ N/m^2, $\rho = 7.83 \times 10^3$ kg/m^3, and $l = 1$ m, determine the maximum bending moment in the beam.

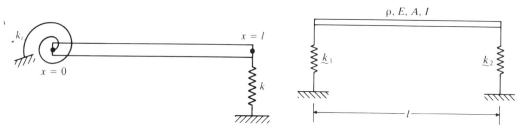

Figure 8.28   **Figure 8.29**

8.34. Derive the frequency equation for the transverse vibration of a uniform beam resting on springs at both ends, as shown in Fig. 8.29. The springs can deflect vertically only and the beam is horizontal in the equilibrium position.

8.35. A simply supported uniform beam of length l carries a mass M at the center of the beam. Assuming the mass M to be a point mass, obtain the frequency equation of the system.

8.36. A uniform fixed-fixed beam of length $2l$ is simply supported at the middle point. Derive the frequency equation for the transverse vibration of the beam.

8.37. A simply supported beam carries initially a uniformly distributed load of intensity f_0. Find the vibration response of the beam if the load is suddenly removed.

8.38. Estimate the fundamental frequency of a cantilever beam whose cross-sectional area and moment of inertia vary as

$$A(x) = A_0 \frac{x}{l} \quad \text{and} \quad I(x) = I_0 \frac{x}{l}$$

where x is measured from the free end.

8.39. (a) Derive a general expression for the response of a uniform beam subjected to an arbitrary force. (b) Use the result of part (a) to find the response of a uniform simply supported beam under the harmonic force $F_0 \sin \omega t$ applied at $x = a$. Assume the initial conditions as $w(x, 0) = (\partial w / \partial t)(x, 0) = 0$.

8.40. Derive Eqs. (E.5) and (E.6) of Example 8.9.

8.41. Derive Eqs. (E.7) and (E.8) of Example 8.9.

8.42. Prove that the constant a in Eq. (8.82) is positive for common boundary conditions. *Hint*: Multiply Eq. (8.83) by $W(x)$ and integrate with respect to x from 0 to l.

8.43. Starting from fundamentals, show that the equation for the lateral vibration of a circular membrane is given by

$$\frac{\partial^2 w}{\partial r^2} + \frac{1}{r}\frac{\partial w}{\partial r} + \frac{1}{r^2}\frac{\partial^2 w}{\partial \theta^2} = \frac{\rho}{P}\frac{\partial^2 w}{\partial t^2}$$

8.44. Using the equation of motion given in Problem 8.43, find the natural frequencies of a circular membrane of radius R clamped around the boundary at $r = R$.

8.45. Consider a rectangular membrane of sides a and b supported along all the edges. (a) Derive an expression for the deflection $w(x, y, t)$ under an arbitrary pressure $f(x, y, t)$. (b) Find the response when a uniformly distributed pressure f_0 is applied to a membrane that is initially at rest.

8.46. Find the free vibration solution and the natural frequencies of a rectangular membrane that is clamped along all the sides. The membrane has dimensions a and b along the x and y directions, respectively.

8.47. Find the free vibration response of a rectangular membrane of sides a and b subject to the following initial conditions:

$$w(x, y, 0) = w_0 \sin\frac{\pi x}{a}\sin\frac{\pi y}{b}, \qquad 0 \leqslant x \leqslant a, \qquad 0 \leqslant y \leqslant b$$

$$\frac{\partial w}{\partial t}(x, y, 0) = 0, \qquad 0 \leqslant x \leqslant a, \qquad 0 \leqslant y \leqslant b$$

8.48. Find the free vibration response of a rectangular membrane of sides a and b subjected to the following initial conditions:

$$\left.\begin{aligned} w(x, y, 0) &= 0 \\ \frac{\partial w}{\partial t}(x, y, 0) &= \dot{w}_0 \sin\frac{\pi x}{a}\sin\frac{2\pi y}{b} \end{aligned}\right\}, \qquad \begin{aligned} 0 \leqslant x \leqslant a \\ 0 \leqslant y \leqslant b \end{aligned}$$

Assume that the edges of the membrane are fixed.

8.49. Compare the fundamental natural frequencies of transverse vibration of membranes of the following shapes:

(a) Square

(b) Circular

(c) Rectangular with sides in the ratio of $2:1$

Assume that all the membranes are clamped around their edges and have the same area, material, and tension.

8.50. Consider a railway car moving on a railroad track as shown in Fig. 8.30(a). The track can be modeled as an infinite beam resting on an elastic foundation and the car can be idealized as a moving load $F_0(x, t)$ (see Fig. 8.30b). If the soil stiffness per unit length is k, and the constant velocity of the car is v_0, show that the equation of motion of the beam can be expressed as

$$EI\frac{\partial^4 w(x, t)}{\partial x^4} + \rho A\frac{\partial^2 w(x, t)}{\partial t^2} + kw(x, t) = F_0(x - v_0 t)$$

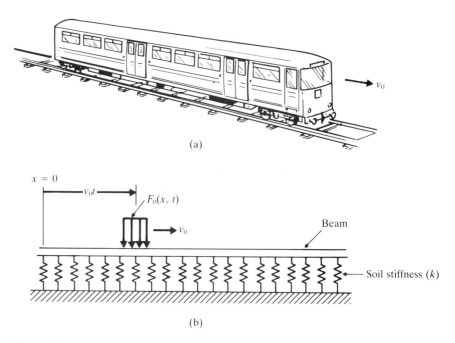

(a)

(b)

Figure 8.30

Indicate a method of solving the equation of motion if the moving load is assumed to be constant in magnitude.

8.51. Find the fundamental natural frequency of a fixed-fixed beam using the static deflection curve

$$W(x) = \frac{c_0 x^2}{24EI}(l - x)^2$$

where c_0 is a constant.

8.52. Solve Problem 8.51 using the deflection shape $W(x) = c_0\left(1 - \cos\frac{2\pi x}{l}\right)$, where c_0 is a constant.

8.53. Find the fundamental natural frequency of vibration of a uniform beam of length l that is fixed at one end and simply supported at the other end. Assume the deflection shape of the beam to be same as the static deflection curve under its self weight. *Hint:* The static deflection of a uniform beam under self weight is governed by

$$EI\frac{d^4 W(x)}{dx^4} = \rho g A$$

where ρ is the density, g is the acceleration due to gravity, and A is the area of cross section of the beam. This equation can be integrated for any known boundary conditions of the beam.

8.54. Determine the fundamental frequency of a uniform fixed-fixed beam carrying a mass M at the middle by applying Rayleigh's method. Use the static deflection curve for $W(x)$.

8.55. Applying Rayleigh's method, determine the fundamental frequency of a cantilever beam (fixed at $x = l$) whose cross-sectional area $A(x)$ and moment of inertia $I(x)$ vary as $A(x) = A_0 x/l$ and $I(x) = I_0 x/l$.

8.56. Using Rayleigh's method, find the fundamental frequency for the lateral vibration of the beam shown in Fig. 8.31. The restoring force in the spring k is proportional to the deflection, and the restoring moment in the spring k_t is proportional to the angular deflection.

8.57. Using Rayleigh's method, estimate the fundamental frequency for the lateral vibration of a uniform beam fixed at both the ends. Assume the deflection curve to be

$$W(x) = c_1\left(1 - \cos\frac{2\pi x}{l}\right)$$

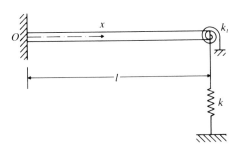

Figure 8.31

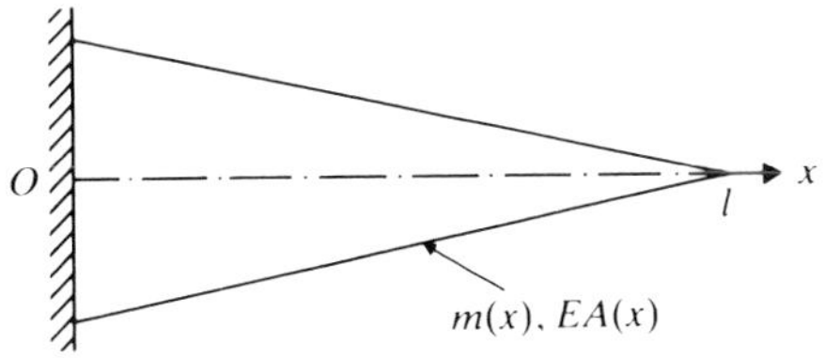

Figure 8.32

8.58. Find the fundamental frequency of longitudinal vibration of the tapered bar shown in Fig. 8.32, using Rayleigh's method with the mode shape

$$U(x) = c_1 \sin\frac{\pi x}{2l}$$

The mass per unit length is given by

$$m(x) = 2m_0\left(1 - \frac{x}{l}\right)$$

and the stiffness by

$$EA(x) = 2EA_0\left(1 - \frac{x}{l}\right)$$

8.59. Approximate the fundamental frequency of a rectangular membrane supported along all the edges by using Rayleigh's method with

$$W(x, y) = c_1 xy(x - a)(y - b).$$

Hint:

$$V = \frac{P}{2}\iint\left[\left(\frac{\partial w}{\partial x}\right)^2 + \left(\frac{\partial w}{\partial y}\right)^2\right]dx\,dy \quad \text{and} \quad T = \frac{\rho}{2}\iint\left(\frac{\partial w}{\partial t}\right)^2 dx\,dy$$

8.60. Estimate the fundamental frequency of a fixed-fixed string, assuming the mode shape

(a) $W(x) = c_1 x(l - x)$

(b) $W(x) = c_1 x(l - x) + c_2 x^2(l - x)^2$

8.61. Estimate the fundamental frequency for the longitudinal vibration of a uniform bar fixed at $x = 0$ and free at $x = l$ by assuming the mode shapes as (a) $U(x) = c_1(x/l)$ and (b) $U(x) = c_1(x/l) + c_2(x/l)^2$.

8.62. A stepped bar, fixed at $x = 0$ and free at $x = l$, has a cross-sectional area of $2A$ for $0 \leqslant x < l/3$ and A for $l/3 \leqslant x \leqslant l$. Assuming the mode shape

$$U(x) = c_1 \sin\frac{\pi x}{2l} + c_2 \sin\frac{3\pi x}{2l}$$

estimate the first two natural frequencies of longitudinal vibration.

8.63. Solve Problem 8.58 using the Rayleigh-Ritz method with the mode shape

$$U(x) = c_1 \sin\frac{\pi x}{2l} + c_2 \sin\frac{3\pi x}{2l}$$

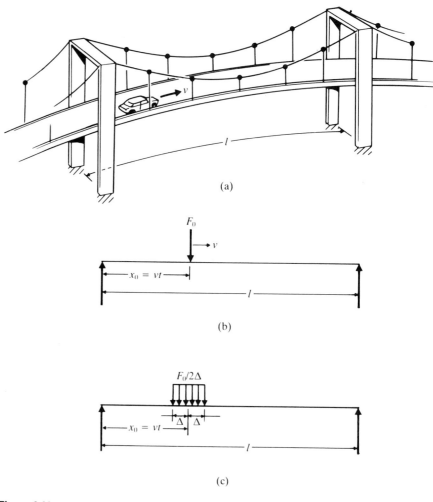

(a)

(b)

(c)

Figure 8.33

8.64. Find the first two natural frequencies of a fixed-fixed uniform string of mass density ρ per unit length stretched between $x = 0$ and $x = l$ with an initial tension P. Assume the deflection functions

$$w_1(x) = x(l - x)$$
$$w_2(x) = x^2(l - x)^2$$

8.65. Find the values of α_1 and α_2 for $\beta = 0.01$ and 1.0 in Example 8.4 using the subroutine NONEQN.

8.66. Find the first five natural frequencies of a thin fixed-fixed beam using subroutine NONEQN.

8.67. Write a computer program for finding numerically the mode shapes of thin fixed-simply supported beams by using the known values of the natural frequencies.

Project:

8.68. A vehicle, of weight F_0, moving at a constant speed on a bridge (Fig. 8.33a) can be modeled as a concentrated load travelling on a simply supported beam as shown in Fig. 8.33(b). The concentrated load F_0 can be considered as a uniformly distributed load over an infinitesimal length 2Δ and can be expressed as a sum of sine terms using Fourier sine series expansion (of the distributed load). Find the transverse displacement of the bridge as a sum of the responses due to each of the moving harmonic load components. Assume the initial conditions of the bridge as $w(x, 0) = \partial w / \partial t (x, 0) = 0$.

CHAPTER 9

Vibration Control

Leonhard Euler (1707–1783) was a Swiss mathematician who became a court mathematician and later a professor of mathematics in St. Petersburg, Russia. He produced many works in algebra and geometry and was interested in the geometrical form of deflection curves in strength of materials. Euler's column buckling load is quite familiar to mechanical and civil engineers, and Euler's constant and Euler's coordinate system are well known to mathematicians. He derived the equation of motion for the bending vibrations of a rod (Euler-Bernoulli theory) and presented a series form of solution, as well as studying the dynamics of a vibrating ring. (By permission of Brown Brothers, Sterling, PA 18463.)

9.1 INTRODUCTION

There are numerous sources of vibration in an industrial environment: impact processes such as pile driving and blasting; rotating or reciprocating machinery such as engines, compressors, and motors; transportation vehicles such as trucks, trains, and aircraft; the flow of fluids; and many others. The presence of vibration often leads to undesirable effects such as structural or mechanical failure, frequent and costly maintenance of machines, and human pain and discomfort. Vibration can sometimes be eliminated on the basis of theoretical analysis. However, the manufacturing costs involved in eliminating the vibration may be too high; a designer must compromise between an acceptable amount of vibration and a reasonable manufacturing cost. In some cases the excitation or shaking force is inherent in the machine. As seen earlier, even a relatively small excitation force can cause an undesirably large response near resonance, especially in lightly damped systems. In these cases, the magnitude of the response can be significantly reduced by the use of isolators and auxiliary mass absorbers [9.1, 9.2]. In this chapter, we shall consider various techniques of vibration control—that is, methods involving the elimination or reduction of vibration.

9.2 REDUCTION OF VIBRATION AT THE SOURCE

The first thing to be explored to control vibrations is to try to alter the source of vibration so that it produces less vibration. This method may not always be feasible. Some examples of the sources of vibration that cannot be altered are earthquake excitation, atmospheric turbulence, road roughness, and engine combustion instability. On the other hand, certain sources of vibration such as unbalance in rotating or reciprocating machines can be altered to reduce the vibrations. This can be achieved, usually, by using either internal balancing or an increase in the precision of machine elements. The use of close tolerances and better surface finish for machine parts (which have relative motion with respect to one another) make the machine less susceptible to vibration. Of course, there may be economic and manufacturing constraints on the degree of balancing that can be achieved or the precision with which the machine parts can be made. We shall consider the analysis of rotating and reciprocating machines in the presence of unbalance as well as the means of controlling the vibrations that result from unbalanced forces.

9.3 BALANCING OF ROTATING MACHINES

The presence of an eccentric or unbalanced mass in a rotating disc causes vibration, which may be acceptable up to a certain level. Figure 9.1 is a vibration-severity chart that can be used to determine acceptable vibration levels at various frequencies or operating speeds [9.3]. If the vibration caused by an unbalanced mass is not acceptable, it can be eliminated either by removing the eccentric mass or by adding an equal mass in such a position that it cancels the effect of the unbalance. In order to use this procedure, we need to determine the amount and location of the eccentric mass experimentally. The unbalance in practical machines can be attributed to such irregularities as machining errors and variations in sizes of bolts, nuts, rivets, and welds. In this section, we shall consider two types of balancing: *single-plane* or *static balancing* and *two-plane or dynamic balancing* [9.4–9.7].

**9.3.1
Single-Plane
Balancing**

Consider a machine element in the form of a thin circular disc such as a fan, flywheel, gear, and a grinding wheel mounted on a shaft. When the center of mass is displaced from the axis of rotation due to manufacturing errors, the machine element is said to be statically unbalanced. To determine whether a disc is balanced or not, mount the shaft on two low friction bearings as shown in Fig. 9.2(a). Rotate the disc and permit it to come to rest. Mark the lowest point on the circumference of the disc by a chalk. Repeat the process several times, each time marking the lowest point on the disc by chalk. If the disc is balanced, the chalk marks will be scattered randomly all over the circumference. On the other hand, if the disc is unbalanced, all the chalk marks coincide.

The unbalance detected by this procedure is known as *static unbalance*. The static unbalance can be corrected by removing (drilling) metal at the chalk mark or by adding a weight at 180° from the chalk mark. Since the magnitude of unbalance is not known, the amount of material to be removed or added must be determined

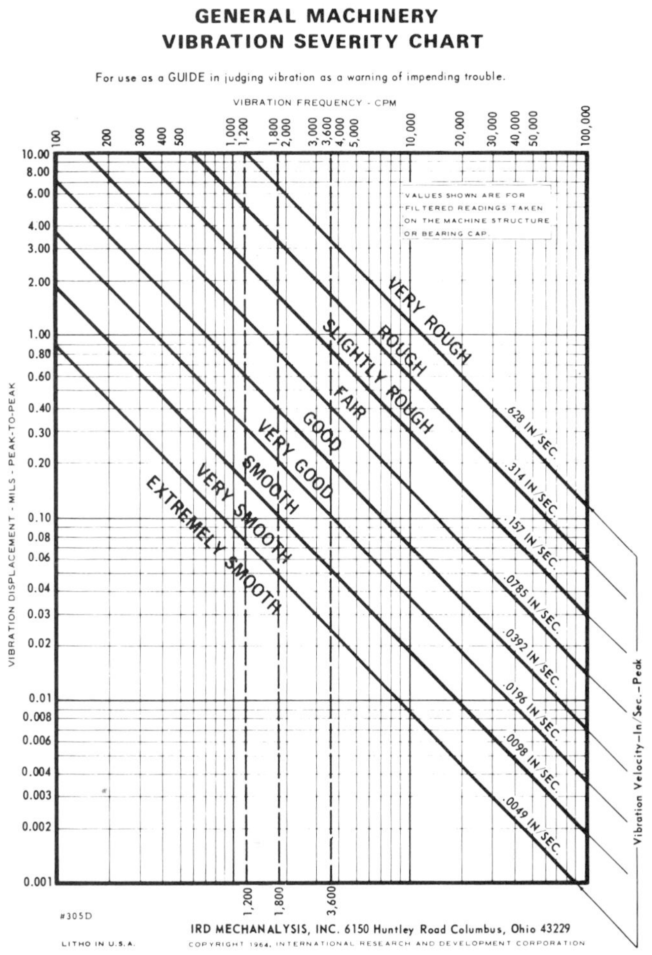

Figure 9.1 (Reprinted with permission of IRD Mechanalysis, Inc. Copyright 1964.)

by trail and error. This procedure is called "single-plane balancing" since all the mass lies practically in a single plane. The amount of unbalance can be found by rotating the disc at a known speed ω and measuring the reactions at the two bearings (see Fig. 9.2b). If an unbalanced mass m is located at a radius r of the disc, the centrifugal force will be $mr\omega^2$. Thus the measured bearing reactions F_1 and F_2 give m and r:

$$F_1 = \frac{a_2}{l}mr\omega^2, \qquad F_2 = \frac{a_1}{l}mr\omega^2 \tag{9.1}$$

Another procedure for single-plane balancing, using a vibration analyzer, is illustrated in Fig. 9.3. Here, a grinding wheel (disc) is attached to a rotating shaft that has bearing at A and is driven by an electric motor rotating at an angular velocity ω.

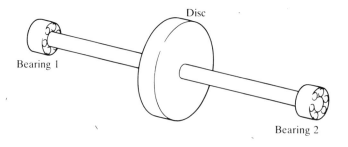

(a)

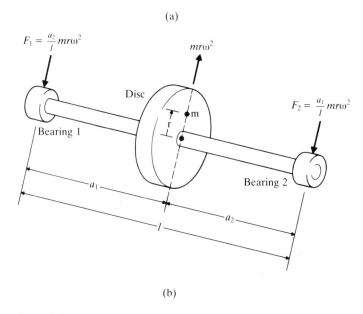

(b)

Figure 9.2

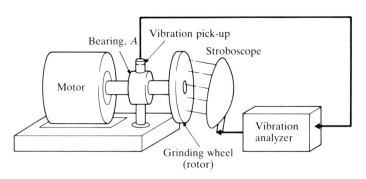

Figure 9.3

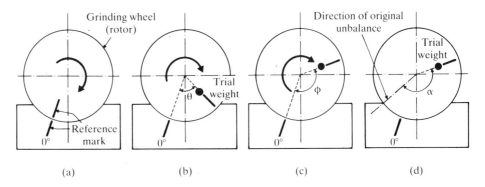

Figure 9.4

Before starting the procedure, *reference marks*, also known as *phase marks*, are placed both on the rotor (wheel) and the stator, as shown in Fig. 9.4(a). A vibration pickup is placed in contact with the bearing, as shown in Fig. 9.3, and the vibration analyzer is set to a frequency corresponding to the angular velocity of the grinding wheel. The vibration signal (the displacement amplitude) produced by the unbalance can be read from the indicating meter of the vibration analyzer. A stroboscopic light is fired by the vibration analyzer at the frequency of the rotating wheel. When the rotor rotates at speed ω, the phase mark on the rotor appears stationary under the stroboscopic light but is positioned at an angle θ from the mark on the stator, as shown in Fig. 9.4(b), due to phase lag in the response. Both the angle θ and the amplitude A_u (read from the vibration analyzer) caused by the original unbalance are noted. The rotor is then stopped, and a known trial weight W is attached to the rotor, as shown in Fig. 9.4(b). When the rotor runs at speed ω, the new angular position of the rotor phase mark ϕ and the vibration amplitude A_{u+w}, caused by the combined unbalance of rotor and trial weight, are noted (see Fig. 9.4c).*

Now we construct a vector diagram to find the magnitude and location of the correction mass for balancing the wheel. The original unbalance vector $\vec{A}_u$ is drawn in an arbitrary direction, with its length equal to A_u, as shown in Fig. 9.5. Then the combined unbalance vector is drawn as $\vec{A}_{u+w}$ at an angle $\phi - \theta$ from the direction of $\vec{A}_u$ with a length of A_{u+w}. The difference vector $\vec{A}_w = \vec{A}_{u+w} - \vec{A}_u$ in Fig. 9.5 then represents the unbalance vector due to the trial weight W. The magnitude of $\vec{A}_w$ can be computed using the law of cosines:

$$A_w = \left[A_u^2 + A_{u+w}^2 - 2 A_u A_{u+w} \cos\left(\phi - \theta \right) \right]^{1/2} \tag{9.2}$$

Since the magnitude of the trial weight W and its direction relative to the original unbalance (α in Fig. 9.5) are known, the original unbalance itself must be at an

---

* Note that if the trial weight is placed in a position that shifts the net unbalance in a clockwise direction, the stationary position of the phase mark will be shifted by exactly the same amount in the counterclockwise direction, and vice versa.

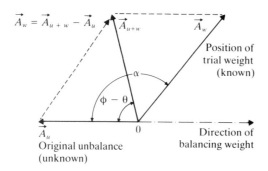

Figure 9.5

angle α away from the position of the trial weight, as shown in Fig. 9.4(d). The angle α can be obtained from the law of cosines:

$$\alpha = \cos^{-1}\left[\frac{A_u^2 + A_w^2 - A_{u+w}^2}{2A_u A_w}\right] \tag{9.3}$$

The magnitude of the original unbalance is $W_0 = (A_u/A_w) \cdot W$, located at the same radial distance from the rotation axis of the rotor as the weight W. Once the location and magnitude of the original unbalance are known, correction weight can be added to balance the wheel properly.

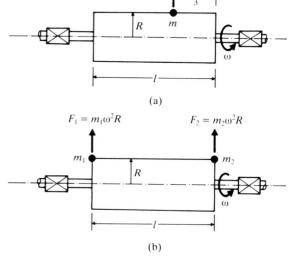

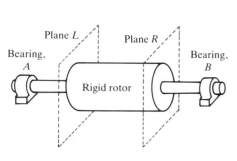

Figure 9.6

Figure 9.7

The single-plane balancing procedure can be used for balancing in one plane—that is, for rotors of the rigid disc type. If the rotor is an elongated rigid body as shown in Fig. 9.6, the unbalance can be anywhere along the length of the rotor. In this case, the rotor can be balanced by adding balancing weights in any two planes [9.7–9.9]. For convenience, the two planes are usually chosen as the end planes of the rotor (shown by dotted lines in Fig. 9.6).

To see that any unbalanced mass in the rotor can be replaced by two equivalent unbalanced masses (in any two planes), consider a rotor with an unbalanced mass m at a distance $l/3$ from the right end, as shown in Fig. 9.7(a). When the rotor rotates at a speed of ω, the force due to the unbalance will be $F = m\omega^2 R$, where R is the radius of the rotor. The unbalanced mass m can be replaced by two masses m_1 and m_2, located at the ends of the rotor, as shown in Fig. 9.7(b). The forces exerted on the rotor by these masses are $F_1 = m_1\omega^2 R$ and $F_2 = m_2\omega^2 R$. For the equivalence of force in Figs. 9.7(a) and (b), we have

$$m\omega^2 R = m_1\omega^2 R + m_2\omega^2 R \qquad \text{or} \qquad m = m_1 + m_2 \qquad (9.4)$$

For the equivalence of moments in the two cases, we consider moments about the right end so that

$$m\omega^2 R \frac{l}{3} = m_1\omega^2 R l \qquad \text{or} \qquad m = 3m_1 \qquad (9.5)$$

Equations (9.4) and (9.5) give $m_1 = m/3$ and $m_2 = 2m/3$. Thus any unbalanced mass can be replaced by two equivalent unbalanced masses in the end planes of the rotor.

We now consider the two-plane balancing procedure using a vibration analyzer. In Fig. 9.8, the total unbalance in the rotor is replaced by two unbalanced weights U_L and U_R in the left and the right planes, respectively. At the rotor's operating speed ω, the vibration amplitude and phase due to the original unbalance are measured at the two bearings A and B, and the results are recorded as vectors $\vec{V}_A$ and $\vec{V}_B$. The magnitude of the vibration vector is taken as the vibration amplitude, while the direction of the vector is taken as the negative of the phase angle observed under stroboscopic light with reference to the stator reference line. The measured vectors $\vec{V}_A$ and $\vec{V}_B$ can be expressed as

$$\vec{V}_A = \vec{A}_{AL}\vec{U}_L + \vec{A}_{AR}\vec{U}_R \qquad (9.6)$$

$$\vec{V}_B = \vec{A}_{BL}\vec{U}_L + \vec{A}_{BR}\vec{U}_R \qquad (9.7)$$

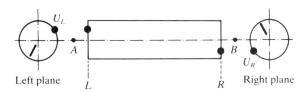

Left plane L $\qquad\qquad$ R Right plane

Figure 9.8

where $\vec{A}_{ij}$ can be considered as a vector, reflecting the effect of the unbalance in plane j $(j = L, R)$ on the vibration at bearing i $(i = A, B)$. Note that $\vec{U}_L, \vec{U}_R$, and all the vectors $\vec{A}_{ij}$ are unknown in Eqs. (9.6) and (9.7).

As in the case of single-plane balancing, we add known trial weights and take measurements to obtain information about the unbalanced masses. First we add a known weight $\vec{W}_L$ in the left plane at a known angular position and measure the displacement and phase of vibration at the two bearings while the rotor is rotating at speed ω. We denote these measured vibrations as vectors as

$$\vec{V}_A' = \vec{A}_{AL}\left(\vec{U}_L + \vec{W}_L\right) + \vec{A}_{AR}\vec{U}_R \tag{9.8}$$

$$\vec{V}_B' = \vec{A}_{BL}\left(\vec{U}_L + \vec{W}_L\right) + \vec{A}_{BR}\vec{U}_R \tag{9.9}$$

By subtracting Eqs. (9.6) and (9.7) from Eqs. (9.8) and (9.9), respectively, and solving, we obtain*

$$\vec{A}_{AL} = \frac{\vec{V}_A' - \vec{V}_A}{\vec{W}_L} \tag{9.10}$$

$$\vec{A}_{BL} = \frac{\vec{V}_B' - \vec{V}_B}{\vec{W}_L} \tag{9.11}$$

We then remove $\vec{W}_L$ and add a known weight $\vec{W}_R$ in the right plane at a known angular position and measure the resulting vibrations while the rotor is running at speed ω. The measured vibrations can be denoted as vectors:

$$\vec{V}_A'' = \vec{A}_{AR}\left(\vec{U}_R + \vec{W}_R\right) + \vec{A}_{AL}\vec{U}_L \tag{9.12}$$

$$\vec{V}_B'' = \vec{A}_{BR}\left(\vec{U}_R + \vec{W}_R\right) + \vec{A}_{BL}\vec{U}_L \tag{9.13}$$

As before, we subtract Eqs. (9.6) and (9.7) from Eqs. (9.12) and (9.13), respectively, to find

$$\vec{A}_{AR} = \frac{\vec{V}_A'' - \vec{V}_A}{\vec{W}_R} \tag{9.14}$$

$$\vec{A}_{BR} = \frac{\vec{V}_B'' - \vec{V}_B}{\vec{W}_R} \tag{9.15}$$

---

* It can be seen that complex subtraction, division, and multiplication are often used in the computation of the balancing weights. If

$$\vec{A} = a\underline{/\theta_A} \qquad \text{and} \qquad \vec{B} = b\underline{/\theta_B}$$

we can rewrite $\vec{A}$ and $\vec{B}$ as $\vec{A} = a_1 + ia_2$ and $\vec{B} = b_1 + ib_2$, where $a_1 = a\cos\theta_A$, $a_2 = a\sin\theta_A$, $b_1 = b\cos\theta_B$, and $b_2 = b\sin\theta_B$. Then the formulas for complex subtraction, division, and multiplication are [9.10]:

$$\vec{A} - \vec{B} = (a_1 - b_1) + i(a_2 - b_2)$$

$$\frac{\vec{A}}{\vec{B}} = \frac{(a_1 b_1 + a_2 b_2) + i(a_2 b_1 - a_1 b_2)}{(b_1^2 + b_2^2)}$$

$$\vec{A} \cdot \vec{B} = (a_1 b_1 - a_2 b_2) + i(a_2 b_1 + a_1 b_2)$$

Once the vector operators $\vec{A}_{ij}$ are known, Eqs. (9.6) and (9.7) can be solved to find the unbalance vectors $\vec{U}_L$ and $\vec{U}_R$:

$$\vec{U}_L = \frac{\vec{A}_{BR}\vec{V}_A - \vec{A}_{AR}\vec{V}_B}{\vec{A}_{BR}\vec{A}_{AL} - \vec{A}_{AR}\vec{A}_{BL}} \tag{9.16}$$

$$\vec{U}_R = \frac{\vec{A}_{BL}\vec{V}_A - \vec{A}_{AL}\vec{V}_B}{\vec{A}_{BL}\vec{A}_{AR} - \vec{A}_{AL}\vec{A}_{BR}} \tag{9.17}$$

The rotor can now be balanced by adding equal and opposite balancing weights in each plane. The balancing weights in the left and right planes can be denoted vectorially as $\vec{B}_L = -\vec{U}_L$ and $\vec{B}_R = -\vec{U}_R$. It can be seen that the two plane balancing procedure is a straightforward extension of the single-plane balancing procedure. Although high-speed rotors are balanced during manufacture, usually it becomes necessary to rebalance them in the field due to slight unbalances introduced due to creep, high temperature operation, and the like. Figure 9.9 shows a practical example of two-plane balancing.

Figure 9.9   (Courtesy of Bruel & Kjaer Instruments, Inc., Marlborough, Mass.)

EXAMPLE 9.1   Two-Plane Balancing of Turbine Rotor

In the two-plane balancing of a turbine rotor, the data obtained from measurement of the original unbalance, the right-plane trial weight, and the left-plane trial weight is shown below. The displacement amplitudes are in mils (1/1000 inch.) Determine the size and location of the balance weights required.

Condition	Vibration (displacement) amplitude		Phase angle	
	At bearing A	At bearing B	At bearing A	At bearing B
Original unbalance	8.5	6.5	60°	205°
$W_L = 10.0$ oz added at 270° from reference mark	6.0	4.5	125°	230°
$W_R = 12.0$ oz added at 180° from reference mark	6.0	10.5	35°	160°

Given: Data from balancing measurements as given in the table.

Find: Size and location of the necessary balance weights.

Approach: Apply the vector equations, Eqs. (9.16) and (9.17), to achieve balancing.

Solution. The given data can be expressed in vector notation as

$$\vec{V}_A = 8.5 \underline{/\,60°\,} = 4.2500 + i7.3612$$
$$\vec{V}_B = 6.5 \underline{/\,205°\,} = -5.8910 - i2.7470$$
$$\vec{V}'_A = 6.0 \underline{/\,125°\,} = -3.4415 + i4.9149$$
$$\vec{V}'_B = 4.5 \underline{/\,230°\,} = -2.8926 - i3.4472$$
$$\vec{V}''_A = 6.0 \underline{/\,35°\,} = 4.9149 + i3.4415$$
$$\vec{V}''_B = 10.5 \underline{/\,160°\,} = -9.8668 + i3.5912$$
$$\vec{W}_L = 10.0 \underline{/\,270°\,} = 0.0000 - i10.0000$$
$$\vec{W}_R = 12.0 \underline{/\,180°\,} = -12.0000 + i0.0000$$

Equations (9.10) and (9.11) give

$$\vec{A}_{AL} = \frac{\vec{V}'_A - \vec{V}_A}{\vec{W}_L} = \frac{-7.6915 - i2.4463}{0.0000 - i10.0000} = 0.2446 - i0.7691$$

$$\vec{A}_{BL} = \frac{\vec{V}'_B - \vec{V}_B}{\vec{W}_L} = \frac{2.9985 - i0.7002}{0.0000 - i10.0000} = 0.0700 + i0.2998$$

The use of Eqs. (9.14) and (9.15) leads to

$$\vec{A}_{AR} = \frac{\vec{V}''_A - \vec{V}_A}{\vec{W}_R} = \frac{0.6649 - i3.9198}{-12.0000 + i0.0000} = -0.0554 + i0.3266$$

$$\vec{A}_{BR} = \frac{\vec{V}''_B - \vec{V}_B}{\vec{W}_R} = \frac{-3.9758 + i6.3382}{-12.0000 + i0.0000} = 0.3313 - i0.5282$$

The unbalance weights can be determined from Eqs. (9.16) and (9.17):

$$\vec{U}_L = \frac{(5.2962 + i0.1941) - (1.2237 - i1.7721)}{(-0.3252 - i0.3840) - (-0.1018 + i0.0063)} = \frac{(4.0725 + i1.9661)}{(-0.2234 - i0.3903)}$$

$$= -8.2930 + i5.6879$$

$$\vec{U}_R = \frac{(-1.9096 + i1.7898) - (-3.5540 + i3.8590)}{(-0.1018 + i0.0063) - (-0.3252 - i0.3840)} = \frac{(1.6443 - i2.0693)}{(0.2234 + i0.3903)}$$

$$= -2.1773 - i5.4592$$

Thus the required balance weights are given by

$$\vec{B}_L = -\vec{U}_L = (8.2930 - i5.6879) = 10.0561 \underline{/145.5548°}$$

$$\vec{B}_R = -\vec{U}_R = (2.1773 + i5.4592) = 5.8774 \underline{/248.2559°}$$

This shows that the addition of a 10.0561 oz weight in the left plane at 145.5548° and a 5.8774 oz weight in the right plane at 248.2559° from the reference position will balance the turbine rotor. It is implied that the balance weights are added at the same radial distance as the trial weights. If a balance weight is to be located at a different radial position, the required balance weight is to be modified in inverse proportion to the radial distance from the axis of rotation.

9.4 CRITICAL SPEEDS OF ROTATING SHAFTS

In the previous section, the rotor system—the shaft as well as the rotating body—was assumed to be rigid. In practice, however, all rotating shafts are flexible. If a flexible shaft carries an unbalanced rotating mass, the unbalance produces bending in the shaft, which in turn alters the effect of the unbalance. At a certain speed, known as the critical speed, the deflection of the shaft becomes very large. We consider the aspects of modeling the rotor system, critical speeds, response of the system, and stability in this section [9.11–9.13].

9.4.1
Equations of
Motion

Consider a shaft supported by two bearings and carrying a rotor or disc of mass m at the middle, as shown in Fig. 9.10. We shall assume that the rotor is subjected to a steady-state excitation due to mass unbalance. The forces acting on the rotor are the inertia force due to the acceleration of the mass center, the spring force due to the elasticity of the shaft, and the external and internal damping forces.*

Let O denote the equilibrium position of the shaft when balanced perfectly (Fig. 9.11). Let the shaft rotate with an angular velocity ω. We use a fixed coordinate system (x and y fixed to the earth) with O as the origin for deriving the equations of motion. The points P and Q denote the geometric center and mass

---

* Any rotating system responds in two different ways to damping or friction forces, depending upon whether the forces rotate with the shaft or not. When the positions at which the forces act remain fixed in space, as in the case of damping forces (which cause energy losses) in the bearing support structure, the damping is called *stationary* or *external damping*. On the other hand, if the positions at which they act rotate with the shaft in space, as in the case of internal friction of the shaft material, the damping is called *rotary* or *internal damping*.

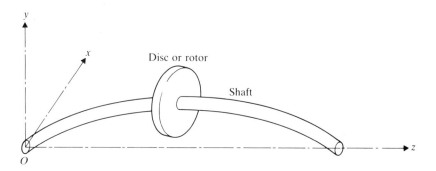

Figure 9.10

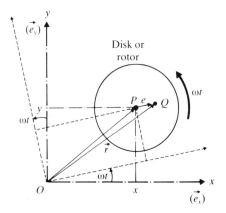

Figure 9.11

center of the disc, respectively. The equation of motion of the system (mass m) can be written as

$$\text{inertia force } \left(\vec{F_i} \right) = \text{elastic force } \left(\vec{F_e} \right) + \text{internal damping force } \left(\vec{F_{di}} \right)$$
$$+ \text{ external damping force } \left(\vec{F_{de}} \right) \tag{9.18}$$

The various forces in Eq. (9.18) can be expressed as follows:

$$\text{Inertia force:} \qquad \vec{F_i} = m\ddot{\vec{r}} \tag{9.19}$$

where $\vec{r}$ denotes the radius vector of the mass center Q given by

$$\vec{r} = (x + e \cos \omega t)\vec{e_x} + (y + e \sin \omega t)\vec{e_y} \tag{9.20}$$

with x and y representing the coordinates of the geometric center and $\vec{e_x}$ and $\vec{e_y}$ denoting the unit vectors along the x and y coordinates, respectively. Equations

(9.19) and (9.20) lead to

$$\vec{F_i} = m\left[\left(\ddot{x} - e\omega^2\cos\omega t\right)\vec{e}_x + \left(\ddot{y} - e\omega^2\sin\omega t\right)\vec{e}_y\right] \qquad (9.21)$$

Elastic force: $\qquad \vec{F}_e = -k\left(x\vec{e}_x + y\vec{e}_y\right) \qquad (9.22)$

where k is the stiffness of the shaft.

Internal damping force: $\qquad \vec{F}_{di} = -c_i\left[\left(\dot{x} + \omega y\right)\vec{e}_x + \left(\dot{y} + \omega x\right)\vec{e}_y\right]$

$$(9.23)$$

where c_i is the internal (rotary) damping coefficient.

External damping force: $\qquad \vec{F}_{de} = -c_e\left(\dot{x}\vec{e}_x + \dot{y}\vec{e}_y\right) \qquad (9.24)$

where c_e is the external damping coefficient. By substituting Eqs. (9.21) to (9.24) into Eq. (9.18), we obtain the equations of motion in scalar form:

$$m\ddot{x} + \left(c_i + c_e\right)\dot{x} + kx + c_i\omega y = m\omega^2 e \cos\omega t \qquad (9.25)$$

$$m\ddot{y} + \left(c_i + c_e\right)\dot{y} + ky - c_i\omega x = m\omega^2 e \sin\omega t \qquad (9.26)$$

These equations indicate that the equations of motion for the lateral vibration of the rotor are coupled and are dependent on the speed of the steady-state rotation. By defining a complex quantity w as

$$w = x + iy \qquad (9.27)$$

where $i = (-1)^{1/2}$ and by adding Eq. (9.25) to Eq. (9.26) multiplied by i, we obtain a single equation of motion:

$$m\ddot{w} + \left(c_i + c_e\right)\dot{w} + kw - i\omega c_i w = m\omega^2 e e^{i\omega t} \qquad (9.28)$$

**9.4.2
Critical Speeds**

A critical speed is said to exist when the frequency of the rotation of a shaft equals one of the natural frequencies of the shaft. The undamped natural frequency of the rotor system can be obtained by solving Eqs. (9.25), (9.26), or (9.28), retaining only the homogeneous part with $c_i = c_e = 0$. This gives the natural frequency of the system (or critical speed of the undamped system):

$$\omega_n = \left(\frac{k}{m}\right)^{1/2} \qquad (9.29)$$

When the rotational speed is equal to this critical speed, the rotor undergoes large deflections, and the force transmitted to the bearings can cause bearing failures. A rapid transition of the rotating shaft through a critical speed is expected to limit the whirl amplitudes, while a slow transition through the critical speed aids the development of large amplitudes. Reference [9.12] investigates the behavior of the rotor during acceleration and deceleration through critical speeds. A Fortran computer program for calculating the critical speeds of rotating shafts is given in Ref. [9.14].

**9.4.3
Response of the
System**

We shall now consider the response of the rotor shown in Fig. 9.11. The excitation is assumed to be harmonic unbalanced force due to the unbalance of the rotor. In addition, the internal damping c_i is taken as zero. For this, we solve Eqs. (9.25) and

(9.26) and find the rotor's dynamic whirl amplitudes resulting from the mass unbalance. With $c_i = 0$, Eqs. (9.25) and (9.26) reduce to

$$m\ddot{x} + c_e\dot{x} + kx = m\omega^2 e \cos \omega t \qquad (9.30)$$

$$m\ddot{y} + c_e\dot{y} + ky = m\omega^2 e \sin \omega t \qquad (9.31)$$

The solution of Eqs. (9.30) and (9.31) can be expressed as

$$x(t) = C_1 e^{-\alpha t} \cos (\beta t + \theta_1) + Dm\omega^2 e \cos (\omega t + \theta_2) \qquad (9.32)$$

$$y(t) = C_2 e^{-\alpha t} \cos (\beta t + \theta_3) + Dm\omega^2 e \cos (\omega t + \theta_4) \qquad (9.33)$$

where C_1, C_2, D, α, β, θ_1, θ_2, θ_3, and θ_4 are constants. The first term in Eqs. (9.32) and (9.33) contains a decaying exponential term representing a transient solution, and the second term denotes a steady-state circular motion (whirl). By substituting the steady-state part of Eq. (9.32) into Eq. (9.30), we can find the amplitude of the circular motion (whirl) as:

$$D = \frac{1}{\left[(k - m\omega^2)^2 + \omega^2 c_e^2 \right]^{1/2}} \qquad (9.34)$$

The phase angle is given by

$$\theta_2 = \theta_4 = -\tan^{-1} \left(\frac{c_e \omega}{k - m\omega^2} \right) \qquad (9.35)$$

By differentiating Eq. (9.34) with respect to ω and setting the result equal to zero, we can find the rotational speed ω at which the whirl amplitude becomes a maximum:

$$\omega = \frac{\omega_n}{\left\{ 1 - \frac{1}{2} \left(\frac{c_e}{\omega_n} \right)^2 \right\}^{1/2}} \qquad (9.36)$$

where ω_n is given by Eq. (9.29). It can be seen that the critical speed corresponds exactly to the natural frequency ω_n only when the damping (c_e) is zero. Furthermore, Eq. (9.36) shows that the presence of damping, in general, increases the value of the critical speed compared to ω_n. A plot of Eqs. (9.34) and (9.35) is shown in Fig. 9.12 [9.13]. Since the forcing function is proportional to ω^2, we normally expect the vibration amplitude to increase with the speed ω. However, the actual amplitude appears as shown in Fig. 9.12. From Eq. (9.34), we note that the amplitude of circular whirl D at low speeds is determined by the spring constant k, since the other two terms, $m\omega^2$ and $c_e^2\omega^2$, are small. Also, the value of the phase angle θ_2 can be seen to be 0° from Eq. (9.35) for small values of ω. As ω increases, the amplitude of the response reaches a peak, since resonance occurs at $k - m\omega^2 = 0$. Around resonance, the response is essentially limited by the damping term. The phase lag is 90° at resonance. As the speed ω increases beyond ω_n, the response is dominated by the mass term $m^2\omega^4$ in Eq. (9.34). Since this term is 180° out of phase with the unbalanced force, the shaft rotates in a direction opposite to that of the unbalanced force and hence the response of the shaft will be limited.

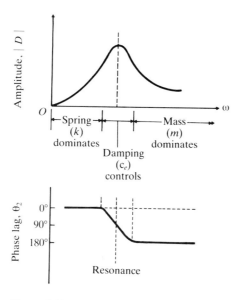

Figure 9.12

The vibration and balancing of unbalanced flexible rotors are presented in Refs. [9.15, 9.16].

9.4.4
Stability Analysis

Instability in a flexible rotor system can occur due to several factors like internal friction, eccentricity of the rotor, and the oil whip in the bearings. As seen earlier, the stability of the system can be investigated by considering the equation governing the dynamics of the system. Assuming $w(t) = e^{st}$, the characteristic equation corresponding to the homogeneous part of Eq. (9.28) can be written as

$$ms^2 + (c_i + c_e)s + k - i\omega c_i = 0 \tag{9.37}$$

With $s = i\lambda$, Eq. (9.37) becomes

$$-m\lambda^2 + (c_i + c_e)i\lambda + k - i\omega c_i = 0 \tag{9.38}$$

This equation is a particular case of the more general equation

$$(p_2 + iq_2)\lambda^2 + (p_1 + iq_1)\lambda + (p_0 + iq_0) = 0 \tag{9.39}$$

A necessary and sufficient condition for the system governed by Eq. (9.39) to be stable, according to Routh-Hurvitz criterion, is that the following inequalities are satisfied.

$$-\begin{vmatrix} p_2 & p_1 \\ q_2 & q_1 \end{vmatrix} > 0 \tag{9.40}$$

and

$$
\begin{vmatrix}
p_2 & p_1 & p_0 & 0 \\
q_2 & q_1 & q_0 & 0 \\
0 & p_2 & p_1 & p_0 \\
0 & q_2 & q_1 & q_0
\end{vmatrix} > 0 \tag{9.41}
$$

Noting that $p_2 = -m$, $q_2 = 0$, $p_1 = 0$, $q_1 = c_i + c_e$, $p_0 = k$, and $q_0 = -\omega c_i$, from Eq. (9.38), the application of Eqs. (9.40) and (9.41) leads to

$$
m(c_i + c_e) > 0 \tag{9.42}
$$

and

$$
km(c_i + c_e)^2 - m^2(\omega^2 c_i^2) > 0 \tag{9.43}
$$

Equation (9.42) is automatically satisfied, while Eq. (9.43) yields the condition

$$
\sqrt{\frac{k}{m}}\left(1 + \frac{c_e}{c_i}\right) - \omega > 0 \tag{9.44}
$$

This equation also shows that internal and external friction can cause instability at rotating speeds above the first critical speed of $\omega = \sqrt{\dfrac{k}{m}}$.

9.5 BALANCING OF RECIPROCATING ENGINES

The essential moving elements of a reciprocating engine are the piston, the crank, and the connecting rod. Vibrations in reciprocating engines arise due to (1) periodic variations of the gas pressure in the cylinder and (2) inertia forces associated with the moving parts [9.17]. We shall now analyze a reciprocating engine and find the unbalanced forces caused by these factors.

**9.5.1
Unbalanced
Forces Due to
Fluctuations in
Gas Pressure**

Fig 9.13(a) is a schematic diagram of a cylinder of a reciprocating engine. The engine is driven by the expanding gas in the cylinder. The expanding gas exerts on the piston a pressure force F, which is transmitted to the crankshaft through the connecting rod. The reaction to the force F can be resolved into two components: one of magnitude $F/\cos\phi$, acting along the connecting rod, and the other of magnitude $F\tan\phi$, acting in a horizontal direction. The force $F/\cos\phi$ induces a torque M_t, which tends to rotate the crankshaft. (In Fig. 9.13b, M_t acts about an axis perpendicular to the plane of the paper and passes through point Q.)

$$
M_t = \left(\frac{F}{\cos\phi}\right) r \cos\theta \tag{9.45}
$$

For force equilibrium of the overall system, the forces at the bearings of the crankshaft will be F in the vertical direction and $F\tan\phi$ in the horizontal direction.

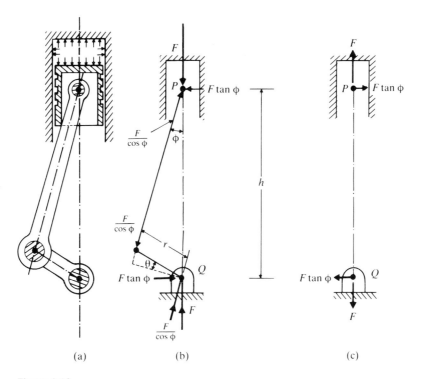

Figure 9.13

Thus the forces transmitted to the stationary parts of the engine are:

1. force F acting upwards at the cylinder head
2. force $F \tan \phi$ acting toward the right at the cylinder head
3. force F acting downwards at the crankshaft bearing Q
4. force $F \tan \phi$ acting toward the left at the crankshaft bearing

These forces are shown in Fig. 9.13(c). Although the total resultant force is zero, there is a resultant torque $M_Q = Fh \tan \phi$ on the body of the engine, where h can be found from the geometry of the system:

$$h = \frac{r \cos \theta}{\sin \phi} \tag{9.46}$$

Thus the resultant torque is given by

$$M_Q = \frac{Fr \cos \theta}{\cos \phi} \tag{9.47}$$

As to be expected, M_t and M_Q given by Eqs. (9.45) and (9.47) can be seen to be identical, which indicates that the torque induced on the crankshaft due to the gas pressure on the piston is felt at the support of the engine. Since the magnitude of the

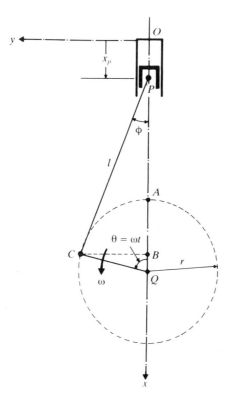

Figure 9.14

gas force F varies with time, the torque M_Q also varies with time. The magnitude of force F changes from a maximum to a minimum at a frequency governed by the number of cylinders in the engine, the type of the operating cycle, and the rotating speed of the engine.

**9.5.2
Unbalanced
Forces Due to
Inertia of the
Moving Parts**

Acceleration of the Piston. Figure 9.14 shows the crank (of length r), the connecting rod (of length l), and the piston of a reciprocating engine. The crank is assumed to rotate in an anticlockwise direction at a constant angular speed of ω, as shown in Fig. 9.14. If we consider the origin of the x axis (O) as the uppermost position of the piston, the displacement of the piston P corresponding to an angular displacement of the crank of $\theta = \omega t$ can be expressed as in Fig. 9.14. The displacement of the piston P corresponding to an angular displacement of the crank $\theta = \omega t$ from its topmost position (origin O) can be expressed as

$$x_p = r + l - r\cos\theta - l\cos\phi = r + l - r\cos\omega t - l\sqrt{1 - \sin^2\phi} \qquad (9.48)$$

But

$$l\sin\phi = r\sin\theta = r\sin\omega t \qquad (9.49)$$

and hence

$$\cos \phi = \left(1 - \frac{r^2}{l^2} \sin^2 \omega t\right)^{1/2} \tag{9.50}$$

By substituting Eq. (9.50) into Eq. (9.48), we obtain

$$x_p = r + l - r \cos \omega t - l\sqrt{1 - \frac{r^2}{l^2} \sin^2 \omega t} \tag{9.51}$$

Due to the presence of the term involving the square root, Eq. (9.51) is not very convenient in further calculation. Equation (9.51) can be simplified by noting that, in general, $r/l < 1/4$ and by using the expansion relation

$$\sqrt{1 - \varepsilon} \simeq 1 - \frac{\varepsilon}{2} \tag{9.52}$$

Hence Eq. (9.51) can be approximated as

$$x_p \simeq r(1 - \cos \omega t) + \frac{r^2}{2l} \sin^2 \omega t \tag{9.53}$$

or, equivalently,

$$x_p \simeq r\left(1 + \frac{r}{2l}\right) - r\left(\cos \omega t + \frac{r}{4l} \cos 2\omega t\right) \tag{9.54}$$

Equation (9.54) can be differentiated with respect to time to obtain expressions for the velocity and the acceleration of the piston:

$$\dot{x}_p = r\omega\left(\sin \omega t + \frac{r}{2l} \sin 2\omega t\right) \tag{9.55}$$

$$\ddot{x}_p = r\omega^2\left(\cos \omega t + \frac{r}{l} \cos 2\omega t\right) \tag{9.56}$$

Acceleration of the Crankpin. With respect to the xy coordinate axes shown in Fig. 9.14, the vertical and horizontal displacements of the crankpin C are given by

$$x_c = OA + AB = l + r(1 - \cos \omega t) \tag{9.57}$$
$$y_c = CB = r \sin \omega t \tag{9.58}$$

Differentiation of Eqs. (9.57) and (9.58) with respect to time gives the velocity and acceleration components of the crankpin as

$$\dot{x}_c = r\omega \sin \omega t \tag{9.59}$$
$$\dot{y}_c = r\omega \cos \omega t \tag{9.60}$$
$$\ddot{x}_c = r\omega^2 \cos \omega t \tag{9.61}$$
$$\ddot{y}_c = -r\omega^2 \sin \omega t \tag{9.62}$$

Inertia Forces. Although the mass of the connecting rod is distributed throughout its length, it is generally idealized as a massless link with two masses concentrated at its ends—the piston end and the crankpin end. If m_p and m_c denote the total mass of the piston and of the crankpin (including the concentrated mass of the connecting rod) respectively, the vertical component of the inertia force (F_x) for one cylinder is given by

$$F_x = m_p \ddot{x}_p + m_c \ddot{x}_c \tag{9.63}$$

By substituting Eqs. (9.56) and (9.61) for the accelerations of P and C, Eq. (9.63) becomes

$$F_x = (m_p + m_c) r\omega^2 \cos \omega t + m_p \frac{r^2 \omega^2}{l} \cos 2\omega t \qquad (9.64)$$

It can be observed that the vertical component of the inertia force consists of two parts. One part, known as the *primary part*, has a frequency equal to the rotational frequency of the crank ω. The other part, known as the *secondary part*, has a frequency equal to twice the rotational frequency of the crank.

Similarly, the horizontal component of inertia force for a cylinder can be obtained:

$$F_y = m_p \ddot{y}_p + m_c \ddot{y}_c \qquad (9.65)$$

where $\ddot{y}_p = 0$ and $\ddot{y}_c$ is given by Eq. (9.62). Thus

$$F_y = -m_c r\omega^2 \sin \omega t \qquad (9.66)$$

The horizontal component of the inertia force can be observed to have only a primary part.

**9.5.3
Balancing of
Reciprocating
Engines**

The unbalanced or inertia forces on a single cylinder are given by Eqs. (9.64) and (9.66). In these equations, m_p and m_c represent the equivalent reciprocating and rotating masses, respectively. The mass m_p is always positive, but m_c can be made zero by counterbalancing the crank. It is therefore possible to reduce the horizontal inertia force F_y to zero, but the vertical unbalanced force always exists. Thus a single cylinder engine is inherently unbalanced.

In a multicylinder engine, it is possible to balance some or all of the inertia forces and torques by proper arrangement of the cranks. Figure 9.15(a) shows the general arrangement of an N-cylinder engine (only six cylinders, $N = 6$, are shown in the figure). The lengths of all the cranks and connecting rods are assumed to be r and l, respectively, and the angular velocity of all the cranks is taken to be a constant, ω. The axial displacement and angular orientation of ith cylinder from those of the first cylinder are assumed to be α_i and l_i, respectively; $i = 2, 3, \ldots, N$. For force balance, the total inertia force in the x and y directions must be zero. Thus

$$(F_x)_{\text{total}} = \sum_{i=1}^{N} (F_x)_i = 0 \qquad (9.67)$$

$$(F_y)_{\text{total}} = \sum_{i=1}^{N} (F_y)_i = 0 \qquad (9.68)$$

where $(F_x)_i$ and $(F_y)_i$ are the vertical and horizontal components of inertia force of

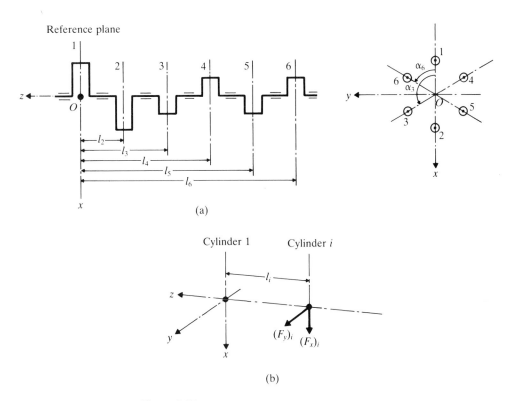

Figure 9.15

cylinder i given by (see Eqs. (9.64) and (9.66)):

$$(F_x)_i = (m_p + m_c)_i r\omega^2 \cos(\omega t + \alpha_i)$$
$$+ (m_p)_i \frac{r^2\omega^2}{l} \cos(2\omega t + 2\alpha_i) \qquad (9.69)$$

$$(F_y)_i = -(m_c)_i r\omega^2 \sin(\omega t + \alpha_i) \qquad (9.70)$$

For simplicity, we assume the reciprocating and rotating masses for each cylinder to be same, that is, $(m_p)_i = m_p$ and $(m_c)_i = m_c$ for $i = 1, 2, \ldots, N$. Without loss of generality, Eqs. (9.67) and (9.68) can be applied at time $t = 0$. Thus the conditions necessary for the total force balance are given by

$$\sum_{i=1}^{N} \cos \alpha_i = 0 \quad \text{and} \quad \sum_{i=1}^{N} \cos 2\alpha_i = 0 \qquad (9.71)$$

$$\sum_{i=1}^{N} \sin \alpha_i = 0 \qquad (9.72)$$

The inertia forces $(F_x)_i$ and $(F_y)_i$ of the ith cylinder induce moments about the y and x axes, respectively, as shown in Fig. 9.15(b). The moments about the z and x

axes are given by

$$M_z = \sum_{i=2}^{N} (F_x)_i l_i = 0 \tag{9.73}$$

$$M_x = \sum_{i=2}^{N} (F_y)_i l_i = 0 \tag{9.74}$$

By substituting Eqs. (9.69) and (9.70) into Eqs. (9.73) and (9.74) and assuming $t = 0$, we obtain the necessary conditions to be satisfied for the balancing of moments about z and x axes as:

$$\sum_{i=2}^{N} l_i \cos \alpha_i = 0 \quad \text{and} \quad \sum_{i=2}^{N} l_i \cos 2\alpha_i = 0 \tag{9.75}$$

$$\sum_{i=2}^{N} l_i \sin \alpha_i = 0 \tag{9.76}$$

Thus we can arrange the cylinders of a multicylinder reciprocating engine so as to satisfy Eqs. (9.71), (9.72), (9.75), and (9.76); it will be completely balanced against the inertia forces and moments.

9.6 CONTROL OF VIBRATION

In many practical situations, it is possible to reduce but not eliminate the dynamic forces that cause vibrations. Several methods can be used to control vibrations. Among them, the following are important.

1. By controlling the natural frequencies of the system and avoiding resonance under external excitations.

2. By preventing excessive response of the system, even at resonance by introducing a damping or energy-dissipating mechanism.

3. By reducing the transmission of the excitation forces from one part of the machine to another, by the use of vibration isolators.

4. By reducing the response of the system, by the addition of an auxiliary mass neutralizer or vibration absorber.

We shall now consider the details of these methods.

9.7 CONTROL OF NATURAL FREQUENCIES

It is well known that whenever the frequency of excitation coincides with one of the natural frequencies of the system, resonance occurs. The most prominent feature of resonance is a large displacement. In most mechanical and structural systems, large displacements indicate undesirably large strains and stresses, which can lead to the failure of the system. Hence resonance conditions must be avoided in any system. In most cases, the excitation frequency cannot be controlled, because it is imposed by the functional requirements of the system or machine. We must concentrate on controlling the natural frequencies of the system to avoid resonance.

As indicated by Eq. (2.11), the natural frequency of a system can be changed either by varying the mass m or the stiffness k.* In many practical cases, however, the mass cannot be changed easily, since its value is determined by the functional requirements of the system. For example, the mass of a flywheel on a shaft is determined by the amount of energy it must store in one cycle. Therefore, the stiffness of the system is the factor that is most often changed to alter its natural frequencies. For example, the stiffness of a rotating shaft can be altered by varying one or more of its parameters, such as materials or number and location of support points (bearings).

9.8 INTRODUCTION OF DAMPING

Although damping is disregarded so as to simplify the analysis, especially in finding the natural frequencies, most systems possess damping to some extent. The presence of damping is helpful in many cases. In systems such as automobile shock absorbers and many vibration-measuring instruments, damping must be introduced to fulfill the functional requirements [9.18–9.20].

If the system undergoes forced vibration, its response or amplitude of vibration tends to become large near resonance if there is no damping. The presence of damping always limits the amplitude of vibration. If the forcing frequency is known, it may be possible to avoid resonance by changing the natural frequency of the system. However, the system or the machine may be required to operate over a range of speeds, as in the case of a variable-speed electric motor or an internal combustion engine. It may not be possible to avoid resonance under all operating conditions. In such cases, we can introduce damping into the system to control its response, by the use of structural materials having high internal damping, such as cast iron or laminated or sandwich materials.

Use of Viscoelastic Materials. The equation of motion of a single degree of freedom system with internal damping, under harmonic excitation $F(t) = F_0 e^{i\omega t}$, can be expressed as

$$m\ddot{x} + k(1 + i\eta)x = F_0 e^{i\omega t} \tag{9.77}$$

where η is called the loss factor (or loss coefficient), which is defined as (see

---

* Although this statement is made with reference to a single degree of freedom system, it is generally true even for multidegree of freedom and continuous systems.

Section 2.6.4)

$$\eta = \frac{(\Delta W/2\pi)}{W} = \left(\frac{\substack{\text{Energy dissipated during one cycle of harmonic} \\ \text{displacement per radian}}}{\text{Maximum strain energy in the cycle}} \right)$$

The amplitude of the response of the system at resonance ($\omega = \omega_n$) is given by

$$\frac{F_0}{k\eta} = \frac{F_0}{aE\eta} \qquad (9.78)$$

since the stiffness is proportional to the Young's modulus ($k = aE$; $a = \text{constant}$).

The viscoelastic materials have larger values of the loss factor and hence are used to provide internal damping. When viscoelastic materials are used for vibration control, they are subjected to shear or direct strains. In the simplest arrangement, a layer of viscoelastic material is attached to an elastic one. In another arrangement, a viscoelastic layer is sandwiched between the elastic layers. Damping tapes, consisting of thin metal foil covered with a viscoelastic adhesive, are used on existing vibrating structures. A disadvantage with the use of viscoelastic materials is that their properties change with temperature, frequency and strain. Equation (9.78) shows that a material with the highest value of ($E\eta$) gives the smallest resonance amplitude. Since the strain is proportional to the displacement x and the stress is proportional to Ex, the material with the largest value of the loss factor will be subjected to the smallest stresses. The values of loss coefficient for some materials are given below [9.40]:

Material	Loss factor (η)
Polystyrene	2.0
Hard rubber	1.0
Fiber mats with matrix	0.1
Cork	0.13–0.17
Aluminum	1×10^{-4}
Iron and steel	$2-6 \times 10^{-4}$

9.9 USE OF VIBRATION ISOLATORS

Vibration isolation methods are used to reduce the undesirable effects of vibration [9.21–9.24]. Basically, vibration isolation involves the insertion of a resilient member (or isolator) between the vibrating mass (or equipment or payload) and the source of vibration so that a reduction in the dynamic response of the system is achieved under specified conditions of vibration excitation. An isolation system is said to be active or passive depending on whether or not external power is required for the isolator to perform its function. A passive isolator consists of a resilient member

(a)

(b)

(c)

Figure 9.16   (a) Undamped spring mount; (b) damped spring mount; (c) pneumatic rubber mount. (Courtesy of *Sound and Vibration*.)

(stiffness) and an energy dissipator (damping). Examples of passive isolators include metal springs, cork, felt, pneumatic springs, and elastomer (rubber) springs. Figure 9.16 shows typical spring and pneumatic mounts that can be used as passive isolators, and Fig. 9.17 illustrates the use of passive isolators in the mounting of a high-speed punch press [9.25]. The optimal synthesis of vibration isolators is presented in Refs. [9.26–9.30].

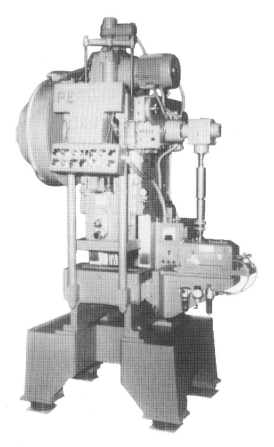

Figure 9.17 High-speed punch press mounted on pneumatic rubber mounts. (Courtesy of *Sound and Vibration*.)

An active isolator is comprised of a servomechanism with a sensor, signal processor and an actuator. The effectiveness of an isolator is stated in terms of its transmissibility. The transmissibility (T_r) is defined as the ratio of the amplitude of the force transmitted to that of the exciting force.

9.9.1 Vibration Isolation System with Rigid Foundation

When a machine or equipment is placed on a resilient member resting on a rigid foundation or support, the system can be idealized as a single degree of freedom system, as shown in Fig. 9.18(a). The resilient member is assumed to have both elasticity and damping and is modeled as a spring k and a dashpot c, as shown in Fig. 9.18(b). It is assumed that the operation of the machine gives rise to a harmonically varying force $F(t) = F_0 \cos \omega t$. The equation of motion of the machine (of mass m) is given by

$$m\ddot{x} + c\dot{x} + kx = F_0 \cos \omega t \tag{9.79}$$

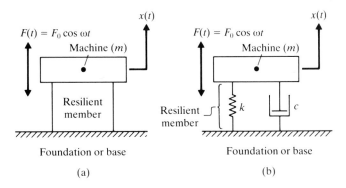

Figure 9.18 Machine and resilient member on rigid foundation.

Since the transient solution dies out after some time, only the steady-state solution will be left. The steady-state solution of Eq. (9.79) is given by (see Eq. (3.25))

$$x(t) = X\cos(\omega t - \phi) \tag{9.80}$$

where

$$X = \frac{F_0}{\left[(k - m\omega^2)^2 + \omega^2 c^2\right]^{1/2}} \tag{9.81}$$

and

$$\phi = \tan^{-1}\left(\frac{\omega c}{k - m\omega^2}\right) \tag{9.82}$$

The force transmitted through the spring and the dashpot are, respectively,

$$kx(t) = kX\cos(\omega t - \phi) \tag{9.83}$$
$$c\dot{x}(t) = -c\omega X\sin(\omega t - \phi) \tag{9.84}$$

The magnitude of the total transmitted force (F_t) is given by

$$F_t = \left[(kx)^2 + (c\dot{x})^2\right]^{1/2} = X\sqrt{k^2 + \omega^2 c^2}$$
$$= \frac{F_0(k^2 + \omega^2 c^2)^{1/2}}{\left[(k - m\omega^2)^2 + \omega^2 c^2\right]^{1/2}} \tag{9.85}$$

The transmissibility or transmission ratio of the isolator (T_r) is defined as the ratio of the magnitude of the force transmitted to that of the exciting force:

$$T_r = \frac{F_t}{F_0} = \left\{\frac{k^2 + \omega^2 c^2}{(k - m\omega^2)^2 + \omega^2 c^2}\right\}^{1/2}$$
$$= \left\{\frac{1 + \left(2\zeta\dfrac{\omega}{\omega_n}\right)^2}{\left[1 - \left(\dfrac{\omega}{\omega_n}\right)^2\right]^2 + \left(2\zeta\dfrac{\omega}{\omega_n}\right)^2}\right\}^{1/2} \tag{9.86}$$

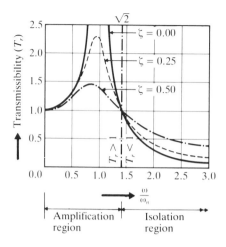

Figure 9.19 Variation of transmission ratio (T_r) with ω.

The variation of T_r with the frequency ratio ω/ω_n is shown in Fig. 9.19. It can be seen that for forcing frequencies greater than 1.414 times the natural frequency of the system, isolation of vibration occurs so that the transmitted force is less than the exciting force.

**9.9.2
Vibration
Isolation System
with Flexible
Foundation**

In many practical situations, the structure or foundation to which the isolator is connected moves when the machine mounted on the isolator operates. For example, in the case of a turbine supported on the hull of a ship or an aircraft engine mounted on the wing of an airplane, the area surrounding the point of support also moves with the isolator. In such cases, the system can be represented as having two degrees of freedom. In Fig. 9.20, m_1 and m_2 denote the masses of the machine and the supporting structure that moves with the isolator, respectively. The isolator is

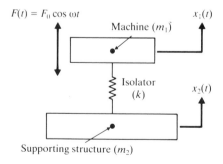

Figure 9.20 Machine with isolator on a flexible foundation.

represented by a spring k, and the damping is disregarded for the sake of simplicity. The equations of motion of the masses m_1 and m_2 are

$$m_1\ddot{x}_1 + k(x_1 - x_2) = F_0\cos \omega t \tag{9.87}$$
$$m_2\ddot{x}_2 + k(x_2 - x_1) = 0 \tag{9.88}$$

By assuming a harmonic solution of the form

$$x_j = X_j\cos \omega t, \qquad j = 1, 2 \tag{9.89}$$

Eqs. (9.87) and (9.88) give

$$\left.\begin{array}{l} X_1(k - m_1\omega^2) - X_2 k = F_0 \\ - X_1 k + X_2(k - m_2\omega^2) = 0 \end{array}\right\} \tag{9.90}$$

The natural frequencies of the system are given by the roots of the equation

$$\begin{vmatrix} (k - m_1\omega^2) & -k \\ -k & (k - m_2\omega^2) \end{vmatrix} = 0 \tag{9.91}$$

The roots of Eq. (9.91) are given by

$$\omega_1^2 = 0, \qquad \omega_2^2 = \frac{(m_1 + m_2)k}{m_1 m_2} \tag{9.92}$$

The value $\omega_1 = 0$ corresponds to rigid-body motion since the system is unconstrained. In the steady state, the amplitudes of m_1 and m_2 are governed by Eq. (9.90), whose solution yields

$$X_1 = \frac{(k - m_2\omega^2) F_0}{[(k - m_1\omega^2)(k - m_2\omega^2) - k^2]} \tag{9.93}$$

$$X_2 = \frac{kF_0}{[(k - m_1\omega^2)(k - m_2\omega^2) - k^2]} \tag{9.94}$$

The force transmitted to the supporting structure (F_t) is given by the amplitude of $m_2\ddot{x}_2$:

$$F_t = -m_2\omega^2 X_2 = \frac{-m_2 k\omega^2 F_0}{[(k - m_1\omega^2)(k - m_2\omega^2) - k^2]} \tag{9.95}$$

The transmissibility of the isolator (T_r) is given by

$$T_r = \frac{F_t}{F_0} = \frac{-m_2 k\omega^2}{[(k - m_1\omega^2)(k - m_2\omega^2) - k^2]}$$

$$= \frac{1}{\left(\dfrac{m_1 + m_2}{m_2} - \dfrac{m_1\omega^2}{k}\right)} = \frac{m_2}{(m_1 + m_2)}\left(\frac{1}{1 - \dfrac{\omega^2}{\omega_2^2}}\right) \tag{9.96}$$

where ω_2 is the natural frequency of the system given by Eq. (9.92). Equation (9.96) shows, as in the case of an isolator on a rigid base, that the force transmitted to the foundation becomes less as the natural frequency of the system ω_2 is reduced.

The value of T_r of an isolator with rigid foundation is plotted in Fig. 9.19. It can be seen that the force transmitted to the foundation cannot become infinite at resonance unless damping is zero. If the force transmitted to the foundation is to be small, ω/ω_n should be large—that is, the isolator should have a natural frequency much smaller than the operating speed of the machine it supports. Although damping reduces the amplitude of motion (X) for all frequencies, it reduces the maximum force transmitted (F_t) only if $\omega/\omega_n > \sqrt{2}$. Above that value, the addition of damping increases the force transmitted. Furthermore, the force transmitted to the foundation is larger than the force applied by the machine ($T_r > 1$) when $\omega/\omega_n < \sqrt{2}$. These observations suggest that the damping of the isolator should be made as small as possible, so as to reduce the force transmitted. If the speed of the machine (forcing frequency) varies, we must compromise in choosing the amount of damping to minimize the force transmitted. The amount of damping should be sufficient to limit the amplitude X and the force transmitted F_t while passing through the resonance, but not so much as to increase unnecessarily the force transmitted at the operating speeds.

EXAMPLE 9.2   Spring Support for Exhaust Fan

An exhaust fan, rotating at 1000 rpm, is to be supported by four springs, each having a stiffness of K. If only 10% of the unbalanced force of the fan is to be transmitted to the base, what should be the value of K? Assume the mass of the exhaust fan to be 40 kg.

Given: Exhaust fan with mass = 40 kg, rotational speed = 1000 rpm, and permissible shaking force to be transmitted to base = 10%.

Find: Stiffness (K) of each of the four supporting springs.

Approach: Use transmissibility equation.

Solution. Since the transmissibility has to be 0.1, we have, from Eq. (9.86),

$$0.1 = \left[\frac{1 + \left(2\zeta\dfrac{\omega}{\omega_n}\right)^2}{\left\{1 - \left(\dfrac{\omega}{\omega_n}\right)^2\right\}^2 + \left(2\zeta\dfrac{\omega}{\omega_n}\right)^2} \right]^{1/2} \tag{E.1}$$

where the forcing frequency is given by

$$\omega = \frac{1000 \times 2\pi}{60} = 104.72 \text{ rad/sec} \tag{E.2}$$

and the natural frequency of the system by

$$\omega_n = \left(\frac{k}{m}\right)^{1/2} = \left(\frac{4K}{40}\right)^{1/2} = \frac{\sqrt{K}}{3.1623} \tag{E.3}$$

By assuming the damping ratio to be $\zeta = 0$, we obtain from Eq. (E.1),

$$0.1 = \frac{\pm 1}{\left\{ 1 - \left(\dfrac{104.72 \times 3.1623}{\sqrt{K}} \right)^2 \right\}} \qquad \text{(E.4)}$$

To avoid imaginary values, we need to consider the negative sign on the right-hand side of Eq. (E.4). This leads to

$$\frac{331.1561}{\sqrt{K}} = 3.3166$$

or

$$K = 9969.6365 \text{ N/m}$$

EXAMPLE 9.3   **Isolation of Vibrating System**

A vibrating system is to be isolated from its supporting base. Find the required damping ratio that must be achieved by the isolator to limit the transmissibility at resonance to $T_r = 4$. Assume the system to have a single degree of freedom.

Given: Transmissibility at resonance = 4.

Find: Damping ratio of the isolator.

Approach: Find the equation for the transmissibility at resonance.

Solution. By setting $\omega = \omega_n$, Eq. (9.86) gives

$$T_r = \frac{\sqrt{1 + (2\zeta)^2}}{2\zeta}$$

or

$$\zeta = \frac{1}{2\sqrt{T_r^2 - 1}} = \frac{1}{2\sqrt{15}} = 0.1291$$

9.9.3 Vibration Isolation System with Partially Flexible Foundation

We shall now consider a more realistic situation. Figure 9.21 shows an isolator whose base, instead of being completely rigid or completely flexible, is partially flexible [9.34]. We can define the mechanical impedance of the base structure, $Z(\omega)$, as the force at frequency ω required to produce a unit displacement of the base (as in Section 3.5):

$$Z(\omega) = \frac{\text{applied force of frequency } \omega}{\text{displacement}}$$

The equations of motion are given by*

$$m_1 \ddot{x}_1 + k(x_1 - x_2) = F_0 \cos \omega t \qquad (9.97)$$
$$k(x_2 - x_1) = -x_2 Z(\omega) \qquad (9.98)$$

---

* If the base is completely flexible with an unconstrained mass of m_2, $Z(\omega) = -\omega^2 m_2$, and Eqs. (9.98) and (9.99) lead to Eq. (9.88).

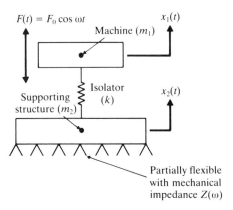

Figure 9.21   Machine with isolator on a partially flexible foundation.

By substituting the harmonic solution

$$x_j(t) = X_j \cos \omega t, \qquad j = 1, 2 \tag{9.99}$$

into Eqs. (9.97) and (9.98), X_1 and X_2 can be obtained as in the previous case:

$$X_1 = \frac{[k + Z(\omega)] X_2}{k} = \frac{[k + Z(\omega)] F_0}{[Z(\omega)(k - m_1\omega^2) - km_1\omega^2]}$$

$$X_2 = \frac{kF_0}{[Z(\omega)(k - m_1\omega^2) - km_1\omega^2]} \tag{9.100}$$

The amplitude of the force transmitted is given by

$$F_t = X_2 Z(\omega) = \frac{kZ(\omega) F_0}{[Z(\omega)(k - m_1\omega^2) - km_1\omega^2]} \tag{9.101}$$

and the transmissibility of the isolator by

$$T_r = \frac{F_t}{F_0} = \frac{kZ(\omega)}{[Z(\omega)(k - m_1\omega^2) - km_1\omega^2]} \tag{9.102}$$

In practice, the mechanical impedance $Z(\omega)$ depends on the nature of the base structure. It can be found experimentally by measuring the displacement produced by a vibrator that applies a harmonic force on the base structure. In some cases, such as in the case of an isolator resting on a concrete raft on soil, the mechanical impedance at any frequency ω can be found in terms of the spring-mass-dashpot model of the soil.

**9.9.4
Active Vibration
Control**

A vibration isolation system is called active if it uses external power to perform its function. It consists of a servomechanism with a sensor, signal processor and an actuator as shown schematically in Fig. 9.22 [9.31–9.33]. This system maintains a

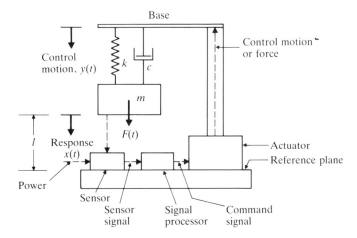

Figure 9.22

constant distance (l) between the vibrating mass and the reference plane. As the force $F(t)$ applied to the system (mass) varies, the distance l tends to vary. This change in l is sensed by the sensor and a signal, proportional to the magnitude of the excitation (or response) of the vibrating body, is produced. The signal processor produces a command signal to the actuator based on the sensor signal it receives. The actuator develops a motion or force proportional to the command signal. The actuator motion or force will control the base displacement such that the distance l is maintained at the desired constant value.

Different types of sensors are available to create feedback signals based on the displacement, velocity, acceleration, jerk, or force. The signal processor may consist of a passive mechanism such as a mechanical linkage or an active electronic or fluidic network that can perform functions such as addition, integration, differentiation, attenuation, or amplification. The actuator may be a mechanical system such as a rack-and-pinion or ball screw mechanism, fluidic system or piezoelectric and electromagnetic force generating systems. Depending on the types of sensor, signal processor, and actuator used, an active vibration control system can be called an electromechanical, electrofluidic, electromagnetic, piezoelectric, or a fluidic system.

9.10 USE OF VIBRATION ABSORBERS

A machine or system may experience excessive vibration if it is acted upon by a force whose excitation frequency nearly coincides with a natural frequency of the machine or system. In such cases, the vibration of the machine or system can be reduced by using a *vibration neutralizer* or *dynamic vibration absorber*. This is simply another spring-mass system. We shall consider the analysis of a dynamic vibration absorber by idealizing the machine as a single degree of freedom system.

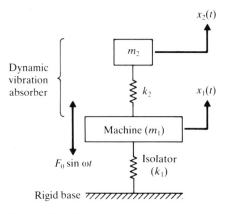

Figure 9.23   Dynamic vibration absorber.

**9.10.1
Dynamic
Vibration
Absorber**

When we couple an auxiliary mass m_2 to a machine of mass m_1 through a spring of stiffness k_2, the resulting two degree of freedom system will look as shown in Fig. 9.23. The equations of motion of the masses m_1 and m_2 are

$$m_1\ddot{x}_1 + k_1 x_1 + k_2(x_1 - x_2) = F_0 \sin \omega t$$
$$m_2\ddot{x}_2 + k_2(x_2 - x_1) = 0 \tag{9.103}$$

By assuming harmonic solution,

$$x_j(t) = X_j \sin \omega t, \qquad j = 1, 2 \tag{9.104}$$

we can obtain

$$X_1 = \frac{(k_2 - m_2\omega^2)F_0}{(k_1 + k_2 - m_1\omega^2)(k_2 - m_2\omega^2) - k_2^2} \tag{9.105}$$

$$X_2 = \frac{k_2 F_0}{(k_1 + k_2 - m_1\omega^2)(k_2 - m_2\omega^2) - k_2^2} \tag{9.106}$$

We are primarily interested in reducing the amplitude of the machine (X_1). In order to make the amplitude of m_1 zero, the numerator of Eq. (9.105) should be set equal to zero. This gives

$$\omega^2 = \frac{k_2}{m_2} \tag{9.107}$$

If the machine, before the addition of the dynamic vibration absorber, operates near its resonance, $\omega^2 \simeq \omega_1^2 = k_1/m_1$. Thus if the absorber is designed such that

$$\omega^2 = \frac{k_2}{m_2} = \frac{k_1}{m_1} \tag{9.108}$$

the amplitude of vibration of the machine, while operating at its original resonant frequency, will be zero. By defining

$$\delta_{\text{st}} = \frac{F_0}{k_1}, \qquad \omega_1 = \left(\frac{k_1}{m_1}\right)^{1/2}$$

as the natural frequency of the machine or main system, and

$$\omega_2 = \left(\frac{k_2}{m_2}\right)^{1/2}$$

(9.109)

as the natural frequency of the absorber or auxiliary system, Eqs. (9.105) and (9.106) can be rewritten as

$$\frac{X_1}{\delta_{st}} = \frac{1 - \left(\dfrac{\omega}{\omega_2}\right)^2}{\left[1 + \dfrac{k_2}{k_1} - \left(\dfrac{\omega}{\omega_1}\right)^2\right]\left[1 - \left(\dfrac{\omega}{\omega_2}\right)^2\right] - \dfrac{k_2}{k_1}}$$

(9.110)

$$\frac{X_2}{\delta_{st}} = \frac{1}{\left[1 + \dfrac{k_2}{k_1} - \left(\dfrac{\omega}{\omega_1}\right)^2\right]\left[1 - \left(\dfrac{\omega}{\omega_2}\right)^2\right] - \dfrac{k_2}{k_1}}$$

(9.111)

Figure 9.24 shows the variation of the amplitude of vibration of the machine (X_1/δ_{st}) with the machine speed (ω/ω_1). The two peaks correspond to the two natural frequencies of the composite system. As seen before, $X_1 = 0$ at $\omega = \omega_1$. At this frequency, Eq. (9.111) gives

$$X_2 = -\frac{k_1}{k_2}\delta_{st} = -\frac{F_0}{k_2}$$

(9.112)

This shows that the force exerted by the auxiliary spring is opposite to the impressed force ($k_2 X_2 = -F_0$) and neutralizes it, thus reducing X_1 to zero. The size of the dynamic vibration absorber can be found from Eqs. (9.112) and (9.108):

$$k_2 X_2 = m_2 \omega^2 X_2 = -F_0$$

(9.113)

Thus the values of k_2 and m_2 depend on the allowable value of X_2.

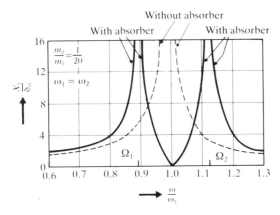

Figure 9.24   Effect of undamped vibration absorber on the response of machine.

It can be seen from Fig. 9.24 that the dynamic vibration absorber, while eliminating vibration at the known impressed frequency ω, introduces two resonant frequencies Ω_1 and Ω_2, at which the amplitude of the machine is infinite. In practice, the operating frequency ω must therefore be kept away from the frequencies Ω_1 and Ω_2. The values of Ω_1 and Ω_2 can be found by equating the denominator of Eq. (9.110) to zero. Noting that

$$\frac{k_2}{k_1} = \frac{k_2}{m_2}\frac{m_2}{m_1}\frac{m_1}{k_1} = \frac{m_2}{m_1}\left(\frac{\omega_2}{\omega_1}\right)^2 \tag{9.114}$$

and setting the denominator of Eq. (9.110) to zero leads to

$$\left(\frac{\omega}{\omega_2}\right)^4\left(\frac{\omega_2}{\omega_1}\right)^2 - \left(\frac{\omega}{\omega_2}\right)^2\left[1 + \left(1 + \frac{m_2}{m_1}\right)\left(\frac{\omega_2}{\omega_1}\right)^2\right] + 1 = 0 \tag{9.115}$$

The two roots of this equation are given by

$$\left.\begin{array}{l}\left(\dfrac{\Omega_1}{\omega_2}\right)^2 \\[4mm] \left(\dfrac{\Omega_2}{\omega_2}\right)^2\end{array}\right\} = \frac{\left[1 + \left(1 + \dfrac{m_2}{m_1}\right)\left(\dfrac{\omega_2}{\omega_1}\right)^2\right] \mp \left\{\left[1 + \left(1 + \dfrac{m_2}{m_1}\right)\left(\dfrac{\omega_2}{\omega_1}\right)^2\right]^2 - 4\left(\dfrac{\omega_2}{\omega_1}\right)^2\right\}^{1/2}}{2\left(\dfrac{\omega_2}{\omega_1}\right)^2} \tag{9.116}$$

which can be seen to be functions of (m_2/m_1) and (ω_2/ω_1).

EXAMPLE 9.4   **Vibration Absorber for Diesel Engine**

A diesel engine, weighing 3000 N, is supported on a pedestal mount. It has been observed that the engine induces vibration into the surrounding area through its pedestal mount at an operating speed of 6000 rpm. Determine the parameters of the vibration absorber that will reduce the vibration when mounted on the pedestal. The magnitude of the exciting force is 250 N, and the amplitude of motion of the auxiliary mass is to be limited to 2 mm.

Given: Diesel engine with weight = 3000 N, speed = 6000 rpm, magnitude of exciting force = 250 N, and permissible amplitude of motion of the auxiliary mass = 2 mm.

Find: Parameters of the vibration absorber (mass and stiffness).

Approach: Set the motion of the pedestal equal to zero. Equate the natural frequency of the auxiliary mass to the resonance frequency of original system.

Solution. The frequency of vibration of the machine is

$$f = \frac{6000}{60} = 100 \text{ Hz} \qquad \text{or} \qquad \omega = 628.32 \text{ rad/sec}$$

Since the motion of the pedestal is to be made equal to zero, the amplitude of motion of the auxiliary mass should be equal and opposite to that of the exciting force. Thus from Eq. (9.113), we obtain

$$|F_0| = m_2\omega^2 X_2 \tag{E.1}$$

Substitution of the given data yields

$$250 = m_2(628.32)^2(0.002)$$

Therefore $m_2 = 0.31665$ kg. The spring stiffness k_2 can be determined from Eq. (9.108):

$$\omega^2 = \frac{k_2}{m_2}$$

Therefore, $k_2 = (628.32)^2(0.31665) = 125009$ N/m.

EXAMPLE 9.5   **Absorber for Motor-Generator Set**

A motor-generator set, shown in Fig. 9.25, is designed to operate in the speed range of 2000 to 4000 rpm. However, the set is found to vibrate violently at a speed of 3000 rpm due to a slight unbalance in the rotor. It is proposed to attach a cantilever mounted lumped mass absorber system to eliminate the problem. When a cantilever carrying a trial mass of 2 kg, tuned to 3000 rpm, is attached to the set, the resulting natural frequencies of the system are found to be 2500 rpm and 3500 rpm. Design the absorber to be attached (by specifying its mass and stiffness) so that the natural frequencies of the total system fall outside the operating speed range of the motor-generator set.

Given: Motor-generator set that vibrates violently at 3000 rpm and has natural frequencies = 2500 rpm and 3500 rpm when trial absorber is attached. Trial absorber of mass = 2 kg and tuned frequency = 3000 rpm.

Find: m and k of absorber so that Ω_1 and Ω_2 lie outside the range 2000 rpm to 4000 rpm.

Approach: Use expressions for the resonant frequencies of the combined system.

Solution. The natural frequencies ω_1 of the motor-generator set and ω_2 of the absorber are given by

$$\omega_1 = \sqrt{\frac{k_1}{m_1}}, \qquad \omega_2 = \sqrt{\frac{k_2}{m_2}} \tag{E.1}$$

The resonant frequencies Ω_1 and Ω_2 of the combined system are given by Eq. (9.116). Since

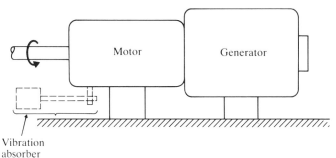

Vibration
absorber

Figure 9.25

the absorber ($m = 2$ kg) is tuned, $\omega_1 = \omega_2 = 314.16$ rad/sec (corresponding to 3000 rpm). Using the notation

$$\mu = \frac{m_2}{m_1}, \qquad r_1 = \frac{\Omega_1}{\omega_2}, \qquad \text{and} \qquad r_2 = \frac{\Omega_2}{\omega_2}$$

Eq. (9.116) becomes

$$r_1^2, r_2^2 = \left(1 + \frac{\mu}{2}\right) \mp \sqrt{\left(1 + \frac{\mu}{2}\right)^2 - 1} \qquad (\text{E.2})$$

Since Ω_1 and Ω_2 are known to be 261.80 rad/sec (or 2500 rpm) and 366.52 rad/sec (or 3500 rpm), respectively, we find that

$$r_1 = \frac{\Omega_1}{\omega_2} = \frac{261.80}{314.16} = 0.8333$$

$$r_2 = \frac{\Omega_2}{\omega_2} = \frac{366.52}{314.16} = 1.1667$$

Hence

$$r_1^2 = \left(1 + \frac{\mu}{2}\right) - \sqrt{\left(1 + \frac{\mu}{2}\right)^2 - 1}$$

or

$$\mu = \left(\frac{r_1^4 + 1}{r_1^2}\right) - 2 \qquad (\text{E.3})$$

Since $r_1 = 0.8333$, Eq. (E.3) gives $\mu = m_2/m_1 = 0.1345$ and $m_1 = m_2/0.1345 = 14.8699$ kg. The specified lower limit of Ω_1 is 2000 rpm or 209.44 rad/sec, and so

$$r_1 = \frac{\Omega_1}{\omega_2} = \frac{209.44}{314.16} = 0.6667$$

With this value of r_1, (E.3) gives $\mu = m_2/m_1 = 0.6942$ and $m_2 = m_1(0.6942) = 10.3227$ kg. With these values, the second resonant frequency can be found from

$$r_2^2 = \left(1 + \frac{\mu}{2}\right) + \sqrt{\left(1 + \frac{\mu}{2}\right)^2 - 1} = 2.2497$$

which gives $\Omega_2 \simeq 4499.4$ rpm, larger than the specified upper limit of 4000 rpm. The spring stiffness of the absorber is given by

$$k_2 = \omega_2^2 m_2 = (314.16)^2 (10.3227) = 1.0188 \times 10^6 \text{ N/m}$$

**9.10.2
Damped Dynamic
Vibration
Absorber**

The dynamic vibration absorber described in the previous section removes the original resonance peak in the response curve of the machine but introduces two new peaks. If it is necessary to reduce the amplitude of vibration of the machine over a range of frequencies, we can use a damped dynamic vibration absorber, as shown in Fig. 9.26. The equations of motion of the two masses are given by

$$m_1 \ddot{x}_1 + k_1 x_1 + k_2(x_1 - x_2) + c_2(\dot{x}_1 - \dot{x}_2) = F_0 \sin \omega t \qquad (9.117)$$

$$m_2 \ddot{x}_2 + k_2(x_2 - x_1) + c_2(\dot{x}_2 - \dot{x}_1) = 0 \qquad (9.118)$$

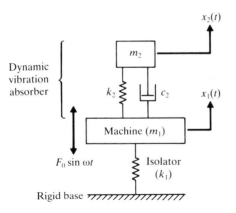

Figure 9.26   Damped dynamic vibration absorber.

By assuming the solution to be

$$x_j(t) = X_j e^{i\omega t}, \qquad j = 1, 2 \tag{9.119}$$

the solution of Eqs. (9.117) and (9.118) can be obtained:

$$X_1 = \frac{F_0(k_2 - m_2\omega^2 + ic_2\omega)}{\left[(k_1 - m_1\omega^2)(k_2 - m_2\omega^2) - m_2 k_2 \omega^2\right] + i\omega c_2(k_1 - m_1\omega^2 - m_2\omega^2)} \tag{9.120}$$

$$X_2 = \frac{X_1(k_2 + i\omega c_2)}{(k_2 - m_2\omega^2 + i\omega c_2)} \tag{9.121}$$

By defining

$$\mu = m_2/m_1 = \text{mass ratio} = \text{absorber mass/main mass} \tag{9.122}$$

$$\delta_{st} = F_0/k_1 = \text{static deflection of the system} \tag{9.123}$$

$$\omega_a^2 = k_2/m_2 = \text{square of natural frequency of the absorber} \tag{9.124}$$

$$\omega_n^2 = k_1/m_1 = \text{square of natural frequency of main mass} \tag{9.125}$$

$$f = \omega_a/\omega_n = \text{ratio of natural frequencies} \tag{9.126}$$

$$g = \omega/\omega_n = \text{forced frequency ratio} \tag{9.127}$$

$$c_c = 2m_2\omega_n = \text{critical damping constant} \tag{9.128}$$

$$\zeta = c_2/c_c = \text{damping ratio} \tag{9.129}$$

the magnitude of X_1 can be expressed as

$$\frac{X_1}{\delta_{st}} = \left[\frac{(2\zeta g)^2 + (g^2 - f^2)^2}{(2\zeta g)^2(g^2 - 1 + \mu g^2)^2 + \{\mu f^2 g^2 - (g^2 - 1)(g^2 - f^2)\}^2}\right]^{1/2} \tag{9.130}$$

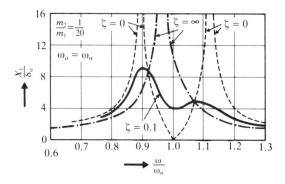

Figure 9.27   Effect of damped vibration absorber on the response of the machine.

This equation shows that the amplitude of vibration of the main mass is a function of μ, f, g, and ζ. The graph of

$$\left| \frac{X_1}{\delta_{st}} \right|$$

against the forced frequency ratio $g = \omega/\omega_n$ is shown in Fig. 9.27 for $f = 1$ and $\mu = 1/20$ for a few different values of ζ. For $c_2 = 0$ ($\zeta = 0$), we have the same case as Fig. 9.24, a known result. When the damping becomes infinite ($\zeta = \infty$), the two masses m_1 and m_2 are virtually clamped together, and the system behaves essentially as a single degree of freedom system with a mass of $(21/20)m_1$. Thus the peak of X_1 is infinite for $c_2 = 0$ as well as for $c_2 = \infty$. Somewhere in between these limits, the peak of X_1 is a minimum. It is possible to find this optimum value of c_2 [9.35]. Additional work relating to the optimum design of vibration absorbers can be found in Refs. [9.36–9.39].

9.11 COMPUTER PROGRAM

A Fortran subroutine BALAN is given for two-plane balancing. The arrays VA, VB, VAP, VBP, VAPP, VBPP, WL, WR, BL, and BR, each of dimension 2, are used to represent the vectors $\vec{V}_A$, $\vec{V}_B$, $\vec{V}_A'$, $\vec{V}_B'$, $\vec{V}_A''$, $\vec{V}_B''$, $\vec{W}_L$, $\vec{W}_R$, $\vec{B}_L$, and $\vec{B}_R$, respectively. The arrays BL and BR give the magnitude and position of the balancing weights in the left and the right planes.

For illustration, Example 9.1 is solved using this subroutine. The main program for this example, subroutine BALAN, and the output of the program are given on the following page.

```
C ================================================================
C
C PROGRAM 18
C TWO-PLANE BALANCING
C
C ================================================================
      DIMENSION VA(2),VB(2),VAP(2),VBP(2),VAPP(2),VBPP(2),WL(2),WR(2),
     2   BL(2),BR(2)
C FOLLOWING 8 LINES CONTAIN PROBLEM-DEPENDENT DATA
      DATA VA/8.5,60.0/
      DATA VAP/6.0,125.0/
      DATA WL/10.0,270.0/
      DATA VB/6.5,205.0/
      DATA VBP/4.5,230.0/
      DATA VAPP/6.0,35.0/
      DATA VBPP/10.5,160.0/
      DATA WR/12.0,180.0/
C END OF PROBLEM-DEPENDENT DATA
      CALL BALAN (VA,VB,VAP,VBP,VAPP,VBPP,WL,WR,BL,BR)
      PRINT 10
   10 FORMAT (//,31H RESULTS OF TWO-PLANE BALANCING)
      PRINT 20,BL(1),BL(2)
   20 FORMAT (//,28H LEFT-PLANE BALANCING WEIGHT,//,11H MAGNITUDE=,
     2   E15.8,/,7H ANGLE=,4X,E15.8)
      PRINT 30,BR(1),BR(2)
   30 FORMAT (//,29H RIGHT-PLANE BALANCING WEIGHT,//,
     2   11H MAGNITUDE=,E15.8,/,7H ANGLE=,4X,E15.8)
      STOP
      END
C ================================================================
C
C SUBROUTINE BALAN
C
C ================================================================
      SUBROUTINE BALAN (VA,VB,VAP,VBP,VAPP,VBPP,WL,WR,BL,BR)
      DIMENSION VA(2),VB(2),VAP(2),VBP(2),VAPP(2),VBPP(2),WL(2),WR(2),
     2   BL(2),BR(2),P(2),Q(2),R(2),S(2),AAL(2),ABL(2),AAR(2),ABR(2),
     3   UL(2),UR(2)
      PI=180.0/3.1415926
      VA(2)=VA(2)/PI
      P(1)=VA(1)
      P(2)=VA(2)
      VA(1)=P(1)*COS(P(2))
      VA(2)=P(1)*SIN(P(2))
      VB(2)=VB(2)/PI
      P(1)=VB(1)
      P(2)=VB(2)
      VB(1)=P(1)*COS(P(2))
      VB(2)=P(1)*SIN(P(2))
      VAP(2)=VAP(2)/PI
      P(1)=VAP(1)
      P(2)=VAP(2)
      VAP(1)=P(1)*COS(P(2))
      VAP(2)=P(1)*SIN(P(2))
```

```
VBP(2)=VBP(2)/PI
P(1)=VBP(1)
P(2)=VBP(2)
VBP(1)=P(1)*COS(P(2))
VBP(2)=P(1)*SIN(P(2))
VAPP(2)=VAPP(2)/PI
P(1)=VAPP(1)
P(2)=VAPP(2)
VAPP(1)=P(1)*COS(P(2))
VAPP(2)=P(1)*SIN(P(2))
VBPP(2)=VBPP(2)/PI
P(1)=VBPP(1)
P(2)=VBPP(2)
VBPP(1)=P(1)*COS(P(2))
VBPP(2)=P(1)*SIN(P(2))
WL(2)=WL(2)/PI
P(1)=WL(1)
P(2)=WL(2)
WL(1)=P(1)*COS(P(2))
WL(2)=P(1)*SIN(P(2))
WR(2)=WR(2)/PI
P(1)=WR(1)
P(2)=WR(2)
WR(1)=P(1)*COS(P(2))
WR(2)=P(1)*SIN(P(2))
CALL VSUB (VAP,VA,R)
CALL VDIV (R,WL,AAL)
CALL VSUB (VBP,VB,S)
CALL VDIV (S,WL,ABL)
CALL VSUB (VAPP,VA,P)
CALL VDIV (P,WR,AAR)
CALL VSUB (VBPP,VB,Q)
CALL VDIV (Q,WR,ABR)
AR1=SQRT(AAR(1)**2+AAR(2)**2)
AR2=ATAN(AAR(2)/AAR(1))*PI
AL1=SQRT(AAL(1)**2+AAL(2)**2)
AL2=ATAN(AAL(2)/AAL(1))*PI
CALL VMULT (ABL,VA,R)
CALL VMULT (AAL,VB,S)
CALL VSUB (R,S,VAP)
CALL VMULT (AAR,ABL,R)
CALL VMULT (AAL,ABR,S)
CALL VSUB (R,S,VBP)
CALL VDIV (VAP,VBP,UR)
CALL VMULT (ABR,VA,R)
CALL VMULT (AAR,VB,S)
CALL VSUB (R,S,VAP)
CALL VMULT (ABR,AAL,R)
CALL VMULT (AAR,ABL,S)
CALL VSUB (R,S,VBP)
CALL VDIV (VAP,VBP,UL)
BL(1)=SQRT(UL(1)**2+UL(2)**2)
A1=UL(2)/UL(1)
BL(2)=ATAN(UL(2)/UL(1))
BR(1)=SQRT(UR(1)**2+UR(2)**2)
```

```
      A2=UR(2)/UR(1)
      BR(2)=ATAN(UR(2)/UR(1))
      BL(2)=BL(2)*PI
      BR(2)=BR(2)*PI
      BL(2)=BL(2)+180.0
      BR(2)=BR(2)+180.0
      RETURN
      END
C ============================================================
C
C SUBROUTINE VSUB (A,B,C)
C
C ============================================================
      SUBROUTINE VSUB (A,B,C)
      DIMENSION A(2),B(2),C(2)
      C(1)=A(1)-B(1)
      C(2)=A(2)-B(2)
      RETURN
      END
C ============================================================
C
C SUBROUTINE VDIV
C
C ============================================================
      SUBROUTINE VDIV (A,B,C)
      DIMENSION A(2),B(2),C(2)
      C(1)=(A(1)*B(1)+A(2)*B(2))/(B(1)**2+B(2)**2)
      C(2)=(A(2)*B(1)-A(1)*B(2))/(B(1)**2+B(2)**2)
      RETURN
      END
C ============================================================
C
C SUBROUTINE VMULT
C
C ============================================================
      SUBROUTINE VMULT (A,B,C)
      DIMENSION A(2),B(2),C(2)
      C(1)=A(1)*B(1)-A(2)*B(2)
      C(2)=A(2)*B(1)+A(1)*B(2)
      RETURN
      END
```

```
                    RESULTS OF TWO-PLANE BALANCING

  LEFT-PLANE BALANCING WEIGHT        RIGHT-PLANE BALANCING WEIGHT

  MAGNITUDE= 0.10056140E+02          MAGNITUDE= 0.58773613E+01
  ANGLE=     0.14555478E+03          ANGLE=     0.24825594E+03
```

REFERENCES

9.1. R. L. Eshleman, "Identification and correction of machinery vibration problems," *Sound and Vibration*, Vol. 15, April 1981, pp. 12–18.

9.2. J. E. Ruzicka, "Fundamental concepts of vibration control," *Sound and Vibration*, Vol. 5, July 1971, pp. 16–22.

9.3. R. L. Fox, "Machinery vibration monitoring and analysis techniques," *Sound and Vibration*, Vol. 5, November 1971, pp. 35–40.

9.4. J. N. Macduff, "A procedure for field balancing rotating machinery," *Sound and Vibration*, Vol. 1, 1967, pp. 16–20.

9.5. J. F. G. Wort, "Industrial balancing machines," *Sound and Vibration*, Vol. 13, October 1979, pp. 14–19.

9.6. D. G. Stadelbauer, "Dynamic balancing with microprocessors," *Shock and Vibration Digest*, Vol. 14, December 1982, pp. 3–7.

9.7. J. Vaughan, *Static and Dynamic Balancing* (2nd Ed.), Bruel and Kjaer Application Notes, Naerum, Denmark.

9.8. D. V. Hutton, *Applied Mechanical Vibrations*, McGraw-Hill, New York, 1981.

9.9. R. L. Baxter, "Dynamic balancing," *Sound and Vibration*, Vol. 6, April 1972, pp. 30–33.

9.10. J. H. Harter and W. D. Beitzel, *Mathematics Applied to Electronics*, Reston Publishing Co., Reston, Virginia, 1980.

9.11. R. G. Loewy and V. J. Piarulli, "Dynamics of rotating shafts," *Shock and Vibration Monograph SVM-4*, Shock and Vibration Information Center, Naval Research Laboratory, Washington, D.C., 1969.

9.12. T. Iwatsuba, "Vibration of rotors through critical speeds," *Shock and Vibration Digest*, Vol. 8, No. 2, February 1976, pp. 89–98.

9.13. J. D. Irwin and E. R. Graf, *Industrial Noise and Vibration Control*, Prentice-Hall, Englewood Cliffs, N.J., 1979.

9.14. R. J. Trivisonno, "Fortran IV computer program for calculating critical speeds of rotating shafts," NASA TN D-7385, 1973.

9.15. R. E. D. Bishop and G. M. L. Gladwell, "The vibration and balancing of an unbalanced flexible rotor," *Journal of Mechanical Engineering Science*, Vol. 1, 1959, pp. 66–77.

9.16. A. G. Parkinson, "The vibration and balancing of shafts rotating in asymmetric bearings," *Journal of Sound and Vibration*, Vol. 2, 1965, pp. 477–501.

9.17. C. E. Crede, *Vibration and Shock Isolation*, Wiley, New York, 1951.

9.18. W. E. Purcell, "Materials for noise and vibration control," *Sound and Vibration*, Vol. 16, July 1982, pp. 6–31.

9.19. J. C. Snowdon, "Rubberlike materials, their internal damping and role in vibration isolation," *Journal of Sound and Vibration*, Vol. 2, 1965, pp. 175–193.

9.20. B. C. Nakra, "Vibration control with viscoelastic materials," *Shock and Vibration Digest*, Vol. 8, No. 6, June 1976, pp. 3–12.

9.21. D. E. Baxa and R. A. Dykstra, "Pneumatic isolation systems control forging hammer vibration," *Sound and Vibration*, Vol. 14, May 1980, pp. 22–25.

9.22. E. I. Rivin, "Vibration isolation of industrial machinery—Basic considerations," *Sound and Vibration*, Vol. 12, November 1978, pp. 14–19.

9.23. E. Rivin, "Vibration isolation of precision machinery," *Sound and Vibration*, Vol. 13, August 1979, pp. 18–23.

9.24. R. A. Waller, "Building on springs," *Sound and Vibration*, Vol. 9, March 1975, pp. 22–25.

9.25. C. M. Salerno and R. M. Hochheiser, "How to select vibration isolators for use as machinery mounts," *Sound and Vibration*, Vol. 7, August 1973, pp. 22–28.

9.26. C. A. Mercer and P. L. Rees, "An optimum shock isolator," *Journal of Sound and Vibration*, Vol. 18, 1971, pp. 511–520.

9.27. M. L. Munjal, "A rational synthesis of vibration isolators," *Journal of Sound and Vibration*, Vol. 39, 1975, pp. 247–263.

9.28. C. Ng and P. F. Cunniff, "Optimization of mechanical vibration isolation systems with multi-degrees of freedom," *Journal of Sound and Vibration*, Vol. 36, 1974, pp. 105–117.

9.29. S. K. Hati and S. S. Rao, "Cooperative solution in the synthesis of multidegree of freedom shock isolation systems," *Journal of Vibration, Acoustics, Stress, and Reliability in Design*, Vol. 105, 1983, pp. 101–103.

9.30. S. S. Rao and S. K. Hati, "Optimum design of shock and vibration isolation systems using game theory," *Journal of Engineering Optimization*, Vol. 4, 1980, pp. 1–8.

9.31. J. E. Ruzicka, "Active vibration and shock isolation," Paper no. 680747, *SAE Transactions*, Vol. 77, 1969, pp. 2872–2886.

9.32. R. W. Horning and D. W. Schubert, "Air suspension and active vibration-isolation systems," Ch. 33, in *Shock and Vibration Handbook* (ed. C. M. Harris) (3rd Ed.), McGraw-Hill, New York, 1988.

9.33. O. Vilnay, "Active control of machinery foundation," *Journal of Engineering Mechanics, ASCE*, Vol. 110, 1984, pp. 273–281.

9.34. J. I. Soliman and M. G. Hallam, "Vibration isolation between non-rigid machines and non-rigid foundations," *Journal of Sound and Vibration*, Vol. 8, 1968, pp. 329–351.

9.35. J. Ormondroyd and J. P. Den Hartog, "The theory of the dynamic vibration absorber," *Transactions of ASME*, Vol. 50, 1928, p. APM-241.

9.36. H. Puksand, "Optimum conditions for dynamic vibration absorbers for variable speed systems with rotating and reciprocating unbalance," *International Journal of Mechanical Engineering Education*, Vol. 3, April 1975, pp. 145–152.

9.37. A. Soom and M.-S. Lee, "Optimal design of linear and nonlinear absorbers for damped systems," *Journal of Vibration, Acoustics, Stress, and Reliability in Design*, Vol. 105, 1983, pp. 112–119.

9.38. J. B. Hunt, *Dynamic Vibration Absorbers*, Mechanical Engineering Publications, London, 1979.

9.39. J. A. Macinante, *Seismic Mountings for Vibration Isolation*, Wiley, New York, 1984.

9.40. L. Cremer and M. Heckl, *Structure-Borne Sound. Structural Vibrations and Sound Radiation at Audio Frequencies* (translation from German by E. E. Ungar), Springer-Verlag, New York, 1973.

REVIEW QUESTIONS

9.1. Name some sources of industrial vibrations.

9.2. What are the various methods available for vibration control?

9.3. What is single-plane balancing?

9.4. Describe the two-plane balancing procedure.

9.5. What is whirling?

9.6. What is the difference between stationary damping and rotary damping?

9.7. How is the critical speed of a shaft determined?

9.8. What causes instability in a rotor system?

9.9. What considerations are to be taken into account for the balancing of a reciprocating engine?

9.10. What is the function of a vibration isolator?

9.11. What is a vibration absorber?

9.12. What is the difference between a vibration isolator and a vibration absorber?

9.13. Does spring mounting always reduce the vibration of the foundation of a machine?

9.14. Is it better to use a soft spring in the flexible mounting of a machine? Why?

9.15. Is the shaking force proportional to the square of the speed of a machine? Does the vibratory force transmitted to the foundation increase with the speed of the machine?

9.16. Why does dynamic balancing imply static balancing?

9.17. Explain why dynamic balancing can never be achieved by a static test alone.

9.18. Why does a rotating shaft always vibrate? What is the source of the shaking force?

9.19. Is it always advantageous to include a damper in the secondary system of a dynamic vibration absorber?

9.20. What is active vibration isolation?

9.21. Explain the difference between passive and active isolations.

PROBLEMS

The problem assignments are organized as follows:

Problems	Section covered	Topic covered
9.1–9.12	9.3	Balancing of rotating machines
9.13, 9.14	9.4	Critical speeds of rotating shafts
9.15–9.19	9.5	Balancing of reciprocating engines
9.20–9.23	9.9	Vibration isolation
9.24–9.32	9.10	Vibration absorbers
9.33–9.35	9.11	Computer program
9.36	—	Project

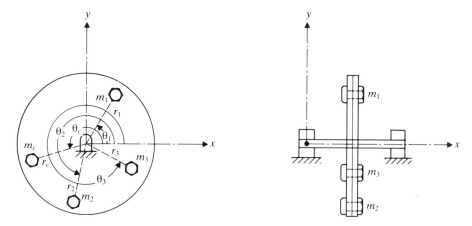

Figure 9.28

9.1. Two identical discs are connected by four bolts of different sizes and mounted on a shaft as shown in Fig. 9.28. The masses and locations of three bolts are given by: $m_1 = 35$ grams, $r_1 = 110$ mm, and $\theta_1 = 40°$; $m_2 = 15$ grams, $r_2 = 90$ mm, and $\theta_2 = 220°$; and $m_3 = 25$ grams, $r_2 = 130$ mm, $\theta_3 = 290°$. Find the mass and location of the fourth bolt (m_c, r_c, and θ_c), which results in the static balance of the discs.

9.2. Four holes are drilled in a uniform circular disc at a radius of 4 in. and angles of 0°, 60°, 120°, and 180°. The weight removed at holes 1 and 2 is 4 oz each and the weight removed at holes 3 and 4 is 5 oz each. If the disc is to be balanced statically by drilling a fifth hole at a radius of 5 in., find the weight to be removed and the angular location of the fifth hole.

9.3. Three masses, weighing 0.5 lb, 0.7 lb, and 1.2 lb, are attached around the rim, of diameter 30 in., of a flywheel at the angular locations $\theta = 10°$, $100°$, and $190°$, respectively. Find the weight and the angular location of the fourth mass to be attached on the rim that leads to the dynamic balance of the flywheel.

9.4. The amplitude and phase angle due to original unbalance in a grinding wheel operating at 1200 rpm are found to be 10 mils and 40° counterclockwise from the phase mark. When a trial weight $W = 6$ oz is added at 65° clockwise from the phase mark and at a radial distance 2.5 in. from the center of rotation, the amplitude and phase angle are observed to be 19 mils and 150° counterclockwise. Find the magnitude and angular position of the balancing weight if it is to be located 2.5 in. radially from the center of rotation.

9.5. An unbalanced flywheel shows an amplitude of 6.5 mils and a phase angle of 15° clockwise from the phase mark. When a trial weight of magnitude 2 oz is added at an angular position 45° counterclockwise from the phase mark, the amplitude and the phase angle become 8.8 mils and 35° counterclockwise, respectively. Find the magnitude and angular position of the balancing weight required. Assume that the weights are added at the same radius.

9.6. In order to determine the unbalance in a grinding wheel, rotating clockwise at 2400 rpm, a vibration analyzer is used and an amplitude of 4 mils and a phase angle of 45°

are observed with the original unbalance. When a trial weight $W = 4$ oz is added at 20° clockwise from the phase mark, the amplitude becomes 8 mils and the phase angle 145°. If the phase angles are measured counterclockwise from the right-hand horizontal, calculate the magnitude and location of the necessary balancing weight.

9.7. A turbine rotor is run at the natural frequency of the system. A stroboscope indicates that the maximum displacement of the rotor occurs at an angle 229° in the direction of rotation. At what angular position must mass be removed from the rotor in order to improve its balancing?

9.8. A rotor, having three eccentric masses in different planes, is shown in Fig. 9.29. The axial, radial, and angular locations of mass m_i are given by l_i, r_i, and θ_i, respectively, for $i = 1, 2, 3$. If the rotor is to be dynamically balanced by locating two masses m_{b1} and m_{b2} at radii r_{b1} and r_{b2} at the angular locations θ_{b1} and θ_{b2}, as shown in Fig. 9.29, derive expressions for $m_{b1}r_{b1}$, $m_{b2}r_{b2}$, θ_{b1}, and θ_{b2}.

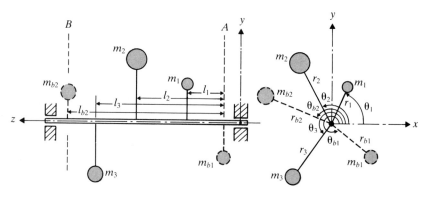

Figure 9.29

9.9. The rotor shown in Fig. 9.30(a) is balanced temporarily in a balancing machine by adding the weights $W_1 = W_2 = 0.2$ lb in the plane A and $W_3 = W_4 = 0.2$ lb in the plane D at a radius of 3 in., as shown in Fig. 9.30(b). If the rotor is permanently

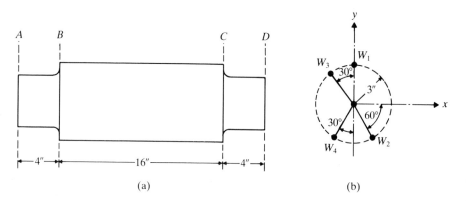

(a)   (b)

Figure 9.30

balanced by drilling holes at a radius of 4 in. in planes B and C, determine the position and amount of material to be removed from the rotor. Assume that the adjustable weights W_1 to W_4 will be removed from the planes A and D.

9.10. Weights of 2 lb, 4 lb and 3 lb are located at radii 2 in., 3 in., and 1 in. in the planes C, D, and E, respectively, on a shaft supported at the bearings B and F, as shown in Fig. 9.31. Find the weights and angular locations of the two balancing weights to be placed in the end planes A and G so that the dynamic load on the bearings will be zero.

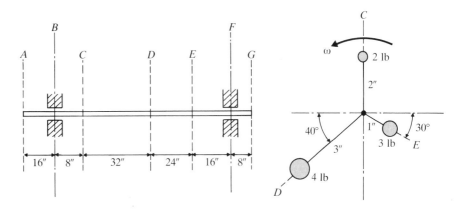

Figure 9.31

9.11. The data obtained in a two-plane balancing procedure is given below. Determine the magnitude and angular position of the balancing weights, assuming that all angles are measured from an arbitrary phase mark and all weights are added at the same radius.

| | Amplitude | | Phase angle | |
| | Bearing *A* | Bearing *B* | Bearing *A* | Bearing *B* |
Condition				
Original unbalance	5	4	100°	180°
W_L = 2 oz added at 30° in the left plane	6.5	4.5	120°	140°
W_R = 2 oz added at 0° in the right plane	6	7	90°	60°

9.12. Figure 9.32 shows a rotating system in which the shaft is supported in bearings at A and B. The three masses m_1, m_2, and m_3 are connected to the shaft as indicated in the figure. (a) Find the bearing reactions at A and B if the speed of the shaft is 1000 rpm. (b) Determine the locations and magnitudes of the balancing masses to be placed at a radius of 0.25 m in the planes L and R, which can be assumed to pass through the bearings A and B.

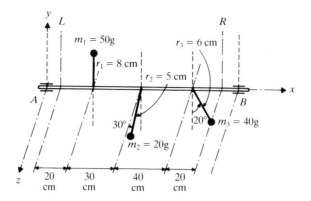

Figure 9.32

9.13. A flywheel, with a weight of 100 lb and an eccentricity of 0.5 in., is mounted at the center of a steel shaft of diameter 1 in. If the length of the shaft between the bearings is 30 in. and the rotational speed of the flywheeel is 1200 rpm, find (a) the critical speed, (b) the vibration amplitude of the rotor, and (c) the force transmitted to the bearing supports.

9.14. Derive the expression for the stress induced in a shaft with an unbalanced concentrated mass located midway between two bearings.

9.15. The cylinders of a four-cylinder in-line engine are placed at intervals of 12 in. in the axial direction. The cranks have the same length, 4 in., and their angular positions are given by 0°, 180°, 180°, and 0°. If the length of the connecting rod is 10 in. and the reciprocating weight is 2 lb for each cylinder, find the unbalanced forces and moments at a speed of 3000 rpm, using the center line through cylinder 1 as the reference plane.

9.16. The reciprocating mass, crank radius, and the connecting rod length of each of the cylinders in a two-cylinder in-line engine is given by m, r, and l, respectively. The crank angles of the two cylinders are separated by 180°. Find the unbalanced forces and moments in the engine.

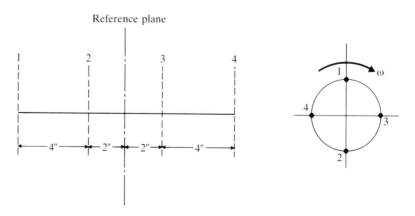

Figure 9.33

9.17. A 4-cylinder in-line engine has a reciprocating weight of 3 lb, stroke of 6 in., and a connecting rod of length 10 in. in each cylinder. The cranks are separated by 4 in. axially and 90° radially, as shown in Fig. 9.33. Find the unbalanced primary and secondary forces and moments with respect to the reference plane shown in Fig. 9.33 at an engine speed of 1500 rpm.

9.18. The arrangement of cranks in a 6-cylinder in-line engine is shown in Fig. 9.34. The cylinders are separated by a distance a in the axial direction and the angular positions of the cranks are given by $\alpha_1 = \alpha_6 = 0°$, $\alpha_2 = \alpha_5 = 120°$, and $\alpha_3 = \alpha_4 = 240°$. If the crank length, connecting rod length, and the reciprocating mass of each cylinder is r, l, and m, respectively, find the primary and secondary unbalanced forces and moments with respect to the reference plane indicated in Fig. 9.34.

Reference plane

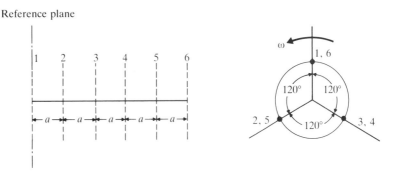

Figure 9.34

9.19. A single-cylinder engine has a total mass of 150 kg. Its reciprocating mass is 5 kg, and the rotating mass is 2.5 kg. The stroke ($2r$) is 15 cm, and the speed is 600 rpm. (a) If the engine is mounted floating on very weak springs, what is the amplitude of vertical vibration of the engine? (b) If the engine is mounted solidly on a rigid foundation, what is the alternating force amplitude transmitted? Assume the connecting rod to be of infinite length.

9.20. An electronic instrument is to be isolated from a panel that vibrates at frequencies ranging from 25 Hz to 35 Hz. It is estimated that at least 80% vibration isolation must be achieved to prevent damage to the instrument. If the instrument weighs 85 N, find the necessary static deflection of the isolator.

9.21.* An exhaust fan, having a small unbalance, weighs 800 N and operates at a speed of 600 rpm. It is desired to limit the response to a transmissibility of 2.5 as the fan passes through resonance during startup. In addition, an isolation of 90% is to be achieved at the operating speed of the fan. Design a suitable isolator for the fan.

9.22.* An air compressor of mass 500 kg has an eccentricity of 50 kg-cm, and operates at a speed of 300 rpm. The compressor is to be mounted on one of the following mountings: (a) An isolator consisting of a spring with negligible damping, and (b) a shock absorber having a damping ratio of 0.1 and negligible stiffness. Select a suitable mounting and specify the design details by considering the static deflection of the compressor, the transmission ratio, and the amplitude of vibration of the compressor.

9.23. The armature of a variable speed electric motor, of mass 200 kg, has an unbalance due to manufacturing errors. The motor is mounted on an isolator having a stiffness of 10 kN/m, and a dashpot having a damping ratio of 0.15. (a) Find the speed range over which the amplitude of the fluctuating force transmitted to the foundation will be larger than the exciting force. (b) Find the speed range over which the transmitted force amplitude will be less than 10% of the exciting force amplitude.

9.24. An air compressor of mass 200 kg, with an unbalance of 0.01 kg-m, is found to have a large amplitude of vibration while running at 1200 rpm. Determine the mass and spring constant of the absorber to be added if the natural frequencies of the system are to be at least 20% from the impressed frequency.

9.25. An electric motor, having an unbalance of 2 kg-cm, is mounted at the end of a steel cantilever beam as shown in Fig. 9.35. The beam is observed to vibrate with large amplitudes at the operating speed of 1500 rpm of the motor. It is proposed to add a vibration absorber to reduce the vibration of the beam. Determine the ratio of the absorber mass to the mass of the motor needed in order to have the lower frequency of the resulting system equal to 75% of the operating speed of the motor. If the mass of the motor is 300 kg, determine the stiffness and mass of the absorber. Also find the amplitude of vibration of the absorber mass.

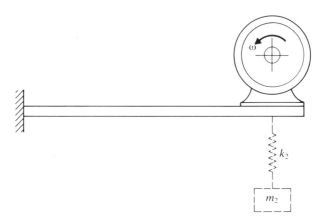

Figure 9.35

9.26.* The pipe carrying feedwater to a boiler in a thermal power plant has been found to vibrate violently at a pump speed of 800 rpm. In order to reduce the vibrations, an absorber consisting of a spring of stiffness k_2 and a trial mass m_2' of 1 kg is attached to the pipe. This arrangement is found to give the natural frequencies of the system as 750 rpm and 1000 rpm. It is desired to keep the natural frequencies of the system outside the operating speed range of the pump, which is 700 rpm to 1040 rpm. Determine the values of k_2 and m_2 that satisfy this requirement.

9.27. A reciprocating engine is installed on the first floor of a building, which can be modeled as a rigid rectangular plate resting on four elastic columns. The equivalent weight of the engine and the floor is 2000 lb. At the rated speed of the engine, which is 600 rpm, the operators experience large vibration of the floor. It has been decided to reduce these vibrations by suspending a spring-mass system from the bottom surface of the floor. If the spring stiffness is $k_2 = 5000$ lb/in., (a) Find the weight of the mass to be attached to absorb the vibrations, and (b) What will be the natural frequencies of the system after the absorber is added?

9.28.* Find the values of k_2 and m_2 in Problem 9.27 in order to have the natural frequencies of the system at least 30% away from the forcing frequency.

9.29.* A hollow steel shaft of outer diameter 2 in., inner diameter 1.5 in. and length 30 in. carries a solid disc of diameter 15 in. and weight 100 lb. Another hollow steel shaft of length 20 in., carrying a solid disc of diameter 6 in. and weight 20 lb, is attached to the first disc as shown in Fig. 9.36. Find the inner and outer diameters of the shaft such that the attached shaft-disc system acts as an absorber.

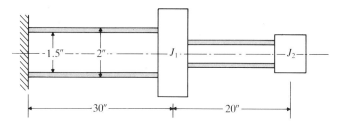

Figure 9.36

9.30.* A rotor, having a mass moment of inertia $J_1 = 15$ kg–m^2, is mounted at the end of a steel shaft having a torsional stiffness of 0.6 MN-m/rad. The rotor is found to vibrate violently when subjected to a harmonic torque of $300 \cos 200t$ N-m. A tuned absorber, consisting of a torsional spring and a mass moment of inertia (k_{t2} and J_2), is to be attached to the first rotor to absorb the vibrations. Find the values of k_{t2} and J_2 such that the natural frequencies of the system are away from the forcing frequency by at least 20%.

9.31. Plot the graphs of (Ω_1/ω_2) against (m_2/m_1) and (Ω_2/ω_2) against (m_2/m_1) as (m_2/m_1) varies from 0 to 1.0 when $\omega_2/\omega_1 = 1$.

9.32. Determine the operating range of the frequency ratio ω/ω_2 for an undamped vibration absorber to limit the value of $|X_1/\delta_{st}|$ to 0.5. Assume that $\omega_1 = \omega_2$ and $m_2 = 0.1m_1$.

9.33. Solve Example 9.1 using subroutine BALAN.

9.34. Solve Problem 9.11 using subroutine BALAN.

9.35. Write a computer program to find the displacements of the main mass and the auxiliary mass of a damped dynamic vibration absorber. Use this program to generate the results of Fig. 9.27.

Project:

9.36. Ground vibrations from a crane operation, a forging press and an air compressor are transmitted to a nearby milling machine and are found to be detrimental to achieving specified accuracies during precision milling operations. The ground vibrations at the locations of the crane, forging press, and air compressor are given by $x_c(t) = A_c e^{-\omega_c \zeta_c t} \sin \omega_c t$, $x_f(t) = A_f \sin \omega_f t$, and $x_a(t) = A_a \sin \omega_a t$, respectively, where $A_c = 20$ μm, $A_f = 30$ μm, $A_a = 25$ μm, $\omega_c = 10$ Hz, $\omega_f = 15$ Hz, $\omega_a = 20$ Hz, and $\zeta_c = 0.1$. The ground vibrations travel at the shear wave velocity of the soil, which is equal to 980 ft/sec, and the amplitudes attenuate according to the relation $A_r = A_0 e^{-0.005r}$, where A_0 is the amplitude at the source and A_r is the amplitude at a

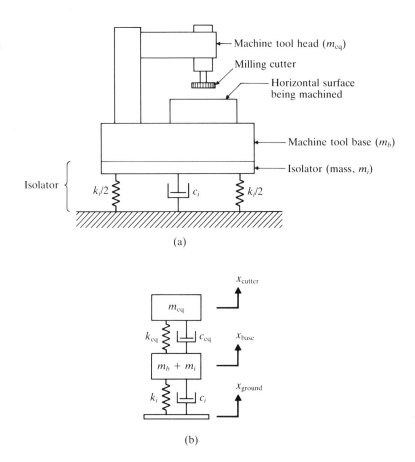

(a)

(b)

Figure 9.37

distance of r ft from the source. The crane, forging press, and air compressor are located at a distance of 60 ft, 80 ft, and 40 ft, respectively, from the milling machine. The equivalent mass, stiffness, and damping ratio of the machine tool head in vertical vibration (at the location of the cutter) are experimentally determined to be 500 kg, 480 kN/m, and 0.15, respectively. The equivalent mass of the machine tool base is 1000 kg. It is proposed to use an isolator for the machine tool as shown in Fig. 9.37(a) to improve the cutting accuracies [9.39]. Design a suitable vibration isolator, consisting of a mass, spring, and damper as shown in Fig. 9.37(b), for the milling machine such that the maximum vertical displacement of the milling cutter, relative to the horizontal surface being machined, due to ground vibration from all the three sources does not exceed 5 μm peak-to-peak.

CHAPTER 10

Vibration Measurement

Ernst Florens Friedrich Chladni (1756–1827) was a German physicist who studied sound and invented musical instruments. He first studied law at the University of Leipzig at the insistence of his father; but changed to the study of science after his father's death circa 1785. Chladni's love of music led him to study the transmission of sound waves, conduct experiments in acoustics, and measure the velocity of sound in various gases. His best known experiment involves the use of a thin metal plate covered with fine sand. When the plate is made to vibrate, the sand collected along the nodal lines of vibration, creating patterns known as Chladni's figures. He demonstrated this technique in Paris to a group of scientists in 1809.

10.1 INTRODUCTION

In practice the measurement of vibration becomes necessary due to the following reasons:

1. The increasing demands of higher productivity and economical design lead to higher operating speeds of machinery[†] and efficient use of materials through lightweight structures. These trends make the occurrence of resonant conditions more frequent during the operation of machinery and reduce the reliability of the system. Hence the periodic measurement of vibration characteristics of machinery and structures becomes essential to ensure adequate safety margins (preventive maintenance).

[†] According to Eshleman, in Ref. [10.12], the average speed of rotating machines doubled—from 1800 rpm to 3600 rpm—during the period between 1940 and 1980.

2. The theoretically computed vibration characteristics of a machine or structure may be different from the actual values due to the assumptions made in the analysis.

3. The measurement of frequencies of vibration and the forces developed is necessary in the design and operation of active vibration isolation systems.

4. The measurement of input and the resulting output vibration characteristics of a system helps in identifying the system in terms of its mass, stiffness, and damping.

5. The information about ground vibrations due to earthquakes, fluctuating wind velocities on structures, random variation of ocean waves, and road surface roughness are important in the design of structures, machines, oil platforms, and vehicle suspension systems.

Vibration Measurement Scheme. Figure 10.1 illustrates the basic features of a vibration measurement scheme. In this figure, the motion (or dynamic force) of the vibrating body is converted into an electrical signal by the vibration transducer or pickup. In general, a transducer is a device that transforms changes in mechanical quantities (such as displacement, velocity, acceleration, or force) into changes in electrical quantities (such as voltage or current). Since the output signal (voltage or current) of a transducer is too small to be recorded directly, a signal conversion instrument is used to amplify the signal to the required value. The output from the signal conversion instrument can be displayed on a display unit for visual inspection, or recorded by a recording unit or stored in a computer for later use. The data can then be analyzed to determine the desired vibration characteristics of the machine or structure.

Depending on the quantity measured, a vibration measuring instrument is called a vibrometer, a velocity meter, an accelerometer, a phase meter, or a frequency meter. If the instrument is designed to record the measured quantity, then the suffix "meter" is to be replaced by "graph" [10.1]. In some applications, we need to vibrate a machine or structure to find its resonance characteristics. For this, electrodynamic vibrators, electrohydraulic vibrators, and signal generators (oscillators) are used.

The following considerations often dictate the type of vibration measuring instruments to be used in a vibration test: (1) Expected ranges of the frequencies and amplitudes, (2) sizes of the machine/structure involved, (3) conditions of operation of the machine/equipment/structure, and (4) type of data processing used (such as graphical display or graphical recording or storing the record in digital form for computer processing).

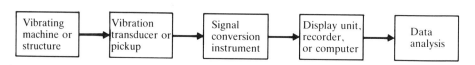

Figure 10.1 Basic vibration measurement scheme.

10.2 TRANSDUCERS

A transducer is a device that transforms values of physical variables into equivalent electrical signals. Several types of transducers are available; some of them are less useful than others due to their nonlinearity or slow response. Some of the transducers commonly used for vibration measurement are given below.

10.2.1 Variable Resistance Transducers

In these transducers, a mechanical motion produces a change in electrical resistance (of a rheostat, a strain gage, or a semiconductor), which, in turn, causes a change in the output voltage or current. The schematic diagram of an electrical resistance strain gage is shown in Fig. 10.2. An electrical resistance strain gage consists of a fine wire whose resistance changes when it is subjected to mechanical deformation. When the strain gage is bonded to a structure, it experiences the same motion (strain) as the structure and hence its resistance change gives the strain applied to the structure. The wire is sandwiched between two sheets of thin paper. The strain gage is bonded to the surface where the strain is to be measured. The most common gage material is a copper-nickel alloy known as Advance. When the surface undergoes a normal strain (ϵ), the strain gage also undergoes the same strain and the resulting change in its resistance is given by [10.6]

$$K = \frac{\Delta R/R}{\Delta L/L} = 1 + 2\nu + \frac{\Delta r}{r}\frac{L}{\Delta L} \approx 1 + 2\nu \qquad (10.1)$$

where

K = gage factor of the wire
R = initial resistance
ΔR = change in resistance
L = initial length of the wire
ΔL = change in length of the wire
ν = Poisson's ratio of the wire
r = resistivity of the wire
Δr = change in resistivity of the wire ≈ 0 for Advance

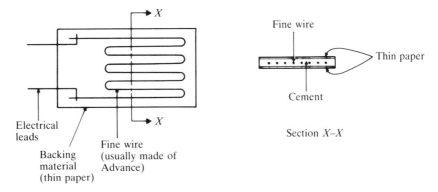

Figure 10.2 Electric resistance strain gage.

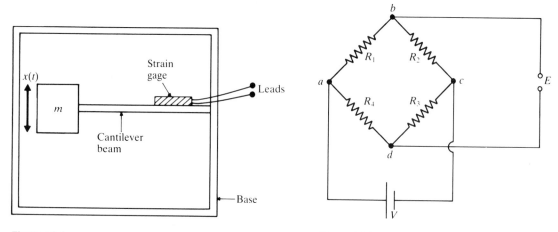

Figure 10.3

Figure 10.4

The value of the gage factor K is given by the manufacturer of the strain gage and hence the value of ϵ can be determined, once ΔR and R are measured, as

$$\epsilon = \frac{\Delta L}{L} = \frac{\Delta R}{RK} \tag{10.2}$$

In a vibration pickup,[†] the strain gage is mounted on an elastic element of a spring mass system as shown in Fig. 10.3. The strain at any point on the cantilever (elastic member) is proportional to the deflection of the mass, $x(t)$, to be measured. Hence the strain indicated by the strain gage can be used to find $x(t)$. The change in resistance of the wire ΔR can be measured using a Wheatstone bridge, potentiometer circuit or voltage divider. A typical Wheatstone bridge, representing a circuit which is sensitive to small changes in the resistance, is shown in Fig. 10.4. A d.c. voltage V is applied across the points a and c. The resulting voltage across the points b and d is given by [10.6]:

$$E = \left[\frac{R_1 R_3 - R_2 R_4}{(R_1 + R_2)(R_3 + R_4)} \right] V \tag{10.3}$$

Initially the resistances are balanced (adjusted) so that the output voltage E is zero. Thus, for initial balance, Eq. (10.3) gives

$$R_1 R_3 = R_2 R_4 \tag{10.4}$$

When the resistances (R_i) change by small amounts (ΔR_i), the change in the output voltage ΔE can be expressed as

$$\Delta E \approx V r_0 \left(\frac{\Delta R_1}{R_1} - \frac{\Delta R_2}{R_2} + \frac{\Delta R_3}{R_3} - \frac{\Delta R_4}{R_4} \right) \tag{10.5}$$

[†] When a transducer is used in conjunction with other components that permit the processing and transmission of the signal, the device is called a *pickup*.

where

$$r_0 = \frac{R_1 R_2}{(R_1 + R_2)^2} = \frac{R_3 R_4}{(R_3 + R_4)^2} \tag{10.6}$$

If the strain gage leads are connected between the points a and b, $R_1 = R_g$, $\Delta R_1 = \Delta R_g$, and $\Delta R_2 = \Delta R_3 = \Delta R_4 = 0$, and Eq. (10.5) gives

$$\frac{\Delta R_g}{R_g} = \frac{\Delta E}{V r_0} \tag{10.7}$$

where R_g is the initial resistance of the gage. Equations (10.2) and (10.7) yield

$$\frac{\Delta R_g}{R_g} = \epsilon K = \frac{\Delta E}{V r_0}$$

or

$$\Delta E = K V r_0 \epsilon \tag{10.8}$$

Since the output voltage is proportional to the strain, it can be calibrated to read the strain directly.

10.2.2 Piezoelectric Transducers

Certain natural and man-made materials like quartz, tourmaline, lithium sulfate, and Rochelle salt generate electrical charge when subjected to a deformation or mechanical stress (see Fig. 10.5a). The electrical charge disappears when the mechanical loading is removed. Such materials are called piezoelectric materials and the transducers, which take advantage of the piezoelectric effect, are known as piezoelectric transducers. The charge generated in the crystal due to a force F_x is given by

$$Q_x = k F_x = k A p_x \tag{10.9}$$

where k is called the piezoelectric constant, A is the area on which the force F_x acts, and p_x is the pressure due to F_x. The output voltage of the crystal is given by

$$E = v t p_x \tag{10.10}$$

where v is called the voltage sensitivity and t is the thickness of the crystal. The values of the piezoelectric constant and voltage sensitivity for quartz are 2.25×10^{-12} Coulomb/Newton and 0.055 volt-meter/Newton, respectively [10.6]. These values are valid only when the perpendicular to the largest face is along the x-axis of the crystal. The electric charge developed and the voltage output will be different if the crystal slab is cut in a different direction.

A typical piezoelectric transducer (accelerometer) is shown in Fig. 10.5(b). In this figure, a small mass is spring loaded against a piezoelectric crystal. When the base vibrates, the load exerted by the mass on the crystal changes with acceleration and hence the output voltage generated by the crystal will be proportional to the acceleration. The main advantages of the piezoelectric accelerometer include compactness, ruggedness, high sensitivity, and high frequency range [10.5, 10.8].

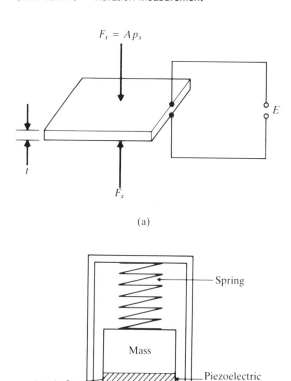

$$F_x = A p_x$$

E

t

$$F_x$$

(a)

Spring

Mass

Piezoelectric
discs

Leads

Base

(b)

Figure 10.5   Piezoelectric accelerometer.

EXAMPLE 10.1	**Output Voltage of a Piezoelectric Transducer**

A quartz crystal having a thickness of 0.1 inch is subjected to a pressure of 50 psi. Find the output voltage if the voltage sensitivity is 0.055 V-m/N.

Given: Quartz crystal with $t = 0.1$ in., $p_x = 50$ psi, and $v = 0.055$ V-m/N.

Find: Output voltage.

Approach: Relationship for voltage developed, Eq. (10.10).

Solution. With $t = 0.1$ in. $= 0.00254$ m, $p_x = 50$ psi $= 344{,}738$ N/m^2, and $v = 0.055$ V-m/N, Eq. (10.10) gives

$$E = (0.055)(0.00254)(344{,}738) = 48.1599 \text{ volts}$$

**10.2.3
Electrodynamic
Transducers**

When an electrical conductor, in the form of a coil, moves in a magnetic field as shown in Fig. 10.6, a voltage E is generated in the conductor. The value of E in volts is given by

$$E = Dlv \qquad (10.11)$$

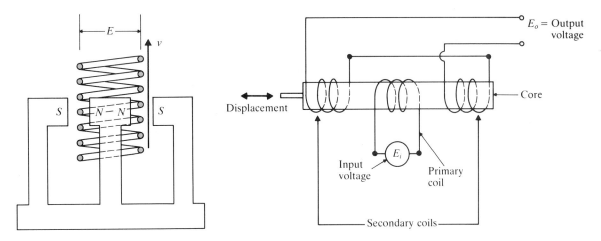

Figure 10.6   **Figure 10.7**

where D is the magnetic flux density (Telsas), l is the length of the conductor (meters), and v is the velocity of the conductor relative to the magnetic field (meters/second). The magnetic field may be produced by either a permanent magnet or an electromagnet. Sometimes, the coil is kept stationary and the magnet is made to move. Since the voltage output of an electromagnetic transducer is proportional to the relative velocity of the coil, they are frequently used in "velocity pickups." Equation (10.11) can be rewritten as

$$Dl = \frac{E}{v} = \frac{F}{I} \qquad (10.12)$$

where F denotes the force (Newtons) acting on the coil while carrying a current I (amperes). Equation (10.12) shows that the performance of an electrodynamic transducer can be reversed. In fact, Eq. (10.12) forms the basis for using an electrodynamic transducer as a "vibration exciter" (see Section 10.5.2).

10.2.4
Linear Variable
Differential
Transformer
(LVDT)
Transducer

The schematic diagram of a LVDT transducer is shown in Fig. 10.7. It consists of a primary coil at the center, two secondary coils at the ends, and a magnetic core that can move freely inside the coils in the axial direction. When an a.c. input voltage is applied to the primary coil, the output voltage will be equal to the difference of the voltages induced in the secondary coils. This output voltage depends on the magnetic coupling between the coils and the core, which in turn, depends on the axial displacement of the core. The secondary coils are connected in phase opposition so that, when the magnetic core is in the exact middle position, the voltages in the two coils will be equal and 180° out of phase. This makes the output voltage of the LVDT as zero. When the core is moved to either side of the middle (zero) position, the magnetic coupling will be increased in one secondary coil and decreased in the other coil. The output polarity depends on the direction of the movement of the magnetic core.

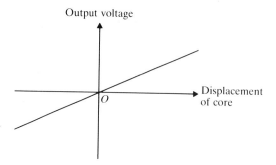

Figure 10.8

The range of displacement for many LVDTs on the market is from 0.0002 cm to 40 cm. The advantages of an LVDT over other displacement transducers include insensitivity to temperature and high output. The mass of the magnetic core restricts the use of the LVDT for high frequency applications [10.4].

As long as the core is not moved very far from the center of the coil, the output voltage varies linearly with the displacement of the core as shown in Fig. 10.8; hence the name linear variable differential transformer.

10.3 VIBRATION PICKUPS

When a transducer is used in conjunction with another device to measure vibrations, it is called a *vibration pickup*. The commonly used vibration pickups are known as seismic instruments. A seismic instrument consists of a mass-spring-damper system mounted on the vibrating body as shown in Fig. 10.9. Then the vibratory motion is measured by finding the displacement of the mass relative to the base on which it is mounted.

The instrument consists of a mass m, a spring k, and a damper c inside a cage, which is fastened to the vibrating body. With this arrangement, the bottom ends of

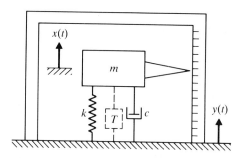

Figure 10.9

the spring and the dashpot will have the same motion as the cage (which is to be measured, y) and their vibration excites the suspended mass into motion. Then the displacement of the mass relative to the cage, $z = x - y$, where x denotes the vertical displacement of the suspended mass, can be measured if we attach a pointer to the mass and a scale to the cage, as shown in Fig. 10.9.[†]

The vibrating body is assumed to have a harmonic motion:

$$y(t) = Y \sin \omega t \qquad (10.13)$$

The equation of motion of the mass m can be written as

$$m\ddot{x} + c(\dot{x} - \dot{y}) + k(x - y) = 0 \qquad (10.14)$$

By defining the relative displacement z as

$$z = x - y \qquad (10.15)$$

Eq. (10.14) can be written as

$$m\ddot{z} + c\dot{z} + kz = -m\ddot{y} \qquad (10.16)$$

Equations (10.13) and (10.16) lead to

$$m\ddot{z} + c\dot{z} + kz = m\omega^2 Y \sin \omega t \qquad (10.17)$$

This equation is identical to Eq. (3.75); hence the steady-state solution is given by

$$z(t) = Z \sin(\omega t - \phi) \qquad (10.18)$$

where Z and ϕ are given by (see Eqs. (3.77) and (3.69)):

$$Z = \frac{Y\omega^2}{\left[(k - m\omega^2)^2 + c^2\omega^2\right]^{1/2}} = \frac{r^2 Y}{\left[(1 - r^2)^2 + (2\zeta r)^2\right]^{1/2}} \qquad (10.19)$$

$$\phi = \tan^{-1}\left(\frac{c\omega}{k - m\omega^2}\right) = \tan^{-1}\left(\frac{2\zeta r}{1 - r^2}\right) \qquad (10.20)$$

$$r = \frac{\omega}{\omega_n} \qquad (10.21)$$

and

$$\zeta = \frac{c}{2m\omega_n} \qquad (10.22)$$

The variations of Z and ϕ with respect to r are shown in Figs. 10.10 and 10.11. As will be seen later, the type of instrument is determined by the useful range of the frequencies, indicated in Fig. 10.10.

10.3.1
Vibrometer

A vibrometer or a seismometer is an instrument that measures the displacement of a vibrating body. It can be observed from Fig. 10.10 that $Z/Y \approx 1$ when $\omega/\omega_n \geq 3$ (range II). Thus the relative displacement between the mass and the base (sensed by the transducer) is essentially the same as the displacement of the base. For an exact

---

[†] The output of the instrument shown in Fig. 10.9 is the relative mechanical motion of the mass, as shown by the pointer and the graduated scale on the cage. For high speed operation and convenience, the motion is often converted into an electrical signal by a transducer.

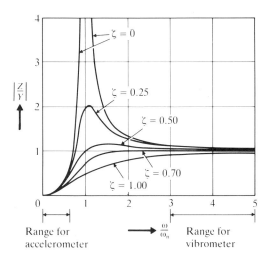

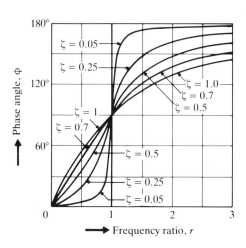

Figure 10.10 Response of a vibration-measuring instrument.

Figure 10.11

analysis, we consider Eq. (10.19). We note that

$$z(t) \simeq Y \sin(\omega t - \phi) \tag{10.23}$$

if

$$\frac{r^2}{\left[(1 - r^2)^2 + (2\zeta r)^2\right]^{1/2}} \approx 1 \tag{10.24}$$

A comparison of Eq. (10.23) with $y(t) = Y \sin \omega t$ shows that $z(t)$ gives directly the motion $y(t)$ except for the phase lag ϕ. This phase lag can be seen to be equal to 180° for $\zeta = 0$. Thus the recorded displacement $z(t)$ lags behind the displacement being measured $y(t)$ by time $t' = \phi/\omega$. This time lag is not important if the base displacement $y(t)$ consists of a single harmonic component.

Since $r = \omega/\omega_n$ has to be large and the value of ω is fixed, the natural frequency $\omega_n = \sqrt{k/m}$ of the mass-spring-damper must be low. This means that the mass must be large and the spring must have a low stiffness. This results in a bulky instrument, which is not desirable in many applications. In practice, the vibrometer may not have a large value of r and hence the value of Z may not be equal to Y exactly. In such a case, the true value of Y can be computed by using Eq. (10.19), as indicated in the following example.

EXAMPLE 10.2   Amplitude by Vibrometer

A vibrometer having a natural frequency of 4 rad/sec and $\zeta = 0.2$ is attached to a structure that performs a harmonic motion. If the difference between the maximum and the minimum recorded values is 8 mm, find the amplitude of motion of the vibrating structure when its frequency is 40 rad/sec.

Given: Vibrometer with $\omega_n = 4$ rad/sec, $\zeta = 0.2$, $Z = 4$ mm, and $\omega = 40$ rad/sec.

Find: Amplitude of the vibrating structure (Y).

Approach: Use Eq. (10.19).

Solution. The amplitude of the recorded motion Z is 4 mm. For $\zeta = 0.2$, $\omega = 40.0$ rad/sec, and $\omega_n = 4$ rad/sec, $r = 10.0$, and Eq. (10.19) gives

$$Z = \frac{Y(10)^2}{\left[\left(1 - 10^2\right)^2 + \{2(0.2)(10)\}^2\right]^{1/2}} = 1.0093Y$$

Thus the amplitude of vibration of the structure is $Y = Z/1.0093 = 3.9631$ mm.

**10.3.2
Accelerometer**

An accelerometer is an instrument that measures the acceleration of a vibrating body (see Fig. 10.12). Accelerometers are widely used for vibration measurements [10.7] and also to record earthquakes. From the accelerometer record, the velocity and displacements are obtained by integration. Differentiation of Eq. (10.13) gives

$$-z(t)\omega_n^2 = \frac{1}{\left[\left(1 - r^2\right)^2 + (2\zeta r)^2\right]^{1/2}}\left\{-Y\omega^2\sin(\omega t - \phi)\right\} \qquad (10.25)$$

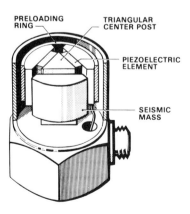

Figure 10.12 Accelerometers. (Courtesy of Bruel & Kjaer Instruments, Inc., Marlborough, Mass.)

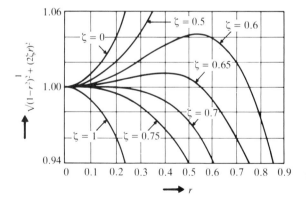

Figure 10.13

This shows that if

$$\frac{1}{\left[(1 - r^2)^2 + (2\zeta r)^2\right]^{1/2}} \simeq 1 \tag{10.26}$$

Eq. (10.25) becomes

$$-z(t)\omega_n^2 \simeq -Y\omega^2 \sin(\omega t - \phi) \tag{10.27}$$

By comparing Eq. (10.27) with $\ddot{y}(t) = -Y\omega^2 \sin \omega t$, we find that the term $z(t)\omega_n^2$ gives the acceleration of the base $\ddot{y}$, except for the phase lag ϕ. Thus the instrument can be made to record (give) directly the value of $\ddot{y} = -z(t)\omega_n^2$. The time by which the record lags the acceleration is given by $t' = \phi/\omega$. If $\ddot{y}$ consists of a single harmonic component, the time lag will not be of importance.

The value of the expression on the left-hand side of Eq. (10.26) is shown plotted in Fig. 10.13. It can be seen that the left-hand side of Eq. (10.26) lies between 0.96 and 1.04 for $0 \leqslant r \leqslant 0.6$ if the value of ζ lies between 0.65 and 0.7. Since r is small, the natural frequency of the instrument has to be large compared to the frequency of vibration to be measured. From the relation $\omega_n = \sqrt{k/m}$, we find that the mass needs to be small and the spring needs to have a large value of k (i.e., short spring), so the instrument will be small in size. Due to their small size and high sensitivity, accelerometers are preferred in vibration measurements. In practice, Eq. (10.26) may not be satisfied exactly; in such cases the quantity

$$\frac{1}{\left[(1 - r^2)^2 + (2\zeta r)^2\right]^{1/2}}$$

can be used to find the correct value of the acceleration measured as illustrated in the following example.

EXAMPLE 10.3   Design of an Accelerometer

An accelerometer has a suspended mass of 0.01 kg with a damped natural frequency of vibration of 150 Hz. When mounted on an engine undergoing an acceleration of 1 g at an operating speed of 6000 rpm, the acceleration is recorded as 9.5 m/s² by the instrument. Find the damping constant and the spring stiffness of the accelerometer.

Given: Accelerometer of mass = 0.01 kg, damped natural frequency = 150 Hz, and recorded acceleration = 9.5 m/s². Engine with operating speed = 6000 rpm, acceleration = 1 g = 9.81 m/s².

Find: Damping constant (c) and spring stiffness (k) of the accelerometer.

Approach: Use the equation for the ratio of recorded and true accelerations in conjunction with the equation for the damped natural frequency.

Solution. The ratio of measured to true accelerations is given by

$$\frac{1}{\left[(1-r^2)^2+(2\zeta r)^2\right]^{1/2}} = \frac{\text{measured value}}{\text{true value}} = \frac{9.5}{9.81} = 0.9684 \qquad \text{(E.1)}$$

which can be rewritten as

$$\left[(1-r^2)^2+(2\zeta r)^2\right] = (1/0.9684)^2 = 1.0663 \qquad \text{(E.2)}$$

The operating speed of the engine gives

$$\omega = \frac{6000(2\pi)}{60} = 628.32 \text{ rad/sec}$$

The damped natural frequency of vibration of the accelerometer is

$$\omega_d = \sqrt{1-\zeta^2}\,\omega_n = 150(2\pi) = 942.48 \text{ rad/sec}$$

Thus

$$\frac{\omega}{\omega_d} = \frac{\omega}{\sqrt{1-\zeta^2}\,\omega_n} = \frac{r}{\sqrt{1-\zeta^2}} = \frac{628.32}{942.48} = 0.6667 \qquad \text{(E.3)}$$

Equation (E.3) gives

$$r = 0.6667\sqrt{1-\zeta^2} \qquad \text{or} \qquad r^2 = 0.4444(1-\zeta^2) \qquad \text{(E.4)}$$

Substitution of Eq. (E.4) into (E.2) leads to a quadratic equation in ζ^2 as

$$1.5801\zeta^4 - 2.2714\zeta^2 + 0.7576 = 0 \qquad \text{(E.5)}$$

The solution of Eq. (E.5) gives

$$\zeta^2 = 0.5260, 0.9115$$

or

$$\zeta = 0.7253, 0.9547$$

By choosing $\zeta = 0.7253$ arbitrarily, the undamped natural frequency of the accelerometer can be found as

$$\omega_n = \frac{\omega_d}{\sqrt{1-\zeta^2}} = \frac{942.48}{\sqrt{1-0.7253^2}} = 1368.8889 \text{ rad/sec}$$

Since $\omega_n = \sqrt{k/m}$, we have

$$k = m\omega_n^2 = (0.01)(1368.8889)^2 = 18738.5628 \text{ N/m}$$

The damping constant can be determined from

$$c = 2m\omega_n\zeta = 2(0.01)(1368.8889)(0.7253) = 19.8571 \text{ N-s/m}$$

10.3.3 Velometer

A velometer measures the velocity of a vibrating body. Equation (10.13) gives the velocity of the vibrating body:

$$\dot{y}(t) = \omega Y \cos \omega t \qquad \text{(10.28)}$$

and Eq. (10.18) gives

$$\dot{z}(t) = \frac{r^2 \omega Y}{\left[(1-r^2)^2+(2\zeta r)^2\right]^{1/2}} \cos(\omega t - \phi) \qquad \text{(10.29)}$$

If

$$\frac{r^2}{\left[(1 - r^2)^2 + (2\zeta r)^2\right]^{1/2}} \simeq 1 \tag{10.30}$$

then

$$\dot{z}(t) \simeq \omega Y \cos(\omega t - \phi) \tag{10.31}$$

A comparison of Eqs. (10.28) and (10.31) shows that, except for the phase difference ϕ, $\dot{z}(t)$ gives directly $\dot{y}(t)$, provided that Eq. (10.30) holds true. In order to satisfy Eq. (10.30), r must be very large. In case Eq. (10.30) is not satisfied, then the velocity of the vibrating body can be computed using Eq. (10.29).

EXAMPLE 10.4   **Design of a Velometer**

Design a velometer if the maximum error is to be limited to 1% of the true velocity. The natural frequency of the velometer is to be 80 Hz and the suspended mass is to be 0.05 kg.

Given: Velometer with $\omega_n = 80$ Hz, $m = 0.05$ kg, and accuracy $= \pm 1\%$.

Find: k and c of velometer.

Approach: Use expression for maximum error.

Solution. The ratio (R) of the recorded and the true velocities is given by Eq. (10.29):

$$R = \frac{r^2}{\left[(1 - r^2)^2 + (2\zeta r)^2\right]^{1/2}} = \frac{\text{recorded velocity}}{\text{true velocity}} \tag{E.1}$$

The maximum of (E.1) occurs when (see Eq. (3.84))

$$r = r^* = \frac{1}{\sqrt{1 - 2\zeta^2}} \tag{E.2}$$

Substitution of Eq. (E.2) into (E.1) gives

$$\frac{\left(\dfrac{1}{1 - 2\zeta^2}\right)}{\sqrt{\left[1 - \left(\dfrac{1}{1 - 2\zeta^2}\right)\right]^2 + 4\zeta^2\left(\dfrac{1}{1 - 2\zeta^2}\right)}} = R$$

which can be simplified as

$$\frac{1}{\sqrt{4\zeta^2 - 4\zeta^4}} = R \tag{E.3}$$

For an error of 1%, $R = 1.01$ or 0.99, and Eq. (E.3) leads to

$$\zeta^4 - \zeta^2 + 0.245075 = 0 \tag{E.4}$$

and

$$\zeta^4 - \zeta^2 + 0.255075 = 0 \tag{E.5}$$

Equation (E.5) gives imaginary roots and Eq. (E.4) gives

$$\zeta^2 = 0.570178, 0.429821$$

or

$$\zeta = 0.755101, 0.655607$$

We choose the value $\zeta = 0.755101$ arbitrarily. The spring stiffness can be found as

$$k = m\omega_n^2 = 0.05(502.656)^2 = 12633.1527 \text{ N/m}$$

since

$$\omega_n = 80(2\pi) = 502.656 \text{ rad/sec}$$

The damping constant can be determined from

$$c = 2\zeta\omega_n m = 2(0.755101)(502.656)(0.05) = 37.9556 \text{ N-s/m}$$

10.3.4
Phase Distortion

As shown by Eq. (10.18), all vibration-measuring instruments exhibit phase lag. Thus the response or output of the instrument lags behind the motion or input it measures. The time lag is given by the phase angle divided by the frequency ω. The time lag is not important if we measure a single harmonic component. But, occasionally, the vibration to be recorded is not harmonic but consists of the sum of two or more harmonic components. In such a case, the recorded graph may not give an accurate picture of the vibration because different harmonics may be amplified by different amounts and their phase shifts also may be different. The distortion in the wave form of the recorded signal is called the phase distortion or phase-shift error. To illustrate the nature of the phase-shift error, we consider a vibration signal of the form shown in Fig. 10.14(a) [10.10]:

$$y(t) = a_1\sin \omega t + a_3\sin 3\omega t \tag{10.32}$$

Let the phase shift be $90°$ for the first harmonic and $180°$ for the third harmonic of Eq. (10.32). The corresponding time lags are given by $t_1 = \theta_1/\omega = 90°/\omega$ and $t_2 = \theta_2/(3\omega) = 180°/(3\omega)$. The output signal is shown in Fig. 10.14(b). It can be seen that the output signal is quite different from the input signal due to phase distortion.

As a general case, let the complex wave being measured be given by the sum of several harmonics as

$$y(t) = a_1\sin \omega t + a_2\sin 2\omega t + \cdots \tag{10.33}$$

If the displacement is measured using a vibrometer, its response to each component of the series is given by an equation similar to Eq. (10.18) so that the output of the vibrometer becomes

$$z(t) = a_1\sin(\omega t - \phi_1) + a_2\sin(2\omega t - \phi_2) + \cdots \tag{10.34}$$

where

$$\tan \phi_j = \frac{2\zeta\left(j\dfrac{\omega}{\omega_n}\right)}{1 - \left(j\dfrac{\omega}{\omega_n}\right)^2}, \qquad j = 1, 2, \ldots \tag{10.35}$$

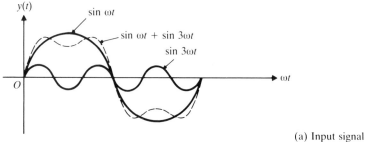

(a) Input signal

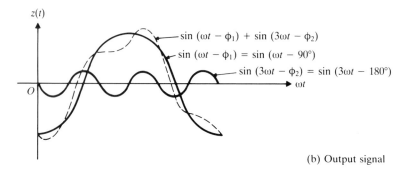

(b) Output signal

Figure 10.14

Since ω/ω_n is large for this instrument, we can find from Fig. 10.11 that $\phi_j \simeq \pi$, $j = 1, 2, \ldots$ and Eq. (10.34) becomes

$$z(t) \simeq -[a_1\sin \omega t + a_2\sin 2\omega t + \cdots] \simeq -y(t) \qquad (10.36)$$

Thus the output record will be simply opposite to the motion being measured. This is unimportant and can easily be corrected.

By using a similar reasoning, we can show, in the case of a velometer, that

$$\dot{z}(t) \simeq -\dot{y}(t) \qquad (10.37)$$

for an input signal consisting of several harmonics. Next we consider the phase distortion for an accelerometer. Let the acceleration curve to be measured be expressed, using Eq. (10.33), as

$$\ddot{y}(t) = -a_1\omega^2\sin \omega t - a_2(2\omega)^2\sin 2\omega t - \cdots \qquad (10.38)$$

The response or output of the instrument to each component can be found as in Eq. (10.34), and so

$$\ddot{z}(t) = -a_1\omega^2\sin(\omega t - \phi_1) - a_2(2\omega)^2\sin(2\omega t - \phi_2) - \cdots \qquad (10.39)$$

where the phase lags ϕ_j are different for different components of the series in Eq. 10.39. Since the phase lag ϕ varies almost linearly from $0°$ at $r = 0$ to $90°$ at $r = 1$ for $\zeta = 0.7$ (see Fig. 10.11), we can express ϕ as

$$\phi \simeq \alpha r = \alpha\frac{\omega}{\omega_n} = \beta\omega \qquad (10.40)$$

where α and $\beta = \alpha/\omega_n$ are constants. The time lag is given by

$$t' = \frac{\phi}{\omega} = \frac{\beta\omega}{\omega} = \beta \qquad (10.41)$$

This shows that the time lag of the accelerometer is independent of the frequency for any component, provided that the frequency lies in the range $0 \leqslant r \leqslant 1$. Since each component of the signal has the same time delay or phase lag, we have, from Eq. (10.39),

$$
\begin{aligned}
-\omega^2 \ddot{z}(t) &= -a_1\omega^2\sin(\omega t - \omega\beta) - a_2(2\omega)^2\sin(2\omega t - 2\omega\beta) - \cdots \\
&= -a_1\omega^2\sin\omega\tau - a_2(2\omega)^2\sin 2\omega\tau - \cdots \qquad (10.42)
\end{aligned}
$$

where $\tau = t - \beta$. Note that Eq. (10.42) assumes that $0 \leqslant r \leqslant 1$, that is, even the highest frequency involved, $n\omega$, is less than ω_n. This may not be true in practice. Fortunately, no significant phase distortion occurs in the output signal even when some of the higher order frequencies are larger than ω_n. The reason is that, generally, only the first few components are important to approximate even a complex wave form; the amplitudes of the higher harmonics are small and contribute very little to the total wave form. Thus the output record of the accelerometer represents a reasonably true acceleration being measured [10.7, 10.11].

10.4 FREQUENCY MEASURING INSTRUMENTS

Most frequency-measuring instruments are of the mechanical type and are based on the principle of resonance. Two kinds of instruments are discussed in the following paragraphs: the Fullarton tachometer and the Frahm tachometer.

Single-Reed Instrument or Fullarton Tachometer. This instrument consists of a variable-length cantilever strip with a mass attached at one of its ends. The other end of the strip is clamped, and its free length can be changed by means of a screw mechanism (see Fig. 10.15a). Since each length of the strip corresponds to a different natural frequency, the reed is marked along its length in terms of its

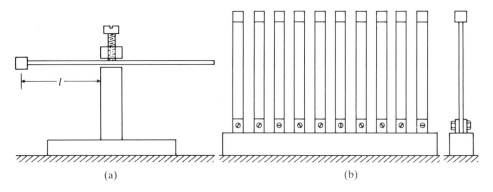

(a)   (b)

Figure 10.15 Frequency-measuring instruments.

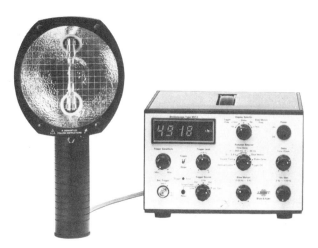

Figure 10.16   A Stroboscope. (Courtesy of Bruel & Kjaer Instruments, Inc., Marlborough, Mass.)

natural frequency. In practice, the clamped end of the strip is pressed against the vibrating body, and the screw mechanism is manipulated to alter its free length until the free end shows the largest amplitude of vibration. At that instant, the excitation frequency is equal to the natural frequency of the cantilever; it can be read directly from the strip.

Multireed-Instrument or Frahm Tachometer. This instrument consists of a number of cantilevered reeds carrying small masses at their free ends (see Fig. 10.15b). Each reed has a different natural frequency and is marked accordingly. Using a number of reeds makes it possible to cover a wide frequency range. When the instrument is mounted on a vibrating body, the reed whose natural frequency is nearest the unknown frequency of the body vibrates with the largest amplitude. The frequency of the vibrating body can be found from the known frequency of the vibrating reed.

Stroboscope. A stroboscope is an instrument which produces light pulses intermittently. The frequency at which the light pulses are produced can be altered and read from the instrument. When a specific point on a rotating (vibrating) object is viewed with the stroboscope, it will appear to be stationary only when the frequency of the pulsating light is equal to the speed of the rotating (vibrating) object. The main advantage of the stroboscope is that it does not make contact with the rotating (vibrating) body. Due to the persistence of vision, the lowest frequency that can be measured with a stroboscope is approximately 15 Hz. A typical stroboscope is shown in Fig. 10.16.

10.5 VIBRATION EXCITERS

The vibration exciters or shakers can be used in several applications such as determination of the dynamic characteristics of machines and structures and fatigue testing of materials. The vibration exciters can be mechanical, electromagnetic, electrodynamic, or hydraulic type. The working principles of mechanical and electromagnetic exciters are described in this section.

**10.5.1
Mechanical
Exciters**

As indicated in Section 1.10 (Fig. 1.25), a Scotch yoke mechanism can be used to produce harmonic vibrations. The crank of the mechanism can be driven either by a constant- or a variable-speed motor. When a structure is to be vibrated, the harmonic force can be applied either as an inertia force as shown in Fig. 10.17(a) or as an elastic spring force as shown in Fig. 10.17(b). These vibrators are generally used for frequencies less than 30 Hz and loads less than 700 N [10.1].

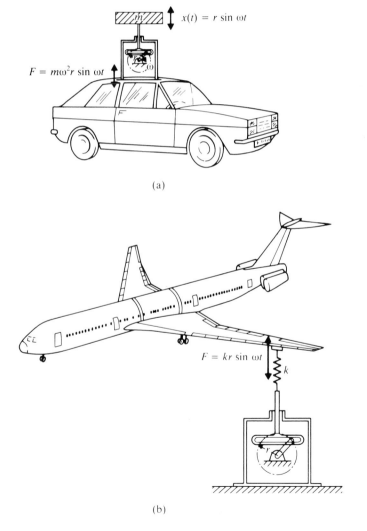

(a)

(b)

Figure 10.17   (a) Vibration of structure through inertia force (b) Vibration of structure through spring force.

The unbalance created by two masses rotating at the same speed in opposite directions (see Fig. 10.18) can be used as a mechanical exciter. This type of shaker can be used to generate relatively large loads between 250 N and 25,000 N. If the two masses, of magnitude m each, rotate at an angular velocity ω at a radius R, the vertical force $F(t)$ generated is given by

$$F(t) = 2mR\omega^2\cos \omega t \qquad (10.43)$$

The horizontal components of the two masses cancel and hence the resultant horizontal force will be zero. The force $F(t)$ will be applied to the structure to which the exciter is attached.

10.5.2 Electrodynamic Shaker

The schematic diagram of an electrodynamic shaker, also known as the electromagnetic exciter, is shown in Fig. 10.19(a). As stated in Section 10.2.3, the electrodynamic shaker can be considered as the reverse of an electrodynamic transducer. When current passes through a coil placed in a magnetic field, a force F (in Newtons) proportional to the current I (in amperes) and the magnetic flux intensity D (in Telsas), is produced which accelerates the component placed on the shaker table:

$$F = DIl \qquad (10.44)$$

where l is the length of the coil (in meters). The magnetic field is produced by a permanent magnet in small shakers while an electromagnet is used in large shakers. The magnitude of acceleration of the table or component depends on the maximum current and the masses of the component and the moving element of the shaker. If the current flowing through the coil varies harmonically with time (a.c. current), the force produced also varies harmonically. On the other hand, if direct current is used to energize the coil, a constant force is generated at the exciter table. The

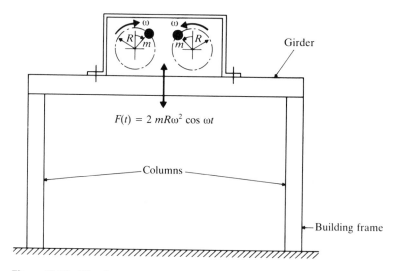

Figure 10.18   Vibration excitation due to unbalanced force.

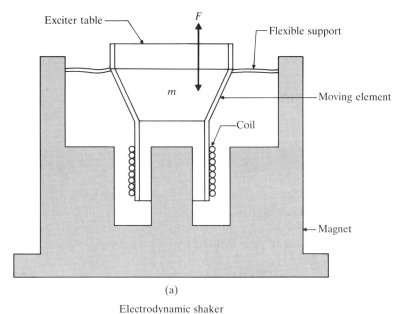

(a)

Electrodynamic shaker

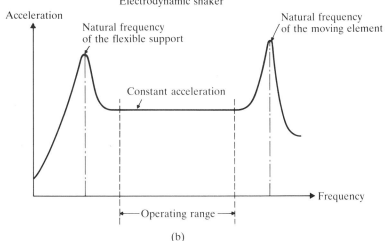

(b)

Typical resonance characteristics of an
electrodynamic exciter

Figure 10.19 (a) Electrodynamic shaker. (b) Typical resonance
characteristics of an electrodynamic exciter.

electrodynamic exciters can be used in conjunction with an inertia or a spring as in
the case of Figs. 10.17(a) and (b) to vibrate a structure.

Since the coil and the moving element should have a linear motion, they are
suspended from a flexible support (having a very small stiffness) as shown in Fig.
10.19(a). Thus the electromagnetic exciter has two natural frequencies; one corresponding to the natural frequency of the flexible support and the other corresponding to the natural frequency of the moving element, which can be made very large.

Figure 10.20 An exciter with a general purpose head. (Courtesy of Bruel & Kjaer Instruments, Inc., Marlborough, Mass.)

These two resonant frequencies are shown in Fig. 10.19(b). The operating frequency range of the exciter lies between these two resonant frequencies as indicated in Fig. 10.19(b) [10.7].

The electrodynamic exciters are used to generate forces up to 30,000 N, displacements up to 25 mm, and frequencies in the range of 5 Hz to 20 kHz [10.1]. A practical electrodynamic exciter is shown in Fig. 10.20.

10.6 SIGNAL ANALYSIS

In signal analysis, we determine the response of a system under a known excitation and present it in a convenient form. Often, the time-response of a system will not give much useful information. However, the frequency-response will show one or more discrete frequencies around which the energy is concentrated. Since the dynamic characteristics of individual components of the system are usually known, we can relate the distinct frequency components (of the frequency-response) to specific components [10.3].

For example, the acceleration-time history of a machine frame that is subjected to excessive vibration might appear as shown in Fig. 10.21(a). This figure cannot be used to identify the cause of vibration. If the acceleration-time history is trans-

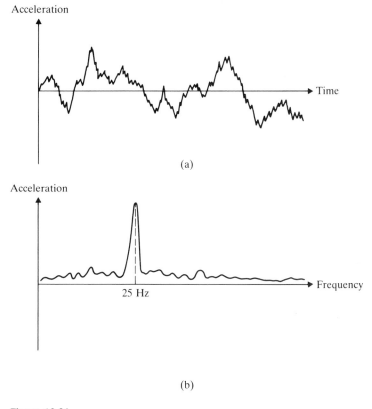

Figure 10.21

formed to the frequency domain, the resulting frequency spectrum might appear as shown in Fig. 10.21(b), where, for specificness, the energy is shown concentrated around 25 Hz. This frequency can easily be related, for example, to the rotational speed of a particular motor. Thus the acceleration spectrum shows a strong evidence that the motor might be the cause of vibration. If the motor is causing the excessive vibrations, changing either the motor or its speed of operation might avoid resonance and hence the problem of excessive vibrations.

**10.6.1
Spectrum
Analyzers**

Spectrum or frequency analyzers can be used for signal analysis. A spectrum or frequency analyzer is a device that analyzes a signal in the frequency domain by separating the energy of the signal into various frequency bands. The separation of signal energy into frequency bands is accomplished through a set of filters. The analyzers are usually classified according to the type of filter employed. For example, if an octave band filter is used, the spectrum analyzer is called an octave band analyzer.

In recent years, digital analyzers have become quite popular for real-time signal analysis. In a real-time frequency analysis, the signal is continuously analyzed over

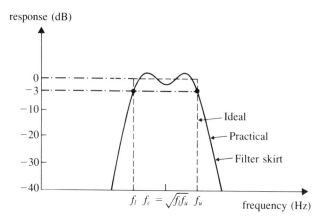

Figure 10.22   Response of a filter.

all the frequency bands. Thus the calculation process must not take more time than the time taken to collect the signal data. Real-time analyzers are especially useful for machinery health monitoring since a change in the noise or vibration spectrum can be observed at the same time that change in the machine occurs. There are two types of real-time analysis procedures, namely, the digital filtering method and the Fast Fourier Transform (FFT) method [10.13]. The digital filtering method is best suited for constant percent bandwidth analysis while the FFT method is most suitable for constant bandwidth analysis. Before we consider the difference between the constant percent bandwidth and constant bandwidth analyses, we first discuss the basic component of a spectrum analyzer, namely, the bandpass filter.

10.6.2
Bandpass Filter

A bandpass filter is a circuit which permits the passage of frequency components of a signal over a frequency band and rejects all other frequency components of the signal. A filter can be built by using, for example, resistors, inductors and capacitors. Figure 10.22 illustrates the response characteristics of a filter whose lower and upper cutoff frequencies are f_l and f_u, respectively. A practical filter will have a response characteristic deviating from the ideal rectangle as shown by the full line in Fig. 10.22. For a good bandpass filter, the ripples within the band will be minimum and the slopes of the filter skirts will be steep to maintain the actual bandwidth close to the ideal value, $B = f_u - f_l$. For a practical filter, the frequencies f_l and f_u at which the response is 3 db[†] below its mean bandpass response are called the cutoff frequencies.

[†]   A decibel (db) of a quantity (such as power, P) is defined as:

$$\text{Quantity in db} = 10 \log_{10}\left(\frac{P}{P_{\text{ref}}}\right)$$

where P is the power and P_{ref} is a reference value of the power.

TABLE 10.1

Lower cutoff limit (Hz)	5.63	11.2	22.4	44.7	89.2	178	355	709	1410
Center frequency (Hz)	8.0	16.0	31.5	63.0	125	250	500	1000	2000
Upper cutoff limit (Hz)	11.2	22.4	44.7	89.2	178	355	709	1410	2820

There are two types of bandpass filters used in signal analysis, namely, the constant percent bandwidth filters and constant bandwidth filters. For a constant percent bandwidth filter, the ratio of the bandwidth to the center (tuned) frequency, $(f_u - f_l)/f_c$, is a constant. The octave[†], one-half-octave and one-third-octave band filters are examples of constant percent bandwidth filters. Some of the cutoff limits and center frequencies of octave bands used in signal analysis are shown in Table 10.1. For a constant bandwidth filter, the bandwidth, $f_u - f_l$, is independent of the center (tuned) frequency, f_c.

10.6.3 Constant Percent Bandwidth and Constant Bandwidth Analyzers

The primary difference between the constant percent bandwidth and constant bandwidth analyzers lies in the detail provided by the various bandwidths. The octave band filters, whose upper cutoff frequency is twice the lower cutoff frequency, give a less detailed (too coarse) analysis for practical vibration and noise encountered in machines. The one-half-octave band filter gives twice the information, but requires twice the amount of time to obtain the data. A spectrum analyzer with a set of octave and one-third octave filters can be used for noise (signal) analysis. Each filter is tuned to a different center frequency to cover the entire frequency range of interest. Since the lower cutoff frequency of a filter is equal to the upper cutoff frequency of the previous filter, the composite filter characteristic will appear as shown in Fig. 10.23. Figure 10.24 shows a real time octave and fractional octave digital frequency analyzer. A constant bandwidth analyzer is used to obtain a more detailed analysis than in the case of a constant percent bandwidth analyzer, especially in the high frequency range of the signal. The constant bandwidth filter, when used with a continuously varying center frequency, is called a wave or heterodyne analyzer. Heterodyne analyzers are available with constant filter bandwidths ranging from one to several hundred Hertz. A practical heterodyne analyzer is shown in Fig. 10.25.

[†] An octave is the interval between any two frequencies $(f_2 - f_1)$, whose frequency ratio (f_2/f_1), is 2. Two frequencies f_1 and f_2 are said to be separated by a number of octaves N when

$$\frac{f_2}{f_1} = 2^N \qquad \text{or} \qquad N \text{ (in octaves)} = \log_2\left(\frac{f_2}{f_1}\right)$$

where N can be an integer or a fraction. If $N = 1$, we have an octave, if $N = \frac{1}{3}$, we get a one-third octave, and so on.

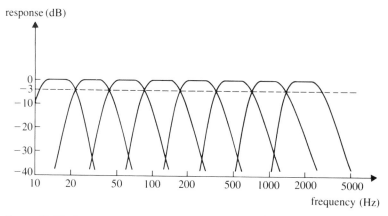

Figure 10.23   Response characteristic of a typical octave-band filter set.

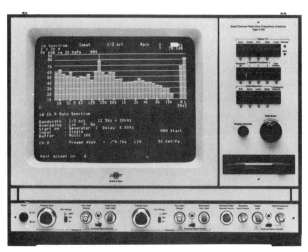

Figure 10.24   Octave and fractional-octave digital frequency analyzer. (Courtesy of Bruel & Kjaer Instruments, Inc., Marlborough, Mass.)

Figure 10.25   Heterodyne analyzer. (Courtesy of Bruel & Kjaer Instruments, Inc., Marlborough, Mass.)

10.7 DYNAMIC TESTING OF MACHINES AND STRUCTURES

The dynamic testing of machines (structures) involves finding the deformation of the machines (structures) at a critical frequency. This can be done using the following two approaches [10.3].

**10.7.1
Using
Operational
Deflection Shape
Measurements**

In this method, the forced dynamic deflection shape is measured under the steady state (operating) frequency of the system. For the measurement, an accelerometer is mounted at some point on the machine (structure) as a reference, and another moving accelerometer is placed at several other points, and in different directions, if necessary (see Fig. 10.26). Then the magnitudes and the phase differences between

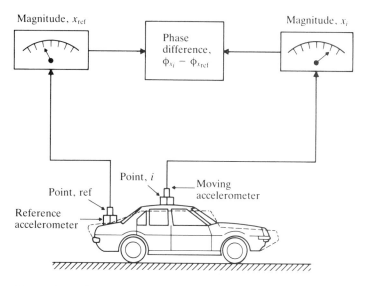

Figure 10.26 Operational deflection shape measurement.

the moving and reference accelerometers at all the points under steady-state operation of the system are measured. By plotting these measurements, we can find how the various parts of the machine (structure) move relative to one another and also absolutely.

The deflection shape measured is valid only for the forces/frequency associated with the operating conditions; as such, we cannot get information about deflections under other forces and/or frequencies. However, the measured deflection shape can be quite useful. For example, if a particular part or location is found to have excessive deflection, we can stiffen that part or location. This, in effect, increases the natural frequency beyond the operational frequency range of the system.

**10.7.2
Using Modal
Testing**

Since any dynamic response of a machine (structure) can be obtained as a combination of its modes, a knowledge of the mode shapes, modal frequencies and modal damping ratios constitute a complete dynamic description of the machine (structure). The experimental modal analysis procedure is described in the following section.

10.8   MODAL ANALYSIS

**10.8.1
Introduction**

As stated in Chapter 6, any dynamic response of a machine/structure can be obtained by superposing its natural modes of vibration when the amplitudes of motion are small. A complete dynamic description of the machine/structure requires the determination of the modal frequencies, mode shapes, and the system parameters—the equivalent mass, stiffness, and the damping ratio.

The frequency response function plays an important role in the experimental modal analysis. The frequency response function is first determined experimentally and then analyzed to find the natural frequencies, mode shapes, and the system parameters—the equivalent mass, stiffness, and the damping ratio. The system parameters can be used to predict the responses to various excitations or to improve the dynamic behavior of the system by design modifications. In the modal analysis, the structure is assumed to be linear and the parameters to be time-invariant.

10.8.2
Types of Forcing
Functions

The following types of forcing functions can be used to determine the frequency response function of a structure: (i) steady-state harmonic excitation, (ii) quasi-steady-state excitation, (iii) transient excitation, and (iv) continuous random excitation [10.9]. For a linear system with time-invariant system parameters, the frequency response measurement will be independent of the testing technique used.

In the case of steady-state harmonic excitation, the system is excited harmonically at a constant frequency and the response is measured. This procedure is repeated at several discrete frequencies to obtain the complete frequency response function. Since the procedure has to be repeated several times, it is time consuming and hence is not used frequently. However, in situations where the expected environment of the system is dominated by one or a few discrete frequencies, this method is quite useful.

The quasi-steady-state excitation method involves a slow frequency sweep and has become popular because of the availability of the transfer function analysis equipment. A sinusoidal force is swept through the frequency range of interest at a sufficiently slow rate to measure the response of the system at all frequencies.

In the transient excitation method, the frequency response function is computed from the Fourier transforms of the excitation and the response time histories. Digital computers and real time analyzers permit the on-line computation of the response of the system. The random excitation method is preferable as it simulates the real environment better. In harmonic excitation, only a single resonance will be excited at a time and any interaction between the resonances will not be detected. The random excitation, on the other hand, acts on all resonances at the same time.

The frequency response data are either displayed and recorded for graphical analysis or stored for computer processing at a later time.

10.8.3
Representation
of Frequency
Response Data

The frequency response data can be represented to obtain plots of (i) modulus and phase angle against frequency, (ii) real and imaginary components of response with varying frequency, or (iii) vector diagram of the real component versus the imaginary component of the response. Since the normal mode method permits the representation of a multidegree of freedom system (with n degrees of freedom) as an equivalent n single degrees of freedom systems, we consider a single degree of freedom system shown in Fig. 10.27. The equation of motion of the system, under the harmonic force $F(t) = F_0 e^{i\omega t}$, is given by

$$m\ddot{x} + c\dot{x} + kx = F_0 e^{i\omega t} \qquad (10.45)$$

By assuming a solution

$$x(t) = X e^{i\omega t} \qquad (10.46)$$

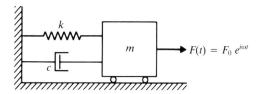

Figure 10.27

the amplitude of the response can be found as (see Section 3.5):

$$X = \frac{H(\omega)F_0}{k} = \frac{F_0}{k}\left\{\frac{1}{1 - \left(\frac{\omega}{\omega_n}\right)^2 + i\left(\frac{c\omega}{k}\right)}\right\} = \frac{F_0}{k}\left\{\frac{1}{1 - r^2 + i2\zeta r}\right\} \tag{10.47}$$

where $\omega_n = \sqrt{k/m}$, $r = \omega/\omega_n$, $c\omega/k = 2\zeta r$, and ζ is the viscous damping ratio. The amplitude of the response X can be expressed as

$$X = X_M e^{i\phi} = X_M\cos\phi + iX_M\sin\phi \equiv X_R + iX_I \tag{10.48}$$

where X_M, X_R, and X_I denote the modulus, real part, and the imaginary part of the amplitude of the response, respectively, with

$$X_R = \frac{F_0}{k}\left\{\frac{1 - r^2}{\left(1 - r^2\right)^2 + (2\zeta r)^2}\right\} \tag{10.49}$$

$$X_I = \frac{F_0}{k}\left\{\frac{-2\zeta r}{\left(1 - r^2\right)^2 + (2\zeta r)^2}\right\} \tag{10.50}$$

$$X_M = \left(X_R^2 + X_I^2\right)^{1/2} = \frac{F_0}{k}\frac{1}{\left[\left(1 - r^2\right)^2 + (2\zeta r)^2\right]^{1/2}} \tag{10.51}$$

The phase angle ϕ can be obtained as

$$\tan\phi = \frac{X_I}{X_R} = \left(\frac{-2\zeta r}{1 - r^2}\right) \tag{10.52}$$

Plots of Modulus and Phase Angle. The variations of the modulus and the phase angle with the frequency are shown in Figs. 10.28 and 10.29. If ω_1 and ω_2 are the frequencies at which the response amplitude is $X_{Mr}/\sqrt{2}$ (half power points) where X_{Mr} is the resonance amplitude (value of X_M when $r = 1$), the damping ratio can be found as follows (see Section 3.4.2):

$$\zeta \simeq \frac{\omega_2^2 - \omega_1^2}{4\omega_n^2} \simeq \frac{\omega_2 - \omega_1}{2\omega_n} \tag{10.53}$$

In the experimental testing, the frequency corresponding to the peak amplitude is

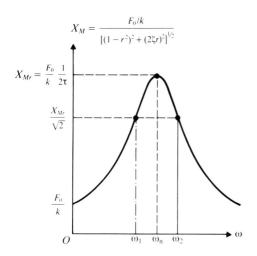

Figure 10.28

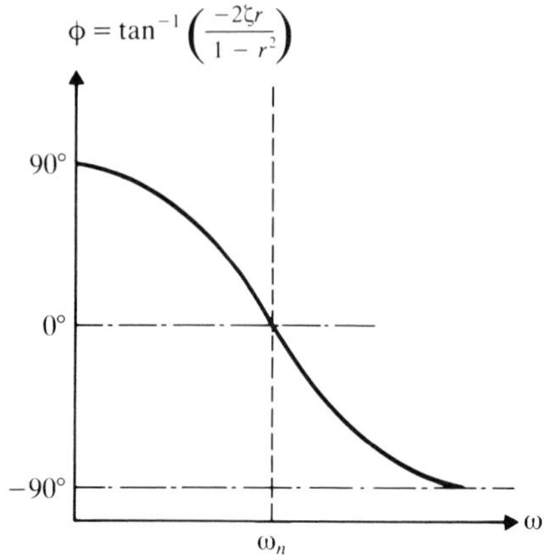

$$\phi = \tan^{-1}\left(\frac{-2\zeta r}{1 - r^2}\right)$$

Figure 10.29

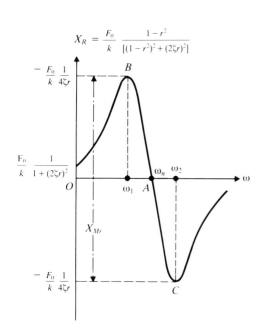

Figure 10.30

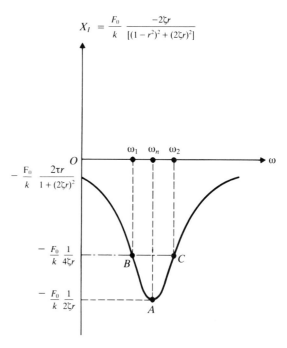

Figure 10.31

identified as ω_n. The peak amplitude (X_{Mr}) gives

$$X_{Mr} = \frac{1}{2\zeta}\frac{F_0}{k} \tag{10.54}$$

The identification of half power points permit the use of Eq. (10.53). Equations (10.53) and (10.54), along with the relation $\omega_n = \sqrt{k/m}$, yield the values of m, k, and c of the system.

Plots of Real and Imaginary Components of Response. The variations of X_R and X_I, given by Eqs. (10.49) and (10.50), with frequency are shown in Figs. 10.30 and 10.31. Resonance can be identified as the value of ω, where X_R is zero or X_I is a maximum. The half power points correspond to $|X_R|_{max}$ and hence Eq. (10.53) can be used.

Plot of Real Component versus Imaginary Component of Response (Vector Diagram). The frequency ω can be eliminated from Eqs. (10.49) and (10.50) to obtain the equation:

$$\left[X_I + \frac{F_0}{k}\frac{1}{(4\zeta r)}\right]^2 + X_R^2 = \left[\frac{F_0}{k}\frac{1}{(4\zeta r)}\right]^2 \tag{10.55}$$

Equation (10.55) denotes the equation of a circle, which is shown in Fig. 10.32. The resonance condition can be identified as the point A. The half-power points correspond to points B and C in Fig. 10.32.

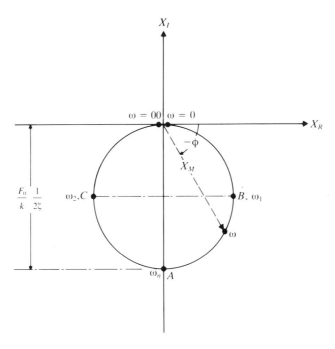

Figure 10.32

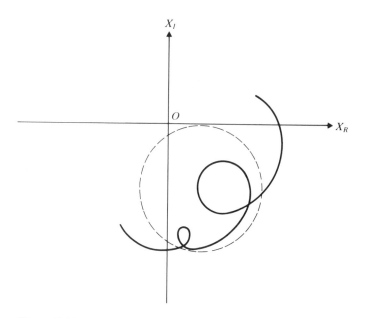

Figure 10.33

10.8.4 Multidegree of Freedom System

The vector diagrams of multidegree of freedom systems may not yield exact circles; usually they lead to curves with many loops, one loop corresponding to each resonance (Fig. 10.33). The system parameters in such cases can be computed using established procedures [10.1, 10.2].

REFERENCES

10.1. G. Buzdugan, E. Mihailescu, and M. Rades, *Vibration Measurement*, Martinus Nijhoff, Dordrecht, The Netherlands, 1986.

10.2. *Vibration Testing*, Bruel & Kjaer, Naerum, Denmark, 1983.

10.3. O. Dossing, *Structural Testing. Part 1. Mechanical Mobility Measurements*, Bruel & Kjaer, Naerum, Denmark, 1987.

10.4. D. N. Keast, *Measurements in Mechanical Dynamics*, McGraw-Hill, New York, 1967.

10.5. B. W. Mitchell (Ed.), *Instrumentation and Measurement for Environmental Sciences* (2nd Ed.), American Society of Agricultural Engineers, St. Joseph, Mich., 1983.

10.6. J. P. Holman, *Experimental Methods for Engineers* (4th Ed.), McGraw-Hill, New York, 1984.

10.7. J. T. Broch, *Mechanical Vibration and Shock Measurements*, Bruel & Kjaer, Naerum, Denmark, 1976.

10.8. R. R. Bouche, *Calibration of Shock and Vibration Measuring Transducers*, The Shock and Vibration Information Center, Washington, D.C., SVM-11, 1979.

10.9. M. Rades, "Methods for the analysis of structural frequency-response measurement data," *Shock and Vibration Digest*, Vol. 8, No. 2, February 1976, pp. 73–88.

10.10. J. D. Irwin and E. R. Graf, *Industrial Noise and Vibration Control*, Prentice-Hall, Englewood Cliffs, N.J., 1979.

10.11. R. K. Vierck, *Vibration Analysis*, Harper & Row, New York, 1979.

10.12. J. A. Macinante, *Seismic Mountings for Vibration Isolation*, Wiley, New York, 1984.

10.13. R. B. Randall and R. Upton, "Digital filters and FFT technique in real-time analysis," pp. 45–67, in *Digital Signal Analysis Using Digital Filters and FFT Techniques*, Bruel & Kjaer, Denmark, 1985.

REVIEW QUESTIONS

10.1. What is the importance of vibration measurement?

10.2. What is the difference between a vibrometer and a vibrograph?

10.3. What is a transducer?

10.4. Discuss the basic principle on which a strain gage works.

10.5. Define the gage factor of a strain gage.

10.6. What is the difference between a transducer and a pickup?

10.7. What is a piezoelectric material? Give two examples of piezoelectric material.

10.8. What is the working principle of an electrodynamic transducer?

10.9. What is an LVDT? How does it work?

10.10. What is a seismic instrument?

10.11. What is the frequency range of a seismometer?

10.12. What is an accelerometer?

10.13. What is phase-shift error? When does it become important?

10.14. Give two examples of a mechanical vibration exciter.

10.15. What is an electromagnetic shaker?

10.16. Discuss the advantage of using operational deflection shape measurement.

10.17. What is modal analysis?

10.18. Describe the use of frequency response function in modal analysis.

10.19. Name two frequency measuring instruments.

10.20. State three methods of representing the frequency response data.

PROBLEMS

The problem assignments are organized as follows:

Problems	Section covered	Topic covered
10.1	10.2	Transducers
10.2–10.18	10.3	Vibration pickups
10.19	10.4	Frequency measuring instruments
10.20	—	Project

10.1. A Rochelle salt crystal, having a voltage sensitivity of 0.098 V-m/N and thickness 2 mm, produced an output voltage of 200 volts under pressure. Find the pressure applied to the crystal.

10.2. A spring-mass system with $m = 0.5$ kg and $k = 10{,}000$ N/m, with negligible damping, is used as a vibration pickup. When mounted on a structure vibrating with an amplitude of 4 mm, the total displacement of the mass of the pickup is observed to be 12 mm. Find the frequency of the vibrating structure.

10.3. The vertical motion of a machine is measured by using the arrangement shown in Fig. 10.34. The motion of the mass m relative to the machine body is recorded on a drum. If the damping constant c is equal to $c_{cri}/\sqrt{2}$, and the vertical vibration of the machine body is given by $y(t) = Y \sin \omega t$, find the amplitude of motion recorded on the drum.

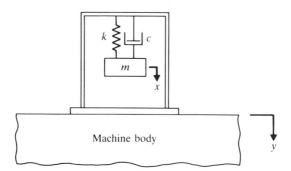

Figure 10.34

10.4. It is proposed to measure the vibration of the foundation of an internal combustion engine over the speed range 500 rpm to 1500 rpm using a vibrometer. The vibration is composed of two harmonics, the first one caused by the primary inertia forces and the second one caused by the secondary inertia forces in the engine. Determine the maximum natural frequency of the vibrometer in order to have an amplitude distortion less than 2%.

10.5. Determine the maximum percent error of a vibrometer in the frequency ratio range $4 \leqslant r < \infty$ with a damping ratio of $\zeta = 0$.

10.6. Solve Problem 10.5 with a damping ratio of $\zeta = 0.67$.

10.7. A vibrometer is used to measure the vibration of an engine whose operating speed range is from 500 to 2000 rpm. The vibration consists of two harmonics. The amplitude distortion must be less than 3%. Find the natural frequency of the vibrometer if (a) the damping is negligible and (b) the damping ratio is 0.6.

10.8. A spring-mass system, having a static deflection of 10 mm and negligible damping, is used as a vibrometer. When mounted on a machine operating at 4000 rpm, the relative amplitude is recorded as 1 mm. Find the maximum values of displacement, velocity, and acceleration of the machine.

10.9. A vibration pickup has a natural frequency of 5 Hz and a damping ratio of $\zeta = 0.5$. Find the lowest frequency that can be measured with a 1% error.

10.10. A vibration pickup has been designed for operation above a frequency level of 100 Hz without exceeding an error of 2%. When mounted on a structure vibrating at a frequency of 100 Hz, the relative amplitude of the mass is found to be 1 mm. Find the suspended mass of the pickup if the stiffness of the spring is 4000 N/m and damping is negligible.

10.11. A vibrometer has an undamped natural frequency of 10 Hz and a damped natural frequency of 8 Hz. Find the lowest frequency in the range to infinity at which the amplitude can be directly read from the vibrometer with less than 2% error.

10.12. Determine the maximum percent error of an accelerometer in the frequency ratio range $0 < r \leqslant 0.65$ with a damping ratio of $\zeta = 0$.

10.13. Solve Problem 10.12 with a damping ratio of 0.75.

10.14. Determine the necessary stiffness and the damping constant of an accelerometer if the maximum error is to be limited to 3% for measurements in the frequency range of 0 to 100 Hz. Assume the suspended mass as 0.05 kg.

10.15. An accelerometer is constructed by suspending a mass of 0.1 kg from a spring of stiffness 10,000 N/m with negligible damping. When mounted on the foundation of an engine, the peak-to-peak travel of the mass of the accelerometer has been found to be 10 mm at an engine speed of 1000 rpm. Determine the maximum displacement, maximum velocity and maximum acceleration of the foundation.

10.16. A spring-mass-damper system, having an undamped natural frequency of 100 Hz and a damping constant of 20 N-s/m, is used as an accelerometer to measure the vibration of a machine operating at a speed of 3000 rpm. If the actual acceleration is 10 m/s^2 and the recorded acceleration is 9 m/s^2, find the mass and the spring constant of the accelerometer.

10.17. A machine shop floor is subjected to the following vibration due to electric motors running at different speeds:

$$x(t) = 20 \sin 4\pi t + 10 \sin 8\pi t + 5 \sin 12\pi t \text{ mm}$$

If a vibrometer having an undamped natural frequency of 0.5 Hz and a damped natural frequency of 0.48 Hz is used to record the vibration of the machine shop floor, what will be the accuracy of the recorded vibration?

10.18. A machine is subjected to the vibration

$$x(t) = 20 \sin 50t + 5 \sin 150t \text{ mm} \qquad (t \text{ in sec})$$

An accelerometer having a damped natural frequency of 80 rad/sec and an undamped natural frequency of 100 rad/sec is mounted on the machine to read the acceleration directly in mm/sec^2. Discuss the accuracy of the recorded acceleration.

10.19. A variable-length cantilever beam of rectangular cross-section $\frac{1}{16}$ in. $\times$ 1 in., made of spring steel, is used to measure the frequency of vibration. The length of the cantilever can be varied between 2 in. and 10 in. Find the range of frequencies that can be measured with this device.

Project:

10.20. Design a vibration exciter to satisfy the following requirements:

(1) Maximum weight of the test specimen = 10 N

(2) Range of operating frequency = 10 to 50 Hz

(3) Maximum acceleration level = 20 g

(4) Maximum vibration amplitude = 0.5 cm peak to peak.

CHAPTER 11

Numerical Integration Methods in Vibration Analysis

Nathan Newmark (1910–1981) was an American engineer and a professor of civil engineering at the University of Illinois at Champaign-Urbana. His research in earthquake resistant structures and structural dynamics is widely known. The numerical method he presented in 1959 for the dynamic response computation of linear and nonlinear systems is quite famous as the "Newmark β-method."

11.1 INTRODUCTION

When the differential equation of motion of a vibrating system cannot be integrated in closed form, a numerical approach must be used. Several numerical methods are available for the solution of vibration problems [11.1–11.3].* Numerical integration methods have two fundamental characteristics. First, they are not intended to satisfy the governing differential equation(s) at all time t, but only at discrete time intervals Δt apart. Second, a suitable type of variation of the displacement x, velocity $\dot{x}$, and acceleration $\ddot{x}$ is assumed within each time interval Δt. Different numerical integration methods can be obtained, depending on the type of variation assumed for the displacement, velocity, and acceleration within each time interval Δt. We shall assume that the values of x and $\dot{x}$ are known to be x_0 and $\dot{x}_0$, respectively, at time $t = 0$ and that the solution of the problem is required from $t = 0$ to $t = T$. In the following, we subdivide the time duration T into n equal steps Δt so that $\Delta t = T/n$ and seek the solution at $t_0 = 0$, $t_1 = \Delta t$, $t_2 = 2\Delta t, \ldots,$ $t_n = n\Delta t = T$. We shall derive formulas for finding the solution at $t_i = i\Delta t$ from the known solution at $t_{i-1} = (i - 1)\Delta t$ according to five different numerical integration schemes: the finite difference method, the Runge-Kutta method, the Houbolt

---

* A numerical procedure using different types of interpolation functions for approximating the forcing function $F(t)$ was presented in Section 4.8.

method, the Wilson method, and the Newmark method. In the finite difference and Runge-Kutta methods, the current displacement (solution) is expressed in terms of the previously determined values of displacement, velocity, and acceleration and the resulting equations are solved to find the current displacement. These methods fall under the category of explicit integration methods. In the Houbolt, Wilson, and Newmark methods, the temporal difference equations are combined with the current equations of motion and the resulting equations are solved to find the current displacement. These methods belong to the category of implicit integration methods.

11.2 FINITE DIFFERENCE METHOD

The main idea in the finite difference method is to use approximations to derivatives. Thus the governing differential equation of motion and the associated boundary conditions, if applicable, are replaced by the corresponding finite difference equations. Three types of formulas—forward, backward, and central difference formulas—can be used to derive the finite difference equations [11.4–11.6]. We shall consider only the central difference formulas in this chapter, since they are most accurate.

In the finite difference method, we replace the solution domain (over which the solution of the given differential equation is required) with a finite number of points, referred to as *mesh* or *grid points*, and seek to determine the values of the desired solution at these points. The grid points are usually considered to be equally spaced along each of the independent coordinates (see Fig. 11.1). By using Taylor's series expansion, x_{i+1} and x_{i-1} can be expressed about the grid point i as

$$x_{i+1} = x_i + h\dot{x}_i + \frac{h^2}{2}\ddot{x}_i + \frac{h^3}{6}\dddot{x}_i + \cdots \tag{11.1}$$

$$x_{i-1} = x_i - h\dot{x}_i + \frac{h^2}{2}\ddot{x}_i - \frac{h^3}{6}\dddot{x}_i + \cdots \tag{11.2}$$

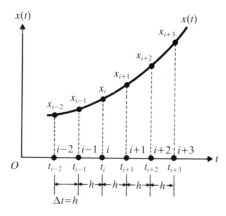

Figure 11.1

where $x_i = x(t = t_i)$ and $h = t_{i+1} - t_i = \Delta t$. By taking two terms only and subtracting Eq. (11.2) from Eq. (11.1), we obtain the central difference approximation to the first derivative of x at $t = t_i$:

$$\dot{x}_i = \frac{dx}{dt}\bigg|_{t_i} = \frac{1}{2h}(x_{i+1} - x_{i-1}) \tag{11.3}$$

By taking terms up to the second derivative and adding Eqs. (11.1) and (11.2), we obtain the central difference formula for the second derivative:

$$\ddot{x}_i = \frac{d^2x}{dt^2}\bigg|_{t_i} = \frac{1}{h^2}(x_{i+1} - 2x_i + x_{i-1}) \tag{11.4}$$

11.3 CENTRAL DIFFERENCE METHOD FOR SINGLE DEGREE OF FREEDOM SYSTEMS

The governing equation of a viscously damped single degree of freedom system is

$$m\frac{d^2x}{dt^2} + c\frac{dx}{dt} + kx = F(t) \tag{11.5}$$

Let the duration over which the solution of Eq. (11.5) is required be divided into n equal parts of interval $h = \Delta t$ each. To obtain a satisfactory solution, we must select a time step Δt that is smaller than a critical time step Δt_{cri}.* Let the initial conditions be given by $x(t = 0) = x_0$ and $\dot{x}(t = 0) = \dot{x}_0$.

Replacing the derivatives by the central differences and writing Eq. (11.5) at grid point i gives

$$m\left\{\frac{x_{i+1} - 2x_i + x_{i-1}}{(\Delta t)^2}\right\} + c\left\{\frac{x_{i+1} - x_{i-1}}{2\,\Delta t}\right\} + kx_i = F_i \tag{11.6}$$

where $x_i = x(t_i)$ and $F_i = F(t_i)$. Solution of Eq. (11.6) for x_{i+1} yields

$$x_{i+1} = \left\{\frac{1}{\dfrac{m}{(\Delta t)^2} + \dfrac{c}{2\,\Delta t}}\right\}\left[\left\{\frac{2m}{(\Delta t)^2} - k\right\}x_i + \left\{\frac{c}{2\,\Delta t} - \frac{m}{(\Delta t)^2}\right\}x_{i-1} + F_i\right] \tag{11.7}$$

---

* Numerical methods that require the use of a time step (Δt) smaller than a critical time step (Δt_{cri}) are said to be *conditionally stable* [11.7]. If Δt is taken to be larger than Δt_{cri}, the method becomes unstable. This means that the truncation of higher-order terms in the derivation of Eqs. (11.3) and (11.4) (or rounding-off in the computer) causes errors that grow and make the response computations worthless in most cases. The critical time step is given by $\Delta t_{\text{cri}} = \tau_n/\pi$, where τ_n is the natural period of the system or the smallest such period in the case of a multidegree of freedom system [11.8]. Naturally, the accuracy of the solution always depends on the size of the time step. By using an unconditionally stable method, we can choose the time step with regard to accuracy only, not with regard to stability. This usually allows a much larger time step to be used for any given accuracy.

This is called the *recurrence formula*. It permits us to calculate the displacement of the mass (x_{i+1}) if we know the previous history of displacements at t_i and t_{i-1}, as well as the present external force F_i. Repeated application of Eq. (11.7) yields the complete time history of the behavior of the system. Note that the solution of x_{i+1} is based on the use of the equilibrium equation at time t_i—that is, Eq. (11.6). For this reason, this integration procedure is called an *explicit integration method*. Certain care has to be exercised in applying Eq. (11.7) for $i = 0$. Since both x_0 and x_{-1} are needed in finding x_1, and the initial conditions provide only the values of x_0 and $\dot{x}_0$, we need to find the value of x_{-1}. Thus the method is not self-starting. However, we can generate the value of x_{-1} by using Eqs. (11.3) and (11.4) as follows. By substituting the known values of x_0 and $\dot{x}_0$ into Eq. (11.5), $\ddot{x}_0$ can be found:

$$\ddot{x}_0 = \frac{1}{m}\left[F(t = 0) - c\dot{x}_0 - kx_0 \right] \tag{11.8}$$

Application of Eqs. (11.3) and (11.4) at $i = 0$ yields the value of x_{-1}:

$$x_{-1} = x_0 - \Delta t\dot{x}_0 + \frac{(\Delta t)^2}{2}\ddot{x}_0 \tag{11.9}$$

EXAMPLE 11.1   **Response of Single d.o.f. System**

Find the response of a viscously damped single degree of freedom system subjected to a force

$$F(t) = F_0\left(1 - \sin\frac{\pi t}{2t_0}\right)$$

with the following data: $F_0 = 1$, $t_0 = \pi$, $m = 1$, $c = 0.2$, and $k = 1$. Assume the values of the displacement and velocity of the mass at $t = 0$ to be zero.

Given: Damped single degree of freedom system with $m = 1$, $c = 0.2$, $k = 1$,

$$F(t) = \left(1 - \sin\frac{t}{2}\right)$$

Initial conditions are $x(t = 0) = \dot{x}(t = 0) = 0$.

Find: Response, $x(t)$.

Approach: Use central difference method with different time steps.

Solution. The governing differential equation is

$$m\ddot{x} + c\dot{x} + kx = F(t) = F_0\left(1 - \sin\frac{\pi t}{2t_0}\right) \tag{E.1}$$

The finite difference solution of Eq. (E.1) is given by Eq. (11.7). Since the initial conditions are $x_0 = \dot{x}_0 = 0$, Eq. (11.8) yields $\ddot{x}_0 = 1$; hence Eq. (11.9) gives $x_{-1} = (\Delta t)^2/2$. Thus the solution of Eq. (E.1) can be found from the recurrence relation

$$x_{i+1} = \frac{1}{\left[\dfrac{m}{(\Delta t)^2} + \dfrac{c}{2\,\Delta t} \right]} \left[\left\{ \dfrac{2m}{(\Delta t)^2} - k \right\} x_i + \left\{ \dfrac{c}{2\,\Delta t} - \dfrac{m}{(\Delta t)^2} \right\} x_{i-1} + F_i \right],$$

$$i = 0, 1, 2, \ldots \qquad (E.2)$$

with $x_0 = 0$, $x_{-1} = (\Delta t)^2/2$, $x_i = x(t_i) = x(i\,\Delta t)$, and

$$F_i = F(t_i) = F_0 \left(1 - \sin\frac{i\pi\,\Delta t}{2t_0} \right)$$

The undamped natural frequency and the natural period of the system are given by

$$\omega_n = \left(\frac{k}{m} \right)^{1/2} = 1 \qquad (E.3)$$

and

$$\tau_n = \frac{2\pi}{\omega_n} = 2\pi \qquad (E.4)$$

Thus the time step Δt must be less than $\tau_n/\pi = 2.0$. We shall find the solution of Eq. (E.1) by using the time steps $\Delta t = \tau_n/40$, $\tau_n/20$, and $\tau_n/2$. The time step $\Delta \tau = \tau_n/2 > \Delta t_{cri}$ is used to illustrate the unstable (diverging) behavior of the solution. The values of the response x_i obtained at different instants of time t_i are shown in Table 11.1.

TABLE 11.1 Comparison of Solutions of Example 11.1

Time (t_i)	Value of $x_i = x(t_i)$ obtained with			Value of x_i given by idealization 4 of Example 4.11
	$\Delta t = \dfrac{\tau_n}{40}$	$\Delta t = \dfrac{\tau_n}{20}$	$\Delta t = \dfrac{\tau_n}{2}$	
0	0.00000	0.00000	0.00000	0.00000
$\pi/10$	0.04638	0.04935	—	0.04541
$2\pi/10$	0.16569	0.17169	—	0.16377
$3\pi/10$	0.32767	0.33627	—	0.32499
$4\pi/10$	0.50056	0.51089	—	0.49746
$5\pi/10$	0.65456	0.66543	—	0.65151
$6\pi/10$	0.76485	0.77491	—	0.76238
$7\pi/10$	0.81395	0.82185	—	0.81255
$8\pi/10$	0.79314	0.79771	—	0.79323
$9\pi/10$	0.70297	0.70340	—	0.70482
π	0.55275	0.54869	4.9348	0.55647
2π	0.19208	0.19898	−29.551	—
3π	2.7750	2.7679	181.90	—
4π	0.83299	0.83852	−1058.8	—
5π	−0.05926	−0.06431	6253.1	—

This example can be seen to be identical to Example 4.11. The results obtained by idealization 4 (piecewise linear type interpolation) of Example 4.11 are shown in Table 11.1 up to time $t_i = \pi$ in the last column of Table 11.1. It can be observed that the finite difference method gives reasonably accurate results with time steps $\Delta t = \tau_n/40$ and $\tau_n/20$ (which are smaller than Δt_{cri}) but gives diverging results with $\Delta \tau = \tau_n/2$ (which is larger than Δt_{cri}).

11.4 RUNGE-KUTTA METHOD FOR SINGLE DEGREE OF FREEDOM SYSTEMS

In the Runge-Kutta method, the approximate formula used for obtaining x_{i+1} from x_i is made to coincide with the Taylor's series expansion of x at x_{i+1} up to terms of order $(\Delta t)^n$. The Taylor's series expansion of $x(t)$ at $t + \Delta t$ is given by

$$x(t + \Delta t) = x(t) + \dot{x}\,\Delta t + \ddot{x}\frac{(\Delta t)^2}{2!} + \dddot{x}\frac{(\Delta t)^3}{3!} + \ddddot{x}\frac{(\Delta t)^4}{4!} + \cdots$$

(11.10)

In contrast to Eq. (11.10), which requires higher order derivatives, the Runge-Kutta method does not require explicitly derivatives beyond the first [11.9–11.11]. For the solution of a second order differential equation, we first reduce it to two first order equations. For example, Eq. (11.5) can be rewritten as

$$\ddot{x} = \frac{1}{m}[F(t) - c\dot{x} - kx] = f(x, \dot{x}, t)$$

(11.11)

By defining $x_1 = x$ and $x_2 = \dot{x}$, Eq. (11.11) can be written as two first order equations:

$$\dot{x}_1 = x_2$$
$$\dot{x}_2 = f(x_1, x_2, t)$$

(11.12)

By defining

$$\vec{X}(t) = \begin{Bmatrix} x_1(t) \\ x_2(t) \end{Bmatrix} \quad \text{and} \quad \vec{F}(t) = \begin{Bmatrix} x_2 \\ f(x_1, x_2, t) \end{Bmatrix}$$

the following recurrence formula is used to find the values of $\vec{X}(t)$ at different grid points t_i according to the fourth order Runge-Kutta method:

$$\vec{X}_{i+1} = \vec{X}_i + \frac{1}{6}\left[\vec{K}_1 + 2\vec{K}_2 + 2\vec{K}_3 + \vec{K}_4\right]$$

(11.13)

where

$$\vec{K}_1 = h\vec{F}\left(\vec{X}_i, t_i\right)$$

(11.14)

$$\vec{K}_2 = h\vec{F}\left(\vec{X}_i + \tfrac{1}{2}\vec{K}_1, t_i + \tfrac{1}{2}h\right)$$

(11.15)

$$\vec{K}_3 = h\vec{F}\left(\vec{X}_i + \tfrac{1}{2}\vec{K}_2, t_i + \tfrac{1}{2}h\right)$$

(11.16)

$$\vec{K}_4 = h\vec{F}\left(\vec{X}_i + \vec{K}_3, t_{i+1}\right)$$

(11.17)

Equation (11.23) gives the initial acceleration vector as

$$\ddot{\vec{x}}_0 = [m]^{-1}\left(\vec{F}_0 - [c]\dot{\vec{x}}_0 - [k]\vec{x}_0\right) \tag{11.26}$$

and Eq. (11.24) gives the displacement vector at t_1 as

$$\vec{x}_1 = \vec{x}_{-1} + 2\,\Delta t\,\dot{\vec{x}}_0 \tag{11.27}$$

Substituting Eq. (11.27) for $\vec{x}_1$, Eq. (11.25) yields

$$\ddot{\vec{x}}_0 = \frac{2}{(\Delta t)^2}\left[\Delta t\,\dot{\vec{x}}_0 - \vec{x}_0 + \vec{x}_{-1}\right]$$

or

$$\vec{x}_{-1} = \vec{x}_0 - \Delta t\,\dot{\vec{x}}_0 + \frac{(\Delta t)^2}{2}\ddot{\vec{x}}_0 \tag{11.28}$$

where $\ddot{\vec{x}}_0$ is given by Eq. (11.26). Thus $\vec{x}_{-1}$ needed for applying Eq. (11.22) at $i = 1$ is given by Eq. (11.28). The computational procedure can be described by the following steps.

1. From the known initial conditions $\vec{x}(t = 0) = \vec{x}_0$ and $\dot{\vec{x}}(t = 0) = \dot{\vec{x}}_0$, compute $\ddot{\vec{x}}(t = 0) = \ddot{\vec{x}}_0$ using Eq. (11.26).
2. Select a time step Δt such that $\Delta t < \Delta t_{\mathrm{cri}}$.
3. Compute $\vec{x}_{-1}$ using Eq. (11.28).

4. Find $\vec{x}_{i+1} = \vec{x}(t = t_{i+1})$, starting with $i = 0$; from Eq. (11.22), as

$$\vec{x}_{i+1} = \left[\frac{1}{(\Delta t)^2}[m] + \frac{1}{2\,\Delta t}[c]\right]^{-1}\left[\vec{F}_i - \left([k] - \frac{2}{(\Delta t)^2}[m]\right)\vec{x}_i\right.$$
$$\left. - \left(\frac{1}{(\Delta t)^2}[m] - \frac{1}{2\,\Delta t}[c]\right)\vec{x}_{i-1}\right] \tag{11.29}$$

where

$$\vec{F}_i = \vec{F}(t = t_i) \tag{11.30}$$

If required, evaluate accelerations and velocities at t_i:

$$\ddot{\vec{x}}_i = \frac{1}{(\Delta t)^2}\left[\vec{x}_{i+1} - 2\vec{x}_i + \vec{x}_{i-1}\right] \tag{11.31}$$

and

$$\dot{\vec{x}}_i = \frac{1}{2\,\Delta t}\left[\vec{x}_{i+1} - \vec{x}_{i-1}\right] \tag{11.32}$$

Repeat step 4 until $\vec{x}_{n+1}$ (with $i = n$) is determined. The stability of the finite difference scheme for solving matrix equations is discussed in Ref. [11.14].

EXAMPLE 11.3   Central Difference Method for a Two d.o.f. System

Find the response of the two degree of freedom system shown in Fig. 11.2 when the forcing functions are given by $F_1(t) = 0$ and $F_2(t) = 10$. Assume the value of c as zero and the initial conditions as $\vec{x}(t = 0) = \dot{\vec{x}}(t = 0) = \vec{0}$.

Given: Two degree of freedom system shown in Fig. 11.2.

Find: Response, $\vec{x}(t)$ at t_i; $i = 1, 2, \ldots$.

Approach: Use $\Delta t = \tau/10$, where τ is the smallest time period in the central difference method.

Solution. The equations of motion are given by

$$[m]\ddot{\vec{x}}(t) + [c]\dot{\vec{x}}(t) + [k]\vec{x}(t) = \vec{F}(t) \tag{E.1}$$

where

$$[m] = \begin{bmatrix} m_1 & 0 \\ 0 & m_2 \end{bmatrix} = \begin{bmatrix} 1 & 0 \\ 0 & 2 \end{bmatrix} \tag{E.2}$$

$$[c] = \begin{bmatrix} c & -c \\ -c & c \end{bmatrix} = \begin{bmatrix} 0 & 0 \\ 0 & 0 \end{bmatrix} \tag{E.3}$$

$$[k] = \begin{bmatrix} k_1 + k & -k \\ -k & k + k_2 \end{bmatrix} = \begin{bmatrix} 6 & -2 \\ -2 & 8 \end{bmatrix} \tag{E.4}$$

$$\vec{F}(t) = \begin{Bmatrix} F_1(t) \\ F_2(t) \end{Bmatrix} = \begin{Bmatrix} 0 \\ 10 \end{Bmatrix} \tag{E.5}$$

and

$$\vec{x}(t) = \begin{Bmatrix} x_1(t) \\ x_2(t) \end{Bmatrix} \tag{E.6}$$

The undamped natural frequencies and the mode shapes of the system can be found by solving the eigenvalue problem

$$\left[-\omega^2 \begin{bmatrix} 1 & 0 \\ 0 & 2 \end{bmatrix} + \begin{bmatrix} 6 & -2 \\ -2 & 8 \end{bmatrix} \right] \begin{Bmatrix} X_1 \\ X_2 \end{Bmatrix} = \begin{Bmatrix} 0 \\ 0 \end{Bmatrix} \tag{E.7}$$

The solution of Eq. (E.7) is given by

$$\omega_1 = 1.807747, \qquad \vec{X}^{(1)} = \begin{Bmatrix} 1.0000 \\ 1.3661 \end{Bmatrix} \tag{E.8}$$

$$\omega_2 = 2.594620, \qquad \vec{X}^{(2)} = \begin{Bmatrix} 1.0000 \\ -0.3661 \end{Bmatrix} \tag{E.9}$$

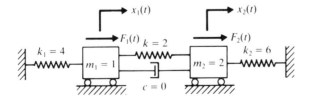

Figure 11.2

TABLE 11.3

Time $(t_i = i\Delta t)$	$\vec{x}_i = \vec{x}(t = t_i)$
t_1	$\begin{Bmatrix} 0 \\ 0.1466 \end{Bmatrix}$
t_2	$\begin{Bmatrix} 0.0172 \\ 0.5520 \end{Bmatrix}$
t_3	$\begin{Bmatrix} 0.0931 \\ 1.1222 \end{Bmatrix}$
t_4	$\begin{Bmatrix} 0.2678 \\ 1.7278 \end{Bmatrix}$
t_5	$\begin{Bmatrix} 0.5510 \\ 2.2370 \end{Bmatrix}$
t_6	$\begin{Bmatrix} 0.9027 \\ 2.5470 \end{Bmatrix}$
t_7	$\begin{Bmatrix} 1.2354 \\ 2.6057 \end{Bmatrix}$
t_8	$\begin{Bmatrix} 1.4391 \\ 2.4189 \end{Bmatrix}$
t_9	$\begin{Bmatrix} 1.4202 \\ 2.0422 \end{Bmatrix}$
t_{10}	$\begin{Bmatrix} 1.1410 \\ 1.5630 \end{Bmatrix}$
t_{11}	$\begin{Bmatrix} 0.6437 \\ 1.0773 \end{Bmatrix}$
t_{12}	$\begin{Bmatrix} 0.0463 \\ 0.6698 \end{Bmatrix}$

Thus the natural periods of the system are

$$\tau_1 = \frac{2\pi}{\omega_1} = 3.4757 \quad \text{and} \quad \tau_2 = \frac{2\pi}{\omega_2} = 2.4216$$

We shall select the time step (Δt) as $\tau_2/10 = 0.24216$. The initial value of $\ddot{\vec{x}}$ can be found as follows:

$$\ddot{\vec{x}}_0 = [m]^{-1}\{\vec{F} - [k]\vec{x}_0\} = \begin{bmatrix} 1 & 0 \\ 0 & 2 \end{bmatrix}^{-1} \begin{Bmatrix} 0 \\ 10 \end{Bmatrix} = \frac{1}{2}\begin{bmatrix} 2 & 0 \\ 0 & 1 \end{bmatrix}\begin{Bmatrix} 0 \\ 10 \end{Bmatrix} = \begin{Bmatrix} 0 \\ 5 \end{Bmatrix}$$

(E.10)

and the value of $\vec{x}_{-1}$ as follows:

$$\vec{x}_{-1} = \vec{x}_0 - \Delta t\,\dot{\vec{x}}_0 + \frac{(\Delta t)^2}{2}\ddot{\vec{x}}_0 = \begin{Bmatrix} 0 \\ 0.1466 \end{Bmatrix}$$

(E.11)

Now Eq. (11.29) can be applied recursively to obtain $\vec{x}_1, \vec{x}_2, \ldots$. The results are shown in Table 11.3.

11.6 FINITE DIFFERENCE METHOD FOR CONTINUOUS SYSTEMS

**11.6.1
Longitudinal
Vibration of Bars**

Equation of Motion. The equation of motion governing the free longitudinal vibration of a uniform bar (see Eqs. (8.49) and (8.20)) can be expressed as

$$\frac{d^2 U}{dx^2} + \alpha^2 U = 0 \tag{11.33}$$

where

$$\alpha^2 = \frac{\omega^2}{c^2} = \frac{\rho \omega^2}{E} \tag{11.34}$$

To obtain the finite difference approximation of Eq. (11.33), we first divide the bar of length l into $n - 1$ equal parts each of length $h = l/(n - 1)$ and denote the mesh points as $1, 2, 3, \ldots, i, \ldots, n$, as shown in Fig. 11.3. Then, by denoting the value of U at mesh point i as U_i and using a formula for the second derivative similar to Eq. (11.4), Eq. (11.33) for mesh point i can be written as

$$\frac{1}{h^2}(U_{i+1} - 2U_i + U_{i-1}) + \alpha^2 U_i = 0$$

or

$$U_{i+1} - (2 - \lambda)U_i + U_{i-1} = 0 \tag{11.35}$$

where $\lambda = h^2 \alpha^2$. The application of Eq. (11.35) at mesh points $i = 2, 3, \ldots, n - 1$ leads to the equations

$$U_3 - (2 - \lambda)U_2 + U_1 = 0$$
$$U_4 - (2 - \lambda)U_3 + U_2 = 0$$
$$\vdots$$
$$U_n - (2 - \lambda)U_{n-1} + U_{n-2} = 0 \tag{11.36}$$

which can be stated in matrix form as

$$\begin{bmatrix} -1 & (2-\lambda) & -1 & 0 & 0 & \cdots & 0 & 0 & 0 \\ 0 & -1 & (2-\lambda) & -1 & 0 & \cdots & 0 & 0 & 0 \\ 0 & 0 & -1 & (2-\lambda) & -1 & \cdots & 0 & 0 & 0 \\ \vdots & \vdots & \vdots & \vdots & \vdots & \cdots & \vdots & \vdots & \vdots \\ 0 & 0 & 0 & 0 & 0 & \cdots & -1 & (2-\lambda) & -1 \end{bmatrix} \begin{Bmatrix} U_1 \\ U_2 \\ U_3 \\ \vdots \\ U_n \end{Bmatrix} = \begin{Bmatrix} 0 \\ 0 \\ 0 \\ \vdots \\ 0 \end{Bmatrix}$$

$$\tag{11.37}$$

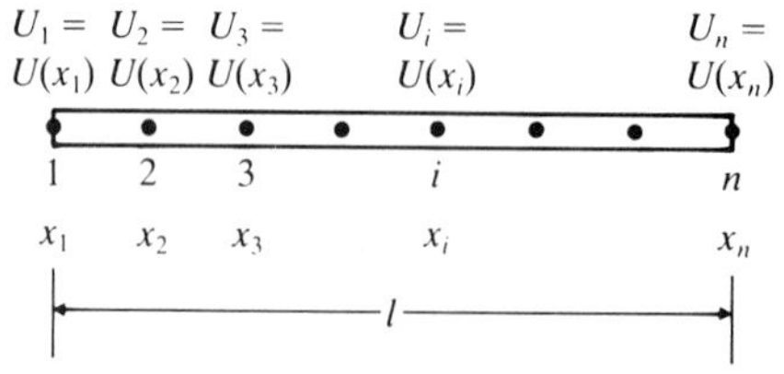

$U_1 = U_2 = U_3 = \qquad U_i = \qquad\qquad U_n =$
$U(x_1)\ U(x_2)\ U(x_3) \qquad U(x_i) \qquad\qquad U(x_n)$

Figure 11.3 Division of a bar for finite difference approximation.

Boundary Conditions

Fixed end. The deflection is zero at a fixed end. Assuming that the bar is fixed at $x = 0$ and $x = l$, we set $U_1 = U_n = 0$ in Eq. (11.37) and obtain the equation

$$[[A] - \lambda[I]]\vec{U} = \vec{0} \tag{11.38}$$

where

$$[A] = \begin{bmatrix} 2 & -1 & 0 & 0 & \cdots & 0 & 0 & 0 \\ -1 & 2 & -1 & 0 & \cdots & 0 & 0 & 0 \\ 0 & -1 & 2 & -1 & \cdots & 0 & 0 & 0 \\ \cdot & \cdot & \cdot & \cdot & \cdots & \cdot & \cdot & \cdot \\ \cdot & \cdot & \cdot & \cdot & \cdots & \cdot & \cdot & \cdot \\ \cdot & \cdot & \cdot & \cdot & \cdots & \cdot & \cdot & \cdot \\ 0 & 0 & 0 & 0 & \cdots & 0 & -1 & 2 \end{bmatrix} \tag{11.39}$$

$$\vec{U} = \begin{Bmatrix} U_2 \\ U_3 \\ \vdots \\ U_{n-1} \end{Bmatrix} \tag{11.40}$$

and $[I]$ = identity matrix of order $n - 2$.

Note that the eigenvalue problem of Eq. (11.38) can be solved easily, since the matrix $[A]$ is a tridiagonal matrix [11.15–11.17].

Free end. The stress is zero at a free end, so $(dU)/(dx) = 0$. We can use a formula for the first derivative similar to Eq. (11.3). To illustrate the procedure, let the bar be free at $x = 0$ and fixed at $x = l$. The boundary conditions can then be stated as

$$\left.\frac{dU}{dx}\right|_1 \simeq \frac{U_2 - U_{-1}}{2h} = 0 \quad \text{or} \quad U_{-1} = U_2 \tag{11.41}$$

$$U_n = 0 \tag{11.42}$$

In order to apply Eq. (11.41), we need to imagine the function $U(x)$ to be continuous beyond the length of the bar and create a fictitious mesh point -1 so that U_{-1} becomes the fictitious displacement of the point x_{-1}. The application of Eq. (11.35) at mesh point $i = 1$ yields

$$U_2 - (2 - \lambda)U_1 + U_{-1} = 0 \tag{11.43}$$

By incorporating the condition $U_{-1} = U_2$ (Eq. (11.41)), Eq. (11.43) can be written as

$$(2 - \lambda)U_1 - 2U_2 = 0 \tag{11.44}$$

By adding Eqs. (11.44) and (11.37), we obtain the final equations:

$$[[A] - \lambda[I]]\vec{U} = \vec{0} \tag{11.45}$$

where

$$[A] = \begin{bmatrix} 2 & -2 & 0 & 0 & \cdots & 0 & 0 & 0 \\ -1 & 2 & -1 & 0 & \cdots & 0 & 0 & 0 \\ 0 & -1 & 2 & -1 & \cdots & 0 & 0 & 0 \\ \vdots & & & & & & & \\ 0 & 0 & 0 & 0 & \cdots & -1 & 2 & -1 \\ 0 & 0 & 0 & 0 & \cdots & 0 & -1 & 2 \end{bmatrix}$$ (11.46)

and

$$\vec{U} = \begin{Bmatrix} U_1 \\ U_2 \\ \vdots \\ U_{n-1} \end{Bmatrix}$$ (11.47)

11.6.2 Transverse Vibration of Beams

Equation of Motion. The governing differential equation for the transverse vibration of a uniform beam is given by Eq. (8.83):

$$\frac{d^4W}{dx^4} - \beta^4 W = 0$$ (11.48)

where

$$\beta^4 = \frac{\rho A \omega^2}{EI}$$ (11.49)

By using the central difference formula for the fourth derivative,* Eq. (11.48) can be written at any mesh point i as

$$W_{i+2} - 4W_{i+1} + (6 - \lambda)W_i - 4W_{i-1} + W_{i-2} = 0$$ (11.50)

where

$$\lambda = h^4 \beta^4$$ (11.51)

Let the beam be divided into $n - 1$ equal parts with n mesh points and $h = l/(n - 1)$. The application of Eq. (11.50) at the mesh points $i = 3, 4, \ldots, n - 2$ leads to the equations

$$\begin{bmatrix} 1 & -4 & (6-\lambda) & -4 & 1 & 0 & 0 & \cdots & 0 & 0 & 0 & 0 & 0 \\ 0 & 1 & -4 & (6-\lambda) & -4 & 1 & 0 & \cdots & 0 & 0 & 0 & 0 & 0 \\ 0 & 0 & 1 & -4 & (6-\lambda) & -4 & 1 & \cdots & 0 & 0 & 0 & 0 & 0 \\ \vdots & & & & & & & & & & & \\ 0 & 0 & 0 & 0 & 0 & 0 & 0 & \cdots & 1 & -4 & (6-\lambda) & -4 & 1 \end{bmatrix} \begin{Bmatrix} W_1 \\ W_2 \\ W_3 \\ \vdots \\ W_n \end{Bmatrix} = \begin{Bmatrix} 0 \\ 0 \\ 0 \\ \vdots \\ 0 \end{Bmatrix}$$ (11.52)

---

* The central difference formula for the fourth derivative (see Problem 11.3) is given by

$$\left. \frac{d^4 f}{dx^4} \right|_i \approx \frac{1}{h^4} (f_{i+2} - 4f_{i+1} + 6f_i - 4f_{i-1} + f_{i-2})$$

Boundary Conditions

Fixed end. The deflection W and the slope $(dW)/(dx)$ are zero at a fixed end. If the end $x = 0$ is fixed, we introduce a fictitious node -1 on the left-hand side of the beam, as shown in Fig. 11.4, and state the boundary conditions, using the central difference formula for $(dW)/(dx)$, as

$$W_1 = 0$$

$$\frac{dW}{dx}\bigg|_1 = \frac{1}{2h}(W_2 - W_{-1}) = 0 \quad \text{or} \quad W_{-1} = W_2 \tag{11.53}$$

where W_i denotes the value of W at node i. If the end $x = l$ is fixed, we introduce the fictitious node $n + 1$ on the right side of the beam, as shown in Fig. 11.4, and state the boundary conditions as

$$W_n = 0$$

$$\frac{dW}{dx}\bigg|_n = \frac{1}{2h}(W_{n+1} - W_{n-1}) = 0, \quad \text{or} \quad W_{n+1} = W_{n-1} \tag{11.54}$$

Simply supported end. If the end $x = 0$ is simply supported (see Fig. 11.5), we have

$$W_1 = 0$$

$$\frac{d^2W}{dx^2}\bigg|_1 = \frac{1}{h^2}(W_2 - 2W_1 + W_{-1}) = 0, \quad \text{or} \quad W_{-1} = -W_2$$

$$\tag{11.55}$$

Similar equations can be written if the end $x = l$ is simply supported.

Free end. Since bending moment and shear force are zero at a free end, we introduce two fictitious nodes outside the beam, as shown in Fig. 11.6, and use central difference formulas for approximating the second and the third derivatives of the deflection W. For example, if the end $x = 0$ is free, we have

$$\frac{d^2W}{dx^2}\bigg|_1 = \frac{1}{h^2}(W_2 - 2W_1 + W_{-1}) = 0$$

$$\frac{d^3W}{dx^3}\bigg|_1 = \frac{1}{2h^3}(W_3 - 2W_2 + 2W_{-1} - W_{-2}) = 0 \tag{11.56}$$

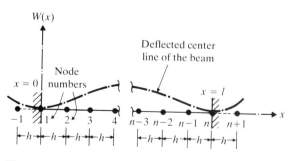

Figure 11.4 Beam with fixed ends.

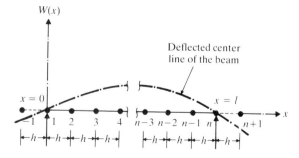

Figure 11.5 Beam with simply supported ends.

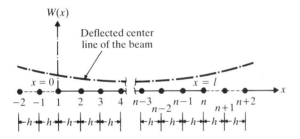

Figure 11.6   Beam with free ends.

EXAMPLE 11.4   Pinned-Fixed Beam

Find the natural frequencies of the simply supported-fixed beam shown in Fig. 11.7. Assume that the cross section of the beam is constant along its length.

Given: Pinned-fixed beam shown in Fig. 11.7.

Find: Natural frequencies and mode shapes.

Approach: Derive and solve central difference equations using four beam segments.

Solution. We shall divide the beam into four segments and express the governing equation

$$\frac{d^4W}{dx^4} - \beta^4 W = 0 \tag{E.1}$$

in finite difference form at each of the interior mesh points. This yields the following equations:

$$W_0 - 4W_1 + (6 - \lambda)W_2 - 4W_3 + W_4 = 0 \tag{E.2}$$
$$W_1 - 4W_2 + (6 - \lambda)W_3 - 4W_4 + W_5 = 0 \tag{E.3}$$
$$W_2 - 4W_3 + (6 - \lambda)W_4 - 4W_5 + W_6 = 0 \tag{E.4}$$

where W_0 and W_6 denote the values of W at the fictitious nodes 0 and 6, respectively, and

$$\lambda = h^4\beta^4 = \frac{h^4\rho A\omega^2}{EI} \tag{E.5}$$

The boundary conditions at the simply supported end (mesh point 1) are

$$W_1 = 0$$
$$W_0 = -W_2 \tag{E.6}$$

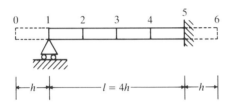

Figure 11.7

At the fixed end (mesh point 5) the boundary conditions are

$$W_5 = 0$$
$$W_6 = W_4 \tag{E.7}$$

With the help of Eqs. (E.6) and (E.7), Eqs. (E.2) to (E.4) can be reduced to

$$(5 - \lambda) W_2 - 4W_3 + W_4 = 0 \tag{E.8}$$
$$- 4W_2 + (6 - \lambda) W_3 - 4W_4 = 0 \tag{E.9}$$
$$W_2 - 4W_3 + (7 - \lambda) W_4 = 0 \tag{E.10}$$

Equations (E.8) to (E.10) can be written in matrix form as

$$\begin{bmatrix} (5 - \lambda) & -4 & 1 \\ -4 & (6 - \lambda) & -4 \\ 1 & -4 & (7 - \lambda) \end{bmatrix} \begin{Bmatrix} W_2 \\ W_3 \\ W_4 \end{Bmatrix} = \begin{Bmatrix} 0 \\ 0 \\ 0 \end{Bmatrix} \tag{E.11}$$

The solution of the eigenvalue problem (Eq. (E.11)) gives the following results:

$$\lambda_1 = 0.7135, \qquad \omega_1 = \frac{0.8447}{h^2} \sqrt{\frac{EI}{\rho A}}, \qquad \begin{Bmatrix} W_2 \\ W_3 \\ W_4 \end{Bmatrix}^{(1)} = \begin{Bmatrix} 0.5880 \\ 0.7215 \\ 0.3656 \end{Bmatrix} \tag{E.12}$$

$$\lambda_2 = 5.0322, \qquad \omega_2 = \frac{2.2433}{h^2} \sqrt{\frac{EI}{\rho A}}, \qquad \begin{Bmatrix} W_2 \\ W_3 \\ W_4 \end{Bmatrix}^{(2)} = \begin{Bmatrix} 0.6723 \\ -0.1846 \\ -0.7169 \end{Bmatrix} \tag{E.13}$$

$$\lambda_3 = 12.2543, \qquad \omega_3 = \frac{3.5006}{h^2} \sqrt{\frac{EI}{\rho A}}, \qquad \begin{Bmatrix} W_2 \\ W_3 \\ W_4 \end{Bmatrix}^{(3)} = \begin{Bmatrix} 0.4498 \\ -0.6673 \\ 0.5936 \end{Bmatrix} \tag{E.14}$$

11.7 RUNGE-KUTTA METHOD FOR MULTIDEGREE OF FREEDOM SYSTEMS

In the Runge-Kutta method, the matrix equations of motion, Eq. (11.18), are used to express the acceleration vector as

$$\ddot{\vec{x}}(t) = [m]^{-1} \left(\vec{F}(t) - [c] \dot{\vec{x}}(t) - [k] \vec{x}(t) \right) \tag{11.57}$$

By treating the displacements as well as velocities as unknowns, a new vector, $\vec{X}(t)$, is defined as $\vec{X}(t) = \begin{Bmatrix} \vec{x}(t) \\ \dot{\vec{x}}(t) \end{Bmatrix}$ so that

$$\dot{\vec{X}} = \begin{Bmatrix} \dot{\vec{x}} \\ \ddot{\vec{x}} \end{Bmatrix} = \begin{Bmatrix} \dot{\vec{x}} \\ [m]^{-1} \left(\vec{F} - [c] \dot{\vec{x}} - [k] \vec{x} \right) \end{Bmatrix} \tag{11.58}$$

Equation (11.58) can be rearranged to obtain

$$\dot{\vec{x}}(t) = \begin{bmatrix} [0] & [I] \\ -[m]^{-1}[k] & -[m]^{-1}[c] \end{bmatrix} \begin{Bmatrix} \vec{x}(t) \\ \dot{\vec{x}}(t) \end{Bmatrix} + \begin{Bmatrix} 0 \\ [m]^{-1}\vec{F}(t) \end{Bmatrix}$$

that is,

$$\dot{\vec{X}}(t) = \vec{f}(\vec{X}, t) \tag{11.59}$$

where

$$\vec{f}(\vec{X}, t) = [A]\vec{X}(t) + \underset{\sim}{\vec{F}}(t) \tag{11.60}$$

$$[A] = \begin{bmatrix} [0] & [I] \\ -[m]^{-1}[k] & -[m]^{-1}[c] \end{bmatrix} \tag{11.61}$$

and

$$\underset{\sim}{\vec{F}}(t) = \begin{Bmatrix} \vec{0} \\ [m]^{-1}\vec{F}(t) \end{Bmatrix} \tag{11.62}$$

With this, the recurrence formula to evaluate $\vec{X}(t)$ at different grid points t_i according to the fourth order Runge-Kutta method becomes [11.10]:

$$\vec{X}_{i+1} = \vec{X}_i + \frac{1}{6}\left[\vec{K}_1 + 2\vec{K}_2 + 2\vec{K}_3 + \vec{K}_4\right] \tag{11.63}$$

where

$$\vec{K}_1 = h\vec{f}\left(\vec{X}_i, t_i\right) \tag{11.64}$$

$$\vec{K}_2 = h\vec{f}\left(\vec{X}_i + \tfrac{1}{2}\vec{K}_1, t_i + \tfrac{1}{2}h\right) \tag{11.65}$$

$$\vec{K}_3 = h\vec{f}\left(\vec{X}_i + \tfrac{1}{2}\vec{K}_2, t_i + \tfrac{1}{2}h\right) \tag{11.66}$$

$$\vec{K}_4 = h\vec{f}\left(\vec{X}_i + \vec{K}_3, t_{i+1}\right) \tag{11.67}$$

EXAMPLE 11.5   Runge-Kutta Method for a Two d.o.f. System

Find the response of the two degree of freedom system considered in Example 11.3 using the fourth order Runge-Kutta method.

Given: Two degree of freedom system shown in Fig. 11.2.

Find: Response, $\vec{x}(t)$ at t_i; $i = 1, 2, \ldots$.

Approach: Use Runge-Kutta method with $\Delta t = 0.2416$.

Solution. Using the initial conditions $\vec{x}(t = 0) = \dot{\vec{x}}(t = 0) = \vec{0}$, Eq. (11.63) is sequentially applied with $\Delta t = 0.2416$ to obtain the results shown in Table 11.4.

TABLE 11.4

Time $t_i = i\,\Delta t$	$\vec{x}_i = \vec{x}(t = t_i)$
t_1	$\left\{ \begin{matrix} 0.0014 \\ 0.1437 \end{matrix} \right\}$
t_2	$\left\{ \begin{matrix} 0.0215 \\ 0.5418 \end{matrix} \right\}$
t_3	$\left\{ \begin{matrix} 0.0978 \\ 1.1041 \end{matrix} \right\}$
t_4	$\left\{ \begin{matrix} 0.2668 \\ 1.7059 \end{matrix} \right\}$
t_5	$\left\{ \begin{matrix} 0.5379 \\ 2.2187 \end{matrix} \right\}$
t_6	$\left\{ \begin{matrix} 0.8756 \\ 2.5401 \end{matrix} \right\}$
t_7	$\left\{ \begin{matrix} 1.2008 \\ 2.6153 \end{matrix} \right\}$
t_8	$\left\{ \begin{matrix} 1.4109 \\ 2.4452 \end{matrix} \right\}$
t_9	$\left\{ \begin{matrix} 1.4156 \\ 2.0805 \end{matrix} \right\}$
t_{10}	$\left\{ \begin{matrix} 1.1727 \\ 1.6050 \end{matrix} \right\}$
t_{11}	$\left\{ \begin{matrix} 0.7123 \\ 1.1141 \end{matrix} \right\}$
t_{12}	$\left\{ \begin{matrix} 0.1365 \\ 0.6948 \end{matrix} \right\}$

11.8 HOUBOLT METHOD

We shall consider the Houbolt method with reference to a multidegree of freedom system. In this method, the following finite difference expansions are employed:

$$\dot{\vec{x}}_{i+1} = \frac{1}{6\,\Delta t}\left(11\vec{x}_{i+1} - 18\vec{x}_i + 9\vec{x}_{i-1} - 2\vec{x}_{i-2}\right) \qquad (11.68)$$

$$\ddot{\vec{x}}_{i+1} = \frac{1}{(\Delta t)^2}\left(2\vec{x}_{i+1} - 5\vec{x}_i + 4\vec{x}_{i-1} - \vec{x}_{i-2}\right) \qquad (11.69)$$

To derive Eqs. (11.68) and (11.69), consider the function $x(t)$. Let the values of x at the equally spaced grid points $t_{i-2} = t_i - 2\,\Delta t$, $t_{i-1} = t_i - \Delta t$, t_i, and $t_{i+1} = t_i + \Delta t$ be given by x_{i-2}, x_{i-1}, x_i, and x_{i+1}, respectively, as shown in Fig. 11.8 [11.18]. The

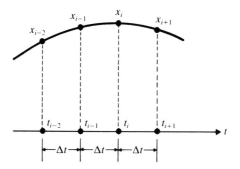

Figure 11.8

Taylor's series expansion, with backward step, gives:
With step size $= \Delta t$:

$$x(t) = x(t + \Delta t) - \Delta t\, \dot{x}(t + \Delta t) + \frac{(\Delta t)^2}{2!}\ddot{x}(t + \Delta t) - \frac{(\Delta t)^3}{3!}\dddot{x}(t + \Delta t) + \cdots$$

or

$$x_i = x_{i+1} - \Delta t\, \dot{x}_{i+1} + \frac{(\Delta t)^2}{2}\ddot{x}_{i+1} - \frac{(\Delta t)^3}{6}\dddot{x}_{i+1} + \cdots \tag{11.70}$$

With step size $= 2\,\Delta t$:

$$x(t - \Delta t) = x(t + \Delta t) - (2\,\Delta t)\dot{x}(t + \Delta t)$$
$$+ \frac{(2\,\Delta t)^2}{2!}\ddot{x}(t + \Delta t) - \frac{(2\,\Delta t)^3}{3!}\dddot{x}(t + \Delta t) + \cdots$$

or

$$x_{i-1} = x_{i+1} - 2\,\Delta t\, \dot{x}_{i+1} + 2(\Delta t)^2\ddot{x}_{i+1} - \tfrac{4}{3}(\Delta t)^3\dddot{x}_{i+1} + \cdots \tag{11.71}$$

With step size $= 3\,\Delta t$:

$$x(t - 2\,\Delta t) = x(t + \Delta t) - (3\,\Delta t)\dot{x}(t + \Delta t)$$
$$+ \frac{(3\,\Delta t)^2}{2!}\ddot{x}(t + \Delta t) - \frac{(3\,\Delta t)^3}{3!}\dddot{x}(t + \Delta t) + \cdots$$

or

$$x_{i-2} = x_{i+1} - 3\,\Delta t\, \dot{x}_{i+1} + \tfrac{9}{2}(\Delta t)^2\ddot{x}_{i+1} - \tfrac{9}{2}(\Delta t)^3\dddot{x}_{i+1} + \cdots \tag{11.72}$$

By considering terms up to $(\Delta t)^3$ only, Eqs. (11.70) to (11.72) can be solved to express $\dot{x}_{i+1}$, $\ddot{x}_{i+1}$, and $\dddot{x}_{i+1}$ in terms of x_{i-2}, x_{i-1}, x_i, and x_{i+1}. This gives $\dot{x}_{i+1}$ and $\ddot{x}_{i+1}$ as [11.18]:

$$\dot{x}_{i+1} = \frac{1}{6(\Delta t)}(11x_{i+1} - 18x_i + 9x_{i-1} - 2x_{i-2}) \tag{11.73}$$

$$\ddot{x}_{i+1} = \frac{1}{(\Delta t)^2}(2x_{i+1} - 5x_i + 4x_{i-1} - x_{i-2}) \tag{11.74}$$

Equations (11.68) and (11.69) represent the vector form of these equations.

To find the solution at step $i + 1 (\vec{x}_{i+1})$, we consider Eq. (11.18) at t_{i+1}, so that

$$[m]\ddot{\vec{x}}_{i+1} + [c]\dot{\vec{x}}_{i+1} + [k]\vec{x}_{i+1} = \vec{F}_{i+1} \equiv \vec{F}(t = t_{i+1}) \qquad (11.75)$$

By substituting Eqs. (11.68) and (11.69) into Eq. (11.75), we obtain

$$\left(\frac{2}{(\Delta t)^2}[m] + \frac{11}{6\Delta t}[c] + [k] \right)\vec{x}_{i+1}$$

$$= \vec{F}_{i+1} + \left(\frac{5}{(\Delta t)^2}[m] + \frac{3}{\Delta t}[c] \right)\vec{x}_i$$

$$- \left(\frac{4}{(\Delta t)^2}[m] + \frac{3[c]}{2\Delta t} \right)\vec{x}_{i-1} + \left(\frac{1}{(\Delta t)^2}[m] + \frac{[c]}{3\Delta t} \right)\vec{x}_{i-2} \qquad (11.76)$$

Note that the equilibrium equation at time t_{i+1}, Eq. (11.75), is used in finding the solution $\vec{X}_{i+1}$ through Eq. (11.76). This is also true of the Wilson and Newmark methods. For this reason, these methods are called *implicit integration methods*.

It can be seen from Eq. (11.76) that a knowledge of $\vec{x}_i$, $\vec{x}_{i-1}$, and $\vec{x}_{i-2}$ is required to find the solution $\vec{x}_{i+1}$. Thus the values of $\vec{x}_{-1}$ and $\vec{x}_{-2}$ are to be found before attempting to find the vector $\vec{x}_1$ using Eq. (11.76). Since there is no direct method to find $\vec{x}_{-1}$ and $\vec{x}_{-2}$, we can not use Eq. (11.76) to find $\vec{x}_1$ and $\vec{x}_2$. This makes the method non-self-starting. To start the method, we can use the central difference method described in Section 11.5 to find $\vec{x}_1$ and $\vec{x}_2$. Once $\vec{x}_0$ is known from the given initial conditions of the problem and $\vec{x}_1$ and $\vec{x}_2$ are known from the central difference method, the subsequent solutions $\vec{x}_3, \vec{x}_4, \ldots$ can be found by using Eq. (11.76).

The step-by-step procedure to be used in the Houbolt method is given below:

1. From the known initial conditions $\vec{x}(t = 0) = \vec{x}_0$ and $\dot{\vec{x}}(t = 0) = \dot{\vec{x}}_0$, find $\ddot{\vec{x}}_0 = \ddot{\vec{x}}(t = 0)$ using Eq. (11.26).
2. Select a suitable time step Δt.
3. Determine $\vec{x}_{-1}$ using Eq. (11.28).
4. Find $\vec{x}_1$ and $\vec{x}_2$ using the central difference equation (11.29).
5. Compute $\vec{x}_{i+1}$, starting with $i = 2$ and using Eq. (11.76):

$$\vec{x}_{i+1} = \left[\frac{2}{(\Delta t)^2}[m] + \frac{11}{6\Delta t}[c] + [k] \right]^{-1}$$

$$\times \left\{ \vec{F}_{i+1} + \left(\frac{5}{(\Delta t)^2}[m] + \frac{3}{\Delta t}[c] \right)\vec{x}_i - \left(\frac{4}{(\Delta t)^2}[m] + \frac{3}{2\Delta t}[c] \right)\vec{x}_{i-1} \right.$$

$$\left. + \left(\frac{1}{(\Delta t)^2}[m] + \frac{1}{3\Delta t}[c] \right)\vec{x}_{i-2} \right\} \qquad (11.77)$$

If required, evaluate the velocity and acceleration vectors $\dot{\vec{x}}_{i+1}$ and $\ddot{\vec{x}}_{i+1}$ using Eqs. (11.68) and (11.69).

EXAMPLE 11.6   **Houbolt Method for a Two d.o.f. System**

Find the response of the two degree of freedom system considered in Example 11.3 using the Houbolt method.

Given: Two degree of freedom system shown in Fig. 11.2.

Find: Response, $\vec{x}(t)$ at t_i; $i = 1, 2, \ldots$.

Approach: Use Houbolt method with $\Delta t = 0.2416$.

Solution. The value of $\ddot{\vec{x}}_0$ can be found using Eq. (11.26):

$$\ddot{\vec{x}} = \begin{Bmatrix} 0 \\ 5 \end{Bmatrix}$$

By using a value of $\Delta t = 0.24216$, Eq. (11.29) can be used to find $\vec{x}_1$ and $\vec{x}_2$, and then Eq. (11.77) can be used recursively to obtain $\vec{x}_3, \vec{x}_4, \ldots$, as shown in Table 11.5.

TABLE 11.5

Time $t_i = i\Delta t$	$\vec{x}_i = \vec{x}(t = t_i)$
t_1	$\begin{Bmatrix} 0.0000 \\ 0.1466 \end{Bmatrix}$
t_2	$\begin{Bmatrix} 0.0172 \\ 0.5520 \end{Bmatrix}$
t_3	$\begin{Bmatrix} 0.0917 \\ 1.1064 \end{Bmatrix}$
t_4	$\begin{Bmatrix} 0.2501 \\ 1.6909 \end{Bmatrix}$
t_5	$\begin{Bmatrix} 0.4924 \\ 2.1941 \end{Bmatrix}$
t_6	$\begin{Bmatrix} 0.7867 \\ 2.5297 \end{Bmatrix}$
t_7	$\begin{Bmatrix} 1.0734 \\ 2.6489 \end{Bmatrix}$
t_8	$\begin{Bmatrix} 1.2803 \\ 2.5454 \end{Bmatrix}$
t_9	$\begin{Bmatrix} 1.3432 \\ 2.2525 \end{Bmatrix}$
t_{10}	$\begin{Bmatrix} 1.2258 \\ 1.8325 \end{Bmatrix}$
t_{11}	$\begin{Bmatrix} 0.9340 \\ 1.3630 \end{Bmatrix}$
t_{12}	$\begin{Bmatrix} 0.5178 \\ 0.9224 \end{Bmatrix}$

11.9 WILSON METHOD

The Wilson method assumes that the acceleration of the system varies linearly between two instants of time. In particular, the two instants of time are taken as indicated in Fig. 11.9. Thus the acceleration is assumed to be linear from time $t_i = i\,\Delta t$ to time $t_{i+\theta} = t_i + \theta\,\Delta t$, where $\theta \geq 1.0$ [11.19]. For this reason, this method is also called the *Wilson θ method*. If $\theta = 1.0$, this method reduces to the linear acceleration scheme [11.20].

A stability analysis of the Wilson method shows that it is unconditionally stable provided that $\theta \geq 1.37$. In this section, we shall consider the Wilson method for a multidegree of freedom system.

Since $\ddot{\vec{x}}(t)$ is assumed to vary linearly between t_i and $t_{i+\theta}$, we can predict the value of $\ddot{\vec{x}}$ at any time $t_i + \tau$, $0 \leq \tau \leq \theta\,\Delta t$:

$$\ddot{\vec{x}}(t_i + \tau) = \ddot{\vec{x}}_i + \frac{\tau}{\theta\,\Delta t}\left(\ddot{\vec{x}}_{i+\theta} - \ddot{\vec{x}}_i\right) \tag{11.78}$$

By integrating Eq. (11.78), we obtain*

$$\dot{\vec{x}}(t_i + \tau) = \dot{\vec{x}}_i + \ddot{\vec{x}}_i\tau + \frac{\tau^2}{2\theta\,\Delta t}\left(\ddot{\vec{x}}_{i+\theta} - \ddot{\vec{x}}_i\right) \tag{11.79}$$

and

$$\vec{x}(t_i + \tau) = \vec{x}_i + \dot{\vec{x}}_i\tau + \frac{1}{2}\ddot{\vec{x}}_i\tau^2 + \frac{\tau^3}{6\theta\,\Delta t}\left(\ddot{\vec{x}}_{i+\theta} - \ddot{\vec{x}}_i\right) \tag{11.80}$$

By substituting $\tau = \theta\,\Delta t$ into Eqs. (11.79) and (11.80), we obtain

$$\dot{\vec{x}}_{i+\theta} = \dot{\vec{x}}(t_i + \theta\,\Delta t) = \dot{\vec{x}}_i + \frac{\theta\,\Delta t}{2}\left(\ddot{\vec{x}}_{i+\theta} + \ddot{\vec{x}}_i\right) \tag{11.81}$$

$$\vec{x}_{i+\theta} = \vec{x}(t_i + \theta\,\Delta t) = \vec{x}_i + \theta\,\Delta t\,\dot{\vec{x}}_i + \frac{\theta^2(\Delta t)^2}{6}\left(\ddot{\vec{x}}_{i+\theta} + 2\ddot{\vec{x}}_i\right) \tag{11.82}$$

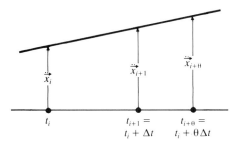

Figure 11.9 Linear acceleration assumption of the Wilson method.

---

* $\dot{\vec{x}}_i$ and $\vec{x}_i$ have been substituted in place of the integration constants in Eqs. (11.79) and (11.80), respectively.

Equation (11.82) can be solved to obtain

$$\ddot{\vec{x}}_{i+\theta} = \frac{6}{\theta^2(\Delta t)^2}(\vec{x}_{i+\theta} - \vec{x}_i) - \frac{6}{\theta \Delta t}\dot{\vec{x}}_i - 2\ddot{\vec{x}}_i \qquad (11.83)$$

By substituting Eq. (11.83) into Eq. (11.81), we obtain

$$\dot{\vec{x}}_{i+\theta} = \frac{3}{\theta \Delta t}(\vec{x}_{i+\theta} - \vec{x}_i) - 2\dot{\vec{x}}_i - \frac{\theta \Delta t}{2}\ddot{\vec{x}}_i \qquad (11.84)$$

To obtain the value of $x_{i+\theta}$, we consider the equilibrium equation (11.18) at time $t_{i+\theta} = t_i + \theta \Delta t$ and write

$$[m]\ddot{\vec{x}}_{i+\theta} + [c]\dot{\vec{x}}_{i+\theta} + [k]\vec{x}_{i+\theta} = \vec{F}_{i+\theta} \qquad (11.85)$$

where the force vector $\vec{\underset{\sim}{F}}_{i+\theta}$ is also obtained by using the linear assumption:

$$\vec{F}_{i+\theta} = \vec{F}_i + \theta(\vec{F}_{i+1} - \vec{F}_i) \qquad (11.86)$$

Substituting Eqs. (11.83), (11.84), and (11.86) for $\ddot{\vec{x}}_{i+\theta}$, $\dot{\vec{x}}_{i+\theta}$, and $\vec{F}_{i+\theta}$, Eq. (11.85) gives

$$\left\{ \frac{6}{\theta^2(\Delta t)^2}[m] + \frac{3}{\theta \Delta t}[c] + [k] \right\}\vec{x}_{i+1}$$

$$= \vec{F}_i + \theta(\vec{F}_{i+1} - \vec{F}_i) + \left\{ \frac{6}{\theta^2(\Delta t)^2}[m] + \frac{3}{\theta \Delta t}[c] \right\}\vec{x}_i$$

$$+ \left\{ \frac{6}{\theta \Delta t}[m] + 2[c] \right\}\dot{\vec{x}}_i + \left\{ 2[m] + \frac{\theta \Delta t}{2}[c] \right\}\ddot{\vec{x}}_i \qquad (11.87)$$

which can be solved for $\vec{x}_{i+1}$.

The Wilson method can be described by the following steps:

1. From the known initial conditions $\vec{x}_0$ and $\dot{\vec{x}}_0$, find $\ddot{\vec{x}}_0$ using Eq. (11.26).
2. Select a suitable time step Δt and a suitable value of θ (θ is usually taken as 1.4).
3. Compute the effective load vector $\vec{\underset{\sim}{F}}_{i+\theta}$, starting with $i = 0$:

$$\vec{\underset{\sim}{F}}_{i+\theta} = \vec{F}_i + \theta(\vec{F}_{i+1} - \vec{F}_i) + [m]\left(\frac{6}{\theta^2(\Delta t)^2}\vec{x}_i + \frac{6}{\theta \Delta t}\dot{\vec{x}}_i + 2\ddot{\vec{x}}_i \right)$$

$$+ [c]\left(\frac{3}{\theta \Delta t}\vec{x}_i + 2\dot{\vec{x}}_i + \frac{\theta \Delta t}{2}\ddot{\vec{x}}_i \right) \qquad (11.88)$$

4. Find the displacement vector at time $t_{i+\theta}$:

$$\vec{x}_{i+\theta} = \left[\frac{6}{\theta^2(\Delta t)^2}[m] + \frac{3}{\theta \Delta t}[c] + [k] \right]^{-1}\vec{\underset{\sim}{F}}_{i+\theta} \qquad (11.89)$$

5. Calculate the acceleration, velocity, and displacement vectors at time t_{i+1}:

$$\ddot{\vec{x}}_{i+1} = \frac{6}{\theta^3(\Delta t)^2}(\vec{x}_{i+\theta} - \vec{x}_i) - \frac{6}{\theta^2 \Delta t}\dot{\vec{x}}_i + \left(1 - \frac{3}{\theta} \right)\ddot{\vec{x}}_i \qquad (11.90)$$

$$\dot{\vec{x}}_{i+1} = \dot{\vec{x}}_i + \frac{\Delta t}{2}(\ddot{\vec{x}}_{i+1} + \ddot{\vec{x}}_i) \qquad (11.91)$$

$$\vec{x}_{i+1} = \vec{x}_i + \Delta t\dot{\vec{x}}_i + \frac{(\Delta t)^2}{6}(\ddot{\vec{x}}_{i+1} + 2\ddot{\vec{x}}_i) \qquad (11.92)$$

EXAMPLE 11.7	Wilson Method for a Two d.o.f. System

Find the response of the system considered in Example 11.3, using the Wilson θ method with $\theta = 1.4$.

Given: Two degree of freedom system shown in Fig. 11.2.

Find: Response, $\vec{x}(t)$ at t_i; $i = 1, 2, \ldots$.

Approach: Use Wilson method with $\Delta t = 0.2416$.

Solution. The value of $\ddot{\vec{x}}_0$ can be obtained as in the case of Example 11.3:

$$\ddot{\vec{x}}_0 = \begin{Bmatrix} 0 \\ 5 \end{Bmatrix}$$

Then, by using Eqs. (11.90) to (11.92) with a time step of $\Delta t = 0.24216$, we obtain the results indicated in Table 11.6.

TABLE 11.6

Time $t_i = i\Delta t$	$\vec{x}_i = \vec{x}(t = t_i)$
t_1	$\begin{Bmatrix} 0.0033 \\ 0.1392 \end{Bmatrix}$
t_2	$\begin{Bmatrix} 0.0289 \\ 0.5201 \end{Bmatrix}$
t_3	$\begin{Bmatrix} 0.1072 \\ 1.0579 \end{Bmatrix}$
t_4	$\begin{Bmatrix} 0.2649 \\ 1.6408 \end{Bmatrix}$
t_5	$\begin{Bmatrix} 0.5076 \\ 2.1529 \end{Bmatrix}$
t_6	$\begin{Bmatrix} 0.8074 \\ 2.4981 \end{Bmatrix}$
t_7	$\begin{Bmatrix} 1.1035 \\ 2.6191 \end{Bmatrix}$
t_8	$\begin{Bmatrix} 1.3158 \\ 2.5056 \end{Bmatrix}$
t_9	$\begin{Bmatrix} 1.3688 \\ 2.1929 \end{Bmatrix}$
t_{10}	$\begin{Bmatrix} 1.2183 \\ 1.7503 \end{Bmatrix}$
t_{11}	$\begin{Bmatrix} 0.8710 \\ 1.2642 \end{Bmatrix}$
t_{12}	$\begin{Bmatrix} 0.3897 \\ 0.8208 \end{Bmatrix}$

11.10 NEWMARK METHOD

The Newmark integration method is also based on the assumption that the acceleration varies linearly between two instants of time. The resulting expressions for the velocity and displacement vectors $\dot{\vec{x}}_{i+1}$ and $\vec{x}_{i+1}$, for a multidegree of freedom system [11.21], are written as in Eqs. (11.79) and (11.80):

$$\dot{\vec{x}}_{i+1} = \dot{\vec{x}}_i + \left[(1 - \beta)\ddot{\vec{x}}_i + \beta\ddot{\vec{x}}_{i+1}\right]\Delta t \tag{11.93}$$

$$\vec{x}_{i+1} = \vec{x}_i + \Delta t\,\dot{\vec{x}}_i + \left[(\tfrac{1}{2} - \alpha)\ddot{\vec{x}}_i + \alpha\ddot{\vec{x}}_{i+1}\right](\Delta t)^2 \tag{11.94}$$

where the parameters α and β indicate how much the acceleration at the end of the interval enters into the velocity and displacement equations at the end of the interval Δt. In fact, α and β can be chosen to obtain the desired accuracy and stability characteristics [11.22]. When $\beta = \tfrac{1}{2}$ and $\alpha = \tfrac{1}{6}$, Eqs. (11.93) and (11.94) correspond to the linear acceleration method (which can also be obtained using $\theta = 1$ in the Wilson method). When $\beta = \tfrac{1}{2}$ and $\alpha = \tfrac{1}{4}$, Eqs. (11.93) and (11.94) correspond to the assumption of constant acceleration between t_i and t_{i+1}. To find the value of $\ddot{\vec{x}}_{i+1}$, the equilibrium equation (11.18) is considered at $t = t_{i+1}$ so that

$$[m]\ddot{\vec{x}}_{i+1} + [c]\dot{\vec{x}}_{i+1} + [k]\vec{x}_{i+1} = \vec{F}_{i+1} \tag{11.95}$$

Equation (11.94) can be used to express $\ddot{\vec{x}}_{i+1}$ in terms of $\vec{x}_{i+1}$, and the resulting expression can be substituted into Eq. (11.93) to express $\dot{\vec{x}}_{i+1}$ in terms of $\vec{x}_{i+1}$. By substituting these expressions for $\ddot{\vec{x}}_{i+1}$ and $\dot{\vec{x}}_{i+1}$ into Eq. (11.95), we can obtain a relation for finding $\vec{x}_{i+1}$:

$$\vec{x}_{i+1} = \left[\frac{1}{\alpha(\Delta t)^2}[m] + \frac{\beta}{\alpha\,\Delta t}[c] + [k]\right]^{-1}$$

$$\times \left\{\vec{F}_{i+1} + [m]\left(\frac{1}{\alpha(\Delta t)^2}\vec{x}_i + \frac{1}{\alpha\,\Delta t}\dot{\vec{x}}_i + \left(\frac{1}{2\alpha} - 1\right)\ddot{\vec{x}}_i\right)\right.$$

$$\left. + [c]\left(\frac{\beta}{\alpha\,\Delta t}\vec{x}_i + \left(\frac{\beta}{\alpha} - 1\right)\dot{\vec{x}}_i + \left(\frac{\beta}{\alpha} - 2\right)\frac{\Delta t}{2}\ddot{\vec{x}}_i\right)\right\} \tag{11.96}$$

The Newmark method can be summarized in the following steps:

1. From the known values of $\vec{x}_0$ and $\dot{\vec{x}}_0$, find $\ddot{\vec{x}}_0$ using Eq. (11.26).
2. Select suitable values of Δt, α, and β.
3. Calculate the displacement vector $\vec{x}_{i+1}$, starting with $i = 0$ and using Eq. (11.96).
4. Find the acceleration and velocity vectors at time t_{i+1}:

$$\ddot{\vec{x}}_{i+1} = \frac{1}{\alpha(\Delta t)^2}(\vec{x}_{i+1} - \vec{x}_i) - \frac{1}{\alpha\,\Delta t}\dot{\vec{x}}_i - \left(\frac{1}{2\alpha} - 1\right)\ddot{\vec{x}}_i \tag{11.97}$$

$$\dot{\vec{x}}_{i+1} = \dot{\vec{x}}_i + (1 - \beta)\Delta t\,\ddot{\vec{x}}_i + \beta\Delta t\,\ddot{\vec{x}}_{i+1} \tag{11.98}$$

It is important to note that unless β is taken as $\tfrac{1}{2}$, there is a spurious damping introduced, proportional to $(\beta - \tfrac{1}{2})$. If β is taken as zero, a negative damping results; this involves a self-excited vibration arising solely from the numerical procedure. Similarly, if β is greater than $\frac{1}{2}$, a positive damping is introduced. This reduces the magnitude of response even without real damping in the problem [11.21]. The method is unconditionally stable for $\alpha \geqslant \frac{1}{4}(\beta + \frac{1}{2})^2$ and $\beta \geqslant \frac{1}{2}$.

EXAMPLE 11.8 **Newmark Method for a Two d.o.f. System**

Find the response of the system considered in Example 11.3, using the Newmark method with $\alpha = \frac{1}{6}$ and $\beta = \frac{1}{2}$.

Given: Two degree of freedom system shown in Fig. 11.2.

Find: Response, $\vec{x}(t)$ at t_i; $i = 1, 2, \ldots$.

Approach: Use Newmark method with $\Delta t = 0.2416$.

TABLE 11.7

Time $t_i = i\Delta t$	$\vec{x}_i = \vec{x}(t = t_i)$
t_1	$\left\{ \begin{matrix} 0.0026 \\ 0.1411 \end{matrix} \right\}$
t_2	$\left\{ \begin{matrix} 0.0246 \\ 0.5329 \end{matrix} \right\}$
t_3	$\left\{ \begin{matrix} 0.1005 \\ 1.0884 \end{matrix} \right\}$
t_4	$\left\{ \begin{matrix} 0.2644 \\ 1.6870 \end{matrix} \right\}$
t_5	$\left\{ \begin{matrix} 0.5257 \\ 2.2027 \end{matrix} \right\}$
t_6	$\left\{ \begin{matrix} 0.8530 \\ 2.5336 \end{matrix} \right\}$
t_7	$\left\{ \begin{matrix} 1.1730 \\ 2.6229 \end{matrix} \right\}$
t_8	$\left\{ \begin{matrix} 1.3892 \\ 2.4674 \end{matrix} \right\}$
t_9	$\left\{ \begin{matrix} 1.4134 \\ 2.1137 \end{matrix} \right\}$
t_{10}	$\left\{ \begin{matrix} 1.1998 \\ 1.6426 \end{matrix} \right\}$
t_{11}	$\left\{ \begin{matrix} 0.7690 \\ 1.1485 \end{matrix} \right\}$
t_{12}	$\left\{ \begin{matrix} 0.2111 \\ 0.7195 \end{matrix} \right\}$

Solution. The value of $\ddot{\vec{x}}_0$ can be found using Eq. (11.26):

$$\ddot{\vec{x}}_0 = \begin{Bmatrix} 0 \\ 5 \end{Bmatrix}$$

With the values of $\alpha = \frac{1}{6}$, $\beta = 0.5$, and $\Delta t = 0.24216$, Eq. (11.96) gives the values of $\vec{x}_i = \vec{x}(t = t_i)$ as shown in Table 11.7.

11.11 COMPUTER PROGRAMS

**11.11.1
Fourth Order
Runge-Kutta
Method**

A Fortran program in the form of subroutine RK4 is given for solving a system of N first order differential equations of the form

$$\frac{d\vec{X}}{dt} = \vec{F}(\vec{X}, t)$$

Subroutine RK4 is to be called *NSTEP* times to generate the solution one time step at a time. The following arguments are used for the subroutine:

T   =   Current value of time. Input data.

DT   =   Time increment. Input data.

N   =   Number of differential equations. Input data.

XX   =   Array of size N, containing the initial conditions $x_i(0)$. Input data.

F, XI, XJ, XK, XL, UU = Arrays of size N each.

The program requires a user-supplied subroutine FUN(X, F, N, T). This subroutine must evaluate the forcing functions $F(1)$, $F(2)$, ..., $F(N)$ at any specified vectors $\vec{X}$ (argument X) and time t (argument T).

Example 11.2 is solved by the use of subroutine RK4. The main program, subroutine RK4, and the results given by the program are shown below.

```
C ==================================================================
C
C PROGRAM 19
C MAIN PROGRAM FOR CALLING THE SUBROUTINE RK4
C
C ==================================================================
C FOLLOWING 7 LINES CONTAIN PROBLEM-DEPENDENT DATA
      DIMENSION TIME(40),X(40,2),XX(2),F(2),YI(2),YJ(2),YK(2),YL(2),
     2   UU(2)
      XX(1)=0.0
      XX(2)=0.0
      NEQ=2
      NSTEP=40
```

```
          DT=0.31416/2
          T=0.0
          PRINT 10
    10    FORMAT (//,3X,5H  I  ,10H   TIME(I),7X,5H X(1),12X,5H X(2),/)
          DO 40 I=1,NSTEP
          CALL RK4 (T,DT,NEQ,XX,F,YI,YJ,YK,YL,UU)
          TIME(I)=T
          DO 20 J=1,NEQ
    20    X(I,J)=XX(J)
          PRINT 30, I, TIME(I),(X(I,J),J=1,NEQ)
    30    FORMAT (2X,I5,F10.4,2X,E15.8,2X,E15.8)
    40    CONTINUE
          STOP
          END
```

```
C ======================================================================
C
C SUBROUTINE RK4
C
C ======================================================================
          SUBROUTINE RK4 (T,DT,N,XX,F,XI,XJ,XK,XL,UU)
          DIMENSION XI(N),XJ(N),XK(N),XL(N),UU(N),XX(N),F(N)
          CALL FUN (XX,XI,N,T)
          DO 10 I=1,N
    10    UU(I)=XX(I)+0.5*DT*XI(I)
          TN=T+0.5*DT
          CALL FUN (UU,XJ,N,TN)
          DO 20 I=1,N
    20    UU(I)=XX(I)+0.5*DT*XJ(I)
          CALL FUN (UU,XK,N,TN)
          DO 30 I=1,N
    30    UU(I)=XX(I)+DT*XK(I)
          TN=T+DT
          CALL FUN (UU,XL,N,TN)
          DO 40 I=1,N
          F(I)=XL(I)
    40    XX(I)=XX(I)+(XI(I)+2.0*XJ(I)+2.0*XK(I)+XL(I))*DT/6.0
          T=T+DT
          RETURN
          END
```

```
C ======================================================================
C
C SUBROUTINE FUN FOR USE IN THE SUBROUTINE RK4
C THIS SUBROUTINE CHANGES FROM PROBLEM TO PROBLEM
C
C ======================================================================
          SUBROUTINE FUN (X,F,N,T)
          DIMENSION X(N),F(N)
          F(1)=X(2)
          F0=1.0
          T0=3.1416
```

```
XM=1.0
XC=0.2
XK=1.0
FT=F0*(1.0-SIN(3.1416*T/(2.0*T0)))
F(2)=(FT-XC*X(2)-XK*X(1))
RETURN
END
```

I	TIME(I)	X(1)	X(2)
1	0.1571	0.11863151E-01	0.14791380E+00
2	0.3142	0.45406420E-01	0.27559111E+00
3	0.4712	0.97257055E-01	0.38067484E+00
4	0.6283	0.16372624E+00	0.46150222E+00
5	0.7854	0.24091978E+00	0.51712251E+00
6	0.9425	0.32485014E+00	0.54729646E+00
7	1.0996	0.41154701E+00	0.55247802E+00
8	1.2566	0.49716362E+00	0.53377944E+00
9	1.4137	0.57807648E+00	0.49292117E+00
10	1.5708	0.65097594E+00	0.43216801E+00
11	1.7279	0.71294606E+00	0.35425371E+00
12	1.8850	0.76153159E+00	0.26229632E+00
13	2.0420	0.79479134E+00	0.15970632E+00
14	2.1991	0.81133646E+00	0.50090402E-01
15	2.3562	0.81035364E+00	-0.62847026E-01
16	2.5133	0.79161268E+00	-0.17540169E+00
17	2.6704	0.75545907E+00	-0.28396457E+00
18	2.8274	0.70279163E+00	-0.38511199E+00
19	2.9845	0.63502711E+00	-0.47568849E+00
20	3.1416	0.55405176E+00	-0.55288076E+00
21	3.2987	0.46216279E+00	-0.61428112E+00
22	3.4558	0.36200023E+00	-0.65793908E+00
23	3.6128	0.25647193E+00	-0.68240052E+00
24	3.7699	0.14867344E+00	-0.68673331E+00
25	3.9270	0.41804813E-01	-0.67053956E+00
26	4.0841	-0.60913362E-01	-0.63395429E+00
27	4.2412	-0.15632270E+00	-0.57763124E+00
28	4.3982	-0.24140659E+00	-0.50271636E+00
29	4.5553	-0.31336465E+00	-0.41080984E+00
30	4.7124	-0.36968049E+00	-0.30391818E+00
31	4.8695	-0.40818110E+00	-0.18439761E+00
32	5.0266	-0.42708698E+00	-0.54890379E-01
33	5.1836	-0.42505166E+00	0.81744030E-01
34	5.3407	-0.40119010E+00	0.22250143E+00
35	5.4978	-0.35509574E+00	0.36430690E+00
36	5.6549	-0.28684574E+00	0.50408858E+00
37	5.8120	-0.19699481E+00	0.63884296E+00
38	5.9690	-0.86557955E-01	0.76573688E+00
39	6.1261	0.43017015E-01	0.88210326E+00
40	6.2832	0.18988648E+00	0.98556501E+00

**11.11.2
Central
Difference
Method**

A Fortran subroutine CDIFF is written to find the dynamic response of a multidegree of freedom system by the use of the central difference method. The following arguments are used:

M	=	Array of size $N \times N$, denoting the mass matrix. Input data.
C	=	Array of size $N \times N$, denoting the damping matrix. Input data.
K	=	Array of size $N \times N$, denoting the stiffness matrix. Input data.
XI	=	Array of size N, containing the initial values $x_i(0)$. Input data.
XDI	=	Array of size N, containing the initial values $\dot{x}_i(0)$. Input data.
XDDI	=	Array of size N.
N	=	Number of degrees of freedom. Input data.
NSTEP	=	Number of time steps at which the solution is to be found. Input data.
DELT	=	Time increment between steps. Input data.
F	=	Array of size N, containing the values of the forcing function at any specified time.

R, RR, XM1, XM2, XP1, XMK, XMI, ZA, ZB, ZC, LA, S = arrays of size N each.

MC, MK, MCI, MMC, MI = arrays of size $N \times N$ each.

X, XD, XDD = arrays of size $NSTEP1 \times N$ each. $X(I, J) = x_j(t_i)$, $XD(I, J) = \dot{x}_j(t_i)$, $XDD(I, J) = \ddot{x}_j(t_i)$. Output.

LB	=	Array of size $N \times 2$.
NSTEP1	=	NSTEP + 1. Input data.

The program requires a user-supplied subroutine EXTFUN(F, TIME, N). This must evaluate the forcing functions $F(1), \ldots, F(N)$ at a given time, TIME.

For illustration, subroutine CDIFF is used to solve Example 11.3. The main program that calls CDIFF, subroutine CDIFF, and the output of the program are given below.

```
C ==========================================================================
C
C PROGRAM 20
C MAIN PROGRAM WHICH CALLS CDIFF
C
C ==========================================================================
C FOLLOWING 10 LINES CONTAIN PROBLEM-DEPENDENT DATA
      REAL M(2,2),K(2,2),MC(2,2),MK(2,2),MCI(2,2),MMC(2,2),MI(2,2)
      DIMENSION C(2,2),XI(2),XDI(2),XDDI(2),XM1(2),F(2),R(2),RR(2),
     2    XMK(2),XMI(2),XM2(2),XP1(2),ZA(2),ZB(2),ZC(2),LA(2),LB(2,2),
     3     S(2),X(25,2),XD(25,2),XDD(25,2)
      DATA N,NSTEP,NSTEP1,DELT/2,24,25,0.24216267/
      DATA XI/0.0,0.0/
      DATA XDI/0.0,0.0/
```

```
         DATA M/1.0,0.0,0.0,2.0/
         DATA C/0.0,0.0,0.0,0.0/
         DATA K/6.0,-2.0,-2.0,8.0/
C END OF PROBLEM-DEPENDENT DATA
         CALL CDIFF (M,C,K,XI,XDI,XDDI,N,NSTEP,DELT,F,R,RR,XM1,XM2,XP1,
        2    MC,MK,MCI,XMK,MMC,XMI,ZA,ZB,ZC,LA,LB,S,X,XD,XDD,NSTEP1,MI)
         PRINT 10
   10    FORMAT (//,38H SOLUTION BY CENTRAL DIFFERENCE METHOD,/)
         PRINT 20, N,NSTEP,DELT
   20    FORMAT (12H GIVEN DATA:,/,3H N=,I5,4X,7H NSTEP=,I5,4X,6H DELT=,
        2    E15.8,/)
         PRINT 30
   30    FORMAT (10H SOLUTION:,//,5H STEP,3X,5H TIME,3X,7H X(I,1),3X,
        2    8H XD(I,1),2X,9H XDD(I,1),4X,7H X(I,2),3X,8H XD(I,2),2X,
        3    9H XDD(I,2),/)
         DO 40  I=1,NSTEP1
         TIME=REAL(I-1)*DELT
   40    PRINT 50, I,TIME,X(I,1),XD(I,1),XDD(I,1),X(I,2),XD(I,2),XDD(I,2)
   50    FORMAT (1X,I4,F8.4,6(1X,E10.4))
         STOP
         END
C =======================================================================
C
C SUBROUTINE CDIFF
C
C =======================================================================
         SUBROUTINE CDIFF (M,C,K,XI,XDI,XDDI,N,NSTEP,DELT,F,R,RR,XM1,
        2    XM2,XP1,MC,MK,MCI,XMK,MMC,XMI,ZA,ZB,ZC,LA,LB,S,X,XD,XDD,NSTEP1
        3    ,MI)
         REAL M(N,N),K(N,N),MC(N,N),MK(N,N),MCI(N,N),MMC(N,N),MI(N,N)
         DIMENSION C(N,N),XI(N),XDI(N),XDDI(N),F(N),R(N),RR(N),XM1(N),
        2    XM2(N),XP1(N),XMK(N),XMI(N),ZA(N),ZB(N),ZC(N),LA(N),LB(N,2),S(N
        3    ,X(NSTEP1,N),XD(NSTEP1,N),XDD(NSTEP1,N)
         DO 5 I=1,N
         DO 5 J=1,N
    5    MI(I,J)=M(I,J)
         CALL SIMUL (MI,ZA,N,0,LA,LB,S)
         CALL EXTFUN (F,0.0,N)
         DO 20 I=1,N
         R(I)=F(I)
         DO 10 J=1,N
   10    R(I)=R(I)-C(I,J)*XDI(J)-K(I,J)*XI(J)
   20    CONTINUE
         CALL XMULT (MI,R,XDDI,N)
         DO 25 J=1,N
         X(1,J)=XI(J)
         XD(1,J)=XDI(J)
   25    XDD(1,J)=XDDI(J)
         DO 30 I=1,N
   30    XM1(I)=XI(I)-DELT*XDI(I)+(DELT**2)*XDDI(I)/2.0
         DO 40 I=1,N
         DO 40 J=1,N
         MC(I,J)=(M(I,J)/(DELT**2))+(C(I,J)/(2.0*DELT))
         MK(I,J)=K(I,J)-2.0*M(I,J)/(DELT**2)
```

```
40     MMC(I,J)=(M(I,J)/(DELT**2))-(C(I,J)/(2.0*DELT))
       DO 45 I=1,N
       DO 45 J=1,N
45     MCI(I,J)=MC(I,J)
       CALL SIMUL (MCI,ZA,N,0,LA,LB,S)
       TIME=-DELT
       DO 90 I=1,NSTEP
       CALL XMULT (MK,XI,XMK,N)
       CALL XMULT (MMC,XM1,XMI,N)
       TIME=TIME+DELT
       CALL EXTFUN (F,TIME,N)
       DO 50 J=1,N
50     RR(J)=F(J)-XMK(J)-XMI(J)
       CALL XMULT (MCI,RR,XP1,N)
       DO 60 J=1,N
       XDI(J)=(XP1(J)-XM1(J))/(2.0*DELT)
60     XDDI(J)=(XP1(J)-2.0*XI(J)+XM1(J))/(DELT**2)
       DO 70 J=1,N
       XM1(J)=XI(J)
70     XI(J)=XP1(J)
       DO 80 J=1,N
       X(I+1,J)=XP1(J)
       XD(I+1,J)=XDI(J)
80     XDD(I+1,J)=XDDI(J)
90     CONTINUE
       RETURN
       END
C ==========================================================
C
C SUBROUTINE EXTFUN
C THIS SUBROUTINE IS PROBLEM-DEPENDENT
C
C ==========================================================
       SUBROUTINE EXTFUN (F,TIME,N)
       DIMENSION F(N)
       F(1)=0.0
       F(2)=10.0
       RETURN
       END
C ==========================================================
C
C SUBROUTINE XMULT
C
C ==========================================================
       SUBROUTINE XMULT (A,B,BB,N)
       DIMENSION A(N,N),B(N),BB(N)
       DO 10 I=1,N
       BB(I)=0.0
       DO 10 J=1,N
10     BB(I)=BB(I)+A(I,J)*B(J)
       RETURN
       END
```

SOLUTION BY CENTRAL DIFFERENCE METHOD

GIVEN DATA:

N=   2   NSTEP=   24   DELT= 0.24216267E+00

SOLUTION:

STEP	TIME	X(I,1)	XD(I,1)	XDD(I,1)	X(I,2)	XD(I,2)	XDD(I,2)
1	0.0000	0.0000E+00	0.0000E+00	0.0000E+00	0.0000E+00	0.0000E+00	0.5000E+0
2	0.2422	0.0000E+00	0.0000E+00	0.0000E+00	0.1466E+00	0.0000E+00	0.5000E+0
3	0.4843	0.1719E-01	0.3550E-01	0.2932E+00	0.5520E+00	0.1140E+01	0.4414E+0
4	0.7265	0.9309E-01	0.1922E+00	0.1001E+01	0.1122E+01	0.2014E+01	0.2809E+0
5	0.9687	0.2678E+00	0.5175E+00	0.1686E+01	0.1728E+01	0.2428E+01	0.6043E+0
6	1.2108	0.5510E+00	0.9455E+00	0.1849E+01	0.2237E+01	0.2302E+01	-.1643E+0
7	1.4530	0.9027E+00	0.1311E+01	0.1168E+01	0.2547E+01	0.1692E+01	-.3397E+0
8	1.6951	0.1235E+01	0.1413E+01	-.3219E+00	0.2606E+01	0.7613E+00	-.4285E+0
9	1.9373	0.1439E+01	0.1108E+01	-.2201E+01	0.2419E+01	-.2646E+00	-.4188E+0
10	2.1795	0.1420E+01	0.3814E+00	-.3797E+01	0.2042E+01	-.1164E+01	-.3236E+0
11	2.4216	0.1141E+01	-.6155E+00	-.4437E+01	0.1563E+01	-.1767E+01	-.1749E+0
12	2.6638	0.6437E+00	-.1603E+01	-.3720E+01	0.1077E+01	-.1992E+01	-.1110E+0
13	2.9060	0.4627E-01	-.2260E+01	-.1708E+01	0.6698E+00	-.1844E+01	0.1335E+0
14	3.1481	-.4889E+00	-.2339E+01	0.1062E+01	0.4012E+00	-.1396E+01	0.2367E+0
15	3.3903	-.8050E+00	-.1758E+01	0.3736E+01	0.3030E+00	-.7575E+00	0.2906E+0
16	3.6324	-.8023E+00	-.6471E+00	0.5436E+01	0.3797E+00	-.4438E-01	0.2983E+0
17	3.8746	-.4728E+00	0.6859E+00	0.5573E+01	0.6135E+00	0.6412E+00	0.2679E+0
18	4.1168	0.9503E-01	0.1853E+01	0.4064E+01	0.9689E+00	0.1217E+01	0.2073E+0
19	4.3589	0.7431E+00	0.2510E+01	0.1368E+01	0.1396E+01	0.1615E+01	0.1219E+0
20	4.6011	0.1293E+01	0.2474E+01	-.1667E+01	0.1832E+01	0.1782E+01	0.1598E+0
21	4.8433	0.1603E+01	0.1776E+01	-.4096E+01	0.2208E+01	0.1676E+01	-.1035E+0
22	5.0854	0.1608E+01	0.6502E+00	-.5205E+01	0.2453E+01	0.1281E+01	-.2227E+0
23	5.3276	0.1335E+01	-.5545E+00	-.4744E+01	0.2510E+01	0.6238E+00	-.3202E+0
24	5.5697	0.8862E+00	-.1491E+01	-.2990E+01	0.2350E+01	-.2124E+00	-.3704E+0
25	5.8119	0.4013E+00	-.1928E+01	-.6176E+00	0.1984E+01	-.1086E+01	-.3513E+0

11.11.3
Houbolt Method

A Fortran subroutine HOBOLT implements the Houbolt method. The following arguments are used for this subroutine:

M	=	Mass matrix of size $N \times N$. Input data.
C	=	Damping matrix of size $N \times N$. Input data.
K	=	Stiffness matrix of size $N \times N$. Input data.
XI, XDI	=	Vectors of size N each, containing the initial values of x_i and $\dot{x}_i$. Input data.
XDDI	=	Vector of size N.
N	=	Order of the matrix $[m]$. Input data.
NSTEP	=	Number of time steps at which the solution is to be found. Input data.
DELT	=	Time increment between steps. Input data.
F	=	Vector of size N, containing the values of the forcing function at any specified time.

R, RR, XM1, XM2, XP1, XMK, XMI, ZA, ZB, ZC, LA, S = Vectors of size N each.

MC, MK, MCI, MMC, MI = Matrices of size $N \times N$ each.

X, XD, XDD = Matrices of size $NSTEP1 \times N$ each. $X(I, J) = x_j(t_i)$, $XD(I, J) = \dot{x}_j(t_i)$, $XDD(I, J) = \ddot{x}_j(t_i)$. Output.

NSTEP1 = NSTEP + 1.

LB = Matrix of size $N \times 2$.

The program requires a user-supplied subroutine EXTFUN(F, TIME, N). This subroutine must evaluate the forcing functions $F(1), F(2), \ldots, F(N)$ at a given time, *TIME*.

For illustration, Example 11.6 is solved by the use of subroutine HOBOLT. The main program, subroutine HOBOLT, and the results of the program are given below.

```
C ================================================================
C
C PROGRAM 21
C MAIN PROGRAM WHICH CALLS HOBOLT
C
C ================================================================
C FOLLOWING 10 LINES CONTAIN PROBLEM-DEPENDENT DATA
      REAL M(2,2),K(2,2),MC(2,2),MK(2,2),MCI(2,2),MMC(2,2),MI(2,2)
      DIMENSION C(2,2),XI(2),XDI(2),XDDI(2),XM1(2),F(2),R(2),RR(2),
     2   XMK(2),XMI(2),XM2(2),XP1(2),ZA(2),ZB(2),ZC(2),LA(2),LB(2,2),
     3   S(2),X(25,2),XD(25,2),XDD(25,2)
      DATA N,NSTEP,NSTEP1,DELT/2,24,25,0.24216267/
      DATA XI/0.0,0.0/
      DATA XDI/0.0,0.0/
      DATA M/1.0,0.0,0.0,2.0/
      DATA C/0.0,0.0,0.0,0.0/
      DATA K/6.0,-2.0,-2.0,8.0/
C END OF PROBLEM-DEPENDENT DATA
      CALL HOBOLT (M,C,K,XI,XDI,XDDI,N,NSTEP,DELT,F,R,RR,XM1,XM2,XP1,
     2   MC,MK,MCI,XMK,MMC,XMI,ZA,ZB,ZC,LA,LB,S,X,XD,XDD,NSTEP1,MI)
      PRINT 10
10    FORMAT (//,27H SOLUTION BY HOUBOLT METHOD,/)
      PRINT 20, N,NSTEP,DELT
20    FORMAT (12H GIVEN DATA:,/,3H N=,I5,4X,7H NSTEP=,I5,4X,6H DELT=,
     2   E15.8,/)
      PRINT 30
30    FORMAT (10H SOLUTION:,//,5H STEP,3X,5H TIME,3X,7H X(I,1),3X,
     2   8H XD(I,1),2X,9H XDD(I,1),4X,7H X(I,2),3X,8H XD(I,2),2X,
     3   9H XDD(I,2),/)
      DO 40  I=1,NSTEP1
      TIME=REAL(I-1)*DELT
40    PRINT 50, I,TIME,X(I,1),XD(I,1),XDD(I,1),X(I,2),XD(I,2),XDD(I,2)
50    FORMAT (1X,I4,F8.4,6(1X,E10.4))
      STOP
      END
```

```
C =====================================================================
C
C SUBROUTINE HOBOLT
C
C =====================================================================
      SUBROUTINE HOBOLT (M,C,K,XI,XDI,XDDI,N,NSTEP,DELT,F,R,RR,XM1,
     2   XM2,XP1,MC,MK,MCI,XMK,MMC,XMI,ZA,ZB,ZC,LA,LB,S,X,XD,XDD,NSTEP1
     3   ,MI)
      REAL M(N,N),K(N,N),MC(N,N),MK(N,N),MCI(N,N),MMC(N,N),MI( N,N)
      DIMENSION C(N,N),XI(N),XDI(N),XDDI(N),F(N),R(N),RR(N),XM1(N),
     2   XM2(N),XP1(N),XMK(N),XMI(N),ZA(N),ZB(N),ZC(N),X(NSTEP1,N),
     3   XD(NSTEP1,N),XDD(NSTEP1,N),LA(N),LB(N,2),S(N)
      DO 5 I=1,N
      DO 5 J=1,N
5     MI(I,J)=M(I,J)
      CALL SIMUL (MI,ZA,N,0,LA,LB,S)
      CALL EXTFUN (F,0.0,N)
      DO 20 I=1,N
      R(I)=F(I)
      DO 10 J=1,N
10    R(I)=R(I)-C(I,J)*XDI(J)-K(I,J)*XI(J)
20    CONTINUE
      CALL XMULT (MI,R,XDDI,N)
      DO 25 J=1,N
      X(1,J)=XI(J)
      XD(1,J)=XDI(J)
25    XDD(1,J)=XDDI(J)
      DO 30 I=1,N
30    XM1(I)=XI(I)-DELT*XDI(I)+(DELT**2)*XDDI(I)/2.0
      DO 40 I=1,N
      DO 40 J=1,N
      MC(I,J)=(M(I,J)/(DELT**2))+(C(I,J)/(2.0*DELT))
      MK(I,J)=K(I,J)-2.0*M(I,J)/(DELT**2)
40    MMC(I,J)=(M(I,J)/(DELT**2))-(C(I,J)/(2.0*DELT))
      DO 45 I=1,N
      DO 45 J=1,N
45    MCI(I,J)=MC(I,J)
      CALL SIMUL (MCI,ZA,N,0,LA,LB,S)
      TIME=-DELT
      DO 90 I=1,2
      CALL XMULT (MK,XI,XMK,N)
      CALL XMULT (MMC,XM1,XMI,N)
      TIME=TIME+DELT
      CALL EXTFUN (F,TIME,N)
      DO 50 J=1,N
50    RR(J)=F(J)-XMK(J)-XMI(J)
      CALL XMULT (MCI,RR,XP1,N)
      DO 60 J=1,N
      XDI(J)=(XP1(J)-XM1(J))/(2.0*DELT)
60    XDDI(J)=(XP1(J)-2.0*XI(J)+XM1(J))/(DELT**2)
      DO 70 J=1,N
      XM1(J)=XI(J)
70    XI(J)=XP1(J)
      DO 80 J=1,N
```

```
          X(I+1,J)=XP1(J)
          XD(I+1,J)=XDI(J)
   80     XDD(I+1,J)=XDDI(J)
   90     CONTINUE
          DO 160 II=3,NSTEP
          DO 110 I=1,N
          DO 110 J=1,N
          MC(I,J)=M(I,J)*2.0/(DELT**2)+11.0*C(I,J)/(6.0*DELT)+K(I,J)
          MK(I,J)=5.0*M(I,J)/(DELT**2)+3.0*C(I,J)/DELT
          MI(I,J)=4.0*M(I,J)/(DELT**2)+3.0*C(I,J)/(2.0*DELT)
  110     MMC(I,J)=M(I,J)/(DELT**2)+C(I,J)/(3.0*DELT)
          DO 120 I=1,N
          XI(I)=X(II,I)
          XM1(I)=X(II-1,I)
  120     XM2(I)=X(II-2,I)
          CALL XMULT (MK,XI,ZA,N)
          CALL XMULT (MI,XM1,ZB,N)
          CALL XMULT (MMC,XM2,ZC,N)
          TIME=REAL(II)*DELT
          CALL EXTFUN (F,TIME,N)
          DO 130 I=1,N
  130     R(I)=F(I)+ZA(I)-ZB(I)+ZC(I)
          DO 135 I=1,N
          DO 135 J=1,N
  135     MCI(I,J)=MC(I,J)
          CALL SIMUL (MCI,ZA,N,0,LA,LB,S)
          CALL XMULT (MCI,R,XP1,N)
          DO 140 I=1,N
          XDI(I)=(11.0*XP1(I)-18.0*XI(I)+9.0*XM1(I)-2.0*XM2(I))/(6.0*DELT)
  140     XDDI(I)=(2.0*XP1(I)-5.0*XI(I)+4.0*XM1(I)-XM2(I))/(DELT**2)
          DO 150 I=1,N
          X(II+1,I)=XP1(I)
          XD(II+1,I)=XDI(I)
  150     XDD(II+1,I)=XDDI(I)
  160     CONTINUE
          RETURN
          END
C ================================================================
C
C SUBROUTINE EXTFUN
C
C ================================================================
          SUBROUTINE EXTFUN (F,TIME,N)
          DIMENSION F(N)
          F(1)=0.0
          F(2)=10.0
          RETURN
          END
C ================================================================
C
C SUBROUTINE XMULT
C
C ================================================================
          SUBROUTINE XMULT (A,B,BB,N)
          DIMENSION A(N,N),B(N),BB(N)
```

```
          DO 10 I=1,N
          BB(I)=0.0
          DO 10 J=1,N
     10   BB(I)=BB(I)+A(I,J)*B(J)
          RETURN
          END
```

```
SOLUTION BY HOUBOLT METHOD

GIVEN DATA:
N=   2    NSTEP=   24    DELT= 0.24216267E+00

SOLUTION:
```

STEP	TIME	X(I,1)	XD(I,1)	XDD(I,1)	X(I,2)	XD(I,2)	XDD(I,2)
1	0.0000	0.0000E+00	0.0000E+00	0.0000E+00	0.0000E+00	0.0000E+00	0.5000E+01
2	0.2422	0.0000E+00	0.0000E+00	0.0000E+00	0.1466E+00	0.0000E+00	0.5000E+01
3	0.4843	0.1719E-01	0.3550E-01	0.2932E+00	0.5520E+00	0.1140E+01	0.4414E+01
4	0.7265	0.9173E-01	0.4815E+00	0.1662E+01	0.1106E+01	0.2446E+01	0.6661E+00
5	0.9687	0.2501E+00	0.8635E+00	0.1881E+01	0.1691E+01	0.2312E+01	-.1513E+01
6	1.2108	0.4924E+00	0.1174E+01	0.1434E+01	0.2194E+01	0.1757E+01	-.3284E+01
7	1.4530	0.7867E+00	0.1278E+01	0.3394E+00	0.2530E+01	0.9207E+00	-.4332E+01
8	1.6951	0.1073E+01	0.1087E+01	-.1143E+01	0.2649E+01	-.2187E-01	-.4522E+01
9	1.9373	0.1280E+01	0.5906E+00	-.2591E+01	0.2545E+01	-.8952E+00	-.3901E+01
10	2.1795	0.1343E+01	-.1264E+00	-.3554E+01	0.2252E+01	-.1555E+01	-.2667E+01
11	2.4216	0.1226E+01	-.9066E+00	-.3690E+01	0.1832E+01	-.1911E+01	-.1104E+01
12	2.6638	0.9340E+00	-.1558E+01	-.2878E+01	0.1363E+01	-.1934E+01	0.4820E+00
13	2.9060	0.5178E+00	-.1906E+01	-.1262E+01	0.9224E+00	-.1652E+01	0.1828E+01
14	3.1481	0.6242E-01	-.1845E+01	0.7795E+00	0.5770E+00	-.1138E+01	0.2754E+01
15	3.3903	-.3323E+00	-.1367E+01	0.2739E+01	0.3725E+00	-.4914E+00	0.3178E+01
16	3.6324	-.5761E+00	-.5704E+00	0.4115E+01	0.3293E+00	0.1839E+00	0.3107E+01
17	3.8746	-.6108E+00	0.3677E+00	0.4552E+01	0.4436E+00	0.7914E+00	0.2615E+01
18	4.1168	-.4258E+00	0.1233E+01	0.3934E+01	0.6899E+00	0.1254E+01	0.1815E+01
19	4.3589	-.6001E-01	0.1829E+01	0.2413E+01	0.1026E+01	0.1520E+01	0.8341E+00
20	4.6011	0.4068E+00	0.2026E+01	0.3642E+00	0.1402E+01	0.1563E+01	-.2027E+00
21	4.8433	0.8737E+00	0.1790E+01	-.1716E+01	0.1763E+01	0.1385E+01	-.1179E+01
22	5.0854	0.1243E+01	0.1187E+01	-.3340E+01	0.2058E+01	0.1013E+01	-.1991E+01
23	5.3276	0.1441E+01	0.3662E+00	-.4155E+01	0.2246E+01	0.4955E+00	-.2543E+01
24	5.5697	0.1436E+01	-.4846E+00	-.4020E+01	0.2299E+01	-.9674E-01	-.2760E+01
25	5.8119	0.1241E+01	-.1182E+01	-.3029E+01	0.2209E+01	-.6813E+00	-.2593E+01

REFERENCES

11.1. G. L. Goudreau and R. L. Taylor, "Evaluation of numerical integration methods in elastodynamics," *Computational Methods in Applied Mechanics and Engineering*, Vol. 2, 1973, pp. 69–97.

11.2. S. W. Key, "Transient response by time integration: Review of implicit and explicit operations," in J. Donéa (Ed.), *Advanced Structural Dynamics*, Applied Science Publishers, London, 1980.

11.3. R. E. Cornwell, R. R. Craig, Jr., and C. P. Johnson, "On the application of the mode-acceleration method to structural engineering problems," *Earthquake Engineering and Structural Dynamics*, Vol. 11, 1983, pp. 679–688.

11.4. T. Wah and L. R. Calcote, *Structural Analysis by Finite Difference Calculus*, Van Nostrand Reinhold, New York, 1970.

11.5. R. Ali, "Finite difference methods in vibration analysis," *Shock and Vibration Digest*, Vol. 15, March 1983, pp. 3–7.

11.6. P. C. M. Lau, "Finite difference approximation for ordinary derivatives," *International Journal for Numerical Methods in Engineering*, Vol. 17, 1981, pp. 663–678.

11.7. R. D. Krieg, "Unconditional stability in numerical time integration methods," *Journal of Applied Mechanics*, Vol. 40, 1973, pp. 417–421.

11.8. S. Levy and W. D. Kroll, "Errors introduced by finite space and time increments in dynamic response computation," *Proceedings, First U.S. National Congress of Applied Mechanics*, 1951, pp. 1–8.

11.9. A. F. D'Souza and V. K. Garg, *Advanced Dynamics. Modeling and Analysis*, Prentice-Hall, Englewood Cliffs, N.J., 1984.

11.10. A Ralston and H. S. Wilf (Eds.), *Mathematical Methods for Digital Computers*, Wiley, New York, 1960.

11.11. S. Nakamura, *Computational Methods in Engineering and Science*, John Wiley, New York, 1977.

11.12. T. Belytschko, "Explicit time integration of structure-mechanical systems," in J. Donéa (Ed.), *Advanced Structural Dynamics*, Applied Science Publishers, London, 1980, pp. 97–122.

11.13. S. Levy and J. P. D. Wilkinson, *The Component Element Method in Dynamics with Application to Earthquake and Vehicle Engineering*, McGraw-Hill, New York, 1976.

11.14. J. W. Leech, P. T. Hsu, and E. W. Mack, "Stability of a finite-difference method for solving matrix equations," *AIAA Journal*, Vol. 3, 1965, pp. 2172–2173.

11.15. S. D. Conte and C. W. DeBoor, *Elementary Numerical Analysis: An Algorithmic Approach* (2nd Ed.), McGraw-Hill, New York, 1972.

11.16. C. F. Gerald and P. O. Wheatley, *Applied Numerical Analysis* (3d Ed.), Addison-Wesley, Reading, Mass., 1984.

11.17. L. V. Atkinson and P. J. Harley, *Introduction to Numerical Methods with PASCAL*, Addison-Wesley, Reading, Mass., 1984.

11.18. J. C. Houbolt, "A recurrence matrix solution for the dynamic response of elastic aircraft," *Journal of Aeronautical Sciences*, Vol. 17, 1950, pp. 540–550, 594.

11.19. E. L. Wilson, I. Farhoomand, and K. J. Bathe, "Nonlinear dynamic analysis of complex structures," *International Journal of Earthquake Engineering and Structural Dynamics*, Vol. 1, 1973, pp. 241–252.

11.20. S. P. Timoshenko, D. H. Young, and W. Weaver, Jr., *Vibration Problems in Engineering* (4th Ed.), John Wiley, New York, 1974.

11.21. N. M. Newmark, "A method of computation for structural dynamics," *ASCE Journal of Engineering Mechanics Division*, Vol. 85, 1959, pp. 67–94.

11.22. T. J. R. Hughes, "A note on the stability of Newmark's algorithm in nonlinear structural dynamics," *International Journal for Numerical Methods in Engineering*, Vol. 11, 1976, pp. 383–386.

REVIEW QUESTIONS

11.1. Describe the procedure of the finite difference method.

11.2. Derive the central difference formulas for the first and the second derivatives of a function, using Taylor's series expansion.

11.3. What is a conditionally stable method?

11.4. What is the main difference between the central difference method and the Runge-Kutta method?

11.5. Why is it necessary to introduce fictitious mesh points in the finite difference method of solution?

11.6. Define a tridiagonal matrix.

11.7. What is the basic assumption of the Wilson method?

11.8. What is a linear acceleration method?

11.9. What is the difference between explicit and implicit integration methods?

11.10. Can we use the numerical integration methods discussed in this chapter to solve nonlinear vibration problems?

PROBLEMS

The problem assignments are organized as follows:

Problems	Section covered	Topic covered
11.1–11.3	11.2	Finite difference approach
11.4–11.10	11.3	Central difference method for single degree of freedom systems
11.11–11.17, 11.22	11.4	Runge-Kutta method for single degree of freedom systems
11.18–11.20	11.5	Central difference method for multidegree of freedom system
11.21, 11.23–11.28	11.6	Central difference method for continuous systems
11.29–11.31	11.7	Runge-Kutta method for multidegree of freedom systems
11.32–11.34	11.8, 11.11	Houbolt method
11.35–11.38	11.9, 11.11	Wilson method
11.39–11.42	11.10, 11.11	Newmark method

11.1. The forward difference formulas make use of the values of the function to the right of the base grid point. Thus the first derivative at point $i(t = t_i)$ is defined as

$$\frac{dx}{dt} = \frac{x(t + \Delta t) - x(t)}{\Delta t} = \frac{x_{i+1} - x_i}{\Delta t}$$

Derive the forward difference formulas for $(d^2x)/(dt^2)$, $(d^3x)/(dt^3)$, and $(d^4x)/(dt^4)$ at t_i.

11.2. The backward difference formulas make use of the values of the function to the left of the base grid point. Accordingly, the first derivative at point $i(t = t_i)$ is defined as

$$\frac{dx}{dt} = \frac{x(t) - x(t - \Delta t)}{\Delta t} = \frac{x_i - x_{i-1}}{\Delta t}$$

Derive the backward difference formulas for $(d^2x)/(dt^2)$, $(d^3x)/(dt^3)$, and $(d^4x)/(dt^4)$ at t_i.

11.3. Derive the formula for the fourth derivative, $(d^4x)/(dt^4)$, according to the central difference method.

11.4. Find the free vibratory response of an undamped single degree of freedom system with $m = 1$ and $k = 1$, using the central difference method. Assume $x_0 = 0$ and $\dot{x}_0 = 1$. Compare the results obtained with $\Delta t = 1$ and $\Delta t = 0.5$ with the exact solution $x(t) = \sin t$.

11.5. Integrate the differential equation

$$-\frac{d^2x}{dt^2} + 0.1x = 0 \qquad \text{for} \qquad 0 \leqslant t \leqslant 10$$

using the backward difference formula with $\Delta t = 1$. Assume the initial conditions as $x_0 = 1$ and $\dot{x}_0 = 0$.

11.6. Find the free vibration response of a viscously damped single degree of freedom system with $m = k = c = 1$, using the central difference method. Assume that $x_0 = 0$, $\dot{x}_0 = 1$, and $\Delta t = 0.5$.

11.7. Solve Problem 11.6 by changing c to 2.

11.8. Solve Problem 11.6 by taking the value of c as 4.

11.9. Find the solution of the equation $4\ddot{x} + 2\dot{x} + 3000x = F(t)$, where $F(t)$ is as shown in Fig. 11.10 for the duration $0 \leqslant t \leqslant 1$. Assume that $x_0 = \dot{x}_0 = 0$ and $\Delta t = 0.05$.

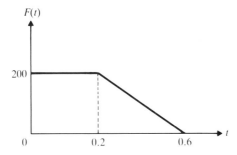

Figure 11.10

11.10. Find the solution of a spring-mass-damper system governed by the equation $m\ddot{x} + c\dot{x} + kx = F(t) = \delta F \cdot t$ with $m = c = k = 1$ and $\delta F = 1$. Assume the initial values of x and $\dot{x}$ to be zero and $\Delta t = 0.5$. Compare the central difference solution with the exact solution given in Example 4.6.

11.11. Express the following nth order differential equation as a system of n first order differential equations:

$$a_n \frac{d^n x}{dt^n} + a_{n-1} \frac{d^{n-1} x}{dt^{n-1}} + \cdots + a_1 \frac{dx}{dt} = g(x, t)$$

11.12. Find the solution of the following equations by using the fourth order Runge-Kutta method with $\Delta t = 0.1$:

(a) $\dot{x} = x - 1.5e^{-0.5t}$; $x_0 = 1$

(b) $\dot{x} = -tx^2$; $x_0 = 1$.

11.13. The second order Runge-Kutta formula is given by

$$\vec{X}_{i+1} = \vec{X}_i + \tfrac{1}{2}\left(\vec{K}_1 + \vec{K}_2 \right)$$

where

$$\vec{K}_1 = h\vec{F}\left(\vec{X}_i, t_i \right) \quad \text{and} \quad \vec{K}_2 = h\vec{F}\left(\vec{X}_i + \vec{K}_1, t_i + h \right)$$

Using this formula, solve the problem considered in Example 11.2.

11.14. The third order Runge-Kutta formula is given by

$$\vec{X}_{i+1} = \vec{X}_i + \tfrac{1}{6}\left(\vec{K}_1 + 4\vec{K}_2 + \vec{K}_3 \right)$$

where

$$\vec{K}_1 = h\vec{F}\left(\vec{X}_i, t_i \right)$$
$$\vec{K}_2 = h\vec{F}\left(\vec{X}_i + \tfrac{1}{2}\vec{K}_1, t_i + \tfrac{1}{2}h \right)$$

and

$$\vec{K}_3 = h\vec{F}\left(\vec{X}_i - \vec{K}_1 + 2\vec{K}_2, t_i + h \right)$$

Using this formula, solve the problem considered in Example 11.2.

11.15. Solve the differential equation $\ddot{x} + 1000x = 0$ with the initial conditions $x_0 = 5$ and $\dot{x}_0 = 0$, using the second order Runge-Kutta method. Use $\Delta t = 0.01$.

11.16. Solve Problem 11.15 using the third order Runge-Kutta method.

11.17. Solve Problem 11.15 using the fourth order Runge-Kutta method.

11.18. Find the response of the two degree of freedom system shown in Fig. 11.2 when $c = 2$, $F_1(t) = 0$, $F_2(t) = 10$, using the central difference method.

11.19. Find the response of the system shown in Fig. 11.2 when $F_1(t) = 10 \sin 5t$ and $F_2(t) = 0$, using the central difference method.

11.20. The equations of motion of a two degree of freedom system are given by $2\ddot{x}_1 + 6x_1 - 2x_2 = 5$ and $\ddot{x}_2 - 2x_1 + 4x_2 = 20 \sin 5t$. Assuming the initial conditions as $x_1(0) = \dot{x}_1(0) = x_2(0) = \dot{x}_2(0) = 0$, find the response of the system, using the central difference method with $\Delta t = 0.25$.

11.21. The ends of a beam are elastically restrained by linear and torsional springs, as shown in Fig. 11.11. Express the boundary conditions using the finite difference method.

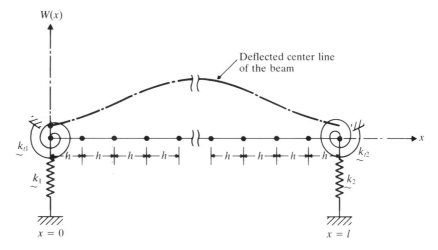

Figure 11.11

11.22. Solve Problem 11.20 using the fourth order Runge-Kutta method.

11.23. Find the natural frequencies of a fixed-fixed bar undergoing longitudinal vibration, using 3 mesh points in the range $0 < x < l$.

11.24. Derive the finite difference equations governing the forced longitudinal vibration of a fixed-free uniform bar, using a total of n mesh points. Find the natural frequencies of the bar, using $n = 4$.

11.25. Derive the finite difference equations for the forced vibration of a fixed-fixed uniform shaft under torsion, using a total of n mesh points.

11.26. Find the first three natural frequencies of a uniform fixed-fixed beam.

11.27. Derive the finite difference equations for the forced vibration of a cantilever beam subjected to a transverse force $f(x, t) = f_0 \cos \omega t$ at the free end.

11.28. Derive the finite difference equations for the forced vibration analysis of a rectangular membrane, using m and n mesh points in the x and y directions, respectively. Assume the membrane to be fixed along all the edges. Use the central difference formula.

11.29. Solve Problem 11.18 using the fourth order Runge-Kutta method with $c = 1$.

11.30. Solve Problem 11.19 using the fourth order Runge-Kutta method.

11.31. Solve Problem 11.20 using the fourth order Runge-Kutta method.

11.32. Solve Problem 11.18 using the Houbolt method.

11.33. Solve Problem 11.19 using the Houbolt method.

11.34. Solve Problem 11.20 using the Houbolt method.

11.35. Solve Problem 11.18 using the Wilson method, with $\theta = 1.4$.

11.36. Solve Problem 11.19 using the Wilson method with $\theta = 1.4$.

11.37. Solve Problem 11.20 using the Wilson method with $\theta = 1.4$.

11.38. Write a subroutine WILSON for implementing the Wilson method. Use this program to find the solution of Example 11.7.

11.39. Solve Problem 11.18 using the Newmark method, with $\alpha = \frac{1}{6}$ and $\beta = \frac{1}{2}$.

11.40. Solve Problem 11.19 using the Newmark method, with $\alpha = \frac{1}{6}$ and $\beta = \frac{1}{2}$.

11.41. Solve Problem 11.20 using the Newmark method, with $\alpha = \frac{1}{6}$ and $\beta = \frac{1}{2}$.

11.42. Write a subroutine NUMARK for implementing the Newmark method. Use this subroutine to find the solution of Example 11.8.

CHAPTER 12

Finite Element Method

Thomas Young (1773–1829) was a British physicist and physician who introduced Young's modulus and principle of interference of light. He studied medicine and received his MD in 1796. He was appointed Professor of Natural Philosophy at the Royal Institution in 1801, but he resigned in 1803 as his lectures were disappointing to popular audience. He joined St. George's hospital in London as a physician in 1811 and continued there until his death. Young made many contributions to mechanics. He was the first to use the terms "energy" and "labor expended" (i.e. work done) for the quantities mv^2 and Fx, respectively, where m is the mass of the body, v is its velocity, F is a force and x is the distance by which F is moved, and to state that the two terms are proportional to one another. He defined the term modulus (which has become known as Young's modulus) as the weight which would double the length of a rod of unit cross section.

12.1 INTRODUCTION

The finite element method is a numerical method that can be used for the accurate solution of complex mechanical and structural vibration problems [12.1, 12.2]. In this method, the actual structure is replaced by several pieces or elements, each of which is assumed to behave as a continuous structural member called a *finite element*. The elements are assumed to be interconnected at certain points known as *joints* or *nodes*. Since it is very difficult to find the exact solution (such as the displacements) of the original structure under the specified loads, a convenient approximate solution is assumed in each finite element. The idea is that if the solutions of the various elements are selected properly, they can be made to converge to the exact solution of the total structure as the element size is reduced. During the solution process, the equilibrium of forces at the joints and the compatibility of displacements between the elements are satisfied so that the entire structure (assemblage of elements) is made to behave as a single entity.

The basic procedure of the finite element method, with application to simple vibration problems, is presented in this chapter. The element stiffness and mass matrices and force vectors are derived for a bar element, a torsion element, and a beam element. The transformation of element matrices and vectors from the local to the global coordinate system is presented. The equations of motion of the complete system of finite elements and the incorporation of the boundary conditions are discussed. The concepts of consistent and lumped mass matrices are presented along with a numerical example. Finally, a computer program for the eigenvalue analysis of stepped beams, is presented. Although the techniques presented in this chapter can be applied to more complex problems involving two- and three-dimensional finite elements, only the use of one-dimensional elements is considered in the numerical treatment.

12.2 EQUATIONS OF MOTION OF AN ELEMENT

For illustration, the finite element model of a plano-milling machine structure (Fig. 12.1(a)) is shown in Fig. 12.1(b). In this model, the columns and the overarm are represented by triangular plate elements and the cross slide and the tool holder are represented by beam elements [12.3]. The elements are assumed to be connected to each other only at the joints. The displacement within an element is expressed in terms of the displacements at the corners or joints of the element. In Fig. 12.1(b), the transverse displacement within a typical element e is assumed to be $w(x, y, t)$. The values of w, $(\partial w)/(\partial x)$, and $(\partial w)/(\partial y)$ at joints 1, 2, and 3, namely $w(x_1, y_1, t)$, $(\partial w)/(\partial x)(x_1, y_1, t)$, $(\partial w)/(\partial y)(x_1, y_1, t), \ldots, (\partial w)/(\partial y)(x_3, y_3, t)$, are treated as unknowns and are denoted as $w_1(t), w_2(t), w_3(t), \ldots, w_9(t)$. The displacement $w(x, y, t)$ can be expressed in terms of the unknown joint displacements $w_i(t)$ in the form

$$w(x, y, t) = \sum_{i=1}^{n} N_i(x, y) w_i(t) \tag{12.1}$$

where $N_i(x, y)$ is called the *shape function* corresponding to the joint displacement $w_i(t)$ and n is the number of unknown joint displacements ($n = 9$ in Fig. 12.1(b)). If a distributed load $f(x, y, t)$ acts on the element, it can be converted into equivalent joint forces $f_i(t)$ ($i = 1, 2, \ldots, 9$). If concentrated forces act at the joints, they can also be added to the appropriate joint force $f_i(t)$. We shall now derive the equations of motion for determining the joint displacements $w_i(t)$ under the prescribed joint forces $f_i(t)$. By using Eq. (12.1), the kinetic energy T and the strain energy V of the element can be expressed as

$$T = \frac{1}{2} \dot{\vec{W}}^T [m] \dot{\vec{W}} \tag{12.2}$$

$$V = \frac{1}{2} \vec{W}^T [k] \vec{W} \tag{12.3}$$

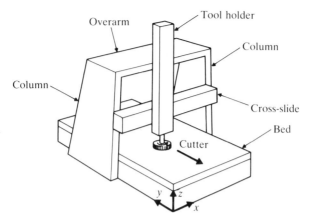

(a) Plano-milling machine structure

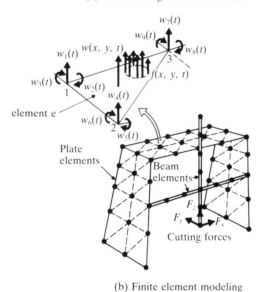

(b) Finite element modeling

Figure 12.1   (a) Plano-milling machine structure. (b) Finite element modeling.

where

$$\vec{W} = \begin{Bmatrix} w_1(t) \\ w_2(t) \\ \vdots \\ w_n(t) \end{Bmatrix}, \qquad \dot{\vec{W}} = \begin{Bmatrix} \dot{w}_1(t) \\ \dot{w}(t) \\ \vdots \\ \dot{w}_n(t) \end{Bmatrix} = \begin{Bmatrix} dw_1/dt \\ dw_2/dt \\ \vdots \\ dw_n/dt \end{Bmatrix}$$

and $[m]$ and $[k]$ are the mass and stiffness matrices of the element. By substituting

Eqs. (12.2) and (12.3) into Lagrange's equations, Eq. (6.38), the equations of motion of the finite element can be obtained as

$$[m]\ddot{\vec{W}} + [k]\vec{W} = \vec{f} \tag{12.4}$$

where $\vec{f}$ is the vector of joint forces and $\ddot{\vec{W}}$ is the vector of joint accelerations given by

$$\ddot{\vec{W}} = \begin{Bmatrix} \ddot{w}_1 \\ \ddot{w}_2 \\ \vdots \\ \ddot{w}_n \end{Bmatrix} = \begin{Bmatrix} d^2w_1/dt^2 \\ d^2w_2/dt^2 \\ \vdots \\ d^2w_n/dt^2 \end{Bmatrix}$$

Note that the shape of the finite elements and the number of unknown joint displacements may differ for different applications. Although the equations of motion of a single element, Eq. (12.4), are not useful directly (as our interest lies in the dynamic response of the assemblage of elements), the mass matrix $[m]$, the stiffness matrix $[k]$, and the joint force vector $\vec{f}$ of individual elements are necessary for the final solution. We shall derive the element mass and stiffness matrices and the joint force vectors for some simple one-dimensional elements in the next section.

12.3 MASS MATRIX, STIFFNESS MATRIX, AND FORCE VECTOR

12.3.1
Bar Element

Consider the uniform bar element shown in Fig. 12.2. For this one-dimensional element, the two end points form the joints (nodes). When the element is subjected to axial loads $f_1(t)$ and $f_2(t)$, the axial displacement within the element is assumed to be linear in x as

$$u(x, t) = a(t) + b(t)x \tag{12.5}$$

When the joint displacements $u_1(t)$ and $u_2(t)$ are treated as unknowns, Eq. (12.5) should satisfy the conditions:

$$u(0, t) = u_1(t), \qquad u(l, t) = u_2(t) \tag{12.6}$$

Equations (12.5) and (12.6) lead to

$$a(t) = u_1(t)$$

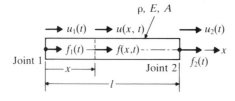

Figure 12.2

and

$$a(t) + b(t)l = u_2(t) \qquad \text{or} \qquad b(t) = \frac{u_2(t) - u_1(t)}{l} \qquad (12.7)$$

Substitution for $a(t)$ and $b(t)$ from Eq. (12.7) into Eq. (12.5) gives

$$u(x, t) = \left(1 - \frac{x}{l}\right) u_1(t) + \frac{x}{l} u_2(t) \qquad (12.8)$$

or

$$u(x, t) = N_1(x) u_1(t) + N_2(x) u_2(t) \qquad (12.9)$$

where

$$N_1(x) = \left(1 - \frac{x}{l}\right), \qquad N_2(x) = \frac{x}{l} \qquad (12.10)$$

are the shape functions.

The kinetic energy of the bar element can be expressed as

$$T(t) = \frac{1}{2} \int_0^l \rho A \left\{ \frac{\partial u(x, t)}{\partial t} \right\}^2 dx = \frac{1}{2} \int_0^l \rho A \left\{ \left(1 - \frac{x}{l}\right) \frac{du_1(t)}{dt} + \left(\frac{x}{l}\right) \frac{du_2(t)}{dt} \right\}^2 dx$$

$$= \frac{1}{2} \frac{\rho A l}{3} \left(\dot{u}_1^2 + \dot{u}_1 \dot{u}_2 + \dot{u}_2^2 \right) \qquad (12.11)$$

where

$$\dot{u}_1 = \frac{du_1(t)}{dt}, \qquad \dot{u}_2 = \frac{du_2(t)}{dt},$$

ρ is the density of the material and A is the cross sectional area of the element.

By expressing Eq. (12.11) in matrix form,

$$T(t) = \frac{1}{2} \dot{\vec{u}}(t)^T [m] \dot{\vec{u}}(t) \qquad (12.12)$$

where

$$\dot{\vec{u}}(t) = \left\{ \begin{array}{c} \dot{u}_1(t) \\ \dot{u}_2(t) \end{array} \right\}$$

and the superscript T indicates the transpose, the mass matrix $[m]$ can be identified as

$$[m] = \frac{\rho A l}{6} \begin{bmatrix} 2 & 1 \\ 1 & 2 \end{bmatrix} \qquad (12.13)$$

The strain energy of the element can be written as

$$V(t) = \frac{1}{2} \int_0^l EA \left\{ \frac{\partial u(x, t)}{\partial x} \right\}^2 dx = \frac{1}{2} \int_0^l EA \left\{ -\frac{1}{l} u_1(t) + \frac{1}{l} u_2(t) \right\}^2 dx$$

$$= \frac{1}{2} \frac{EA}{l} \left(u_1^2 - 2u_1 u_2 + u_2^2 \right) \qquad (12.14)$$

where $u_1 = u_1(t)$, $u_2 = u_2(t)$, and E is Young's modulus. By expressing Eq. (12.14) in matrix form as

$$V(t) = \frac{1}{2} \vec{u}(t)^T [k] \vec{u}(t) \qquad (12.15)$$

where

$$\vec{u}(t) = \begin{Bmatrix} u_1(t) \\ u_2(t) \end{Bmatrix} \qquad \text{and} \qquad \vec{u}(t)^T = \{ u_1(t) \quad u_2(t) \}$$

the stiffness matrix $[k]$ can be identified as

$$[k] = \frac{EA}{l} \begin{bmatrix} 1 & -1 \\ -1 & 1 \end{bmatrix} \tag{12.16}$$

The force vector

$$\vec{f} = \begin{Bmatrix} f_1(t) \\ f_2(t) \end{Bmatrix}$$

can be derived from the virtual work expression. If the bar is subjected to the distributed force $f(x, t)$, the virtual work δW can be expressed as

$$\delta W(t) = \int_0^l f(x, t)\,\delta u(x, t)\,dx = \int_0^l f(x, t)\left\{\left(1 - \frac{x}{l}\right)\delta u_1(t) + \left(\frac{x}{l}\right)\delta u_2(t)\right\} dx$$

$$= \left(\int_0^l f(x, t)\left(1 - \frac{x}{l}\right) dx\right)\delta u_1(t) + \left(\int_0^l f(x, t)\left(\frac{x}{l}\right) dx\right)\delta u_2(t) \tag{12.17}$$

By expressing Eq. (12.17) in matrix form as

$$\delta W(t) = \delta \vec{u}(t)^T \vec{f}(t) \equiv f_1(t)\,\delta u_1(t) + f_2(t)\,\delta u_2(t) \tag{12.18}$$

the equivalent joint forces can be identified as

$$\left. \begin{aligned} f_1(t) &= \int_0^l f(x, t)\left(1 - \frac{x}{l}\right) dx \\ f_2(t) &= \int_0^l f(x, t)\left(\frac{x}{l}\right) dx \end{aligned} \right\} \tag{12.19}$$

12.3.2
Torsion Element

Consider a uniform torsion element with the x axis taken along the centroidal axis, as shown in Fig. 12.3. Let I_p denote the polar moment of inertia about the centroidal axis and GJ represent the torsional stiffness ($J = I_p$ for a circular cross section). When the torsional displacement (rotation) within the element is assumed to be linear in x as

$$\theta(x, t) = a(t) + b(t)x \tag{12.20}$$

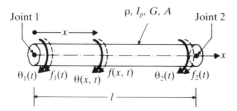

Figure 12.3

and the joint rotations $\theta_1(t)$ and $\theta_2(t)$ are treated as unknowns, Eq. (12.20) can be expressed, by proceeding as in the case of a bar element, as

$$\theta(x, t) = N_1(x)\theta_1(t) + N_2(x)\theta_2(t) \tag{12.21}$$

where $N_1(x)$ and $N_2(x)$ are the same as in Eq. (12.10). The kinetic energy, the strain energy, and the virtual work for pure torsion are given by

$$T(t) = \frac{1}{2}\int_0^l \rho I_p\left\{\frac{\partial\theta(x, t)}{\partial t}\right\}^2 dx \tag{12.22}$$

$$V(t) = \frac{1}{2}\int_0^l GJ\left\{\frac{\partial\theta(x, t)}{\partial x}\right\}^2 dx \tag{12.23}$$

$$\delta W(t) = \int_0^l f(x, t)\delta\theta(x, t)\, dx \tag{12.24}$$

where ρ is the mass density and $f(x, t)$ is the distributed torque per unit length. Using the procedures employed in Section 12.3.1, we can derive the element mass and stiffness matrices and the force vector:

$$[m] = \frac{\rho I_p l}{6}\begin{bmatrix} 2 & 1 \\ 1 & 2 \end{bmatrix} \tag{12.25}$$

$$[k] = \frac{GJ}{l}\begin{bmatrix} 1 & -1 \\ -1 & 1 \end{bmatrix} \tag{12.26}$$

$$\vec{f} = \begin{Bmatrix} f_1(t) \\ f_2(t) \end{Bmatrix} = \begin{Bmatrix} \int_0^l f(x, t)\left(1 - \frac{x}{l}\right) dx \\ \int_0^l f(x, t)\left(\frac{x}{l}\right) dx \end{Bmatrix} \tag{12.27}$$

12.3.3
Beam Element

We now consider a beam element according to the Euler-Bernoulli theory.* Figure 12.4 shows a uniform beam element subjected to the transverse force distribution $f(x, t)$. In this case, the joints undergo both translational and rotational displacements, so the unknown joint displacements are labeled as $w_1(t)$, $w_2(t)$, $w_3(t)$, and $w_4(t)$. Thus there will be linear joint forces $f_1(t)$ and $f_3(t)$ corresponding to the linear joint displacements $w_1(t)$ and $w_3(t)$ and rotational joint forces (bending moments) $f_2(t)$ and $f_4(t)$ corresponding to the rotational joint displacements $w_2(t)$ and $w_4(t)$, respectively. The transverse displacement within the element is assumed to be a cubic equation in x (as in the case of static deflection of a beam):

$$w(x, t) = a(t) + b(t)x + c(t)x^2 + d(t)x^3 \tag{12.28}$$

---

*   The beam element, according to the Timoshenko theory, was considered in Refs. [12.4–12.7].

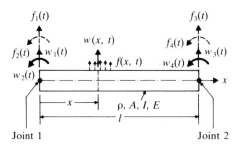

Joint 1   Joint 2

Figure 12.4

The unknown joint displacements must satisfy the conditions

$$w(0, t) = w_1(t), \qquad \frac{\partial w}{\partial x}(0, t) = w_2(t) \left.\right\}$$

$$w(l, t) = w_3(t), \qquad \frac{\partial w}{\partial x}(l, t) = w_4(t) \left.\right\}$$

(12.29)

Equations (12.28) and (12.29) yield

$$a(t) = w_1(t)$$
$$b(t) = w_2(t)$$
$$c(t) = \frac{1}{l^2}\left[-3w_1(t) - 2w_2(t)l + 3w_3(t) - w_4(t)l\right]$$
$$d(t) = \frac{1}{l^3}\left[2w_1(t) + w_2(t)l - 2w_3(t) + w_4(t)l\right]$$

(12.30)

By substituting Eqs. (12.30) into Eq. (12.28), we can express $w(x, t)$ as

$$w(x, t) = \left(1 - 3\frac{x^2}{l^2} + 2\frac{x^3}{l^3}\right)w_1(t) + \left(\frac{x}{l} - 2\frac{x^2}{l^2} + \frac{x^3}{l^3}\right)lw_2(t)$$
$$+ \left(3\frac{x^2}{l^2} - 2\frac{x^3}{l^3}\right)w_3(t) + \left(-\frac{x^2}{l^2} + \frac{x^3}{l^3}\right)lw_4(t)$$

(12.31)

This equation can be rewritten as

$$w(x, t) = \sum_{i=1}^{4} N_i(x)w_i(t).$$

(12.32)

where $N_i(x)$ are the shape functions given by

$$N_1(x) = 1 - 3\left(\frac{x}{l}\right)^2 + 2\left(\frac{x}{l}\right)^3$$

(12.33)

$$N_2(x) = x - 2l\left(\frac{x}{l}\right)^2 + l\left(\frac{x}{l}\right)^3$$

(12.34)

$$N_3(x) = 3\left(\frac{x}{l}\right)^2 - 2\left(\frac{x}{l}\right)^3$$

(12.35)

$$N_4(x) = -l\left(\frac{x}{l}\right)^2 + l\left(\frac{x}{l}\right)^3$$

(12.36)

The kinetic energy, bending strain energy, and virtual work of the element can be expressed as

$$T(t) = \frac{1}{2} \int_0^l \rho A \left\{ \frac{\partial w(x,t)}{\partial t} \right\}^2 dx \equiv \frac{1}{2} \dot{\vec{w}}(t)^T [m] \dot{\vec{w}}(t) \qquad (12.37)$$

$$V(t) = \frac{1}{2} \int_0^l EI \left\{ \frac{\partial^2 w(x,t)}{\partial x^2} \right\}^2 dx \equiv \frac{1}{2} \vec{w}(t)^T [k] \vec{w}(t) \qquad (12.38)$$

$$\delta W(t) = \int_0^l f(x,t) \delta w(x,t) \, dx \equiv \delta \vec{w}(t)^T \vec{f}(t) \qquad (12.39)$$

where ρ is the density of the beam, E is Young's modulus, I is the moment of inertia of the cross section, A is the area of cross section, and

$$\vec{w}(t) = \begin{Bmatrix} w_1(t) \\ w_2(t) \\ w_3(t) \\ w_4(t) \end{Bmatrix}, \quad \dot{\vec{w}}(t) = \begin{Bmatrix} dw_1/dt \\ dw_2/dt \\ dw_3/dt \\ dw_4/dt \end{Bmatrix}, \quad \delta \vec{w}(t) = \begin{Bmatrix} \delta w_1(t) \\ \delta w_2(t) \\ \delta w_3(t) \\ \delta w_4(t) \end{Bmatrix}, \quad \vec{f}(t) = \begin{Bmatrix} f_1(t) \\ f_2(t) \\ f_3(t) \\ f_4(t) \end{Bmatrix}$$

By substituting Eq. (12.31) into Eqs. (12.37) to (12.39) and carrying out the necessary integrations, we obtain

$$[m] = \frac{\rho A l}{420} \begin{bmatrix} 156 & 22l & 54 & -13l \\ 22l & 4l^2 & 13l & -3l^2 \\ 54 & 13l & 156 & -22l \\ -13l & -3l^2 & -22l & 4l^2 \end{bmatrix} \qquad (12.40)$$

$$[k] = \frac{EI}{l^3} \begin{bmatrix} 12 & 6l & -12 & 6l \\ 6l & 4l^2 & -6l & 2l^2 \\ -12 & -6l & 12 & -6l \\ 6l & 2l^2 & -6l & 4l^2 \end{bmatrix} \qquad (12.41)$$

$$f_i(t) = \int_0^l f(x,t) N_i(x) \, dx, \qquad i = 1,2,3,4 \qquad (12.42)$$

12.4 TRANSFORMATION OF ELEMENT MATRICES AND VECTORS

As stated earlier, the finite element method considers the given dynamical system as an assemblage of elements. The joint displacements of an individual element are selected in a convenient direction, depending on the nature of the element. For example, for the bar element shown in Fig. 12.2, the joint displacements $u_1(t)$ and $u_2(t)$ are chosen along the axial direction of the element. However, other bar elements can have different orientations in an assemblage, as shown in Fig. 12.5. Here x denotes the axial direction of an individual element and is called a *local coordinate axis*. If we use $u_1(t)$ and $u_2(t)$ to denote the joint displacements of different bar elements, there will be one joint displacement at joint 1, three at joint

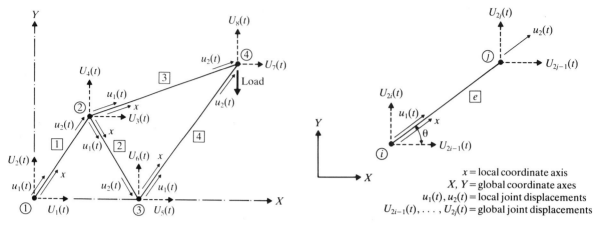

Figure 12.5   A dynamical system (truss) idealized as an assemblage of four bar elements.

Figure 12.6   Local and global joint displacements of ele - ment e.

2, two at joint 3, and 2 at joint 4. However, the displacements of joints can be specified more conveniently using reference or global coordinate axes X and Y. Then the displacement components of joints parallel to the X and Y axes can be used as the joint displacements in the global coordinate system. These are shown as $U_i(t)$, $i = 1, 2, \ldots, 8$ in Fig. 12.5. The joint displacements in the local and the global coordinate system for a typical bar element e are shown in Fig. 12.6. The two sets of joint displacements are related as follows:

$$u_1(t) = U_{2i-1}(t)\cos\theta + U_{2i}(t)\sin\theta$$
$$u_2(t) = U_{2j-1}(t)\cos\theta + U_{2j}(t)\sin\theta \qquad (12.43)$$

which can be rewritten as

$$\vec{u}(t) = [\lambda]\vec{U}(t) \qquad (12.44)$$

where $[\lambda]$ is the coordinate transformation matrix given by

$$[\lambda] = \begin{bmatrix} \cos\theta & \sin\theta & 0 & 0 \\ 0 & 0 & \cos\theta & \sin\theta \end{bmatrix} \qquad (12.45)$$

and $\vec{u}(t)$ and $\vec{U}(t)$ are the vectors of joint displacements in the local and the global coordinate system, respectively, and are given by

$$\vec{u}(t) = \begin{Bmatrix} u_1(t) \\ u_2(t) \end{Bmatrix}, \qquad \vec{U}(t) = \begin{Bmatrix} U_{2i-1}(t) \\ U_{2i}(t) \\ U_{2j-1}(t) \\ U_{2j}(t) \end{Bmatrix}$$

It is useful to express the mass matrix, stiffness matrix, and joint force vector of an element in terms of the global coordinate system while finding the dynamical response of the complete system. Since the kinetic and strain energies of the element must be independent of the coordinate system, we have

$$T(t) = \frac{1}{2}\dot{u}(t)^T[m]\dot{u}(t) = \frac{1}{2}\vec{\dot{U}}(t)^T[\overline{m}]\vec{\dot{U}}(t) \tag{12.46}$$

$$V(t) = \frac{1}{2}\vec{u}(t)^T[k]\vec{u}(t) = \frac{1}{2}\vec{U}(t)^T[\overline{k}]\vec{U}(t) \tag{12.47}$$

where $[\overline{m}]$ and $[\overline{k}]$ denote the element mass and stiffness matrices, respectively, in the global coordinate system and $\vec{\dot{U}}(t)$ is the vector of joint velocities in the global coordinate system, related to $\vec{u}(t)$ as in Eq. (12.44):

$$\dot{\vec{u}}(t) = [\lambda]\vec{\dot{U}}(t) \tag{12.48}$$

By inserting Eqs. (12.44) and (12.48) into (12.46) and (12.47), we obtain

$$T(t) = \frac{1}{2}\dot{\vec{u}}(t)^T[\lambda]^T[m][\lambda]\dot{\vec{u}}(t) \equiv \frac{1}{2}\vec{\dot{U}}(t)^T[\overline{m}]\vec{\dot{U}}(t) \tag{12.49}$$

$$V(t) = \frac{1}{2}\vec{u}(t)^T[\lambda]^T[k][\lambda]\vec{u}(t) \equiv \frac{1}{2}\vec{U}(t)^T[\overline{k}]\vec{U}(t) \tag{12.50}$$

Equations (12.49) and (12.50) yield

$$[\overline{m}] = [\lambda]^T[m][\lambda] \tag{12.51}$$

$$[\overline{k}] = [\lambda]^T[k][\lambda] \tag{12.52}$$

Similarly, by equating the virtual work in the two coordinate systems,

$$\delta W(t) = \delta\vec{u}(t)^T\vec{f}(t) = \delta\vec{U}(t)^T\vec{\overline{f}}(t) \tag{12.53}$$

we find the vector of element joint forces in the global coordinate system $\vec{\overline{f}}(t)$:

$$\vec{\overline{f}}(t) = [\lambda]^T\vec{f}(t) \tag{12.54}$$

Equations (12.51), (12.52), and (12.54) can be used to obtain the equations of motion of a single finite element in the global coordinate system:

$$[\overline{m}]\vec{\ddot{U}}(t) + [\overline{k}]\vec{U}(t) = \vec{\overline{f}}(t) \tag{12.55}$$

Although this equation is not of much use, since our interest lies in the equations of motion of an assemblage of elements, the matrices $[\overline{m}]$ and $[\overline{k}]$ and the vector $\vec{\overline{f}}$ are useful in deriving the equations of motion of the complete system, as indicated in the following section.

12.5   EQUATIONS OF MOTION OF THE COMPLETE SYSTEM OF FINITE ELEMENTS

Since the complete structure is considered an assemblage of several finite elements, we shall now extend the equations of motion obtained for single finite elements in the global system to the complete structure. We shall denote the joint displacements of the complete structure in the global coordinate system as $U_1(t), U_2(t), \ldots, U_M(t)$ or, equivalently, as a column vector:

$$\vec{U}(t) = \begin{Bmatrix} U_1(t) \\ U_2(t) \\ \vdots \\ U_M(t) \end{Bmatrix}$$

For convenience, we shall denote the quantities pertaining to an element e in the assemblage by the superscript e. Since the joint displacements of any element e can be identified in the vector of joint displacements of the complete structure, the vectors $\vec{U}^{(e)}(t)$ and $\vec{U}(t)$ are related:

$$\vec{U}^{(e)}(t) = [A^{(e)}]\vec{U}(t) \tag{12.56}$$

where $[A^{(e)}]$ is a rectangular matrix composed of zeros and ones. For example, for element 1 in Fig. 12.5, Eq. (12.56) becomes

$$\vec{U}^{(1)}(t) \equiv \begin{Bmatrix} U_1(t) \\ U_2(t) \\ U_3(t) \\ U_4(t) \end{Bmatrix} = \begin{bmatrix} 1 & 0 & 0 & 0 & 0 & 0 & 0 & 0 \\ 0 & 1 & 0 & 0 & 0 & 0 & 0 & 0 \\ 0 & 0 & 1 & 0 & 0 & 0 & 0 & 0 \\ 0 & 0 & 0 & 1 & 0 & 0 & 0 & 0 \end{bmatrix} \begin{Bmatrix} U_1(t) \\ U_2(t) \\ \vdots \\ U_8(t) \end{Bmatrix} \tag{12.57}$$

The kinetic energy of the complete structure can be obtained by adding the kinetic energies of individual elements:

$$T = \sum_{e=1}^{E} \frac{1}{2} \dot{\vec{U}}^{(e)^T}[\overline{m}]\dot{\vec{U}}^{(e)} \tag{12.58}$$

where E denotes the number of finite elements in the assemblage. By differentiating Eq. (12.56), the relation between the velocity vectors can be derived:

$$\dot{\vec{U}}^{(e)}(t) = [A^{(e)}]\dot{\vec{U}}(t) \tag{12.59}$$

Substitution of Eq. (12.59) into (12.58) leads to

$$T = \frac{1}{2} \sum_{e=1}^{E} \dot{\vec{U}}^T [A^{(e)}]^T [\overline{m}^{(e)}][A^{(e)}]\dot{\vec{U}} \tag{12.60}$$

The kinetic energy of the complete structure can also be expressed in terms of joint velocities of the complete structure $\vec{U}$:

$$T = \frac{1}{2} \dot{\vec{U}}^T [M]\dot{\vec{U}} \tag{12.61}$$

where $[M]$ is called the mass matrix of the complete structure. A comparison of Eqs.

(12.60) and (12.61) gives the relation[†]

$$[\underset{\sim}{M}] = \sum_{e=1}^{E} [A^{(e)}]^{T} [\overline{m}^{(e)}][A^{(e)}] \qquad (12.62)$$

Similarly, by considering strain energy, the stiffness matrix of the complete structure, $[\underset{\sim}{K}]$, can be expressed as

$$[\underset{\sim}{K}] = \sum_{e=1}^{E} [A^{(e)}]^{T} [\overline{k}^{(e)}][A^{(e)}] \qquad (12.63)$$

Finally the consideration of virtual work yields the vector of joint forces of the complete structure, $\vec{F}$:

$$\vec{F} = \sum_{e=1}^{E} [A^{(e)}]^{T} \vec{f}^{(e)} \qquad (12.64)$$

Once the mass and stiffness matrices and the force vector are known, Lagrange's equations of motion for the complete structure can be expressed as

$$[\underset{\sim}{M}]\ddot{\vec{U}} + [\underset{\sim}{K}]\vec{U} = \vec{F} \qquad (12.65)$$

Note that the joint force vector $\vec{F}$ in Eq. (12.65) was generated by considering only the distributed loads acting on the various elements. If there is any concentrated load acting along the joint displacement $U_i(t)$, it must be added to the ith component of $\vec{F}$.

12.6   INCORPORATION OF BOUNDARY CONDITIONS

In the preceding derivation, no joint was assumed to be fixed. Thus the comple\[structure is capable of undergoing rigid body motion under the joint forces. Th means that $[\underset{\sim}{K}]$ is a singular matrix (see Section 6.11). Usually the structure supported such that the displacements are zero at a number of joints, to avoid rig body motion of the structure. A simple method of incorporating the zero displac ment conditions is to eliminate the corresponding rows and columns from tl matrices $[\underset{\sim}{M}]$ and $[\underset{\sim}{K}]$ and the vector $\vec{F}$. The final equations of motion of tl

---

[†]   An alternative procedure can be used for the assembly of element matrices. In this procedure, ea of the rows and columns of the element (mass or stiffness) matrix is identified by the correspondi degree of freedom in the assembled structure. Then the various entries of the element matrix can placed at their proper locations in the overall (mass or stiffness) matrix of the assembled system. F example, the entry belonging to the ith row (identified by the degree of freedom p) and the jth colu (identified by the degree of freedom q) of the element matrix is to be placed in the pth row and column of the overall matrix. This procedure is illustrated in Example 12.2.

restrained structure can be expressed as

$$\underset{N \times N}{[M]} \; \underset{N \times 1}{\ddot{\vec{U}}} + \underset{N \times N}{[K]} \; \underset{N \times 1}{\vec{U}} = \underset{N \times 1}{\vec{F}} \tag{12.66}$$

where N denotes the number of free joint displacements of the structure.

Note the following points concerning finite element analysis:

1. The approach used in the above presentation is called the *displacement method* of finite element analysis because it is the displacements of elements that are directly approximated. Other methods, such as the force method, the mixed method, and hybrid methods, are also available [12.8, 12.9].

2. The stiffness matrix, mass matrix, and force vector for other finite elements, including two-dimensional and three-dimensional elements, can be derived in a similar manner, provided the shape functions are known [12.1, 12.2].

3. In the Rayleigh-Ritz method discussed in Section 8.8, the displacement of the continuous system is approximated by a sum of assumed functions, where each function denotes a deflection shape of the entire structure. In the finite element method, an approximation using shape functions (similar to the assumed functions) is also used for a finite element instead of the entire structure. Thus the finite element procedure can also be considered a Rayleigh-Ritz method.

4. Error analysis of the finite element method can also be conducted [12.10].

EXAMPLE 12.1   Natural Frequencies of a Simply Supported Beam

Find the natural frequencies of the simply supported beam shown in Fig. 12.7(a) using one finite element.

Given: Simply supported uniform beam of length l.

Find: Natural frequencies of the beam.

Approach: State the eigenvalue problem using the mass and stiffness matrices of the beam. Apply the boundary conditions and solve the problem.

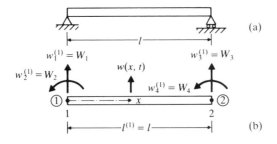

Figure 12.7

Solution. Since the beam is idealized using only one element, the element joint displacements are the same in both local and global systems as indicated in Fig. 12.7(b). The stiffness and mass matrices of the beam are given by

$$[\underset{\sim}{K}] = [K^{(1)}] = \frac{EI}{l^3} \begin{bmatrix} 12 & 6l & -12 & 6l \\ 6l & 4l^2 & -6l & 2l^2 \\ -12 & -6l & 12 & -6l \\ 6l & 2l^2 & -6l & 4l^2 \end{bmatrix} \tag{E.1}$$

$$[\underset{\sim}{M}] = [M^{(1)}] = \frac{\rho A l}{420} \begin{bmatrix} 156 & 22l & 54 & -13l \\ 22l & 4l^2 & 13l & -3l^2 \\ 54 & 13l & 156 & -22l \\ -13l & -3l^2 & -22l & 4l^2 \end{bmatrix} \tag{E.2}$$

and the vector of joint displacements by

$$\vec{W} = \begin{Bmatrix} W_1 \\ W_2 \\ W_3 \\ W_4 \end{Bmatrix} \equiv \begin{Bmatrix} w_1^{(1)} \\ w_2^{(1)} \\ w_3^{(1)} \\ w_4^{(1)} \end{Bmatrix} \tag{E.3}$$

The boundary conditions corresponding to the simply supported ends ($W_1 = 0$ and $W_3 = 0$) can be incorporated* by deleting the rows and columns corresponding to W_1 and W_3 in Eqs. (E.1) and (E.2). This leads to the overall matrices

$$[K] = \frac{2EI}{l} \begin{bmatrix} 2 & 1 \\ 1 & 2 \end{bmatrix} \tag{E.4}$$

$$[M] = \frac{\rho A l^3}{420} \begin{bmatrix} 4 & -3 \\ -3 & 4 \end{bmatrix} \tag{E.5}$$

and the eigenvalue problem can be written as

$$\left[\frac{2EI}{l} \begin{bmatrix} 2 & 1 \\ 1 & 2 \end{bmatrix} - \frac{\rho A l^3 \omega^2}{420} \begin{bmatrix} 4 & -3 \\ -3 & 4 \end{bmatrix} \right] \begin{Bmatrix} W_2 \\ W_4 \end{Bmatrix} = \begin{Bmatrix} 0 \\ 0 \end{Bmatrix} \tag{E.6}$$

By multiplying throughout by $l/(2EI)$, Eq. (E.6) can be expressed as

$$\begin{bmatrix} 2 - 4\lambda & 1 + 3\lambda \\ 1 + 3\lambda & 2 - 4\lambda \end{bmatrix} \begin{Bmatrix} W_2 \\ W_4 \end{Bmatrix} = \begin{Bmatrix} 0 \\ 0 \end{Bmatrix} \tag{E.7}$$

where

$$\lambda = \frac{\rho A l^4 \omega^2}{840 EI} \tag{E.8}$$

By setting the determinant of the coefficient matrix in Eq. (E.7) equal to zero, we obtain the frequency equation

$$\begin{vmatrix} 2 - 4\lambda & 1 + 3\lambda \\ 1 + 3\lambda & 2 - 4\lambda \end{vmatrix} = (2 - 4\lambda)^2 - (1 + 3\lambda)^2 = 0 \tag{E.9}$$

---

*   The bending moment cannot be set equal to zero at the simply supported ends explicitly, since there is no degree of freedom (joint displacement) involving the second derivative of the displacement w.

The roots of Eq. (E.9) give the natural frequencies of the beam as

$$\lambda_1 = \frac{1}{7} \quad \text{or} \quad \omega_1 = \left(\frac{120 EI}{\rho A l^4} \right)^{1/2} \tag{E.10}$$

$$\lambda_2 = 3 \quad \text{or} \quad \omega_2 = \left(\frac{2520 EI}{\rho A l^4} \right)^{1/2} \tag{E.11}$$

These results can be compared with the exact values (see Fig. 8.15):

$$\omega_1 = \left(\frac{97.41 EI}{\rho A l^4} \right)^{1/2}, \quad \omega_2 = \left(\frac{1558.56 EI}{\rho A l^4} \right)^{1/2} \tag{E.12}$$

EXAMPLE 12.2   **Stresses in a Two Bar Truss**

Find the stresses developed in the two members of the truss shown in Fig. 12.8(a) under a vertical load of 200 lb at joint 3. The areas of cross section are 1 in^2 for member 1 and 2 in^2 for member 2, and the Young's modulus is 30×10^6 psi.

Given: Coordinates of joints: $(X_1, Y_1) = (0, 10)$ in.; $(X_2, Y_2) = (0, 0)$ in.; $(X_3, Y_3) = (10, 5)$ in. Areas of cross section of members: $A^{(1)} = 1$ in^2 and $A^{(2)} = 2$ in^2. Young's modulus: 30×10^6 psi. Vertical load at joint 3: 200 lb downwards.

Find: Stresses in the two members.

Approach: Derive the static equilibrium equations and solve them to find the joint displacements. Use the elasticity relations to find the element stresses. Each member is to be treated as a bar element.

Solution. The modeling of the truss as an assemblage of two bar elements and the displacement degrees of freedom of the joints are shown in Fig. 12.8(b). The lengths of the elements can be computed from the coordinates of the ends (joints) as

$$l^{(1)} = \left\{ (X_3 - X_1)^2 + (Y_3 - Y_1)^2 \right\}^{1/2} = \left\{ (10 - 0)^2 + (5 - 10)^2 \right\}^{1/2} = 11.1803 \text{ in.}$$

$$l^{(2)} = \left\{ (X_3 - X_2)^2 + (Y_3 - Y_2)^2 \right\}^{1/2} = \left\{ (10 - 0)^2 + (5 - 0)^2 \right\}^{1/2} = 11.1803 \text{ in.}$$

The element stiffness matrices in the local coordinate system can be obtained as

$$[k^{(1)}] = \frac{A^{(1)} E^{(1)}}{l^{(1)}} \begin{bmatrix} 1 & -1 \\ -1 & 1 \end{bmatrix} = \frac{(1)(30 \times 10^6)}{11.1803} \begin{bmatrix} 1 & -1 \\ -1 & 1 \end{bmatrix} = 2.6833 \times 10^6 \begin{bmatrix} 1 & -1 \\ -1 & 1 \end{bmatrix}$$

$$[k^{(2)}] = \frac{A^{(2)} E^{(2)}}{l^{(2)}} \begin{bmatrix} 1 & -1 \\ -1 & 1 \end{bmatrix} = \frac{(2)(30 \times 10^6)}{11.1803} \begin{bmatrix} 1 & -1 \\ -1 & 1 \end{bmatrix} = 5.3666 \times 10^6 \begin{bmatrix} 1 & -1 \\ -1 & 1 \end{bmatrix}$$

$$\tag{E.2}$$

The angle between the local x-coordinate and the global X-coordinate is given by

$$\left. \begin{aligned} \cos \theta_1 &= \frac{X_3 - X_1}{l^{(1)}} = \frac{10 - 0}{11.1803} = 0.8944 \\ \sin \theta_1 &= \frac{Y_3 - Y_1}{l^{(1)}} = \frac{5 - 10}{11.1803} = -0.4472 \end{aligned} \right\} \text{ for element 1} \tag{E.3}$$

$$\left. \begin{aligned} \cos\theta_2 &= \frac{X_3 - X_2}{l^{(2)}} = \frac{10 - 0}{11.1803} = 0.8944 \\ \sin\theta_2 &= \frac{Y_3 - Y_2}{l^{(2)}} = \frac{5 - 0}{11.1803} = 0.4472 \end{aligned} \right\} \text{for element 2} \qquad \text{(E.4)}$$

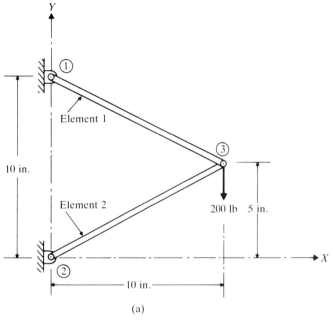

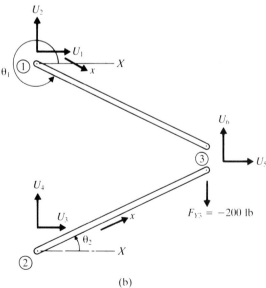

(a)

(b)

Figure 12.8

The stiffness matrices of the elements in the global (X, Y) coordinate system can be derived as

$$[\bar{k}^{(1)}] = [\lambda^{(1)}]^T [k^{(1)}][\lambda^{(1)}]$$

$$= 2.6833 \times 10^6 \begin{array}{c} \\ \end{array} \begin{matrix} 1 & 2 & 5 & 6 \\ \begin{bmatrix} 0.8 & -0.4 & -0.8 & 0.4 \\ -0.4 & 0.2 & 0.4 & -0.2 \\ -0.8 & 0.4 & 0.8 & -0.4 \\ 0.4 & -0.2 & -0.4 & 0.2 \end{bmatrix} & \begin{matrix} 1 \\ 2 \\ 5 \\ 6 \end{matrix} \end{matrix} \qquad (E.5)$$

$$[\bar{k}^{(2)}] = [\lambda^{(2)}]^T [k^{(2)}][\lambda^{(2)}]$$

$$= 5.3666 \times 10^6 \begin{matrix} 3 & 4 & 5 & 6 \\ \begin{bmatrix} 0.8 & 0.4 & -0.8 & -0.4 \\ 0.4 & 0.2 & -0.4 & -0.2 \\ -0.8 & -0.4 & 0.8 & 0.4 \\ -0.4 & -0.2 & 0.4 & 0.2 \end{bmatrix} & \begin{matrix} 3 \\ 4 \\ 5 \\ 6 \end{matrix} \end{matrix} \qquad (E.6)$$

where

$$[\lambda^{(1)}] = \begin{bmatrix} \cos\theta_1 & \sin\theta_1 & 0 & 0 \\ 0 & 0 & \cos\theta_1 & \sin\theta_1 \end{bmatrix} = \begin{bmatrix} 0.8944 & -0.4472 & 0 & 0 \\ 0 & 0 & 0.8944 & -0.4472 \end{bmatrix} \qquad (E.7)$$

$$[\lambda^{(2)}] = \begin{bmatrix} \cos\theta_2 & \sin\theta_2 & 0 & 0 \\ 0 & 0 & \cos\theta_2 & \sin\theta_2 \end{bmatrix} = \begin{bmatrix} 0.8944 & 0.4472 & 0 & 0 \\ 0 & 0 & 0.8944 & 0.4472 \end{bmatrix} \qquad (E.8)$$

Note that the top and right hand sides of Eqs. (E.5) and (E.6) denote the global degrees of freedom corresponding to the rows and columns of the respective stiffness matrices. The assembled stiffness matrix of the system, $[\underset{\sim}{K}]$, can be obtained, by placing the elements of $[\bar{k}^{(1)}]$ and $[\bar{k}^{(2)}]$ at their proper places in $[\underset{\sim}{K}]$, as

$$[\underset{\sim}{K}] = \begin{matrix} 1 & 2 & 3 & 4 & 5 & 6 \\ \begin{bmatrix} 0.8 & -0.4 & & & -0.8 & 0.4 \\ -0.4 & 0.2 & & & 0.4 & -0.2 \\ & & 1.6 & 0.8 & -1.6 & -0.8 \\ & & 0.8 & 0.4 & -0.8 & -0.4 \\ -0.8 & 0.4 & -1.6 & -0.8 & 0.8 & -0.4 \\ & & & & +1.6 & +0.8 \\ 0.4 & -0.2 & -0.8 & -0.4 & -0.4 & 0.2 \\ & & & & +0.8 & +0.4 \end{bmatrix} & \begin{matrix} 1 \\ 2 \\ 3 \\ 4 \\ 5 \\ \\ 6 \end{matrix} \end{matrix} \qquad (E.9)$$

The assembled force vector can be written as

$$\vec{F} = \begin{Bmatrix} F_{X1} \\ F_{Y1} \\ F_{X2} \\ F_{Y2} \\ F_{X3} \\ F_{Y3} \end{Bmatrix} \qquad (E.10)$$

where, in general, (F_{Xi}, F_{Yi}) denote the forces applied at joint i along (X, Y) directions. Specifically, (F_{X1}, F_{Y1}) and (F_{X2}, F_{Y2}) represent the reactions at joints 1 and 2, while

$(F_{X3}, F_{Y3}) = (0, -200)$ lb shows the external forces applied at joint 3. By applying the boundary conditions $U_1 = U_2 = U_3 = U_4 = 0$ (i.e., by deleting the rows and columns 1, 2, 3, and 4 in Eqs. (E.9) and (E.10)), we get the final assembled stiffness matrix and the force vector as

$$[K] = 2.6833 \times 10^6 \begin{matrix} & 5 & 6 \\ & \begin{bmatrix} 2.4 & 0.4 \\ 0.4 & 0.6 \end{bmatrix} & \begin{matrix} 5 \\ 6 \end{matrix} \end{matrix} \tag{E.11}$$

$$\vec{F} = \begin{Bmatrix} 0 \\ -200 \end{Bmatrix} \begin{matrix} 5 \\ 6 \end{matrix} \tag{E.12}$$

The equilibrium equations of the system can be written as

$$[K]\vec{U} = \vec{F} \tag{E.13}$$

where $\vec{U} = \begin{Bmatrix} U_5 \\ U_6 \end{Bmatrix}$. The solution of Eq. (E.13) can be found as

$$U_5 = 23.2922 \times 10^{-6} \text{ in.}, \qquad U_6 = -139.7532 \times 10^{-6} \text{ in.} \tag{E.14}$$

The axial displacements of elements 1 and 2 can be found as

$$\begin{Bmatrix} u_1 \\ u_2 \end{Bmatrix}^{(1)} = [\lambda^{(1)}] \begin{Bmatrix} U_1 \\ U_2 \\ U_5 \\ U_6 \end{Bmatrix} = \begin{bmatrix} 0.8944 & -0.4472 & 0 & 0 \\ 0 & 0 & 0.8944 & -0.4472 \end{bmatrix} \begin{Bmatrix} 0 \\ 0 \\ 23.2922 \times 10^{-6} \\ -139.7532 \times 10^{-6} \end{Bmatrix}$$

$$= \begin{Bmatrix} 0 \\ 83.3301 \times 10^{-6} \end{Bmatrix} \text{ in.} \tag{E.15}$$

$$\begin{Bmatrix} u_1 \\ u_2 \end{Bmatrix}^{(2)} = [\lambda^{(2)}] \begin{Bmatrix} U_3 \\ U_4 \\ U_5 \\ U_6 \end{Bmatrix} = \begin{bmatrix} 0.8944 & 0.4472 & 0 & 0 \\ 0 & 0 & 0.8944 & 0.4472 \end{bmatrix} \begin{Bmatrix} 0 \\ 0 \\ 23.2922 \times 10^{-6} \\ -139.7532 \times 10^{-6} \end{Bmatrix}$$

$$= \begin{Bmatrix} 0 \\ -41.6651 \times 10^{-6} \end{Bmatrix} \text{ in.} \tag{E.16}$$

The stresses in elements 1 and 2 can be determined as

$$\sigma^{(1)} = E^{(1)}\epsilon^{(1)} = E^{(1)}\frac{\Delta l^{(1)}}{l^{(1)}} = \frac{E^{(1)}(u_2 - u_1)^{(1)}}{l^{(1)}}$$

$$= \frac{(30 \times 10^6)(83.3301 \times 10^{-6})}{11.1803} = 223.5989 \text{ psi} \tag{E.17}$$

$$\sigma^{(2)} = E^{(2)}\epsilon^{(2)} = \frac{E^{(2)}\Delta l^{(2)}}{l^{(2)}} = \frac{E^{(2)}(u_2 - u_1)^{(2)}}{l^{(2)}}$$

$$= \frac{(30 \times 10^6)(-41.6651 \times 10^{-6})}{11.1803} = -111.7996 \text{ psi} \tag{E.18}$$

where $\sigma^{(i)}$ denotes the stress, $\epsilon^{(i)}$ represents the strain, and $\Delta l^{(i)}$ indicates the change in length of element i ($i = 1, 2$).

12.7 CONSISTENT AND LUMPED MASS MATRICES

The mass matrices derived in Section 12.3 are called consistent mass matrices. They are called consistent because the same displacement model that is used for deriving the element stiffness matrix is used for the derivation of mass matrix. It is of interest to note that several dynamic problems have been solved with simpler forms of mass matrices. The simplest form of the mass matrix, known as the lumped mass matrix, can be obtained by placing point (concentrated) masses m_i at node points i in the directions of the assumed displacement degrees of freedom. The concentrated masses refer to translational and rotational inertia of the element, and are calculated by assuming that the material within the mean locations on either side of the particular displacement behaves like a rigid body while the remainder of the element does not participate in the motion. Thus this assumption excludes the dynamic coupling that exists between the element displacements and hence the resulting element mass matrix is purely diagonal [12.11].

12.7.1 Lumped Mass Matrix for a Bar Element

By dividing the total mass of the element equally between the two nodes, the lumped mass matrix of a uniform bar element can be obtained as

$$[m] = \frac{\rho A l}{2} \begin{bmatrix} 1 & 0 \\ 0 & 1 \end{bmatrix} \tag{12.67}$$

12.7.2 Lumped Mass Matrix for a Beam Element

In Fig. 12.4, by lumping one half of the total beam mass at each of the two nodes, along the translational degrees of freedom, we obtain the lumped mass matrix of the beam element as

$$[m] = \frac{\rho A l}{2} \begin{bmatrix} 1 & 0 & 0 & 0 \\ 0 & 0 & 0 & 0 \\ 0 & 0 & 1 & 0 \\ 0 & 0 & 0 & 0 \end{bmatrix} \tag{12.68}$$

Note that the inertia effect associated with the rotational degrees of freedom has been assumed to be zero in Eq. (12.68). If the inertia effect is to be included, we compute the mass moment of inertia of half of the beam segment about each end and include it at the diagonal locations corresponding to the rotational degrees of freedom. Thus, for a uniform beam, we have

$$I = \frac{1}{3} \left(\frac{\rho A l}{2} \right) \left(\frac{l}{2} \right)^2 = \frac{\rho A l^3}{24} \tag{12.69}$$

and hence the lumped mass matrix of the beam element becomes

$$[m] = \frac{\rho A l}{2} \begin{bmatrix} 1 & 0 & 0 & 0 \\ 0 & \left(\dfrac{l^2}{12} \right) & 0 & 0 \\ 0 & 0 & 1 & 0 \\ 0 & 0 & 0 & \left(\dfrac{l^2}{12} \right) \end{bmatrix} \tag{12.70}$$

12.7.3
Lumped Mass versus Consistent Mass Matrices

It is not obvious as to whether the lumped mass matrices or consistent mass matrices yield more accurate results for a general dynamic response problem. The lumped mass matrices are approximate in the sense that they do not consider the dynamic coupling present between the various displacement degrees of freedom of the element. However, since the lumped mass matrices are diagonal, they require less storage space during computation. On the other hand, the consistent mass matrices are not diagonal and hence require more storage space. They too are approximate in the sense that the shape functions, which are derived using static displacement patterns, are used even for the solution of dynamics problems. The following example illustrates the application of lumped and consistent mass matrices in a simple vibration problem.

EXAMPLE 12.3   Consistent and Lumped Mass Matrices of a Bar

Find the natural frequencies of the fixed-fixed uniform bar shown in Fig. 12.9 using consistent and lumped mass matrices.

Given: Fixed-fixed uniform bar under longitudinal vibration.

Find: Natural frequencies of the bar using a two-element idealization with consistent and lumped mass matrices.

Approach: Derive the eigenvalue equation using consistent and lumped mass matrices. Solve the equations after applying the boundary conditions.

Solution. The stiffness and mass matrices of a bar element are

$$[k] = \frac{AE}{l}\begin{bmatrix} 1 & -1 \\ -1 & 1 \end{bmatrix} \tag{E.1}$$

$$[m]_c = \frac{\rho A l}{6}\begin{bmatrix} 2 & 1 \\ 1 & 2 \end{bmatrix} \tag{E.2}$$

$$[m]_l = \frac{\rho A l}{2}\begin{bmatrix} 1 & 0 \\ 0 & 1 \end{bmatrix} \tag{E.3}$$

where the subscripts c and l to the mass matrices denote the consistent and lumped matrices, respectively. Since the bar is modeled by two elements, the assembled stiffness and mass

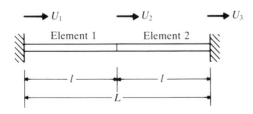

Figure 12.9

matrices are given by

$$[\underset{\sim}{K}] = \frac{AE}{l}\begin{array}{ccc}1 & 2 & 3\end{array}\begin{bmatrix}1 & -1 & 0 \\ -1 & 1+1 & -1 \\ 0 & -1 & 1\end{bmatrix}\begin{array}{c}1 \\ 2 \\ 3\end{array} = \frac{AE}{l}\begin{bmatrix}1 & -1 & 0 \\ -1 & 2 & -1 \\ 0 & -1 & 1\end{bmatrix} \tag{E.4}$$

$$[\underset{\sim}{M}]_c = \frac{\rho A l}{6}\begin{array}{ccc}1 & 2 & 3\end{array}\begin{bmatrix}2 & 1 & 0 \\ 1 & 2+2 & 1 \\ 0 & 1 & 2\end{bmatrix}\begin{array}{c}1 \\ 2 \\ 3\end{array} = \frac{\rho A l}{6}\begin{bmatrix}2 & 1 & 0 \\ 1 & 4 & 1 \\ 0 & 1 & 2\end{bmatrix} \tag{E.5}$$

$$[\underset{\sim}{M}]_l = \frac{\rho A l}{2}\begin{array}{ccc}1 & 2 & 3\end{array}\begin{bmatrix}1 & 0 & 0 \\ 0 & 1+1 & 0 \\ 0 & 0 & 1\end{bmatrix}\begin{array}{c}1 \\ 2 \\ 3\end{array} = \frac{\rho A l}{2}\begin{bmatrix}1 & 0 & 0 \\ 0 & 2 & 0 \\ 0 & 0 & 1\end{bmatrix} \tag{E.6}$$

The dashed boxes in Eqs. (E.4) through (E.6) enclose the contributions of elements 1 and 2. The degrees of freedom corresponding to the columns and rows of the matrices are indicated at the top and the right-hand side of the matrices. The eigenvalue problem, after applying the boundary conditions $U_1 = U_3 = 0$, becomes

$$\big[[K] - \omega^2[M]\big]\{U_2\} = \{0\} \tag{E.7}$$

The eigenvalue ω^2 can be determined by solving the equation

$$\big|[K] - \omega^2[M]\big| = 0 \tag{E.8}$$

which, for the present case, becomes

$$\left|\frac{AE}{l}[2] - \omega^2\frac{\rho A l}{6}[4]\right| = 0 \quad \text{with consistent mass matrices} \tag{E.9}$$

and

$$\left|\frac{AE}{l}[2] - \omega^2\frac{\rho A l}{2}[2]\right| = 0 \quad \text{with lumped mass matrices} \tag{E.10}$$

Equations (E.9) and (E.10) can be solved to obtain

$$\omega_c = \sqrt{\frac{3E}{\rho l^2}} = 3.4641\sqrt{\frac{E}{\rho L^2}} \tag{E.11}$$

$$\omega_l = \sqrt{\frac{2E}{\rho l^2}} = 2.8284\sqrt{\frac{E}{\rho L^2}} \tag{E.12}$$

These values can be compared with the exact value (see Fig. 8.7)

$$\omega_1 = \pi\sqrt{\frac{E}{\rho L^2}} \tag{E.13}$$

12.8 COMPUTER PROGRAM

A Fortran program is given for the eigenvalue analysis of the stepped beam shown in Fig. 12.10, using the finite element method. Although the program is written for the beam of Fig. 12.10, it can be easily generalized to find the natural frequencies

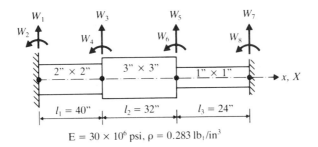

Figure 12.10

and mode shapes of any beam with a specified number of steps (see Problem 12.34). The input data includes XL(I) = length of element I, XI(I) = moment of inertia of element I, A(I) = area of cross section of element I, BJ(I, J) = global degree of freedom number corresponding to the local Jth degree of freedom of element I, E = Young's modulus, and RHO = mass density of the material.

The listing and output of the program are given below.

```
C ===============================================================
. C
C PROGRAM 22
C PROGRAM FOR FINITE ELEMENT VIBRATION ANALYSIS OF STEPPED BEAM
C
C ===============================================================
C DIMENSIONS: XL(NE),XI(NE),A(NE),XMAS(NE),BJ(NE,4),XM(4,4),XK(4,4),
C             AI(4,8),AIT(8,4),XKA(4,8),XMA(4,8),AKA(8,8),AMA(8,8),
C             BIGM(N,N),BIGK(N,N),BM(ND,ND),BK(ND,ND)
C             U(ND,ND),UI(ND,ND),UTI(ND,ND),BMU(ND,ND),UMU(ND,ND),
C             XF(ND,ND),EV(ND,ND)
C
C NE = NUMBER OF ELEMENTS
C N  = TOTAL NUMBER OF DEGREES OF FREEDOM
C ND = TOTAL NUMBER OF DEGREES OF FREEDOM AFTER DELETING ZERO D.O.F.
C
        DIMENSION XL(3),XI(3),A(3),XMAS(3),BJ(3,4),XM(4,4),XK(4,4),
     2    AI(4,8),AIT(8,4),XKA(4,8),XMA(4,8),AKA(8,8),AMA(8,8),BIGM(8,8),
     3    BIGK(8,8),BM(4,4),BK(4,4)
        DIMENSION U(4,4),UI(4,4),UTI(4,4),BMU(4,4),UMU(4,4),XF(4,4),
     2    EV(4,4)
        INTEGER BJ
        DATA XL/40.0,32.0,24.0/
        DATA XI/1.333333,6.75,0.083333/
        DATA A/4.0,9.0,1.0/
        DATA BJ/1,3,5,2,4,6,3,5,7,4,6,8/
```

```
                    E=30.0E+06
                    RHO=0.283/386.4
                    DO 10 I=1,3
         10      XMAS(I)=A(I)*RHO
                    DO 20 I=1,8
                    DO 20 J=1,8
                    BIGM(I,J)=0.0
         20      BIGK(I,J)=0.0
                    DO 100 II=1,3
                    DO 30 I=1,4
                    DO 30 J=1,8
         30      AI(I,J)=0.0
                    I1=BJ(II,1)
                    I2=BJ(II,2)
                    I3=BJ(II,3)
                    I4=BJ(II,4)
                    AI(1,I1)=1.0
                    AI(2,I2)=1.0
                    AI(3,I3)=1.0
                    AI(4,I4)=1.0
                    XM(1,1)=156.0
                    XM(1,2)=22.0*XL(II)
                    XM(1,3)=54.0
                    XM(1,4)=-13.0*XL(II)
                    XM(2,2)=4.0*(XL(II)**2)
                    XM(2,3)=13.0*XL(II)
                    XM(2,4)=-3.0*(XL(II)**2)
                    XM(3,3)=156.0
                    XM(3,4)=-22.0*XL(II)
                    XM(4,4)=4.0*(XL(II)**2)
                    XK(1,1)=12.0
                    XK(1,2)=6.0*XL(II)
                    XK(1,3)=-12.0
                    XK(1,4)=6.0*XL(II)
                    XK(2,2)=4.0*(XL(II)**2)
                    XK(2,3)=-6.0*XL(II)
                    XK(2,4)=2.0*(XL(II)**2)
                    XK(3,3)=12.0
                    XK(3,4)=-6.0*XL(II)
                    XK(4,4)=4.0*(XL(II)**2)
                    DO 40 I=1,4
                    DO 40 J=1,4
                    XM(J,I)=XM(I,J)
         40      XK(J,I)=XK(I,J)
                    DO 50 I=1,4
                    DO 50 J=1,4
                    XM(I,J)=(XMAS(II)*XL(II)/420.0)*XM(I,J)
         50      XK(I,J)=(E*XI(II)/(XL(II)**3))*XK(I,J)
                    DO 60 I=1,8
                    DO 60 J=1,4
         60      AIT(I,J)=AI(J,I)
                    CALL MATMUL (XKA,XK,AI,4,4,8)
                    CALL MATMUL (XMA,XM,AI,4,4,8)
```

```
            CALL MATMUL (AKA,AIT,XKA,8,4,8)
            CALL MATMUL (AMA,AIT,XMA,8,4,8)
            DO 70 I=1,8
            DO 70 J=1,8
            BIGM(I,J)=BIGM(I,J)+AMA(I,J)
  70        BIGK(I,J)=BIGK(I,J)+AKA(I,J)
 100    CONTINUE
C APPLICATION OF BOUNDARY CONDITIONS
C ROWS AND COLUMNS CORRESPONDING TO ZERO DISPLACEMENTS ARE DELETED
            DO 110 I=1,4
            DO 110 J=1,4
            BM(I,J)=BIGM(I+2,J+2)
 110       BK(I,J)=BIGK(I+2,J+2)
C DECOMPOSING THE MATRIX [BK] INTO TRIANGULAR MATRICES
            ND=4
            CALL DECOMP (BK,U,ND)
C FINDING THE INVERSE OF THE UPPER TRIANGULAR MATRIX [U]
            DO 120 I=1,ND
            DO 120 J=1,ND
 120       UI(I,J)=0.0
            DO 130 I=1,ND
 130       UI(I,I)=1.0/U(I,I)
            DO 150 J=1,ND
            DO 150 II=1,ND
            I=ND-II+1
            IF (I .GE. J) GO TO 150
            IP=I+1
            SUM=0.0
            DO 140 K=IP,J
 140       SUM=SUM+U(I,K)*UI(K,J)
            UI(I,J)=-SUM/U(I,I)
 150       CONTINUE
            DO 160 I=1,ND
            DO 160 J=1,ND
 160       UTI(I,J)=UI(J,I)
            CALL MATMUL (BMU,BM,UI,ND,ND,ND)
            CALL MATMUL (UMU,UTI,BMU,ND,ND,ND)
            CALL JACOBI (UMU,ND,EV,1.0E-05,200)
            CALL MATMUL (XF,UI,EV,ND,ND,ND)
            DO 170 I=1,ND
 170       UMU(I,I)=SQRT(1.0/UMU(I,I))
            PRINT 180,(UMU(I,I),I=1,ND)
 180       FORMAT (//,40H NATURAL FREQUENCIES OF THE STEPPED BEAM,//,
          2    4(1X,E15.6))
            PRINT 190
 190       FORMAT (//,12H MODE SHAPES)
            DO 200 J=1,ND
 200       PRINT 210, J, (XF(I,J),I=1,ND)
 210       FORMAT (/,I4,5X,4(1X,E15.6))
            STOP
            END
```

NATURAL FREQUENCIES OF THE STEPPED BEAM

    0.160083E+03   0.617460E+03   0.225198E+04   0.712653E+04

MODE SHAPES

1	0.103333E-01	0.189147E-03	0.141626E-01	0.445258E-04
2	-0.376593E-02	0.202976E-03	0.471097E-02	0.259495E-03
3	0.167902E-03	-0.181687E-03	0.135672E-02	0.207580E-03
4	0.182648E-03	0.607247E-04	0.373775E-03	0.163869E-03

REFERENCES

12.1. O. C. Zienkiewicz, *The Finite Element Method* (4th ed.), McGraw-Hill, London, 1987.

12.2. S. S. Rao, *The Finite Element Method in Engineering* (2nd ed.), Pergamon Press, Oxford, 1989.

12.3. G. V. Ramana and S. S. Rao, "Optimum design of plano-milling machine structure using finite element analysis," *Computers and Structures*, Vol. 18, 1984, pp. 247–253.

12.4. R. Davis, R. D. Henshell, and G. B. Warburton, "A Timoshenko beam element," *Journal of Sound and Vibration*, Vol. 22, 1972, pp. 475–487.

12.5. D. L. Thomas, J. M. Wilson, and R. R. Wilson, "Timoshenko beam finite elements," *Journal of Sound and Vibration*, Vol. 31, 1973, pp. 315–330.

12.6. J. Thomas and B. A. H. Abbas, "Finite element model for dynamic analysis of Timoshenko beams," *Journal of Sound and Vibration*, Vol. 41, 1975, pp. 291–299.

12.7. R. S. Gupta and S. S. Rao, "Finite element eigenvalue analysis of tapered and twisted Timoshenko beams," *Journal of Sound and Vibration*, Vol. 56, 1978, pp. 187–200.

12.8. T. H. H. Pian, "Derivation of element stiffness matrices by assumed stress distribution," *AIAA Journal*, Vol. 2, 1964, pp. 1333–1336.

12.9. H. Alaylioglu and R. Ali, "Analysis of an automotive structure using hybrid stress finite elements," *Computers and Structures*, Vol. 8, 1978, pp. 237–242.

12.10. I. Fried, "Accuracy of finite element eigenproblems," *Journal of Sound and Vibration*, Vol. 18, 1971, pp. 289–295.

12.11. P. Tong, T. H. H. Pian, and L. L. Bucciarelli, "Mode shapes and frequencies by the finite element method using consistent and lumped matrices," *Computers and Structures*, Vol. 1, 1971, pp. 623–638.

REVIEW QUESTIONS

12.1. What is the basic idea behind the finite element method?

12.2. What is a shape function?

12.3. What is the role of transformation matrices in the finite element method?

12.4. What is the basis for the derivation of transformation matrices?

12.5. How are fixed boundary conditions incorporated in the finite element equations?

12.6. How do you solve a finite element problem having symmetry in geometry and loading by modeling only half of the problem?

12.7. Why is the finite element approach presented in this chapter called the displacement method?

12.8. What is a consistent mass matrix?

12.9. What is a lumped mass matrix?

12.10. What is the difference between the finite element method and the Rayleigh-Ritz method?

12.11. How is the distributed load converted into equivalent joint force vector in the finite element method?

PROBLEMS

The problem assignments are organized as follows:

Problems	Section covered	Topic covered
12.1, 12.2, 12.4	12.3	Derivation of element matrices and vectors
12.5, 12.7	12.4	Transformation matrix
12.6, 12.9	12.5	Assembly of matrices and vectors
12.4, 12.8, 12.10–12.30	12.6	Application of boundary conditions and solution of problem
12.31, 12.32	12.7	Consistent and lumped mass matrices
12.3, 12.33–12.35	12.8	Computer program
12.36, 12.37	—	Projects

12.1. Derive the stiffness matrix and the consistent and lumped mass matrices of the tapered bar element shown in Fig. 12.11. The diameter of the bar decreases from D to d over its length.

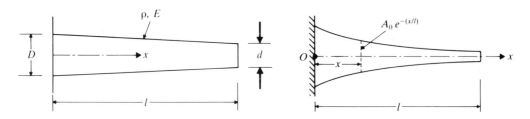

Figure 12.11   **Figure 12.12**

12.2.   Derive the stiffness matrix of the bar element in longitudinal vibration whose cross-sectional area varies as $A(x) = A_0 e^{-(x/l)}$, where A_0 is the area at the root (see Fig. 12.12).

12.3.   Write a computer program for finding the stresses in a planar truss.

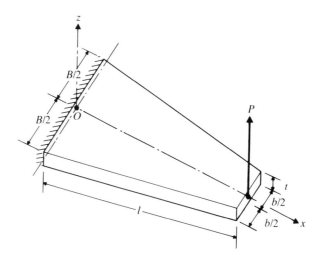

Figure 12.13

12.4.   The tapered cantilever beam shown in Fig. 12.13 is used as a spring to carry a load P. (i) Derive the stiffness matrix of the beam element. (ii) Use the result of (i) to find the stress induced in the beam when $B = 25$ cm, $b = 10$ cm, $t = 2.5$ cm, $l = 2$ m, $E = 2.07 \times 10^{11}$ N/m^2, and $P = 1000$ N. Use one beam element for idealization.

12.5.   Find the global stiffness matrix of each of the four elements of the truss shown in Fig. 12.5 using the following data:
Nodal coordinates: $(X_1, Y_1) = (0, 0)$, $(X_2, Y_2) = (50, 100)$ in., $(X_3, Y_3) = (100, 0)$ in., $(X_4, Y_4) = (200, 150)$ in.
Cross-sectional areas: $A_1 = A_2 = A_3 = A_4 = 2$ in.^{2}.
Young's modulus of all members: 30×10^6 lb/in.^{2}.

12.6. Using the result of Problem 12.5, find the assembled stiffness matrix of the truss and formulate the equilibrium equations if the vertical downward load applied at node 4 is 1000 lb.

12.7. Derive the stiffness and mass matrices of the planar frame element (general beam element) shown in Fig. 12.14 in the global XY-coordinate system.

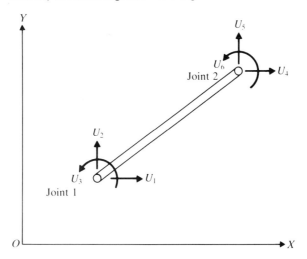

Figure 12.14 A frame element in global system.

12.8. A multiple-leaf spring used in automobiles is shown in Fig. 12.15 It consists of five leaves, each of thickness $t = 0.25$ in. and width $w = 1.5$ in. Find the deflection of the leaves under a load of $P = 2000$ lb. Model only a half of the spring for the finite element analysis. The Young's modulus of the material is 30×10^6 psi.

12.9. Derive the assembled stiffness and mass matrices of the multiple-leaf spring of Problem 12.8 assuming a specific weight of 0.283 lb/in^3 for the material.

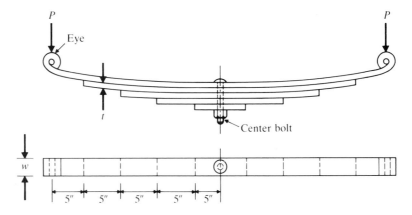

Figure 12.15 A multi-leaf spring.

12.10. Find the nodal displacements of the crane shown in Fig. 12.16 when a vertically downward load of 1000 lb is applied at node 4. The Young's modulus is 30×10^6 psi and the cross-sectional area is 2 in^2 for elements 1 and 2 and 1 in^2 for elements 3 and 4.

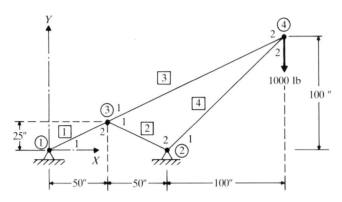

Figure 12.16

12.11. Find the tip deflection of the cantilever beam shown in Fig. 12.17 when a vertical load of $P = 500$ N is applied at point Q using (i) a one element approximation and (ii) a two element approximation. Assume $l = 0.25$ m, $h = 25$ mm, $b = 50$ mm, $E = 2.07 \times 10^{11}$ Pa, and $k = 10^5$ N/m.

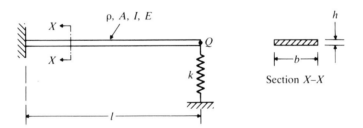

Figure 12.17

12.12. Find the stresses in the stepped beam shown in Fig. 12.18 when a moment of 1000 N-m is applied at node 2 using a two-element idealization. The beam has a square cross section 50 × 50 mm between nodes 1 and 2 and 25 × 25 mm between nodes 2 and 3. Assume the Young's modulus as 2.1×10^{11} Pa.

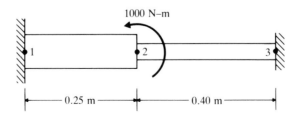

Figure 12.18

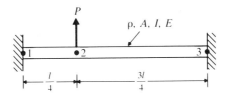

Figure 12.19

12.13. Find the transverse deflection and slope of node 2 of the beam shown in Fig. 12.19 using a two element idealization. Compare the solution with that of simple beam theory.

12.14. Find the natural frequencies of a cantilever beam of length l, cross-sectional area A, moment of inertia I, Young's modulus E, and density ρ, using one finite element.

12.15. Find the natural frequencies of the uniform pinned-free beam shown in Fig. 12.20 using one beam element.

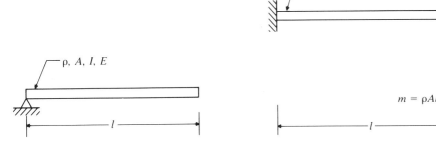

Figure 12.20   **Figure 12.21**

12.16. Find the natural frequencies of the uniform, spring-supported cantilever beam shown in Fig. 12.17 using one beam element and one spring element.

12.17. Find the natural frequencies of the system shown in Fig. 12.21 using one beam element and one spring element.

12.18. Find the natural frequencies and mode shapes of the uniform fixed-fixed beam shown in Fig. 12.22 using two beam elements.

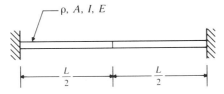

Figure 12.22

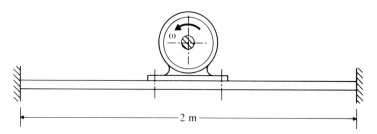

Figure 12.23

12.19.* An electric motor, of mass $m = 100$ kg and operating speed $= 1800$ rpm, is fixed at the middle of a clamped-clamped steel beam of rectangular cross section as shown in Fig. 12.23. Design the beam such that the natural frequency of the system exceeds the operating speed of the motor.

12.20. Find the natural frequencies of the beam shown in Fig. 12.24 using three finite elements of length l each.

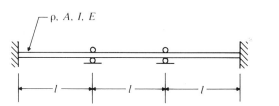

Figure 12.24

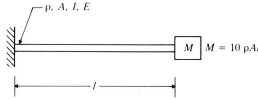

Figure 12.25

12.21. Find the natural frequencies of the cantilever beam carrying an end mass M shown in Fig. 12.25 using a one beam element idealization.

12.22. Find the natural frequencies of vibration of the beam shown in Fig. 12.26 using two beam elements. Also find the load vector if a uniformly distributed transverse load p is applied to element 1.

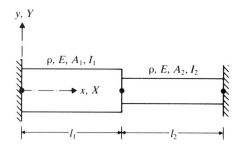

Figure 12.26

12.23. Find the natural frequencies of a beam of length l, which is pin connected at $x = 0$ and fixed at $x = l$, using one beam element.

12.24. Find the natural frequencies of torsional vibration of the stepped shaft shown in Fig. 12.27. Assume that $\rho_1 = \rho_2 = \rho$, $G_1 = G_2 = G$, $I_{p_1} = 2I_{p_2} = 2I_p$, $J_1 = 2J_2 = 2J$, and $l_1 = l_2 = l$.

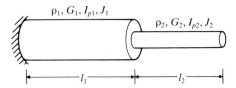

ρ_1, G_1, I_{p1}, J_1

ρ_2, G_2, I_{p2}, J_2

l_1 l_2

Figure 12.27

12.25. Find the dynamic response of the stepped bar shown in Fig. 12.28(a) when its free end is subjected to the load given in Fig. 12.28(b).

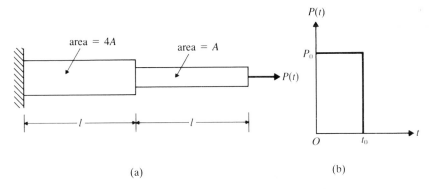

area = 4A

area = A

$P(t)$

$P(t)$

P_0

O t_0 t

(a)

(b)

Figure 12.28

12.26. Find the displacement of node 3 and the stresses in the two members of the truss shown in Fig. 12.29. Assume that the Young's modulus and the cross sectional areas of the two members are same with $E = 30 \times 10^6$ psi and $A = 1$ in^2.

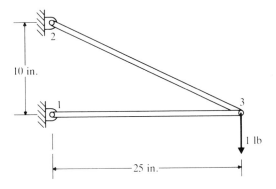

10 in.

2

1

3

1 lb

25 in.

Figure 12.29

12.27. The simplified model of a radial drilling machine structure is shown in Fig. 12.30. Using two beam elements for the column and one beam element for the arm, find the natural frequencies and mode shapes of the machine. Assume the material of the structure as steel.

12.28. If a vertical force of 5000 N along the z-direction and a bending moment of 500 N-m in the xz-plane are developed at point A during a metal cutting operation, find the stresses developed in the machine tool structure shown in Fig. 12.30.

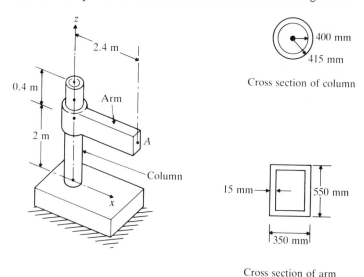

Cross section of column

Cross section of arm

Figure 12.30 A radial drilling machine structure.

12.29. The crank in the slider-crank mechanism shown in Fig. 12.31 rotates at a constant clockwise angular speed of 1000 rpm. Find the stresses in the connecting rod and the

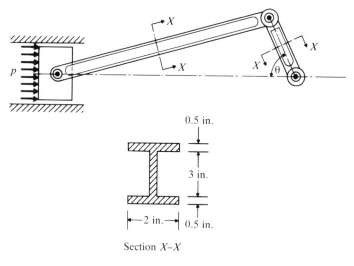

Section X–X

Figure 12.31 A slider-crank mechanism.

crank when the pressure acting on the piston is 200 psi and $\theta = 30°$. The diameter of the piston is 12 in. and the material of the mechanism is steel. Model the connecting rod and the crank by one beam element each. The lengths of the crank and connecting rod are 12 in. and 48 in., respectively.

12.30. A water tank of weight W is supported by a hollow circular steel column of inner diameter d, wall thickness t, and height l. The wind pressure acting on the column can be assumed to vary linearly from 0 to p_{max} as shown in Fig. 12.32. Find (i) the bending stress induced in the column under the loads, and (ii) the natural frequencies of the water tank using a one beam element idealization. Data: $W = 10,000$ lb, $l = 40$ ft, $d = 2$ ft, $t = 1$ in., and $p_{max} = 100$ psi.

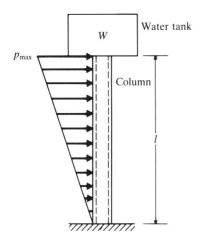

Figure 12.32

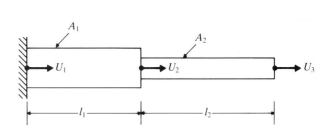

Figure 12.33

12.31. Find the natural frequencies of the stepped bar shown in Fig. 12.33 with the following data using consistent and lumped mass matrices: $A_1 = 2$ in^2, $A_2 = 1$ in^2, $E = 30 \times 10^6$ psi, $\rho = 0.283$ lb/in^3, and $l_1 = l_2 = 50$ in.

12.32. Find the undamped natural frequencies of longitudinal vibration of the stepped bar shown in Fig. 12.34 with the following data using consistent and lumped mass matrices: $l_1 = l_2 = l_3 = 0.2$ m, $A_1 = 2A_2 = 4A_3 = 0.4 \times 10^{-3}$ m^2, $E = 2.1 \times 10^{11}$ N/m^2, and $\rho = 7.8 \times 10^3$ kg/m^3.

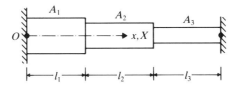

Figure 12.34

12.33. Write a computer program for finding the assembled stiffness matrix of a general planar truss.

12.34. Generalize the computer program of Section 12.8 to make it applicable to the solution of any stepped beam having a specified number of steps.

12.35. Find the natural frequencies and mode shapes of the beam shown in Fig. 12.10 with $l_1 = l_2 = l_3 = 10$ in. and a uniform cross section of 1 in. $\times$ 1 in. throughout the length, using the computer program of Section 12.8. Compare your results with those given in Chapter 8. (Hint: Only the data XL, XI, and A need to be changed.)

Projects:

12.36. Derive the stiffness and mass matrices of a uniform beam element in transverse vibration rotating at an angular velocity of Ω rad/sec about a vertical axis as shown in Fig. 12.35(a). Using these matrices, find the natural frequencies of transverse vibration of the rotor blade of a helicopter (see Fig. 12.35(b)) rotating at a speed of 300 rpm. Assume a uniform rectangular cross section $1'' \times 12''$ and a length $48''$ for the blade. The material of the blade is aluminum.

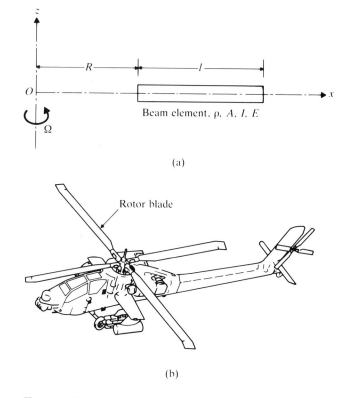

(a)

(b)

Figure 12.35

12.37. An electric motor weighing 1000 lb operates on the first floor of a building frame that can be modeled by a steel girder supported by two reinforced concrete columns as

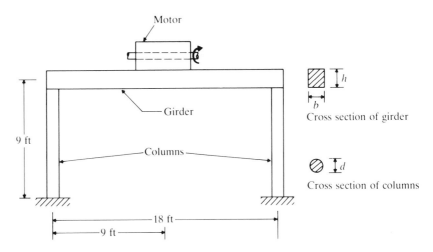

Figure 12.36

shown in Fig. 12.36. If the operating speed of the motor is 1500 rpm, design the girder and the columns such that the fundamental frequency of vibration of the building frame is greater than the operating speed of the motor. Use two beam and two bar elements for the idealization. Assume the following data:

Girder: $E = 30 \times 10^6$ psi, $\rho = 8.8 \times 10^{-3}$ lbm/in^3, $h/b = 2$.

Columns: $E = 4 \times 10^6$ psi, $\rho = 2.7 \times 10^{-3}$ lbm/in^3.

CHAPTER 13

Nonlinear Vibration

Jules Henri Poincaré (1854–1912) was a French mathematician and professor of celestial mechanics at the University of Paris and of mechanics at the École Polytechnique. His contributions to pure and applied mathematics, particularly to celestial mechanics and electrodynamics, are outstanding. His classification of singular points of nonlinear autonomous systems is important in the study of nonlinear vibrations. (Courtesy Culver Pictures)

13.1 INTRODUCTION

In the preceding chapters, the equation of motion contained displacement or its derivatives only to the first degree, and no square or higher powers of displacement or velocity were involved. For this reason, the governing differential equations of motion and the corresponding systems were called *linear*. For convenience of analysis, most systems are modeled as linear systems, but real systems are actually more often nonlinear than linear [13.1–13.6]. Whenever finite amplitudes of motion are encountered, nonlinear analysis becomes necessary. The superposition principle, which is very useful in linear analysis, does not hold true in the case of nonlinear analysis. Since mass, damper, and spring are the basic components of a vibratory system, nonlinearity into the governing differential equation may be introduced through any of these components. In many cases, linear analysis is insufficient to describe the behavior of the physical system adequately. One of the main reasons for modeling a physical system as a nonlinear one is that totally unexpected phenomena sometimes occur in nonlinear systems—phenomena that are not predicted or even hinted at by linear theory. Several methods are available for the solution of nonlinear vibration problems. Some of the exact methods, approximate analytical techniques, graphical procedures, and numerical methods are presented in this chapter.

13.2 EXAMPLES OF NONLINEAR VIBRATION PROBLEMS

The following examples are given to illustrate the nature of nonlinearity in some physical systems.

**13.2.1
Simple Pendulum**

Consider a simple pendulum of length l, having a bob of mass m, as shown in Fig. 13.1(a). The differential equation governing the free vibration of the pendulum can be derived from Fig. 13.1(b):

$$ml^2\ddot{\theta} + mgl\sin\theta = 0 \tag{13.1}$$

For small angles, $\sin\theta$ may be approximated by θ and Eq. (13.1) reduces to a linear equation:

$$\ddot{\theta} + \omega_0^2\theta = 0 \tag{13.2}$$

where

$$\omega_0 = (g/l)^{1/2} \tag{13.3}$$

The solution of Eq. (13.2) can be expressed as

$$\theta(t) = A_0\sin(\omega_0 t + \phi) \tag{13.4}$$

where A_0 is the amplitude of oscillation, ϕ is the phase angle, and ω_0 is the angular frequency. The values of A_0 and ϕ are determined by the initial conditions and the angular frequency ω_0 is independent of the amplitude A_0. Equation (13.4) denotes an approximate solution of the simple pendulum. A better approximate solution can be obtained by using a two-term approximation for $\sin\theta$ near $\theta = 0$ as $\theta - \theta^3/6$ in Eq. (13.1):

$$ml^2\ddot{\theta} + mgl\left(\theta - \frac{\theta^3}{6}\right) = 0$$

or

$$\ddot{\theta} + \omega_0^2\left(\theta - \tfrac{1}{6}\theta^3\right) = 0 \tag{13.5}$$

It can be seen that Eq. (13.5) is nonlinear because of the term involving θ^3 (due to geometric nonlinearity). Equation (13.5) is similar to the equation of motion of a

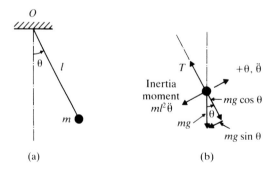

(a)   (b)

Figure 13.1

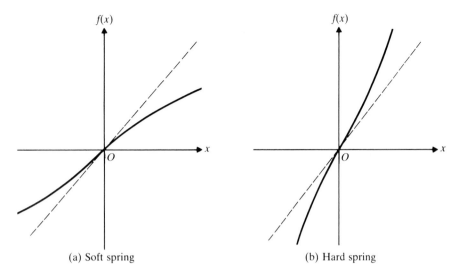

(a) Soft spring   (b) Hard spring

Figure 13.2

spring-mass system with a nonlinear spring. If the spring is nonlinear (due to material nonlinearity), the restoring force can be expressed as $f(x)$, where x is the deformation of the spring and the equation of motion of the spring-mass system becomes

$$m\ddot{x} + f(x) = 0 \qquad (13.6)$$

If $df/dx(x) = k = $ constant, the spring is linear. If df/dx is a strictly increasing function of x, the spring is called a hard spring, and if df/dx is a strictly decreasing function of x, the spring is called a soft spring (Fig. 13.2). Due to the similarity of Eqs. (13.5) and (13.6), a pendulum with large amplitudes is considered, in a loose sense, as a system with nonlinear elastic (spring) component.

13.2.2
Mechanical
Chatter, Belt
Friction System

Nonlinearity may be reflected in the damping term as in the case of Fig. 13.3(a). The system behaves nonlinearly because of the dry friction between the mass m and the moving belt. For this system, there are two friction coefficients: the static coefficient of friction (μ_s), corresponding to the force required to initiate the motion of the

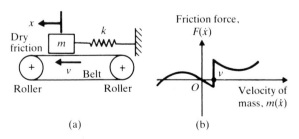

(a)   (b)

Figure 13.3

body held by dry friction; and the kinetic coefficient of friction (μ_k), corresponding to the force required to maintain the body in motion. In either case, the component of the applied force tangent to the friction surface (F) is the product of the appropriate friction coefficient and the force normal to the surface.

The sequence of motion of the system shown in Fig. 13.3(a) is as follows [13.7]. The mass is initially at rest on the belt. Due to the displacement of the mass m along with the belt, the spring elongates. As the spring extends, the spring force on the mass increases until the static friction force is overcome and the mass begins to slide. It slides rapidly towards the right, thereby relieving the spring force until the kinetic friction force halts it. The spring then begins to build up the spring force again. The variation of the damping force with the velocity of the mass is shown in Fig. 13.3(b). The equation of motion can be expressed as

$$m\ddot{x} + F(\dot{x}) + kx = 0 \tag{13.7}$$

where the friction force F is a nonlinear function of $\dot{x}$, as shown in Fig. 13.3(b).

For large values of $\dot{x}$, the damping force is positive (the curve has a positive slope) and energy is removed from the system. On the other hand, for small values of $\dot{x}$, the damping force is negative (the curve has a negative slope) and energy is put into the system. Although there is no external stimulus, the system can have an oscillatory motion; it corresponds to a nonlinear self-excited system. This phenomenon of self-excited vibration is called *mechanical chatter*.

13.2.3
Variable Mass
System

Nonlinearity may appear in the mass term as in the case of Fig. 13.4 [13.8]. For large deflections, the mass of the system depends on the displacement x, and so the equation of motion becomes

$$\frac{d}{dt}(m\dot{x}) + kx = 0 \tag{13.8}$$

Note that this is a nonlinear differential equation due to the nonlinearity of the first term.

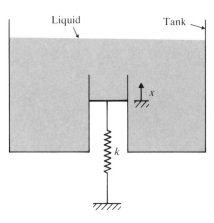

Figure 13.4

13.3 EXACT METHODS

An exact solution is possible only for a relatively few nonlinear systems whose motion is governed by specific types of second order nonlinear differential equations. The solutions are exact in the sense that they are given either in closed form or in the form of an expression that can be numerically evaluated to any degree of accuracy. In this section, we shall consider a simple nonlinear system for which the exact solution is available. For a single degree of freedom system with a general restoring (spring) force $F(x)$, the free vibration equation can be expressed as

$$\ddot{x} + a^2 F(x) = 0 \tag{13.9}$$

where a^2 is a constant. Equation (13.9) can be rewritten as

$$\frac{d}{dx}(\dot{x}^2) + 2a^2 F(x) = 0 \tag{13.10}$$

Assuming the initial displacement as x_0 and the velocity as zero at $t = t_0$, Eq. (13.10) can be integrated to obtain

$$\dot{x}^2 = 2a^2 \int_x^{x_0} F(\eta) \, d\eta \quad \text{or} \quad |\dot{x}| = \sqrt{2}\, a \left\{ \int_x^{x_0} F(\eta) \, d\eta \right\}^{1/2} \tag{13.11}$$

where η is the integration variable. Equation (13.11), when integrated again, gives

$$t - t_0 = \frac{1}{\sqrt{2}\, a} \int_0^x \frac{d\xi}{\left\{ \int_\xi^{x_0} F(\eta) \, d\eta \right\}^{1/2}} \tag{13.12}$$

where ξ is the new integration variable and t_0 corresponds to the time when $x = 0$. Equation (13.12) thus gives the exact solution of Eq. (13.9) in all those situations where the integrals of Eq. (13.12) can be evaluated in closed form. After evaluating the integrals of Eq. (13.12), one can invert the result and obtain the displacement-time relation. If $F(x)$ is an odd function,

$$F(-x) = -F(x) \tag{13.13}$$

By considering Eq. (13.12) from zero displacement to maximum displacement, the period of vibration τ can be obtained:

$$\tau = \frac{4}{\sqrt{2}\, a} \int_0^{x_0} \frac{d\xi}{\left\{ \int_\xi^{x_0} F(\eta) \, d\eta \right\}^{1/2}} \tag{13.14}$$

For illustration, let $F(x) = x^n$. In this case Eqs. (13.12) and (13.14) become

$$t - t_0 = \frac{1}{a} \sqrt{\frac{n+1}{2}} \int_0^{x_0} \frac{d\xi}{\left(x_0^{n+1} - \xi^{n+1} \right)^{1/2}} \tag{13.15}$$

and

$$\tau = \frac{4}{a} \sqrt{\frac{n+1}{2}} \int_0^{x_0} \frac{d\xi}{\left(x_0^{n+1} - \xi^{n+1} \right)^{1/2}} \tag{13.16}$$

By setting $y = \xi/x_0$, Eq. (13.16) can be written as

$$\tau = \frac{4}{a} \frac{1}{\left(x_0^{n-1}\right)^{1/2}} \sqrt{\frac{n+1}{2}} \int_0^1 \frac{dy}{\left(1 - y^{n+1}\right)^{1/2}} \tag{13.17}$$

This expression can be evaluated numerically to any desired level of accuracy.

13.4 APPROXIMATE ANALYTICAL METHODS

In the absence of an exact analytical solution to a nonlinear vibration problem, we wish to find at least an approximate solution. Although both analytical and numerical methods are available for approximate solution of nonlinear vibration problems, the analytical methods are more desirable [13.6, 13.9]. The reason is that once the analytical solution is obtained, any desired numerical values can be substituted and the entire possible range of solutions can be investigated. We shall now consider four analytical techniques in the following subsections.

**13.4.1
Basic Philosophy**

Let the equations governing the vibration of a nonlinear system be represented by a system of n first order differential equations:[†]

$$\dot{\vec{x}}(t) = \vec{f}\,(\vec{x}, t) + \alpha \vec{g}(\vec{x}, t) \tag{13.18}$$

where the nonlinear terms are assumed to appear only in $\vec{g}(\vec{x}, t)$ and α is a small parameter. In Eq. (13.18),

$$\vec{x} = \begin{Bmatrix} x_1 \\ x_2 \\ \vdots \\ x_n \end{Bmatrix}, \qquad \dot{\vec{x}} = \begin{Bmatrix} dx_1/dt \\ dx_2/dt \\ \vdots \\ dx_n/dt \end{Bmatrix}, \qquad \vec{f}\,(\vec{x}, t) = \begin{Bmatrix} f_1(x_1, x_2, \ldots, x_n, t) \\ f_2(x_1, x_2, \ldots, x_n, t) \\ \vdots \\ f_n(x_1, x_2, \ldots, x_n, t) \end{Bmatrix}$$

and

$$\vec{g}(\vec{x}, t) = \begin{Bmatrix} g_1(x_1, x_2, \ldots, x_n, t) \\ g_2(x_1, x_2, \ldots, x_n, t) \\ \vdots \\ g_n(x_1, x_2, \ldots, x_n, t) \end{Bmatrix}$$

The solution of differential equations having nonlinear terms associated with a small parameter was studied by Poincaré [13.6]. Basically, he assumed the solution of Eq.

[†] Systems governed by Eq. (13.18), in which the time appears explicitly, are known as *nonautonomous* systems. On the other hand, systems for which the governing equations are of the type

$$\dot{\vec{x}}(t) = \vec{f}(x) + \alpha \vec{g}(x)$$

where time does not appear explicitly are called *autonomous* systems.

(13.18) in series form as

$$\vec{x}(t) = \vec{x}_0(t) + \alpha\vec{x}_1(t) + \alpha^2\vec{x}_2(t) + \alpha^3\vec{x}_3(t) + \cdots \qquad (13.19)$$

The series solution of Eq. (13.19) has two basic characteristics:

1. As $\alpha \to 0$, Eq. (13.19) reduces to the exact solution of the linear equations $\dot{\vec{x}} = \vec{f}(\vec{x}, t)$.
2. For small values of α, the series converges fast so that even the first two or three terms in the series of Eq. (13.19) yields a reasonably accurate solution.

The various approximate analytical methods presented in this section can be considered to be modifications of the basic idea contained in Eq. (13.19). Although Poincaré's solution, Eq. (13.19), is valid for only small values of α, the method can still be applied to systems with large values of α. The solution of the pendulum equation, Eq. (13.5), is presented to illustrate the Poincaré's method.

Solution of Pendulum Equations. Equation (13.5) can be rewritten as

$$\ddot{x} + \omega_0^2 x + \alpha x^3 = 0 \qquad (13.20)$$

where $x = \theta$, $\omega_0 = (g/l)^{1/2}$, and $\alpha = -\omega_0^2/6$. Equation (13.20) is known as the free Duffing's equation. Assuming weak nonlinearity (i.e., α is small), the solution of Eq. (13.20) is expressed as

$$x(t) = x_0(t) + \alpha x_1(t) + \alpha^2 x_2(t) + \cdots + \alpha^n x_n(t) + \cdots \qquad (13.21)$$

where $x_i(t)$, $i = 0, 1, 2, \ldots, n$, are functions to be determined. By using a two-term approximation in Eq. (13.21), Eq. (13.20) can be written as

$$(\ddot{x}_0 + \alpha\ddot{x}_1) + \omega_0^2(x_0 + \alpha x_1) + \alpha(x_0 + \alpha x_1)^3 = 0$$

that is,

$$(\ddot{x}_0 + \omega_0^2 x_0) + \alpha(\ddot{x}_1 + \omega_0^2 x_1 + x_0^3) + \alpha^2(3x_0^2 x_1) + \alpha^3(3x_0 x_1^2) + \alpha^4 x_1^3 = 0 \qquad (13.22)$$

If terms involving α^2, α^3, and α^4 are neglected (since α is assumed to be small), Eq. (13.22) will be satisfied if the following equations are satisfied:

$$\ddot{x}_0 + \omega_0^2 x_0 = 0 \qquad (13.23)$$

$$\ddot{x}_1 + \omega_0^2 x_1 = -x_0^3 \qquad (13.24)$$

The solution of Eq. (13.23) can be expressed as

$$x_0(t) = A_0 \sin(\omega_0 t + \phi) \qquad (13.25)$$

In view of Eq. (13.25), Eq. (13.24) becomes

$$\ddot{x}_1 + \omega_0^2 x_1 = -A_0^3 \sin^3(\omega_0 t + \phi)$$

$$= -A_0^3 \left[\tfrac{3}{4}\sin(\omega_0 t + \phi) - \tfrac{1}{4}\sin 3(\omega_0 t + \phi) \right] \qquad (13.26)$$

The particular solution of Eq. (13.26) is (and can be verified by substitution)

$$x_1(t) = \frac{3}{8\omega_0} t A_0^3 \cos(\omega_0 t + \phi) - \frac{A_0^3}{32\omega_0^2} \sin 3(\omega_0 t + \phi) \qquad (13.27)$$

Thus the approximate solution of Eq. (13.20) becomes

$$x(t) = x_0(t) + \alpha x_1(t)$$

$$= A_0 \sin(\omega_0 t + \phi) + \frac{3\alpha t}{8\omega_0} A_0^3 \cos(\omega_0 t + \phi) - \frac{A_0^3 \alpha}{32\omega_0^2} \sin 3(\omega_0 t + \phi) \quad (13.28)$$

The initial conditions on $x(t)$ can be used to evaluate the constants A_0 and ϕ.

Notes: 1. It can be seen that even a weak nonlinearity (i.e., small value of α) leads to a nonperiodic solution since Eq. (13.28) is not periodic due to the second term on the right-hand side of Eq. (13.28). In general, the solution given by Eq. (13.21) will not be periodic if we retain only a finite number of terms.

2. In Eq. (13.28), the second term, and hence the total solution, can be seen to approach infinity as t tends to infinity. However, the exact solution of Eq. (13.20) is known to be bounded for all values of t. The reason for the unboundedness of the solution, Eq. (13.28), is that only two terms are considered in Eq. (13.21). The second term in Eq. (13.28) is called a *secular term*. The infinite series in Eq. (13.21) leads to a bounded solution of Eq. (13.20) because the process is a convergent one. To illustrate this point, consider the Taylor's series expansion of the function $\sin(\omega t + \alpha t)$:

$$\sin(\omega + \alpha)t = \sin \omega t + \alpha t \cos \omega t - \frac{\alpha^2 t^2}{2!} \sin \omega t - \frac{\alpha^3 t^3}{3!} \cos \omega t + \cdots \quad (13.29)$$

If only two terms are considered on the right-hand side of Eq. (13.29), the solution approaches infinity as $t \to \infty$. However, the function itself and hence its infinite series expansion can be seen to be a bounded one.

**13.4.2
Lindstedt's
Perturbation
Method**

This method assumes that the angular frequency along with the solution varies as a function of the amplitude A_0. This method eliminates the secular terms in each step of the approximation [13.5] by requiring the solution to be periodic in each step. The solution and the angular frequency are assumed as

$$x(t) = x_0(t) + \alpha x_1(t) + \alpha^2 x_2(t) + \cdots \qquad (13.30)$$

$$\omega^2 = \omega_0^2 + \alpha \omega_1(A_0) + \alpha^2 \omega_2(A_0) + \cdots \qquad (13.31)$$

We consider the solution of the pendulum equation, Eq. (13.20), to illustrate the perturbation method. We use only linear terms in α in Eqs. (13.30) and (13.31):

$$x(t) = x_0(t) + \alpha x_1(t) \qquad (13.32)$$

$$\omega^2 = \omega_0^2 + \alpha \omega_1(A_0) \qquad \text{or} \qquad \omega_0^2 = \omega^2 - \alpha \omega_1(A_0) \qquad (13.33)$$

Substituting Eqs. (13.32) and (13.33) into Eq. (13.20), we get

$$\ddot{x}_0 + \alpha\ddot{x}_1 + \left[\omega^2 - \alpha\omega_1(A_0)\right]\left[x_0 + \alpha x_1\right] + \alpha\left[x_0 + \alpha x_1\right]^3 = 0$$

that is,

$$\ddot{x}_0 + \omega_0^2 x_0 + \alpha\left(\omega^2 x_1 + x_0^3 - \omega_1 x_0 + \ddot{x}_1\right)$$
$$+ \alpha^2\left(3x_1 x_0^2 - \omega_1 x_1\right) + \alpha^3\left(3x_1^2 x_0\right) + \alpha^4\left(x_1^3\right) = 0 \qquad (13.34)$$

Setting the coefficients of various powers of α to zero and neglecting the terms involving α^2, α^3, and α^4 in Eq. (13.34), we obtain

$$\ddot{x}_0 + \omega^2 x_0 = 0 \qquad (13.35)$$
$$\ddot{x}_1 + \omega^2 x_1 = -x_0^3 + \omega_1 x_0 \qquad (13.36)$$

Using the solution of Eq. (13.35),

$$x_0(t) = A_0 \sin(\omega t + \phi) \qquad (13.37)$$

into Eq. (13.36), we obtain

$$\ddot{x}_1 + \omega^2 x_1 = -\left[A_0 \sin(\omega t + \phi)\right]^3 + \omega_1\left[A_0 \sin(\omega t + \phi)\right]$$
$$= -\tfrac{3}{4}A_0^3 \sin(\omega t + \phi) + \tfrac{1}{4}A_0^3 \sin 3(\omega t + \phi) + \omega_1 A_0 \sin(\omega t + \phi) \qquad (13.38)$$

It can be seen that the first and the last terms on the right-hand side of Eq. (13.38) lead to secular terms. They can be eliminated by taking ω_1 as

$$\omega_1 = \tfrac{3}{4}A_0^2, \qquad A_0 \neq 0 \qquad (13.39)$$

With this, Eq. (13.38) becomes

$$\ddot{x}_1 + \omega^2 x_1 = \tfrac{1}{4}A_0^3 \sin 3(\omega t + \phi) \qquad (13.40)$$

The solution of Eq. (13.40) is

$$x_1(t) = A_1 \sin(\omega t + \phi_1) - \frac{A_0^3}{32\omega^2} \sin 3(\omega t + \phi) \qquad (13.41)$$

Let the initial conditions be $x(t = 0) = A$ and $\dot{x}(t = 0) = 0$. In the Lindstedt's method, we force the solution $x_0(t)$ given by Eq. (13.37) to satisfy the initial conditions so that

$$x(0) = A = A_0 \sin\phi, \qquad \dot{x}(0) = 0 = A_0\omega \cos\phi$$

or

$$A_0 = A \qquad \text{and} \qquad \phi = \frac{\pi}{2}$$

Since the initial conditions are satisfied by $x_0(t)$ itself, the solution $x_1(t)$ given by Eq. (13.41) must satisfy zero initial conditions.[†] Thus

$$x_1(0) = 0 = A_1 \sin\phi_1 - \frac{A_0^3}{32\omega^2} \sin 3\phi$$

$$\dot{x}_1(0) = 0 = A_1\omega \cos\phi_1 - \frac{A_0^3}{32\omega^2}(3\omega)\cos 3\phi$$

[†] If $x_0(t)$ satisfies the initial conditions, each of the solutions $x_1(t), x_2(t), \ldots$ appearing in Eq. (13.30) must satisfy zero initial conditions.

In view of the known relations $A_0 = A$ and $\phi = \pi/2$, the above equations yield

$$A_1 = -\left(\frac{A^3}{32\omega^2}\right) \quad \text{and} \quad \phi_1 = \frac{\pi}{2}.$$

Thus the total solution of Eq. (13.20) becomes

$$x(t) = A_0 \sin(\omega t + \phi) - \frac{\alpha A_0^3}{32\omega^2} \sin 3(\omega t + \phi) \tag{13.42}$$

with

$$\omega^2 = \omega_0^2 + \alpha \tfrac{3}{4} A_0^2 \tag{13.43}$$

For the solution obtained by considering three terms in the expansion of Eq. (13.30), see Problem 13.9. It is to be noted that the Lindstedt's method gives only the periodic solutions of Eq. (13.20); it cannot give any nonperiodic solutions, even if they exist.

13.4.3
Iterative Method

In the basic iterative method, first the equation is solved by neglecting certain terms. The resulting solution is then inserted in the terms that were neglected at first to obtain a second, improved, solution. We shall illustrate the iterative method to find the solution of Duffing's equation, which represents the equation of motion of a damped, harmonically excited, single degree of freedom system with a nonlinear spring. First we consider the solution of the undamped equation.

Solution of the Undamped Equation. If damping is disregarded, Duffing's equation becomes

$$\ddot{x} + \omega_0^2 x \pm \alpha x^3 = F \cos \omega t$$

or

$$\ddot{x} = -\omega_0^2 x \mp \alpha x^3 + F \cos \omega t \tag{13.44}$$

As a first approximation, we assume the solution to be

$$x_1(t) = A \cos \omega t \tag{13.45}$$

where A is an unknown. By substituting Eq. (13.45) into Eq. (13.44), we obtain the differential equation for the second approximation:

$$\ddot{x}_2 = -A\omega_0^2 \cos \omega t \mp A^3 \alpha \cos^3 \omega t + F \cos \omega t \tag{13.46}$$

By using the identity

$$\cos^3 \omega t = \tfrac{3}{4} \cos \omega t + \tfrac{1}{4} \cos 3\omega t \tag{13.47}$$

Eq. (13.46) can be expressed as

$$\ddot{x}_2 = -\left(A\omega_0^2 \pm \tfrac{3}{4} A^3 \alpha - F\right) \cos \omega t \mp \tfrac{1}{4} A^3 \alpha \cos 3\omega t \tag{13.48}$$

By integrating this equation and setting the constants of integration to zero (so as to make the solution harmonic with period $\tau = 2\pi/\omega$) we obtain the second approximation:

$$x_2(t) = \frac{1}{\omega^2}\left(A\omega_0^2 \pm \tfrac{3}{4}A^3\alpha - F \right)\cos \omega t \pm \frac{A^3\alpha}{36\omega^2}\cos 3\omega t \qquad (13.49)$$

Duffing [13.7] reasoned at this point that if $x_1(t)$ and $x_2(t)$ are good approximations to the solution $x(t)$, the coefficients of $\cos \omega t$ in the two equations (13.45) and (13.49) should not be very different. Thus by equating these coefficients, we obtain

$$A = \frac{1}{\omega^2}\left(A\omega_0^2 \pm \frac{3}{4}A^3\alpha - F \right)$$

or

$$\omega^2 = \omega_0^2 \pm \frac{3}{4}A^2\alpha - \frac{F}{A} \qquad (13.50)$$

For present purposes, we will stop the procedure with the second approximation. It can be verified that this procedure yields the exact solution for the case of a linear spring (with $\alpha = 0$):

$$A = \frac{F}{\omega_0^2 - \omega^2} \qquad (13.51)$$

where A denotes the amplitude of the harmonic response of the linear system.

For a nonlinear system (with $\alpha \neq 0$), Eq. (13.50) shows that the frequency ω is a function of α, A, and F. Note that the quantity A, in the case of a nonlinear system, is not the amplitude of the harmonic response but only the coefficient of the first term of its solution. However, it is commonly taken as the amplitude of the harmonic response of the system.[†] For the free vibration of the nonlinear system, $F = 0$ and Eq. (13.50) reduces to

$$\omega^2 = \omega_0^2 \pm \frac{3}{4}A^2\alpha \qquad (13.52)$$

This equation shows that the frequency of the response increases with the amplitude A for the hardening spring and decreases for the softening spring. Also, the solution, Eq. (13.52), can be seen to be same as the one given by the Lindstedt's method, Eq. (13.43).

For both linear and nonlinear systems, when $F \neq 0$ (forced vibration), there are two values of the frequency ω for any given amplitude $|A|$. One of these values of ω is smaller and the other larger than the corresponding frequency of free vibration at that amplitude. For the smaller value of ω, $A > 0$ and the harmonic response of the system is in phase with the external force. For the larger value of ω, $A < 0$ and the response is 180° out of phase with the external force. Note that only the harmonic solutions of Duffing's equation—that is, solutions for which the frequency is the same as that of the external force $F\cos \omega t$—have been considered in the present analysis. It has been observed [13.2] that oscillations whose frequency is a fraction,

---

[†]   The first approximate solution, Eq. (13.45), can be seen to satisfy the initial conditions $x(0) = A$ and $\dot{x}(0) = 0$.

such as $\frac{1}{2}, \frac{1}{3}, \ldots, \frac{1}{n}$, of that of the applied force, are also possible for Duffing's equation. Such oscillations, known as subharmonic oscillations, are considered in Section 13.5.

Solution of the Damped Equation. If we consider viscous damping, we obtain Duffing's equation:

$$\ddot{x} + c\dot{x} + \omega_0^2 x \pm \alpha x^3 = F \cos \omega t \qquad (13.53)$$

For a damped system, it was observed in earlier chapters that there is a phase difference between the applied force and the response or solution. The usual procedure is to prescribe the applied force first and then determine the phase of the solution. In the present case, however, it is more convenient to fix the phase of the solution and keep the phase of the applied force as a quantity to be determined. We take the differential equation, Eq. (13.53), in the form

$$\ddot{x} + c\dot{x} + \omega_0^2 x \pm \alpha x^3 = F \cos(\omega t + \phi)$$
$$= A_1 \cos \omega t - A_2 \sin \omega t \qquad (13.54)$$

in which the amplitude $F = (A_1^2 + A_2^2)^{1/2}$ of the applied force is considered fixed, but the ratio $A_1/A_2 = \tan^{-1} \phi$ is left to be determined. We assume that c, A_1, and A_2 are all small, of order α. As with Eq. (13.44), we assume the first approximation to the solution to be

$$x_1 = A \cos \omega t \qquad (13.55)$$

where A is assumed fixed and ω to be determined. By substituting Eq. (13.55) into Eq. (13.54) and making use of the relation (13.47), we obtain

$$\left[(\omega_0^2 - \omega^2) A \pm \frac{3}{4} \alpha A^3 \right] \cos \omega t - c\omega A \sin \omega t \pm \frac{\alpha A^3}{4} \cos 3\omega t = A_1 \cos \omega t - A_2 \sin \omega t \qquad (13.56)$$

By disregarding the term involving $\cos 3\omega t$ and equating the coefficients of $\cos \omega t$ and $\sin \omega t$ on both sides of Eq. (13.56), we obtain the following relations:

$$\left(\omega_0^2 - \omega^2 \right) A \pm \frac{3}{4} \alpha A^3 = A_1$$
$$c\omega A = A_2 \qquad (13.57)$$

The relation between the amplitude of the applied force and the quantities A and ω can be obtained by squaring and adding the equations (13.57):

$$\left[(\omega_0^2 - \omega^2) A \pm \frac{3}{4} \alpha A^3 \right]^2 + (c\omega A)^2 = A_1^2 + A_2^2 = F^2 \qquad (13.58)$$

Equation (13.58) can be rewritten as

$$S^2(\omega, A) + c^2 \omega^2 A^2 = F^2 \qquad (13.59)$$

where

$$S(\omega, A) = (\omega_0^2 - \omega^2) A \pm \frac{3}{4} \alpha A^3 \qquad (13.60)$$

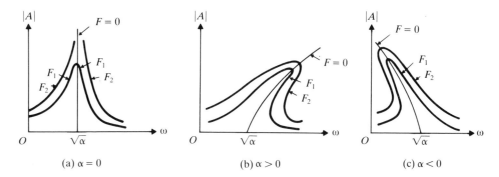

Figure 13.5   Response curves of Duffing's equation.

It can be seen that for $c = 0$, Eq. (13.59) reduces to $S(\omega, A) = F$, which is the same as Eq. (13.50). The response curves given by Eq. (13.59) are shown in Fig. 13.5.

Jump Phenomenon. As mentioned earlier, nonlinear systems exhibit phenomena that cannot occur in linear systems. For example, the amplitude of vibration of the system described by Eq. (13.54) has been found to increase or decrease suddenly as the excitation frequency ω is increased or decreased, as shown in Fig. 13.6. For a constant magnitude of F, the amplitude of vibration will increase along the points $1, 2, 3, 4, 5$ on the curve when the excitation frequency ω is slowly increased. The amplitude of vibration jumps from point 3 to 4 on the curve. Similarly, when the forcing frequency ω is slowly decreased, the amplitude of vibration follows the curve along the points $5, 4, 6, 7, 2, 1$ and makes a jump from point 6 to 7. This behavior is known as the *jump phenomenon*. It is evident that there exist two amplitudes of vibration for a given forcing frequency, as shown in the shaded

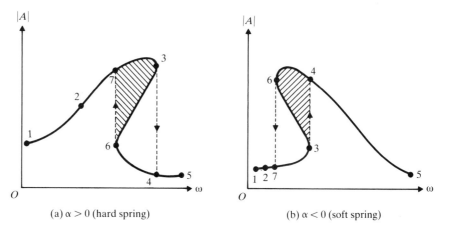

Figure 13.6

regions of the curves of Fig. 13.6. The shaded region can be thought of as unstable in some sense. Thus an understanding of the jump phenomena requires a knowledge of the mathematically involved stability analysis of periodic solutions [13.24, 13.25]. The jump phenomena was also observed experimentally by several investigators [13.26, 13.27].

**13.4.4
Ritz-Galerkin
Method**

In the Ritz-Galerkin method, an approximate solution of the problem is found by satisfying the governing nonlinear equation in the average. To see how the method works, let the nonlinear differential equation be represented as

$$E[x] = 0 \tag{13.61}$$

An approximate solution of Eq. (13.61) is assumed as

$$\underset{\sim}{x}(t) = a_1\phi_1(t) + a_2\phi_2(t) + \cdots + a_n\phi_n(t) \tag{13.62}$$

where $\phi_1(t), \phi_2(t), \ldots, \phi_n(t)$ are prescribed functions of time and $a_1, a_2, \ldots, a_n$ are weighting factors to be determined. If Eq. (13.62) is substituted in Eq. (13.61), we get a function $E[\underset{\sim}{x}(t)]$. Since $\underset{\sim}{x}(t)$ is not, in general, the exact solution of Eq. (13.61), $\underset{\sim}{E}(t) = E[\underset{\sim}{x}(t)]$ will not be zero. However, the value of $\underset{\sim}{E}[t]$ will serve as a measure of the accuracy of the approximation; in fact, $\underset{\sim}{E}[t] \to 0$ as $\underset{\sim}{x} \to x$.

The weighting factors a_i are determined by minimizing the integral

$$\int_0^\tau \underset{\sim}{E}^2[t] \, dt \tag{13.63}$$

where τ denotes the period of the motion. The minimization of the function of Eq. (13.63) requires

$$\frac{\partial}{\partial a_i}\left(\int_0^\tau \underset{\sim}{E}^2[t] \, dt\right) = 2\int_0^\tau \underset{\sim}{E}[t]\frac{\partial \underset{\sim}{E}[t]}{\partial a_i} \, dt = 0, \qquad i = 1, 2, \ldots, n \tag{13.64}$$

Equation (13.64) represents a system of n algebraic equations that can be solved simultaneously to find the values of $a_1, a_2, \ldots, a_n$. The procedure is illustrated with the following example.

EXAMPLE 13.1   Solution of Pendulum Equation Using Ritz-Galerkin Method

Find the solution of the pendulum equation

$$E[x] = \ddot{x} + \omega_0^2 x - \frac{\omega_0^2}{6}x^3 = 0 \tag{E.1}$$

by the Ritz-Galerkin method.

Given: Nonlinear pendulum equation.

Find: Approximate solution using Ritz-Galerkin method.

Approach: Use a one-term approximate solution.

Solution. By using a one-term approximation for $x(t)$ as

$$\underset{\sim}{x}(t) = A_0 \sin \omega t, \tag{E.2}$$

Eqs. (E.1) and (E.2) lead to

$$E\left[\underset{\sim}{x}(t)\right] = -\omega^2 A_0 \sin \omega t + \omega_0^2 \left[A_0 \sin \omega t - \frac{1}{6} \sin^3 \omega t \right]$$

$$= \left(\omega_0^2 - \omega^2 - \frac{1}{8} \omega_0^2 A_0^2 \right) A_0 \sin \omega t + \frac{\omega_0^2}{24} A_0^3 \sin 3\omega t \tag{E.3}$$

The Ritz-Galerkin method requires the minimization of

$$\int_0^\tau E^2\left[\underset{\sim}{x}(t)\right] dt \tag{E.4}$$

for finding A_0. The application of Eq. (13.64) gives

$$\int_0^\tau \underset{\sim}{E} \frac{\partial E}{\partial A_0} \, dt = \int_0^\tau \left[\left(\omega_0^2 - \omega^2 - \frac{1}{8} \omega_0^2 A_0^2 \right) A_0 \sin \omega t + \frac{\omega_0^2}{24} A_0^3 \sin 3\omega t \right]$$

$$\times \left[\left(\omega_0^2 - \omega^2 - \frac{3}{8} \omega_0^2 A_0^2 \right) \sin \omega t + \frac{1}{8} \omega_0^2 A_0^2 \sin 3\omega t \right] dt = 0$$

that is,

$$A_0 \left(\omega_0^2 - \omega^2 - \frac{1}{8} \omega_0^2 A_0^2 \right) \left(\omega_0^2 - \omega^2 - \frac{3}{8} \omega_0^2 A_0^2 \right) \int_0^\tau \sin^2 \omega t \, dt$$

$$+ \frac{\omega_0^2 A_0^3}{24} \left(\omega_0^2 - \omega^2 - \frac{3}{8} \omega_0^2 A_0^2 \right) \int_0^\tau \sin \omega t \sin 3\omega t \, dt$$

$$+ \frac{1}{8} \omega_0^2 A_0^2 \left(\omega_0^2 - \omega^2 - \frac{1}{8} \omega_0^2 A_0^2 \right) \int_0^\tau \sin \omega t \sin 3\omega t \, dt$$

$$+ \frac{\omega_0^4 A_0^5}{192} \int_0^\tau \sin^2 3\omega t \, dt = 0$$

that is,

$$A_0 \left[\left(\omega_0^2 - \omega^2 - \frac{1}{8} \omega_0^2 A_0^2 \right) \left(\omega_0^2 - \omega^2 - \frac{3}{8} \omega_0^2 A_0^2 \right) + \frac{\omega_0^4 A_0^4}{192} \right] = 0 \tag{E.5}$$

For a nontrivial solution, $A_0 \neq 0$, and Eq. (E.5) leads to

$$\omega^4 + \omega^2 \omega_0^2 \left(\tfrac{1}{2} A_0^2 - 2 \right) + \omega_0^4 \left(1 - \tfrac{1}{2} A_0^2 + \tfrac{5}{96} A_0^4 \right) = 0 \tag{E.6}$$

The roots of the quadratic equation in ω^2, Eq. (E.5), can be found as

$$\omega^2 = \omega_0^2 \left(1 - 0.147938 \, A_0^2 \right) \tag{E.7}$$

$$\omega^2 = \omega_0^2 \left(1 - 0.352062 \, A_0^2 \right) \tag{E.8}$$

It can be verified that ω^2 given by Eq. (E.7) minimizes the quantity of (E.4), while the one given by Eq. (E.8) maximizes it. Thus the solution of Eq. (E.1) is given by Eq. (E.2) with

$$\omega^2 = \omega_0^2 \left(1 - 0.147936 \, A_0^2 \right) \tag{E.9}$$

This expression can be compared with the solution given by the Lindstedt's and the iteration methods (Eqs. (13.43) and (13.52)):

$$\omega^2 = \omega_0^2 \left(1 - 0.125 \, A_0^2 \right) \tag{E.10}$$

The solution can be improved by using a two-term approximation for $x(t)$ as

$$\underline{x}(t) = A_0 \sin \omega t + A_3 \sin 3\omega t \qquad (E.11)$$

The application of Eq. (13.64) with the solution of Eq. (E.11) leads to two simultaneous algebraic equations that must be numerically solved for A_0 and A_3.

Other approximate methods, such as the equivalent linearization scheme, and the harmonic balance procedure, are also available for solving nonlinear vibration problems [13.10–13.12]. Specific solutions found using these techniques include the free vibration response of single degree of freedom oscillators [13.13, 13.14], two degree of freedom systems [13.15], and elastic beams [13.16, 13.17], and the transient response of forced systems [13.18, 13.19]. Several nonlinear problems of structural dynamics have been discussed by Crandall [13.30].

13.5 SUBHARMONIC AND SUPERHARMONIC OSCILLATIONS

We noted in Chapter 3 that for a linear system, when the applied force has a certain frequency of oscillation, the steady-state response will have the same frequency of oscillation. However, a nonlinear system will exhibit subharmonic and superharmonic oscillations. Subharmonic response involves oscillations whose frequencies (ω_n) are related to the forcing frequency (ω) as

$$\omega_n = \frac{\omega}{n} \qquad (13.65)$$

where n is an integer ($n = 2, 3, 4, \dots$). Similarly, superharmonic response involves oscillations whose frequencies (ω_n) are related to the forcing frequency (ω) as

$$\omega_n = n\omega \qquad (13.66)$$

where $n = 2, 3, 4, \dots$.

13.5.1
Subharmonic
Oscillations

In this section, we consider the subharmonic oscillations of order $\frac{1}{3}$ of an undamped pendulum whose equation of motion is given by (undamped Duffing's equation):

$$\ddot{x} + \omega_0^2 x + \alpha x^3 = F \cos 3\omega t \qquad (13.67)$$

where α is assumed to be small. We find the response using the perturbation method [13.4, 13.6]. Accordingly, we seek a solution of the form

$$x(t) = x_0(t) + \alpha x_1(t) \qquad (13.68)$$

$$\omega^2 = \omega_0^2 + \alpha \omega_1 \qquad \text{or} \qquad \omega_0^2 = \omega^2 - \alpha \omega_1 \qquad (13.69)$$

where ω denotes the fundamental frequency of the solution (equal to the third subharmonic frequency of the forcing frequency). Substitution of Eqs. (13.68) and (13.69) into Eq. (13.67) gives

$$\ddot{x}_0 + \alpha \ddot{x}_1 + \omega^2 x_0 + \omega^2 \alpha x_1 - \alpha \omega_1 x_0 - \alpha^2 x_1 \omega_1 + \alpha (x_0 + \alpha x_1)^3 = F \cos 3\omega t \quad (13.70)$$

If terms involving α^2, α^3, and α^4 are neglected, Eq. (13.70) reduces to

$$\ddot{x}_0 + \omega^2 x_0 + \alpha \ddot{x}_1 + \alpha \omega^2 x_1 - \alpha \omega_1 x_0 + \alpha x_0^3 = F \cos 3\omega t \qquad (13.71)$$

We first consider the linear equation (by setting $\alpha = 0$):

$$\ddot{x}_0 + \omega^2 x_0 = F \cos 3\omega t \qquad (13.72)$$

The solution of Eq. (13.72) can be expressed as

$$x_0(t) = A_1 \cos \omega t + B_1 \sin \omega t + C \cos 3\omega t \qquad (13.73)$$

If the initial conditions are assumed as $x(t = 0) = A$ and $\dot{x}(t = 0) = 0$, we obtain $A_1 = A$ and $B_1 = 0$ so that Eq. (13.73) reduces to

$$x_0(t) = A \cos \omega t + C \cos 3\omega t \qquad (13.74)$$

where C denotes the amplitude of the forced vibration. The value of C can be determined by substituting Eq. (13.74) into Eq. (13.72) and equating the coefficients of $\cos 3\omega t$ on both sides of the resulting equation, which yields

$$C = -\frac{F}{8\omega^2} \qquad (13.75)$$

Now we consider the terms involving α in Eq. (13.71) and set them equal to zero:

$$\alpha \left(\ddot{x}_1 + \omega^2 x_1 - \omega_1 x_0 + x_0^3 \right) = 0$$

or

$$\ddot{x}_1 + \omega^2 x_1 = \omega_1 x_0 - x_0^3 \qquad (13.76)$$

The substitution of Eq. (13.74) into Eq. (13.76) results in

$$\begin{aligned}
\ddot{x}_1 + \omega^2 x_1 = {} & \omega_1 A \cos \omega t + \omega_1 C \cos 3\omega t - A^3 \cos^3 \omega t - C^3 \cos^3 3\omega t \\
& - 3A^2 C \cos^2 \omega t \cos 3\omega t - 3AC^2 \cos \omega t \cos^2 3\omega t \qquad (13.77)
\end{aligned}$$

By using the trigonometric relations

$$\left. \begin{aligned}
\cos^2 \theta &= \tfrac{1}{2} + \tfrac{1}{2} \cos 2\theta \\
\cos^3 \theta &= \tfrac{3}{4} \cos \theta + \tfrac{1}{4} \cos 3\theta \\
\cos \theta \cos \phi &= \tfrac{1}{2} \cos(\theta - \phi) + \tfrac{1}{2} \cos(\theta + \phi)
\end{aligned} \right\} \qquad (13.78)$$

Eq. (13.77) can be expressed as

$$\begin{aligned}
\ddot{x}_1 + \omega^2 x_1 = {} & A \left(\omega_1 - \frac{3}{4} A^2 - \frac{3}{2} C^2 - \frac{3}{4} AC \right) \cos \omega t \\
& + \left(\omega_1 C - \frac{A^3}{4} - \frac{3}{4} C^3 - \frac{3}{2} A^2 C \right) \cos 3\omega t \\
& - \frac{3}{4} AC(A + C) \cos 5\omega t - \frac{3AC^2}{4} \cos 7\omega t - \frac{C^3}{4} \cos 9\omega t \qquad (13.79)
\end{aligned}$$

The condition to avoid a secular term in the solution is that the coefficient of $\cos \omega t$ in Eq. (13.79) must be zero. Since $A \neq 0$ in order to have a subharmonic response,

$$\omega_1 = \tfrac{3}{4}(A^2 + AC + 2C^2) \qquad (13.80)$$

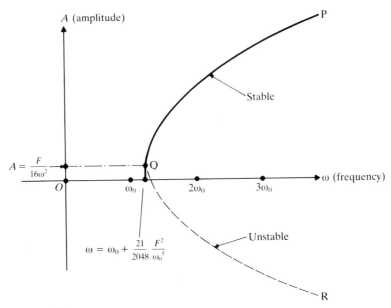

Figure 13.7

Equations (13.80) and (13.75) give

$$\omega_1 = \frac{3}{4}\left(A^2 - \frac{AF}{8\omega^2} + \frac{2F^2}{64\omega^4} \right) \tag{13.81}$$

Substituting Eq. (13.81) into Eq. (13.69) and rearranging the terms, we obtain the equation to be satisfied by A and ω in order to have subharmonic oscillation as

$$\omega^6 - \omega_0^2\omega^4 - \frac{3\alpha}{256}(64A^2\omega^4 - 8AF\omega^2 + 2F^2) = 0 \tag{13.82}$$

Equation (13.82) can be seen to be a cubic equation in ω^2 and a quadratic in A. The relationship between the amplitude (A) and the subharmonic frequency (ω), given by Eq. (13.82), is shown graphically in Fig. 13.7. It has been observed that the curve PQ, where the slope is positive, represents stable solutions while the curve QR, where the slope is negative, denotes unstable solutions [13.4, 13.6]. The minimum value of amplitude for the existence of stable subharmonic oscillations can be found by setting $d\omega^2/dA = 0$ as $A = (F/16\omega^2)$.[†]

[†] Equation (13.82) can be rewritten as

$$(\omega^2)^3 - \omega_0^2(\omega^2)^2 - \frac{3\alpha}{4}A^2(\omega^2)^2 + \frac{3\alpha F}{32}A(\omega^2) - \frac{3\alpha F^2}{128} = 0$$

which, upon differentiation, gives

$$3(\omega^2)^2\, d\omega^2 - 2\omega_0^2\omega^2\, d\omega^2 - \frac{3\alpha}{4}(2A\, dA)(\omega^2)^2 - \frac{3\alpha}{2}A^2\omega^2\, d\omega^2 + \frac{3\alpha F}{32}\omega^2\, dA + \frac{3\alpha F}{32}A\, d\omega^2 = 0$$

By setting $d\omega^2/dA = 0$, we obtain $A = (F/16\omega^2)$.

**13.5.2
Superharmonic
Solution**

Consider the undamped Duffing's equation

$$\ddot{x} + \omega_0^2 x + \alpha x^3 = F \cos \omega t \tag{13.83}$$

The solution of this equation is assumed as

$$x(t) = A \cos \omega t + C \cos 3\omega t \tag{13.84}$$

where the amplitudes of the harmonic and superharmonic components, A and C, are to be determined. The substitution of Eq. (13.84) into Eq. (13.83) gives, with the use of the trigonometric relations of Eq. (13.78),

$$\cos \omega t \left[-\omega^2 A + \omega_0^2 A + \tfrac{3}{4}\alpha A^3 + \tfrac{3}{4}\alpha A^2 C + \tfrac{3}{2}\alpha A C^2 \right]$$
$$+ \cos 3\omega t \left[-9\omega^2 C + \omega_0^2 C + \tfrac{1}{4}\alpha A^3 + \tfrac{3}{4}\alpha C^3 + \tfrac{3}{2}\alpha A^2 C \right]$$
$$+ \cos 5\omega t \left[\tfrac{3}{4}\alpha A^2 C + \tfrac{3}{4}\alpha A C^2 \right] + \cos 7\omega t \left[\tfrac{3}{4}\alpha A C^2 \right]$$
$$+ \cos 9\omega t \left[\tfrac{1}{4}\alpha C^3 \right] = F \cos \omega t \tag{13.85}$$

Neglecting the terms involving $\cos 5\omega t$, $\cos 7\omega t$, and $\cos 9\omega t$, and equating the coefficients of $\cos \omega t$ and $\cos 3\omega t$ on both sides of Eq. (13.85), we obtain

$$\omega_0^2 A - \omega^2 A + \tfrac{3}{4}\alpha A^3 + \tfrac{3}{4}\alpha A^2 C + \tfrac{3}{2}\alpha A C^2 = F \tag{13.86}$$

$$\omega_0^2 C - 9\omega^2 C + \tfrac{1}{4}\alpha A^3 + \tfrac{3}{4}\alpha C^3 + \tfrac{3}{2}\alpha A^2 C = 0 \tag{13.87}$$

Equations (13.86) and (13.87) represent a set of simultaneous nonlinear equations that can be solved numerically for A and C.

As a particular case, if C is assumed to be small compared to A, the terms involving C^2 and C^3 can be neglected and Eq. (13.87) gives

$$C \approx \frac{-\tfrac{1}{4}\alpha A^3}{\left(\tfrac{3}{2}\alpha A^2 + \omega_0^2 - 9\omega^2 \right)} \tag{13.88}$$

and Eq. (13.86) gives

$$C \approx \frac{F - \omega_0^2 A + \omega^2 A - \tfrac{3}{4}\alpha A^3}{\tfrac{3}{4}\alpha A^2} \tag{13.89}$$

Equating C from Eqs. (13.88) and (13.89) leads to

$$\left(-\tfrac{1}{4}\alpha A^3 \right)\left(\tfrac{3}{4}\alpha A^2 \right) = \left(\tfrac{3}{2}\alpha A^2 + \omega_0^2 - 9\omega^2 \right)\left(F - \omega_0^2 A + \omega^2 A - \tfrac{3}{4}\alpha A^3 \right) \tag{13.90}$$

which can be rewritten as

$$-A^5 \left(\tfrac{15}{16}\alpha^2 \right) + A^3 \left(\tfrac{33}{4}\alpha\omega^2 - \tfrac{9}{4}\alpha\omega_0^2 \right) + A^2 \left(\tfrac{3}{2}\alpha F \right)$$
$$+ A \left(10\omega^2\omega_0^2 - 9\omega^4 - \omega_0^4 \right) + \left(\omega_0^2 F - 9\omega^2 F \right) = 0 \tag{13.91}$$

Equation (13.88), in conjunction with Eq. (13.91), gives the relationship between the amplitude of superharmonic oscillations (C) and the corresponding frequency (3ω).

13.6 SYSTEMS WITH TIME-DEPENDENT COEFFICIENTS (MATHIEU EQUATION)

Consider the simple pendulum shown in Fig. 13.8(a). The pivot point of the pendulum is made to vibrate in the vertical direction as

$$y(t) = Y \cos \omega t \tag{13.92}$$

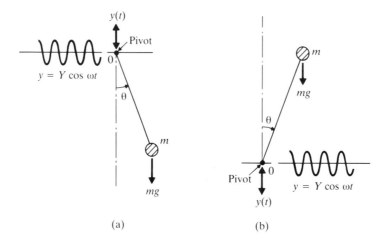

Figure 13.8

where Y is the amplitude and ω is the frequency of oscillation. Since the entire pendulum accelerates in the vertical direction, the net acceleration is given by $g - \ddot{y}(t) = g - \omega^2 Y \cos \omega t$. The equation of motion of the pendulum can be derived as

$$ml^2\ddot{\theta} + m(g - \ddot{y})l \sin \theta = 0 \tag{13.93}$$

For small deflections near $\theta = 0$, $\sin \theta \approx \theta$ and Eq. (13.93) reduces to

$$\ddot{\theta} + \left(\frac{g}{l} - \frac{\omega^2 Y}{l} \cos \omega t \right) \theta = 0 \tag{13.94}$$

If the pendulum is inverted as shown in Fig. 13.8(b), the equation of motion becomes

$$ml^2\ddot{\theta} - mgl \sin \theta = 0$$

or

$$\ddot{\theta} - \frac{g}{l} \sin \theta = 0 \tag{13.95}$$

where θ is the angle measured from the vertical (unstable equilibrium) point. If the pivot point 0 vibrates as $y(t) = Y \cos \omega t$, the equation of motion becomes

$$\ddot{\theta} + \left(-\frac{g}{l} + \frac{\omega^2 Y}{l} \cos \omega t \right) \sin \theta = 0 \tag{13.96}$$

For small angular displacements around $\theta = 0$, Eq. (13.96) reduces to

$$\ddot{\theta} + \left(-\frac{g}{l} + \frac{\omega^2 Y}{l} \cos \omega t \right) \theta = 0 \tag{13.97}$$

Equations (13.94) and (13.97) are particular forms of an equation called the *Mathieu equation* for which the coefficient of θ in the differential equation varies with time (nonautonomous equation). We shall study the periodic solutions and their stability characteristics of the system for small values of Y in this section.

Periodic Solutions Using Perturbation Method [13.7]. Consider the Mathieu equation in the form

$$\frac{d^2y}{dt^2} + (a + \epsilon \cos t)y = 0 \tag{13.98}$$

where ϵ is assumed to be small. We approximate the solution of Eq. (13.98) as

$$y(t) = y_0(t) + \epsilon y_1(t) + \epsilon^2 y_2(t) + \cdots \tag{13.99}$$

$$a = a_0 + \epsilon a_1 + \epsilon^2 a_2 + \cdots \tag{13.100}$$

where $a_0, a_1, \ldots$ are constants. Since the periodic coefficient $\cos t$ in Eq. (13.98) varies with a period of 2π, it was found that only two types of solutions exist—one with period 2π and the other with period 4π [13.7, 13.28]. Thus we seek the functions $y_0(t), y_1(t), \ldots$ in Eq. (13.99) in such a way that $y(t)$ is a solution of Eq. (13.98) with period 2π or 4π. Substituting Eqs. (13.99) and (13.100) into Eq. (13.98) results in

$$(\ddot{y}_0 + a_0 y_0) + \epsilon(\ddot{y}_1 + a_1 y_0 + y_0 \cos t + a_0 y_1)$$
$$+ \epsilon^2(\ddot{y}_2 + a_2 y_0 + a_1 y_1 + y_1 \cos t + a_0 y_2) + \cdots = 0 \tag{13.101}$$

where $\ddot{y}_i = d^2y_i/dt^2$, $i = 0, 1, 2, \ldots$. Setting the coefficients of various powers of ϵ in Eq. (13.101) equal to zero, we obtain

$$\epsilon^0: \quad \ddot{y}_0 + a_0 y_0 = 0 \tag{13.102}$$

$$\epsilon^1: \quad \ddot{y}_1 + a_0 y_1 + a_1 y_0 + y_0 \cos t = 0 \tag{13.103}$$

$$\epsilon^2: \quad \ddot{y}_2 + a_0 y_2 + a_2 y_0 + a_1 y_1 + y_1 \cos t = 0 \tag{13.104}$$
$$\vdots$$

where each of the functions y_i is required to have a period of 2π or 4π. The solution of Eq. (13.102) can be expressed as

$$y_0(t) = \begin{cases} \cos\sqrt{a_0}\,t \\ \sin\sqrt{a_0}\,t \end{cases} \equiv \begin{cases} \cos\dfrac{n}{2}t \\ \sin\dfrac{n}{2}t \end{cases}, \quad n = 0, 1, 2, \ldots \tag{13.105}$$

and

$$a_0 = \frac{n^2}{4}, \quad n = 0, 1, 2, \ldots$$

Now, we consider the following specific values of n.

When $n = 0$: Equation (13.105) gives $a_0 = 0$ and $y_0 = 1$ and Eq. (13.103) yields

$$\ddot{y}_1 + a_1 + \cos t = 0 \quad \text{or} \quad \ddot{y}_1 = -a_1 - \cos t \tag{13.106}$$

In order to have y_1 as a periodic function, a_1 must be zero. When Eq. (13.106) is integrated twice, the resulting periodic solution can be expressed as

$$y_1(t) = \cos t + \alpha \tag{13.107}$$

where α is a constant. With the known values of $a_0 = 0$, $a_1 = 0$, $y_0 = 1$ and $y_1 = \cos t + \alpha$, Eq. (13.104) can be rewritten as

$$\ddot{y}_2 + a_2 + (\cos t + \alpha)\cos t = 0$$

or

$$\ddot{y}_2 = -\tfrac{1}{2} - a_2 - \alpha \cos t - \tfrac{1}{2}\cos 2t \tag{13.108}$$

In order to have y_2 as a periodic function, $(-\tfrac{1}{2} - a_2)$ must be zero (i.e., $a_2 = -\tfrac{1}{2}$). Thus, for $n = 0$, Eq. (13.100) gives

$$a = -\tfrac{1}{2}\epsilon^2 + \cdots \tag{13.109}$$

When n = 1: For this case, Eq. (13.105) gives $a_0 = \tfrac{1}{4}$ and $y_0 = \cos(t/2)$ or $\sin(t/2)$. With $y_0 = \cos(t/2)$, Eq. (13.103) gives

$$\ddot{y}_1 + \frac{1}{4}y_1 = \left(-a_1 - \frac{1}{2}\right)\cos\frac{t}{2} - \frac{1}{2}\cos\frac{3t}{2} \tag{13.110}$$

The homogeneous solution of Eq. (13.110) is given by

$$y_1(t) = A_1 \cos\frac{t}{2} + A_2 \sin\frac{t}{2}$$

where A_1 and A_2 are constants of integration. Since the term involving $\cos(t/2)$ appears in the homogeneous solution as well as the forcing function, the particular solution will contain a term of the form $t\cos(t/2)$, which is not periodic. Thus the coefficient of $\cos(t/2)$, namely $(-a_1 - 1/2)$, must be zero in the forcing function to ensure periodicity of $y_1(t)$. This gives $a_1 = -1/2$ and Eq. (13.110) becomes

$$\ddot{y}_1 + \frac{1}{4}y_1 = -\frac{1}{2}\cos\frac{3t}{2}. \tag{13.111}$$

By substituting the particular solution $y_1(t) = A_2 \cos(3t/2)$ into Eq. (13.111), we obtain $A_2 = \tfrac{1}{4}$, and hence $y_1(t) = \tfrac{1}{4}\cos(3t/2)$. Using $a_0 = \tfrac{1}{4}$, $a_1 = -\tfrac{1}{2}$ and $y_1 = \tfrac{1}{4}\cos(3t/2)$, Eq. (13.104) can be expressed as

$$\ddot{y}_2 + \frac{1}{4}y_2 = -a_2\cos\frac{t}{2} + \frac{1}{2}\left(\frac{1}{4}\cos\frac{3t}{2}\right) - \left(\frac{1}{4}\cos\frac{3t}{2}\right)\cos t$$

$$= \left(-a_2 - \frac{1}{8}\right)\cos\frac{t}{2} + \frac{1}{8}\cos\frac{3t}{2} - \frac{1}{8}\cos\frac{5t}{2} \tag{13.112}$$

Again, since the homogeneous solution of Eq. (13.112) contains the term $\cos(t/2)$, the coefficient of the term $\cos(t/2)$ on the right hand side of Eq. (13.112) must be zero. This leads to $a_2 = -\tfrac{1}{8}$ and hence Eq. (13.100) becomes

$$a = \frac{1}{4} - \frac{\epsilon}{2} - \frac{\epsilon^2}{8} + \cdots \tag{13.113a}$$

Similarly, by starting with the solution $y_0 = \sin(t/2)$, we obtain the relation (see Problem 13.13)

$$a = \frac{1}{4} + \frac{\epsilon}{2} - \frac{\epsilon^2}{8} + \cdots \tag{13.113b}$$

When n = 2: Equation (13.105) gives $a_0 = 1$ and $y_0 = \cos t$ or $\sin t$. With $a_0 = 1$ and $y_0 = \cos t$, Eq. (13.103) can be written as

$$\ddot{y}_1 + y_1 = -a_1 \cos t - \frac{1}{2} - \frac{1}{2}\cos 2t \tag{13.114}$$

Since $\cos t$ is a solution of the homogeneous equation, the term involving $\cos t$ in Eq. (13.114) gives rise to $t \cos t$ in the solution of y_1. Thus, to impose periodicity of y_1, we set $a_1 = 0$. With this, the particular solution of $y_1(t)$ can be assumed as $y_1(t) = A_3 + B_3 \cos 2t$. When this solution is substituted into Eq. (13.114), we obtain $A_3 = -\frac{1}{2}$ and $B_3 = \frac{1}{6}$. Thus Eq. (13.104) becomes

$$\ddot{y}_2 + y_2 + a_2 \cos t + y_1 \cos t = 0$$

or

$$\ddot{y}_2 + y_2 = -a_2 \cos t - \left(-\tfrac{1}{2} + \tfrac{1}{6}\cos 2t \right)\cos t$$
$$= \cos t \left(-a_1 + \tfrac{1}{2} - \tfrac{1}{12} \right) + \tfrac{1}{2}\cos 3t \qquad (13.115)$$

For periodicity of $y_2(t)$, we set the coefficient of $\cos t$ equal to zero in the forcing function of Eq. (13.115). This gives $a_2 = \frac{5}{12}$, and hence

$$a = 1 + \tfrac{5}{12}\epsilon^2 + \cdots \qquad (13.116a)$$

Similarly, by proceeding with $y_0 = \sin t$, we obtain (see Problem 13.13)

$$a = 1 - \tfrac{1}{12}\epsilon^2 + \cdots \qquad (13.116b)$$

To observe the stability characteristics of the system, Eqs. (13.109), (13.113), and (13.116) are plotted in the (a, ϵ) plane as indicated in Fig. 13.9. These equations

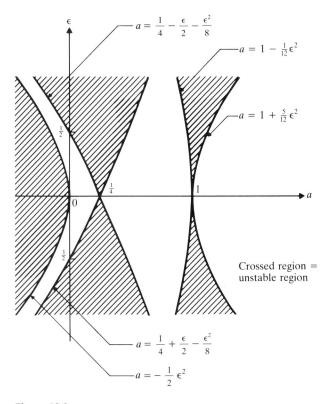

Figure 13.9

represent curves that are known as the *boundary* or *transition curves* that divide the (a, ϵ) plane into regions of stability and instability. These boundary curves are such that a point belonging to any one curve represents a periodic solution of Eq. (13.98). The stability of these periodic solutions can be investigated [13.7, 13.25, 13.28]. In Fig. 13.9, the points inside the shaded region denote unstable motion. It can be noticed from Fig. 13.9 that stability is also possible for negative values of a, which correspond to the equilibrium position $\theta = 180°$. Thus with the right choice of the parameters, the pendulum can be made to be stable in the upright position by moving its support harmonically.

13.7 GRAPHICAL METHODS

**13.7.1
Phase Plane
Representation**

Graphical methods can be used to obtain qualitative information about the behavior of the nonlinear system and also to integrate the equations of motion. We shall first consider a basic concept known as the *phase plane*. For a single degree of freedom system, two parameters are needed to describe the state of motion completely. These parameters are usually taken as the displacement and velocity of the system. When the parameters are used as coordinate axes, the resulting graphical representation of the motion is called the *phase plane representation*. Thus each point in the phase plane represents a possible state of the system. As time changes, the state of the system changes. A typical or representative point in the phase plane (such as the point representing the state of the system at time $t = 0$) moves and traces a curve known as the *trajectory*. The trajectory shows how the solution of the system varies with time.

EXAMPLE 13.2   Trajectories of a Simple Harmonic Oscillator

Find the trajectories of a simple harmonic oscillator.

Given: Simple harmonic oscillator.

Find: Trajectories.

Approach: Find the equation of the trajectories.

Solution. The equation of motion of an undamped linear system is given by

$$\ddot{x} + \omega_n^2 x = 0 \qquad \text{(E.1)}$$

By setting $y = \dot{x}$, Eq. (E.1) can be written as

$$\frac{dy}{dt} = -\omega_n^2 x$$
$$\frac{dx}{dt} = y \qquad \text{(E.2)}$$

from which we can obtain

$$\frac{dy}{dx} = -\frac{\omega_n^2 x}{y} \qquad \text{(E.3)}$$

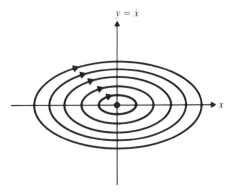

Figure 13.10

Integration of Eq. (E.3) leads to

$$y^2 + \omega_n^2 x^2 = c^2 \tag{E.4}$$

where c is a constant. The value of c is determined by the initial conditions of the system. Equation (E.4) shows that the trajectories of the system in the phase plane (x-y plane) are a family of ellipses, as shown in Fig. 13.10. It can be observed that the point ($x = 0$, $y = 0$) is surrounded by closed trajectories. Such a point is called a *center*. The direction of motion of the trajectories can be determined from Eq. (E.2). For instance, if $x > 0$ and $y > 0$, Eq. (E.2) implies that $dx/dt > 0$ and $dy/dt < 0$; therefore, the motion is clockwise.

EXAMPLE 13.3   **Phase-plane of an Undamped Pendulum**

Find the trajectories of an undamped pendulum.

Given: Undamped pendulum.

Find: Trajectories.

Approach: Find the equation of the trajectories.

Solution. The equation of motion is given by Eq. (13.1):

$$\ddot{\theta} = -\omega_0^2 \sin \theta \tag{E.1}$$

where $\omega_0^2 = g/l$. Introducing $x = \theta$ and $y = \dot{x} = \dot{\theta}$, Eq. (E.1) can be rewritten as

$$\frac{dx}{dt} = y, \qquad \frac{dy}{dt} = -\omega_0^2 \sin x$$

or

$$\frac{dy}{dx} = -\frac{\omega_0^2 \sin x}{y}$$

or

$$y \, dy = -\omega_0^2 \sin x \, dx \tag{E.2}$$

Integrating Eq. (E.2) and using the condition that $\dot{x} = 0$ when $x = x_0$ (at the end of the swing), we obtain

$$y^2 = 2\omega_0^2 (\cos x - \cos x_0) \tag{E.3}$$

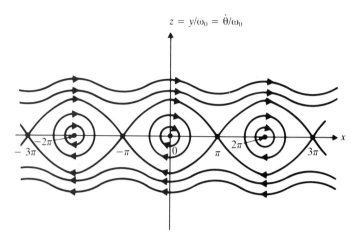

$$z = y/\omega_0 = \dot{\theta}/\omega_0$$

Figure 13.11

Introducing $z = y/\omega_0$ Eq. (E.3) can be expressed as

$$z^2 = 2(\cos x - \cos x_0) \tag{E.4}$$

The trajectories given by Eq. (E.4) are shown in Fig. 13.11.

EXAMPLE 13.4   **Phase-plane of an Undamped Nonlinear System**

Find the trajectories of a nonlinear spring-mass system governed by the equation

$$\ddot{x} + \omega_0^2(x - 2\alpha x^3) = 0 \tag{E.1}$$

Given: Free undamped nonlinear spring-mass system.

Find: Trajectories.

Approach: Find the equation of the trajectories.

Solution. The nonlinear pendulum equation can be considered as a special case of Eq. (E.1). To see this, we use the approximation $\sin \theta \simeq \theta - \theta^3/6$ in the neighborhood of $\theta = 0$ in Eq. (E.1) of Example 13.3 to obtain

$$\ddot{\theta} + \omega_0^2\left(\theta - \frac{\theta^3}{6}\right) = 0$$

which can be seen to be a special case of Eq. (E.1). Equation (E.1) can be rewritten as

$$\frac{dx}{dt} = y, \qquad \frac{dy}{dt} = -\omega_0^2(x - 2\alpha x^3)$$

or

$$\frac{dy}{dx} = -\frac{\omega_0^2(x - 2\alpha x^3)}{y}$$

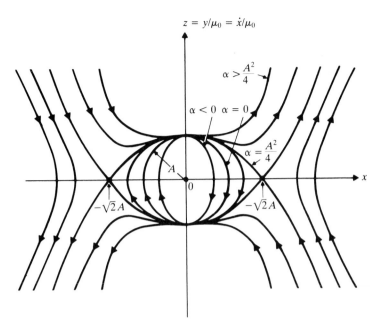

$$z = y/\mu_0 = \dot{x}/\mu_0$$

Figure 13.12

or

$$y \, dy = -\omega_0^2 (x - 2\alpha x^3) \, dx \tag{E.2}$$

Integration of Eq. (E.2), with the condition $\dot{x} = 0$ when $x = x_0$ (at the end of the swing in the case of a pendulum), gives

$$z^2 + x^2 - \alpha x^4 = A^2 \tag{E.3}$$

where $z = y/\omega_0$ and $A^2 = x_0^2(1 - \alpha x_0^2)$ is a constant. The trajectories, in the phase-plane, given by Eq. (E.3), are shown in Fig. 13.12 for several values of α.

It can be observed that for $\alpha = 0$, Eq. (E.3) denotes a circle of radius A and corresponds to a simple harmonic motion. When $\alpha < 0$, Eq. (E.3) represents ovals within the circle given by $\alpha = 0$ and the ovals touch the circle at the points $(0, \pm A)$. When $\alpha = (1/4A^2)$, Eq. (E.3) becomes

$$y^2 + x^2 - \frac{x^4}{4A^2} - A^2 = \left[y - \left(A - \frac{x^2}{2A} \right) \right]\left[y + \left(A - \frac{x^2}{2A} \right) \right] = 0 \tag{E.4}$$

Equation (E.4) indicates that the trajectories are given by the parabolas

$$y = \pm \left(A - \frac{x^2}{2A} \right) \tag{E.5}$$

These two parabolas intersect at the points ($x = \pm \sqrt{2} A$, $y = 0$), which correspond to points of unstable equilibrium.

When $(1/4A^2) \geqslant \alpha \geqslant 0$, the trajectories given by Eq. (E.3) will be closed ovals lying between the circle given by $\alpha = 0$ and the two parabolas given by $\alpha = (1/4A^2)$. Since these trajectories are closed curves, they represent periodic vibrations. When $\alpha > (1/4A^2)$, the trajectories given by Eq. (E.3) lie outside the region between the parabolas and extend to infinity. These trajectories correspond to the conditions that permit the body to escape from the center of force.

To see some of the characteristics of trajectories, consider a single degree of freedom nonlinear oscillatory system whose governing equation is of the form

$$\ddot{x} + f(x, \dot{x}) = 0 \tag{13.117}$$

By defining

$$\frac{dx}{dt} = \dot{x} = y \tag{13.118}$$

and

$$\frac{dy}{dt} = \dot{y} = -f(x, y) \tag{13.119}$$

we obtain

$$\frac{dy}{dx} = \frac{(dy/dt)}{(dx/dt)} = -\frac{f(x, y)}{y} = \phi(x, y), \text{ say.} \tag{13.120}$$

Thus there is a unique slope of the trajectory at every point (x, y) in the phase plane, provided that $\phi(x, y)$ is not indeterminate. If $y = 0$ and $f \neq 0$ (that is, if the point lies on the x axis), the slope of the trajectory is infinite. This means that all trajectories must cross the x axis at right angles. If $y = 0$ and $f = 0$, the point is called a *singular point*, and the slope is indeterminate at such points. A singular point corresponds to a state of equilibrium of the system—the velocity $y = \dot{x}$ and the force $\ddot{x} = -f$ are zero at a singular point. Further investigation is necessary to establish whether the equilibrium represented by a singular point is stable or unstable.

13.7.2
Phase Velocity

The velocity $\vec{v}$ with which a representative point moves along a trajectory is called the *phase velocity*. The components of phase velocity parallel to the x and y axes are

$$v_x = \dot{x}, \qquad v_y = \dot{y} \tag{13.121}$$

and the magnitude of $\vec{v}$ is given by

$$|\vec{v}| = \sqrt{v_x^2 + v_y^2} = \sqrt{\left(\frac{dx}{dt}\right)^2 + \left(\frac{dy}{dt}\right)^2} \tag{13.122}$$

We can note that if the system has a periodic motion, its trajectory in the phase plane is a closed curve. This follows from the fact that the representative point, having started its motion along a closed trajectory at an arbitrary point (x, y), will return to the same point after one period. The time required to go around the closed trajectory (the period of oscillation of the system) is finite because the phase velocity is bounded away from zero at all points of the trajectory.

**13.7.3
Method of
Constructing
Trajectories**

We shall now consider a method known as the *method of isoclines* for constructing the trajectories of a dynamical system with one degree of freedom. By writing the equations of motion of the system as

$$\frac{dx}{dt} = f_1(x, y)$$
$$\frac{dy}{dt} = f_2(x, y) \tag{13.123}$$

where f_1 and f_2 are nonlinear functions of x and $y = \dot{x}$, the equation for the integral curves can be obtained as

$$\frac{dy}{dx} = \frac{f_2(x, y)}{f_1(x, y)} = \phi(x, y), \text{ say.} \tag{13.124}$$

The curve

$$\phi(x, y) = c \tag{13.125}$$

for a fixed value of c is called an *isocline*. An isocline can be defined as the locus of points at which the trajectories passing through them have the constant slope c. In the method of isoclines we fix the slope $(dy)/(dx)$ by giving it a definite number c_1 and solve Eq. (13.125) for the trajectory. The curve $\phi(x, y) - c_1 = 0$ thus represents an isocline in the phase plane. We plot several isoclines by giving different values $c_1, c_2, \ldots$ to the slope $\phi = (dy)/(dx)$. Let $h_1, h_2, \ldots$ denote these isoclines in Fig. 13.13(a). Suppose that we are interested in constructing the trajectory passing through the point R_1 on the isocline h_1. We draw two straight line segments from R_1: one with a slope c_1, meeting h_2 at R_2', and the other with a slope c_2 meeting h_2 at R_2''. The middle point between R_2' and R_2'' lying on h_2 is denoted as R_2. Starting at R_2, this construction is repeated, and the point R_3 is determined on h_3. This procedure is continued until the polygonal trajectory with sides $R_1 R_2$, $R_2 R_3$, $R_3 R_4, \ldots$ is taken as an approximation to the actual trajectory passing through the point R_1. Obviously, the larger the number of isoclines, the better is the approximation obtained by this graphical method. A typical final trajectory looks like that in Fig. 13.13(b).

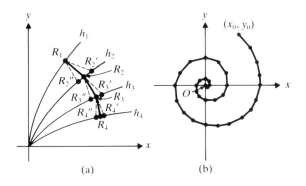

(a) (b)

Figure 13.13 Method of isoclines.

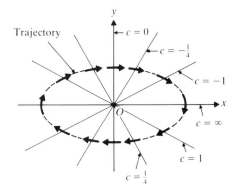

Figure 13.14 Isoclines of a simple harmonic oscillator.

EXAMPLE 13.5   Trajectories Using the Method of Isoclines

Construct the trajectories of a simple harmonic oscillator by the method of isoclines.

Given: Simple harmonic oscillator.

Find: Trajectories.

Approach: Use the method of isoclines.

Solution. The differential equation defining the trajectories of a simple harmonic oscillator is given by Eq. (E.3) of Example 13.2. Hence the family of isoclines is given by

$$c = -\frac{\omega_n^2 x}{y} \qquad \text{or} \qquad y = \frac{-\omega_n^2}{c} x \qquad (E.1)$$

This equation represents a family of straight lines passing through the origin, with c representing the slope of the trajectories on each isocline. The isoclines given by Eq. (E.1) are shown in Fig. 13.14. Once the isoclines are known, the trajectory can be plotted as indicated above.

**13.7.4
Obtaining Time
Solution from
Phase Plane
Trajectories**

The trajectory plotted in the phase plane is a plot of $\dot{x}$ as a function of x, and time (t) does not appear explicitly in the plot. For the qualitative analysis of the system, the trajectories are enough, but in some cases we may need the variation of x with time t. In such cases, it is possible to obtain the time solution $x(t)$ from the phase plane diagram, although the original differential equation cannot be solved for x and $\dot{x}$ as functions of time. The method of obtaining a time solution is essentially a step-by-step procedure; several different schemes may be used for this purpose. In this section, we shall present a method based on the relation $\dot{x} = (\Delta x)/(\Delta t)$.

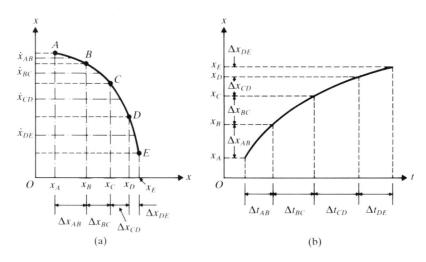

Figure 13.15   Obtaining time solution from a phase plane plot.

For small increments of displacement and time (Δx and Δt), the average velocity can be taken as $\dot{x}_{av} = (\Delta x)/(\Delta t)$, so that

$$\Delta t = \frac{\Delta x}{\dot{x}_{av}} \tag{13.126}$$

In the phase plane trajectory shown in Fig. 13.15, the incremental time needed for the representative point to traverse the incremental displacement Δx_{AB} is shown as Δt_{AB}. If $\dot{x}_{AB}$ denotes the average velocity during Δt_{AB}, we have $\Delta t_{AB} = \Delta x_{AB}/\dot{x}_{AB}$. Similarly, $\Delta t_{BC} = \Delta x_{BC}/\dot{x}_{BC}$, etc. Once $\Delta t_{AB}, \Delta t_{BC}, \ldots$ are known, the time solution $x(t)$ can be plotted easily, as shown in Fig. 13.15(b). It is evident that for good accuracy, the incremental displacements $\Delta x_{AB}, \Delta x_{BC}, \ldots$ must be chosen small enough that the corresponding incremental changes in $\dot{x}$ and t are reasonably small. Note that Δx need not be constant; it can be changed depending on the nature of the trajectories.

13.8 STABILITY OF EQUILIBRIUM STATES

**13.8.1
Stability Analysis**

Consider a single degree of freedom nonlinear vibratory system described by two first order differential equations

$$\frac{dx}{dt} = f_1(x, y)$$
$$\frac{dy}{dt} = f_2(x, y) \tag{13.127}$$

where f_1 and f_2 are nonlinear functions of x and $y = \dot{x} = dx/dt$. For this system, the slope of the trajectories in the phase plane is given by

$$\frac{dy}{dx} = \frac{\dot{y}}{\dot{x}} = \frac{f_2(x, y)}{f_1(x, y)} \tag{13.128}$$

Let (x_0, y_0) be a singular point or an equilibrium point so that $(dy)/(dx)$ has the form $0/0$:

$$f_1(x_0, y_0) = f_2(x_0, y_0) = 0 \tag{13.129}$$

A study of Eqs. (13.127) in the neighborhood of the singular point provides us with answers as to the stability of equilibrium. We first note that there is no loss of generality if we assume that the singular point is located at the origin $(0, 0)$. This is because the slope $(dy)/(dx)$ of the trajectories does not vary with a translation of the coordinate axes x and y to x' and y':

$$x' = x - x_0$$
$$y' = y - y_0$$
$$\frac{dy}{dx} = \frac{dy'}{dx'} \tag{13.130}$$

Thus we assume that $(x = 0, y = 0)$ is a singular point, so that

$$f_1(0, 0) = f_2(0, 0) = 0$$

If f_1 and f_2 are expanded in terms of Taylor's series about the singular point $(0, 0)$, we obtain

$$\dot{x} = f_1(x, y) = a_{11}x + a_{12}y + \text{higher order terms}$$
$$\dot{y} = f_2(x, y) = a_{21}x + a_{22}y + \text{higher order terms} \tag{13.131}$$

where

$$a_{11} = \frac{\partial f_1}{\partial x}\bigg|_{(0,0)}, \quad a_{12} = \frac{\partial f_1}{\partial y}\bigg|_{(0,0)}, \quad a_{21} = \frac{\partial f_2}{\partial x}\bigg|_{(0,0)}, \quad a_{22} = \frac{\partial f_2}{\partial y}\bigg|_{(0,0)}$$

In the neighborhood of $(0, 0)$, x and y are small; f_1 and f_2 can be approximated by linear terms only, so that Eqs. (13.131) can be written as

$$\left\{ \begin{array}{c} \dot{x} \\ \dot{y} \end{array} \right\} = \begin{bmatrix} a_{11} & a_{12} \\ a_{21} & a_{22} \end{bmatrix} \left\{ \begin{array}{c} x \\ y \end{array} \right\} \tag{13.132}$$

The solutions of Eq. (13.132) are expected to be geometrically similar to those of Eq. (13.127). We assume the solution of Eq. (13.132) in the form

$$\left\{ \begin{array}{c} x \\ y \end{array} \right\} = \left\{ \begin{array}{c} X \\ Y \end{array} \right\} e^{\lambda t} \tag{13.133}$$

where X, Y, and λ are constants. Substitution of Eq. (13.133) into Eq. (13.132) leads to the eigenvalue problem

$$\begin{bmatrix} a_{11} - \lambda & a_{12} \\ a_{21} & a_{22} - \lambda \end{bmatrix} \left\{ \begin{array}{c} X \\ Y \end{array} \right\} = \left\{ \begin{array}{c} 0 \\ 0 \end{array} \right\} \tag{13.134}$$

The eigenvalues λ_1 and λ_2 can be found by solving the characteristic equation

$$\begin{vmatrix} a_{11} - \lambda & a_{12} \\ a_{21} & a_{22} - \lambda \end{vmatrix} = 0$$

as

$$\lambda_1, \lambda_2 = \tfrac{1}{2}\left(p \pm \sqrt{p^2 - 4q} \right) \tag{13.135}$$

where $p = a_{11} + a_{22}$ and $q = a_{11}a_{22} - a_{12}a_{21}$. If

$$\left\{ \begin{array}{c} X_1 \\ Y_1 \end{array} \right\} \quad \text{and} \quad \left\{ \begin{array}{c} X_2 \\ Y_2 \end{array} \right\}$$

denote the eigenvectors corresponding to λ_1 and λ_2, respectively, the general solution of Eqs. (13.127) can be expressed as (assuming $\lambda_1 \neq 0$, $\lambda_2 \neq 0$, and $\lambda_1 \neq \lambda_2$):

$$\left\{ \begin{array}{c} x \\ y \end{array} \right\} = C_1 \left\{ \begin{array}{c} X_1 \\ Y_1 \end{array} \right\} e^{\lambda_1 t} + C_2 \left\{ \begin{array}{c} X_2 \\ Y_2 \end{array} \right\} e^{\lambda_2 t} \tag{13.136}$$

where C_1 and C_2 are arbitrary constants. We can notice that

if $(p^2 - 4q) < 0$, the motion is oscillatory;

if $(p^2 - 4q) > 0$, the motion is aperiodic;

if $p > 0$, the system is unstable;

if $p < 0$, the system is stable.

If we use the transformation

$$\begin{Bmatrix} x \\ y \end{Bmatrix} = \begin{bmatrix} X_1 & X_2 \\ Y_1 & Y_2 \end{bmatrix} \begin{Bmatrix} \alpha \\ \beta \end{Bmatrix} = [T] \begin{Bmatrix} \alpha \\ \beta \end{Bmatrix}$$

where $[T]$ is the modal matrix and α and β are the generalized coordinates, Eqs. (13.132) will be uncoupled:

$$\begin{Bmatrix} \dot{\alpha} \\ \dot{\beta} \end{Bmatrix} = \begin{bmatrix} \lambda_1 & 0 \\ 0 & \lambda_2 \end{bmatrix} \begin{Bmatrix} \alpha \\ \beta \end{Bmatrix} \qquad \text{or} \qquad \begin{matrix} \dot{\alpha} = \lambda_1 \alpha \\ \dot{\beta} = \lambda_2 \beta \end{matrix} \tag{13.137}$$

The solution of Eqs. (13.137) can be expressed as

$$\alpha(t) = e^{\lambda_1 t}$$
$$\beta(t) = e^{\lambda_2 t} \tag{13.138}$$

13.8.2
Classification of
Singular Points

Depending on the values of λ_1 and λ_2 in Eq. (13.135), the singular or equilibrium points can be classified as follows [13.20, 13.23].

Case (i)—λ_1 and λ_2 Are Real and Distinct ($p^2 > 4q$). In this case, Eq. (13.138) gives

$$\alpha(t) = \alpha_0 e^{\lambda_1 t} \qquad \text{and} \qquad \beta(t) = \beta_0 e^{\lambda_2 t} \tag{13.139}$$

where α_0 and β_0 are the initial values of α and β, respectively. The type of motion depends on whether λ_1 and λ_2 are of the same sign or of opposite signs. If λ_1 and λ_2 have the same sign ($q > 0$) the equilibrium point is called a *node*. The phase plane diagram for the case $\lambda_2 < \lambda_1 < 0$ (when λ_1 and λ_2 are real and negative or $p < 0$) is shown in Fig. 13.16(a). In this case, Eq. (13.139) shows that all the trajectories tend to the origin as $t \to \infty$ and hence the origin is called a *stable node*. On the other hand, if $\lambda_2 > \lambda_1 > 0$ ($p > 0$), the arrow heads change in direction and the origin is called an *unstable node* (see Fig. 13.16b). If λ_1 and λ_2 are real but of opposite signs ($q < 0$ irrespective of the sign of p), one solution tends to the origin while the other tends to infinity. In this case the origin is called a *saddle point* and it corresponds to unstable equilibrium (see Fig. 13.16d).

Case (ii)—λ_1 and λ_2 Are Real and Equal ($p^2 = 4q$). In this case Eq. (13.138) gives

$$\alpha(t) = \alpha_0 e^{\lambda_1 t} \qquad \text{and} \qquad \beta(t) = \beta_0 e^{\lambda_1 t} \tag{13.140}$$

The trajectories will be straight lines passing through the origin and the equilibrium point (origin) will be a *stable node* if $\lambda_1 < 0$ (see Fig. 13.16c) and an *unstable node* if $\lambda_1 > 0$.

Case (iii)—λ_1 and λ_2 Are Complex Conjugates ($p^2 < 4q$). Let $\lambda_1 = \theta_1 + i\theta_2$ and $\lambda_2 = \theta_1 - i\theta_2$, where θ_1 and θ_2 are real. Then Eq. (13.137) gives

$$\dot{\alpha} = (\theta_1 + i\theta_2)\alpha \qquad \text{and} \qquad \dot{\beta} = (\theta_1 - i\theta_2)\beta \tag{13.141}$$

This shows that α and β must also be complex conjugates. We can rewrite Eq. (13.138) as

$$\alpha(t) = \left(\alpha_0 e^{\theta_1 t}\right) e^{i\theta_2 t}, \qquad \beta(t) = \left(\beta_0 e^{\theta_1 t}\right) e^{-i\theta_2 t} \tag{13.142}$$

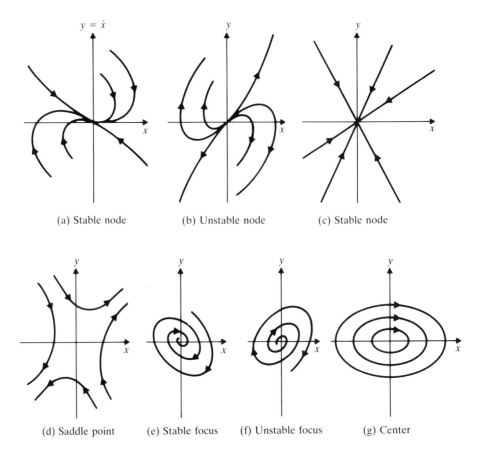

(a) Stable node   (b) Unstable node   (c) Stable node

(d) Saddle point   (e) Stable focus   (f) Unstable focus   (g) Center

Figure 13.16

These equations represent logarithmic spirals. In this case the equilibrium point (origin) is called a *focus* or a *spiral point*. Since the factor $e^{i\theta_2 t}$ in $\alpha(t)$ represents a vector of unit magnitude rotating with angular velocity θ_2 in the complex plane, the magnitude of the complex vector $\alpha(t)$ and hence the stability of motion, is determined by $e^{\theta_1 t}$. If $\theta_1 < 0$, the motion will be asymptotically stable and the focal point will be stable ($p < 0$ and $q > 0$). If $\theta_1 > 0$, the focal point will be unstable ($p > 0$ and $q > 0$). The sign of θ_2 merely gives the direction of rotation of the complex vector, counterclockwise for $\theta_2 > 0$ and clockwise for $\theta_2 < 0$.

If $\theta_1 = 0$ ($p = 0$), the magnitude of the complex radius vector $\alpha(t)$ will be constant and the trajectories reduce to circles with the center as the equilibrium point (origin). The motion will be periodic and hence will be stable. The equilibrium point in this case is called a *center* or *vertex point* and the motion is simply stable and not asymptotically stable. The stable focus, unstable focus and center are shown in Figs. 13.16(e) to (g).

13.9 LIMIT CYCLES

In certain vibration problems involving nonlinear damping, the trajectories, starting either very close to the origin or far away from the origin, tend to a single closed curve, which corresponds to a periodic solution of the system. This means that every solution of the system tends to a periodic solution as $t \to \infty$. The closed curve to which all the solutions approach is called a limit cycle.

For illustration, we consider the following equation, known as the *van der Pol equation*:

$$\ddot{x} - \alpha(1 - x^2)\dot{x} + x = 0 \qquad \alpha > 0 \qquad (13.143)$$

This equation exhibits, mathematically, the essential features of some vibratory systems like certain electrical feedback circuits controlled by valves where there is a source of power that increases with the amplitude of vibration. Van der Pol invented Eq. (13.143) by introducing a type of damping that is negative for small amplitudes but becomes positive for large amplitudes. In this equation, he assumed the damping term to be a multiple of $-(1 - x^2)\dot{x}$ in order to make the magnitude of the damping term independent of the sign of x.

The qualitative nature of the solution depends upon the value of the parameter α. Although the analytical solution of the equation is not known, it can be represented using the method of isoclines, in the phase plane. Equation (13.143) can be rewritten as

$$y = \dot{x} = \frac{dx}{dt} \qquad (13.144)$$

$$\dot{y} = \frac{dy}{dt} = \alpha(1 - x^2)y - x \qquad (13.145)$$

Thus the isocline corresponding to a specified value of the slope $dy/dx = c$ is given by

$$\frac{dy}{dx} = \frac{dy/dt}{dx/dt} = \frac{\alpha(1 - x^2)y - x}{y} = c$$

or

$$y + \left[\frac{x}{-\alpha(1 - x^2) + c} \right] = 0 \qquad (13.146)$$

By drawing the curves represented by Eq. (13.146) for a set of values of c as shown in Fig. 13.17, the trajectories can be sketched in with fair accuracy as shown in the same figure. The isoclines will be curves since the equation, Eq. (13.146), is nonlinear. An infinity of isoclines pass through the origin, which is a singularity.

An interesting property of the solution can be observed from Fig. 13.17. All the trajectories, irrespective of the initial conditions, approach asymptotically a particular closed curve, known as the limit cycle, which represents a steady state periodic (but not harmonic) oscillation. This is a phenomenon that can be observed only with certain nonlinear vibration problems and not in any linear problem. If the initial point is inside the limit cycle, the ensuing solution curve spirals outward. On the other hand, if the initial point falls outside the limit cycle, the solution curve

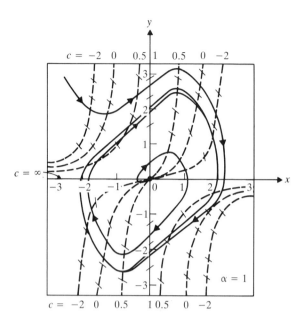

Figure 13.17   Trajectories and limit cycle for van der Pol's equation.

spirals inward. As stated above, in both the cases, the limit cycle is attained finally. An important characteristic of the limit cycle is that the maximum value of x is always close to 2 irrespective of the value of α. This result can be seen by solving Eq. (13.143) using the perturbation method (see Problem 13.30).

13.10 NUMERICAL METHODS

Most of the numerical methods described in the earlier chapters can be used for finding the response of nonlinear systems. The Runge-Kutta method described in Section 11.4 is directly applicable for nonlinear systems and is illustrated in Section 13.11. The central difference, Houbolt, Wilson, and Newmark methods considered in Chapter 11 can aiso be used for solving nonlinear multidegree of freedom vibration problems with slight modification. Let a multidegree of freedom system be governed by the equation

$$[m]\ddot{\vec{x}}(t) + [c]\dot{\vec{x}}(t) + \vec{P}(\vec{x}(t)) = \vec{F}(t) \qquad (13.147)$$

where the internal set of forces opposing the displacements $\vec{P}$ are assumed to be nonlinear functions of $\vec{x}$. For the linear case, $\vec{P} = [k]\vec{x}$. In order to find the displacement vector $\vec{x}$ that satisfies the nonlinear equilibrium in Eq. (13.147), it is necessary to perform an equilibrium iteration sequence in each time step. In implicit methods (Houbolt, Wilson, and Newmark methods), the equilibrium conditions are considered at the same time for which solution is sought. If the solution is known for time t_i and we wish to find the solution for time t_{i+1}, then the following equilibrium equations are considered

$$[m]\ddot{\vec{x}}_{i+1} + [c]\dot{\vec{x}}_{i+1} + \vec{P}_{i+1} = \vec{F}_{i+1} \tag{13.148}$$

where $\vec{F}_{i+1} = \vec{F}(t = t_{i+1})$ and $\vec{P}_{i+1}$ is computed as

$$\vec{P}_{i+1} = \vec{P}_i + [k_i]\Delta\vec{x} = \vec{P}_i + [k_i](\vec{x}_{i+1} - \vec{x}_i) \tag{13.149}$$

where $[k_i]$ is the linearized or tangent stiffness matrix computed at time t_i. Substitution of Eq. (13.149) in Eq. (13.148) gives

$$[m]\ddot{\vec{x}}_{i+1} + [c]\dot{\vec{x}}_{i+1} + [k_i]\vec{x}_{i+1} = \hat{\vec{F}}_{i+1} = \vec{F}_{i+1} - \vec{P}_i + [k_i]\vec{x}_i \tag{13.150}$$

Since the right hand side of Eq. (13.150) is completely known, it can be solved for $\vec{x}_{i+1}$ using any of the implicit methods directly. The $\vec{x}_{i+1}$ found is only an approximate vector due to the linearization process used in Eq. (13.149). To improve the accuracy of the solution and to avoid the development of numerical instabilities, an iterative process has to be used within the current time step [13.21].

13.11 COMPUTER PROGRAM

A Fortran program is given for numerically solving the following free nonlinear vibration problem, using the fourth order Runge-Kutta method:

$$m\ddot{x} + c\dot{x} + kx + k^*x^3 = 0 \tag{E.1}$$

with $m = 0.01$, $c = 0.1$, $k = 2.0$, $k^* = 0.5$, and the initial conditions $x(0) = 7.5$ and $\dot{x}(0) = 0$. Equation (E.1) is written as a system of two first order differential equations:

$$\dot{x}_1 = f_1 = x_2$$
$$\dot{x}_2 = f_2 = -\frac{c}{m}x_2 - \frac{k}{m}x_1 - \frac{k^*}{m}x_1^3 \tag{E.2}$$

with the initial conditions $x_1(0) = 7.5$ and $x_2(0) = 0$. The time step Δt is taken as 0.0025, the number of steps NSTEP as 400, and the number of equations N as 2. The values of m, c, k, and k^* are denoted as YM, YC, YK and YKS and are transferred to subroutine FUN through a common statement. Subroutine FUN provides the values of $f_1 = $ F(1) and $f_2 = $ F(2) at specified values of array $X = \begin{Bmatrix} x_1 \\ x_2 \end{Bmatrix}$ and T = time t. The listing of the program and the results follow.

```
C =================================================================
C
C PROGRAM 23
C MAIN PROGRAM FOR SOLVING A NONLINEAR VIBRATION PROBLEM USING THE
C SUBROUTINE RK4
C
C =================================================================
C DIMENSIONS: TIME(NSTEP),X(NSTEP,N),XX(N),F(N),XI(N),XJ(N),XK(N),XL(N),
C             UU(N)
C             XX(1),...,XX(N) CONTAIN INITIAL VALUES
C FOLLOWING 9 LINES CONTAIN PROBLEM-DEPENDENT DATA
      DIMENSION TIME(400),X(400,2),XX(2),F(2),XI(2),XJ(2),XK(2),XL(2),
     2   UU(2)
      COMMON /BLOCK1/ YM,YC,YK,YKS
      XX(1)=7.5
      XX(2)=0.0
      N=2
      NSTEP=400
      DT=0.0025
      DATA YM,YC,YK,YKS/0.01,0.1,2.0,0.5/
C END OF PROBLEM-DEPENDENT DATA
      T=0.0
      PRINT 10
10    FORMAT (//,40H SOLUTION OF NONLINEAR VIBRATION PROBLEM,/,
     2   35H BY FOURTH ORDER RUNGE-KUTTA METHOD)
      PRINT 20,YM,YC,YK,YKS
20    FORMAT (//,6H DATA:,/,6H YM  =,E15.6,/,6H YC  =,E15.6,/,
     2   6H YK  =,E15.6,/,6H YKS =,E15.6,/)
      PRINT 60
      DO 50 I=1,NSTEP
      CALL RK4 (T,DT,N,XX,F,XI,XJ,XK,XL,UU)
      TIME(I)=T
      DO 30 J=1,N
30    X(I,J)=XX(J)
      PRINT 40, I, TIME(I),(X(I,J),J=1,N)
40    FORMAT (2X,I5,F12.6,2E15.6)
50    CONTINUE
60    FORMAT (/,9H RESULTS:,//,5X,2H I,4X,8H TIME(I),5X,7H X(I,1),
     2   5X,7H X(I,2),/)
      STOP
      END
C =================================================================
C
C SUBROUTINE FUN FOR USE IN SUBROUTINE RK4
C THIS SUBROUTINE IS PROBLEM-DEPENDENT
C
C =================================================================
      SUBROUTINE FUN (X,F,N,T)
      DIMENSION X(N),F(N)
      COMMON /BLOCK1/ YM,YC,YK,YKS
      F(1)=X(2)
      F(2)=-(YC/YM)*X(2)-(YK/YM)*X(1)-(YKS/YM)*(X(1)**3)
      RETURN
      END
```

```
SOLUTION OF NONLINEAR VIBRATION PROBLEM
BY FOURTH ORDER RUNGE-KUTTA METHOD

DATA:
YM  =   0.100000E-01
YC  =   0.100000E+00
YK  =   0.200000E+01
YKS =   0.500000E+00

RESULTS:

     I      TIME(I)       X(I,1)        X(I,2)

     1     0.002500    0.743030E+01   -0.552857E+02
     2     0.005000    0.722711E+01   -0.106350E+03
     3     0.007500    0.690398E+01   -0.150944E+03
     4     0.010000    0.647900E+01   -0.187662E+03
     5     0.012500    0.597268E+01   -0.216014E+03
     6     0.015000    0.540567E+01   -0.236317E+03
     7     0.017500    0.479708E+01   -0.249463E+03
     8     0.020000    0.416331E+01   -0.256674E+03
     9     0.022500    0.351755E+01   -0.259270E+03
    10     0.025000    0.286976E+01   -0.258502E+03
     .
     .
     .
   395     0.987511    0.906302E-01   -0.951725E+00
   396     0.990011    0.882246E-01   -0.972474E+00
   397     0.992511    0.857693E-01   -0.991504E+00
   398     0.995011    0.832685E-01   -0.100883E+01
   399     0.997511    0.807265E-01   -0.102448E+01
   400     1.000011    0.781475E-01   -0.103847E+01
```

REFERENCES

13.1. C. Hayashi, *Nonlinear Oscillations in Physical Systems*, McGraw-Hill, New York, 1964.

13.2. A. A. Andronow and C. E. Chaikin, *Theory of Oscillations* (English language edition), Princeton University Press, Princeton, N.J., 1949.

13.3. N. V. Butenin, *Elements of the Theory of Nonlinear Oscillations*, Blaisdell Publishing Co., New York, 1965.

13.4. A. Blaquiere, *Nonlinear System Analysis*, Academic Press, New York, 1966.

13.5. Y. H. Ku, *Analysis and Control of Nonlinear Systems*, Ronald Press, New York, 1958.

13.6. J. N. J. Cunningham, *Introduction to Nonlinear Analysis*, McGraw-Hill, New York, 1958.

13.7. J. J. Stoker, *Nonlinear Vibrations in Mechanical and Electrical Systems*, Interscience Publishers, New York, 1950.

13.8. J. P. Den Hartog, *Mechanical Vibrations* (4th Ed.), McGraw-Hill, New York, 1956.

13.9. N. Minorsky, *Nonlinear Oscillations*, D. Van Nostrand Co., Princeton, N.J., 1962.

13.10. R. E. Mickens, "Perturbation solution of a highly non-linear oscillation equation," *Journal of Sound and Vibration*, Vol. 68, 1980, pp. 153–155.

13.11. B. V. Dasarathy and P. Srinivasan, "Study of a class of non-linear systems reducible to equivalent linear systems," *Journal of Sound and Vibration*, Vol. 7, 1968, pp. 27–30.

13.12. G. L. Anderson, "A modified perturbation method for treating non-linear oscillation problems," *Journal of Sound and Vibration*, Vol. 38, 1975, pp. 451–464.

13.13. B. L. Ly, "A note on the free vibration of a non-linear oscillator," *Journal of Sound and Vibration*, Vol. 68, 1980, pp. 307–309.

13.14. V. A. Bapat and P. Srinivasan, "Free vibrations of non-linear cubic spring mass systems in the presence of Coulomb damping," *Journal of Sound and Vibration*, Vol. 11, 1970, pp. 121–137.

13.15. H. R. Srirangarajan, P. Srinivasan, and B. V. Dasarathy, "Analysis of two degrees of freedom systems through weighted mean square linearization approach," *Journal of Sound and Vibration*, Vol. 36, 1974 pp. 119–131.

13.16. S. R. Woodall, "On the large amplitude oscillations of a thin elastic beam," *International Journal of Nonlinear Mechanics*, Vol. 1, 1966, pp. 217–238.

13.17. D. A. Evenson, "Nonlinear vibrations of beams with various boundary conditions," *AIAA Journal*, Vol. 6, 1968, pp. 370–372.

13.18. M. E. Beshai and M. A. Dokainish, "The transient response of a forced non-linear system," *Journal of Sound and Vibration*, Vol. 41, 1975, pp. 53–62.

13.19. V. A. Bapat and P. Srinivasan, "Response of undamped non-linear spring mass systems subjected to constant force excitation," *Journal of Sound and Vibration*, Vol. 9, 1969, Part I: pp. 53–58 and Part II: pp. 438–446.

13.20. W. E. Boyce and R. C. DiPrima, *Elementary Differential Equations and Boundary Value Problems* (4th Ed.), Wiley, New York, 1986.

13.21. D. R. J. Owen, "Implicit finite element methods for the dynamic transient analysis of solids with particular reference to nonlinear situations," pp. 123–152 in *Advanced Structural Dynamics*, J. Donéa (ed.), Applied Science Publishers, London, 1980.

13.22. B. van der Pol, "Relaxation oscillations," *Philosophical Magazine*, Vol. 2, pp. 978–992, 1926.

13.23. L. A. Pipes and L. R. Harvill, *Applied Mathematics for Engineers and Physicists* (3rd Ed.), McGraw-Hill, New York, 1970.

13.24. N. N. Bogoliubov and Y. A. Mitropolsky, *Asymptotic Methods in the Theory of Nonlinear Oscillations*, Hindustan Publishing, Delhi, 1961.

13.25. A. H. Nayfeh and D. T. Mook, *Nonlinear Oscillations*, John Wiley, New York, 1979.

13.26. G. Duffing, "Erzwungene Schwingungen bei veranderlicher Eigenfrequenz und ihre technische Bedeutung," Ph.D. thesis (Sammlung Vieweg, Braunschweig, 1918).

13.27. C. A. Ludeke, "An experimental investigation of forced vibrations in a mechanical system having a non-linear restoring force," *Journal of Applied Physics*, Vol. 17, pp. 603–609, 1946.

13.28. D. W. Jordan and P. Smith, *Nonlinear Ordinary Differential Equations* (2nd Ed.), Clarendon Press, Oxford, 1987.

13.29. R. E. Mickens, *An Introduction to Nonlinear Oscillations*, Cambridge University Press, Cambridge, 1981.

13.30. S. H. Crandall, "Nonlinearities in structural dynamics," *The Shock and Vibration Digest*, Vol. 6, No. 8, August 1974, pp. 2–14.

REVIEW QUESTIONS

13.1. How do you recognize a nonlinear vibration problem?

13.2. What are the various sources of nonlinearity in a vibration problem?

13.3. What is the source of nonlinearity in Duffing's equation?

13.4. How is the frequency of the solution of Duffing's equation affected by the nature of the spring?

13.5. What are subharmonic oscillations?

13.6. Explain the jump phenomenon.

13.7. What principle is used in the Ritz-Galerkin method?

13.8. Define these terms: phase plane, trajectory, singular point, phase velocity.

13.9. What is the method of isoclines?

13.10. What is the difference between a hard spring and a soft spring?

13.11. Explain the difference between subharmonic and superharmonic oscillations.

13.12. What is a secular term?

13.13. Give an example of a system that leads to an equation of motion with time dependent coefficients.

13.14. Explain the significance of the following: Stable node, unstable node, saddle point, focus, and center.

13.15. What is a limit cycle?

13.16. Give two examples of physical phenomena that can be represented by the van der Pol's equation.

PROBLEMS

The problem assignments are organized as follows:

Problems	Section covered	Topic covered
13.1	13.1	Introduction
13.2–13.4	13.2	Examples of nonlinear vibration
13.5–13.7	13.3	Exact methods of solution
13.8, 13.9	13.4	Approximate analytical methods

13.1. The equation of motion of a simple pendulum, subjected to a constant torque, $M_t = ml^2 f$, is given by

$$\ddot{\theta} + \omega_0^2 \sin \theta = f \tag{E.1}$$

If $\sin \theta$ is replaced by its two-term approximation, $\theta - (\theta^3/6)$, the equation of motion becomes

$$\ddot{\theta} + \omega_0^2 \theta = f + \frac{\omega_0^2}{6} \theta^3 \tag{E.2}$$

Let the solution of the linearized equation

$$\ddot{\theta} + \omega_0^2 \theta = f \tag{E.3}$$

be denoted as $\theta_1(t)$, and the solution of the equation

$$\ddot{\theta} + \omega_0^2 \theta = \frac{\omega_0^2}{6} \theta^3 \tag{E.4}$$

be denoted as $\theta_2(t)$. Discuss the validity of the total solution, $\theta(t)$, given by $\theta(t) = \theta_1(t) + \theta_2(t)$, for Eq. (E.2).

13.2. Two springs, having different stiffnesses k_1 and k_2 with $k_2 > k_1$, are placed on either side of a mass m, as shown in Fig. 13.18. When the mass is in its equilibrium position, no spring is in contact with the mass. However, when the mass is displaced from its equilibrium position, only one spring will be compressed. If the mass is given an initial velocity $\dot{x}_0$ at $t = 0$, determine (a) the maximum deflection and (b) the period of vibration of the mass.

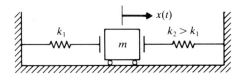

Figure 13.18

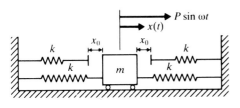

Figure 13.19   **Figure 13.20**

13.3. Find the equation of motion of the mass shown in Fig. 13.19. Draw the spring force versus x diagram.

13.4. Two masses m_1 and m_2 are attached to a stretched wire as shown in Fig. 13.20. If the initial tension in the wire is P, derive the equations of motion for large transverse displacements of the masses.

13.5. Find the natural time period of oscillation of the pendulum shown in Fig. 13.1(a) when it oscillates between the limits $\theta = -\pi/2$ and $\theta = \pi/2$, using Eqs. (13.1) and (13.12).

13.6. A simple pendulum of length 30 in. is released from the initial position of $80°$ from the vertical. How long does it take to reach the position $\theta = 0°$?

13.7. Find the exact solution of the nonlinear pendulum equation

$$\ddot{\theta} + \omega_0^2 \left(\theta - \frac{\theta^3}{6} \right) = 0$$

with $\dot{\theta} = 0$ when $\theta = \theta_0$, where θ_0 denotes the maximum angular displacement.

13.8. Find the solution of Example 13.1 using the following two-term approximation for $x(t)$:

$$\underset{\sim}{x}(t) = A_0 \sin \omega t + A_3 \sin 3\omega t$$

13.9. Find the solution of the pendulum equation, Eq. (13.20), using a three-term expansion in the Lindstedt's perturbation method [Eq. (13.30)] referred in text on p. 614.

13.10. The equation of motion for the forced vibration of a single degree of freedom nonlinear system can be expressed as

$$\ddot{x} + c\dot{x} + k_1 x + k_2 x^3 = a_1 \cos 3\omega t - a_2 \sin 3\omega t$$

Derive the conditions for the existence of subharmonics of order 3 for this system.

13.11. The equation of motion of a nonlinear system is given by

$$\ddot{x} + c\dot{x} + k_1 x + k_2 x^2 = a \cos 2\omega t$$

Investigate the subharmonic solution of order 2 for this system.

13.12. Prove that, for the system considered in Section 13.5.1, the minimum value of ω^2 for which the amplitude of subharmonic oscillations A will have a real value is given by

$$\omega_{\min} = \omega_0 + \frac{21}{2048} \frac{F^2}{\omega_0^5}$$

Also, show that the minimum value of the amplitude, for stable subharmonic oscillations, is given by

$$A_{\min} = \frac{F}{16\omega^2}$$

13.13. Derive Eqs. (13.113b) and (13.116b) for the Mathieu equation.

13.14. The equation of motion of a single degree of freedom system is given by

$$2\ddot{x} + 0.8\dot{x} + 1.6x = 0$$

with initial conditions $x(0) = -1$ and $\dot{x}(0) = 2$. (a) Plot the graph $x(t)$ versus t for $0 \leqslant t \leqslant 10$. (b) Plot a trajectory in the phase plane.

13.15. Find the equilibrium position and plot the trajectories in the neighborhood of the equilibrium position corresponding to the following equation:

$$\ddot{x} + 0.1(x^2 - 1)\dot{x} + x = 0$$

13.16. Obtain the phase trajectories for a system governed by the equation

$$\ddot{x} + 0.4\dot{x} + 0.8x = 0$$

with the initial conditions $x(0) = 2$ and $\dot{x}(0) = 1$ using the method of isoclines.

13.17. Plot the phase-plane trajectories for the following system:

$$\ddot{x} + 0.1\dot{x} + x = 5$$

The initial conditions are $x(0) = \dot{x}(0) = 0$.

13.18. A single degree of freedom system is subjected to Coulomb friction so that the equation of motion is given by

$$\ddot{x} + f\frac{\dot{x}}{|\dot{x}|} + \omega_n^2 x = 0$$

Construct the phase plane trajectories of the system using the initial conditions $x(0) = 10(f/\omega_n^2)$ and $\dot{x}(0) = 0$.

13.19. The equation of motion of a simple pendulum subject to viscous damping can be expressed as

$$\ddot{\theta} + c\dot{\theta} + \sin\theta = 0$$

If the initial conditions are $\theta(0) = \theta_0$ and $\dot{\theta}(0) = 0$, show that the origin in the phase plane diagram represents (i) a stable focus for $c > 0$ and (ii) an unstable focus for $c < 0$.

13.20. The equation of motion of a simple pendulum, subjected to external force, is given by

$$\ddot{\theta} + 0.5\dot{\theta} + \sin\theta = 0.8$$

Find the nature of singularity at $\theta = \sin^{-1}(0.8)$.

13.21. The phase plane equation of a single degree of freedom system is given by

$$\frac{dy}{dx} = \frac{-cy - (x - 0.1x^3)}{y}$$

Investigate the nature of singularity at $(x, y) = (0, 0)$ for $c > 0$.

13.22. Identify the singularity and find the nature of solution near the singularity for the van der Pol's equation:

$$\ddot{x} - \alpha(1 - x^2)\dot{x} + x = 0$$

13.23. Identify the singularity and investigate the nature of solution near the singularity for an undamped system with a hard spring:

$$\ddot{x} + \omega_n^2(1 + k^2 x^2)x = 0$$

13.24. Solve Problem 13.23 for an undamped system with a soft spring:

$$\ddot{x} + \omega_n^2(1 - k^2x^2)x = 0$$

13.25. Solve Problem 13.23 for a simple pendulum:

$$\ddot{\theta} + \omega_n^2 \sin \theta = 0$$

13.26. Determine the eigenvalues and eigenvectors of the following equations:

(i) $\quad \dot{x} = x - y, \qquad \dot{y} = x + 3y$

(ii) $\quad \dot{x} = x + y, \qquad \dot{y} = 4x + y$

13.27. Find the trajectories of the system governed by the equations:

$$\dot{x} = x - 2y, \qquad \dot{y} = 4x - 5y$$

13.28. Find the trajectories of the system governed by the equations:

$$\dot{x} = x - y, \qquad \dot{y} = x + 3y$$

13.29. Find the trajectories of the system governed by the equations:

$$\dot{x} = 2x + y, \qquad \dot{y} = -3x - 2y$$

13.30. Find the solution of the van der Pol's equation, Eq. (13.143), using the Lindstedt's perturbation method.

13.31. Solve the equation of motion $\ddot{x} + 0.5\dot{x} + x + 1.2x^3 = 1.8 \cos 0.4t$ using the Runge-Kutta method with $\Delta t = 0.05$, $t_{max} = 5.0$, and $x_0 = \dot{x}_0 = 0$. Plot the variation of x with t.

13.32. Find the time variation of the angular displacement of a simple pendulum (i.e., the solution of Eq. (13.5)) for $g/l = 0.5$, using the initial conditions $\theta_0 = 45°$ and $\dot{\theta}_0 = 0$. Use the Runge-Kutta method.

13.33. Write a computer program for finding the period of vibration corresponding to Eq. (13.14). Use a suitable numerical integration procedure. Using this program, find the solution of Problem 13.32.

13.34. In the static firing test of a rocket, the rocket is anchored to a rigid wall by a nonlinear spring-damper system and fuel is burnt to develop a thrust, as shown in Fig. 13.21. The thrust acting on the rocket during the time period $0 \leqslant t \leqslant t_0$ is given by $F = m_0 v$, where m_0 is the constant rate at which fuel is burnt and v is the velocity of the jet stream. The initial mass of the rocket is M, so that its mass at any time t is given by $m = M - m_0 t$, $0 \leqslant t \leqslant t_0$. The data is: spring force $= 8 \times 10^5 x + 6 \times 10^3 x^3$ N, damping force $= 10\dot{x} + 20\dot{x}^2$ N, $m_0 = 10$ kg/s, $v = 2000$ m/s, $M = 2000$ kg, and

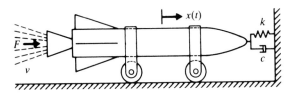

Figure 13.21

$t_0 = 100$ s. (i) Derive the equation of motion of the rocket and (ii) Find the variation of the displacement of the rocket using the Runge-Kutta method of numerical integration.

Projects:

13.35. In some periodic vibratory systems, external energy is supplied to the system over part of a period and dissipated within the system in another part of the period. Such periodic oscillations are known as *relaxation oscillations*. Van der Pol [13.22] indicated several instances of occurrence of relaxation oscillations such as a pneumatic hammer, the scratching noise of a knife on a plate, the squeaking of a door, and the fluctuation of populations of animal species. Many relaxation oscillations are governed by the van der Pol's equation:

$$\ddot{x} - \alpha(1 - x^2)\dot{x} + x = 0 \qquad (\text{E.1})$$

(i) Plot the phase plane trajectories for three values of α: $\alpha = 0.1$, $\alpha = 1$, and $\alpha = 10$. Use the initial conditions (a) $x(0) = 0.5$, $\dot{x}(0) = 0$ and (b) $x(0) = 0$ and $\dot{x}(0) = 5$.
(ii) Solve Eq. (E.1) using the fourth-order Runge-Kutta method using the initial conditions stated in (i) for $\alpha = 0.1$, $\alpha = 1$, and $\alpha = 10$.

13.36. A machine tool is mounted on two nonlinear elastic mounts as shown in Fig. 13.22. The equations of motion, in terms of the coordinates $x(t)$ and $\theta(t)$, are given by

$$m\ddot{x} + k_{11}(x - l_1\theta) + k_{12}(x - l_1\theta)^3 + k_{21}(x + l_2\theta) + k_{22}(x + l_2\theta)^3 = 0 \quad (\text{E.1})$$
$$J_0\ddot{\theta} - k_{11}(x - l_1\theta)l_1 - k_{12}(x - l_1\theta)^3 l_1$$
$$+ k_{21}(x + l_2\theta)l_2 + k_{22}(x + l_2\theta)^3 l_2 = 0 \qquad (\text{E.2})$$

where m is the mass and J_0 is the mass moment of inertia about G of the machine tool. Use Runge-Kutta method and find $x(t)$ and $\theta(t)$ for the following data: $m = 1000$ kg, $J_0 = 2500$ kg-m^2, $l_1 = 1$ m, $l_2 = 1.5$ m, $k_1 = 40x_1 + 10x_1^3$ kN/m, and $k_2 = 50x_2 + 5x_2^3$ kN/m.

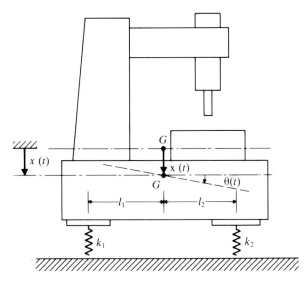

Figure 13.22

CHAPTER 14

Random Vibration

Karl Friedrich Gauss (1777 – 1855) was a German mathematician, astronomer and physicist. The trio, Gauss, Archimedes and Newton, are considered to be in a class by themselves among the great mathematicians. Although Gauss was born in a poor family, his extraordinary intelligence and genius in childhood inspired the Duke of Brunswick to pay all the expenses of Gauss during his entire education. In 1795, Gauss entered the University of Göttingen to study mathematics and began to keep his scientific diary. It was found after his death that his diary contained theories which were rediscovered and published by others. He moved to the University of Helmstedt in 1798 from which he received his doctor's degree in 1799. He published his most famous work, Disquisitiones Arithmeticae (Arithmetical Researches), in 1801. After that, Gauss was made the Director of Göttingen Observatory and also broadened his activities to include the mathematical and practical aspects of astronomy, geodesy and electromagnetism. The instrument used for measuring magnetic field strength is called "Gaussmeter." He invented the method of least squares and the law of normal distribution (Gaussian distribution) which is widely used in probability and random vibration. (Courtesy of Simon & Schuster, Inc.)

14.1 INTRODUCTION

If vibrational response characteristics such as displacement, acceleration, and stress are known precisely as functions of time, the vibration is known as *deterministic vibration*. This implies a deterministic system (or structure) and a deterministic loading (or excitation); deterministic vibration exists only if there is perfect control over all the variables that influence the structural properties and the loading. In practice, however, there are many processes and phenomena whose parameters cannot be precisely predicted. Such processes are therefore called *random processes*

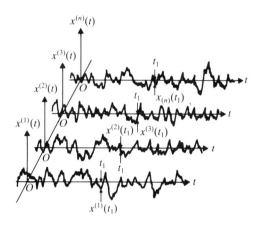

Figure 14.1 Ensemble of a random process ($x^{(i)}(t)$ is the ith sample function of the ensemble).

[14.1–14.4]. An example of a random process is pressure fluctuation at a particular point on the surface of an aircraft flying in air. If several records of these pressure fluctuations are taken under the same flight speed, altitude, and load factor, they might look as indicated in Fig. 14.1. The records are not identical even though the measurements are taken under seemingly identical conditions. Similarly, a building subjected to ground acceleration due to an earthquake, a water tank under wind loading, and a car running on a rough road represent random processes. An elementary treatment of random vibration is presented in this chapter.

14.2 RANDOM VARIABLES AND RANDOM PROCESSES

Most phenomena in real life are nondeterministic. For example, the tensile strength of steel and the dimensions of a machined part are nondeterministic. If many samples of steel are tested, their tensile strengths will not be the same—they will fluctuate about its mean or average value. Any quantity, like the tensile strength of steel, whose magnitude cannot be precisely predicted, is known as a *random variable* or a *probabilistic quantity*. If experiments are conducted to find the value of a random variable x, each experiment will give a number that is not a function of any parameter. For example, if 20 samples of steel are tested, their tensile strengths might be $x^{(1)} = 28.4$, $x^{(2)} = 30.2$, $x^{(3)} = 26.9, \ldots, x^{(20)} = 29.8$ kg/mm^2. Each of these outcomes is called a *sample point*. If n experiments are conducted, all the n possible outcomes of the random variable constitute what is known as the *sample space* of the random variable.

There are other types of probabilistic phenomena for which the outcome of an experiment is a function of some parameter such as time or spatial coordinate. Quantities such as the pressure fluctuations shown in Fig. 14.1 are called *random processes*. Each outcome of an experiment, in the case of a random process, is called a *sample function*. If n experiments are conducted, all the n possible outcomes of a random process constitute what is known as the *ensemble* of the process [14.5]. Notice that if the parameter t is fixed at a particular value t_1, $x(t_1)$ is a random variable whose sample points are given by $x^{(1)}(t_1), x^{(2)}(t_1), \ldots, x^{(n)}(t_1)$.

14.3 PROBABILITY DISTRIBUTION

Consider a random variable x such as the tensile strength of steel. If n experimental values of x are available as $x_1, x_2, \ldots, x_n$, the probability of realizing the value of x smaller than some specified value $\underset{\sim}{x}$ can be found as

$$\text{Prob}(x \leqslant \underset{\sim}{x}) = \frac{\underset{\sim}{n}}{n} \tag{14.1}$$

where $\underset{\sim}{n}$ denotes the number of x_i values which are less than or equal to $\underset{\sim}{x}$. As the number of experiments $n \to \infty$, Eq (14.1) defines the probability distribution function of x, $P(x)$:

$$P(x) = \lim_{n \to \infty} \frac{\underset{\sim}{n}}{n} \tag{14.2}$$

The probability distribution function can also be defined for a random time function. For this, we consider the random time function shown in Fig. 14.2. During a fixed time span t, the time intervals for which the value of $x(t)$ is less than $\underset{\sim}{x}$ are denoted as Δt_1, Δt_2, Δt_3, and Δt_4. Thus the probability of realizing $x(t)$ less than or equal to $\underset{\sim}{x}$ is given by

$$\text{Prob}\big[x(t) \leqslant \underset{\sim}{x}\big] = \frac{1}{t}\sum_i \Delta t_i \tag{14.3}$$

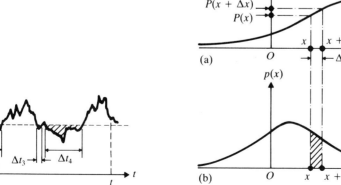

Figure 14.2

Figure 14.3

As $t \to \infty$, Eq. (14.3) gives the probability distribution function of $x(t)$:

$$P(x) = \lim_{t \to \infty} \frac{1}{t} \sum_i \Delta t_i \qquad (14.4)$$

If $x(t)$ denotes a physical quantity, the magnitude of $x(t)$ will always be a finite number, so Prob$[x(t) < -\infty] = P(-\infty) = 0$ (impossible event), and Prob$[x(t) < \infty] = P(\infty) = 1$ (certain event). The typical variation of $P(x)$ with x is shown in Fig. 14.3(a). The function $P(x)$ is called the *probability distribution function* of x. The derivative of $P(x)$ with respect to x is known as the *probability density function* and is denoted as $p(x)$. Thus

$$p(x) = \frac{dP(x)}{dx} = \lim_{\Delta x \to 0} \frac{P(x + \Delta x) - P(x)}{\Delta x} \qquad (14.5)$$

where the quantity $P(x + \Delta x) - P(x)$ denotes the probability of realizing $x(t)$ between the values x and $x + \Delta x$. Since $p(x)$ is the derivative of $P(x)$, we have

$$P(x) = \int_{-\infty}^{x} p(x')\, dx' \qquad (14.6)$$

As $P(\infty) = 1$, Eq (14.6) gives

$$P(\infty) = \int_{-\infty}^{\infty} p(x')\, dx' = 1 \qquad (14.7)$$

which means that the total area under the curve of $p(x)$ is unity.

14.4 MEAN VALUE AND STANDARD DEVIATION

If $f(x)$ denotes a function of the random variable x, the expected value of $f(x)$, denoted as μ_f or $E[f(x)]$ or $\overline{f(x)}$, is defined as

$$\mu_f = E[f(x)] = \overline{f(x)} = \int_{-\infty}^{\infty} f(x)p(x)\, dx \qquad (14.8)$$

If $f(x) = x$, Eq. (14.8) gives the expected value, also known as the *mean value* of x:

$$\mu_x = E[x] = \bar{x} = \int_{-\infty}^{\infty} xp(x)\, dx \qquad (14.9)$$

Similarly, if $f(x) = x^2$, we get the mean square value of x:

$$\mu_{x^2} = E[x^2] = \overline{x^2} = \int_{-\infty}^{\infty} x^2 p(x)\, dx \qquad (14.10)$$

The variance of x, denoted as σ_x^2, is defined as the mean square value of x about the mean,

$$\sigma_x^2 = E\left[(x - \bar{x})^2\right] = \int_{-\infty}^{\infty} (x - \bar{x})^2 p(x)\, dx = \overline{(x^2)} - (\bar{x})^2 \qquad (14.11)$$

The positive square root of the variance, $\sigma(x)$, is called the *standard deviation* of x.

EXAMPLE 14.1   **Probabilistic Characteristics of Eccentricity of a Rotor**

The eccentricity of a rotor (x), due to manufacturing errors, is found to have the following distribution:

$$p(x) = \begin{cases} kx^2, & 0 \leqslant x \leqslant 5 \text{ mm} \\ 0, & \text{elsewhere} \end{cases} \tag{E.1}$$

where k is a constant. Find (i) the mean, standard deviation, and the mean square value of the eccentricity and (ii) the probability of realizing x less than or equal to 2 mm.

Given: Probability distribution of the eccentricity of a rotor (x).

Find: $\bar{x}$, σ_x, $\overline{x^2}$, and Prob$[x \leqslant 2]$.

Approach: Use basic definitions.

Solution. The value of k in Eq. (E.1) can be found by normalizing the probability density function:

$$\int_{-\infty}^{\infty} p(x)\, dx = \int_0^5 kx^2\, dx = 1$$

that is,

$$k \left(\frac{x^3}{3} \right)_0^5 = 1$$

that is,

$$k = \tfrac{3}{125} \tag{E.2}$$

(i) The mean value of x is given by Eq. (14.9):

$$\bar{x} = \int_0^5 p(x) x\, dx = k \left(\frac{x^4}{4} \right)_0^5 = 3.75 \text{ mm} \tag{E.3}$$

The standard deviation of x is given by Eq. (14.11):

$$\sigma_x^2 = \int_0^5 (x - \bar{x})^2 p(x)\, dx = \int_0^5 (x^2 + \bar{x}^2 - 2\bar{x}x) p(x)\, dx$$

$$= \int_0^5 kx^4\, dx - (\bar{x})^2$$

$$= k \left(\frac{x^5}{5} \right)_0^5 - (\bar{x})^2 = k \left(\frac{3125}{5} \right) - (3.75)^2 = 0.9375$$

$$\therefore \quad \sigma_x = 0.9682 \text{ mm} \tag{E.4}$$

The mean square value of x is

$$\overline{x^2} = k \left(\frac{3125}{5} \right) = 15 \text{ mm}^2 \tag{E.5}$$

(ii)

$$\text{Prob}[x \leqslant 2] = \int_0^2 p(x)\, dx = k \int_0^2 x^2\, dx$$

$$= k \left(\frac{x^3}{3} \right)_0^2 = \frac{9}{125} = 0.072 \tag{E.6}$$

14.5 JOINT PROBABILITY DISTRIBUTION OF SEVERAL RANDOM VARIABLES

When two or more random variables are being considered simultaneously, their joint behavior is determined by a *joint probability distribution function*. For example, while testing the tensile strength of steel specimens, we can obtain the values of yield strength and ultimate strength in each experiment. If we are interested in knowing the relation between these two random variables, we must know the joint probability density function of yield strength and ultimate strength. The probability distributions of single random variables are called *univariate distributions*; the distributions that involve two random variables are called *bivariate distributions*. In general, if a distribution involves more than one random variable, it is called a *multivariate distribution*.

The bivariate density function of the random variables x_1 and x_2 is defined by

$$p(x_1, x_2) \, dx_1 \, dx_2 = \text{Prob}[x_1 < x_1' < x_1 + dx_1, \, x_2 < x_2' < x_2 + dx_2] \quad (14.12)$$

that is, the probability of realizing the value of the first random variable between x_1 and $x_1 + dx_1$ and the value of the second random variable between x_2 and $x_2 + dx_2$. The joint probability density function has the property that

$$\int_{-\infty}^{\infty} \int_{-\infty}^{\infty} p(x_1, x_2) \, dx_1 \, dx_2 = 1 \quad (14.13)$$

The joint distribution function of x_1 and x_2 is

$$P(x_1, x_2) = \text{Prob}[x_1' < x_1, x_2' < x_2]$$

$$= \int_{-\infty}^{x_1} \int_{-\infty}^{x_2} p(x_1', x_2') \, dx_1' \, dx_2' \quad (14.14)$$

The marginal or individual density functions can be obtained from the joint probability density function as

$$p(x) = \int_{-\infty}^{\infty} p(x, y) \, dy \quad (14.15)$$

$$p(y) = \int_{-\infty}^{\infty} p(x, y) \, dx \quad (14.16)$$

The variances of x and y can be determined as

$$\sigma_x^2 = E\left[(x - \mu_x)^2\right] = \int_{-\infty}^{\infty} (x - \mu_x)^2 p(x) \, dx \quad (14.17)$$

$$\sigma_y^2 = E\left[(y - \mu_y)^2\right] = \int_{-\infty}^{\infty} (y - \mu_y)^2 p(y) \, dy \quad (14.18)$$

The covariance of x and y, σ_{xy}, is defined as the expected value or average of the product of the deviations from the respective mean values of x and y. It is given by

$$\sigma_{xy} = E\big[(x - \mu_x)(y - \mu_y)\big] = \int_{-\infty}^{\infty}\int_{-\infty}^{\infty}(x - \mu_x)(y - \mu_y)p(x, y)\, dx\, dy$$

$$= \int_{-\infty}^{\infty}\int_{-\infty}^{\infty}(xy - x\mu_y - y\mu_x + \mu_x\mu_y)p(x, y)\, dx\, dy$$

$$= \int_{-\infty}^{\infty}\int_{-\infty}^{\infty}xyp(x, y)\, dx\, dy - \mu_y\int_{-\infty}^{\infty}\int_{-\infty}^{\infty}xp(x, y)\, dx\, dy$$

$$- \mu_x\int_{-\infty}^{\infty}\int_{-\infty}^{\infty}yp(x, y)\, dx\, dy$$

$$+ \mu_x\mu_y\int_{-\infty}^{\infty}\int_{-\infty}^{\infty}p(x, y)\, dx\, dy = E[xy] - \mu_x\mu_y \qquad (14.19)$$

The correlation coefficient between x and y, ρ_{xy}, is defined as the normalized covariance:

$$\rho_{xy} = \frac{\sigma_{xy}}{\sigma_x\sigma_y} \qquad (14.20)$$

It can be seen that the correlation coefficient satisfies the relation $-1 \leqslant \rho_{xy} \leqslant 1$.

14.6 CORRELATION FUNCTIONS OF A RANDOM PROCESS

If $t_1, t_2, \ldots$ are fixed values of t, we use the abbreviations $x_1, x_2, \ldots$ to denote the values of $x(t)$ at $t_1, t_2, \ldots$, respectively. Since there are several random variables $x_1, x_2, \ldots$, we form the products of the random variables $x_1, x_2, \ldots$ (values of $x(t)$ at different times) and average the products over the set of all possibilities to obtain a sequence of functions:

$$K(t_1, t_2) = E\big[x(t_1)x(t_2)\big] = E[x_1 x_2]$$
$$K(t_1, t_2, t_3) = E\big[x(t_1)x(t_2)x(t_3)\big] = E[x_1 x_2 x_3] \qquad (14.21)$$

and so on. These functions describe the statistical connection between the values of $x(t)$ at different times $t_1, t_2, \ldots$ and are called *correlation functions* [14.6, 14.7].

Autocorrelation Function. The mathematical expectation of $x_1 x_2$—the correlation function $K(t_1, t_2)$—is also known as the *autocorrelation function*, designated as $R(t_1, t_2)$. Thus

$$R(t_1, t_2) = E[x_1 x_2] \qquad (14.22)$$

If the joint probability density function of x_1 and x_2 is known to be $p(x_1, x_2)$, the autocorrelation function can be expressed as

$$R(t_1, t_2) = \int_{-\infty}^{\infty}\int_{-\infty}^{\infty}x_1 x_2 p(x_1, x_2)\, dx_1\, dx_2 \qquad (14.23)$$

Experimentally, we can find $R(t_1, t_2)$ by taking the product of $x^{(i)}(t_1)$ and $x^{(i)}(t_2)$

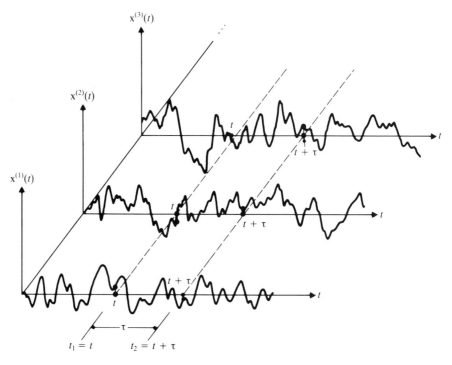

Figure 14.4

in the ith sample function and averaging over the ensemble:

$$R(t_1, t_2) = \frac{1}{n} \sum_{i=1}^{n} x^{(i)}(t_1) x^{(i)}(t_2) \qquad (14.24)$$

where n denotes the number of sample functions in the ensemble (see Fig. 14.4). If t_1 and t_2 are separated by τ (with $t_1 = t$ and $t_2 = t + \tau$), we have $R(t + \tau) = E[x(t)x(t + \tau)]$.

14.7 STATIONARY RANDOM PROCESS

A stationary random process is one for which the probability distributions remain invariant under a shift of the time scale; the family of probability density functions applicable now also applies 5 hours from now or 500 hours from now. Thus the probability density function $p(x_1)$ becomes a universal density function $p(x)$, independent of time. Similarly, the joint density function $p(x_1, x_2)$, to be invariant under a shift of the time scale, becomes a function of $\tau = t_2 - t_1$, but not a function of t_1 or t_2 individually. Thus $p(x_1, x_2)$ can be written as $p(t, t + \tau)$. The expected

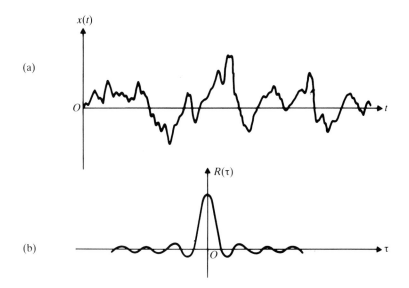

Figure 14.5

value of a stationary random process $x(t)$ can be written as

$$E[x(t_1)] = E[x(t_1 + t)] \quad \text{for any } t \qquad (14.25)$$

and the autocorrelation function becomes independent of the absolute time t and will depend only on the time separation τ:

$$R(t_1, t_2) = E[x_1 x_2] = E[x(t)x(t + \tau)] = R(\tau) \quad \text{for any } t \quad (14.26)$$

where $\tau = t_2 - t_1$. We shall use subscripts to R to identify the random process when more than one random process is involved. For example, we shall use $R_x(\tau)$ and $R_y(\tau)$ to denote the autocorrelation functions of the random processes $x(t)$ and $y(t)$, respectively. The autocorrelation function has the following characteristics [14.2, 14.4]:

1. If $\tau = 0$, $R(\tau)$ gives the mean square value of $x(t)$:
$$R(0) = E[x^2] \qquad (14.27)$$

2. If the process $x(t)$ has a zero mean and is extremely irregular as shown in Fig. 14.5(a), its autocorrelation function $R(\tau)$ will have small values except at $\tau = 0$ as indicated in Fig. 14.5(b).

3. If $x(t) \simeq x(t + \tau)$, the autocorrelation function $R(\tau)$ will have a constant value as shown in Fig. 14.6.

4. If $x(t)$ is stationary, its mean and standard deviations will be independent of t:
$$E[x(t)] = E[x(t + \tau)] = \mu \qquad (14.28)$$

and

$$\sigma_{x(t)} = \sigma_{x(t+\tau)} = \sigma \qquad (14.29)$$

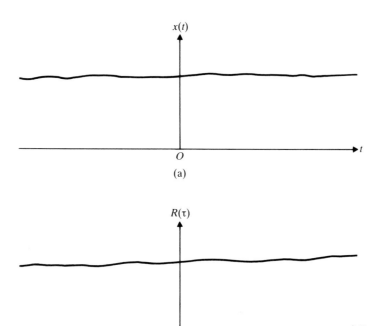

$x(t)$

(a)

$R(\tau)$

(b)

Figure 14.6

The correlation coefficient, ρ, of $x(t)$ and $x(t + \tau)$ can be found as

$$\rho = \frac{E\left[\{x(t) - \mu\}\{x(t + \tau) - \mu\}\right]}{\sigma^2}$$

$$= \frac{E[x(t)x(t + \tau)] - \mu E[x(t + \tau)] - \mu E[x(t)] + \mu^2}{\sigma^2}$$

$$= \frac{R(\tau) - \mu^2}{\sigma^2} \tag{14.30}$$

that is,

$$R(\tau) = \rho\sigma^2 + \mu^2 \tag{14.31}$$

Since $|\rho| \leqslant 1$, Eq. (14.31) shows that

$$-\sigma^2 + \mu^2 \leqslant R(\tau) \leqslant \sigma^2 + \mu^2 \tag{14.32}$$

This shows that the autocorrelation function will not be greater than the mean square value, $E[x^2] = \sigma^2 + \mu^2$.

5. Since $R(\tau)$ depends only on the separation time τ and not on the absolute time t for a stationary process,

$$R(\tau) = E[x(t)x(t + \tau)] = E[x(t)x(t - \tau)] = R(-\tau) \tag{14.33}$$

Thus $R(\tau)$ is an even function of τ.

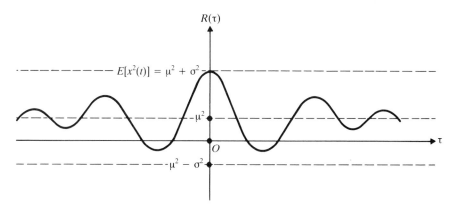

Figure 14.7

6. When τ is large ($\tau \to \infty$), there will not be a coherent relationship between the two values $x(t)$ and $x(t + \tau)$. Hence the correlation coefficient $\rho \to 0$ and Eq. (14.31) gives

$$R(\tau \to \infty) \to \mu^2 \tag{14.34}$$

A typical autocorrelation function is shown in Fig. 14.7.

Ergodic Process. An ergodic process is a stationary random process for which we can obtain all the probability information from a single sample function and assume that it is applicable to the entire ensemble. If $x^{(i)}(t)$ represents a typical sample function of duration T, the averages can be computed by averaging with respect to time along $x^{(i)}(t)$. Such averages are called *temporal averages*. By using the notation $\langle x(t) \rangle$ to represent the temporal average of $x(t)$ (the time average of x), we can write

$$E[x] = \langle x(t) \rangle = \lim_{T \to \infty} \frac{1}{T} \int_{-T/2}^{T/2} x^{(i)}(t) \, dt \tag{14.35}$$

where $x^{(i)}(t)$ has been assumed to be defined from $t = -T/2$ to $t = T/2$ with $T \to \infty$ (T very large). Similarly,

$$E[x^2] = \langle x^2(t) \rangle = \lim_{T \to \infty} \frac{1}{T} \int_{-T/2}^{T/2} [x^{(i)}(t)]^2 \, dt \tag{14.36}$$

and

$$R(\tau) = \langle x(t)x(t + \tau) \rangle = \lim_{T \to \infty} \frac{1}{T} \int_{-T/2}^{T/2} x^{(i)}(t)x^{(i)}(t + \tau) \, dt \tag{14.37}$$

14.8 GAUSSIAN RANDOM PROCESS

The most commonly used distribution for modeling physical random processes is called the *Gaussian* or *normal random process*. The Gaussian process has a number of remarkable properties that permit the computation of the random vibration characteristics in a simple manner. The probability density function of a Gaussian process $x(t)$ is given by

$$p(x) = \frac{1}{\sqrt{2\pi}\,\sigma_x} e^{-\frac{1}{2}\left(\frac{x-\bar{x}}{\sigma_x}\right)^2} \tag{14.38}$$

where $\bar{x}$ and σ_x denote the mean value and standard deviation of x. The mean ($\bar{x}$) and standard deviation (σ_x) of $x(t)$ vary with t for a nonstationary process but are constants (independent of t) for a stationary process. A very important property of the Gaussian process is that the forms of its probability distributions are invariant with respect to linear operations. This means that if the excitation of a linear system is a Gaussian process, the response is generally a different random process, but still a normal one. The only changes are that the magnitude of the mean and standard deviations of the response are different from those of the excitation.

The graph of a Gaussian probability density function is a bell-shaped curve, symmetric about the mean value; its spread is governed by the value of the standard deviation, as shown in Fig. 14.8. By defining a standard normal variable z as

$$z = \frac{x - \bar{x}}{\sigma_x} \tag{14.39}$$

Eq. (14.38) can be expressed as

$$p(x) = \frac{1}{\sqrt{2\pi}} e^{-\frac{1}{2}z^2} \tag{14.40}$$

The probability of $x(t)$ lying in the interval $-c\sigma$ and $+c\sigma$ where c is any positive number can be found, assuming $\bar{x} = 0$:

$$\text{Prob}[-c\sigma \leqslant x(t) \leqslant c\sigma] = \int_{-c\sigma}^{c\sigma} \frac{1}{\sqrt{2\pi}\,\sigma} e^{-\frac{1}{2}\frac{x^2}{\sigma^2}} \, dx \tag{14.41}$$

The probability of $x(t)$ lying outside the range $\mp c\sigma$ is one minus the value given by Eq. (14.41). This can also be expressed as

$$\text{Prob}[|x(t)| > c\sigma] = \frac{2}{\sqrt{2\pi}\,\sigma} \int_{c\sigma}^{\infty} e^{-\frac{1}{2}\frac{x^2}{\sigma^2}} \, dx \tag{14.42}$$

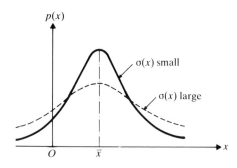

Figure 14.8

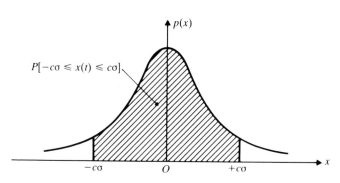

Figure 14.9

The integrals in Eqs. (14.41) and (14.42) have been evaluated numerically and tabulated [14.5]; some typical values are indicated in the following table (see Fig. 14.9 also).

Value of c	1	2	3	4
Prob$[-c\sigma \leqslant x(t) \leqslant c\sigma]$	0.6827	0.9545	0.9973	0.999937
Prob$[|x(t)| > c\sigma]$	0.3173	0.0455	0.0027	0.000063

14.9 FOURIER ANALYSIS

14.9.1
Fourier Series

We saw in Chapter 1 that any periodic function $x(t)$, of period τ, can be expressed in the form of a complex Fourier series

$$x(t) = \sum_{n=-\infty}^{\infty} c_n e^{in\omega_0 t} \tag{14.43}$$

where ω_0 is the fundamental frequency given by

$$\omega_0 = \frac{2\pi}{\tau} \tag{14.44}$$

and the complex Fourier coefficients c_n can be determined by multiplying both sides of Eq. (14.43) with $e^{-im\omega_0 t}$ and integrating over one time period:

$$\int_{-\tau/2}^{\tau/2} x(t) e^{-im\omega_0 t} \, dt = \sum_{n=-\infty}^{\infty} \int_{-\tau/2}^{\tau/2} c_n e^{i(n-m)\omega_0 t} \, dt$$

$$= \sum_{n=-\infty}^{\infty} c_n \int_{-\tau/2}^{\tau/2} \left[\cos(n-m)\omega_0 t \right.$$

$$\left. + i \sin(n-m)\omega_0 t \right] \, dt \tag{14.45}$$

Equation (14.45) can be simplified to obtain (see Problem 14.27)

$$c_n = \frac{1}{\tau} \int_{-\tau/2}^{\tau/2} x(t) e^{-in\omega_0 t} \, dt \tag{14.46}$$

Equation (14.43) shows that the function $x(t)$ of period τ can be expressed as a sum of an infinite number of harmonics. The harmonics have amplitudes given by Eq. (14.46) and frequencies which are multiples of the fundamental frequency ω_0. The difference between any two consecutive frequencies is given by

$$\omega_{n+1} - \omega_n = (n+1)\omega_0 - n\omega_0 = \Delta\omega = \frac{2\pi}{\tau} = \omega_0 \tag{14.47}$$

Thus the larger the period τ, the denser the frequency spectrum becomes. Equation (14.46) shows that the Fourier coefficients c_n are, in general, complex numbers. However, if $x(t)$ is a real and even function, then c_n will be real. If $x(t)$ is real, the integrand of c_n in Eq. (14.46) can also be identified as the complex conjugate of that of c_{-n}. Thus

$$c_n = c^*_{-n} \tag{14.48}$$

The mean square value of $x(t)$, that is, the time average of the square of the function $x(t)$, can be determined as

$$
\overline{x^2(t)} = \frac{1}{\tau} \int_{-\tau/2}^{\tau/2} x^2(t)\, dt = \frac{1}{\tau} \int_{-\tau/2}^{\tau/2} \left(\sum_{n=-\infty}^{\infty} c_n e^{in\omega_0 t} \right)^2 dt
$$

$$
= \frac{1}{\tau} \int_{-\tau/2}^{\tau/2} \left(\sum_{n=-\infty}^{-1} c_n e^{in\omega_0 t} + c_0 + \sum_{n=1}^{\infty} c_n e^{in\omega_0 t} \right)^2 dt
$$

$$
= \frac{1}{\tau} \int_{-\tau/2}^{\tau/2} \left\{ \sum_{n=1}^{\infty} \left(c_n e^{in\omega_0 t} + c^*_n e^{-in\omega_0 t} \right) + c_0 \right\}^2 dt
$$

$$
= \frac{1}{\tau} \int_{-\tau/2}^{\tau/2} \left\{ \sum_{n=1}^{\infty} 2 c_n c^*_n + c_0^2 \right\} dt
$$

$$
= c_0^2 + \sum_{n=1}^{\infty} 2|c_n|^2 = \sum_{n=-\infty}^{\infty} |c_n|^2 \tag{14.49}
$$

Thus the mean square value of $x(t)$ is given by the sum of the squares of the absolute values of the Fourier coefficients. Equation (14.49) is known as Parseval's formula for periodic functions [14.1].

EXAMPLE 14.2   Complex Fourier Series Expansion

Find the complex Fourier series expansion of the function shown in Fig. 14.10(a).

Given: Periodic function shown in Fig. 14.10(a).

Find: Fourier series expansion.

Approach: Application of Eqs. (14.43) and (14.46).

Solution. The given function can be expressed as

$$
x(t) = \begin{cases} A\left(1 + \dfrac{t}{a}\right), & -\dfrac{\tau}{2} \leqslant t \leqslant 0 \\[2mm] A\left(1 - \dfrac{t}{a}\right), & 0 \leqslant t \leqslant \dfrac{\tau}{2} \end{cases} \tag{E.1}
$$

where the period (τ) and the fundamental frequency (ω_0) are given by

$$
\tau = 2a \quad \text{and} \quad \omega_0 = \frac{2\pi}{\tau} = \frac{\pi}{a} \tag{E.2}
$$

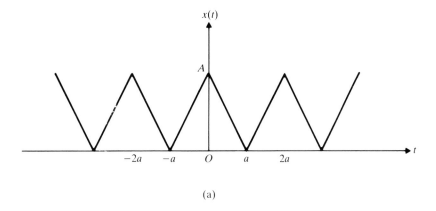

(a)

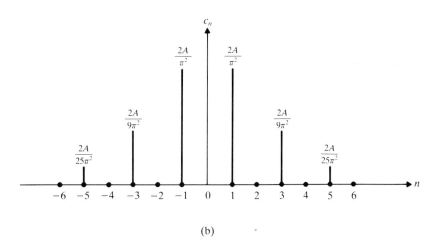

(b)

Figure 14.10

The Fourier coefficients can be determined as

$$c_n = \frac{1}{\tau} \int_{-\tau/2}^{\tau/2} x(t) e^{-in\omega_0 t} \, dt$$

$$= \frac{1}{\tau} \left[\int_{-\tau/2}^{0} A\left(1 + \frac{t}{a}\right) e^{-in\omega_0 t} \, dt + \int_{0}^{\tau/2} A\left(1 - \frac{t}{a}\right) e^{-in\omega_0 t} \, dt \right] \qquad (E.3)$$

Using the relation

$$\int t e^{kt} \, dt = \frac{e^{kt}}{k^2} (kt - 1) \qquad (E.4)$$

c_n can be evaluated as

$$c_n = \frac{1}{\tau}\left[\frac{A}{-in\omega_0}e^{-in\omega_0 t}\Big|_{-\tau/2}^{0} + \frac{A}{a}\left\{ \frac{e^{-in\omega_0 t}}{(-in\omega_0)^2}[-in\omega_0 t - 1]\right\}\Big|_{-\tau/2}^{0} + \frac{A}{-in\omega_0}e^{-in\omega_0 t}\Big|_{0}^{\tau/2}\right.$$

$$\left. - \frac{A}{a}\left\{ \frac{e^{-in\omega_0 t}}{(-in\omega_0)^2}[-in\omega_0 t - 1]\right\}\Big|_{0}^{\tau/2}\right] \qquad (\text{E.5})$$

This equation can be reduced to

$$c_n = \frac{1}{\tau}\left[\frac{A}{in\omega_0}e^{in\pi} + \frac{2A}{a}\frac{1}{n^2\omega_0^2} - \frac{A}{in\omega_0}e^{-in\pi} - \frac{A}{a}\frac{1}{n^2\omega_0^2}e^{in\pi} - \frac{A}{a}\frac{1}{n^2\omega_0^2}e^{-in\pi}\right.$$

$$\left. + \frac{A}{a}\frac{1}{n^2\omega_0^2}(in\pi)e^{in\pi} - \frac{A}{a}\frac{1}{n^2\omega_0^2}(in\pi)e^{-in\pi}\right] \qquad (\text{E.6})$$

Noting that

$$e^{in\pi} \quad \text{or} \quad e^{-in\pi} = \begin{cases} 1, & n = 0 \\ -1, & n = 1,3,5,\ldots \\ 1, & n = 2,4,6,\ldots \end{cases} \qquad (\text{E.7})$$

Eq. (E.6) can be simplified to obtain

$$c_n = \begin{cases} 0, & n = 0 \\ \left(\dfrac{4A}{a\tau n^2\omega_0^2}\right) = \dfrac{2A}{n^2\pi^2}, & n = 1,3,5,\ldots \\ 0, & n = 2,4,6,\ldots \end{cases} \qquad (\text{E.8})$$

The frequency spectrum is shown in Fig. 14.10(b).

14.9.2
Fourier Integral

A nonperiodic function, such as the one shown by the solid curve in Fig. 14.11, can be treated as a periodic function having an infinite period ($\tau \to \infty$). The Fourier series expansion of a periodic function is given by Eqs. (14.43), (14.44) and (14.46):

$$x(t) = \sum_{n=-\infty}^{\infty} c_n e^{in\omega_0 t} \qquad (14.50)$$

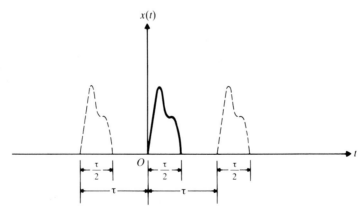

Figure 14.11

with

$$\omega_0 = \frac{2\pi}{\tau} \tag{14.51}$$

and

$$c_n = \frac{1}{\tau} \int_{-\tau/2}^{\tau/2} x(t) e^{-in\omega_0 t} \, dt \tag{14.52}$$

As $\tau \to \infty$, the frequency spectrum becomes continuous and the fundamental frequency becomes infinitesimal. Since the fundamental frequency ω_0 is very small, we can denote it as $\Delta\omega$, $n\omega_0$ as ω, and rewrite Eq. (14.52) as

$$\lim_{\tau \to \infty} \tau c_n = \lim_{\tau \to \infty} \int_{-\tau/2}^{\tau/2} x(t) e^{-i\omega t} \, dt = \int_{-\infty}^{\infty} x(t) e^{-i\omega t} \, dt \tag{14.53}$$

By defining $X(\omega)$ as

$$X(\omega) = \lim_{\tau \to \infty} (\tau c_n) = \int_{-\infty}^{\infty} x(t) e^{-i\omega t} \, dt \tag{14.54}$$

we can express $x(t)$ from Eq. (14.50) as

$$
\begin{aligned}
x(t) &= \lim_{\tau \to \infty} \sum_{n=-\infty}^{\infty} c_n e^{i\omega t} \frac{2\pi\tau}{2\pi\tau} \\
&= \lim_{\tau \to \infty} \sum_{n=-\infty}^{\infty} (c_n \tau) e^{i\omega t} \left(\frac{2\pi}{\tau} \right) \frac{1}{2\pi} \\
&= \frac{1}{2\pi} \int_{-\infty}^{\infty} X(\omega) e^{i\omega t} \, d\omega
\end{aligned}
\tag{14.55}
$$

This equation indicates the frequency decomposition of the nonperiodic function $x(t)$ in a continuous frequency domain, similar to Eq. (14.50) for a periodic function in a discrete frequency domain. The equations

$$x(t) = \frac{1}{2\pi} \int_{-\infty}^{\infty} X(\omega) e^{i\omega t} \, d\omega \tag{14.56}$$

and

$$X(\omega) = \int_{-\infty}^{\infty} x(t) e^{-i\omega t} \, dt \tag{14.57}$$

are known as the (integral) Fourier transform pair for a nonperiodic function $x(t)$, similar to Eqs. (14.50) and (14.52) for a periodic function $x(t)$ [14.9, 14.10].

The mean square value of a nonperiodic function $x(t)$ can be determined from Eq. (14.49):

$$
\begin{aligned}
\frac{1}{\tau} \int_{-\tau/2}^{\tau/2} x^2(t) \, dt &= \sum_{n=-\infty}^{\infty} |c_n|^2 \\
&= \sum_{n=-\infty}^{\infty} c_n c_n^* \frac{\tau\omega_0}{\tau\omega_0} = \sum_{n=-\infty}^{\infty} c_n c_n^* \frac{\tau\omega_0}{\tau\left(\frac{2\pi}{\tau}\right)} \\
&= \frac{1}{\tau} \sum_{n=-\infty}^{\infty} (\tau c_n)(c_n^* \tau) \frac{\omega_0}{2\pi}
\end{aligned}
\tag{14.58}
$$

Since $\tau c_n \to X(\omega)$, $\tau c_n^* \to X^*(\omega)$, and $\omega_0 \to d\omega$ as $\tau \to \infty$, Eq. (14.58) gives the mean square value of $x(t)$ as

$$\overline{x^2(t)} = \lim_{\tau \to \infty} \frac{1}{\tau} \int_{-\tau/2}^{\tau/2} x^2(t)\, dt = \int_{-\infty}^{\infty} \frac{|X(\omega)|^2}{2\pi\tau}\, d\omega \qquad (14.59)$$

Equation (14.59) is known as Parseval's formula for nonperiodic functions [14.1].

EXAMPLE 14.3   **Fourier Transform of a Triangular Pulse**

Find the Fourier transform of the triangular pulse shown in Fig. 14.12(a).

Given: Triangular pulse shown in Fig. 14.12(a).

Find: Fourier transform, $F(\omega)$.

Approach: Application of Eq. (14.57).

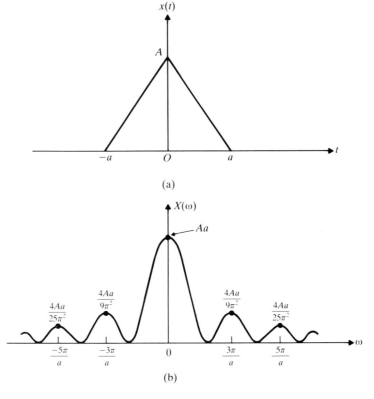

(a)

(b)

Figure 14.12

Solution. The triangular pulse can be expressed as

$$x(t) = \begin{cases} A\left(1 - \dfrac{|t|}{a}\right), & |t| \leqslant a \\ 0, & \text{otherwise} \end{cases} \tag{E.1}$$

The Fourier transform of $x(t)$ can be found, using Eq. (14.57), as

$$\begin{aligned} X(\omega) &= \int_{-\infty}^{\infty} A\left(1 - \frac{|t|}{a}\right) e^{-i\omega t}\, dt \\ &= \int_{-\infty}^{0} A\left(1 + \frac{t}{a}\right) e^{-i\omega t}\, dt + \int_{0}^{\infty} A\left(1 - \frac{t}{a}\right) e^{-i\omega t}\, dt \end{aligned} \tag{E.2}$$

Since $x(t) = 0$ for $|t| > 0$, Eq. (E.2) can be expressed as

$$\begin{aligned} X(\omega) &= \int_{-a}^{0} A\left(1 + \frac{t}{a}\right) e^{-i\omega t}\, dt + \int_{0}^{a} A\left(1 - \frac{t}{a}\right) e^{-i\omega t}\, dt \\ &= \left(\frac{A}{-i\omega}\right) e^{-i\omega t}\bigg|_{-a}^{0} + \frac{A}{a}\left\{\frac{e^{-i\omega t}}{(-i\omega)^2}[-i\omega t - 1]\right\}\bigg|_{-a}^{0} \\ &\quad + \left(\frac{A}{-i\omega}\right) e^{-i\omega t}\bigg|_{0}^{a} - \frac{A}{a}\left\{\frac{e^{-i\omega t}}{(-i\omega)^2}[-i\omega t - 1]\right\}\bigg|_{0}^{a} \end{aligned} \tag{E.3}$$

Equation (E.3) can be simplified to obtain

$$\begin{aligned} X(\omega) &= \frac{2A}{a\omega^2} + e^{i\omega a}\left(-\frac{A}{a\omega^2}\right) + e^{-i\omega a}\left(-\frac{A}{a\omega^2}\right) \\ &= \frac{2A}{a\omega^2} - \frac{A}{a\omega^2}(\cos\omega a + i\sin\omega a) - \frac{A}{a\omega^2}(\cos\omega a - i\sin\omega a) \\ &= \frac{2A}{a\omega^2}(1 - \cos\omega a) = \frac{4A}{a\omega^2}\sin^2\left(\frac{\omega a}{2}\right) \end{aligned} \tag{E.4}$$

Equation (E.4) is plotted in Fig. 14.12(b). Notice the similarity of this figure with the discrete Fourier spectrum shown in Fig. 14.10(b).

14.10   POWER SPECTRAL DENSITY

The power spectral density $S(\omega)$ of a stationary random process is defined as the Fourier transform of $R(\tau)/2\pi$:

$$S(\omega) = \frac{1}{2\pi}\int_{-\infty}^{\infty} R(\tau) e^{-i\omega\tau}\, d\tau \tag{14.60}$$

so that

$$R(\tau) = \int_{-\infty}^{\infty} S(\omega) e^{i\omega\tau}\, d\omega \tag{14.61}$$

Equations (14.60) and (14.61) are known as the Wiener-Khintchine formulas [14.1]. The power spectral density is more often used in random vibration analysis than the

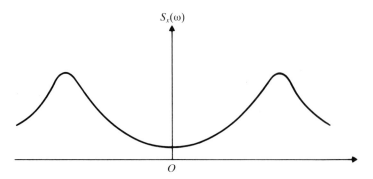

Figure 14.13

autocorrelation function. The following properties of power spectral density can be observed:

1. From Eqs. (14.27) and (14.61), we obtain

$$R(0) = E[x^2] = \int_{-\infty}^{\infty} S(\omega)\,d\omega \qquad (14.62)$$

If the mean is zero, the variance of $x(t)$ is given by

$$\sigma_x^2 = R(0) = \int_{-\infty}^{\infty} S(\omega)\,d\omega \qquad (14.63)$$

If $x(t)$ denotes the displacement, $R(0)$ represents the average energy. From Eq. (14.62), it is clear that $S(\omega)$ represents the energy density associated with the frequency ω. Thus $S(\omega)$ indicates the spectral distribution of energy in a system. Also, in electrical circuits, if $x(t)$ denotes random current, then the mean square value indicates the power of the system (when the resistance is unity). This is the origin of the term *power spectral density*.

2. Since $R(\tau)$ is an even function of τ and real, $S(\omega)$ is also an even and real function of ω. Thus $S(-\omega) = S(\omega)$. A typical power spectral density function is shown in Fig. 14.13.

3. From Eq. (14.62), the units of $S(\omega)$ can be identified as those of $x^2/$unit of angular frequency. It can be noted that both negative and positive frequencies are counted in Eq. (14.62). In experimental work, for convenience, an equivalent one-sided spectrum $W_x(f)$ is widely used [14.1, 14.2].[†]

The spectrum $W_x(f)$ is defined in terms of linear frequency (i.e., cycles per unit time) and only the positive frequencies are counted. The relationship between $S_x(\omega)$ and $W_x(f)$ can be seen with reference to Fig. 14.14. The

---

[†] When several random processes are involved, a subscript is used to identify the power spectral density function (or simply the spectrum) of a particular random process. Thus $S_x(\omega)$ denotes the spectrum of $x(t)$.

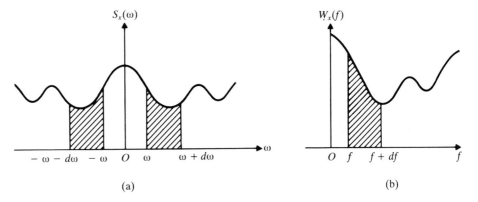

Figure 14.14

differential frequency $d\omega$ in Fig. 14.14(a) corresponds to the differential frequency $df = d\omega/2\pi$ in Fig. 14.14(b). Since $W_x(f)$ is the equivalent spectrum defined over positive values of f only, we have

$$E[x^2] = \int_{-\infty}^{\infty} S_x(\omega)\, d\omega \equiv \int_0^{\infty} W_x(f)\, df \qquad (14.64)$$

In order to have the contributions of the frequency bands $d\omega$ and df to the mean square value to be same, the shaded areas in both Figs. 14.14(a) and (b) must be the same. Thus

$$2S_x(\omega)\, d\omega = W_x(f)\, df \qquad (14.65)$$

which gives

$$W_x(f) = 2S_x(\omega)\frac{d\omega}{df} = 2S_x(\omega)\frac{d\omega}{(d\omega/2\pi)} = 4\pi S_x(\omega) \qquad (14.66)$$

14.11 WIDE-BAND AND NARROW-BAND PROCESSES

A wide-band process is a stationary random process whose spectral density function $S(\omega)$ has significant values over a range or band of frequencies which is approximately the same order of magnitude as the center frequency of the band. An example of a wide-band random process is shown in Fig. 14.15. The pressure fluctuations on the surface of a rocket due to acoustically transmitted jet noise or due to supersonic boundary layer turbulence are examples of physical processes that are typically wide-band. A narrow-band random process is a stationary process whose spectral density function $S(\omega)$ has significant values only in a range or band of frequencies whose width is small compared to the magnitude of the center frequency of the process. Figure 14.16 shows the sample function and the corresponding spectral density and autocorrelation functions of a narrow-band process.

A random process whose power spectral density is constant over a frequency range is called *white noise*, by analogy with the white light which spans the visible

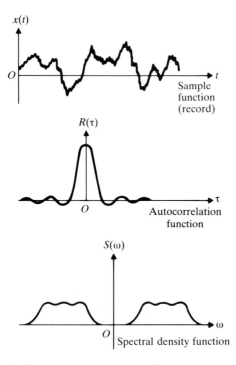

Figure 14.15   Wide-band stationary random process.

spectrum more or less uniformly. It is called *ideal white noise* if the band of frequencies $\omega_2 - \omega_1$ is infinitely wide. Ideal white noise is a physically unrealizable concept, since the mean square value of such a random process would be infinite because the area under the spectrum would be infinite. It is called *band-limited white noise* if the band of frequencies has finite cut-off frequencies ω_1 and ω_2 [14.8]. The mean square value of a band-limited white noise is given by the total area under the

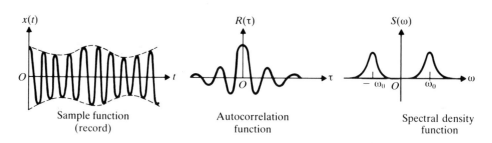

Figure 14.16   Narrow-band stationary random process.

spectrum, namely, $2S_0(\omega_2 - \omega_1)$, where S_0 denotes the constant value of the spectral density function.

EXAMPLE 14.4 **Autocorrelation and Mean Square Value of a Stationary Process**

The power spectral density of a stationary random process $x(t)$ is shown in Fig. 14.17(a). Find its autocorrelation function and the mean square value.

Given: Power spectral density function, $S_x(\omega) = S_0$.

Find: Autocorrelation function $R_x(\tau)$ and the mean square value $E[x^2]$.

Approach: Use Wiener-Khintchine relationships and Eq. (14.62).

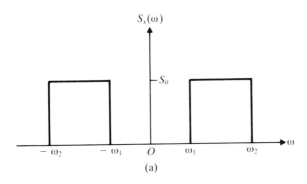

(a)

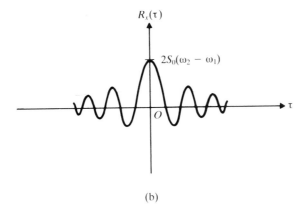

(b)

Figure 14.17

Solution. (i) Since $S_x(\omega)$ is real and even in ω, Eq. (14.61) can be rewritten as

$$R_x(\tau) = 2\int_0^\infty S_x(\omega)\cos \omega\tau \, d\omega = 2S_0 \int_{\omega_1}^{\omega_2} \cos \omega\tau \, d\omega$$

$$= 2S_0\left(\frac{1}{\tau}\sin \omega\tau\right)\Bigg|_{\omega_1}^{\omega_2} = \frac{2S_0}{\tau}(\sin \omega_2\tau - \sin \omega_1\tau)$$

$$= \frac{4S_0}{\tau}\cos \frac{\omega_1 + \omega_2}{2}\tau \sin \frac{\omega_2 - \omega_1}{2}\tau$$

This function is shown graphically in Fig. 14.17(b).
(ii) The mean square value of the random process is given by

$$E[x^2] = \int_{-\infty}^{\infty} S_x(\omega) \, d\omega = 2S_0\int_{\omega_1}^{\omega_2} d\omega = 2S_0(\omega_2 - \omega_1)$$

14.12 RESPONSE OF A SINGLE DEGREE OF FREEDOM SYSTEM

The equation of motion for the system shown in Fig. 14.18 is

$$\ddot{y} + 2\zeta\omega_n\dot{y} + \omega_n^2 y = x(t) \tag{14.67}$$

where

$$x(t) = \frac{F(t)}{m}, \qquad \omega_n = \sqrt{\frac{k}{m}}, \qquad \zeta = \frac{c}{c_c}, \qquad \text{and} \qquad c_c = 2km.$$

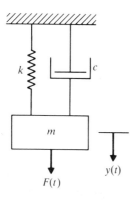

Figure 14.18

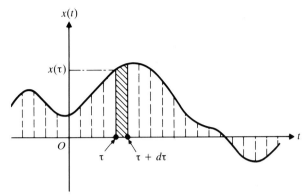

(a) Forcing function in the form of a series of impulses

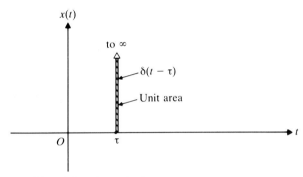

(b) Unit impulse excitation at $t = \tau$

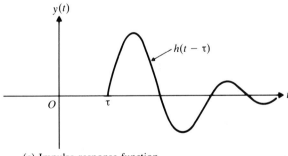

(c) Impulse response function

Figure 14.19

The solution of Eq. (14.67) can be obtained by using either the impulse response approach or the frequency response approach.

14.12.1 Impulse Response Approach

Here we consider the forcing function $x(t)$ to be made up of a series of impulses of varying magnitude as shown in Fig. 14.19(a) (see Section 4.5.2). Let the impulse applied at time τ be denoted as $x(\tau)\,d\tau$. If $y(t) = H(t - \tau)$ denotes the response the unit impulse[†] excitation $\delta(t - \tau)$, it is called the impulse response function. The total response of the system at time t can be found by superposing the responses to impulses of magnitude $x(\tau)\,d\tau$ applied at different values of $t = \tau$. The response to the excitation $x(\tau)\,d\tau$ will be $[x(\tau)\,d\tau]h(t - \tau)$, and the response to the total excitation will be given by the superposition or convolution integral:

$$y(t) = \int_{-\infty}^{t} x(\tau)h(t - \tau)\,d\tau \tag{14.68}$$

**14.12.2
Frequency
Response
Approach**

The transient function $x(t)$ can be expressed in terms of its Fourier transform $X(\omega)$ as

$$x(t) = \frac{1}{2\pi} \int_{\omega = -\infty}^{\infty} X(\omega)e^{i\omega t}\,d\omega \tag{14.69}$$

Thus $x(t)$ can be considered as the superposition of components of different frequencies ω. If we consider the forcing function of unit modulus as

$$\underset{\sim}{x}(t) = e^{i\omega t} \tag{14.70}$$

its response can be denoted as

$$\underset{\sim}{y}(t) = H(\omega)e^{i\omega t} \tag{14.71}$$

where $H(\omega)$ is called the complex frequency response function (see Section 3.5). Since the actual excitation is given by the superposition of components of different frequencies (Eq. (14.69)), the total response of the system also can be obtained by superposition as

$$y(t) = H(\omega)x(t) = \int_{-\infty}^{\infty} H(\omega)\frac{1}{2\pi}X(\omega)e^{i\omega t}\,d\omega$$

$$= \frac{1}{2\pi} \int_{-\infty}^{\infty} H(\omega)X(\omega)e^{i\omega t}\,d\omega \tag{14.72}$$

If $Y(\omega)$ denotes the Fourier transform of the response function $y(t)$, we can express $y(t)$ in terms of $Y(\omega)$ as

$$y(t) = \frac{1}{2\pi} \int_{-\infty}^{\infty} Y(\omega)e^{i\omega t}\,d\omega \tag{14.73}$$

Comparison of Eqs. (14.72) and (14.73) yields

$$Y(\omega) = H(\omega)X(\omega) \tag{14.74}$$

[†]   The unit impulse applied at $t = \tau$ is denoted as

$$x(t) = \delta(t - \tau)$$

where $\delta(t - \tau)$ is the Dirac delta function with [see Fig. 14.19(b)]

$$\delta(t - \tau) \to \infty \qquad \text{as } t \to \tau$$
$$\delta(t - \tau) = 0 \qquad \text{for all } t \text{ except at } t = \tau$$
$$\int_{-\infty}^{\infty} \delta(t - \tau)\,dt = 1 \text{ (area under the curve is unity)}$$

14.12.3
Characteristics of
the Response
Function

(i) Since $h(t - \tau) = 0$ when $t < \tau$ or $\tau > t$ (i.e., the response before the application of the impulse is zero), the upper limit of integration in Eq. (14.68) can be replaced by ∞ so that

$$y(t) = \int_{-\infty}^{\infty} x(\tau) h(t - \tau) \, d\tau \tag{14.75}$$

(ii) By changing the variable from τ to $\theta = t - \tau$, Eq. (14.75) can be rewritten as

$$y(t) = \int_{-\infty}^{\infty} x(t - \theta) h(\theta) \, d\theta \tag{14.76}$$

(iii) The superposition integral, Eq. (14.68) or (14.75) or (14.76), can be used to find the response of the system $y(t)$ for any arbitrary excitation $x(t)$ once the impulse-response function of the system $h(t)$ is known. The Fourier integral, Eq. (14.72), can also be used to find the response of the system once the complex frequency response of the system, $H(\omega)$, is known. Although the two approaches appear to be different, they are intimately related to one another. To see their inter-relationship, consider the excitation of the system to be a unit impulse $\delta(\tau)$ in Eq. (14.72). By definition, the response is $h(t)$ and Eq. (14.72) gives

$$y(t) = h(t) = \frac{1}{2\pi} \int_{-\infty}^{\infty} X(\omega) H(\omega) e^{i\omega t} \, d\omega \tag{14.77}$$

where $X(\omega)$ is the Fourier transform of $x(t) = \delta(t)$:

$$X(\omega) = \int_{-\infty}^{\infty} x(t) e^{-i\omega t} \, dt = \int_{-\infty}^{\infty} \delta(t) e^{-i\omega t} \, dt = 1 \tag{14.78}$$

since $\delta(t) = 0$ everywhere except at $t = 0$ where it has a unit area and $e^{-i\omega t} = 1$ at $t = 0$. Equations (14.77) and (14.78) give

$$h(t) = \frac{1}{2\pi} \int_{-\infty}^{\infty} H(\omega) e^{i\omega t} \, d\omega \tag{14.79}$$

which can be recognized as the Fourier integral representation of $h(t)$ in which $H(\omega)$ is the Fourier transformation of $h(t)$:

$$H(\omega) = \int_{-\infty}^{\infty} h(t) e^{-i\omega t} \, dt \tag{14.80}$$

14.13 RESPONSE DUE TO STATIONARY RANDOM EXCITATIONS

In the previous section, the relationships between excitation and response were derived for arbitrary known excitations $x(t)$. In this section, we consider similar relationships when the excitation is a stationary random process. When the excitation is a stationary random process, the response will also be a stationary random process [14.15, 14.16]. We consider the relation between the excitation and the response using the impulse response (time domain) as well as the frequency response (frequency domain) approaches.

**14.13.1
Impulse
Response
Approach**

Mean Values. The response for any particular sample excitation is given by Eq. (14.76):

$$y(t) = \int_{-\infty}^{\infty} x(t - \theta) h(\theta) \, d\theta \qquad (14.81)$$

For ensemble average, we write Eq. (14.81) for every (x, y) pair in the ensemble and then take the average to obtain[†]

$$E[y(t)] = E\left[\int_{-\infty}^{\infty} x(t - \theta) h(\theta) \, d\theta\right] = \int_{-\infty}^{\infty} E[x(t - \theta)] h(\theta) \, d\theta \quad (14.82)$$

Since the excitation is assumed to be stationary, $E[x(\tau)]$ is a constant independent of τ, Eq. (14.82) becomes

$$E[y(t)] = E[x(t)] \int_{-\infty}^{\infty} h(\theta) \, d\theta \qquad (14.83)$$

The integral in Eq. (14.83) can be obtained by setting $\omega = 0$ in Eq. (14.80) so that

$$H(0) = \int_{-\infty}^{\infty} h(t) \, dt \qquad (14.84)$$

Thus a knowledge of either the impulse response function $h(t)$ or the frequency response function $H(\omega)$ can be used to find the relationship between the mean values of the excitation and the response. It is to be noted that both $E[x(t)]$ and $E[y(t)]$ are independent of t.

Autocorrelation. We can use a similar procedure to find the relationship between the autocorrelation functions of the excitation and the response. For this, we first write

$$y(t) y(t + \tau) = \int_{-\infty}^{\infty} x(t - \theta_1) h(\theta_1) \, d\theta_1 \cdot \int_{-\infty}^{\infty} x(t + \tau - \theta_2) h(\theta_2) \, d\theta_2$$

$$= \int_{-\infty}^{\infty} \int_{-\infty}^{\infty} x(t - \theta_1) x(t + \tau - \theta_2) h(\theta_1) h(\theta_2) \, d\theta_1 \, d\theta_2 \quad (14.85)$$

where θ_1 and θ_2 are used instead of θ to avoid confusion. The autocorrelation function of $y(t)$ can be found as

$$R_y(\tau) = E[y(t) y(t + \tau)]$$

$$= \int_{-\infty}^{\infty} \int_{-\infty}^{\infty} E[x(t - \theta_1) x(t + \tau - \theta_2)] h(\theta_1) h(\theta_2) \, d\theta_1 \, d\theta_2$$

$$= \int_{-\infty}^{\infty} \int_{-\infty}^{\infty} R_x(\tau + \theta_1 - \theta_2) h(\theta_1) h(\theta_2) \, d\theta_1 \, d\theta_2 \qquad (14.86)$$

[†] In deriving Eq. (14.82), the integral is considered as a limiting case of a summation and hence the average of a sum is treated to be same as the sum of the averages, that is,

$$E[x_1 + x_2 + \cdots] = E[x_1] + E[x_2] + \cdots$$

**14.13.2
Frequency
Response
Approach**

Power Spectral Density. The response of the system can also be described by its power spectral density, which by definition, is (see Eq. (14.60)):

$$S_y(\omega) = \frac{1}{2\pi} \int_{-\infty}^{\infty} R_y(\tau) e^{-i\omega\tau} \, d\tau \qquad (14.87)$$

Substitution of Eq. (14.86) into Eq. (14.87) gives

$$S_y(\omega) = \frac{1}{2\pi} \int_{-\infty}^{\infty} e^{-i\omega\tau} \, d\tau \int_{-\infty}^{\infty} \int_{-\infty}^{\infty} R_x(\tau + \theta_1 - \theta_2) h(\theta_1) h(\theta_2) \, d\theta_1 \, d\theta_2 \qquad (14.88)$$

Introduction of

$$e^{i\omega\theta_1} e^{-i\omega\theta_2} e^{-i\omega(\theta_1 - \theta_2)} = 1 \qquad (14.89)$$

into Eq. (14.88) results in

$$S_y(\omega) = \int_{-\infty}^{\infty} h(\theta_1) e^{i\omega\theta_1} \, d\theta_1 \int_{-\infty}^{\infty} h(\theta_2) e^{-i\omega\theta_2} \, d\theta_2$$
$$\times \frac{1}{2\pi} \int_{-\infty}^{\infty} R_x(\tau + \theta_1 - \theta_2) e^{-i\omega(\theta_1 - \theta_2)} \, d\tau \qquad (14.90)$$

In the third integral on the right-hand side of Eq. (14.90), θ_1 and θ_2 are constants and the introduction of a new variable of integration η as

$$\eta = \tau + \theta_1 - \theta_2 \qquad (14.91)$$

leads to

$$\frac{1}{2\pi} \int_{-\infty}^{\infty} R_x(\tau + \theta_1 - \theta_2) e^{-i\omega(\tau + \theta_1 - \theta_2)} \, d\tau = \frac{1}{2\pi} \int_{-\infty}^{\infty} R_x(\eta) e^{-i\omega\eta} \, d\eta \equiv S_x(\omega) \qquad (14.92)$$

The first and the second integrals on the right-hand side of Eq. (14.90) can be recognized as the complex frequency response functions $H(\omega)$ and $H(-\omega)$, respectively. Since $H(-\omega)$ is the complex conjugate of $H(\omega)$, Eq. (14.90) gives

$$S_y(\omega) = |H(\omega)|^2 S_x(\omega) \qquad (14.93)$$

This equation gives the relationship between the power spectral densities of the excitation and the response.

Mean Square Response. The mean square response of the stationary random process $y(t)$ can be determined either from the autocorrelation function $R_y(\tau)$ or from the power spectral density $S_y(\omega)$:

$$E[y^2] = R_y(0) = \int_{-\infty}^{\infty} \int_{-\infty}^{\infty} R_x(\theta_1 - \theta_2) h(\theta_1) h(\theta_2) \, d\theta_1 \, d\theta_2 \qquad (14.94)$$

and

$$E[y^2] = \int_{-\infty}^{\infty} S_y(\omega) \, d\omega = \int_{-\infty}^{\infty} |H(\omega)|^2 S_x(\omega) \, d\omega \qquad (14.95)$$

Note: Equations (14.93) and (14.95) form the basis for the random vibration analysis of single- and multi-degree of freedom systems [14.11, 14.12]. The random vibration analysis of road vehicles is given in Refs. [14.13, 14.14].

EXAMPLE 14.5   **Mean Square Value of Response**

A single degree of freedom system (Fig. 14.20a) is subjected to a force whose spectral density is a white noise $S_x(\omega) = S_0$. Find:

(i) the complex frequency response function of the system,

(ii) the power spectral density of the response, and

(iii) the mean square value of the response.

Given: Single degree of freedom system of Fig. 14.20(a) with $S_x(\omega) = S_0$.

Find: $H(\omega)$, $S_y(\omega)$, and $E[y^2]$.

Approach: Assume harmonic response to find the frequency response function. Use the relationship between the power spectral densities of input and output.

Solution.

(i) To find the complex frequency response function $H(\omega)$, we substitute the input as $e^{i\omega t}$ and the corresponding response as $y(t) = H(\omega)e^{i\omega t}$ in the equation of motion:

$$m\ddot{y} + c\dot{y} + ky = x(t)$$

to obtain

$$(-m\omega^2 + ic\omega + k)H(\omega)e^{i\omega t} = e^{i\omega t}$$

and

$$H(\omega) = \frac{1}{-m\omega^2 + ic\omega + k} \tag{E.1}$$

(ii) The power spectral density of the output can be found as

$$S_y(\omega) = |H(\omega)|^2 S_x(\omega) = S_0 \left| \frac{1}{-m\omega^2 + ic\omega + k} \right|^2 \tag{E.2}$$

(iii) The mean square value of the output is given by[†]

$$E[y^2] = \int_{-\infty}^{\infty} S_y(\omega)\, d\omega$$

$$= S_0 \int \left| \frac{1}{-m\omega^2 + k + ic\omega} \right|^2 d\omega = \frac{\pi S_0}{kc} \tag{E.3}$$

which can be seen to be independent of the magnitude of the mass m. The functions $H(\omega)$ and $S_y(\omega)$ are shown graphically in Fig. 14.20(b).

EXAMPLE 14.6   **Design of Columns of a Building**

A single-story building is modeled by four identical columns of Young's modulus E and height h and a rigid floor of weight W as shown in Fig. 14.21(a). The columns act as

---

[†]   The values of this and other similar integrals can be found in the literature [14.1]. For example, if

$$H(\omega) = \frac{i\omega B_1 + B_0}{-\omega^2 A_2 + i\omega A_1 + A_0}, \qquad \int_{-\infty}^{\infty} |H(\omega)|^2 d\omega = \pi \left\{ \frac{(B_0^2/A_0)A_2 + B_1^2}{A_1 A_2} \right\} \tag{E.4}$$

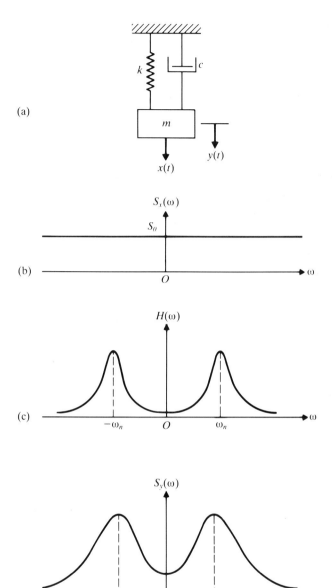

(a)

(b)

(c)

(d)

Figure 14.20

cantilevers fixed at the ground. The damping in the structure can be approximated by an equivalent viscous damping constant c. The ground acceleration due to an earthquake is approximated by a constant spectrum S_0. If each column has a tubular cross section with mean diameter d and wall thickness $t = d/10$, find the mean diameter of the columns such that the standard deviation of the displacement of the floor relative to the ground does not exceed a specified value δ.

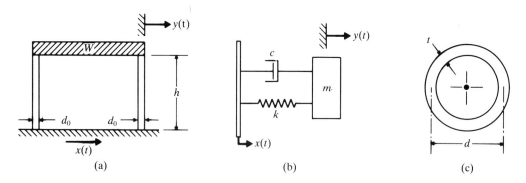

Figure 14.21

Given: Single story building: weight of floor = W. Columns: four columns, Young's modulus = E, mean diameter = d, wall thickness = $t = d/10$, and damping constant = c. Power spectral density of ground acceleration = $S_x(\omega) = S_0$. Permissible value of the standard deviation of displacement of floor relative to ground = δ.

Find: Mean diameter (d) of the columns.

Approach: Model the building as a single degree of freedom system. Use the relation between the power spectral densities of excitation and output.

Solution. The building can be modeled as a single degree of freedom system as shown in Fig. 14.21(b) with

$$m = W/g \tag{E.1}$$

and

$$k = 4\left(\frac{3EI}{h^3}\right) \tag{E.2}$$

since the stiffness of one cantilever beam (column) is equal to $(3EI/h^3)$, where E is the Young's modulus, h is the height, and I is the moment of inertia of the cross section of the columns given by (see Fig. 14.19(c)):

$$I = \frac{\pi}{64}\left(d_0^4 - d_i^4\right) \tag{E.3}$$

Equation (E.3) can be simplified, using $d_0 = d + t$ and $d_i = d - t$, as

$$
\begin{aligned}
I &= \frac{\pi}{64}\left(d_0^2 + d_i^2\right)\left(d_0 + d_i\right)\left(d_0 - d_i\right) \\
&= \frac{\pi}{64}\left[(d+t)^2 + (d-t)^2\right]\left[(d+t) + (d-t)\right]\left[(d+t) - (d-t)\right] \\
&= \frac{\pi}{8}\,dt\left(d^2 + t^2\right)
\end{aligned}
\tag{E.4}
$$

With $t = d/10$, Eq. (E.4) becomes

$$I = \frac{101\pi}{8000}d^4 = 0.03966d^4 \tag{E.5}$$

and hence Eq. (E.2) gives

$$k = \frac{12E(0.03966d^4)}{h^3} = \frac{0.47592\,Ed^4}{h^3} \tag{E.6}$$

When the base of the system moves, the equation of motion is given by (see Section 3.6):

$$m\ddot{z} + c\dot{z} + kz = -m\ddot{x} \tag{E.7}$$

where $z = y - x$ is the displacement of the mass (floor) relative to the ground. Equation (E.7) can be rewritten as

$$\ddot{z} + \frac{c}{m}\dot{z} + \frac{k}{m}z = -\ddot{x} \tag{E.8}$$

The complex frequency response function $H(\omega)$ can be obtained by making the substitution

$$\ddot{x} = e^{i\omega t} \qquad \text{and} \qquad z(t) = H(\omega)e^{i\omega t} \tag{E.9}$$

so that

$$\left[-\omega^2 + i\omega\frac{c}{m} + \frac{k}{m}\right]H(\omega)e^{i\omega t} = -e^{i\omega t}$$

which gives

$$H(\omega) = \frac{-1}{\left(-\omega^2 + i\omega\frac{c}{m} + \frac{k}{m}\right)} \tag{E.10}$$

The power spectral density of the response $z(t)$ is given by

$$S_z(\omega) = |H(\omega)|^2 S_{\ddot{x}}(\omega)$$

$$= S_0\left|\frac{-1}{\left(-\omega^2 + i\omega\frac{c}{m} + \frac{k}{m}\right)}\right|^2 \tag{E.11}$$

The mean square value of the response $z(t)$ can be determined, using Eq. (E.4) of Example 14.5, as

$$E[z^2] = \int_{-\infty}^{\infty} S_z(\omega)\, d\omega$$

$$= S_0\int_{-\infty}^{\infty} \left|\frac{-1}{\left(-\omega^2 + i\omega\frac{c}{m} + \frac{k}{m}\right)}\right|^2 d\omega$$

$$= S_0\left(\frac{\pi m^2}{kc}\right) \tag{E.12}$$

Substitution of the relations (E.1) and (E.6) into Eq. (E.12) gives

$$E[z^2] = \pi S_0 \frac{W^2 h^3}{g^2 c (0.47592\, E d^4)} \tag{E.13}$$

Assuming the mean value of $z(t)$ to be zero, the standard deviation of z can be found as

$$\sigma_z = \sqrt{E[z^2]} = \sqrt{\frac{\pi S_0 W^2 h^3}{0.47592\, g^2 c E d^4}} \tag{E.14}$$

Since $\sigma_z \leqslant \delta$, we find that

$$\frac{\pi S_0 W^2 h^3}{0.47592\, g^2 c E d^4} \leqslant \delta^2$$

or

$$d^4 \geqslant \frac{\pi S_0 W^2 h^3}{0.47592\, g^2 c E \delta^2} \tag{E.15}$$

Thus the required mean diameter of the columns is given by

$$d \geqslant \left\{ \frac{\pi S_0 W^2 h^3}{0.47592\, g^2 c E \delta^2} \right\}^{1/4} \tag{E.16}$$

REFERENCES

14.1. S. H. Crandall and W. D. Mark, *Random Vibration in Mechanical Systems*, Academic Press, New York, 1963.

14.2. D. E. Newland, *An Introduction to Random Vibrations and Spectral Analysis*, Longman, London, 1975.

14.3. J. D. Robson, *An Introduction to Random Vibration*, Edinburgh University Press, Edinburgh, 1963.

14.4. C. Y. Yang, *Random Vibration of Structures*, Wiley, New York, 1986.

14.5. A. Papoulis, *Probability, Random Variables and Stochastic Processes*, McGraw-Hill, New York, 1965.

14.6. J. S. Bendat and A. G. Piersol, *Engineering Applications of Correlation and Spectral Analysis*, John Wiley, New York, 1980.

14.7. P. Z. Peebles, Jr., *Probability, Random Variables, and Random Signal Principles*, McGraw-Hill, New York, 1980.

14.8. J. B. Roberts, "The response of a simple oscillator to band-limited white noise," *Journal of Sound and Vibration*, Vol. 3, 1966, pp. 115–126.

14.9. M. H. Richardson, "Fundamentals of the discrete Fourier transform," *Sound and Vibration*, Vol. 12, March 1978, pp. 40–46.

14.10. E. O. Brigham, *The Fast Fourier Transform*, Prentice-Hall, Englewood Cliffs, N.J., 1974.

14.11. J. K. Hammond, "On the response of single and multidegree of freedom systems to non-stationary random excitations," *Journal of Sound and Vibration*, Vol. 7, 1968, pp. 393–416.

14.12. S. H. Crandall, G. R. Khabbaz, and J. E. Manning, "Random vibration of an oscillator with nonlinear damping," *Journal of the Acoustical Society of America*, Vol. 36, 1964, pp. 1330–1334.

14.13. S. Kaufman, W. Lapinski, and R. C. McCaa, "Response of a single-degree-of-freedom isolator to a random disturbance," *Journal of the Acoustical Society of America*, Vol. 33, 1961, pp. 1108–1112.

14.14. C. J. Chisholm, "Random vibration techniques applied to motor vehicle structures," *Journal of Sound and Vibration*, Vol. 4, 1966, pp. 129–135.

14.15. Y. K. Lin, *Probabilistic Theory of Structural Dynamics*, McGraw-Hill, New York, 1967.

14.16. I. Elishakoff, *Probabilistic Methods in the Theory of Structures*, Wiley, New York, 1983.

14.17. H. W. Liepmann, "On the application of statistical concepts to the buffeting problem," *Journal of the Aeronautical Sciences*, Vol. 19, No. 12, 1952, pp. 793–800, 822.

REVIEW QUESTIONS

14.1. What is the difference between a sample space and an ensemble?

14.2. Define probability density function and probability distribution function.

14.3. How are the mean value and variance of a random variable defined?

14.4. What is a bivariate distribution function?

14.5. What is the covariance between two random variables X and Y?

14.6. Define the correlation coefficient, ρ_{XY}.

14.7. What are the bounds on the correlation coefficient?

14.8. What is a marginal density function?

14.9. What is autocorrelation function?

14.10. Explain the difference between a stationary process and a nonstationary process.

14.11. What are the bounds on the autocorrelation function of a stationary random process?

14.12. Define an ergodic process.

14.13. What are temporal averages?

14.14. What is a Gaussian random process? Why is it frequently used in vibration analysis?

14.15. What is Parseval's formula?

14.16. Define the following terms: power spectral density function, white noise, band-limited white noise, wide-band process, and narrow-band process.

14.17. How are the mean square value, autocorrelation function, and the power spectral density function of a stationary random process related?

14.18. What is an impulse response function?

14.19. Express the response of a single degree of freedom system using Duhamel integral.

14.20. What is complex frequency response function?

14.21. How are the power spectral density functions of input and output of a single degree of freedom system related?

14.22. What are Wiener-Khintchine relations?

PROBLEMS

The problem assignments are organized as follows:

Problems	Section covered	Topic covered
14.1, 14.10	14.3	Probability distribution
14.3, 14.11	14.4	Mean value and standard deviation
14.2, 14.4	14.5	Joint probability distribution
14.7, 14.9	14.6	Correlation functions
14.5, 14.26	14.7	Stationary random process
14.12	14.8	Gaussian random process
14.8, 14.13–14.16, 14.27	14.9	Fourier analysis
14.6, 14.17–14.22 14.23–14.25, 14.28– 14.30	14.10	Power spectral density
	14.12	Response of a single degree of freedom system
14.31	—	Project

14.1. The strength of the foundation of a reciprocating machine (x) has been found to vary between 20 and 30 kips/ft^2 according to the probability density function:

$$p(x) = \begin{cases} k\left(1 - \dfrac{x}{30}\right), & 20 \leqslant x \leqslant 30 \\ 0, & \text{elsewhere} \end{cases}$$

What is the probability of the foundation carrying a load greater than 28 kips/ft^2.

14.2. The joint density function of two random variables X and Y is given by

$$p_{X,Y}(x, y) = \begin{cases} \dfrac{xy}{9}, & 0 \leqslant x \leqslant 2, 0 \leqslant y \leqslant 3 \\ 0, & \text{elsewhere} \end{cases}$$

(i) Find the marginal density functions of X and Y. (ii) Find the means and standard deviations of X and Y. (iii) Find the correlation coefficient, $\rho_{X,Y}$.

14.3. The probability density function of a random variable x is given by

$$p(x) = \begin{cases} 0 & \text{for } x < 0 \\ 0.5 & \text{for } 0 \leqslant x \leqslant 2 \\ 0 & \text{for } x > 2 \end{cases}$$

Determine $E[x]$, $E[x^2]$, and σ_x.

14.4. If x and y are statistically independent, then $E[xy] = E[x]E[y]$. That is, the expected value of the product xy is equal to the product of the separate mean values. If $z = x + y$, where x and y are statistically independent, show that $E[z^2] = E[x^2] + E[y^2] + 2E[x]E[y]$.

14.5. The autocorrelation function of a random process $x(t)$ is given by

$$R_x(\tau) = 20 + \frac{5}{1 + 3\tau^2}$$

Find the mean square value of $x(t)$.

14.6. The autocorrelation function of a random process is given by

$$R_x(\tau) = A \cos \omega\tau; \quad -\frac{\pi}{2\omega} \leqslant \tau \leqslant \frac{\pi}{2\omega}$$

where A and ω are constants. Find the power spectral density of the random process.

14.7. Find the autocorrelation functions of the periodic functions shown in Fig. 14.22.

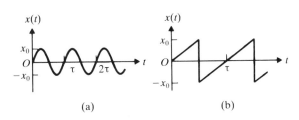

(a)   (b)

Figure 14.22

14.8. Find the complex form of the Fourier series for the wave shown in Fig. 14.22(b).

14.9. Compute the autocorrelation function of a periodic square wave with zero mean value and compare this result with that of a sinusoidal wave of the same period. Assume the amplitudes to be the same for both the waves.

14.10. The life, T in hours, of a vibration transducer is found to follow exponential distribution:

$$p_T(t) = \begin{cases} \lambda e^{-\lambda t}, & t \geqslant 0 \\ 0, & t < 0 \end{cases}$$

where λ is a constant. Find: (i) the probability distribution function of T, (ii) mean value of T, and (iii) standard deviation of T.

14.11. Find the temporal mean value and the mean square value of the function $x(t) = x_0 \sin(\pi t/2)$.

14.12. An air compressor of mass 100 kg is mounted on an undamped isolator and operates at an angular speed of 1800 rpm. The stiffness of the isolator is found to be a random variable with mean value $\bar{k} = 2.25 \times 10^6$ N/m and standard deviation $\sigma_k = 0.225 \times 10^6$ N/m following normal distribution. Find the probability of the natural frequency of the system exceeding the forcing frequency.

14.13–14.16. Find the Fourier transform of the functions shown in Figs. 14.23–14.26 and plot the corresponding spectrum.

14.17. A periodic function $F(t)$ is shown in Fig. 14.27. Use the values of the function $F(t)$ at ten equally spaced time stations t_i to find: (i) the spectrum of $F(t)$ and (ii) the mean square value of $F(t)$.

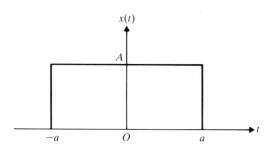

Figure 14.23

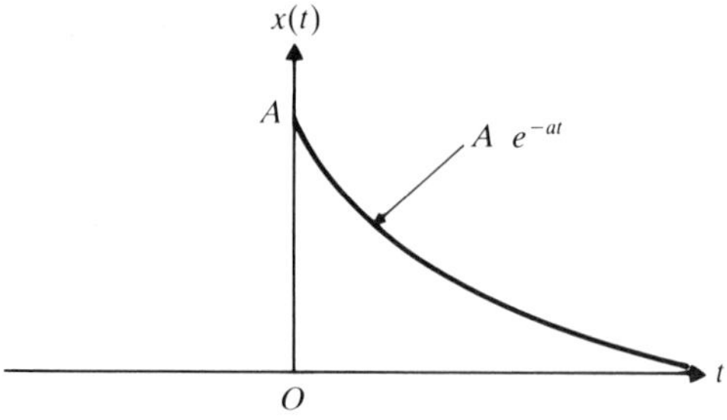

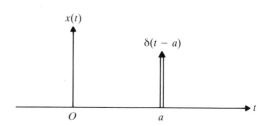

Figure 14.24

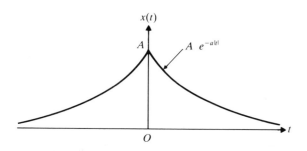

Figure 14.25

Figure 14.26

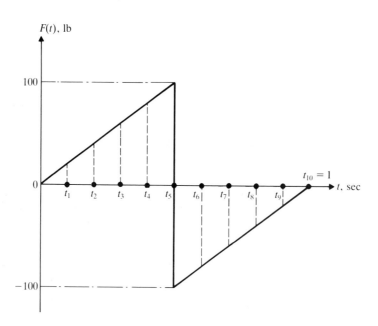

Figure 14.27

14.18. The autocorrelation function of a stationary random process $x(t)$ is given by

$$R_x(\tau) = ae^{-b|\tau|}$$

where a and b are constants. Find the power spectral density of $x(t)$.

14.19. Find the autocorrelation function of a random process whose power spectral density is given by $S(\omega) = S_0 = $ constant between the frequencies ω_1 and ω_2.

14.20. The autocorrelation function of a Gaussian random process representing the unevenness of a road surface is given by

$$R_x(\tau) = \sigma_x^2 e^{-\alpha|v\tau|} \cos \beta v \tau$$

where σ_x^2 is the variance of the random process and v is the velocity of the vehicle. The values of σ_x, α, and β for different types of road are as follows:

Type of road	σ_x	α	β
Asphalt	1.1	0.2	0.4
Paved	1.6	0.3	0.6
Gravel	1.8	0.5	0.9

Compute the spectral density of the road surface for the different types of road.

14.21. Compute the autocorrelation function corresponding to the ideal white noise spectral density.

14.22. Starting from Eqs. (14.60) and (14.61), derive the relations

$$R(\tau) = \int_0^\infty S(f) \cos 2\pi f \tau \cdot df$$

$$S(f) = 4 \int_0^\infty R(\tau) \cos 2\pi f \tau \cdot d\tau$$

14.23. Write a computer program to find the mean square value of the response of a single degree of freedom system subjected to a random excitation whose power spectral density function is given as $S_x(\omega)$.

14.24. A machine, modeled as a single degree of freedom system, has the following parameters: $mg = 2000$ lb, $k = 4 \times 10^4$ lb/in., and $c = 1200$ lb-in./sec. It is subjected to the force shown in Fig. 14.27. Find the mean square value of the response of the machine (mass).

14.25. A mass, connected to a damper as shown in Fig. 14.28, is subjected to a force $F(t)$. Find the frequency response function $H(\omega)$ for the velocity of the mass.

14.26. The spectral density of a random signal is given by

$$S(f) = \begin{cases} 0.0001 \text{ m}^2/\text{cycle}/\text{sec}, & 10 \text{ Hz} \leqslant f \leqslant 1000 \text{ Hz} \\ 0, & \text{elsewhere} \end{cases}$$

Find the standard deviation and the root mean square value of the signal by assuming its mean value to be 0.05 m.

14.27. Derive Eq. (14.46) from Eq. (14.45).

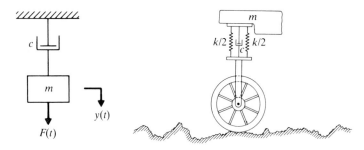

Figure 14.28   **Figure 14.29**

14.28. A simplified model of a motor cycle traveling over a rough road is shown in Fig. 14.29. It is assumed that the wheel is rigid, the wheel does not leave the road surface, and the cycle moves at a constant speed v. The cycle has a mass m and the suspension system has a spring constant k and a damping constant c. If the power spectral density of the rough road surface is taken as S_0, find the mean square value of the vertical displacement of the cycle (mass, m).

14.29. The motion of a lifting surface about the steady flight path due to atmospheric turbulence can be represented by the equation

$$\ddot{x}(t) + 2\zeta\omega_n\dot{x}(t) + \omega_n^2 x(t) = \frac{1}{m}F(t)$$

where ω_n is the natural frequency, m is the mass, and ζ is the damping coefficient of the system. The forcing function $F(t)$ denotes the random lift due to the air turbulence and its spectral density is given by [14.17]

$$S_F(\omega) = \frac{S_T(\omega)}{\left(1 + \dfrac{\pi\omega c}{v}\right)}$$

where c is the chord length and v is the forward velocity of the lifting surface and $S_T(\omega)$ is the spectral density of the upward velocity of air due to turbulence, given by

$$S_T(\omega) = A^2\frac{1 + \left(\dfrac{L\omega}{v}\right)^2}{\left\{1 + \left(\dfrac{L\omega}{v}\right)^2\right\}^2}$$

where A is a constant and L is the scale of turbulence (constant). Find the mean square value of the response $x(t)$ of the lifting surface.

14.30. The wing of an airplane flying in gusty wind has been modeled as a spring-mass-damper system as shown in Fig. 14.30. The undamped and damped natural frequencies of the wing are found to be ω_1 and ω_2, respectively. The mean square value of the displacement of m_{eq} (i.e., the wing) is observed to be δ under the action of the random wind force whose power spectral density is given by $S(\omega) = S_0$. Derive expressions for the system parameters m_{eq}, k_{eq}, and c_{eq} in terms of ω_1, ω_2, δ, and S_0.

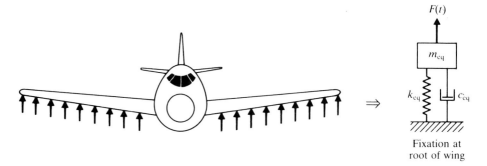

Fixation at
root of wing

Figure 14.30

Project:

14.31. The water tank shown in Fig. 14.31 is supported by a hollow circular steel column. The tank, made of steel, is in the form of a thin-walled pressure vessel and has a capacity of 10,000 gallons. Design the column to satisfy the following specifications:
(1) The undamped natural frequency of vibration of the tank, either empty or full, must exceed a value of 1 Hz.
(2) The mean square value of the displacement of the tank, either empty or full, must not exceed a value of 16 in^2 when subjected to an earthquake ground acceleration whose power spectral density is given by $S(\omega) = 0.0002$ m^2/cycle/sec. Assume damping to be 10 percent of the critical value.

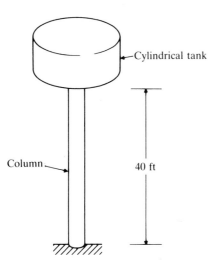

Cylindrical tank

Column

40 ft

Figure 14.31

Matrices

Arthur Cayley (1821–1895) was a British mathematician and professor of mathematics at Cambridge University. His greatest work, produced with James Joseph Sylvester, was the development of the theory of invariants, which played a crucial role in the theory of relativity. He made many important contributions to n-dimensional geometry and invented and developed the theory of matrices. (Courtesy The Granger Collection)

A.1 DEFINITIONS

Matrix. A matrix is a rectangular array of numbers. An array having m rows and n columns enclosed in brackets is called an m-by-n matrix. If $[A]$ is an $m \times n$ matrix, it is denoted as

$$[A] = [a_{ij}] = \begin{bmatrix} a_{11} & a_{12} & \cdots & a_{1n} \\ a_{21} & a_{22} & \cdots & a_{2n} \\ \vdots & & & \\ a_{m1} & a_{m2} & \cdots & a_{mn} \end{bmatrix} \tag{A.1}$$

where the numbers a_{ij} are called the *elements* of the matrix. The first subscript i denotes the row and the second subscript j specifies the column in which the element a_{ij} appears.

Square Matrix. When the number of rows (m) is equal to the number of columns (n), the matrix is called a *square matrix of order n*.

Column Matrix. A matrix consisting of only one column—that is, an $m \times 1$ matrix —is called a *column matrix* or more commonly a *column vector*. Thus if $\vec{a}$ is a column vector having m elements, it can be represented as

$$\vec{a} = \begin{Bmatrix} a_1 \\ a_2 \\ \vdots \\ a_m \end{Bmatrix} \tag{A.2}$$

Row Matrix. A matrix consisting of only one row—that is, a $1 \times n$ matrix—is called a *row matrix* or a *row vector*. If $\lfloor b \rfloor$ is a row vector, it can be denoted as

$$\lfloor b \rfloor = \begin{bmatrix} b_1 & b_2 & \cdots & b_n \end{bmatrix} \tag{A.3}$$

Diagonal Matrix. A square matrix in which all the elements are zero except those on the principal diagonal is called a *diagonal matrix*. For example, if $[A]$ is a diagonal matrix of order n, it is given by

$$[A] = \begin{bmatrix} a_{11} & 0 & 0 & \cdots & 0 \\ 0 & a_{22} & 0 & \cdots & 0 \\ 0 & 0 & a_{33} & \cdots & 0 \\ \vdots & & & & \\ 0 & 0 & 0 & \cdots & a_{nn} \end{bmatrix} \tag{A.4}$$

Identity Matrix. If all the elements of a diagonal matrix have a value 1, then the matrix is called an *identity matrix* or *unit matrix* and is usually denoted as $[I]$.

Zero Matrix. If all the elements of a matrix are zero, it is called a *zero* or *null matrix* and is denoted as [0]. If [0] is of order 2×4, it is given by

$$[0] = \begin{bmatrix} 0 & 0 & 0 & 0 \\ 0 & 0 & 0 & 0 \end{bmatrix} \tag{A.5}$$

Symmetric Matrix. If the element in ith row and jth column is same as the one in jth row and ith column in a square matrix, it is called a *symmetric matrix*. This means that if $[A]$ is a symmetric matrix, we have $a_{ji} = a_{ij}$. For example,

$$[A] = \begin{bmatrix} 4 & -1 & -3 \\ -1 & 0 & 7 \\ -3 & 7 & 5 \end{bmatrix} \tag{A.6}$$

is a symmetric matrix of order 3.

Transpose of a Matrix. The transpose of an $m \times n$ matrix $[A]$ is the $n \times m$ matrix obtained by interchanging the rows and columns of $[A]$ and is denoted as $[A]^T$. Thus if

$$[A] = \begin{bmatrix} 2 & 4 & 5 \\ 3 & 1 & 8 \end{bmatrix} \tag{A.7}$$

then $[A]^T$ is given by

$$[A]^T = \begin{bmatrix} 2 & 3 \\ 4 & 1 \\ 5 & 8 \end{bmatrix} \tag{A.8}$$

Note that the transpose of a column matrix (vector) is a row matrix (vector), and vice versa.

Trace. The sum of the main diagonal elements of a square matrix $[A] = [a_{ij}]$ is called the *trace* of $[A]$ and is given by

$$\text{Trace}[A] = a_{11} + a_{22} + \cdots + a_{nn} \qquad (A.9)$$

Determinant. If $[A]$ denotes a square matrix of order n, then the determinant of $[A]$ is denoted as $|[A]|$. Thus

$$|[A]| = \begin{vmatrix} a_{11} & a_{12} & \cdots & a_{1n} \\ a_{21} & a_{22} & \cdots & a_{2n} \\ \vdots & & & \\ a_{n1} & a_{n2} & \cdots & a_{nn} \end{vmatrix} \qquad (A.10)$$

The value of a determinant can be found by obtaining the minors and cofactors of the determinant.

The *minor* of the element a_{ij} of the determinant $|[A]|$ of order n is a determinant of order $(n - 1)$ obtained by deleting the row i and the column j of the original determinant. The minor of a_{ij} is denoted as M_{ij}.

The *cofactor* of the element a_{ij} of the determinant $|[A]|$ of order n is the minor of the element a_{ij}, with either a plus or a minus sign attached; it is defined as

$$\text{cofactor of } a_{ij} = \beta_{ij} = (-1)^{i+j} M_{ij} \qquad (A.11)$$

where M_{ij} is the minor of a_{ij}. For example, the cofactor of the element a_{32} of

$$\det[A] = \begin{vmatrix} a_{11} & a_{12} & a_{13} \\ a_{21} & a_{22} & a_{23} \\ a_{31} & a_{32} & a_{33} \end{vmatrix} \qquad (A.12)$$

is given by

$$\beta_{32} = (-1)^5 M_{32} = -\begin{vmatrix} a_{11} & a_{13} \\ a_{21} & a_{23} \end{vmatrix} \qquad (A.13)$$

The value of a second order determinant $|[A]|$ is defined as

$$\det[A] = \begin{vmatrix} a_{11} & a_{12} \\ a_{21} & a_{22} \end{vmatrix} = a_{11}a_{22} - a_{12}a_{21} \qquad (A.14)$$

The value of an nth order determinant $|[A]|$ is defined as

$$\det[A] = \sum_{j=1}^{n} a_{ij}\beta_{ij} \quad \text{for any specific row } i$$

or

$$\det[A] = \sum_{i=1}^{n} a_{ij}\beta_{ij} \quad \text{for any specific column } j \qquad (A.15)$$

For example, if

$$det[A] = |[A]| = \begin{vmatrix} 2 & 2 & 3 \\ 4 & 5 & 6 \\ 7 & 8 & 9 \end{vmatrix} \qquad (A.16)$$

then, by selecting the first column for expansion, we obtain

$$det[A] = 2\begin{vmatrix} 5 & 6 \\ 8 & 9 \end{vmatrix} - 4\begin{vmatrix} 2 & 3 \\ 8 & 9 \end{vmatrix} + 7\begin{vmatrix} 2 & 3 \\ 5 & 6 \end{vmatrix}$$
$$= 2(45 - 48) - 4(18 - 24) + 7(12 - 15) = -3 \qquad (A.17)$$

Properties of Determinants

1. The value of a determinant is not affected if rows (or columns) are written as columns (or rows) in the same order.

2. If all the elements of a row (or a column) are zero, the value of the determinant is zero.

3. If any two rows (or two columns) are interchanged, the value of the determinant is multiplied by -1.

4. If all the elements of one row (or one column) are multiplied by the same constant a, the value of the new determinant is a times the value of the original determinant.

5. If the corresponding elements of two rows (or two columns) of a determinant are proportional, the value of the determinant is zero. For example,

$$det[A] = \begin{vmatrix} 4 & 7 & -8 \\ 2 & 5 & -4 \\ -1 & 3 & 2 \end{vmatrix} = 0 \qquad (A.18)$$

Adjoint Matrix. The adjoint matrix of a square matrix $[A] = [a_{ij}]$ is defined as the matrix obtained by replacing each element a_{ij} by its cofactor β_{ij} and then transposing. Thus

$$\text{Adjoint}[A] = \begin{bmatrix} \beta_{11} & \beta_{12} & \cdots & \beta_{1n} \\ \beta_{21} & \beta_{22} & \cdots & \beta_{2n} \\ \vdots & & & \\ \beta_{n1} & \beta_{n2} & \cdots & \beta_{nn} \end{bmatrix}^T = \begin{bmatrix} \beta_{11} & \beta_{21} & \cdots & \beta_{n1} \\ \beta_{12} & \beta_{22} & \cdots & \beta_{n2} \\ \vdots & & & \\ \beta_{1n} & \beta_{2n} & \cdots & \beta_{nn} \end{bmatrix} \qquad (A.19)$$

Inverse Matrix. The inverse of a square matrix $[A]$ is written as $[A]^{-1}$ and is defined by the following relationship:

$$[A]^{-1}[A] = [A][A]^{-1} = [I] \qquad (A.20)$$

where $[A]^{-1}[A]$, for example, denotes the product of the matrix $[A]^{-1}$ and $[A]$. The inverse matrix of $[A]$ can be determined (see Ref. [A.1]):

$$[A]^{-1} = \frac{\text{adjoint}[A]}{det[A]} \qquad (A.21)$$

when $\det[A]$ is not equal to zero. For example, if

$$[A] = \begin{bmatrix} 2 & 2 & 3 \\ 4 & 5 & 6 \\ 7 & 8 & 9 \end{bmatrix} \tag{A.22}$$

its determinant has a value $\det[A] = -3$. The cofactor of a_{11} is

$$\beta_{11} = (-1)^2 \begin{vmatrix} 5 & 6 \\ 8 & 9 \end{vmatrix} = -3 \tag{A.23}$$

In a similar manner, we can find the other cofactors and determine

$$[A]^{-1} = \frac{\text{adjoint}[A]}{\det[A]} = \frac{1}{-3} \begin{bmatrix} -3 & 6 & -3 \\ 6 & -3 & 0 \\ -3 & -2 & 2 \end{bmatrix} = \begin{bmatrix} 1 & -2 & 1 \\ -2 & 1 & 0 \\ 1 & 2/3 & -2/3 \end{bmatrix} \tag{A.24}$$

Singular Matrix. A square matrix is said to be singular if its determinant is zero.

A.2   BASIC MATRIX OPERATIONS

Equality of Matrices. Two matrices $[A]$ and $[B]$, having the same order, are equal if and only if $a_{ij} = b_{ij}$ for every i and j.

Addition and Subtraction of Matrices. The sum of the two matrices $[A]$ and $[B]$, having the same order, is given by the sum of the corresponding elements. Thus if $[C] = [A] + [B] = [B] + [A]$, we have $c_{ij} = a_{ij} + b_{ij}$ for every i and j. Similarly, the difference between two matrices $[A]$ and $[B]$ of the same order, $[D]$, is given by $[D] = [A] - [B]$ with $d_{ij} = a_{ij} - b_{ij}$ for every i and j.

Multiplication of Matrices. The product of two matrices $[A]$ and $[B]$ is defined only if they are conformable—that is, if the number of columns of $[A]$ is equal to the number of rows of $[B]$. If $[A]$ is of order $m \times n$ and $[B]$ is of order $n \times p$, then the product $[C] = [A][B]$ is of order $m \times p$ and is defined by $[C] = [c_{ij}]$, with

$$c_{ij} = \sum_{k=1}^{n} a_{ik} b_{kj} \tag{A.25}$$

This means that c_{ij} is the quantity obtained by multiplying the ith row of $[A]$ and the jth column of $[B]$ and summing these products. For example, if

$$[A] = \begin{bmatrix} 2 & 3 & 4 \\ 1 & -5 & 6 \end{bmatrix} \quad \text{and} \quad [B] = \begin{bmatrix} 8 & 0 \\ 2 & 7 \\ -1 & 4 \end{bmatrix} \tag{A.26}$$

then

$$[C] = [A][B] = \begin{bmatrix} 2 & 3 & 4 \\ 1 & -5 & 6 \end{bmatrix} \begin{bmatrix} 8 & 0 \\ 2 & 7 \\ -1 & 4 \end{bmatrix}$$

$$= \begin{bmatrix} 2 \times 8 + 3 \times 2 + 4 \times (-1) & 2 \times 0 + 3 \times 7 + 4 \times 4 \\ 1 \times 8 + (-5) \times 2 + 6 \times (-1) & 1 \times 0 + (-5) \times 7 + 6 \times 4 \end{bmatrix}$$

$$= \begin{bmatrix} 18 & 37 \\ -8 & -11 \end{bmatrix} \tag{A.27}$$

If the matrices are conformable, the matrix multiplication process is associative:

$$([A][B])[C] = [A]([B][C]) \tag{A.28}$$

and is distributive:

$$([A] + [B])[C] = [A][C] + [B][C] \tag{A.29}$$

The product $[A][B]$ denotes the premultiplication of $[B]$ by $[A]$ or the postmultiplication of $[A]$ by $[B]$. It is to be noted that the product $[A][B]$ is not necessarily equal to $[B][A]$.

The transpose of a matrix product can be found to be the product of the transposes of the separate matrices in reverse order. Thus, if $[C] = [A][B]$,

$$[C]^T = ([A][B])^T = [B]^T[A]^T \tag{A.30}$$

The inverse of a matrix product can be determined from the product of the inverses of the separate matrices in reverse order. Thus if $[C] = [A][B]$,

$$[C]^{-1} = ([A][B])^{-1} = [B]^{-1}[A]^{-1} \tag{A.31}$$

REFERENCE

A.1. S. Barnett, *Matrix Methods for Engineers and Scientists*, McGraw-Hill, New York, 1982.

Laplace Transform Pairs

Pierre Simon Laplace (1749–1827) was a French
mathematician, remembered for his fundamental
contributions to probability theory, mathematical physics,
and celestial mechanics; the name Laplace occurs in both
mechanical and electrical engineering. Much use is made
of Laplace transforms in vibrations and applied mechanics,
and the Laplace equation is applied extensively in the
study of electric and magnetic fields. (Courtesy Brown
Brothers)

Laplace Domain $\bar{f}(s) = \int_0^\infty f(t)e^{-st}\, dt$	Time Domain $f(t)$
1. $c_1 \bar{f}(s) + c_2 \bar{g}(s)$	$c_1 f(t) + c_2 g(t)$
2. $\bar{f}\left(\dfrac{s}{a}\right)$	$f(a \cdot t)a$
3. $\bar{f}(s)\bar{g}(s)$	$\displaystyle\int_0^t f(t-\tau)g(\tau)\,d\tau = \int_0^t f(\tau)g(t-\tau)\,d\tau$
4. $s^n \bar{f}(s) - \displaystyle\sum_{j=1}^n s^{n-j}\dfrac{d^{j-1}f}{dt^{j-1}}(0)$	$\dfrac{d^n f}{dt^n}(t)$
5. $\dfrac{1}{s^n}\bar{f}(s)$	$\underbrace{\displaystyle\int_0^t \cdots \int_0^t f(\tau)\,d\tau \cdots d\tau}_{n}$
6. $\bar{f}(s+a)$	$e^{-at}f(t)$
7. $\dfrac{a}{s(s+a)}$	$1 - e^{-at}$

Laplace Domain $$\bar{f}(s) = \int_0^\infty f(t)e^{-st}\,dt$$	**Time Domain** $$f(t)$$
8. $\dfrac{s+a}{s^2}$	$1 + at$
9. $\dfrac{a^2}{s^2(s+a)}$	$at - (1 - e^{-at})$
10. $\dfrac{s+b}{s(s+a)}$	$\dfrac{b}{a}\left\{1 - \left(1 - \dfrac{a}{b}\right)e^{-at}\right\}$
11. $\dfrac{a}{s^2 + a^2}$	$\sin at$
12. $\dfrac{s}{s^2 + a^2}$	$\cos at$
13. $\dfrac{a^2}{s(s^2 + a^2)}$	$1 - \cos at$
14.* $\dfrac{1}{s^2 + 2\zeta\omega_n s + \omega_n^2}$	$\dfrac{1}{\omega_d}e^{-\zeta\omega_n t}\sin\omega_d t$
15.* $\dfrac{s}{s^2 + 2\zeta\omega_n s + \omega_n^2}$	$-\dfrac{\omega_n}{\omega_d}e^{-\zeta\omega_n t}\sin(\omega_d t - \phi_1)$
16.* $\dfrac{s + 2\zeta\omega_n}{s^2 + 2\zeta\omega_n s + \omega_n^2}$	$\dfrac{\omega_n}{\omega_d}e^{-\zeta\omega_n t}\sin(\omega_d t + \phi_1)$
17.* $\dfrac{\omega_n^2}{s(s^2 + 2\zeta\omega_n s + \omega_n^2)}$	$1 - \dfrac{\omega_n}{\omega_d}e^{-\zeta\omega_n t}\sin(\omega_d t + \phi_1)$
18.* $\dfrac{s + \zeta\omega_n}{s(s^2 + 2\zeta\omega_n s + \omega_n^2)}$	$e^{-\zeta\omega_n t}\sin(\omega_d t + \phi_1)$
19. 1	Unit impulse at $t = 0$
20. $\dfrac{e^{-as}}{s}$	Unit step function at $t = a$

*$\omega_d = \omega_n\sqrt{1 - \zeta^2}\,;\quad \zeta < 1$

$\phi_1 = \cos^{-1}\zeta;\ \zeta < 1$

Units

Heinrich Rudolf Hertz (1857–1894), a German physicist and a professor of physics at the Polytechnic Institute in Karlsruhe and later at the University of Bonn, gained fame through his experiments on radio waves. His investigations in the field of elasticity formed a relatively small part of his achievements, but are of vital importance to engineers. His work on the analysis of elastic bodies in contact is referred to as "Hertzian stresses," and is very important in the design of ball and roller bearings. The unit of frequency of periodic phenomena, measured in cycles per second, is named Hertz in SI units. (Courtesy The Granger Collection)

The English system of units is now being replaced by the International System of units (SI). The SI system is the modernized version of the metric system of units. Its name in French is Système International; hence the abbreviation *SI*. The SI system has seven basic units. All other units can be derived from these seven [C.1–C.2]. The three basic units of concern in the study of vibrations are meter for length, kilogram for mass, and second for time.

The consistent set of units for various quantities is given in Table C.1. The common prefixes for multiples and submultiples of SI units are given in Table C.2. In the SI system, the combined units must be abbreviated with care. For example, a torque of 4 kg × 2 m must be stated as 8 kg m or 8 kg · m with either a space or a dot between kg and m. It should not be written as kgm. Another example is 8 m × 5 s = 40 m s or 40 m · s or 40 meter-seconds. If it is written as 40 ms, it means 40 milliseconds.

Conversion of Units

To convert the units of any given quantity from one system to another, we use the equivalence of units given on the inside back cover. The following examples illustrate the procedure.

EXAMPLE C.1

Mass moment of inertia:

$$\begin{pmatrix} \text{mass moment of} \\ \text{inertia in SI units} \end{pmatrix} = \begin{pmatrix} \text{mass moment of inertia} \\ \text{in English units} \end{pmatrix} \times \begin{pmatrix} \text{multiplication} \\ \text{factor} \end{pmatrix}$$

$$(\text{kg} \cdot \text{m}^2) \equiv (\text{N} \cdot \text{m} \cdot \text{s}^2) = \left(\frac{\text{N}}{\text{lb}_f} \cdot \text{lb}_f \right) \left(\frac{\text{m}}{\text{in.}} \cdot \text{in.} \right) (\text{sec}^2)$$

$$= (\text{N per 1 lb}_f)(\text{m per 1 in.})(\text{lb}_f\text{-in.-sec}^2)$$

$$= (4.448222)(0.0254)(\text{lb}_f\text{-in-sec}^2)$$

$$= 0.1129848(\text{lb}_f\text{-in.-sec}^2)$$

TABLE C.1   Consistent set of units

Quantity	SI Unit Unit	SI Unit Symbol	English Unit* (symbol)
Mass	kilogram	kg	$\text{lb}_f\text{-sec}^2/\text{in.}$
Length	meter	m	in.
Time	second	s	sec
Force	Newton = $\text{kg} \cdot \text{m/s}^2$	N	lb_f
Stress or pressure or elastic modulus	Pascal = N/m^2	Pa	$\text{lb}_f/\text{in}^2 = \text{psi}$
Moment of a force	Newton meter = $\text{kg} \cdot \text{m}^2/\text{s}^2$	$\text{N} \cdot \text{m}$	in.-lb_f
Work or energy	Joule = $\text{N} \cdot \text{m}$	J	in.-lb_f
Power	Watt = J/s	W	$\text{in.-lb}_f/\text{sec}$
Frequency	Hertz = 1/s	Hz	Hz
Mass density	kilogram per cubic meter	kg/m^3	$\text{lb}_f\text{-sec}^2/\text{in}^4$
Velocity or speed	meter per second	m/s	in./sec
Acceleration	meter per second squared	m/s^2	in./sec^2
Angular displacement or plane angle	radian	rad	rad
Angular velocity	radian per second	rad/s	rad/sec
Angular acceleration	radian per second squared	rad/s^2	rad/sec^2
Area moment of inertia	(meter)⁴	m^4	in^4
Mass moment of inertia	kilogram · meter squared	$\text{kg} \cdot \text{m}^2$	$\text{in.-lb}_f\text{-sec}^2$
Stiffness coefficient:			
Rectilinear (translational)	Newton per meter	N/m	$\text{lb}_f/\text{in.}$
Rotational (torsional)	Newton meter per radian	$\text{N} \cdot \text{m/rad}$	$\text{lb}_f\text{-in./rad}$
Damping coefficient:			
Rectilinear (translational)	Newton second per meter	$\text{N} \cdot \text{s/m}$	$\text{lb}_f\text{-sec/in.}$

| Quantity | SI Unit | | English Unit* |
	Unit	Symbol	(symbol)
Rotational (torsional)	Newton meter second per radian	$N \cdot m \cdot s/rad$	lb_f-in.-sec/rad
Mass:			
Rectilinear (translational)	kilogram	kg	lb_f-sec^2/in.
Rotational (torsional)	kilogram meter squared	$kg \cdot m^2$	lb_f-sec^2-in.

*ft (foot) can be interchanged with in. (inch).

EXAMPLE C.2

Stress:

$$(\text{Stress in SI units}) = (\text{stress in English units}) \times \left(\begin{array}{c} \text{multiplication} \\ \text{factor} \end{array}\right)$$

$$(\text{Pa}) \equiv (N/m^2) = \left(\frac{N}{lb_f} \cdot lb_f\right)\frac{1}{\left(\dfrac{m}{in.} \cdot in.\right)^2} = \frac{N}{lb_f}\frac{1}{\left(\dfrac{m}{in.}\right)^2}\left(lb_f/in^2\right)$$

$$= \frac{(N \text{ per } 1\ lb_f)}{(m \text{ per } 1\ in.)^2}\left(lb_f/in^2\right)$$

$$= \frac{(4.448222)}{(0.0254)^2}\left(lb_f/in^2\right)$$

$$= 6894.757\left(lb_f/in^2\right)$$

TABLE C.2   Prefixes for Multiples and Submultiples of SI Units

Multiple	Prefix	Symbol	Submultiple	Prefix	Symbol
10	deka	da	10^{-1}	deci	d
10^2	hecto	h	10^{-2}	centi	c
10^3	kilo	k	10^{-3}	milli	m
10^6	mega	M	10^{-6}	micro	μ
10^9	giga	G	10^{-9}	nano	n
10^{12}	tera	T	10^{-12}	pico	p

REFERENCES

C.1. E. A. Mechtly, "The International System of Units" (Second Revision), NASA SP-7012, 1973.

C.2. C. Wandmacher, "Metric Units in Engineering—Going SI," Industrial Press, New York, 1978.

Answers to Selected Problems

Chapter 1

1.6. $k_{eq} = k_4 + \{k_1 k_2 k_3 /(k_1 k_2 + k_2 k_3 + k_3 k_1)\} + R^2 (k_5 + k_6)$
$\qquad + R^2 \{k_7 k_8 /(k_7 + k_8)\}$

1.8. (a) $k_{eq} = k(4a^2 - b^2)/b^2$, (b) $k_{eq} = k$

1.10. $J_{eq} = m_1 l_1^2 + (m_2 + m) l_3^2$, $k_{eq} = k_1 l_1^2 + \{k_2 k_3 l_2^2 /(k_2 + k_3)\} + k_t + k_4 l_3^2$

1.13. $k = p\gamma A^2 / v$ **1.17.** $m_{eq} = m_1 \left(\dfrac{a}{b}\right)^2 + m_2 + J_0 \left(\dfrac{1}{b^2}\right)$

1.20. (i) $c_{eq} = c_1 + c_2 + c_3$,   (ii) $\dfrac{1}{c_{eq}} = \dfrac{1}{c_1} + \dfrac{1}{c_2} + \dfrac{1}{c_3}$,

     (iii) $c_{eq} = c_1 + c_2 \left(\dfrac{l_2}{l_1}\right)^2 + c_3 \left(\dfrac{l_3}{l_1}\right)^2$,

     (iv) $c_{teq} = c_{t1} + c_{t2} \left(\dfrac{n_1}{n_2}\right)^2 + c_{t3} \left(\dfrac{n_1}{n_3}\right)^2$

1.23. $c_t = \dfrac{\pi\mu D^2 (l - h)}{2d} + \dfrac{\pi\mu D^3}{32h}$ **1.25.** $A = 4.4721$, $\theta = -26.5651°$

1.27. (i) $A = 20.2237$ mm, $\alpha = -1.4219$ rad; (ii) $A_1 = 3$ mm, $A_2 = 20$ mm

1.30. $x_2(t) = 6.1966 \sin(\omega t + 83.7938°)$ **1.33.**   No

1.36. $A = 0.5522$ mm, $\dot{x}_{max} = 52.04$ mm/sec

1.38. $c_n = \dfrac{1}{\tau} \displaystyle\int_0^\tau x(t) e^{-in\omega t}\, dt$

1.41. $x(t) = \dfrac{A}{\pi} + \dfrac{A}{2} \sin \omega t - \dfrac{2A}{\pi} \displaystyle\sum_{n=2,4,6,\dots}^{\infty} \dfrac{\cos n\omega t}{(n^2 - 1)}$

1.43. $x(t) = \dfrac{8A}{\pi^2} \displaystyle\sum_{n=1,3,5,\dots}^{\infty} (-1)^{\frac{n-1}{2}} \dfrac{\sin n\omega t}{n^2}$

1.47. $a_0 = 2275.0$, $a_1 = -414.94$, $a_2 = 28.61$, $a_3 = 35.73$; $b_1 = 150.31$, $b_2 = -146.17$, $b_3 = 55.15$

1.49. $a_0 = 19.92$, $a_1 = -20.16$, $a_2 = 3.31$, $a_3 = 3.77$; $b_1 = 23.52$, $b_2 = 12.26$, $b_3 = -0.41$

1.53. $a_0 = -0.38$, $a_1 = -0.62$, $a_2 = 0.46$, $a_3 = 0.41$; $b_1 = -0.35$, $b_2 = 0.92$, $b_3 = -0.17$

Chapter 2

2.2. (i) 0.1715 sec,   (ii) 0.2970 sec   **2.4.**   0.0993 sec

2.6. $\omega_n = [(k_1 k_2 l_1^2 + k_2 k_3 l_2^2)/\{m(k_1 l_1^2 + k_2 l_2^2 + k_3 l_3^2)\}]^{1/2}$

2.8. $\omega_n = [(k_1 + k_2)/m]^{1/2}$ **2.10.** $\omega_n = [k/(4m)]^{1/2}$

2.13. 0.2204 sec, 0.07015 m   **2.16.** $m\ddot{x} + (k_1 + k_2)x = 0$ **2.19.** $\omega_n = (2k/m)^{1/2}$

2.21. Longitudinal $\omega_n = \left[\dfrac{glAE}{Wa(l-a)}\right]^{1/2}$,

Transverse $\omega_n = \left[\dfrac{3EIl^3 g}{Wa^3(l-a)^3}\right]^{1/2}$,

Torsional $\omega_n = \left[\dfrac{GJ}{J_0}\left(\dfrac{1}{a} + \dfrac{1}{b}\right)\right]^{1/2}$

2.23. 0.4330 sec   **2.26.**   454.7935 N/m

2.29. $\omega_n = \left\{\dfrac{\left(M + \dfrac{m}{2}\right)g}{\left(M + \dfrac{m}{3}\right)l}\right\}^{1/2}$ **2.32.** $\omega_n = \left\{\dfrac{(k_1 + k_2)(R + a)^2}{1.5mR^2}\right\}^{1/2}$

2.34. $\frac{1}{3}ml^2\ddot{\theta} + (k_t + k_1 a^2 + k_2 l^2)\theta = 0$

2.38. $\omega_n = [(k_1 + k_2)/m]^{1/2}$ **2.40.**   45.1547 rad/sec

2.43. $c_t = 8.5013$ N-m-s/rad   **2.46.** $x_{max} = \left(x_0 + \dfrac{\dot{x}_0}{\omega_n}\right)e^{-\{\dot{x}_0/(\dot{x}_0 + \omega_n x_0)\}}$

2.49. (a) 1.0548 m, (b) $t = 0.5736$ sec   **2.52.** $m = 500$ kg, $k = 27066.403$ N/m

2.55. Coulomb, 5N, 14.1421 rad/sec   **2.58.**   12.5 N

2.61. $c_{eq} = \dfrac{4\,\mu N}{\pi \omega X}$ **2.63.** $\zeta_{eq} = 0.1736$ **2.65.** $\beta = 0.001291$

Chapter 3

3.2. 5 sec   **3.4.** 9.1189 kg   **3.6.**
$$\dfrac{F_0}{k\left[1 - \left(\dfrac{\omega}{\omega_n}\right)^2\right]}\cos \omega t$$

3.8. 0.676 sec   **3.10.** $\zeta = 0.1180$ **3.12.** (i) $m\ddot{x} + (c_1 + c_2)\dot{x} + k_2 x = c_1 Y \cos \omega t$
3.14. (i) 64.16 rad/sec, (ii) 967.2 N-m
3.18. 0.3339 sin 25 t mm   **3.20.** $X = 0.106$ m, $s = 246.73$ km/hr
3.22. $c = (k - m\omega^2)/\omega$ **3.25.** 0.4145×10^{-3} m, 1.0400×10^{-3} m
3.27. 1.4195 N-m   **3.29.** $\zeta = 0.1364$ **3.32.** Maximum force = 26.68 1b
3.35. $\mu = 0.1$ **3.38.** (i) 10.2027 lb-in,   (ii) 40.8108 lb-in

3.41. (iii) $\dfrac{1}{\left\{\dfrac{4\,\mu N}{\pi X k} + \dfrac{3}{4k}c\omega^3 X^2\right\}}$

Chapter 4

4.2. $x(t) = \dfrac{F_0}{2k} - \dfrac{4F_0}{\pi^2 k}\displaystyle\sum_{n=1,3,\ldots}^{\infty}\dfrac{1}{n^2}\dfrac{1}{\sqrt{(1 - r^2 n^2)^2 + (2\zeta nr)^2}}\cos(n\omega t - \phi_n)$

with $r = \omega/\omega_n$ and $\phi_n = \tan^{-1}\left(\dfrac{2\zeta nr}{1 - n^2 r^2}\right)$

4.6. $x_p(t) = 6.6389 \times 10^{-4} - 13.7821 \times 10^{-4}\cos(10.472t - 0.0172)$
$\qquad\qquad +15.7965 \times 10^{-4}\sin(10.472t - 0.0172) + \ldots$ m

4.9. $x(t) = \dfrac{F_0}{k}\left\{1 + \dfrac{\sin \omega_n(t - t_0) - \sin \omega_n t}{\omega_n t_0}\right\}$; for $t \geqslant t_0$

4.12. $x(t) = \dfrac{F_0}{2k\left(1 - \dfrac{\omega^2}{\omega_n^2}\right)}\left[2 - \dfrac{\omega^2}{\omega_n^2}\left(1 - \cos\dfrac{\omega_n\pi}{\omega}\right)\right] + \dfrac{F_0}{k}\left[1 - \cos\omega_n\left(t - \dfrac{\pi}{\omega}\right)\right]$

for $t > \pi/\omega$.

4.17. $x(t) = 1.7689 \sin 6.2832\,(t - 0.018) - 0.8845 \sin 6.2832t$
$\qquad\qquad -0.8845 \sin 6.2832\,(t - 0.036)$m; for $t > 0.036$ sec

4.21. $\qquad x_p(t) = 0.002667$ m

4.25. $\qquad x_m = \dfrac{F_0}{k\omega_n t_0}[(1 - \cos \omega_n t_0)^2 + (\omega_n t_0 - \sin \omega_n t_0)^2]^{1/2}$; for $t > t_0$

4.29. $\qquad x(t) = \begin{cases} \dfrac{F_0}{m\omega_n^2}(1 - \cos \omega_n t); & 0 \leqslant t \leqslant t_0 \\[2mm] \dfrac{F_0}{m\omega_n^2}\left[\cos \omega_n(t - t_0) - \cos \omega_n t\right]; & t \geqslant t_0 \end{cases}$

4.32. $\qquad \dot{x}_i(t_i = \pi) = \begin{cases} -0.549289, & \text{Eq. } (4.68) \\ -0.551730, & \text{Eq. } (4.71) \end{cases}$

Chapter 5

5.1. $\omega_1 = 3.6603$ rad/sec, $\omega_2 = 13.6603$ rad/sec

5.3. $\omega_1 = \sqrt{k/m}$, $\omega_2 = \sqrt{2k/m}$ **5.5.** $\omega_1 = 7.3892$ rad/s, $\omega_2 = 58.2701$ rad/s

5.7. $\omega_1 = 0.7654\sqrt{g/l}$, $\omega_2 = 1.8478\sqrt{g/l}$

5.10. $\omega_1 = 12.8817$ rad/s, $\omega_2 = 30.5624$ rad/s

5.13. $x_1(t) = 0.1046 \sin 40.4225t + 0.2719 \sin 58.0175t$,
$x_2(t) = 0.1429 \sin 40.4225t - 0.09952 \sin 58.0175t$

5.15. $\omega_1 = 3.7495\sqrt{\dfrac{EI}{mh^3}}$, $\omega_2 = 9.0524\sqrt{\dfrac{EI}{mh^3}}$

5.17. $\vec{X}^{(1)} = \left\{ \begin{matrix} 1.0 \\ 2.3029 \end{matrix} \right\}$, $\vec{X}^{(2)} = \left\{ \begin{matrix} 1.0 \\ -1.3028 \end{matrix} \right\}$

5.20. $\omega_1 = 0.5176\sqrt{k_t/J_0}$, $\omega_2 = 1.9319\sqrt{k_t/J_0}$

5.23. $\omega_1 = 0.38197\sqrt{k_t/J_0}$, $\omega_2 = 2.61803\sqrt{k_t/J_0}$

5.27. $\omega_{1,2}^2 = \left\{ \dfrac{(J_0 k + mk_t) \pm \sqrt{(J_0 k + mk_t)^2 - 4(J_0 - me^2)mkk_t}}{2m(J_0 - me^2)} \right\}$

5.30. (i) $\omega_1 = 12.2474$ rad/sec, $\omega_2 = 38.7298$ rad/sec   **5.34.** $k_2 = m_2 \omega^2$

5.37. $x_1(t) = (17.2915 \, F_0 \cos \omega t + 6.9444 \, F_0 \sin \omega t)10^{-4}$
$x_2(t) = (17.3165 \, F_0 \cos \omega t + 6.9684 \, F_0 \sin \omega t)10^{-4}$

5.41. $x_2(t) = \left(\frac{1}{60} - \frac{1}{40} \cos 10t + \frac{1}{120} \cos 10\sqrt{3}\,t \right) u(t)$ **5.44.** $b_1 c_2 - c_1 b_2 = 0$

5.46. $\ddot{\alpha} + \left(\dfrac{k_t}{J_1} + \dfrac{k_t}{J_2} \right) \alpha = 0$ where $\alpha = \theta_1 - \theta_2$

Chapter 6

6.2. $[k] = \begin{bmatrix} (k_1 + k_2) & -k_2 & 0 \\ -k_2 & (k_2 + k_3) & -k_3 \\ 0 & -k_3 & (k_3 + k_4) \end{bmatrix}$

6.4. $[a] = \dfrac{l^3}{EI} \begin{bmatrix} 9/64 & 1/6 & 13/192 \\ 1/6 & 1/3 & 1/6 \\ 13/192 & 1/6 & 9/64 \end{bmatrix}$ **6.9.** $2k$

6.12. $2\,m\ddot{x} + kx = 0$, $l\ddot{\theta} + g\theta = 0$

6.17. $\omega_1 = 0.44504\sqrt{k/m}$, $\omega_2 = 1.2471\sqrt{k/m}$, $\omega_3 = 1.8025\sqrt{k/m}$

6.20. $\omega_1 = 0.533399\sqrt{k/m}$, $\omega_2 = 1.122733\sqrt{k/m}$, $\omega_3 = 1.669817\sqrt{k/m}$

6.23. $\lambda_1 = 2.21398$, $\lambda_2 = 4.16929$, $\lambda_3 = 10.6168$

6.26. $\omega_1 = 0.644798\sqrt{g/l}$, $\omega_2 = 1.514698\sqrt{g/l}$, $\omega_3 = 2.507977\sqrt{g/l}$

6.29. $\omega_1 = 0.562587\sqrt{\dfrac{P}{ml}}$, $\omega_2 = 0.915797\sqrt{\dfrac{P}{ml}}$, $\omega_3 = 1.584767\sqrt{\dfrac{P}{ml}}$

6.32. $[X] = \dfrac{1}{2} \begin{bmatrix} 1 & 1 & 0 \\ -1 & 1 & \sqrt{2/3} \\ 1 & 1 & \sqrt{8/3} \end{bmatrix}$

6.35. $\omega_1 = 0$, $\omega_2 = 0.752158\sqrt{k/m}$, $\omega_3 = 1.329508\sqrt{k/m}$

6.38. (a) $\omega_1 = 0.44497\sqrt{k_t/J_0}$, $\omega_2 = 1.24700\sqrt{k_t/J_0}$, $\omega_3 = 1.80194\sqrt{k_t/J_0}$

(b) $\vec{\theta}(t) = \begin{Bmatrix} -0.0000025 \\ 0.0005190 \\ -0.0505115 \end{Bmatrix} \cos 100t$ radians

6.42. $x_3(t) = 0.0256357\cos(\omega t + 0.5874^0)$ m   **6.46.** $x^3 - 17x^2 + 77x - 98 = 0$

Chapter 7

7.1. (a) $\omega_1 \simeq 2.6917\sqrt{\dfrac{EI}{ml^3}}$ (b) $\omega_1 \simeq 2.7994\sqrt{\dfrac{EI}{ml^3}}$

7.3. $3.5987\sqrt{\dfrac{EI}{ml^3}}$ **7.5.** $0.3015\sqrt{k/m}$

7.7. $0.4082\sqrt{k/m}$ **7.9.** $1.0954\sqrt{\dfrac{T}{lm}}$

7.13. $\omega_1 = 0$, $\omega_2 \simeq 6.2220$ rad/s, $\omega_3 \simeq 25.7156$ rad/s   **7.16.** $\omega_1 = \sqrt{k/m}$

7.18. $\omega_1 = 0.3104$, $\omega_2 = 0.4472$, $\omega_3 = 0.6869$ where $\omega_i = 1/\sqrt{\lambda_i}$

7.21. $\omega_1 = 0.765366$, $\omega_2 = 1.414213$, $\omega_3 = 1.847759$ with $\omega_i = \omega_i\sqrt{\dfrac{GJ}{lJ_0}}$

7.24. $\omega_1 = 0.2583$, $\omega_2 = 3.0$, $\omega_3 = 7.7417$

7.27. $[U]^{-1} = \begin{bmatrix} 0.44721359 & 0.083045475 & -0.12379687 \\ 0 & 0.41522738 & 1.1760702 \\ 0 & 0 & 1.7950547 \end{bmatrix}$

7.34. $\omega_1 = 0.2430$, $\omega_2 = 0.5728$, $\omega_3 = 7.1842$

7.37. $\omega_1 = 0.4450$, $\omega_2 = 1.2470$, $\omega_3 = 1.8019$

7.39. $\omega_1 = 5.8694$, $\omega_2 = 85.5832$, $\omega_3 = 293.5470$

Chapter 8

8.1. 28.2843 m/sec   **8.3.** $\omega_3 = 9000$ Hz, both increased by 9.54%

8.6. (a) 0.1248×10^6 N, (b) 3.12×10^6 N

8.11. $w\left(x, \dfrac{l}{c}\right) = -\dfrac{\sqrt{3}\,9h}{2\pi^2}\sin\dfrac{\pi x}{l} + \dfrac{\sqrt{3}\,9h}{8\pi^2}\sin\dfrac{2\pi x}{l} - \dfrac{\sqrt{3}\,9h}{32\pi^2}\sin\dfrac{4\pi x}{l} + \dfrac{\sqrt{3}\,9h}{50\pi^2}\sin\dfrac{5\pi x}{l}$

8.18. $\tan\dfrac{\omega l_1}{c_1}\tan\dfrac{\omega l_2}{c_2} = \dfrac{A_1 E_1 c_2}{A_2 E_2 c_1}$

8.22. $\omega_n = \dfrac{(2n+1)\pi}{2}\sqrt{\dfrac{G}{\rho l^2}}$; $n = 0, 1, 2, \ldots$

8.25. 5030.59 rad/sec   **8.28.** $\cos\beta l\cosh\beta l = -1$

8.31. $\tan\beta l - \tanh\beta l = 0$ **8.33.** 20.2328 N-m

8.36. $\cos\beta l\cosh\beta l = 1$, and $\tan\beta l - \tanh\beta l = 0$

8.44. $\omega_{mn}^2 = \dfrac{\gamma_n P}{\rho}$, where $J_m(\gamma_n R) = 0$; $m = 0, 1, 2, \ldots$; $n = 1, 2, \ldots$

8.48. $w(x, y, t) = \dfrac{\dot{w}_0}{\omega_{12}} \sin \dfrac{\pi x}{a} \sin \dfrac{2\pi y}{b} \sin \omega_{12} t$ **8.51.** $22.4499 \sqrt{\dfrac{EI}{\rho A l^4}}$

8.55. $7.7460 \sqrt{\dfrac{EI_0}{\rho A_0 l^4}}$ **8.58.** $2.4146 \sqrt{\dfrac{EA_0}{m_0 l^2}}$

8.61. (a) $1.73205 \sqrt{\dfrac{E}{\rho l^2}}$, (b) $1.57669 \sqrt{\dfrac{E}{\rho l^2}}$, $5.67280 \sqrt{\dfrac{E}{\rho l^2}}$

8.64. $\omega_1 = 3.142 \sqrt{\dfrac{P}{\rho l^2}}$, $\omega_2 = 10.12 \sqrt{\dfrac{P}{\rho l^2}}$

Chapter 9

9.1. $m_c r_c = 3354.6361$ g-mm, $\theta_c = -25.5525^0$
9.3. $m_4 = 0.99$ oz, $\theta_4 = -35^0$ **9.6.** 1.6762 oz, $\alpha = 75.6261^0$ CW
9.9. Remove 0.1336 lb at 10.8377^0 CCW in plane B and 0.2063 lb at 1.3957^0 CCW in plane C at radii 4 in.
9.12. (a) $\vec{R}_A = -28.4021 \vec{j} - 3.5436 \vec{k}$, $\vec{R}_B = 13.7552 \vec{j} + 4.7749 \vec{k}$
        (b) $m_L = 10.44$ g, $\theta_L = 7.1141^0$
9.15. $F_{xp} = 0$, $F_{xs} = 3269.4495$ 1b, $M_{zp} = M_{zs} = 0$
9.18. Engine completely force and moment balanced
9.20. 0.2385 mm   **9.23** (a) $\omega < 95.4927$ rpm, (b) $\omega > 276.7803$ rpm
9.27. (a) 487.379 1b, (b) $\Omega_1 = 469.65$ rpm, $\Omega_2 = 766.47$ rpm
9.32. $0.9764 \leqslant \dfrac{\omega}{\omega_2} \leqslant 1.05125$

Chapter 10

10.2. 18.3777 Hz   **10.4.** 3.6935 Hz   **10.6.** 0.53%   **10.9.** 35.2635 Hz
10.12. 73.16%   **10.14.** $k = 33623.85$ N/m, $c = 50.55$ N-sec/m
10.16. $m = 19.41$ g, $k = 7622.8$ N/m   **10.19.** 111.20 rad/sec $-$ 2780.02 rad/sec

Chapter 11

11.2. $\left. \dfrac{d^4 x}{dt^4} \right|_i = \dfrac{x_i - 4x_{i-1} + 6x_{i-2} - 4x_{i-3} + x_{i-4}}{(\Delta t)^4}$

11.4. $x(t = 5) = -1$ with $\Delta t = 1$ and -0.9733 with $\Delta t = 0.5$
11.6. $x_{10} = -0.0843078$, $x_{15} = 0.00849639$
11.9. $x(t = 0.1) = 0.131173$, $x(t = 0.4) = -0.0215287$, $x(t = 0.8) = -0.0676142$
11.14. With $\Delta t = 0.07854$, $x_1 = x$ and $x_2 = \dot{x}$, $x_1(t = 0.2356) = 0.100111$,
        $x_2(t = 0.2356) = 0.401132$, $x_1(t = 1.5708) = 1.040726$,
        $x_2 = (t = 1.5708) = -0.378066$

11.20.

t	x_1	x_2
0.25	0.07813	1.1860
1.25	2.3360	−0.2832
3.25	−0.6363	2.3370

11.23. $\omega_1 = 3.06147\sqrt{\dfrac{E}{\rho l^2}}$, $\omega_2 = 5.65685\sqrt{\dfrac{E}{\rho l^2}}$, $\omega_3 = 7.39103\sqrt{\dfrac{E}{\rho l^2}}$

11.26. $\omega_1 = 17.9274\sqrt{\dfrac{EI}{\rho A l^4}}$, $\omega_2 = 39.1918\sqrt{\dfrac{EI}{\rho A l^4}}$, $\omega_3 = 57.1193\sqrt{\dfrac{EI}{\rho A l^4}}$

11.32. With $\Delta t = 0.24216267$,

t	x_1	x_2
0.2422	0.0000	0.1466
2.4216	0.8691	1.9430
4.1168	−0.1154	0.6498

11.39. With $\Delta t = 0.24216267$,

t	x_1	x_2
0.2422	0.01776	0.1335
2.4216	0.7330	1.8020
4.1168	0.1059	0.8573

Chapter 12

12.2. $[k] = \dfrac{EA_0}{l}(0.6321)\begin{bmatrix} 1 & -1 \\ -1 & 1 \end{bmatrix}$

12.4. 3.3392×10^7 N/m²   **12.8.** 5.184 in. under load

12.10. 0.05165 in. under load

12.13. Deflection $= 0.002197\dfrac{Pl^3}{EI}$, slope $= 0.008789\dfrac{Pl^3}{EI}$

12.17. $\omega_1 = 0.8587\sqrt{\dfrac{EI}{\rho A l^4}}$, $\omega_2 = 4.0965\sqrt{\dfrac{EI}{\rho A l^4}}$, $\omega_3 = 34.9210\sqrt{\dfrac{EI}{\rho A l^4}}$

12.20. $\omega_1 = 15.1357\sqrt{\dfrac{EI}{\rho A l^4}}$, $\omega_2 = 28.9828\sqrt{\dfrac{EI}{\rho A l^4}}$

12.23. $\omega_1 = 20.4939\sqrt{\dfrac{EI}{\rho A l^4}}$ **12.26.** $\sigma^{(1)} = -2.5056$ psi, $\sigma^{(2)} = 2.6936$ psi

12.29. Maximum bending stresses: −37218 psi (in both connecting rod and crank), maximum axial stresses: −6411 psi (in connecting rod), −5649 psi (in crank)

12.31. $\omega_1 = 6445$ rad/sec, $\omega_2 = 12451$ rad/sec

Chapter 13

13.2. (a) $\sqrt{\dfrac{m}{k_1}}\,\dot{x}_0$, (b) $\tau_n = \pi\left(\sqrt{\dfrac{m}{k_1}} + \sqrt{\dfrac{m}{k_2}}\right)$

13.5. $\tau = 4\sqrt{\dfrac{l}{g}}\displaystyle\int_0^{\pi/2}\dfrac{d\phi}{\sqrt{1 - k^2\sin^2\phi}}$, where $k = \sin(\theta_0/2)$

13.7. $\dfrac{4}{\omega_0\left(1 - \dfrac{\theta_0^2}{12}\right)}F\left(a, \dfrac{\pi}{2}\right)$ where $F(a, \beta)$ is an incomplete elliptic integral

of the first kind

13.9. $x(t) = A_0\cos\omega t - \dfrac{A_0^3\alpha}{32\omega^2}(\cos\omega t - \cos 3\omega t) - \dfrac{A_0^5\alpha^2}{1024\omega^4}(\cos\omega t - \cos 5\omega t);$

$\omega^2 = \omega_0^2 + \dfrac{3}{4}A_0^2\alpha - \dfrac{3}{128}\dfrac{A_0^4}{\omega^2}\alpha^2$

13.17. $x(t) = 5[1 - 1.0013e^{-0.05t}\{\cos(0.9987t - 2.8681^0)\}]$

13.21. For $0 < c < 2$: stable focus, for $c \geqslant 2$: stable nodal point

13.23. Equilibrium point is a center

13.26. (i) $\lambda_1 = \lambda_2 = 2$, (ii) $\lambda_1 = -1$, $\lambda_2 = 3$

13.28. $x(t) = c_1\begin{Bmatrix}1\\-1\end{Bmatrix}e^{2t} + c_2\begin{Bmatrix}1\\-1\end{Bmatrix}te^{2t} + c_2\begin{Bmatrix}0\\-1\end{Bmatrix}e^{2t}$

13.30. $x(t) = 2\cos\omega t + \dfrac{\alpha}{4\omega}\sin 3\omega t + \dfrac{3\alpha^2}{32\omega^2}\cos 3\omega t + \dfrac{5\alpha^2}{96\omega^2}\cos 5\omega t;$ $\omega^2 = 1 + \dfrac{\alpha^2}{8}$

Chapter 14

14.1. 0.04   **14.3.** 1.0, 1.3333, 0.5773   **14.5.** 25

14.10. (i) $1 - e^{-\lambda t}$, (ii) $\dfrac{1}{\lambda}$, (iii) $\dfrac{1}{\lambda}$ **14.12.** 0.3316×10^{-8}

14.14. $X(\omega) = \left(\dfrac{Aa}{a^2 + \omega^2}\right) - i\left(\dfrac{A\omega}{a^2 + \omega^2}\right)$ **14.17.** (ii) 3400.0

14.19. $\dfrac{2S_0}{\tau}(\sin\omega_2\tau - \sin\omega_1\tau)$ **14.26.** $\sigma = 0.3106$ m

14.28. $E[z^2] = \dfrac{\pi S_0\omega^4}{2\zeta\omega_n^3}$ **14.30.** $m_{eq} = \left\{\dfrac{\pi S_0}{2\delta\omega_1^2(\omega_1^2 - \omega_2^2)^{1/2}}\right\}^{1/2},$

$k_{eq} = \left\{\dfrac{\pi S_0\omega_1^2}{2\delta(\omega_1^2 - \omega_2^2)^{1/2}}\right\}^{1/2},$

$c_{eq} = \left\{\dfrac{2\pi S_0(\omega_1^2 - \omega_2^2)^{1/2}}{\delta\omega_1^2}\right\}^{1/2}$

Index

Equivalent Masses, Springs and Dampers

Equivalent masses

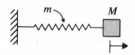

Mass (M) attached at end of spring of mass m

$$m_{eq} = M + \frac{m}{3}$$

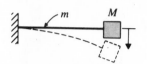

Cantilever beam of mass m carrying an end mass M

$$m_{eq} = M + 0.23\,m$$

Simply supported beam of mass m carrying a mass M at the middle

$$m_{eq} = M + 0.5\,m$$

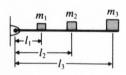

Coupled translational and rotational masses

$$m_{eq} = m + \frac{J_0}{R^2}$$
$$J_{eq} = J_0 + mR^2$$

Masses on a hinged bar

$$m_{eq_1} = m_1 + \left(\frac{l_2}{l_1}\right)^2 m_2 + \left(\frac{l_3}{l_1}\right)^2 m_3$$

Equivalent springs

Rod under axial load
(l = length, A = cross sectional area)

$$k_{eq} = \frac{EA}{l}$$

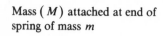

Tapered rod under axial load
(D, d = end diameters)

$$k_{eq} = \frac{\pi E D d}{4l}$$

Helical spring under axial load
(d = wire diameter, D = mean coil diameter, n = number of active turns)

$$k_{eq} = \frac{Gd^4}{8nD^3}$$

Fixed-fixed beam with load at the middle

$$k_{eq} = \frac{192\,EI}{l^3}$$

Cantilever beam with end load

$$k_{eq} = \frac{3EI}{l^3}$$